Microplastics Degradation and Characterization

Microplastics Degradation and Characterization

Microplastics Degradation and Characterization

Editor

Jacopo La Nasa

MDPI • Basel • Beijing • Wuhan • Barcelona • Belgrade • Manchester • Tokyo • Cluj • Tianjin

Editor
Jacopo La Nasa
Department of Chemistry and
Industrial Chemistry
University of Pisa
Pisa
Italy

Editorial Office
MDPI
St. Alban-Anlage 66
4052 Basel, Switzerland

This is a reprint of articles from the Special Issue published online in the open access journal *Polymers* (ISSN 2073-4360) (available at: www.mdpi.com/journal/polymers/special_issues/microplastics_degradation_characterization).

For citation purposes, cite each article independently as indicated on the article page online and as indicated below:

LastName, A.A.; LastName, B.B.; LastName, C.C. Article Title. *Journal Name* **Year**, *Volume Number*, Page Range.

ISBN 978-3-0365-5266-8 (Hbk)
ISBN 978-3-0365-5265-1 (PDF)

Cover image courtesy of Jacopo La Nasa

© 2022 by the authors. Articles in this book are Open Access and distributed under the Creative Commons Attribution (CC BY) license, which allows users to download, copy and build upon published articles, as long as the author and publisher are properly credited, which ensures maximum dissemination and a wider impact of our publications.
The book as a whole is distributed by MDPI under the terms and conditions of the Creative Commons license CC BY-NC-ND.

Contents

Valter Castelvetro, Andrea Corti, Jacopo La Nasa, Francesca Modugno, Alessio Ceccarini and Stefania Giannarelli et al.
Polymer Identification and Specific Analysis (PISA) of Microplastic Total Mass in Sediments of the Protected Marine Area of the Meloria Shoals
Reprinted from: *Polymers* **2021**, *13*, 796, doi:10.3390/polym13050796 1

Sabiqah Tuan Anuar, Raad Shaher Altarawnah, Ahmad Ammarluddin Mohd Ali, Bai Qin Lee, Wan Mohd Afiq Wan Mohd Khalik and Ku Mohd Kalkausar Ku Yusof et al.
Utilizing Pyrolysis–Gas Chromatography/Mass Spectrometry for Monitoring and Analytical Characterization of Microplastics in Polychaete Worms
Reprinted from: *Polymers* **2022**, *14*, 3054, doi:10.3390/polym14153054 19

Solange Magalhães, Luís Alves, Anabela Romano, Bruno Medronho and Maria da Graça Rasteiro
Extraction and Characterization of Microplastics from Portuguese Industrial Effluents
Reprinted from: *Polymers* **2022**, *14*, 2902, doi:10.3390/polym14142902 33

Diego David Pinzon-Moreno, Isabel Rosali Maurate-Fernandez, Yury Flores-Valdeon, Antony Alexander Neciosup-Puican and María Verónica Carranza-Oropeza
Degradation of Hydrogels Based on Potassium and Sodium Polyacrylate by Ionic Interaction and Its Influence on Water
Reprinted from: *Polymers* **2022**, *14*, 2656, doi:10.3390/polym14132656 45

Lalitsuda Phutthimethakul and Nuta Supakata
Partial Replacement of Municipal Incinerated Bottom Ash and PET Pellets as Fine Aggregate in Cement Mortars
Reprinted from: *Polymers* **2022**, *14*, 2597, doi:10.3390/polym14132597 61

Nguyen Thao Nguyen, Nguyen Thi Thanh Nhon, Ho Truong Nam Hai, Nguyen Doan Thien Chi and To Thi Hien
Characteristics of Microplastics and Their Affiliated PAHs in Surface Water in Ho Chi Minh City, Vietnam
Reprinted from: *Polymers* **2022**, *14*, 2450, doi:10.3390/polym14122450 79

Ilaria Savino, Claudia Campanale, Pasquale Trotti, Carmine Massarelli, Giuseppe Corriero and Vito Felice Uricchio
Effects and Impacts of Different Oxidative Digestion Treatments on Virgin and Aged Microplastic Particles
Reprinted from: *Polymers* **2022**, *14*, 1958, doi:10.3390/polym14101958 99

Bárbara Abaroa-Pérez, Sara Ortiz-Montosa, José Joaquín Hernández-Brito and Daura Vega-Moreno
Yellowing, Weathering and Degradation of Marine Pellets and Their Influence on the Adsorption of Chemical Pollutants
Reprinted from: *Polymers* **2022**, *14*, 1305, doi:10.3390/polym14071305 121

Martina Miloloža, Kristina Bule, Viktorija Prevarić, Matija Cvetnić, Šime Ukić and Tomislav Bolanča et al.
Assessment of the Influence of Size and Concentration on the Ecotoxicity of Microplastics to Microalgae *Scenedesmus* sp., Bacterium *Pseudomonas putida* and Yeast *Saccharomyces cerevisiae*
Reprinted from: *Polymers* **2022**, *14*, 1246, doi:10.3390/polym14061246 133

Claudia Cella, Rita La Spina, Dora Mehn, Francesco Fumagalli, Giacomo Ceccone and Andrea Valsesia et al.
Detecting Micro- and Nanoplastics Released from Food Packaging: Challenges and Analytical Strategies
Reprinted from: *Polymers* **2022**, *14*, 1238, doi:10.3390/polym14061238 . . . . **153**

Cristina De Monte, Marina Locritani, Silvia Merlino, Lucia Ricci, Agnese Pistolesi and Simona Bronco
An In Situ Experiment to Evaluate the Aging and Degradation Phenomena Induced by Marine Environment Conditions on Commercial Plastic Granules
Reprinted from: *Polymers* **2022**, *14*, 1111, doi:10.3390/polym14061111 . . . . **167**

Chun-Ting Lin, Ming-Chih Chiu and Mei-Hwa Kuo
A Mini-Review of Strategies for Quantifying Anthropogenic Activities in Microplastic Studies in Aquatic Environments
Reprinted from: *Polymers* **2022**, *14*, 198, doi:10.3390/polym14010198 . . . . **193**

Mehmet Murat Monkul and Hakkı O. Özhan
Microplastic Contamination in Soils: A Review from Geotechnical Engineering View
Reprinted from: *Polymers* **2021**, *13*, 4129, doi:10.3390/polym13234129 . . . . **211**

Franja Prosenc, Pia Leban, Urška Šunta and Mojca Bavcon Kralj
Extraction and Identification of a Wide Range of Microplastic Polymers in Soil and Compost
Reprinted from: *Polymers* **2021**, *13*, 4069, doi:10.3390/polym13234069 . . . . **235**

Sola Choi, Miyeon Kwon, Myung-Ja Park and Juhea Kim
Analysis of Microplastics Released from Plain Woven Classified by Yarn Types during Washing and Drying
Reprinted from: *Polymers* **2021**, *13*, 2988, doi:10.3390/polym13172988 . . . . **251**

Greta Biale, Jacopo La Nasa, Marco Mattonai, Andrea Corti, Virginia Vinciguerra and Valter Castelvetro et al.
A Systematic Study on the Degradation Products Generated from Artificially Aged Microplastics
Reprinted from: *Polymers* **2021**, *13*, 1997, doi:10.3390/polym13121997 . . . . **263**

Valeria Caponetti, Alexandra Mavridi-Printezi, Matteo Cingolani, Enrico Rampazzo, Damiano Genovese and Luca Prodi et al.
A Selective Ratiometric Fluorescent Probe for No-Wash Detection of PVC Microplastic
Reprinted from: *Polymers* **2021**, *13*, 1588, doi:10.3390/polym13101588 . . . . **287**

Aranza Denisse Vital-Grappin, Maria Camila Ariza-Tarazona, Valeria Montserrat Luna-Hernández, Juan Francisco Villarreal-Chiu, Juan Manuel Hernández-López and Cristina Siligardi et al.
The Role of the Reactive Species Involved in the Photocatalytic Degradation of HDPE Microplastics Using C,N-TiO$_2$ Powders
Reprinted from: *Polymers* **2021**, *13*, 999, doi:10.3390/polym13070999 . . . . **303**

Lukas Miksch, Lars Gutow and Reinhard Saborowski
pH-Stat Titration: A Rapid Assay for Enzymatic Degradability of Bio-Based Polymers
Reprinted from: *Polymers* **2021**, *13*, 860, doi:10.3390/polym13060860 . . . . **321**

Benjamin O. Asamoah, Pauliina Salmi, Jukka Räty, Kalle Ryymin, Julia Talvitie and Anna K. Karjalainen et al.
Optical Monitoring of Microplastics Filtrated from Wastewater Sludge and Suspended in Ethanol
Reprinted from: *Polymers* **2021**, *13*, 871, doi:10.3390/polym13060871 335

Benjamin O. Asamoah, Emilia Uurasjärvi, Jukka Räty, Arto Koistinen, Matthieu Roussey and Kai-Erik Peiponen
Towards the Development of Portable and In Situ Optical Devices for Detection of Micro-and Nanoplastics in Water: A Review on the Current Status
Reprinted from: *Polymers* **2021**, *13*, 730, doi:10.3390/polym13050730 349

Hongxia Li, Jianqun Yang, Feng Tian, Xingji Li and Shangli Dong
Study on the Microstructure of Polyether Ether Ketone Films Irradiated with 170 keV Protons by Grazing Incidence Small Angle X-ray Scattering (GISAXS) Technology
Reprinted from: *Polymers* **2020**, *12*, 2717, doi:10.3390/polym12112717 377

Article

Polymer Identification and Specific Analysis (PISA) of Microplastic Total Mass in Sediments of the Protected Marine Area of the Meloria Shoals

Valter Castelvetro [1,2,*], Andrea Corti [1,2], Jacopo La Nasa [1], Francesca Modugno [1,2], Alessio Ceccarini [1], Stefania Giannarelli [1], Virginia Vinciguerra [1,2] and Monica Bertoldo [3,4]

1. Department of Chemistry and Industrial Chemistry, University of Pisa, 56124 Pisa, Italy; andrea.corti@unipi.it (A.C.); jacopo.lanasa@for.unipi.it (J.L.N.); francesca.modugno@unipi.it (F.M.); alessio.ceccarini@unipi.it (A.C.); stefania.giannarelli@unipi.it (S.G.); virginia.vinciguerra@dcci.unipi.it (V.V.)
2. CISUP—Center for the Integration of Scientific Instruments of the University of Pisa, 56124 Pisa, Italy
3. Department of Chemical, Pharmaceutical and Agricultural Sciences, University of Ferrara, via L. Borsari, 45121 Ferrara, Italy; brtmnc@unife.it
4. Institute of Organic Synthesis and Photoreactivity, National Research Council of Italy (ISOF-CNR), via P. Gobetti 101, 40129 Bologna, Italy
* Correspondence: valter.castelvetro@unipi.it; Tel.: +39-0502219256

Abstract: Microplastics (MPs) quantification in benthic marine sediments is typically performed by time-consuming and moderately accurate mechanical separation and microscopy detection. In this paper, we describe the results of our innovative Polymer Identification and Specific Analysis (PISA) of microplastic total mass, previously tested on either less complex sandy beach sediment or less demanding (because of the high MPs content) wastewater treatment plant sludges, applied to the analysis of benthic sediments from a sublittoral area north-west of Leghorn (Tuscany, Italy). Samples were collected from two shallow sites characterized by coarse debris in a mixed seabed of *Posidonia oceanica*, and by a very fine silty-organogenic sediment, respectively. After sieving at <2 mm the sediment was sequentially extracted with selective organic solvents and the two polymer classes polystyrene (PS) and polyolefins (PE and PP) were quantified by pyrolysis-gas chromatography-mass spectrometry (Pyr-GC/MS). A contamination in the 8–65 ppm range by PS could be accurately detected. Acid hydrolysis on the extracted residue to achieve total depolymerization of all natural and synthetic polyamides, tagging of all aminated species in the hydrolysate with a fluorophore, and reversed-phase high performance liquid chromatography (HPLC) (RP-HPLC) analysis, allowed the quantification within the 137–1523 ppm range of the individual mass of contaminating nylon 6 and nylon 6,6, based on the detected amounts of the respective monomeric amines 6-aminohexanoic acid (AHA) and hexamethylenediamine (HMDA). Finally, alkaline hydrolysis of the residue from acid hydrolysis followed by RP-HPLC analysis of the purified hydrolysate showed contamination by polyethylene terephthalate (PET) in the 12.1–2.7 ppm range, based on the content of its comonomer, terephthalic acid.

Keywords: microplastics; marine sediment; pet; nylon 6; nylon 6,6; reversed-phase HPLC; polyolefin; polystyrene; Pyr-GC/MS; polymer degradation

Citation: Castelvetro, V.; Corti, A.; La Nasa, J.; Modugno, F.; Ceccarini, A.; Giannarelli, S.; Vinciguerra, V.; Bertoldo, M. Polymer Identification and Specific Analysis (PISA) of Microplastic Total Mass in Sediments of the Protected Marine Area of the Meloria Shoals. *Polymers* **2021**, *13*, 796. https://doi.org/10.3390/polym13050796

Academic Editor: Victor Tcherdyntsev

Received: 7 February 2021
Accepted: 2 March 2021
Published: 5 March 2021

Publisher's Note: MDPI stays neutral with regard to jurisdictional claims in published maps and institutional affiliations.

Copyright: © 2021 by the authors. Licensee MDPI, Basel, Switzerland. This article is an open access article distributed under the terms and conditions of the Creative Commons Attribution (CC BY) license (https://creativecommons.org/licenses/by/4.0/).

1. Introduction

Plastic microparticles, commonly referred to as microplastics (MPs), either deriving from the environmental degradation of larger plastic waste items [1–3] or directly released as primary microparticles (microbeads, textile microfibers) in wastewaters, are a class of pollutants detected in virtually all natural environments, from oceans to inland waters [4], soils and even as airborne material [5], reaching such remote areas as the Arctic and Antarctica [6,7]. The ubiquitous presence of MPs, and likely so also of their ultimate products of further degradation into sub-micrometer sized particles (nanoplastics) [8], along with incipient evidence of their adverse interaction with living organisms [9], has stimulated increasing research efforts aimed at understanding their transport, distribution and fate [10–12]. Due to their small size microplastic can be ingested by various organisms at all trophic levels, and increasing scientific evidence highlights the possibility of their transfer into animal tissue and up the food chain reaching humans [13,14].

The most common synthetic polymers in plastic waste are polyolefins (polypropylene, PP, high density polyethylene, HDPE, and low-density polyethylene, LDPE) and polystyrene (PS), widely used in packaging and single-use disposable items such as tableware; polyester (mainly polyethylene terephthalate, PET, used for beverage bottles, packaging and as staple textile fiber) and polyamides (often referred to according to the tradename nylons) represent an additional significant fraction of MPs pollution. In the case of polyolefins, the environmental degradation processes are mainly ascribed to photo-oxidation, resulting in oxygen pickup due to free radical reactions with cascade effects eventually leading to polymer chain fragmentation and insertion of oxidized functional groups (carbonyls, carboxyl, hydroxyl, etc.) [15]. Such chemical transformations bear several consequences: (i) the initially high molecular weight is reduced and the polymeric material becomes more brittle, promoting progressive fragmentation into increasingly smaller particles; (ii) its density and hydrophilicity increase along with surface polarity and reactivity, enhancing adsorption/absorption of low molecular weight organic (including toxic polycyclic aromatic hydrocarbons, PAHs, and polychlorinated biphenyls, PCBs) and inorganic (heavy metals) environmental pollutants; (iii) increased wettability and specific surface area facilitate biofouling and adhesion inorganic particulate, all of the above promoting sinking down the water column and deposition in both shore and benthic sediments [16–19]. It has been estimated that less than 1% of the 5–12 million tons per year of plastics entering the oceans stays afloat for a long time, the remaining fraction reaching the seabed either in a very short time (this is the case of larger items of higher density plastics such as e.g., PET or low-density polymers with inorganic fillers) or over longer periods regardless of the initial density because of the abovementioned degradation and fouling phenomena [12,20,21].

Here we report the results of the application of our recently developed analytical protocol to the quantitative determination of the total mass content of a well-defined set of microplastics [22], hereafter Polymer Identification and Specific Analysis (PISA), in benthic marine sediments. The PISA protocol provides accurate quantitative (total mass of the contaminating MPs in the sediment sample, with separate quantification for each polymer type, as specified below) and qualitative (type of polymer) information with sensitivities orders of magnitude higher than those attainable with the general methodology most commonly adopted so far by researchers worldwide. The methods described in the literature, based on MPs separation from the sediment by flotation in a high density saline solution (NaCl, NaI, Na tungstate, etc.) followed by quantification and characterization typically by means of optical microscopy and micro-spectroscopy techniques [23], may suffer from inaccuracy due to the underestimation caused by the missed detection of MPs below the mesh size of the filtering device, and to the overestimation caused by residual biogenic and inorganic contaminating material. In particular, the PISA protocol allows quantification of the total mass of MPs, regardless of their size and morphology, that are constituted by the following polymers: polyolefins, PS, PET, and the two polyamides nylon 6 (polycaprolactame, the homopolymer of AHA) and nylon 6,6 (copolymer of HMDA, with adipic acid) [24]. These are also the main commodity polymers and, not incidentally, are also considered to be the main macro- and microplastic marine pollutants.

Although techniques similar to those comprised in the PISA protocol have been described, they do not include the complete set of relevant polymers; in particular, pressurized solvent extraction [25] is quite effective for polyolefins and other polymers soluble in common solvents (mainly PS and other vinyl polymers) but misses PET and polyamides that are the most abundant microplastics from synthetic textile fibers, while a previously described depolymerization of PET followed by high performance liquid chromatography

(HPLC) coupled with mass spectrometry [26] does not include polyamides. On the other hand, a thermoanalytical method based on the accurate quantification of the total carbon content from synthetic polymers recently proposed by J. Lin et al. [27], besides losing the information on size and shape as in the PISA protocol, does not allow specification of the type of polymeric materials in the contaminating MPs and only provides an estimation of the total amount of microplastics due to the different fractional carbon content in each polymer type.

The benthic sediment samples analyzed in the present work were collected in two close locations of the sublittoral southern Ligurian Sea close to the harbor of Leghorn and the estuary of the Arno river, Italy, and in particular within the shoals of the Meloria protected marine area and in nearby shallow coastal waters, respectively. The sediments were sieved at 2 mm mesh and then submitted to a sequence of fractional solvent extractions with refluxing dichloromethane (DCM) and xylene (Xy) as selective solvents for PS and polyolefins, respectively [28], followed by sequential hydrolytic depolymerization of polyamides, under acidic conditions, and of PET, under alkaline conditions. Although the extracted polyolefins and PS are quantified by gravimetry and pyrolysis-GC/MS, the total content of nylon 6, nylon 6,6, and PET are calculated from the quantitative analysis, by reversed-phase HPLC, of their respective monomers: the two amines AHA and HMDA after tagging with a fluorophore [29], and terephthalic acid (TPA) [30]. Differently from the previously reported examples in which this procedure (or part of it) had been tested on less complex sandy beach sediment or less demanding (because of the high MPs content) wastewater treatment plant sludges, this is the first report on the PISA procedure for the detection and quantification of MPs in benthic sediments.

2. Materials and Methods

2.1. Sediment Sampling

Benthic (bottom) marine sediment samples were collected in four sites of relatively shallow waters of the continental shelf in the Ligurian Sea along the northern coastline of Tuscany, Italy (Table 1). The sampling was performed on 3 July 2018, using a single corer 10 cm in diameter. After collection, the top ~5 cm of the sediment was placed in glass flasks with metal lid and then air-dried in a laminar flow hood in the lab and stored in a fridge at 2 °C. The subsequent analyses were performed in second half of 2020. Care was taken as to avoid contamination from airborne and other environmental MPs; for this purpose, all glassware was rinsed with the given solvent (previously filtered on 0.45 μm pore size membrane) prior to use; all open surfaces of solutions and solid samples and extracts were kept covered with aluminum foils throughout the various manipulations except during the actual operations and transfer; personal protective equipment included cotton protective coats.

Table 1. Sample acronyms and relevant sampling sites coordinates and depth.

Sample	Acronym	Geolocalization		Depth (m)
Meloria 1	MEL1	43°32′50.0″ N 10°13′08.2″ E	43.547219 lat. 10.218944 lon.	3
Meloria 2	MEL2	43°33′1,02″ N 10°13′4,03″ E	43.5502778 lat. 10.2177778 lon.	4
Calambrone 1	CAL1	43°35′5,21″ N 10°17′2,34″ E	43.5847778 lat. 10.2839722 lon.	20
Tirrenia-Calambrone 2	CAL2	43°36′9,67″ N 10°16′7,80″ E	43.6026944 lat. 10.2688333 lon.	17

The sedimentologic features are representative of two distinct benthic zones: the MEL samples (as identified in Table 1) were collected in the shoals of the marine Protected Area "Secche della Meloria", about 3 miles west of the harbor of Leghorn, with the bottom sediments consisting mainly of fragmented organogenic shells and carbonate sand in a seabed partially covered by meadows of *Posidonia oceanica*, a seagrass endemic to the Mediterranean Sea also known as Mediterranean tapeweed; the CAL samples were collected northeast of the MEL area, in the upper shore platform about 0.5–1 mile off of the sandy beaches of Calambrone, characterized by intensive seasonal touristic presence, and 5 miles south of the Arno river estuary, resulting in the bottom sediments consisting of very fine sandy to silty material (Figure 1). The sampling sites were chosen as they are influenced by the currents carrying the estuarine waters of the Arno river, the main river in the Tuscany region collecting wastewaters from about 2,200,000 inhabitants and from several industrial districts (tanning and textile, among others), and by the commercial and touristic harbor of Leghorn. Moreover, the benthic sediments of the Meloria shoals are quite peculiar as they are strongly affected by surface currents while being possibly too far from the coastline to collect high density debris carried by the riverine freshwaters.

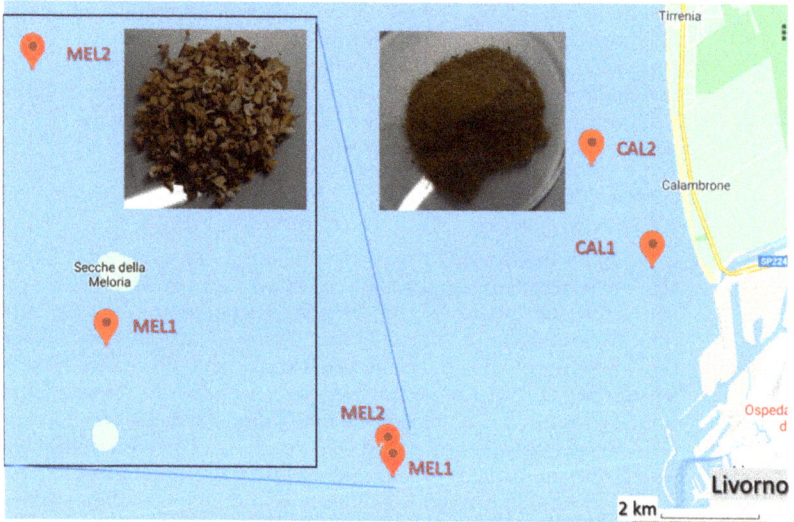

Figure 1. Sampling sites and morphology of the two types of sediment sample: CAL and MEL.

2.2. Chemicals

Dichloromethane (DCM, 99.9%, stabilized with amylene, Romil-SpS, Romil Ltd., Cambridge, UK), xylene (Xy, 98.5%, Sigma-Aldrich, Merck Life Science S.r.l., Milano, Italy, hereafter Sigma-Aldrich), methanol (99.9%, Sigma-Aldrich), acetic acid (99.85% Sigma-Aldrich), sulfuric acid (95–98%, Sigma-Aldrich), hydrogen peroxide (30% w/v, Panreac, Nova Chimica Srl, Cinisello Balsamo, Italy), 6 N aqueous hydrochloric acid (prepared from 37% HCl, Sigma-Aldrich), 1.9 N aqueous sodium hydroxide (from NaOH pellets, 98.0%, Sigma-Aldrich), hexadecyl-tributyl-phosphonium bromide (TBHDPB, 97%, Sigma-Aldrich), HPLC-grade water (Sigma-Aldrich), and reversed-phase Solid Phase Extraction (SPE) cartridges (Chromabond® C18ec loaded with 500 mg stationary phase, Macherey-Nagel GmbH & Co., Düren, Germany) were used for sediment extractions, extracts purifications, and for the acid and alkaline depolymerizations of the hydrolysates. Dansyl chloride (DNS-Cl, 96%, Alfa Aesar Thermo Fisher (Kandel) GmbH, Kandel, Germany), *n*-butyl amine (99.5%, Honeywell Fluka Chemicals, Fisher Scientific Italia, 20053 Rodano, Italy, hereafter Fluka) and potassium carbonate (K_2CO_3, Carlo Erba Reagents S.r.l., 20010

Cornaredo, Italy) were used in the dansylation of AHA and HMDA amino monomers. Chloroform (HPLC-grade stabilized with ethanol, Sigma-Aldrich) was used as mobile phase in size exclusion chromatography (SEC) analysis. Acetonitrile (HPLC-grade, ≥99.9%, Sigma-Aldrich), HPLC-grade water (Sigma-Aldrich), triethyl amine (≥99.9%, Fluka), acetic acid (99.85%, Sigma-Aldrich), methanol (99.9%, Sigma-Aldrich), and phosphoric acid (Sigma-Aldrich) were used in the preparation of the HPLC eluents for determination of dansylated amine monomers in the acid hydrolysates for the quantitative analysis of terephthalic acid in the alkaline hydrolysates.

2.3. Analytical Techniques

SEC analyses were performed with an instrument consisting of a Jasco (Jasco Europe Srl, Cremella, LC, Italy) PU-2089 Plus four-channel pump, a PLgel pre-column packed with polystyrene/divinylbenzene and two in series PLgel MIXED-D columns (Agilent Technologies Italia S.p.A., 20063 Cernusco sul Naviglio, Italy) placed in a Jasco CO_2063 column oven, a Jasco RI 2031 Plus refractive index detector, and a Jasco UV-2077 Plus multi-channel UV spectrometer. Chloroform was used as the eluent at 1 mL/min flow rate.

Pyrolysis-Gas Chromatography/Mass spectrometry (Py-GC/MS). Analyses were performed using a multi-shot pyrolyzer EGA/PY-3030D (Frontier Laboratories Ltd., Koriyama 963-8862, Japan) coupled with a 6890N gas chromatographic system with a split/splitless injection port and combined with a 5973-mass selective single quadrupole mass spectrometer (Agilent Technologies, Inc., Santa Clara, CA 95051, USA). For the analysis of the extracts, 100–150 µL of solution were placed in stainless steel cups together with 2.0 µL of a solution of dibutyl phthalate-3,4,5,6-d_4 (500 ppm), and dried under nitrogen steam prior to the analyses [31]. The temperatures used for the double shot pyrolysis were 350 °C and 600 °C while the interface was set at 280 °C. The GC injection port temperature was 280 °C. The GC injection was operated in split mode with a split ratio of 1:10. For the analysis of particles, the pyrolysis was performed in a single shot at 600 °C with the interface set at 280 °C and split ratio 1:10 [32]. The chromatographic and mass spectrometric conditions were as follows: 5 min isotherm at 50 °C, heating up to 180 °C at 12 °C/min, 2 min isotherm, heating up to 300 °C at 8 °C/min, and 20 min isotherm; 1.2 mL/min He (99.9995%) carrier gas; GC/MS interface temperature 280 °C and MS electron ionization voltage 70 eV. Perfluorotributylamine (PFTBA) was used for mass spectrometer tuning. MSD ChemStation D.02.00.275 software (Agilent Technologies, Inc., Santa Clara, CA 95051, USA) was used for data analysis and the peak assignment was based on a comparison with libraries of mass spectra (NIST 8.0).

Infrared spectra in the mid-IR region (700–4000 cm^{-1}) were recorded with a Perkin Elmer Spectrum Autoimage System microscope equipped with an Attenuated Total Reflectance (ATR) module with germanium crystal; each spectrum is the result of 64 scans accumulation at 4 cm^{-1} spectral resolution. The lateral spatial resolution corresponds to the contact area with the germanium crystal tip (30–40 micron).

Two instrumental setups were used for HPLC. Quantitative determination of TPA was performed using a Jasco PU-1580 isocratic pump connected with a Jones-Genesis Aq 120 reversed-phase column (150 mm × 4.6 mm, 4 µm particle size) operating at room temperature and a Jasco 1575 UV-Vis detector (UVD) set at 242 nm wavelength. Analyses were carried out on 20 µL of the solutions at 0.8 mL/min flow rate of an isocratic 30/70 vol/vol methanol/HPLC-grade water (acidulated with 0.1 wt.% phosphoric acid) eluent. The DNS-Cl derivatives of AHA and HMDA were analyzed with an Agilent 1260 Infinity Binary LC instrument equipped with pre-column, a reversed-phase Phenomenex-Aqua C18 column (250 mm × 4.6 mm, 5 µm particle size) and diode array (DAD VL + 1260/G1315C, set at 335 nm wavelength) plus fluorescence (FLD 1260/G1321B, set at 335/522 nm excitation/emission wavelengths) double detector. Elution was performed at 1.0 mL/min flow rate in gradient mode by combining an aqueous solution of 2.5% acetic acid and 0.83% triethylamine (phase A) with acetonitrile (phase B) according to the program reported in Table 2.

Table 2. Elution program adopted for the analysis of dansylated amine derivatives.

Elution Time (min)	Mobile Phase (A) %	Mobile Phase (B) %	Elution Mode
0 → 20	60	40	isocratic
20 → 25	60 → 30	40 → 70	gradient
25 → 35	30	70	isocratic
35 → 37	30 → 60	70 → 40	gradient
37 → 50	60	40	isocratic

2.4. Synthetyc Polymer Recovery by Sequential Extractions with Selective Solvents

Each sediment sample was submitted to a first extraction with refluxing dichloromethane (DCM), targeting the solubilization of polystyrene, followed by a second step in refluxing xylene to solubilize semicrystalline polyolefins. For this purpose, approximately 100 g of sediment that had been previously sieved at 2 mm and air-dried to nearly constant weight in equilibrium with the atmospheric lab conditions, was loaded into a cellulose thimble that was placed into a kumagawa apparatus and the extraction was performed for 2 h with 250 mL refluxing DCM. Before each extraction, the apparatus was conditioned by refluxing 100 mL DCM for 3 h to remove any contaminant. The DCM extract was reduced to 1–2 mL in a rotatory evaporator, then transferred into a 5 mL glass vial conditioned to constant weight in an oven at 60 °C and weighed. The residue from the DCM extraction was then further extracted in the same apparatus with 250 mL refluxing xylene. The obtained xylene solution was transferred into a two-necked flask fitted with distillation head and condenser, reduced to 5 mL by distilling off the excess solvent, and then added with 20 mL of 1.6 M KOH in methanol to induce the precipitation of polyolefins; the obtained precipitate, mainly consisting of PP, LDPE, HDPE, was then collected by vacuum filtration through a 0.22 µm Durapore PVDF membrane (without PP backing).

2.5. Hydrolytic Depolymerization Procedures

The total content of nylon 6, nylon 6,6, and PET contaminating microparticles (mainly found as microfibers) in the sediment samples was calculated from the content of the corresponding monomers 6-aminocaproic acid (AHA), 1,6-hexanediamine (or hexamethylenediamine, HMDA), and terephthalic acid (TPA), respectively, in the hydrolysates obtained from the acid and alkaline depolymerizations for the polyamides and the polyester, respectively. As nylons and PET are neither soluble in DCM nor in xylene, the solid residue from the organic solvent extraction was analyzed for their quantification. In particular, the dried residue of the organic solvent extraction of each sample was transferred into a 100 mL round-bottomed flask equipped with a reflux condenser and magnetic stirring bar. The selective determination of nylons and PET was based on the sequential selective hydrolysis of all aliphatic polyamides in acidic conditions, and of PET in alkaline conditions.

After addition of approximately 80 mL 6 N HCl the stirred mixture was heated to the reflux temperature of about at 105 °C for 24 h. At the end of the hydrolysis, the reaction mixture was vacuum-filtered on a 0.22 µm PVDF membrane to separate the solid residue from the acid solution. The filter membrane with the solid residue was carefully rinsed with small amounts of HPLC-grade water for the subsequent treatments, while the acid solution was transferred in a 100 mL volumetric flask and taken to volume with 6 N HCl. A given volume (5 mL) of the obtained solution was weighed and neutralized to pH 6.5–7.5 with 5 N NaOH. To enable a highly sensitive quantification of the amino-monomers AHA and HMDA, the solutions were treated with 5-dimethylaminonaphtalene-1-sulfonyl chloride (dansyl chloride, DNS-Cl), a derivatizing fluorophore commonly used in protein sequencing (Figure 2).

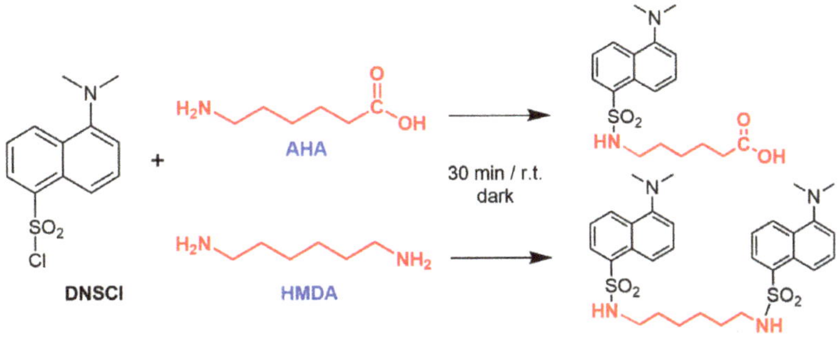

Figure 2. Dansylation of the amino-monomers AHA and HMDA from the depolymerization of nylon 6 and nylon 6,6.

For the derivatization of AHA and HMDA, 1 mL of the neutralized product of acid hydrolysis was loaded in a 5 mL glass vial, added with 1.0 mL aqueous K_2CO_3 solution (80 g/L), to favor the precipitation of calcium carbonate if present in the neutralized solution. After allowing the obtained mixture to settle, 1 mL was taken and added with a further 1.0 mL aqueous K_2CO_3 solution and 1.0 mL of a 5 g/L solution of DNS-Cl in acetone (18.5 µmol). After 30 min stirring at room temperature in the dark, an excess of *n*-butyl amine (5.0 µL, 51 µmol) was added to quantitatively convert the unreacted DNS-Cl. The solution containing the derivatized amines (including those from the hydrolysis of both natural and synthetic polyamides) was then transferred into a 10 mL volumetric flask and taken to volume with a 1:1 (*v/v*) water/acetone mixture before HPLC analysis.

For the determination of the PET content, the solid residues collected at the end of the acid hydrolysis were treated under alkaline hydrolytic conditions to achieve the complete PET depolymerization. For this purpose, each residue was rinsed with deionized water on the same PVDF membrane used for filtration, then transferred into a 100 mL round-bottomed flask equipped with a reflux condenser and magnetic stirring bar, added with 40 mL 1.9 N NaOH and TBHDPB as a phase transfer catalyst, then the mixture was stirred at 85 °C for 6 h. The final solution was vacuum-filtered on a 0.22 µm PVDF membrane, then transferred into 50 mL volumetric flasks and taken to volume with 1.9 N NaOH. For the removal of most of the residual biogenic contaminants that might interfere with the subsequent purification by elution through a SPE cartridge (e.g., by saturating the adsorption capacity of the cartridge), 1 mL of hydrolysate was weighed at 0.1 mg accuracy, transferred into a 10 mL glass vial with 1–2 mL of 30 vol% H_2O_2 until complete discoloration and/or end of visible bubble formation, then added with 1 mL 1.9 M H_2SO_4. The resulting acidic solution was further purified to remove potential interferents before HPLC analysis; for this purpose, the pre-treated hydrolysate was eluted through a reversed-phase SPE cartridge, the adsorbate was then desorbed with 0.8 mL MeOH and the recovered roughly 0.8 mL solution in methanol was weighed at 0.1 mg accuracy. Finally, 0.5 mL of the solution was taken up with a micropipette, placed in a vial and weighed again at 0.1 mg accuracy, then added with 0.75 mL aqueous CH_3COOH (1 wt.% in HPLC-grade water) to obtain a 40/60 vol% methanol/water mixture.

The amounts of nylon 6, nylon 6,6 and PET in each sample (given in ppm, or mg polymer/kg dry sludge) was calculated from the corresponding monomer concentration C_{AHA}, C_{HMDA}, and C TPA (in ppm) as determined by HPLC, based on the calibrated response of both UV and fluorescence detectors (see Figure 3), according to Equations (1)–(3):

$$\text{nylon 6 (ppm)} = C_{AHA} \cdot \frac{MW_{PA6}}{MW_{AHA}} \quad (1)$$

$$\text{nylon 6,6 (ppm)} = C_{HMDA} \cdot \frac{MW_{PA6,6}}{MW_{HMDA}} \quad (2)$$

$$\text{PET (ppm)} = C_{TPA} \cdot \frac{MW_{PET}}{MW_{TPA}} \quad (3)$$

where MW_{PA6} = 113.16 g/mol, $MW_{PA6,6}$ = 226.32 g/mol, and MW_{PET} = 192.2 g/mol, are the molecular weights of the repeating units in the corresponding polymer (Figure 4), and MW_{AHA} = 131.17 g/mol, MW_{HMDA} = 116.21 g/mol, and MW_{TPA} = 166.13 g/mol those of the analytes.

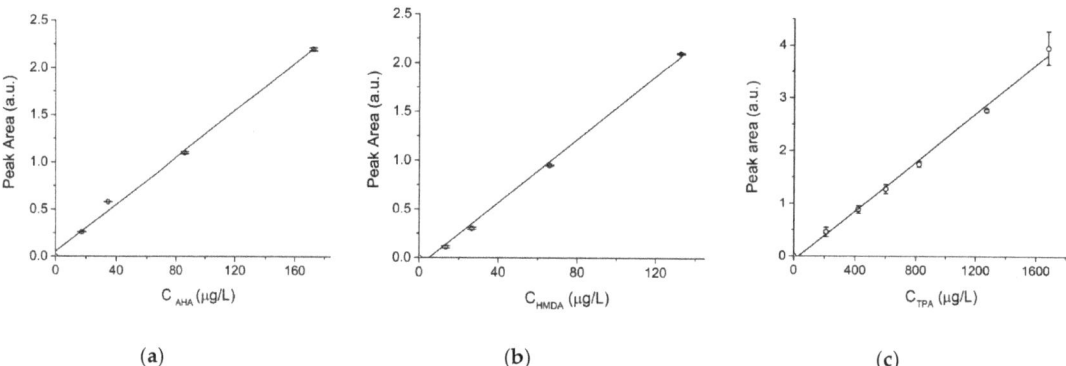

Figure 3. Linear regressions of the calibration dataset for the quantitative determination of the monomeric units: (**a**) AHA, with FLD (fitting parameters: peak area = 0.01246 × C_{AHA} + 0.05411; R2 = 0.99424); (**b**) HMDA, with FLD (fitting parameters: peak area = 0.01622 × C − 0.08077; R2 = 0.99466); (**c**) TPA, with UV detector (fitting parameters: peak area = 231955.9 × C − 7236.6; R2 = 0.99532). Each calibration was performed by running the measurements in triplicate.

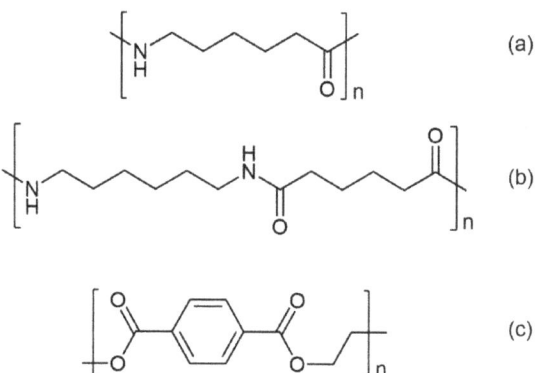

Figure 4. Polymer repeating units: (**a**) nylon 6 (polycaprolactame, the homopolymer of AHA); (**b**) nylon 6,6 (copolymer of adipic acid and HMDA); (**c**) PET (polyethylene terephthalate).

2.6. Calibrations

Calibration of the response of the pyrolysis-gas chromatography-mass spectrometry (Py-GC/MS) system used for PS quantification in the DCM extract was performed by analyzing a set of PS solutions prepared starting from a 100 ppm solution of PS in DCM, then weighed amounts of calibration solution with a total PS content in the 20−150 µg range were loaded in the crucible and dried. The quantification was based on the response (GC/MS peak) for the fragment corresponding to the styrene dimer [32], which gave a linear regression of the experimental calibration (r^2 = 0.9998), from which the following limit of detection (LOD) and limit of quantification (LOQ) were calculated: LOD = 3·SD/m = 0.10 µg; LOQ = 10·SD/m = 0.35 µg (with SD standard deviation of the blank areas, and m slope of the calibration curve).

For the HPLC analyses of the dansylated amines the FLD detector response was calibrated against the concentration of dansylated AHA by recording a 4-point calibration using solutions in the 17.25–172.5 µg/L range plus a blank sample, all in triplicate; the same procedure was followed for dansylated HMDA, in the 13.25–132.5 µg/L range plus the blank sample. From the linear regression (Figure 3a,b) obtained for dansylated AHA (A = $1.246 \cdot 10^{-2} \cdot C_{AHA}$ + $5.4 \cdot 10^{-2}$; r^2 = 0.99784) and dansylated HMDA (A = $1.682 \cdot 10^{-2} \cdot C_{HMDA}$ − $8.195 \cdot 10^{-2}$; r^2 = 0.99791) the following values were calculated: LOD_{AHA} = 0.903 µg/L; LOQ_{AHA}: 3.910 µg/L; LOD_{HMDA} = 0.301 µg/L; LOQ_{HMDA}: 0.758 µg/L. The LOD and LOQ values are given as:

$$LOD = \frac{standard\ deviation\ of\ most\ diluited\ solution}{slope\ of\ linear\ regression} \cdot 3; \quad (4)$$

$$LOQ = \frac{standard\ deviation\ of\ most\ diluited\ solution}{slope\ of\ linear\ regression} \cdot 10. \quad (5)$$

For the HPLC quantification of TPA a linear calibration of the UV detector response was obtained by recording a 6-point calibration based on standard TPA solutions in 2N NaOH in the 0.21–1.68 mg/L range plus a blank, all in triplicate [23]. From the linear regression (Figure 3c, A = $2.32 \cdot 10^5 \cdot C_{TPA}$ − 7237; r^2 > 0.995) the following values were calculated: LOD = 0.117 mg/L; LOQ: 0.391 mg/L. (calculated in this case as the ratio between the concentration, or the calibrated peak area, and the signal-to-noise ratio, times 3 for LOD and times 10 for LOQ; the blank sample gave no detectable peak).

3. Results

The overall procedure for the determination of the total mass of individual polymer types that are present as microparticles and fragments in the sediment involves a preliminary step of sieving to recover the fraction below 2 mm in size and air drying, followed by a first sequence of extractions with boiling solvents that are selective for the hydrocarbon polymers. In particular, extraction with DCM (boiling point b.p. = 39.6 °C) allows recovery of the amorphous polystyrene, along with most low-molecular-weight (MW) organic compounds (both biogenic, such as fats, and synthetic, such as plasticizers and surfactants) and the oligomeric fraction deriving from the extensive photo- and thermal oxidation of polyolefins (PE, PP, and olefin copolymers). In addition, most vinyl polymers such as e.g., acrylics, polyvinyl chloride (PVC) and polyvinyl acetate may be co-extracted in boiling DCM, but their presence as microplastic contaminants in marine sediments is likely to be negligible because they are not (or no longer in the case of PVC) commonly used in the disposable items and packaging materials that are by far the main contributors to the plastic waste reaching the marine environment. The second extraction of the residue from the DCM extraction is performed in boiling xylene (b.p. = 139 °C), to recover the less oxidized and high-MW polyolefins, possibly along with some proteins that may be co-extracted. The extractable fraction may then be further purified to remove the biogenic fraction, and analyzed by one or more techniques for the quantification of the total mass of each polymer type.

The subsequent steps, consisting of a sequence of hydrolytic treatments performed under acid and then alkaline conditions, are performed under optimized conditions to selectively and sequentially achieve the complete depolymerization of all aliphatic polyamides (both synthetic and natural) and all polyesters, respectively. The resulting hydrolysates may then be submitted to further purification (different environmental matrices may require different purification procedures) before performing reversed-phase HPLC analysis that allows the accurate and sensitive quantification of the monomers and to calculate the corresponding amount of the original polymer. In the case of the polyamides, an additional tagging of the amino-monomers with a fluorophore is performed prior to the HPLC analysis to increase of orders of magnitude the sensitivity of the measurement. The overall procedure had been previously validated on different matrices (marine beach and underwater lakebed sediments [28,30], wastewater treatment plant sludges [29]) by performing microplastic spiking and recovery experiments.

3.1. MPs Fractionation by Polymer Type through Selective Solvent Extraction

3.1.1. Polystyrene and Highly Degraded Hydrocarbon Polymers

The DCM extracts are expected to contain not only PS, but also highly oxidized and degraded polyolefin oligomers and other vinyl polymers less frequently found as microplastic pollutants (e.g., polyacrylates, PVC, etc.), in addition to biogenic low-MW species. The total amounts of DCM extractable fraction in the four samples are reported in Table 3. Further extractions with refluxing xylene to collect the DCM-insoluble, less degraded polyolefin fraction gave in most cases very small amounts of dry matter that could be neither weighed with sufficient accuracy nor further purified, and were therefore not further analyzed; the only exception was the xylene extract from CAL2, from which a sizable solid particulate could be recovered (see Section 3.1.2).

Table 3. Extractable fraction in DCM from the sediment samples.

Sediment Sample	Extracted Sediment (g)	Extractables (mg)	Total Extract [1] (ppm)	PS Content [2] (ppm)
MEL1	82.85	6.2	75	8
MEL2	107.96	12.8	119	11
CAL1	100.91	9.1	90	65
CAL2	112.57	10.7	95	16

[1] Total concentration expressed as mg of dry extractable matter per kg dry sediment. [2] Determined by double shot-Py-GC/MS measurements, from the styrene dimer peak and the corresponding instrumental calibration.

The dried DCM extracts were picked up with chloroform to about 5 mg/mL and the obtained solutions were microfiltered (to remove any contamination by inorganic particles) and analyzed by SEC. The chromatographic profiles obtained with UV detectors set at 260 nm and 340 nm always show the presence of a main very broad and structured peak at retention times r.t. > 15 min due to low-MW polymers and oligomers along with other low MW species; an additional weak peak roughly centered at r.t. ≈ 12.5 min could be detected for samples MEL1, MEL2, and CAL1, corresponding to higher MW polymers (Figure 5 and Table 4).

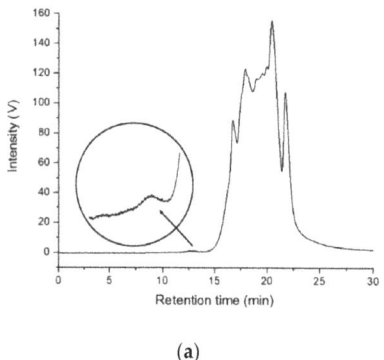

(a)

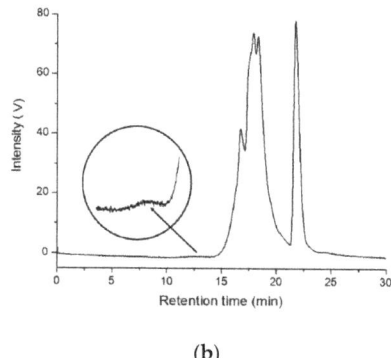

(b)

Figure 5. Representative examples of the SEC traces (the curves displayed are those recorded with UV detector set at 260 nm) for the DCM extractable fractions of: (a) MEL1; (b) MEL2.

Table 4. SEC analysis of the DCM extractable fractions in the sediment samples.

Sample	Retention Time [1] (min)	$\overline{M}_n$ (g·mol^{-1})	$\overline{M}_w$ (g·mol^{-1})	PDI [2]
MEL1	12.83	35,883	42,245	1.18
	20.41	92	367	4.0
MEL2	12.22	36,855	48,336	1.31
	17.89	254	585	2.30
CAL1	12.71	29,653	46,250	1.56
	17.88	270	529	1.96
CAL2	n.d. [3]	n.a. [3]	n.a. [3]	n.a. [3]
	18.04	196	475	2.43

[1] peak value reported here only as an aid for better visualization; [2] Polydispersity Index PDI = $\overline{M}_w / \overline{M}_n$; [3] n.d.= not detectable (below the limit of detection, LOD); n.a. = not applicable.

The SEC-UV detector response for the high MW fractions in MEL1, MEL2, and CAL1 is characterized by a strong absorption at 260 nm that becomes negligible at 340 nm, as one would expect from polystyrene [33], but differently from most nonaromatic polymers.

The overall PS content in the sediment samples (including the low MW oligomeric fraction) was determined by double shot-Py-GC/MS analysis performed on the DCM extracts, according to a calibration based on the MS count corresponding to the PS dimer fragment; the results are reported in Table 3. The double shot technique also allows separate detection of different species; in the first shot at lower temperature (here 350 °C) low MW compounds such as hydrocarbon species deriving from highly degraded polyolefins or plasticizers and other common plastics additives are typically observed, along with some polystyrene oligomers, while the presence of other synthetic polymers could be detected in the second shot (here 600 °C). The main species identified in the DCM extracts of the four samples are listed in Table 5, while the Py-GC/MS chromatograms recorded after each shot are shown in Figures 6 and 7.

Table 5. Most abundant species identified by double shot-Py-GC/MS in the DCM extracts.

Acronym	First Shot (350 °C) [1]	Second Shot (600 °C)
MEL1	TBP, DBP	PS, siloxane
MEL2	TBP, DBP, BEHP, fatty acids	PS, PE, branched hydrocarbons, sterols
CAL1	TBP, DBP, fatty acids	PS, sterols, branched hydrocarbons
CAL2	TBP, DBP, DOA	PS, PE, sterols, branched hydrocarbons

[1] TBP = tributyl phosphate; DBP = dibutyl phthalate; BEHP = bis(2-ethylhexyl) phthalate; DOA = diisooctyl adipate.

The Py-GC/MS chromatogram obtained at 350 °C from the MEL1 extract was mainly characterized by the presence of tributyl phosphate (TBP) and dibutyl phthalate (DBP), two nearly ubiquitous environmental pollutants largely used as plasticizers in many applications. The pyrolysis products from the high temperature shot were mainly the typical markers of PS (styrene and its low oligomers) and of polysiloxane.

Similarly, in all the other MEL and CAL extracts the 350 °C shot resulted in the release of various plasticizers such as TBP, DBP, bis(2-ethylhexyl) phthalate (BEHP) and diisooctyl adipate (DOA), along with naturally occurring fatty acids. At 600 °C the pyrolysis markers of PS were always detected, along with those of PE (only in the case of the extracts from MEL2 and CAL2); finally, various sterols of likely natural origin, and branched hydrocarbons possibly originating from the degradation of synthetic surfactants could also be detected.

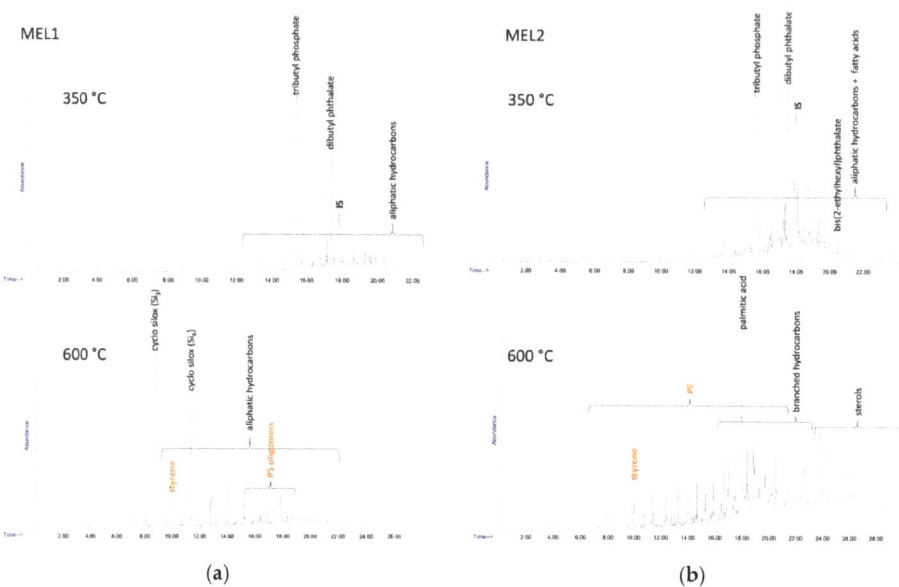

Figure 6. Py-GC-MS chromatograms of sediment DCM extracts: (**a**) MEL1; (**b**) MEL2.

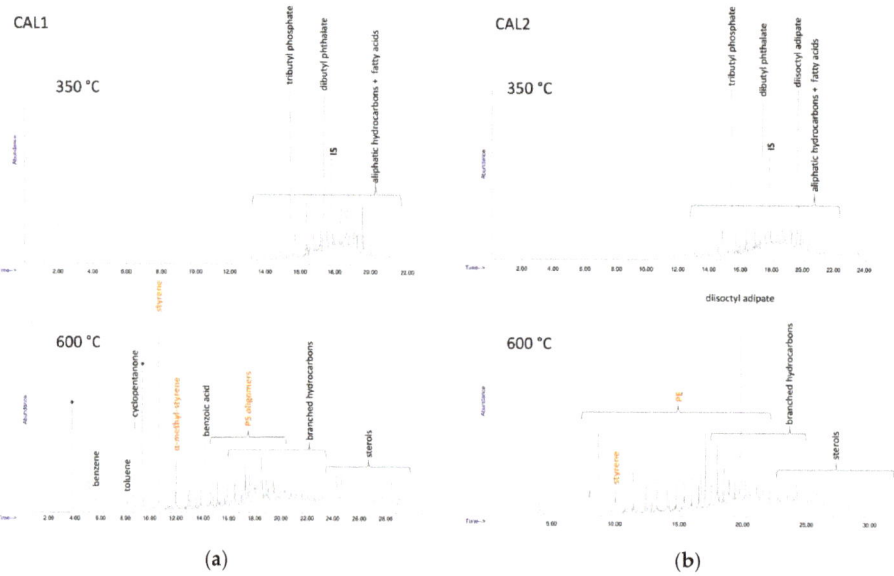

Figure 7. Py-GC-MS chromatograms of sediment DCM extracts: (**a**) CAL1; (**b**) CAL2.

3.1.2. High Molecular Weight Polyethylenes (HDPE, LDPE) and Polypropylene

For the semi-quantitative determination of semi-crystalline polyolefin MPs (polyethylenes and polypropylene) that are insoluble in DCM unless highly oxidized and degraded to low molecular weights, the residues from DCM extraction were further extracted with refluxing xylene. After distilling off most of the xylene from the final extracts they were added with an excess of a solution of KOH in methanol and the precipitate was then collected by filtration, dried, and weighed. A quantifiable number of precipitated particles

(2.9 mg from 81.6 g of sediment) was only recovered from sample CAL2. The micro-ATR FTIR spectrum in Figure 8 clearly indicates that the precipitate mainly consists of oxidized polyethylene (methylene CH asymmetric and symmetric stretching bands at 2917 and 2849 cm^{-1}, respectively; weak methyl CH stretchings at 2960 and 2866 cm^{-1}, methylene bendings at 1452, and 1375 cm^{-1}), with a high oxidation level shown by the intense and broad carbonyl absorption with main peaks at 1710 and 1660 cm^{-1} (isolated and conjugated aldehydes and ketones generated by photo-oxidation and subsequent chain cleavage reactions) and the broad absorption centered at 3400 cm^{-1} from hydroxyl groups. Further absorptions can be ascribed at least partially to polydimethylsiloxane (methyl deformation at 1260 cm^{-1}, symmetric and asymmetric Si–O–Si stretchings at 1088 and 1018 cm^{-1}, and Si–C stretching at 800 cm^{-1}, in addition to a small C–H stretching peak at 2950 cm^{-1}) possibly due to contamination by silicone grease during the lab operations.

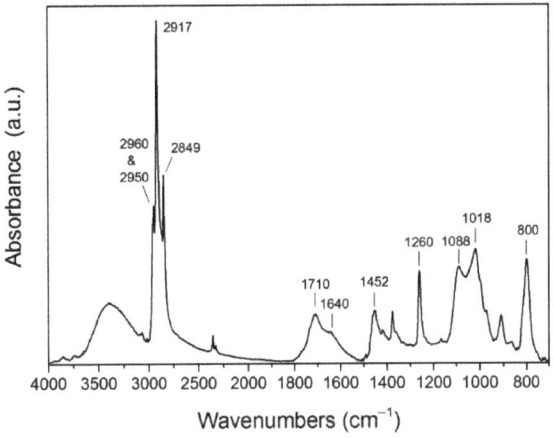

Figure 8. Micro-ATR-FTIR spectrum of the xylene-extractable fraction of CAL2.

3.2. Total Mass Content of Polyamide (nylon 6 and nylon 6,6) and Polyester (PET) Mps by Depolymerization and Quantitative Analysis of the Resulting Comonomers

For the quantification of nylon 6 and nylon 6,6 polyamides the solid residues from the sequential extractions with DCM and Xy were treated with refluxing 6 N HCl to achieve the total depolymerization of both natural and synthetic polyamides. Due to the high carbonate content of the two MEL sediments samples a high volume of HCl had to be slowly added to allow complete evolution of CO_2 upon conversion of the carbonate mineral into the corresponding chlorides. The acid hydrolysate solutions were separated from the residue by filtration on 0.22 μm PVDF membranes, neutralized and treated with the fluorescent tag dansyl chloride (DNS-Cl) before reversed-phase HPLC analysis, as described in detail in Section 2. The solid residues from the acid hydrolysis were then treated with 1.9 N NaOH to achieve quantitative depolymerization of PET MPs, followed by purification of the alkaline hydrolysate and quantification of the TPA content by reversed-phase HPLC analysis, as described in detail in Section 2.

Table 6 reports the detected concentrations of the dansylated AHA and HMDA and of TPA from which the concentration of nylon 6, and nylon 6,6, and PET MPs in the sediment samples (the air-dried sediments were considered to be a starting material) could be calculated.

Table 6. Concentration of PA's monomers and relative polymers in sediment sample.

Acronym	AHA [1] (μg/L)	Nylon 6 [2] (ppm)	HMDA [1] (μg/L)	Nylon 6,6 [2] (ppm)	TPA [1] (mg/L) [1]	PET (ppm) [2]
MEL1	n.d. [3]	n.a. [3]	n.d.	n.a.	0.101	290
MEL2	n.d.	n.a.	n.d.	n.a.	0.0489	137
CAL1	36.6	11.2	8.97	2.7	0.633	1523
CAL2	35.0	12.1	n.d.	n.a.	0.061	174

[1] Concentration of the monomer (or its dansyl derivative in the case of the two amines) in the solution obtained after purification of the corresponding acid (for the two amines) or alkaline (for TPA) hydrolysate. [2] Total concentration in ppm (mg polymer/kg dry sediment) as calculated from the detected amount of the corresponding monomers. [3] n.d.= not detectable (below the limit of detection, LOD); n.a. = not applicable.

3.3. Analysis of Microplastic's Particles Detected on Filter Membrane

The final residue recovered from the filter in last step of the overall procedure was observed under an optical microscope to detect the presence of any microplastic particle resistant to all the extraction and hydrolysis processes. In the case of the MEL1 sample, a few sub-millimeter sized green plastic fragments weighing about 50 μg each could easily be detected in the brown-greyish inorganic residue (Figure 9). The fragments were identified as polytetrafluoroethylene (PTFE) from the presence of a tetrafluoroethene main peak in the Py-GC/MS chromatogram recorded from each individual particle.

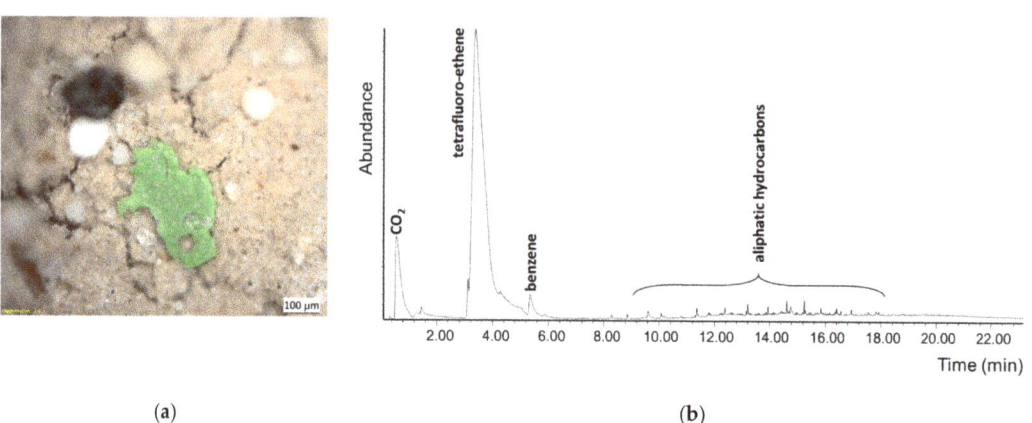

(a) (b)

Figure 9. PTFE microparticles found in the inorganic sediment residue after all extraction and hydrolysis proceduresperformed on the MEL1 sample: (a) micrograph of a green microparticle taken with a stereomicroscope; (b) pyrogram of the microparticle.

4. Discussion

The complete PISA protocol for the separation, purification, and quantification of the total mass contents of PS, polyolefins, PET, nylon 6 and nylon 6,6 MPs in environmental matrices, recently developed and previously applied in the analysis of the contamination level in sandy shore sediments and in wastewater treatment plant sludges, has been successfully applied for the first time to benthic marine sediments, considered to be the final sink of most of the plastic waste inflow in the oceans.

The selection of the polymer types to be investigated, which dictated the design of the overall separation, fractionation, and analysis scheme, was based on the considerations, supported by an increasing number of scientific papers and technical reports, that the most abundant polymer types in benthic marine sediments correspond to those that are also produced globally in larger amounts. These are: polyolefins and polystyrene, largely used in short lifetime applications such as packaging and single use disposable items, and therefore likely to end up as unmanaged plastic waste; the two main synthetic polymer classes used as staple textile fibers or as materials of fishery and aquaculture activities, i.e., is the polyester PET and the two polyamides nylon 6 and nylon 6,6, included among the target polymers because textile fibers released in laundering wastewaters and mismanaged fishing gears are well recognized threats for the marine ecosystems.

The overall procedure, schematically shown in the flowchart of Figure 10, allows the tackling of the two main challenges faced when such polymeric materials end up in sea bottom sediments, either because of their high density or as a result of photo-oxidation and/or biofouling promoting vertical transport down the water column. These are the lengthy (and possibly inaccurate) procedures for the density separation of the MPs from the sediment, and the size threshold for their detection by micro-spectroscopy techniques, a possibly critical issue in particular for the low-density polyolefins and PS. Indeed, the latter hydrocarbon polymers are likely to reach the benthic sediments only once they have undergone significant degradation, which may include fragmentation down to the sub-micrometer size range, well below the detection limit of a few micrometers and up to tens of micrometers typical of the micro-spectroscopy techniques commonly used for MPs in sediments.

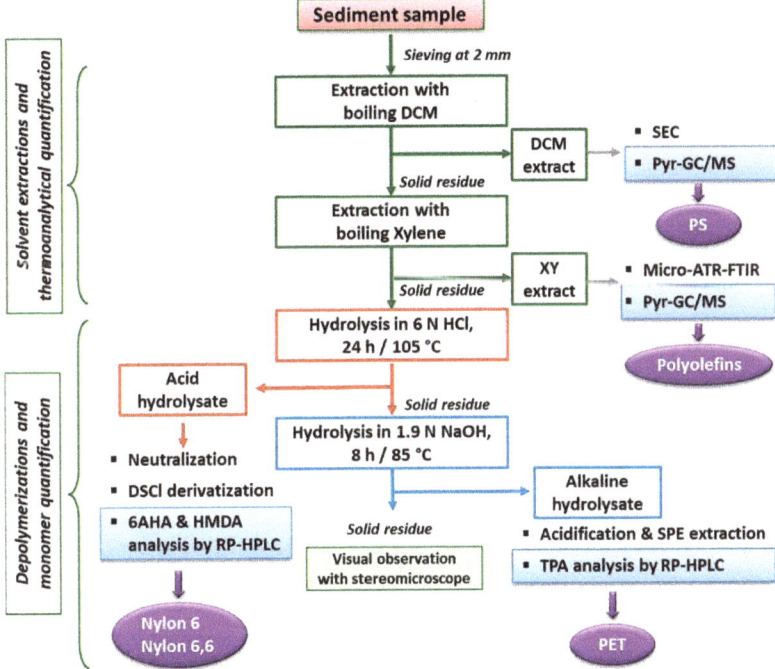

Figure 10. Flowchart of the entire analytical protocol for the separate quantification of the total mass of micro- and nanoparticles of different polymer types.

Among the most noteworthy results presented here are the successful implementation of an improved version of the procedure for the quantification of PS, and the results confirming the presence of both low-MW and high-MW low-density polyolefins in benthic sediments. In particular, the Pyr-GC/MS technique adopted for the quantification of PS allows improvement to improve the accuracy with respect to the previously reported procedure based on FD-SEC [33], in which only the high MW fraction could be evaluated due to the interference by biogenic species and oxidized polyolefin molecular fragments in the low molecular weight fraction. The presence of PS in the 8–65 ppm concentration range could thus be determined with accuracy, while the MW distribution determined by SEC analysis highlighted the presence of a significant fraction of degraded low-MW PS, in agreement with the expected presence in the bottom sediment of low-density hydrocarbon polymers that have undergone significant photo-oxidation. The presence of highly oxidized high-MW polyolefin MPs, although in very small amount, clearly indicates that polymer oxidation contributes to their vertical transport and ultimate deposition due to their increased density and hydrophilicity.

The results obtained in this work also highlight the versatility of the procedure, which deal with sediments of very different compositions (silicate silt and coarse organogenic carbonate debris).

Finally, the double shot Py-GC/MS technique allowed separate detection in the DCM extracts of all samples various phthalates and other low-MW plasticizers (suspect endocrine disruptors), while the high-MW fraction was found to contain, in addition to PS, also polysiloxane (in the DCM extract of MEL1) possibly from silicones used in formulations of personal care products, and degraded fractions of polyethylene (in MEL2 and CAL2). Although not mentioned before, the likely presence of biodegradable (not necessarily so once in the marine sediment) aliphatic-aromatic polyesters could also be identified from the presence of some markers of poly(butylene terephthalate-co-adipate).

The hydrolytic depolymerization of the higher density heteropolymers (PET, nylon 6 and nylon 6,6) followed by HPLC analysis of the resulting monomers allowed accurate quantification of the contamination level presumably associated with the deposition of synthetic textile fibers carried by urban wastewaters. The somewhat surprising higher level of contamination by polyamides (2.7–12.1 ppm) compared to PET (1.5–0.2 ppm) may be the result of different sources of pollution (e.g., fishing gears), although the number of analyzed samples was too small to allow drawing general and accurate considerations. In fact, while PET was detected in all the analyzed samples, polyamides were only detected in some of them. Finally, the presence of PTFE particles isolated from the final residue could be the result of a point source or of a specialized source of pollution, as this material is widely used in technical fishing equipment.

Although the methodology used in this work cannot provide information on important parameters such as the number, size, and shape of the individual plastic particles, it provides important complementary and unique quantitative information allowing the gaining of an accurate picture on the transport, extent, and distribution of MPs in the marine environment, thus clarifying the actual role of the sea bottom as MPs sink.

Author Contributions: Conceptualization, V.C. and A.C. (Andrea Corti); methodology, A.C. (Andrea Corti) and J.L.N.; validation, A.C. (Andrea Corti) and J.L.N.; investigation, A.C. (Andrea Corti), J.L.N., S.G. and V.V.; resources, V.C., F.M. and A.C. (Alessio Ceccarini); writing—original draft preparation, A.C. (Andrea Corti), J.L.N. and V.V.; writing—review and editing, V.C.; supervision, A.C. (Andrea Corti), F.M. and M.B.; funding acquisition, V.C. All authors have read and agreed to the published version of the manuscript.

Funding: This research was funded by Fondazione Cassa di Risparmio di Lucca (Bando Ricerca 2019-21, CISUP project "Micro- e nano-plastiche: metodologie di quantificazione, valutazione dell'impatto in ecosistemi marini e lacustri, strategie di remediation ambientale") and University of Pisa PRA project 2020_27 on *Micro- and nanoplastics*.

Institutional Review Board Statement: Not applicable.

Informed Consent Statement: Not applicable.

Acknowledgments: The authors are thankful to the Harbor Masters Corps—Coast Guard of Leghorn (Capitaneria di Porto—Guardia Costiera di Livorno) for logistic support in the sampling.

Conflicts of Interest: The authors declare no conflict of interest. The funders had no role in the design of the study; in the collection, analyses, or interpretation of data; in the writing of the manuscript, or in the decision to publish the results.

References

1. Da Costa, J.P.; Nunes, A.R.; Santos, P.S.M.; Girão, A.V.; Duarte, A.C.; Rocha-Santos, T. Degradation of polyethylene microplastics in seawater: Insights into the environmental degradation of polymers. *J. Environ. Sci. Health Part A* **2018**, *53*, 866–875. [CrossRef] [PubMed]
2. Pabortsava, K.; Lampitt, R.S. High concentrations of plastic hidden beneath the surface of the Atlantic Ocean. *Nat. Commun.* **2020**, *11*, 1–11. [CrossRef] [PubMed]
3. Alimi, O.S.; Budarz, J.F.; Hernandez, L.M.; Tufenkji, N. Microplastics and Nanoplastics in Aquatic Environments: Aggregation, Deposition, and Enhanced Contaminant Transport. *Environ. Sci. Technol.* **2018**, *52*, 1704–1724. [CrossRef]
4. Eriksen, M.; Lebreton, L.C.M.; Carson, H.S.; Thiel, M.; Moore, C.J.; Borerro, J.C.; Galgani, F.; Ryan, P.G.; Reisser, J. Plastic Pollution in the World's Oceans: More than 5 Trillion Plastic Pieces Weighing over 250,000 Tons Afloat at Sea. *PLoS ONE* **2014**, *9*, e111913. [CrossRef] [PubMed]
5. Evangeliou, N.; Grythe, H.; Klimont, Z.; Heyes, C.; Eckhardt, S.; Lopez-Aparicio, S.; Stohl, A. Atmospheric transport is a major pathway of microplastics to remote regions. *Nat. Commun.* **2020**, *11*, 1–11. [CrossRef] [PubMed]
6. Waller, C.L.; Griffiths, H.J.; Waluda, C.M.; Thorpe, S.E.; Loaiza, I.; Moreno, B.; Pacherres, C.O.; Hughes, K.A. Microplastics in the Antarctic marine system: An emerging area of research. *Sci. Total Environ.* **2017**, *598*, 220–227. [CrossRef] [PubMed]
7. Bergmann, M.; Wirzberger, V.; Krumpen, T.; Lorenz, C.; Primpke, S.; Tekman, M.B.; Gerdts, G. High Quantities of Microplastic in Arctic Deep-Sea Sediments from the HAUSGARTEN Observatory. *Environ. Sci. Technol.* **2017**, *51*, 11000–11010. [CrossRef]
8. Schwaferts, C.; Niessner, R.; Elsner, M.; Ivleva, N.P. Methods for the analysis of submicrometer- and nanoplastic particles in the environment. *TrAC Trends Anal. Chem.* **2019**, *112*, 52–65. [CrossRef]
9. Coyle, R.; Hardiman, G.; Driscoll, K.O. Microplastics in the marine environment: A review of their sources, distribution processes, uptake and exchange in ecosystems. *Case Stud. Chem. Environ. Eng.* **2020**, *2*, 100010. [CrossRef]
10. Kihara, S.; Köper, I.; Mata, J.P.; McGillivray, D.J. Reviewing nanoplastic toxicology: It's an interface problem. *Adv. Colloid Interface Sci.* **2021**, *288*, 102337. [CrossRef] [PubMed]
11. Choy, C.A.; Robison, B.H.; Gagne, T.O.; Erwin, B.; Firl, E.; Halden, R.U.; Hamilton, J.A.; Katija, K.; Lisin, S.E.; Rolsky, C.; et al. The vertical distribution and biological transport of marine microplastics across the epipelagic and mesopelagic water column. *Sci. Rep.* **2019**, *9*, 1–9. [CrossRef] [PubMed]
12. Van Sebille, E.; Aliani, S.; Law, K.L.; Maximenko, N.; Alsina, J.M.; Bagaev, A.; Bergmann, M.; Chapron, B.; Chubarenko, I.; Cózar, A.; et al. The physical oceanography of the transport of floating marine debris. *Environ. Res. Lett.* **2020**, *15*, 023003. [CrossRef]
13. Courtene-Jones, W.; Quinn, B.; Gary, S.F.; Mogg, A.O.; Narayanaswamy, B.E. Microplastic pollution identified in deep-sea water and ingested by benthic invertebrates in the Rockall Trough, North Atlantic Ocean. *Environ. Pollut.* **2017**, *231*, 271–280. [CrossRef]
14. Botterell, Z.L.; Beaumont, N.; Dorrington, T.; Steinke, M.; Thompson, R.C.; Lindeque, P.K. Bioavailability and effects of microplastics on marine zooplankton: A review. *Environ. Pollut.* **2019**, *245*, 98–110. [CrossRef] [PubMed]
15. Waldman, W.R.; Rillig, M.C. Microplastic Research Should Embrace the Complexity of Secondary Particles. *Environ. Sci. Technol.* **2020**, *54*, 7751–7753. [CrossRef]
16. Kooi, M.; Van Nes, E.H.; Scheffer, M.; Koelmans, A.A. Ups and Downs in the Ocean: Effects of Biofouling on Vertical Transport of Microplastics. *Environ. Sci. Technol.* **2017**, *51*, 7963–7971. [CrossRef]
17. Martin, J.; Lusher, A.; Thompson, R.C.; Morley, A. The Deposition and Accumulation of Microplastics in Marine Sediments and Bottom Water from the Irish Continental Shelf. *Sci. Rep.* **2017**, *7*, 1–9. [CrossRef]
18. Peng, X.; Chen, M.; Chen, S.; Dasgupta, S.; Xu, H.; Ta, K.; Du, M.; Li, J.; Guo, Z.; Bai, S. Microplastics contaminate the deepest part of the world's ocean. *Geochem. Perspect. Lett.* **2018**, *9*, 1–5. [CrossRef]
19. Singh, N.; Tiwari, E.; Khandelwal, N.; Darbha, G.K. Understanding the stability of nanoplastics in aqueous environments: Effect of ionic strength, temperature, dissolved organic matter, clay, and heavy metals. *Environ. Sci. Nano* **2019**, *6*, 2968–2976. [CrossRef]
20. Van Sebille, E.; Wilcox, C.; Lebreton, L.C.M.; A Maximenko, N.; Hardesty, B.D.; A Van Franeker, J.; Eriksen, M.; A Siegel, D.; Galgani, F.; Law, K.L. A global inventory of small floating plastic debris. *Environ. Res. Lett.* **2015**, *10*, 124006. [CrossRef]
21. Van Cauwenberghe, L.; Devriese, L.; Galgani, F.; Robbens, J.; Janssen, C.R. Microplastics in sediments: A review of techniques, occurrence and effects. *Mar. Environ. Res.* **2015**, *111*, 5–17. [CrossRef] [PubMed]
22. Corti, A.; Vinciguerra, V.; Iannilli, V.; Pietrelli, L.; Manariti, A.; Bianchi, S.; Petri, A.; Cifelli, M.; Domenici, V.; Castelvetro, V. Thorough Multianalytical Characterization and Quantification of Micro- and Nanoplastics from Bracciano Lake's Sediments. *Sustainability* **2020**, *12*, 878. [CrossRef]
23. Frias, J.; Pagter, E.; Nash, R.; O'Connor, I.; Carretero, O.; Filgueiras, A.; Viñas, L.; Gago, J.; Antunes, J.; Bessa, F.; et al. *Standardised Protocol for Monitoring Microplastics in Sediments*; Technical Report; JPI-Oceans BASEMAN: Brussels, Belgium, 2018. [CrossRef]
24. Castelvetro, V.; Corti, A.; Biale, G.; Ceccarini, A.; Degano, I.; La Nasa, J.; Lomonaco, T.; Manariti, A.; Manco, E.; Modugno, F.; et al. New methodologies for the detection, identification, and quantification of microplastics and their environmental degradation by-products. *Environ. Sci. Pollut. Res.* **2021**, *28*, 1–17. [CrossRef]
25. Okoffo, E.D.; Ribeiro, F.; O'Brien, J.W.; O'Brien, S.; Tscharke, B.J.; Gallen, M.; Samanipour, S.; Mueller, J.F.; Thomas, K.V. Identification and quantification of selected plastics in biosolids by pressurized liquid extraction combined with double-shot pyrolysis gas chromatography–mass spectrometry. *Sci. Total Environ.* **2020**, *715*, 136924. [CrossRef]
26. Zhang, J.; Wang, L.; Halden, R.U.; Kannan, K. Polyethylene Terephthalate and Polycarbonate Microplastics in Sewage Sludge Collected from the United States. *Environ. Sci. Technol. Lett.* **2019**, *6*, 650–655. [CrossRef]

27. Lin, J.; Xu, X.-P.; Yue, B.-Y.; Li, Y.; Zhou, Q.-Z.; Xu, X.-M.; Liu, J.-Z.; Wang, Q.-Q.; Wang, J.-H. A novel thermoanalytical method for quantifying microplastics in marine sediments. *Sci. Total Environ.* **2021**, *760*, 144316. [CrossRef]
28. Ceccarini, A.; Corti, A.; Erba, F.; Modugno, F.; La Nasa, J.; Bianchi, S.; Castelvetro, V. The Hidden Microplastics: New Insights and Figures from the Thorough Separation and Characterization of Microplastics and of Their Degradation Byproducts in Coastal Sediments. *Environ. Sci. Technol.* **2018**, *52*, 5634–5643. [CrossRef] [PubMed]
29. Castelvetro, V.; Corti, A.; Ceccarini, A.; Petri, A.; Vinciguerra, V. Nylon 6 and nylon 6,6 micro- and nanoplastics: A first example of their accurate quantification, along with polyester (PET), in wastewater treatment plant sludges. *J. Hazard. Mater.* **2021**, *407*, 124364. [CrossRef]
30. Castelvetro, V.; Corti, A.; Bianchi, S.; Ceccarini, A.; Manariti, A.; Vinciguerra, V. Quantification of poly(ethylene terephthalate) micro- and nanoparticle contaminants in marine sediments and other environmental matrices. *J. Hazard. Mater.* **2020**, *385*, 121517. [CrossRef]
31. La Nasa, J.; Biale, G.; Mattonai, M.; Modugno, F. Microwave-assisted solvent extraction and double-shot analytical pyrolysis for the quali-quantitation of plasticizers and microplastics in beach sand samples. *J. Hazard. Mater.* **2021**, *401*, 123287. [CrossRef]
32. La Nasa, J.; Biale, G.; Fabbri, D.; Modugno, F. A review on challenges and developments of analytical pyrolysis and other thermoanalytical techniques for the quali-quantitative determination of microplastics. *J. Anal. Appl. Pyrolysis* **2020**, *149*, 104841. [CrossRef]
33. Biver, T.; Bianchi, S.; Carosi, M.R.; Ceccarini, A.; Corti, A.; Manco, E.; Castelvetro, V. Selective determination of poly(styrene) and polyolefin microplastics in sandy beach sediments by gel permeation chromatography coupled with fluorescence detection. *Mar. Pollut. Bull.* **2018**, *136*, 269–275. [CrossRef] [PubMed]

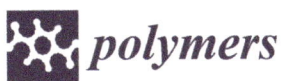

Article

Utilizing Pyrolysis–Gas Chromatography/Mass Spectrometry for Monitoring and Analytical Characterization of Microplastics in Polychaete Worms

Sabiqah Tuan Anuar [1,*], Raad Shaher Altarawnah [2], Ahmad Ammarluddin Mohd Ali [2], Bai Qin Lee [3], Wan Mohd Afiq Wan Mohd Khalik [1], Ku Mohd Kalkausar Ku Yusof [1] and Yusof Shuaib Ibrahim [1,*]

[1] Microplastic Research Interest Group (MRIG), Faculty of Science and Marine Environment, Universiti Malaysia Terengganu, Kuala Nerus 21030, Terengganu, Malaysia; wan.afiq@umt.edu.my (W.M.A.W.M.K.); kukautsar@umt.edu.my (K.M.K.K.Y.)

[2] Faculty of Science and Marine Environment, Universiti Malaysia Terengganu, Kuala Nerus 21030, Terengganu, Malaysia; raedshaher01@gmail.com (R.S.A.); ammarali69.com@gmail.com (A.A.M.A.)

[3] ALS Technichem (M) Sdn. Bhd., Wisma ALS, No. 21, Jalan Astaka U8/84, Bukit Jelutong, Shah Alam 40150, Selangor, Malaysia; lee.baiqin@alsglobal.com

* Correspondence: sabiqahanuar@umt.edu.my (S.T.A.); yusofshuaib@umt.edu.my (Y.S.I.); Tel.: +60-192723430 (S.T.A.); +60-132992926 (Y.S.I.)

Abstract: Microplastics (the term for plastics at sizes of <5 mm) might be introduced into the environment from domestic or agricultural activities or from the breakdown of plastic pieces, particles, and debris that are bigger in size. Their presence in the aquatic environment has caused accumulation problems, as microplastics do not easily break down and can be digested by some aquatic organisms. This study was conducted to screen and monitor the level of microplastic pollution in polychaete worms using pyrolysis–gas chromatography/mass spectrometry (Py-GC/MS). The study was conducted in Setiu Wetlands, Malaysia from November 2015 to January 2017 at five-month intervals and covered all monsoon changes. Results from physical and visual analyses indicated that a total number of 371.4 ± 20.2 items/g microplastics were retrieved from polychaete for all seasons, in which the majority comprised transparent microplastics (49.87%), followed by brown with 138.3 ± 13.6 items/g (37.24%), 21.7 ± 1.9 items/g for blue (5.84%), and 12.9 ± 1.1 items/g for black (3.47%), while the remaining were green and grey-red colors. Statistical analysis using Kruskal–Wallis showed insignificant differences ($p > 0.05$) between the sampling station and period for the presence of a microplastics amount. Most of the microplastics were found in fiber form (81.5%), whereas the remaining comprised fragment (18.31%) and film (0.19%) forms. Further analysis with Py-GC/MS under a selective ion monitoring mode indicated that pyrolytic products and fragment ions for a variety of polymers, such as polyvinyl chloride, polypropylene, polyethylene, polyethylene terephthalate, polyamide, and polymethylmethacrylate, were detected. This study provides an insightful application of Py-GC/MS techniques for microplastics monitoring, especially when dealing with analytical amounts of samples.

Keywords: South China Sea; pollution; Py-GC/MS; fragmentation and degradation; mechanism

Citation: Anuar, S.T.; Altarawnah, R.S.; Mohd Ali, A.A.; Lee, B.Q.; Khalik, W.M.A.W.M.; Yusof, K.M.K.K.; Ibrahim, Y.S. Utilizing Pyrolysis–Gas Chromatography/Mass Spectrometry for Monitoring and Analytical Characterization of Microplastics in Polychaete Worms. *Polymers* **2022**, *14*, 3054. https://doi.org/10.3390/polym14153054

Academic Editor: Jacopo La Nasa

Received: 19 May 2022
Accepted: 26 July 2022
Published: 28 July 2022

Publisher's Note: MDPI stays neutral with regard to jurisdictional claims in published maps and institutional affiliations.

Copyright: © 2022 by the authors. Licensee MDPI, Basel, Switzerland. This article is an open access article distributed under the terms and conditions of the Creative Commons Attribution (CC BY) license (https://creativecommons.org/licenses/by/4.0/).

1. Introduction

The ever-increasing human population has made a high demand for resources and living space, which has resulted in the expansion of terrestrial urban areas and subsequently intruded the coastal territory. The construction of buildings, the constant generation of municipal wastes, and the unearthing of natural resources has not only brought ecological changes, but has also negatively affected many organisms living therein through pollution, food scarcity, poor reproduction, and alternative predation [1,2]. As a result, these organisms are forced to change their routine and perquisites to cater for the rapid changes in their surroundings. As such, hydrocarbon-based contamination is threatening both aquatic and terrestrial life forms in human-infringed areas; ultimately, these contaminations have caused a species-wise population reduction, especially among less tolerant organisms [3,4]. The plastic derived from the industrial synthesis of natural gas and crude oil refining also represents another source of contamination/pollution that requires immediate attention [5,6]. For instance, the plastic wastes resulting from human-used products such as toothpaste, cosmetics, clothing, and plastic packaging are usually identified as microplastics due to their small size (1–5 mm) [7,8]. Among others, polystyrene (PS), polyethylene (PE), polypropylene (PP), polyethylene terephthalate (PET), polyvinyl chloride (PVC), nylon, and polyamides (PA) are some of the examples of the most abundant microplastics found in the coastal water [9].

The biological interaction between microplastics and biota is crucial to understanding the movement, impact, and fate of microplastics in the environment. Several controlled laboratory and/or field studies have been conducted to clarify the impact of this problem. Recent attention has been focused on lower trophic level organisms, as these marine invertebrates have significant effects on the introduction of microplastics into the food chain [10]. Recent studies revealed that an excessive amount of microplastics has been found in the tissue and gastrointestinal tract of small marine deposit feeders and scavengers [11–13] and respiratory and digestive tracts of crustaceans [14,15]. These microplastics were also detected in the body of larger predators after consuming lower trophic animals [16,17]. As a result, these higher trophic predators will suffer from physical blockage at their alimentary and digestive tract and will be poisoned after the digested plastic leaches into their digestive system and adsorbs on the organ interior lining [18]. In addition, the direct ingestion of microplastics by fish in riverine areas [17], as well as in estuaries [19], was also reported. This could then trigger major environmental and health issues [20,21].

Despite microplastic pollution being an emerging field of study, the impacts of microplastics on human health and the environment are yet to be fully discovered. Hence, the monitoring and identification of microplastic pollution is urgently needed prior to the development of efficient treatment strategies. As such, standardized field methods for collecting microplastics in sediment, sand, and surface-water samples and a novel analysis technique for the identification of microplastics have been developed and continue to be optimized. In due course, a global comparison of the amount of microplastics released can be carried out with field and laboratory protocols to elucidate the final distribution, impacts, and fate of microplastics [22]. Two innovative analysis techniques have been developed for the determination of microplastics in complex environmental samples using pyrolysis–gas chromatography/mass spectrometry (Py-GC/MS) and thermal desorption–gas chromatography/mass spectrometry (TD-GC/MS). The scientific and practical challenge of detecting microplastics in the target environment in a rapid manner can be resolved by thermogravimetric analysis with the mass spectrometry method. Using Py-GC/MS and TD-GC/MS, the rapid identification and quantitative determination of most thermoplastic polymers and elastomers is possible [23–25] with a limited sample size; for instance, in polychaete worms, the composition of the plastic material can be obtained via the Py-GC/MS. This study aims to investigate and monitor the pollution level of microplastics in Setiu Wetlands, Terengganu, Malaysia using Py-GC/MS, in which, to our knowledge, this is the first study in Malaysia to utilize the technique for microplastics identification.

2. Materials and Methods

2.1. Sampling Site and Sample Collection

Setiu Wetland comprises 23,000 ha terrestrial and aquatic environments that experience hot and humid climate intervals, with air temperatures ranging between 28 °C and 33 °C, seasonal salinity shifts from 3–33%, and annual rainfall between 2000 and 4000 mm [26,27]. This large ecological system has freshwater and peat swamps that host riparian forests and mangrove–mangrove-associated vegetation along its riverbanks, grass beds, and sandy beaches, which also house diverse macrobenthos and invertebrates [28,29]. Four sampling sites at Setiu Wetlands were selected according to the preliminary investigation of polychaete distribution. Each of the sampling sites were coordinated and illustrated as shown in Table 1 and Figure 1, respectively. The source of fresh water in these areas is streamed from River Setiu and River Ular (Figure 1). These areas are covered by tidal action and wind-driven currents from mangrove trees (i.e., *Nypa fruiticans* and *Avicennia* sp.), which act as a shelter and dominate sand islets. On the other hand, marine water is channeled into the estuary of the South China Sea [30]. The input into the Setiu Wetlands system is usually retained for long periods during low tide because it is an almost closed system with only a single opening to the sea that is <20 m apart.

Table 1. Information of each sampling sites.

Sampling Sites	Latitude	Longitude
S1	5°43′21.48″ N	102°40′13.87″ E
S2	5°41′30.77″ N	102°41′54.98″ E
S3	5°40′54.46″ N	102°42′33.25″ E
S4	5°39′49.27″ N	102°43′57.69″ E

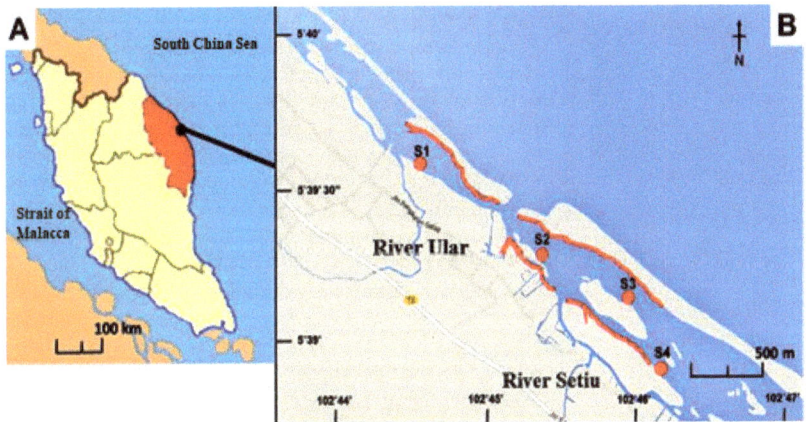

Figure 1. Map showing (**A**) Setiu Wetlands location at the east coast of peninsular Malaysia (facing South China Sea), and (**B**) Sampling stations S1–S4 at Setiu Wetland, Malaysia.

The sampling was performed four times from November 2015 until January 2017 in Setiu Wetlands, Terengganu, Malaysia, following the protocol of Hamzah et al. [11]. The field visits were conducted at four sites (i.e., three in Ular River and marked as S1–S3, and one in Setiu River and marked as S4, Figure 1). Each transect of 400 m^2 (20 m × 20 m) was set and had a 200 m interval before the next station, as illustrated by the red lines in the map. Sampling was conducted during the intermediate tide (1.2–2.0 m) by searching for polychaete worms (*Namalycastis* sp.) at the marked *Nypa fruticans* (nypa palm) vegetated riverbanks. This species is among the most abundant organisms in the benthic community in the mangrove area of Setiu Wetlands. These worms have special ability to adapt to their environment, thus allowing them to structure the dynamics of their surroundings and feed on rotten nypa palm [29,31]. During the sample collection, encrusted decaying pieces of the plants (such as fronds and roots) were carefully sectioned to retrieve the polychaete worms using bare hands. A total of 160 individuals was collected by hand, and the site was marked for re-visiting using a portable handheld global positioning system (Garmin GPSMAP 78S GPS, Garmin Ltd., Olathe, KS, USA). The polychaetes collected from each site were kept in a glass container and stored in an icebox for laboratory analysis.

2.2. Extraction and Isolation of Microplastics

Ten polychaetes with approximate similar size and weights were pooled (n = 10, weight~10 g) from each site and added to 10 mL of 10 M NaOH in a digestion tube. Data from the physical analysis of microplastics (sorting and isolation) were expressed as microplastics item/g of the polychaete, since the polychaete was pooled based on different sampling stations and periods during the digestion process. The digestion was conducted at 60 °C for 48 h. After the digestion process, the brown-translucent solution was filtered using a vacuum pump with cellulose nitrate filter paper (Whatman 0.45 µm pore size, Cytiva, Marlborough, MA, USA) before being placed in a petri dish and dried in a desiccator. Physical identification of microplastics was conducted using the standardized size and color-sorted system according to the characterization protocol provided by Hidalgo-Ruz et al. [32]. The sample was sorted into three groups: fiber, fragment, and film, with various colors of brown (Br), transparent (T), red (R), blue (B), grey (Gy), black (Bl), and green (Gn) using dissecting microscope (Olympus CX21 8X-56X Magnification, Olympus Corp., Tokyo, Japan). All sorted samples were collected using a fine-tip stainless steel forceps and kept in wide-neck glass bottles (McCartney, Fisher Scientific, Waltham, MA, USA) containing 10 mL of filtered Milli-Q water.

2.3. Chemical Analysis for Microplastic Identification Using Py-GC/MS

The microplastics samples were analyzed via Py-GC/MS technique. This technique combines the process of pyrolysis, which is the degradation of high molecular polymer into smaller organic compounds by heating in a pyrolysis chamber (Frontier Lab. Ltd., Koriyama, Japan) in the absence of oxygen. The organic compounds were separated via the gas chromatography technique (GC 6890 N, Agilent Tech., Santa Clara, CA, USA). In brief, the GC is equipped with a pre-column (Ultra Alloy*-5, 30 m × 0.25 mm × 0.5 µm) and column (Ultra Alloy*-50, 2 m × 0.25 mm × 1.0 µm, Frontier Lab. Ltd., Koriyama, Japan) and characterized using mass spectrometer (MS 5973 N, Agilent Tech., CA, USA). For the analysis using Py-GC/MS, 20% of the microplastics item (for each station and period) was considered by adding 4 mg of calcium carbonate powder (Frontier Lab, Japan), and was derivatized with 5 µL 25% tetramethylammonium hydroxide (TMAH) solution in methanol (Sigma Aldrich, Petaling Jaya, Malaysia) after being filtered using an aluminum oxide filter prior to the pyrolysis. The temperature programs were developed and optimized according to the manual from the supplier of pyrolyzer, Frontier Laboratories Ltd. [33], whereas the selection of characteristic ions was following Fisher and Scholz-Bottcher [34].

In brief, after the pyrolysis cup was dropped into the pyrolysis chamber, the temperature of the pyrolysis furnace was maintained at 400 °C for 10 min. The transfer line was heated and maintained at 300 °C. The initial oven temperature of the GC was set at 40 °C and held for 2 min, then ramped to 280 °C at 20 °C/min and held for 10 min. Finally, the oven was ramped to 320 °C at 40 °C/min and held for 15 min. The inlet was set up with a split mode ratio of 50:1 and a constant pressure of 150 kPa by helium gas. Mass spectrometry was set to run with scan mode in the range between 29 and 350 amu. The MS scan rate was set at n = 2, equivalent to 4 scans/s. Subsequently, a set of the calibration curve was prepared based on twelve common polymers, including polyethylene (PE), polypropylene (PP), polystyrene (PS), polycarbonate (PC), polyvinyl chloride (PVC), polyurethane (PU), polyethylene terephthalate (PET), acrylonitrile butadiene styrene (ABS), styrene-butadiene (SBR), and polymethyl methacrylate (PMMA), and polyamides such as nylon-6 (N-6) and nylon-6,6 (N-66). Quantitation of these polymers was performed referring to the selected pyrolytic products (Table 2).

Table 2. Characteristic compounds for 12 polymers, their characteristics ions, and retention time.

Polymer	Characteristic Compound	Main Characteristic Ion (m/z)	Sub Characteristic Ion (m/z)	Retention Time (min)
PE	1,20-Heneicosadiene	82	294	15.427
PP	2,4-Dimethyl-1-heptene	126	70	6.052
PS	Styrene trimer	91	312	19.96
ABS	2-Phenethly-4-phenylpent-4-enentrile	170	91	17.17
SBR	4-phenylcyclohexene	104	158	11.016
PMMA	Methyl methacrylate	100	69	4.328
PC	4-Isopropenylphenol	134	119	10.696
PVC	Naphthalene	128	115	9
PU	4,4'-Methylenedianiline	198	106	17.25
PET	Benzophenone	182	105	13.608
N-6	ε-Caprolactam	113	85	10.677
N-66	Cyclopentanone	84	55	5.637

Additionally, microplastics calibration standard set was purchased from Frontier Lab (Frontier Lab. Ltd., Koriyama, Japan). Three-point calibration curve was created by the microplastics calibration standards containing calcium carbonate. In brief, 0.4 mg, 2.0 mg, and 4 mg of calibration standard were weighted precisely in the pyrolysis cup before 5 μL of 25% TMAH in methanol solution was added into the pyrolysis cup and left at room temperature for 20 min before the analysis. Analysis of calibration curve is as per described in Section 2.3. The R^2 for built calibration is between 0.96 and 0.99 (Supplementary Figure S1).

2.4. Multivariate and Statistical Analysis

The statistical analyses were performed using OriginPro version 9.1 software (Origin Lab Corp., Northampton, MA, USA) and R Software (R development team, Version 1.4.1). The abundance of microplastics was evaluated using Kruskal–Wallis (post hoc: Bonferonni correction) subjected to a non-parametric dataset for various types and colors between the station and period of the study. The significance level was set at $p < 0.05$. The results obtained are presented in the form of a dendrogram graph, with 3 clusters of color following the number of the collected color sample within the sampling time. Box plot analysis was used to compare the shape distribution of the samples.

2.5. Quality Assurance and Control

Precautions were taken to prevent microplastics contamination during sample collection in the field and analysis in the dedicated clean room of the Microplastic Research Interest Group (MRIG) laboratory. Cotton lab gowns and single-use latex gloves were worn during all procedures of the experiment. The work surfaces were carefully washed with 80% filtered ethanol prior to the start of all procedures and during the experiment period. Glassware and instrumentation were washed with filtered Milli-Q water (Milli-Q water underwent double filtration with 1.2 μm Whatman GF/C filter paper prior usage) and rinsed three times with 80% ethanol. In addition, procedural blanks and control filter papers were first inspected by dissecting microscope (Olympus SZX-ZB7, Olympus Corp., Tokyo, Japan) and later thoroughly analyzed to detect any contamination of microplastics from the laboratory surroundings and during sample handling [11].

3. Results and Discussion

Originated from synthetic polymers, microplastics (size < 5 mm) can present in various shapes, sizes, and colors [22,35]. In this study, analyses of microplastics were separated into two parts. Firstly, physical and visual analysis, and secondly, polymer identification by chemical analysis. In the first part, the color of microplastics was sorted, followed by their type (i.e., fiber, fragment, and film). From the physical analysis, a total of 371.4 ± 20.2 items/g microplastic particles was extracted in collected polychaetes from four sampling stations. From that data, 185.2 ± 24.7 items/g are transparent microplastics (49.87%), followed by brown with 138.3 ± 13.6 items/g (37.24%), 21.7 ± 1.9 items/g for blue (5.84%), 12.9 ± 1.1 items/g for black (3.47%), 1.53% for green, 1.08% for grey, and, lastly, around 0.97% for red samples. The Kruskal–Wallis test showed that the total microplastics abundance between sampling stations was insignificantly different: $H(3) = 1.51$, $p = 0.68$. When comparing the sampling stations (Figure 2A), S1 and S2 indicated that brown microplastics were dominantly found in collected worms, followed by transparent, blue, black, and other colors of microplastics. However, in sampling sites of S3 and S4, the transparent color is the most dominant color among the microplastic found in polychaete samples, followed by brown, blue, black, and others.

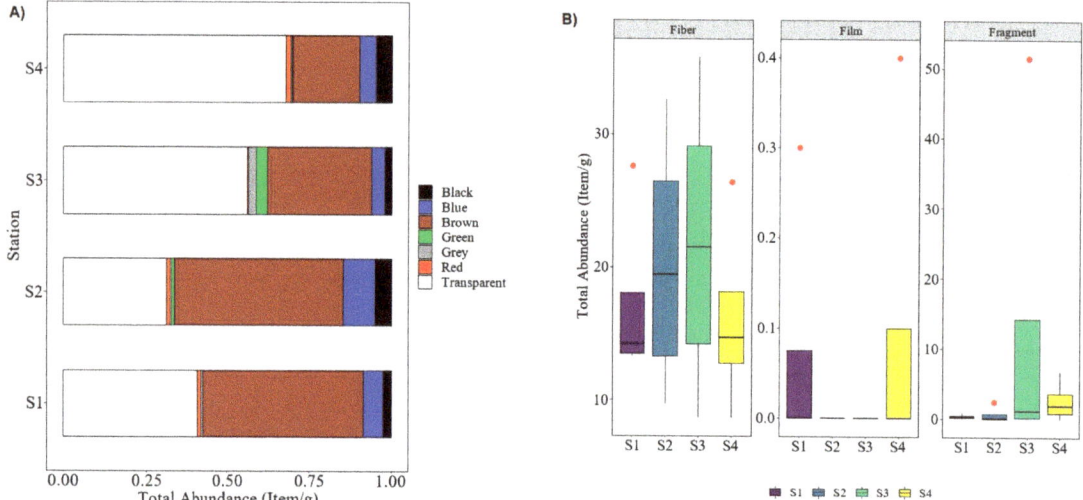

Figure 2. Percentage abundance of microplastic found in polychaete of sampling stations S1–S4: (**A**) based on color, (**B**) based on shape.

Between stations, polychaete samples from S3 recorded the highest number of microplastics (141.2 items/g, 38.02%), while the total abundance for the remaining sampling stations (S1, S2, and S4) was 19.1–22.6%. Since S3 is located on the depositional bank of the terrestrial area (low island) and near the mouth of the Ular River, it is thus susceptible to the wash-over pollution from the river downstream and the terrestrial area, as well as the tidal influx activity of the sea that brings marine debris and microplastics to this area [35–37]. Additionally, the other reason for S3 becoming the most polluted area (in terms of ingested microplastics by polychaetes) could be due to its geographical area that is sheltered by the small island and near the closed inlet; thus, the ocean movement might be relatively slow in that area and subsequently cause an accumulation of microplastics waste. Therefore, the population of polychaetes in this area consumed more microplastics compared to other areas, which corroborated the findings reported by [11]. On the other hand, previous studies have suggested that the different amounts of microplastics collected at the sampling stations might be affected by the movement of the ocean or winds [30,38]. In this study, three types of microplastics were identified by the physical analysis, namely fiber (thread), fragment, and film, as specified by the boxplot analysis in Figure 2B. Of the total microplastics, 81.5% were fiber, followed by 18.31% in the form of fragments, and only 0.19% were film form. This finding was similar to the previous studies conducted by Su et al. [39] and Huang et al. [40]. Both studies indicated that more than 70% microplastics were found in the form of fiber, and the fewest microplastics were found in the form of film, which concurred with the findings obtained in this study. Generally, the fiber-type of microplastics are derived from the decomposition of synthetic fishing gears such as lines, nets, and ropes [18]. Since major economic activities at the Setiu Wetlands were agriculture-based, the pollution might therefore be sourced from these activities.

Meanwhile, Figure 3 indicates the total abundance of microplastics in polychaetes in four different sampling periods of November 2015 (1st), March 2016 (2nd), August 2016 (3rd), and January 2017 (4th) in term of colors (Figure 3A) and shapes (Figure 3B). During the first sampling period, the northeast monsoon was blown in the Setiu Wetlands and caused a heavy rainy season from November to January [41]. Generally, the movement of pollution agents into river streams was positively affected by the amount of rainfall. Therefore, when sampling activity happened in the second sampling period in March (post–northeast-monsoon), the number of microplastics increased by approximately 41.62% compared to the first sampling period. Interestingly, the most abundant microplastics found were transparent during the second and third sampling periods, while the brown microplastics became dominant when it was close to the northeast monsoon season. However, Kruskal–Wallis confirmed that there was an insignificant difference in terms of the total microplastics abundance between the sampling period, even with the variation in specific color observed: $H(3) = 0.86$, $p = 0.834$.

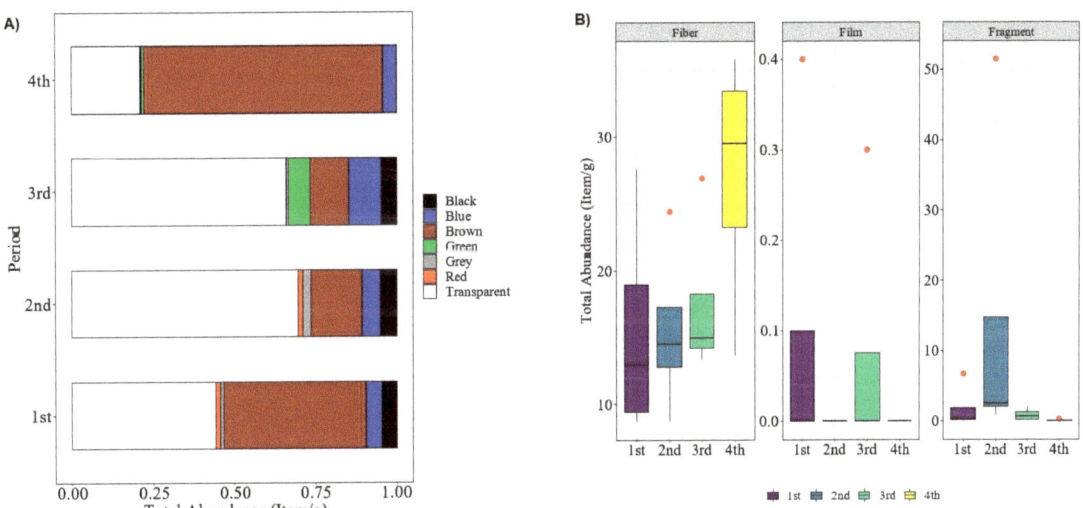

Figure 3. Total abundance of microplastic found in polychaete of sampling period 1 to 4: (**A**) based on color, (**B**) based on shape.

Similar to the previous findings by other researchers, the fiber type of microplastics represents the most common shape found in the polychaete samples [11,42] (Figure 3B). This could be attributed to the feeding behavior of the polychaete itself. This species tends to mistakenly ingest fiber microplastics since it can be similar to their feed, such as *Nypa fruticans* fibers. The discoloration of microplastics could be suggested as being due to the oxidation of microplastic-associated compounds either caused by environmental (e.g., weathering, chemical degradation) or biological (e.g., digestive environment of polychaete gut) factors. Polychaetes usually dominate the bottom community in open water, thus making them important for bioturbation. This organism can break down, deposit, and incorporate organic matter to recycle nutrients in their substrata [43,44]; ruling out pollutants such as microplastics during this process is inevitable. Previously, polychaetes have been used as a bioindicator for environmental pollution, especially for heavy metals and other toxic substances [45–47]. Additionally, the occurrence of the fiber type of microplastics in the estuarine ecosystem can be caused by both terrestrial activities and hydrological factors; since the size of microplastics is very small, they can easily get into the food chain and be consumed directly by the polychaetes in the ecosystem [11,42,47].

The next analysis for the microplastics was the chemical characterization via pyrolysis-GC/MS. Theoretically, this technique required the pyrolysis of microplastics at a high temperature, and the volatiles were separated using gas chromatography, finally detected, or analyzed via mass spectrometry. Table 2 summarizes the fragment ion data used for this technique, which include the pyrolytic products and m/z ratio that were selected for quantitative purposes, while Table 3 shows the polymer found using the analysis.

Table 3. Identification of polymer associated with microplastics found in polychaete samples during the sampling period and stations.

Period and Station	1st				2nd				3rd				4th			
	S1	S2	S3	S4	S1	S2	S3	S4	S1	S2	S3	S4	S1	S2	S3	S4
PE	+	+	-	-	-	+	+	-	+	-	+	+	+	+	+	-
PP	+	+	+	-	+	-	+	+	+	+	+	+	-	+	+	+
PS	-	-	-	+	-	-	+	-	+	-	-	-	-	-	+	-
PVC	++	++	++	++	++	++	++	++	++	++	++	++	++	++	++	++
ABS	-	-	-	-	-	-	-	-	-	-	-	-	-	-	-	-
PET	-	-	-	-	+	-	+	+	-	-	-	-	-	-	-	-
PC	-	-	-	-	-	-	-	-	-	-	-	-	-	-	-	-
PMMA	+	+	-	-	-	+	+	-	+	-	-	-	-	-	+	-
N-6	-	-	-	+	-	+	+	-	+	-	+	-	+	+	-	-
N-66	-	-	-	-	-	-	-	-	-	-	-	-	-	-	-	-
PU	-	-	-	-	-	-	-	-	-	-	-	-	-	-	-	-
SBR	-	-	-	-	-	-	-	-	-	-	-	-	-	-	-	-

"+" refers to detected, "++" refers to much detected", and, "-" refers to undetected.

Py-GC/MS is a well-established separation technique for the investigation of volatile and non-volatile compounds. Py-GC/MS has been frequently applied in the microstructural characterization and identification of synthetic polymers and copolymers in many industrial applications [48,49]. Chemical additives such as organic plastic additives (OPAs) are usually added to modify and improve the physical, chemical, thermal, mechanical, and electrical properties of the polymers during the manufacturing process [50]. This chemical additive is available in small, volatile, and semi-volatile molecules, or large and less or non-volatile compounds. As such, pyrolytic products for PE were successfully identified (i.e., tetradecadiene with characteristic and sub-characteristic ions of 82 m/z and 294 m/z), respectively [51]. For PP, the decomposition fragment that could be used for this polymer identification is 2,4-dimethyl-1-heptene (126 m/z); whereas the presence of styrene trimer (91 m/z), naphthalene (128 m/z), and benzene (77 m/z) could be used for PS and PVC, respectively. Additionally, 2,2-bis(4′methoxy-phenyl) propane and 4-isopropenylphenol were identified as a decomposition fragment for PC [25,51,52].

ε-Caprolactam, a monomer in the production of nylon-6, was used in the polyamide identification, in which, this fragment yielded a molecular weight of 133 m/z. For PET and PMMA identifications, the pyrolytic products benzophenone (182 m/z), primarily used in the manufacturing of PET fiber, film and container plastics, and methyl methacrylate (100 m/z), commonly used in the paint industry and as rheology modifiers, were used to identify the respective polymers. Examples of fragmentation mechanisms for PVC into naphthalene and subsequently into benzene or toluene monomer (a), PS into styrene monomer (b), and PET into dimethyl terephthalate (c) are illustrated in Figure 4 [53–55]. These mechanisms show the possible degradation pathways of the microplastics in the environment. For instance, PVC contains a high abundance of chlorine in the structure, thus leading to the formation and leachate of other chlorinated hydrocarbons when being induced by hydrolysis, photooxidation, and thermal degradation, which continuously happen in the environment. It will later be fragmented into the polyene chain and other PAHs derivatizes (e.g., toluene, benzene, and naphthalene) via random bond scission [54], as observed in the thermoanalysis of this polymeric material. Additionally, the possible fragmentation mechanism of PMMA into methyl methacrylate is illustrated in Figure 5. Similar to the industrial modification of PMMA into the monomers, the physical, biological, chemical, and thermal modifications occurred in the environment when the polymers were subjected to UV light, thermal changes, weathering, mechanical abrasion, and fragmentation. During the degradation process, the polymeric PMMA will lose its H^+, which leads to the bond fragmentation to a smaller unit of monomers, such as methyl acrylate, as observed in the pyrolysis-GC/MS analysis.

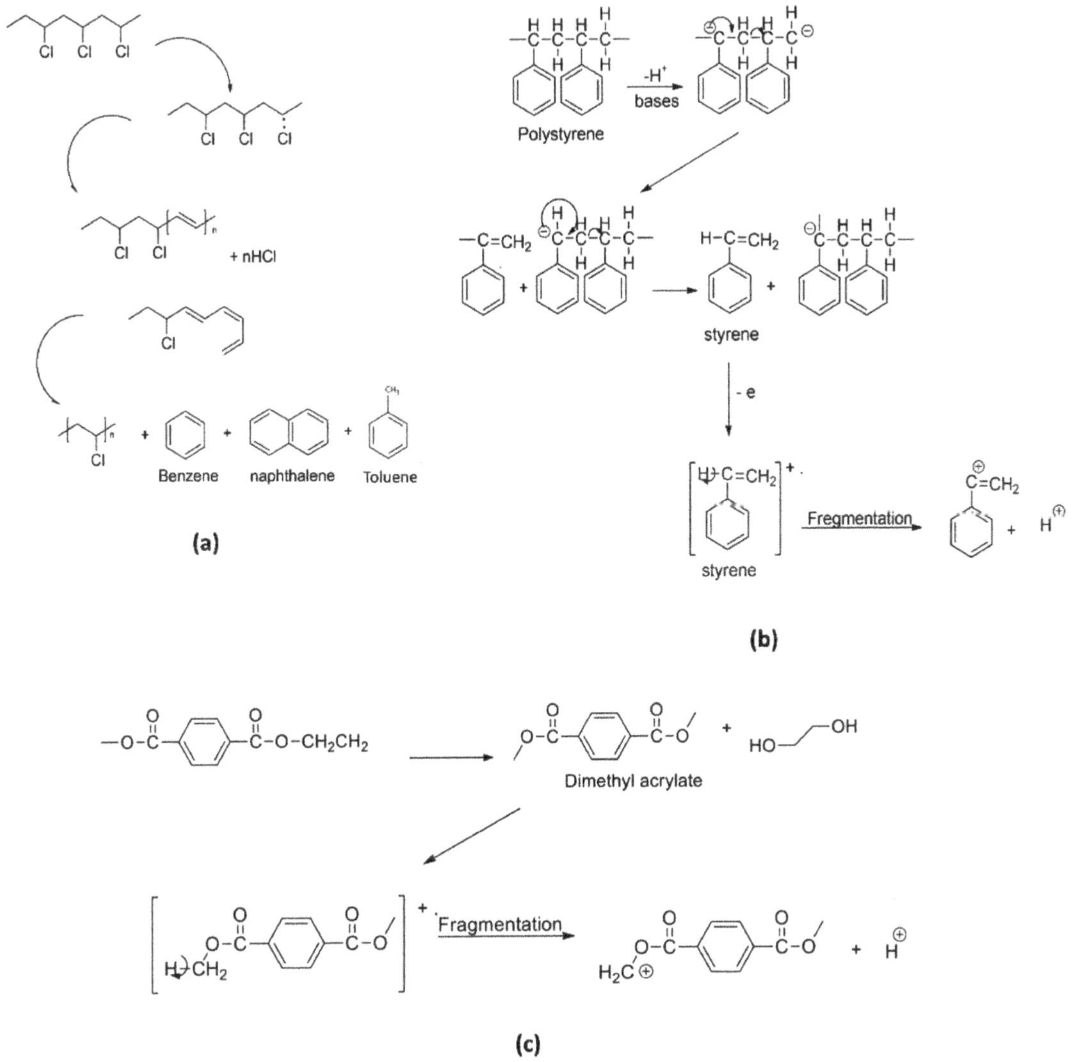

Figure 4. Simplification of degradation mechanism of common polymers into respective monomers: (a) PVC, (b) PS, and (c) PET.

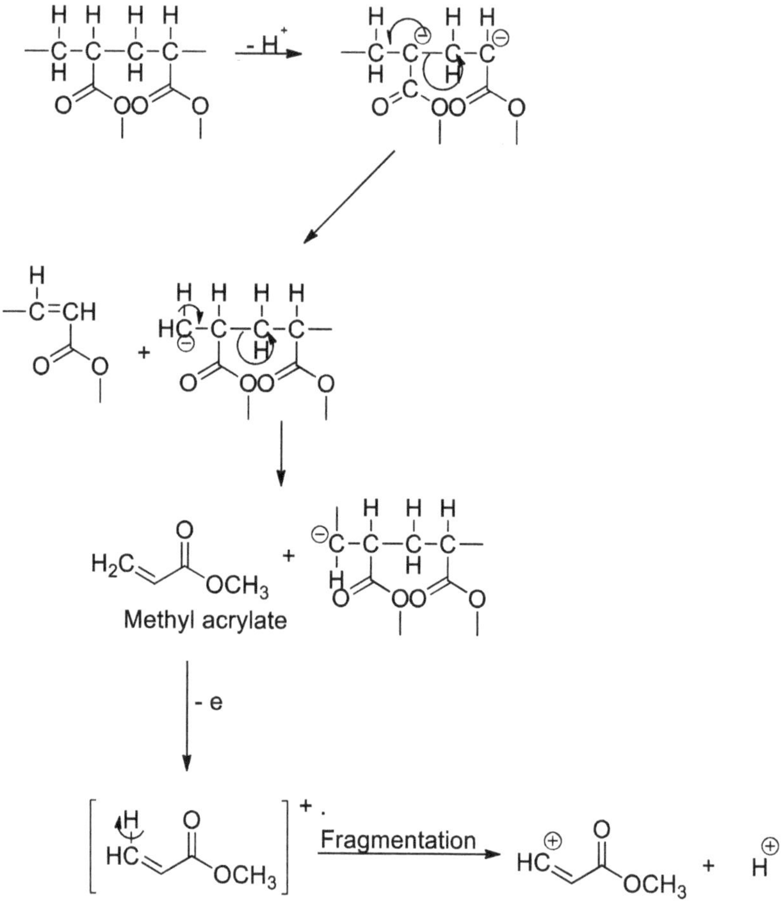

Figure 5. Fragmentation mechanism of PMMA into methyl acrylate monomer.

Py-GC/MS could undoubtedly be used to characterize and quantify the relative mass content of microplastics in environmental samples with >50 µg/L (or ppb when describing the environmental contaminant concentrations) via the determination of specific polymeric decomposition products, independent of the natural organic and inorganic matrix [34]. It is noteworthy to state that, in our case, no additional preparation steps were needed if the homogenous sample size was used. Nonetheless, this technique can be applied to different matrices [56] and can allow for the quantification of the analytical size of samples. In this study, some of the samples were quantitatively measured in the range of 60–90 µg/L, while the rest were also detected despite being below the calibration limit. This, however, limits the information about the concentration of each polymer, which will need to be optimized in future work.

Nevertheless, based on the findings obtained (Table 3), PVC was found in all samples collected from all sampling locations. This contradicts PC, which has not been detected. PVC has a higher density than water (~1.38 g/cm3); thus, this polymer is susceptible to being deposited in the riverbed of the mangrove vegetation area during the tidal changes [11,57]. The high availability of this type of microplastic in such an ecosystem will lead to indirect or direct ingestion by the polychaetes. Additionally, both PE and PP were also identified in most of the polychaete samples, followed by PA, PET, and PMMA.

The abundance of lower-density polymers found among the polychaete samples could be attributed to their ability to be transported by river flow and seawater influx around the estuary area (e.g., Setiu Wetlands) [19,30,40]. Notwithstanding, the molecular information on the polymer, additives, and the additives–polymers mixture will help to extend the knowledge on the state of polymer degradation, establish the link between formulation, properties, and degradation [58], and target toxic aspects; thus, this technique should be optimized and used in microplastics monitoring.

4. Conclusions

The sampling of polychaete was carried out within one year, with a 5-month interval at four different sampling stations. The result indicated that the most abundant color of microplastics that was ingested by the polychaetes during the northeast monsoon season was a brown color, whereas those that are transparent in color were more preferable during the southwest monsoon. The main reason for this observation is that all samples were collected in the digestive system of polychaetes and the original color of microplastics could be decolorized due to the ageing process or photo-degradation, as well as the biological interaction in the polychaete guts. The microplastics were further analyzed by employing chemical characterization via pyrolysis-GC/MS techniques, and the data indicated that several polymers were detected in the sample, including PVC, PE, PP, PS, PA, PMMA, and PET. The application of Py-GC/MS enables information on the composition of the plastic material to be obtained, though without clear data on the formulation. Additionally, the possibility of obtaining false positive data cannot be eliminated since Py-GC/MS will eventually pyrolyze all possible materials, hence producing similar molecular structures to those monomers. Nevertheless, it is proven that this technique can be applied to the qualitative and semi-quantitative analysis of a large variety of additives in polymeric items. Furthermore, analytical pyrolysis provides an extensive benefit in analyzing a complex sample for most compounds. It can determine mixed media and analyze both the polymer and associated organic plastic additives content in both an analytical amount and small size of samples, such as samples of invertebrates and planktonic microorganisms.

Supplementary Materials: The following supporting information can be downloaded at: https://www.mdpi.com/article/10.3390/polym14153054/s1, Figure S1: Prepared calibration curves for the analysis of twelve common polymers using Py-GC/MS.

Author Contributions: Y.S.I. and R.S.A. carried out the sampling and laboratory works, especially the physical analysis, and drafted the manuscript, while A.A.M.A., S.T.A. and B.Q.L. conducted the chemical analysis and interpretation. W.M.A.W.M.K. and K.M.K.K.Y. analyzed the data and performed the statistical analysis; S.T.A. and Y.S.I. designed the study, acquired the funding, and completed the manuscript. All authors have read and agreed to the published version of the manuscript.

Funding: This research was funded by the Ministry of Higher Education Malaysia, grant number FRGS59618 (FRGS/1/2020/STG04/UMT/02/2) and the APC was partly funded by the Research Management Office, Universiti Malaysia Terengganu.

Institutional Review Board Statement: Not applicable.

Informed Consent Statement: Not applicable.

Data Availability Statement: Not applicable.

Acknowledgments: The authors would like to express their gratitude to the Faculty of Science and Marine Environment and Central Laboratory UMT in providing the facilities used in this research. The authors also wish to thank Chin Teen Teen and Yap Chen Loon from ALS Technichem (M) Sdn. Bhd. for their continuous support of microplastic research in Malaysia. The authors are also thankful to Siti Rabaah Hamzah, Yuzwan Mohamad, Syazwani Azmi, Mohd Zan Husin, and all MRIG members for their valuable support and technical assistance throughout the field sampling in this study.

Conflicts of Interest: The authors declare no conflict of interest. The funders had no role in the design of the study; in the collection, analyses, or interpretation of data; in the writing of the manuscript, or in the decision to publish the results.

References

1. Kim, S.G. The evolution of coastal wetland policy in developed countries and Korea. *Ocean Coast. Manag.* **2010**, *53*, 562–569. [CrossRef]
2. Suckall, N.; Tompkins, E.; Stringer, L. Identifying trade-offs between adaptation, mitigation and development in community responses to climate and socio-economic stresses: Evidence from Zanzibar, Tanzania. *Appl. Geogr.* **2014**, *46*, 111–121. [CrossRef]
3. Förstner, U.; Wittmann, G.T. *Metal Pollution in the Aquatic Environment*; Springer Science & Business Media: New York, NY, USA, 2012; pp. 9–21.
4. Drury, B.; Rosi-Marshall, E.; Kelly, J.J. Wastewater treatment effluent reduces the abundance and diversity of benthic bacterial communities in urban and suburban rivers. *Appl. Environ. Microbiol.* **2013**, *79*, 1897–1905. [CrossRef] [PubMed]
5. Ali, M.F.; Siddiqui, M.N. Thermal and catalytic decomposition behaviour of PVC mixed plastic waste with petroleum residue. *J. Anal. Appl. Pyrolysis* **2005**, *74*, 282–289. [CrossRef]
6. Liptow, C.; Tillman, A.M. A comparative life cycle assessment study of polyethylene based on sugarcane and crude oil. *J. Ind. Ecol.* **2012**, *16*, 420–435. [CrossRef]
7. Müller, Y.K.; Wernicke, T.; Pittroff, M.; Witzig, C.S.; Strock, F.R.; Klinger, J.; Zumbülte, N. Microplastic analysis—Are we measuring the same? Results on the first global comparative study for microplastic analysis in a water sample. *Anal. Bioanal. Chem.* **2020**, *412*, 555–560. [CrossRef]
8. Browne, M.A. Sources and pathways of microplastics to habitats. In *Marine Anthropogenic Litter*; Springer International Publishing: New York, NY, USA, 2015; pp. 229–244.
9. Ribic, C.A.; Sheavly, S.B.; Rugg, D.J.; Erdmann, E.S. Trends and drivers of marine debris on the Atlantic coast of the United States 1997–2007. *Mar. Pollut. Bull.* **2010**, *60*, 1231–1242. [CrossRef]
10. Seltenrich, N. New link in the food chain? Marine plastic pollution and seafood safety. *Environ. Health Perspect.* **2015**, *123*, A34. [CrossRef]
11. Hamzah, S.R.; Altrawneh, R.S.; Anuar, S.T.; Khalik, W.M.A.W.M.; Kolandhasamy, P.; Ibrahim, Y.S. Ingestion of microplastics by the estuarine polychaete, *Namalycastis* sp. in the Setiu Wetlands, Malaysia. *Mar. Pollut. Bull.* **2021**, *170*, 112617. [CrossRef]
12. Murray, F.; Cowie, P.R. Plastic contamination in the decapod crustacean *Nephrops norvegicus* (Linnaeus, 1758). *Mar. Pollut. Bull.* **2011**, *62*, 1207–1217. [CrossRef]
13. Wright, S.L.; Thompson, R.C.; Galloway, T.S. The physical impacts of microplastics on marine organisms: A review. *Environ. Pollut.* **2013**, *178*, 483–492. [CrossRef] [PubMed]
14. Taha, Z.D.; Amin, R.M.; Anuar, S.T.; Abdul Nasser, A.A.; Sohaimi, E.S. Microplastics in seawater and zooplankton: A case study from Terengganu estuary and offshore waters, Malaysia. *Sci. Total Environ.* **2021**, *786*, 147466. [CrossRef] [PubMed]
15. Welden, N.A.; Cowie, P.R. Environment and gut morphology influence microplastic retention in langoustine, *Nephrops norvegicus*. *Environ. Pollut.* **2016**, *214*, 859–865. [CrossRef] [PubMed]
16. De Sá, L.C.; Luís, L.G.; Guilhermino, L. Effects of microplastics on juveniles of the common goby (*Pomatoschistus microps*): Confusion with prey, reduction of the predatory performance and efficiency, and possible influence of developmental conditions. *Environ. Pollut.* **2015**, *196*, 359–362. [CrossRef] [PubMed]
17. Peters, C.A.; Bratton, S.P. Urbanization is a major influence on microplastic ingestion by sunfish in the Brazos River Basin, Central Texas, USA. *Environ. Pollut.* **2016**, *210*, 380–387. [CrossRef] [PubMed]
18. Lusher, A.L.; Welden, N.A.; Sobral, P.; Cole, M. Sampling, isolating and identifying microplastics ingested by fish and invertebrates. *Anal. Methods* **2017**, *9*, 1346–1360. [CrossRef]
19. Zaki, M.R.M.; Zaid, S.H.M.; Zainuddin, A.H.; Aris, A.Z. Microplastic pollution in tropical estuary gastropods: Abundance, distribution and potential sources if Klang River estuary, Malaysia. *Mar. Pollut. Bull.* **2021**, *162*, 11866. [CrossRef]
20. Rochman, C.M.; Brookson, C.; Bikker, J.; Djuric, N.; Earn, A.; Bucci, K.; Hung, C. Rethinking microplastics as a diverse contaminant suite. *Environ. Toxicol. Chem.* **2019**, *38*, 703–711. [CrossRef] [PubMed]
21. Ma, Z.F.; Ibrahim, Y.S.; Lee, Y.Y. Microplastic Pollution and Health and Relevance to the Malaysia's Roadmap to Zero Single-Use Plastics 2018–2030. *Malays. J. Med. Sci.* **2020**, *27*, 1–6. [CrossRef]
22. De Frond, H.; Hampton, L.T.; Kotar, S.; Gesulga, K.; Matuch, C.; Lao, W.; Weisberg, S.B.; Wong, C.S.; Rochman, C.M. Monitoring microplastics in drinking water: An interlaboratory study to inform effective methods for quantifying and characterizing microplastics. *Chemosphere* **2022**, *298*, 134282. [CrossRef]
23. Wiesner, Y.; Altmann, K.; Braun, U. Determination of Microplastic Mass Content by Thermal Extraction Desorption Gas Chromatography–Mass Spectrometry. *Column* **2021**, *17*, 2–7.
24. Akoueson, F.; Chbib, C.; Monchy, S.; Paul-Pont, I.; Doyen, P.; Dehaut, A.; Duflos, G. Identification and quantification of plastic additives using Pyrolysis-GC/MS: A review. *Sci. Total Environ.* **2021**, *773*, 145073. [CrossRef] [PubMed]
25. Riess, M.; Thoma, H.; Vierle, O.; Van Eldik, R. Identification of flame retardants in polymers using curie point pyrolysis-gas chromatography/mass spectrometry. *J. Anal. Appl. Pyrolysis* **2000**, *53*, 135–148. [CrossRef]

26. Suratman, S.; Hussein, A.; Tahir, N.M.; Latif, M.T.; Mostapa, R.; Weston, K. Seasonal and spatial variability of selected surface water quality parameters in Setiu wetland, Terengganu, Malaysia. *Sains Malays.* **2016**, *45*, 551–558.
27. Rosli, N.S.; Yahya, N.; Arifin, I.; Bachok, Z. Diversity of polychaeta (Annelida) in the continental shelf of Southern South China Sea. *Middle-East J. Sci. Res.* **2016**, *24*, 2086–2092.
28. WWF (World Wildlife Federation). Sustainable Management of Setiu Wetlands. Available online: http://www.wwf.org.my (accessed on 10 January 2022).
29. Ibrahim, N.F.; Ibrahim, Y.S.; Sato, M. New record of an estuarine polychaete, *Neanthes glandicincta* (Annelida, Nereididae) on the eastern coast of Peninsular Malaysia. *ZooKeys* **2019**, *831*, 81. [CrossRef] [PubMed]
30. Ibrahim, Y.S.; Hamzah, S.R.; Khalik, W.M.A.W.M.; Yusof, K.M.K.K.; Anuar, S.T. Spatiotemporal microplastic occurrence study of Setiu Wetland, South China Sea. *Sci. Total Environ.* **2021**, *788*, 147809. [CrossRef]
31. Junardi; Anggraeni, T.; Ridwan, A.; Yowono, E. Larval development of nypa palm worm *Namalycastis rhodochorde* (Polychaeta: Nereididae). *Nusant. Biosci.* **2020**, *12*, 148–153. [CrossRef]
32. Hidalgo-Ruz, V.; Gutow, L.; Thompson, R.C.; Thiel, M. Microplastics in the marine environment: A review of the methods used for identification and quantification. *Environ. Sci. Technol.* **2012**, *46*, 3060–3075. [CrossRef] [PubMed]
33. Frontier Laboratories Ltd. *Qualitative and Quantitative Analysis if Microplastic Using Pyrolysis-GC/MS (Ver. 1.1)*; Manual Operation; Frontier Laboratories Ltd.: Koriyama, Japan.
34. Fischer, M.; Scholz-Böttcher, B.M. Simultaneous trace identification and quantification of common types of microplastics in environmental samples by pyrolysis-gas chromatography-mass spectrometry. *Environ. Sci. Technol.* **2017**, *51*, 5052–5060. [CrossRef]
35. Khalik, W.M.A.W.; Ibrahim, Y.S.; Anuar, S.T.; Govindasamy, S.; Baharuddin, N.F. Microplastics analysis in Malaysian marine waters: A field study of Kuala Nerus and Kuantan. *Mar. Pollut. Bull.* **2018**, *135*, 451–457. [CrossRef] [PubMed]
36. Koh, H.L.; Teh, S.Y.; Khang, X.Y.; Raja Barizan, R.S. Mangrove Forests: Protection Against and Resilience to Coastal Disturbances. *J. Trop. For. Sci.* **2018**, *30*, 446–460. [CrossRef]
37. Suratman, S.; Latif, M.T. Reassessment of nutrient status in Setiu Wetland, Terengganu, Malaysia. *Asian J. Chem.* **2015**, *271*, 239–242. [CrossRef]
38. Isobe, A.; Iwasaki, S.; Uchida, K.; Tokai, T. Abundance of non-conservative microplastics in the upper ocean from 1957 to 2066. *Nat. Commun.* **2019**, *10*, 417. [CrossRef] [PubMed]
39. Su, Y.; Cai, H.; Kolandhasamy, P.; Wu, C.; Rochman, C.M.; Shi, H. Using the Asian clam as an indicator of microplastic pollution in freshwater ecosystems. *Environ. Pollut.* **2018**, *234*, 347–355. [CrossRef] [PubMed]
40. Huang, D.; Li, X.; Ouyang, Z.; Zhao, X.; Wu, R.; Zhang, C.; Lin, C.; Li, Y.; Guo, X. The occurrence and abundance of microplastics in surface water and sediment of the West River downstream, in the south of China. *Sci. Total Environ.* **2021**, *156*, 143857. [CrossRef]
41. Ariffin, B.E.H. Effect of Monsoons on Beach Morphodynamics in the East Coast of Peninsular Malaysia: Examples from Kuala Terengganu Coast. Ph.D. Thesis, Geomorphology, Universiteé de Bretagne Sud, Lorient, France, 2017.
42. Jaritkhuan, S.; Damrongrojwattana, P.; Chewprecha, B.; Kunbou, V. Diversity of polychaetes in mangrove forest, Prasae Estuary, Rayong Province, Thailand. *Chiang Mai J. Sci.* **2016**, *44*, 816–823.
43. Fadhullah, W.; Syakir, M.I. Polychaetes as Ecosystem Engineers: Agents of Sustainable Technologies. In *Renewable Energy and Sustainable Technologies for Building and Environmental Applications*; Springer International Publishing: New York, NY, USA, 2016; pp. 137–150.
44. Herringshaw, L.G.; Sherwood, O.A.; McIlroy, D. Ecosystem engineering by bioturbating polychaetes in event bed microcosms. *Palaios* **2010**, *25*, 46–58. [CrossRef]
45. Otegui, M.B.; Brauko, K.M.; Pagliosa, P.R. Matching ecological functioning with polychaete morphology: Consistency patterns along sedimentary habitats. *J. Sea Res.* **2016**, *114*, 13–21. [CrossRef]
46. Rodriguez, V.F.; Mesa, M.H.L. Polychaetes (Annelida: Polychaeta) as biological indicators of marine pollution: Cases in Colombia. *Manag. Environ.* **2015**, *18*, 189–204.
47. Jang, M.; Shim, W.J.; Han, G.M.; Song, K.Y.; Hong, S.H. Formation of microplastics by polychaetes *Marphysa sanguinea* inhabiting expanded polystyrene marine debris. *Mar. Pollut. Bull.* **2018**, *131*, 365–369. [CrossRef] [PubMed]
48. Kusch, P. *Pyrolysis-gas Chromatography: Mass Spectrometry of Polymeric Materials*; World Scientific: London, UK, 2018.
49. Fries, E.; Dekiff, J.H.; Willmeyer, J.; Nuelle, M.T.; Ebert, M.; Remy, D. Identification of polymer types and additives in marine microplastic particles using pyrolysis-GC/MS and scanning electron microscopy. *Environ. Sci. Processes Impacts* **2013**, *15*, 1949–1956. [CrossRef]
50. Hahladakis, J.N.; Velis, C.A.; Weber, R.; Iacovidou, E.; Purnell, P. An overview of chemical additives present in plastics: Migration, release, fate and environmental impact during their use, disposal and recycling. *J. Hazard. Mater.* **2018**, *344*, 179–199. [CrossRef]
51. Tsuge, S.; Ohtani, H.; Watanabe, C. *Pyrolysis-GC/MS Data Book of Synthetic Polymers: Pyrograms, Thermograms and MS of Pyrolyzates*; Elsevier Science & Technology: London, UK, 2011.
52. Dümichen, E.; Eisentraut, P.; Celina, M.; Braun, U. Automated thermal extraction-desorption gas chromatography mass spectrometry: A multifunctional tool for comprehensive characterization of polymers and their degradation products. *J. Chromatogr. A* **2019**, *1592*, 133–142. [CrossRef] [PubMed]
53. Pham, D.D.; Cho, J. Low-energy catalytic methanolysis of poly(ethyleneterephthalate). *Green Chem.* **2021**, *23*, 511–525. [CrossRef]

54. Yu, J.; Sun, I.; Ma, C.; Qiao, Y.; Yao, H. Thermal degradation of PVC: A review. *Waste Manag.* **2016**, *48*, 300–314. [CrossRef] [PubMed]
55. Jan, M.R.; Shah, J.; Rahim, A. Recovery of styrene monomer from waste polystyrene using catalytic degradation. *Am. Lab.* **2008**, *40*, 12–14.
56. Jansson, K.D.; Zawodny, C.P.; Wampler, T.P. Determination of polymer additives using analytical pyrolysis. *J. Anal. Appl. Pyrolysis* **2007**, *79*, 353–361. [CrossRef]
57. Tibbetts, J.; Krause, S.; Lynch, I.; Smith, G.H.S. Abundance, distribution, and drivers of microplastic contamination in urban river environments. *Water* **2018**, *10*, 1597. [CrossRef]
58. Nasa, J.L.; Biale, G.; Ferriani, B.; Trevisan, R.; Colombini, M.P.; Modugno, F. Plastics in heritage science: Analytical pyrolysis techniques applied to objects of design. *Molecules* **2020**, *25*, 1705. [CrossRef]

Article

Extraction and Characterization of Microplastics from Portuguese Industrial Effluents

Solange Magalhães [1], Luís Alves [1,*], Anabela Romano [2], Bruno Medronho [2,3,*] and Maria da Graça Rasteiro [1,*]

1 CIEPQPF, Department of Chemical Engineering Pólo II–R. Silvio Lima, University of Coimbra, 3030-790 Coimbra, Portugal; solangemagalhaes@eq.uc.pt
2 MED–Mediterranean Institute for Agriculture, Environment and Development, Faculdade de Ciências e Tecnologia, Campus de Gambelas, Universidade do Algarve, Ed. 8, 8005-139 Faro, Portugal; aromano@ualg.pt
3 FSCN, Surface and Colloid Engineering, Mid Sweden University, SE-851 70 Sundsvall, Sweden
* Correspondence: luisalves@ci.uc.pt (L.A.); bfmedronho@ualg.pt (B.M.); mgr@eq.uc.pt (M.d.G.R.)

Abstract: Microplastics (MPs) are contaminants present in the environment. The current study evaluates the contribution of different well-established industrial sectors in Portugal regarding their release of MPs and potential contamination of the aquifers. For each type of industry, samples were collected from wastewater treatment plants (WWTP), and different parameters were evaluated, such as the potential contamination sources, the concentration, and the composition of the MPs, in both the incoming and outcoming effluents. The procedures to extract and identify MPs in the streams entering or leaving the WWTPs were optimized. All industrial effluents analysed were found to contribute to the increase of MPs in the environment. However, the paint and pharmaceutical activities were the ones showing higher impact. Contrary to many reports, the textile industry contribution to aquifers contamination was not found to be particularly relevant. Its main impact is suggested to come from the numerous washing cycles that textiles suffer during their lifetime, which is expected to strongly contribute to a continuous release of MPs. The predominant chemical composition of the isolated MPs was found to be polyethylene terephthalate (PET). In 2020, the global need for PET was 27 million tons and by 2030, global PET demand is expected to be 42 million tons. Awareness campaigns are recommended to mitigate MPs release to the environment and its potential negative impact on ecosystems and biodiversity.

Keywords: microplastics; Portugal; resin; pharmaceutical; PVC; paint; wastewater treatment plant

Citation: Magalhães, S.; Alves, L.; Romano, A.; Medronho, B.; Rasteiro, M.d.G. Extraction and Characterization of Microplastics from Portuguese Industrial Effluents. *Polymers* **2022**, *14*, 2902. https://doi.org/10.3390/polym14142902

Academic Editors: Zbigniew Bartczak and Graeme Moad

Received: 19 May 2022
Accepted: 13 July 2022
Published: 17 July 2022

Publisher's Note: MDPI stays neutral with regard to jurisdictional claims in published maps and institutional affiliations.

Copyright: © 2022 by the authors. Licensee MDPI, Basel, Switzerland. This article is an open access article distributed under the terms and conditions of the Creative Commons Attribution (CC BY) license (https://creativecommons.org/licenses/by/4.0/).

1. Introduction

Microplastics (MPs) have been gaining increasing awareness after several reports regarding "garbage patches" in the world's oceans [1]. Similarly to climate changes and persistent organic pollutants, plastic residues are also perfect examples of the human capacity to significantly affect ecosystems and biodiversity on a global scale. MPs have been detected and identified in many different environments, such as aquatic media, ground waters, landfill leachate, wastewater, and sewage sludge [2–4]. MPs have also been detected in remote areas, including polar regions, such as Antarctica of the southern ocean and on the deepest parts of the ocean at the Mariana Trench, North Pacific gyre, and south pacific islands, with particularly high concentrations [5–7]. Even in the world's highest hills, the Tibetan plateau, MPs have been detected within its rivers and lakes [8]. Wastewater treatment plants (WWTPs) can mitigate the spreading of MPs by retaining part of them (larger dimensions) with screens and filters [9–11]. In WWTPs, ca. 78–98% of MPs can be removed after primary treatment, while the secondary treatment is responsible for a smaller decrease, ca. 7–20% [12,13]. However, it has been shown that MPs can pass through the WWTP, flowing into the aquatic media and accumulating in the environment [14,15]. Furthermore, the aggregation of MPs with other suspended solids in wastewaters induces their accumulation in sewage sludge [16,17]. The abundance of MPs in wastewaters and sewage sludge varies depending on the time of the day, season, and type of wastewaters (i.e., domestic or industrial). Urbanization and industrialization are among the main reasons for the observed increasing contamination of MPs. In this respect, daily industrial discharges of different manufacturing industries, such as paint, pharmaceutical, textile, resins, cosmetics, etc., can have a significant impact on the total amount of MPs released [18,19]. MPs can then easily enter the food chain. Their chemical composition is complex and includes different polymeric substances and additives (e.g., flame-retardants, dyes, plasticizers, and UV-inhibitors) that may impart toxicity to living beings, or act as adsorbent agents for other harmful organic pollutants [1,19].

It is generally recognized that the most important contamination sources of MPs arise from the textile and paint/coating industries [1,20–22]. Textiles are flexible materials that mainly consist of natural and/or synthetic fibres. Today´s technology allows the manufacturing of several types of fabrics, which can be made of pure fibres (e.g., viscose, polyester, etc.) or mixtures of fibres (e.g., cotton–polyester). Mixing different types of fibres helps to enhance the physical characteristics of the final material (e.g., elasticity, strength, durability) and eventually lowers the price [22]. On the other hand, the paint/coating industries' contamination focuses mainly on alkyd ship paint resins and poly (acrylate/styrene) from fibreglass resins. In this case, MPs are concentrated at the surface microlayer of the ship [1].

In Portugal, there are about sixty-nine thousand manufacturing companies, 17% being chemical-related industries and 5% dealing with textiles. It has been estimated that ca. 72% of the waste found in Portuguese industrial areas and estuaries are MPs [23–26]. Industrial spillages, emissions from road traffic, atmospheric deposition, wind-blown debris from littering or loss during waste disposal, and the degradation of larger plastic debris directly in the aquatic media may further contribute to MPs contamination in aquatic ecosystems [27].

The efficient separation and identification of MPs in the inflow and outflow effluents of WWTPs is very desirable but technically challenging due to the complexity of these streams [15,28]. Typically, the MPs detection involves three main steps but, so far, the methods used in each stage have not been standardized [29].

The MPs separation process is usually performed with a series of sieves or filters of different mesh/pore sizes in which the collected effluent is forced to pass [12]. Afterwards, the material on each sieve can be washed with distilled water and deposited into glass vials. However, the procedure has several problems, such as the clog of the sieves with organic load [30]. Ziajahromi et al. (2017) developed a method for MPs separation from wastewater effluents [31]. This method involves a high-volume sampling device with multiple mesh screens to collect a wide size range of MPs from wastewater effluents. This process is combined with an efficient sample processing procedure using organic matter digestion, density separation, and staining to identify and eliminate the non-plastic particles [31]. The efficiency of the technology is highly satisfactory as the capture of MPs ranged from 92% for the 25 μm mesh screen to 99% for the 500 μm mesh screen. This shows that the sampling device is suitable to capture MP with a wide range of particle sizes. However, the sieve-based separation presents some limitations, also highlighted in the National Ocean and Atmospheric Administration (NOAA) norm, which are related to the MP morphology and size. For example, microfibres, have a high aspect ratio and thus can be either retained horizontally in the sieve or pass longitudinally through the sieve holes.

Another procedure to extract MPs of different nature (i.e., PP, PVC, and PET) was described by Nuelle et al. (2014), where MPs were extracted from sediments using a two-step approach. The authors suggest that the developed method is suitable to monitor MPs in marine sediments [32]. As mentioned, the technique consists of two main steps: (i) fluidisation of sediments in a saturated NaCl solution and (ii) subsequent flotation of MPs in a high-density NaI (aqueous solution). With this two-step approach, it was possible to efficiently extract common synthetic macromolecules, including high-density polymers from contaminated sediments. Compared with other systems based on flotation in high-density salts, this method is more advantageous, due to the availability of materials used, and the simplicity of the equipments, rendering the whole process cost-effective [32].

A slightly different method was proposed by Besley et al. (2017) who used a fully saturated salt solution of NaCl combined with filtration for the extraction of MPs [33]. The extraction of the MPs is accomplished by density separation followed by drying with a saturated salt solution. The supernatant is afterwards submitted to vacuum filtration and the filter membranes containing the MP particles are examined by stereo-microscopy. This method allows for the quantification of MPs in the range of 0.3–5 mm [33].

Wei et al. (2020) used sieves to remove materials larger than 5 mm in length. Then, H_2O_2 is added to the filtrate at 80 °C to degrade the organic matter. After volume reduction, plastic components are separated from non-plastic materials via density separation (40% $CaCl_2$ solution). Finally, the samples are centrifuged at 5000 rpm and, the supernatant particles are vacuum filtered using a glass filter [34]. The drawback of this laborious process is that MPs with a higher density can precipitate during the centrifugation process thus underestimating the global MP quantification.

In general, all the methods used to separate the MPs from the aqueous medium are based on the National Ocean and Atmospheric Administration (NOAA) norms for processing samples. Note that the standard procedure was designed for samples collected in marine environments. However, due to the complex matrix of wastewater samples and the presence of numerous particles of organic nature and other compounds of different sources, this procedure is not considered suitable for other non-marine samples. This method involves the digestion of organic matter using H_2O_2 in the presence of an aqueous ferrous solution (Fe(II)) as a catalyst. The digestion step is usually followed by a separation stage, which uses NaCl (aq) or $ZnCl_2$ (aq) to increase the density of the liquid phase. Through this procedure, the low-density MPs tend to float, while the high-density particles settle at the bottom. Then, the liquid phase is filtered through mesh sizes varying from 0.7 to 125 μm [35].

Despite the low concentrations of MPs typically reported for treated effluents, the number of MPs released to recipient waters can still be very high over relevant temporal scales [36]. Therefore, the present work is focused on the evaluation of the release of MPs in the treated effluents of WWTPs and the main goal is to elucidate the role of different Portuguese industries as potential sources of MPs contamination to ecosystems. To do so, different effluents from the paint, resin, pharmaceutical, textile, and PVC industries were collected, and the MPs were analyzed regarding their concentration and type, after implementing suitable separation and cleaning protocols. The relationship between the different industries and MPs predominancy is discussed. Another relevant point that should be highlighted is that the procedure suitable for extracting MPs was based on a generic method, but it had to be adapted to each effluent, based on the specificities of each sample (i.e., different organic or inorganic fractions, presence of other contaminants, etc).

2. Materials and Methods
2.1. Materials

Five different Portuguese industrial effluents (i.e., paint, resin, textile, pharmaceutical, and PVC) were collected and analysed before and after the company's in-house WWTP. Ultra-pure water (UPW) and ultra-filters (pore size < 2 mm) were used for all extraction and characterization steps. The H_2O_2 (30%) was purchased from Greendet, Portugal. KOH was obtained from LabKem, Spain. HCl (37%) was purchased from Honeywell, Portugal. To minimize the potential contamination by other agents, all reagents were filtered before use with 0.45 μm syringe PTFE filters. The anionic surfactant sodium alkylbenzenesulfonate (99%) was obtained from Sigma-Aldrich (St. Louis, MO, USA).

Sample filtration was performed using a vacuum filtration unit made of glass (FiltresRS par Jean-Pierre D., France). The glass fibre filters (pore size 1–2 μm) were burned at 250 °C for 30 min before use.

2.2. Sample Preparation and Extraction of the Microplastics

The protocols commonly used for MPs extraction and analysis often require long digestion times and different preparation steps [37]. Our approach involved an initial alkaline treatment, 10% KOH (aq) for 12 h at 50 °C, to remove organic matter [38]. After the alkaline treatment, acid-based digestion was employed, consisting of 20% HCl (aq) solution for 15 min (adapted from Nuelle et al., 2014 [32]). This treatment digests any biological residues or inorganic materials, such as calcium carbonate. After the basic and acidic digestions, the samples were filtered with 1–2 μm glass filters. The MPs were then separated via their density differences with a concentrated NaCl (aq) solution, with a density of 1.2 g/cm^3. Finally, after separated, the MPs were cleaned with H_2O_2 and ethanol, to remove the presence of any hypothetical microorganisms from their surfaces. It is important to point out that this protocol was adjusted depending on the type of effluent. That is, depending on the effluent source, the amounts of organic, inorganic compounds and/or other contaminants may vary, thus requiring special care. For instance, in the case of the pharmaceutical effluent, a 1% anionic surfactant solution was used before the alkaline digestion to enhance filtration and MPs separation.

2.3. Characterization and Quantification of Microplastics

The size and shape of MPs were analysed in an Olympus BH-2 KPA optical microscope (Olympus Optical Co., Ltd., Tokyo, Japan) equipped with a high-resolution CCD colour camera (ColorView III, Olympus Optical Co., Ltd., Tokyo, Japan). The size was further accessed by laser diffraction spectroscopy (LDS) in a Malvern Masterziser 2000 (Malvern Instruments, Malvern, UK). The effluent samples were loaded in the equipment dispersion unit until a fixed level of obscuration (i.e., 6–7% guarantees good signal-to-noise quality). The LDS tests were carried out by setting the pump speed to 1500 rpm, corresponding to an average shear rate of 334 s^{-1}.

The charge density of MPs was evaluated by measuring the zeta potential of the effluent samples in a Zetasizer NanoZS equipment (ZN 3500, Malvern Instruments, Malvern, UK), in the electrophoretic light scattering mode. In brief, the samples were gently transferred to a folded capillary zeta cell, the presence of air bubbles was visually checked and then the cell was left to equilibrate at 25 °C before starting the assay. Each sample was scanned 3 times, each time with 12 runs. The results were processed using the Zetasizer Nano Software 7.13 (Malvern Instruments).

The MPs quantification was performed by gravimetry, weighing the filter before starting the filtration process (after burning at 250 °C) and after filtration and being oven-dried at 105 °C for 8 h. The MPs amount was estimated by the mass difference in 100 mL of each effluent type.

2.4. Microplastics Identification

The identification of the polymer types present in the isolated MPs was accessed by Fourier transform infrared spectroscopy (FTIR). The obtained spectra of the MPs before and after the company's in-house WWTP treatment were obtained and compared with an existing FTIR library. The FTIR spectra were obtained using an ATR-FTIR (Perkin–Elmer FT-IR spectrometer, Waltham, MA, USA) with a universal ATR sampling accessory, between 500 cm^{-1} and 4000 cm^{-1}, resolution of 4 cm^{-1} and applying 128 scans.

Fluorescence microscopy was also employed to characterize the isolated MPs, using pyrene as the fluorescent probe. Pyrene has demonstrated excellent sensitivity towards surface-modified carboxyl particles [39]. In brief, 200 μL of the pyrene stock solution (1 mg/mL) were diluted into 20 mL of milli-Q water and poured into the filtration funnel. To achieve a reliable stanning, the solution was allowed to be in contact with the filter disc for ca. 15 min. After this period, the excess solution was removed by vacuum filtration and the filter disc was placed in the filter holder. Samples were imaged at room temperature in a fluorescence microscope (Olympus BX51M), equipped with a 100 × objective lens, a filter set type U-MNU2 (360–370 nm excitation and 400 nm dichromatic mirror) and a UV–

mercury lamp (100 W Ushio Olympus, Olympus Optical Co., Ltd., Tokyo, Japan). Images were obtained through a video camera (Olympus digital camera DP70, Olympus Optical Co., Ltd., Tokyo, Japan) and analysed with an image processor (Olympus DP Controller 2.1.1.176, Olympus DP Manager 2.1.1.158, Olympus Optical Co., Ltd., Tokyo, Japan).

2.5. Statistical Analysis

Statistical analysis was performed using one-way ANOVA ($\alpha = 0.05$) to evaluate significant differences between the zeta potential of the effluents studied.

3. Results

MPs were not readily considered as potential sources of contamination and, for a long time, most samples were analysed with little to no concern about their potential impact. However, with the awareness that MPs are virtually everywhere, contaminating different ecosystems and having potential toxic effects on animals and plants, new strategies must be developed to understand their hypothetical impact and mitigate it. This motto has been the driving force for this study. A total of five effluents from different Portuguese industries (i.e., paint, resin, pharmaceutical, PVC, and textile) were analysed to infer their contribution to the MPs release and pollution of aquifers. Effluents are strongly dependent on the source and thus differ from each other, since these are complex mixtures of several pollutants, including synthetic chemicals, hydrocarbons, acrylic polymers, inorganic compounds, and heavy metals [40]. Effluents from the resin industry are usually composed of acrylic polymers and inorganic compounds, such as calcium carbonate, plasticizers, etc., while PVC effluents are generally composed only of polymers and additives that are incorporated during the polymerization process for PVC morphology control [41]. On the other hand, effluents from the pharmaceutical industry are typically heterogeneous mixtures of polymers, surfactants, antibiotics, and organic contaminants, among others [42]. Similarly, textile-based effluents can contain numerous toxic compounds, such as nonylphenol ethoxylates, benzothiazole, dyes, etc. [43].

The particle size of the effluents was characterized in samples obtained before and after each in-house WWTP treatment (Figure 1).

As can be observed in Figure 1, all effluents have two well-defined size populations of the particles. The exception is the effluent from the PVC industry (Figure 1D), where the size distribution is the same before and after the in-house WWTP. In the effluent from the textile industry (Figure 1E), the in-house WWTP treatment seems to remove the larger particles, thus increasing the volume% of the smaller particles. The effluents with larger particles are the ones coming from the pharmaceutical (Figure 1C) and the PVC (Figure 1D) industries, while the smallest particles were detected in the effluent from the textile industry. Except for the PVC effluent, the in-house WWTPs decrease the particle size. Therefore, it is reasonable to assume that these treatments are efficiently removing the bigger particles shifting the population towards smaller sizes.

The average charge density of the collected effluents was estimated via zeta potential, as summarized in Table 1. Again, samples were analysed before and after each in-house WWTP.

Table 1. Zeta potential of the different effluents, before and after each company in-house WWTP.

Industries	Before Treatment (mV)	After Treatment (mV)
Resin	−13.13 (±0.90) *	−4.93 (±0.24)
Paint	−3.94 (±0.21)	−6.20 (±0.18)
Pharmaceutical	8.47 (±1.07)	−9.90 (±0.70)
PVC	−12.30 (±0.56) *	−7.89 (±0.08)
Textile	−28.30 (±1.47)	−31.13 (±1.40)

* values marked with the same symbol mean no statistical difference ($p < 0.05$).

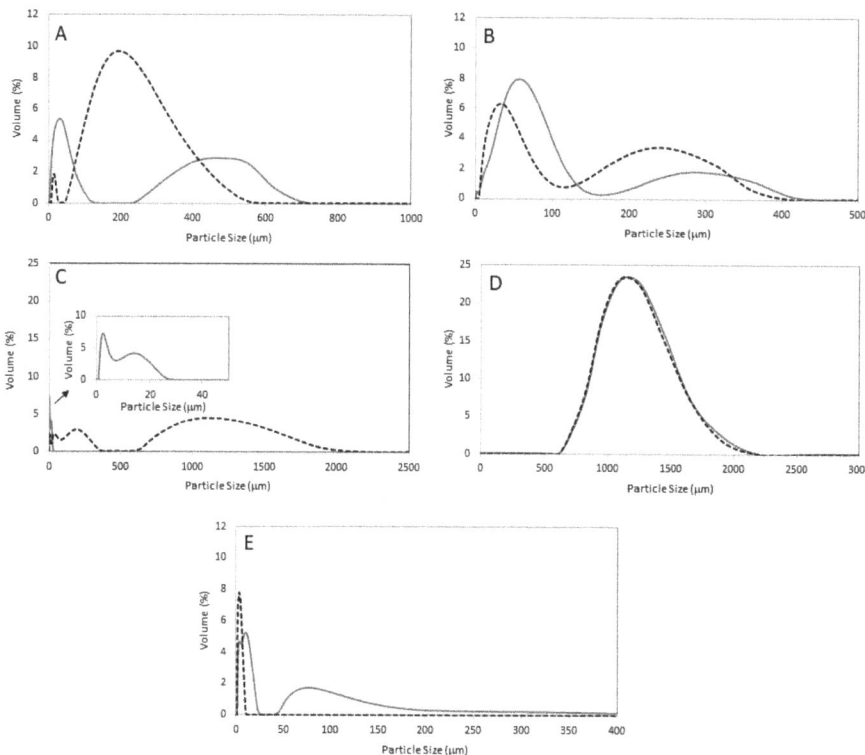

Figure 1. LDS of the different effluents as collected, before (solid grey lines) and after (dashed black lines) the in-house WWTP of each industry: (**A**) resins; (**B**) paint; (**C**) pharmaceutical; (**D**) PVC, and (**E**) textil. The assays were performed at 25 °C.

All effluents showed significant differences in zeta potential, except the effluent of resin and PVC industries, before treatment. For the effluents after treatment, all are significantly different in zeta potential. The results of the one-way ANOVA tests are summarized in Table S1. Larger differences are observed comparing the inflow and outflow effluents for the same industry.

Overall, all the effluents presented a negative charge density before and after treatment, except the pharmaceutical effluent before treatment, probably due to the presence of cationic polymers in the solution (e.g., cationic polyacrylamides). Therefore, at environmentally relevant pH and above, the MPs are generally negatively charged and consequently so is the effluent [44].

Following the procedure described in the experimental section, the MPs from the different effluents (before and after the in-house WWTP) were extracted, cleaned, and quantified (Table 2).

Although all in-house WWTP decreased the % of MPs released, significant amounts of MPs are still liberated and potentially contaminating different ecosystems. Among the different sectors analysed, the textile industry is surprisingly observed to contribute the least. Yet, some studies have shown that synthetic fibres are the dominant type of polyester MPs detected in aqueous media, sediments, and various organisms [43,45,46]. This can be reasoned by the massified and extensive domestic and industrial washing cycles continuously releasing MPs into wastewaters. It should be highlighted that the textile WWTP shows a lesser ability to retain MPs particularly due to their smaller size (see Figure 1).

Table 2. Gravimetric quantification of MPs before and after each company's in-house WWTP. An estimative of the MPs amount released per ton of effluent is also presented.

	Before in-House WWTP (g MPs/100 mL Effluent)	After in-House WWTP (g MPs/100 mL Effluent)	MPs Released after in-House WWTP (g MPs/Ton Effluent)
Resins	0.044 ± 0.020	0.004 ± 0.002	41.66
Paint	0.172 ± 0.095	0.009 ± 0.004	89.01
Pharmaceutical	0.004 ± 0.003	0.002 ± 0.002	24.69
PVC	0.029 ± 0.001	0.007 ± 0.006	70.58
Textile	0.002 ± 0.002	0.002 ± 0.002	15.65

The shapes of MPs recovered from the different effluent samples were mainly elongated fibres and fragments. Fibres constitute the majority as this shape allows an easier diffusion through the in-house WWTP filters. The fragments, in addition to being easily retained in the WWTP filters, can form larger aggregates with the flocculating agents used and thus become more accessible for trapping in the in-house WWTP. In the samples studied, the observed sizes range from 10 µm to 500–600 µm (Figure 2); it has been reported that 80% of the MPs fall within a size range of 125–500 µm [47]. Smaller MPs have also been detected in this work, which can be either due to the specific nature of the effluents tested and/or partial MPs degradation during the cleaning procedure. The alkaline, acidic, and oxidizing treatments used may further fractionate the MPs [48,49]. To infer the effect of our developed "cleaning" treatment on the MPs structure, the samples were analysed before and after applying it. In Figure 2, a typical example of the MPs shape and size distributions is shown for the resin effluent. Similar qualitative data was obtained for the other effluents (Figure S1). Although the trends are not fully clear, three main conclusions can be drawn: (i) the lowest size reported (i.e., 10–50 µm) tends to decrease upon "cleaning" the samples (this is particularly striking in the resins effluent); (ii) the bigger size reported (i.e., 500–600 µm) decreases after "cleaning" the effluents; (iii) the intermediate sizes (i.e., 50–500 µm) tend to increase their number density, most likely as a result of the fractionation of the bigger fibres. Therefore, this analysis shows that the cleaning procedure (applied to remove unwanted inorganic and organic compounds) is not innocuous and can, in fact, affect the MPs shape and size. Although often neglected in the literature, any reliable analysis should not rule out its contribution.

Besides the shape and size, it is also important to study the predominant type of MPs in each effluent. Based on the composition of the plastics, they can be classified into two main categories: (i) thermosetting plastics, such as polyimides and bakelite, and (ii) thermoplastics, such as polyethylene terephthalate, polypropylene, and polyethylene [50]. Typically, thermoplastic materials have a highly cross-linked structure, consisting of a three-dimensional network of covalently bonded atoms, displaying remarkable physical properties, such as high thermal stability, high rigidity, high dimensional stability, resistance to creep or deformation under load and high electrical and thermal insulating capacity [51]. Moreover, thermoplastics, which are far more abundant, can be heated and melted and then cooled and reformed into new shapes, thereby allowing many of them to be recycled [52]. On the other hand, the majority of thermosetting plastics are permanently set and cannot be melted and reformed. This irreversible chemical change does not allow the recycling of thermosetting plastics.

A useful recent technique to infer the chemical composition of the MPs is fluorescence microscopy [43,53]. The fundamentals are fairly easy to understand; samples are stained with suitable dyes, such as pyrene, and, depending on the composition, they will fluoresce (decay after excitation with UV light) at different wavelengths. In practice, this means that the colour observed in the fluorescence microscope can be correlated to the particle composition. Pyrene is particularly sensitive to the different MPs polarities, thus being a suitable dye [34,38]. In Figure 3, typical fluorescent micrographs of the MPs from the PVC effluent are shown. Depending on the effluent type (and MPs composition) different emissions can be observed (Figure S2).

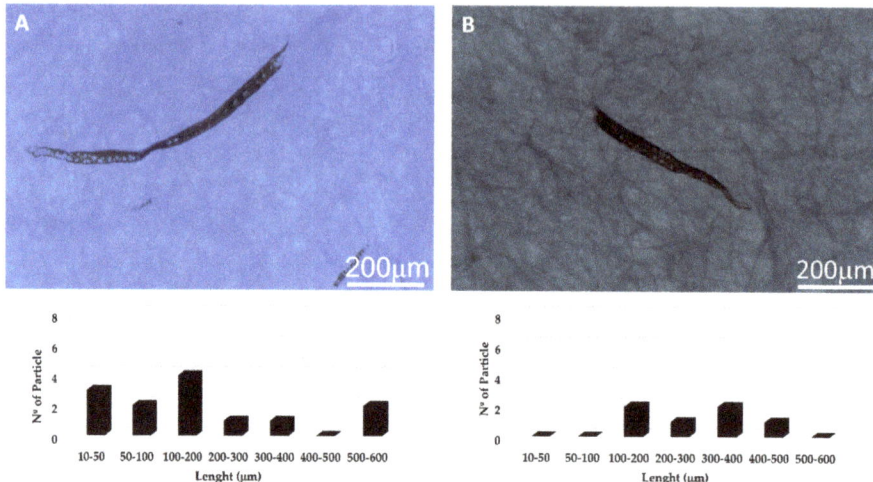

Figure 2. Optical micrographs and size distributions of the MPs before (**A**) and after (**B**) applying the MPs extracting procedure developed in this work to the resin effluent. Note that the data regarding the other effluents can be found in the Supplementary Information section.

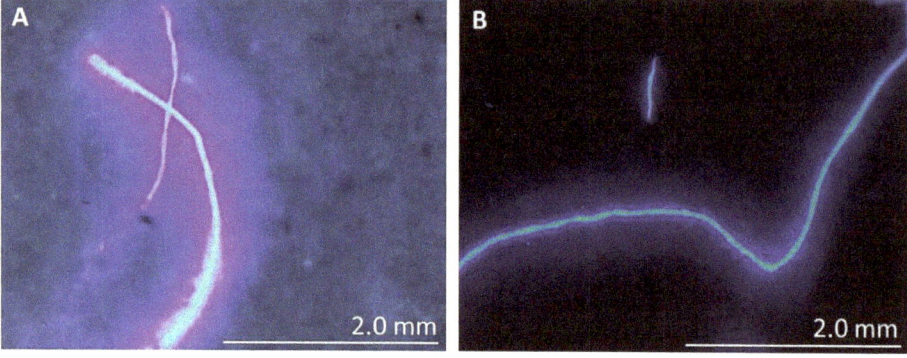

Figure 3. Fluorescence micrographs of the MPs before (**A**) and after (**B**) applying the cleaning procedure developed in this work to the PVC effluent. Samples were stained with pyrene (see the experimental section for details). Note that the data regarding the other effluents can be found in the Supplementary Information section.

As can be seen in Figures Figure 3 and S2, the steady-state fluorescence emission spectra of pyrene-stained samples mostly present MPs with green, blue, and red colours, which can be assigned to high-density polyethylene (HDPE), polyethylene (PE), and polyethylene terephthalate (PET), respectively [43]. In the effluents from the paint, resin, PVC, and textile industries (see Figures Figure 3 and S2), the presence of PET can be mainly observed. Only in the case of the effluent from the pharmaceutical industry, was the presence of HDPE, PE and PET MPs possible to observe through the same measurements (Figure S2E,F). The worldwide production of plastics, according to their polymer composition, is as follows: 36% polyethylene (PE), 21% polypropylene (PP), <10% polyethylene terephthalate (PET), <10% polyurethane (PUR), and <10% polystyrene (PS) [47]. Therefore, the largest used plastic types worldwide are in perfect agreement with most of the MPs observed in this work.

ATR-FTIR was also used to complement the fluorescent data and infer the MPs composition. This technique is usually suitable only for clean isolated particles that are big enough to be placed on the ATR crystal [48]. However, under optimized conditions, it was possible to identify the composition of the most prevalent types of MPs. In Figure 4, typical FTIR spectra are shown for the MPs from pharmaceutical effluent, before and after being submitted to the in-house WWTP procedure. The spectrum of neat PE is also shown for reference. The pharmaceutical effluent shows vibrational modes that are also characteristic of model PE, such as the rocking deformation, wagging deformation, and asymmetric stretching of CH_2 groups. It is also possible to identify vibrational bands that can suggest the presence of PP. In particular, the band at 600 cm^{-1} can be attributed to the CH wagging mode, the band at 844 cm^{-1} is assigned to C-CH stretching, at 961 cm^{-1} is assigned to the trans CH wagging, while the band at 1240–1252 cm^{-1} is assigned to CH rocking vibration and the band at 2890–2950 cm^{-1} can be attributed to the CH_2 asymmetric stretching of PP. Similar assignments can be done for the remaining effluents (see Figure S3 for details). Another striking observation for all samples tested (see Figure S3 for the remaining effluent types) is that after each in-house WWTP treatment, the intensity of the FTIR bands decreases. Although FTIR is not a truly quantitative method, these results suggest that the in-house WWTP reduce the number of MPs in the treated effluent, which agrees with the gravimetric quantifications presented in Table 2. Overall, FTIR complements and supports the fluorescent microscopy data showing that the main compounds present in the MPs are PE and PP.

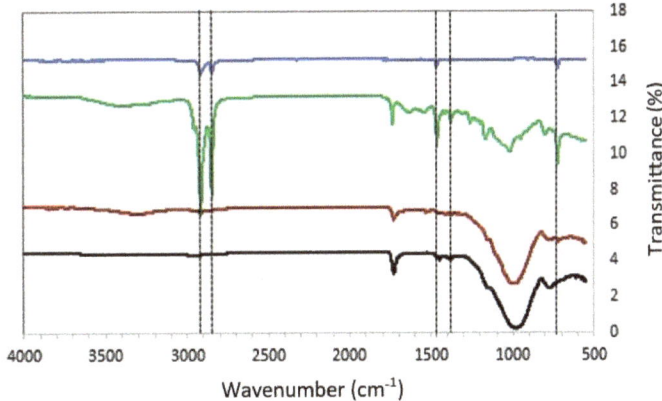

Figure 4. FTIR spectra of the MPs from the pharmaceutical effluent before (red curve) and after (green curve) the in-house WWTP. The FTIR spectra from the filter (black curve) and neat PE (blue curve) are also shown. The main vibrational modes are highlighted with the dashed vertical lines and their assignment are discussed in the main text. The FTIR spectra from the remaining effluents can be found in the Supplementary Information section [54,55].

In brief, it was possible to detect MPs in the studied effluents from different industries being prevalent the MPs from the plastic types widely used, such as PE, PP, and PET. The contamination with MPs is a real problem, and the WVTP treatments used nowadays, at least in the industries analysed, are not sufficiently robust to remove/retain all MPs. Thus, it is urgent to develop methods that allow improved efficiency and higher removal yields of these contaminants from the industrial effluents.

4. Conclusions

Presently, MPs are a well-distributed contaminant virtually all around the world, whose potential negative impact on different ecosystems is still hard to evaluate. It has been shown that the amount of MPs varies significantly depending on the source, but it is particularly relevant in industrial effluents. The inland Portuguese case here explored is no exception. Nevertheless, the predominant MPs type in all the industrial areas studied was PET, which agrees with the fact that it is the most common and widely used thermoplastic polymer. It is also striking, as can be gleaned from this work, that accurate MP extraction and quantitative and qualitative characterizations require special care for a reliable and free-of-objects analysis. This work also revealed that, in the Portuguese case, the textile industry is the sector that does not contaminate (regarding MPs release) the most; the larger number of MPs detected come from pharmaceutical and paint effluents. Further studies are required to understand whether these results reflect the real Portuguese scenario or sporadic technical issues in the in-house WWTP of these industries. This hypothesis should not be ruled out as occasional malfunction problems may compromise an overall appreciation. Regardless of the case, we are confident in stating that the in-house WWTP does not reveal a satisfactory efficiency regarding MPs retention. Serious future efforts should be considered to efficiently mitigate MPs release, especially when, at this stage, little is known about their real impact on the ecosystems. This work expects to bring awareness to this potential threat, as it is important that industries do not neglect their treated effluents as a significant possible MP disseminator source, despite having a WWTP in their facilities. We suggest that this study or other similar analyses should be extended to other countries to understand the trends on a global scale and the real dimensions of the MP threat.

Supplementary Materials: The following supporting information can be downloaded at: https://www.mdpi.com/article/10.3390/polym14142902/s1, Figure S1: Optical micrographs and size distributions of the MPs before (left column) and after (right column) applying the cleaning procedure developed in this work to the resin (A,B), paint (C,D), pharmaceutical (E,F), PVC (G,H) and textile (I,J) effluents. Note that below each representative optical micrograph the correspondent size distributions can be found. Figure S2: Fluorescent micrographs of the MPs before (left column) and after (right column) applying the cleaning procedure developed in this work to the PVC (A,B), resin (C,D), paint (E,F), pharmaceutical (G,H) and textile (I,J) effluents. Figure S3: FTIR spectra of the MPs from the (A) resin, (B) paint, (C) PVC and (D) textile effluents before (red curves) and after (green curves) the in-house WWTP. The FTIR spectra from the filter (black curve) is also shown. The main vibrational modes are highlighted with the dashed vertical lines and its assignment is discussed in the main text and bellow. Table S1: Statistical treatment of charge results. The symbol (-) indicates significant differences in the effluents studied, and (*) indicates no significant differences from one-way ANOVA tests ($p \leq 0.05$).

Author Contributions: Conceptualization: S.M., L.A., B.M. and M.d.G.R.; funding acquisition: L.A, B.M. and M.d.G.R.; investigation: S.M., L.A., B.M. and M.d.G.R.; methodology: S.M., L.A., B.M. and M.d.G.R.; project administration: L.A, B.M. and M.d.G.R.; supervision: L.A, B.M. and M.d.G.R.; writing (original draft preparation): S.M.; writing (preparation, review, and editing): S.M., L.A., B.M., A.R. and M.d.G.R. All authors have read and agreed to the published version of the manuscript.

Funding: This work was financially supported by the Portuguese Foundation for Science and Technology, FCT, through the PhDs grant 2020.07638.BD. L.A. acknowledges FCT for the research grant 2021.00399.CEECIND. B.M. acknowledges FCT for the project PTDC/ASP-SIL/30619/2017 and the researcher grant CEECIND/01014/2018. The MED is supported by FCT through the project UIDB/05183/2020 and COMPETE. The Strategic Research Centre Project UIDB00102/2020, funded by the FCT, is also acknowledged.

Institutional Review Board Statement: Not applicable.

Informed Consent Statement: Not applicable.

Data Availability Statement: The data presented in this study are available on request from the corresponding author.

Conflicts of Interest: The authors declare no conflict of interest.

References

1. Abbasi, S.; Soltani, N.; Keshavarzi, B.; Moore, F.; Turner, A.; Hassanaghaei, M. Microplastics in different tissues of fish and prawn from the Musa Estuary, Persian Gulf. *Chemosphere* **2018**, *205*, 80–87. [CrossRef] [PubMed]

2. Al-Azzawi, M.S.M.; Kefer, S.; Weißer, J.; Reichel, J.; Schwaller, C.; Glas, K.; Knoop, O.; Drewes, J.E. Validation of Sample Preparation Methods for Microplastic Analysis in Wastewater Matrices—Reproducibility and Standardization. *Water* **2020**, *12*, 2445. [CrossRef]
3. Aldalbahi, A.; El-Naggar, M.; El-Newehy, M.; Rahaman, M.; Hatshan, M.; Khattab, T. Effects of Technical Textiles and Synthetic Nanofibers on Environmental Pollution. *Polymers* **2021**, *13*, 155. [CrossRef] [PubMed]
4. Andrady, A.L. Microplastics in the marine environment. *Mar. Pollut. Bull.* **2011**, *62*, 1596–1605. [CrossRef]
5. Antunes, J.; Frias, J.; Sobral, P. Microplastics on the Portuguese coast. *Mar. Pollut. Bull.* **2018**, *131*, 294–302. [CrossRef]
6. Atugoda, T.; Vithanage, M.; Wijesekara, H.; Bolan, N.; Sarmah, A.K.; Bank, M.S.; You, S.; Ok, Y.S. Interactions between microplastics, pharmaceuticals and personal care products: Implications for vector transport. *Environ. Int.* **2021**, *149*, 106367. [CrossRef]
7. Besley, A.; Vijver, M.G.; Behrens, P.; Bosker, T. A standardized method for sampling and extraction methods for quantifying microplastics in beach sand. *Mar. Pollut. Bull.* **2017**, *114*, 77–83. [CrossRef]
8. Blanco, L.; Hermosilla, D.; Negro, C.; Swinnen, N.; Prieto, D.; Blanco, Á. Assessing demineralization treatments for PVC effluent reuse in the resin polymerization step. *Environ. Sci. Pollut. Res.* **2017**, *24*, 16631–16638. [CrossRef]
9. Bretas Alvim, C.; Mendoza-Roca, J.A.; Bes-Piá, A. Wastewater treatment plant as microplastics release source—Quantification and identification techniques. *J. Environ. Manag.* **2020**, *255*, 109739. [CrossRef]
10. Caputo, F.; Vogel, R.; Savage, J.; Vella, G.; Law, A.; Della Camera, G.; Hannon, G.; Peacock, B.; Mehn, D.; Ponti, J.; et al. Measuring particle size distribution and mass concentration of nanoplastics and microplastics: Addressing some analytical challenges in the sub-micron size range. *J. Colloid Interface Sci.* **2021**, *588*, 401–417. [CrossRef]
11. Carney Almroth, B.M.; Åström, L.; Roslund, S.; Petersson, H.; Johansson, M.; Persson, N.-K. Quantifying shedding of synthetic fibers from textiles; a source of microplastics released into the environment. *Environ. Sci. Pollut. Res.* **2018**, *25*, 1191–1199. [CrossRef] [PubMed]
12. Costa, C.Q.V.; Cruz, J.; Martins, J.; Teodósio, M.A.A.; Jockusch, S.; Ramamurthy, V.; Da Silva, J.P. Fluorescence sensing of microplastics on surfaces. *Environ. Chem. Lett.* **2021**, *19*, 1797–1802. [CrossRef]
13. Da Costa, J.; Duarte, A.; Rocha-Santos, T. Microplastics—Occurrence, Fate and Behaviour in the Environment. In *Comprehensive Analytical Chemistry*; Elsevier: Amsterdam, The Netherlands, 2016. [CrossRef]
14. Deng, H.; Wei, R.; Luo, W.; Hu, L.; Li, B.; Di, Y.; Shi, H. Microplastic pollution in water and sediment in a textile industrial area. *Environ. Pollut.* **2020**, *258*, 113658. [CrossRef] [PubMed]
15. Duan, J.; Bolan, N.; Li, Y.; Ding, S.; Atugoda, T.; Vithanage, M.; Sarkar, B.; Tsang, D.C.W.; Kirkham, M.B. Weathering of microplastics and interaction with other coexisting constituents in terrestrial and aquatic environments. *Water Res.* **2021**, *196*, 117011. [CrossRef]
16. Elkhatib, D.; Oyanedel-Craver, V. A Critical Review of Extraction and Identification Methods of Microplastics in Wastewater and Drinking Water. *Environ. Sci. Technol.* **2020**, *54*, 7037–7049. [CrossRef]
17. Fernando, J. Gómez, Rima Simonetta. 2019. Available online: http://www.weforum.org/agenda/2019/10/plastics-what-are-they-explainer/ (accessed on 4 May 2021).
18. Gatidou, G.; Arvaniti, O.S.; Stasinakis, A.S. Review on the occurrence and fate of microplastics in Sewage Treatment Plants. *J. Hazard. Mater.* **2019**, *367*, 504–512. [CrossRef]
19. Ghosh, A. Performance modifying techniques for recycled thermoplastics. *Resour. Conserv. Recycl.* **2021**, *175*, 105887. [CrossRef]
20. Goldschmidt, A.; Streitberger, H.-J. *BASF Handbook on Basics of Coating Technology*, 2nd rev. ed.; Vincentz Network: Hannover, Germany, 2007.
21. Gregory, M.R. Environmental implications of plastic debris in marine settings—Entanglement, ingestion, smothering, hangers-on, hitch-hiking and alien invasions. *Phil. Trans. R. Soc. B* **2009**, *364*, 2013–2025. [CrossRef]
22. Guo, X.; Wang, J. The chemical behaviors of microplastics in marine environment: A review. *Mar. Pollut. Bull.* **2019**, *142*, 1–14. [CrossRef]
23. Hale, R.; Seeley, M.; La Guardia, M.; Mai, L.; Zeng, E. A Global Perspective on Microplastics. *J. Geophys. Res. Ocean.* **2020**, *125*, e2018JC014719. [CrossRef]
24. Hu, K.; Tian, W.; Yang, Y.; Nie, G.; Zhou, P.; Wang, Y.; Duan, X.; Wang, S. Microplastics remediation in aqueous systems: Strategies and technologies. *Water Res.* **2021**, *198*, 117144. [CrossRef] [PubMed]
25. John, J.; Nandhini, A.R.; Velayudhaperumal Chellam, P.; Sillanpää, M. Microplastics in mangroves and coral reef ecosystems: A review. *Environ. Chem. Lett.* **2022**, *20*, 397–416. [CrossRef] [PubMed]
26. Kim, S.W.; An, Y.-J. Edible size of polyethylene microplastics and their effects on springtail behavior. *Environ. Pollut.* **2020**, *266*, 115255. [CrossRef] [PubMed]
27. Koelmans, A.A.; Besseling, E.; Shim, W.J. Nanoplastics in the Aquatic Environment. Critical Review. In *Marine Anthropogenic Litter*; Bergmann, M., Gutow, L., Klages, M., Eds.; Springer International Publishing: Cham, Switzerland, 2015; pp. 325–340. [CrossRef]
28. Kotwani, A.; Joshi, J.; Kaloni, D. Pharmaceutical effluent: A critical link in the interconnected ecosystem promoting antimicrobial resistance. *Environ. Sci. Pollut. Res.* **2021**, *28*, 32111–32124. [CrossRef] [PubMed]

29. Lourenço, P.M.; Serra-Gonçalves, C.; Ferreira, J.L.; Catry, T.; Granadeiro, J.P. Plastic and other microfibers in sediments, macroinvertebrates and shorebirds from three intertidal wetlands of southern Europe and west Africa. *Environ. Pollut.* **2017**, *231*, 123–133. [CrossRef]
30. Magalhães, S. Microplastics in Ecosystems: From Current Trends to Bio-Based Removal Strategies. *Molecules* **2020**, *25*, 3954. [CrossRef]
31. Martins, J.; Sobral, P. Plastic marine debris on the Portuguese coastline: A matter of size? *Mar. Pollut. Bull.* **2011**, *62*, 2649–2653. [CrossRef]
32. Mason, S.A.; Garneau, D.; Sutton, R.; Chu, Y.; Ehmann, K.; Barnes, J.; Fink, P.; Papazissimos, D.; Rogers, D.L. Microplastic pollution is widely detected in US municipal wastewater treatment plant effluent. *Environ. Pollut.* **2016**, *218*, 1045–1054. [CrossRef]
33. Misra, A.; Zambrzycki, C.; Kloker, G.; Kotyrba, A.; Anjass, M.H.; Franco Castillo, I.; Mitchell, S.G.; Güttel, R.; Streb, C. Water Purification and Microplastics Removal Using Magnetic Polyoxometalate-Supported Ionic Liquid Phases (magPOM-SILPs). *Angew. Chem. Int. Ed.* **2020**, *59*, 1601–1605. [CrossRef]
34. Nan, M.I.; Lakatos, E.; Giurgi, G.-I.; Szolga, L.; Po, R.; Terec, A.; Jungsuttiwong, S.; Grosu, I.; Roncali, J. Mono- and di-substituted pyrene-based donor-π-acceptor systems with phenyl and thienyl π-conjugating bridges. *Dye. Pigment.* **2020**, *181*, 108527. [CrossRef]
35. Chias, J.; Hoffman, J. Consumo de Plástico em Portugal: Estamos no Bom Caminho? 2019. National Geographic. Available online: https://www.natgeo.pt/planeta-ou-plastico/2019/02/consumo-de-plastico-em-portugal-estamos-no-bom-caminho (accessed on 13 May 2022).
36. Nuelle, M.-T.; Dekiff, J.H.; Remy, D.; Fries, E. A new analytical approach for monitoring microplastics in marine sediments. *Environ. Pollut.* **2014**, *184*, 161–169. [CrossRef] [PubMed]
37. O'Kelly, B.C.; El-Zein, A.; Liu, X.; Patel, A.; Fei, X.; Sharma, S.; Mohammad, A.; Goli, V.S.N.S.; Wang, J.J.; Li, D.; et al. Microplastics in soils: An environmental geotechnics perspective. *Environ. Geotech.* **2021**, *8*, 586–618. [CrossRef]
38. Park, J.H.; Jang, M.D.; Shin, M.J. Solvatochromic hydrogen bond donor acidity of cyclodextrins and reversed-phase liquid chromatographic retention of small molecules on a β-cyclodextrin-bonded silica stationary phase. *J. Chromatogr. A* **1992**, *595*, 45–52. [CrossRef]
39. Peets, P.; Leito, I.; Pelt, J.; Vahur, S. Identification and classification of textile fibres using ATR-FT-IR spectroscopy with chemometric methods. *Spectrochim. Acta Part A Mol. Biomol. Spectrosc.* **2017**, *173*, 175–181. [CrossRef]
40. Prata, J.C.; da Costa, J.P.; Lopes, I.; Duarte, A.C.; Rocha-Santos, T. Environmental status of (micro)plastics contamination in Portugal. *Ecotoxicol. Environ. Saf.* **2020**, *200*, 110753. [CrossRef]
41. Prata, J.C.; Reis, V.; Matos, J.T.V.; da Costa, J.P.; Duarte, A.C.; Rocha-Santos, T. A new approach for routine quantification of microplastics using Nile Red and automated software (MP-VAT). *Sci. Total Environ.* **2019**, *690*, 1277–1283. [CrossRef]
42. Rodríguez-Narvaez, O.M.; Goonetilleke, A.; Perez, L.; Bandala, E.R. Engineered technologies for the separation and degradation of microplastics in water: A review. *Chem. Eng. J.* **2021**, *414*, 128692. [CrossRef]
43. Sancataldo, G.; Ferrara, V.; Bonomo, F.P.; Martino, D.F.C.; Licciardi, M.; Pignataro, B.G.; Vetri, V. Identification of microplastics using 4-dimethylamino-4′-nitrostilbene solvatochromic fluorescence. *Microsc. Res. Tech.* **2021**, *84*, 2820–2831. [CrossRef]
44. Schell, T.; Hurley, R.; Nizzetto, L.; Rico, A.; Vighi, M. Spatio-temporal distribution of microplastics in a Mediterranean river catchment: The importance of wastewater as an environmental pathway. *J. Hazard. Mater.* **2021**, *420*, 126481. [CrossRef]
45. Thompson, R.C. Lost at Sea: Where Is All the Plastic? *Science* **2004**, *304*, 838. [CrossRef]
46. Uddin, S.; Fowler, S.W.; Saeed, T.; Naji, A.; Al-Jandal, N. Standardized protocols for microplastics determinations in environmental samples from the Gulf and marginal seas. *Mar. Pollut. Bull.* **2020**, *158*, 111374. [CrossRef]
47. Ugwu, K.; Herrera, A.; Gómez, M. Microplastics in marine biota: A review. *Mar. Pollut. Bull.* **2021**, *169*, 112540. [CrossRef] [PubMed]
48. Wagner, M.; Lambert, S. (Eds.) Freshwater Microplastics: Emerging Environmental Contaminants? In *The Handbook of Environmental Chemistry*; Springer International Publishing: Cham, Switzerland, 2018. [CrossRef]
49. Wei, S.; Luo, H.; Zou, J.; Chen, J.; Pan, X.; Rousseau, D.P.L.; Li, J. Characteristics and removal of microplastics in rural domestic wastewater treatment facilities of China. *Sci. Total Environ.* **2020**, *739*, 139935. [CrossRef] [PubMed]
50. Wu, T.; Liang, T.; Hu, W.; Du, M.; Zhang, S.; Zhang, Y.; Anslyn, E.V.; Sun, X. Chemically Triggered Click and Declick Reactions: Application in Synthesis and Degradation of Thermosetting Plastics. *ACS Macro Lett.* **2021**, *10*, 1125–1131. [CrossRef] [PubMed]
51. Zhang, S.; Wang, J.; Liu, X.; Qu, F.; Wang, X.; Wang, X.; Li, Y.; Sun, Y. Microplastics in the environment: A review of analytical methods, distribution, and biological effects. *TrAC Trends Anal. Chem.* **2019**, *111*, 62–72. [CrossRef]
52. Ziajahromi, S.; Neale, P.A.; Rintoul, L.; Leusch, F.D.L. Wastewater treatment plants as a pathway for microplastics: Development of a new approach to sample wastewater-based microplastics. *Water Res.* **2017**, *112*, 93–99. [CrossRef]
53. Prata, J.C.; da Costa, J.P.; Duarte, A.C.; Rocha-Santos, T. Methods for sampling and detection of microplastics in water and sediment: A critical review. *TrAC Trends Anal. Chem.* **2018**, *110*, 150–159. [CrossRef]
54. Gulmine, J.; Janissek, P.; Heise, H.; Akcelrud, L. Polyethylene characterization by FTIR. *Polym. Test.* **2002**, *21*, 557–563. [CrossRef]
55. Pandey, M.; Joshi, G.M.; Mukherjee, A.; Thomas, P. Electrical properties and thermal degradation of poly(vinyl chloride)/polyvinylidene fluoride/ZnO polymer nanocomposites. *Polym. Int.* **2016**, *65*, 1098–1106. [CrossRef]

Article

Degradation of Hydrogels Based on Potassium and Sodium Polyacrylate by Ionic Interaction and Its Influence on Water

Diego David Pinzon-Moreno [1,*], Isabel Rosali Maurate-Fernandez [1], Yury Flores-Valdeon [1], Antony Alexander Neciosup-Puican [2] and María Verónica Carranza-Oropeza [1]

1. Faculty of Chemistry and Chemical Engineering, National University of San Marcos, Lima 15081, Peru; isabel.maurate@unmsm.edu.pe (I.R.M.-F.); yfvaldeon071@gmail.com (Y.F.-V.); mcarranzao@unmsm.edu.pe (M.V.C.-O.)
2. Centro de Investigaciones Tecnológicas, Biomédicas y Medioambientales, National University of San Marcos, Lima 15081, Peru; antony.neciosup@unmsm.edu.pe
* Correspondence: diegodpinzon@gmail.com or diegopinzon@alumni.usp.br

Abstract: Hydrogels are a very useful type of polymeric material in several economic sectors, acquiring great importance due to their potential applications; however, this type of material, similarly to all polymers, is susceptible to degradation, which must be studied to improve its use. In this sense, the present work shows the degradation phenomena of commercial hydrogels based on potassium and sodium polyacrylate caused by the intrinsic content of different types of potable waters and aqueous solutions. In this way, a methodology for the analysis of this type of phenomenon is presented, facilitating the understanding of this type of degradation phenomenon. In this context, the hydrogels were characterized through swelling and FTIR to verify their performance and their structural changes. Likewise, the waters and wastewaters used for the swelling process were characterized by turbidity, pH, hardness, metals, total dissolved solids, electrical conductivity, DLS, Z-potential, and UV-vis to determine the changes generated in the types of waters caused by polymeric degradation and which are the most relevant variables in the degradation of the studied materials. The results obtained suggest a polymeric degradation reducing the swelling capacity and the useful life of the hydrogel; in addition, significant physicochemical changes such as the emergence of polymeric nanoparticles are observed in some types of analyzed waters.

Keywords: hydrogel; polymer degradation; potassium and sodium polyacrylate; swelling; physicochemical changes in the water; polymeric nanoparticles

Citation: Pinzon-Moreno, D.D.; Maurate-Fernandez, I.R.; Flores-Valdeon, Y.; Neciosup-Puican, A.A.; Carranza-Oropeza, M.V. Degradation of Hydrogels Based on Potassium and Sodium Polyacrylate by Ionic Interaction and Its Influence on Water. *Polymers* **2022**, *14*, 2656. https://doi.org/10.3390/polym14132656

Academic Editor: Jacopo La Nasa

Received: 16 May 2022
Accepted: 27 June 2022
Published: 29 June 2022

Publisher's Note: MDPI stays neutral with regard to jurisdictional claims in published maps and institutional affiliations.

Copyright: © 2022 by the authors. Licensee MDPI, Basel, Switzerland. This article is an open access article distributed under the terms and conditions of the Creative Commons Attribution (CC BY) license (https://creativecommons.org/licenses/by/4.0/).

1. Introduction

Currently, the use of hydrogels has gained popularity because of their intrinsic property of retaining significant amounts of water or other fluids, being increasingly used in various sectors, and becoming a material that presents itself as a strategic alternative to many current problems associated with consumption, management, conservation, optimization, and release of water and/or aqueous solutions of interest [1]. Consequently, hydrogels have gained space in several sectors such as agriculture, chemistry, electricity, electronics and magnetism, energy, environment and purification, mechanics, medicine-pharmacy, etc. [2–12], presenting themselves as a versatile and economical solution to facilitate the realization of different processes [1,13–15].

Hydrogels have a complex tangled three-dimensional structure that will traditionally consist of a polymer chain, polar side groups, and crosslinkings; these fractions of the polymeric network can absorb, swell, retain and release controlled amounts of water, liquids, or aqueous solutions [13]. The previously commented parts that constitute the hydrogel provide important characteristics to this type of material; in the case of the polymeric chain, it supports other elements such as polar side groups and multiple crosslinkings, as well as providing flexibility to the polymer and allowing their dilation. Polar lateral groups provide hydrophilic characteristics to hydrogels providing water affinity. On the other hand, crosslinkings, which are bridges that connect the polymer chains with the help of molecules called crosslinkers that contain two or more reactive points capable of attaching to specific functional groups [16,17]. Thus, it creates a network and consequently a porous material that allows for the storage of significant amounts of water or other fluids. Naturally, this set of elements enables the functionality of hydrogels by determining the mass flows of absorption, releasing further the storage and retention capacity of liquids that this material can have during their useful life. However, this ability allows several agents to penetrate the polymeric network during the swelling process, directly interacting; this can have negative effects on the structure, decreasing the efficiency and useful life of this type of polymer [15,18].

In this context, the stability of this type of material must be studied to understand its behavior, efficiency, and useful life when subjected to different conditions. The different types of degradation existing in polymers (mechanical, chemical, photo-oxidative, catalytic, thermal, biological, etc.) [19–22] suggest that, when faced with different types of stimuli, hydrogels can undergo changes that alter their properties, as well as altering aqueous media in contact with this type of polymer [23,24]. The absorption-release mass fluxes of a hydrogel depend on different environmental factors and are also determined by the interactions between the polymeric and aqueous phases. In this way, substrates can be dragged from one phase to another, a phenomenon induced by the different types of degradation, logically generating cross-contamination that should be considered as an analysis criterion depending on the complexity of the hydrogel application [25]. Examples of this type of care can be found in pharmaceutical applications during the release of substrates in the body or in the case of agriculture in the release of substrates in the soil. Therefore, the composition of hydrogels should guarantee that the surrounding environment is not contaminated or harmed by any type of residue. Alternatives to avoid such contamination are through the stabilization of this type of polymer or the reduction in toxic reagents as base materials of hydrogels.

Consequently, several authors have shown that the efficiency concerning swelling can vary depending on the type of salts, solvents, solids, colloids, aggregates, minerals, chemicals, etc. that integrate into the water and that swell the hydrogel [26]; even so, the ionic exchange can change the polymer and the surrounding environment [13,18]. In this context, these agents can interact during the swelling process with different parts of the hydrogel's three-dimensional structural network (polymer chain, lateral polar functional groups, crosslinkings, etc.) modifying it partially or totally during service [15,18,23]. This particularity of hydrogels must be characterized to optimize their use, to know their useful life, properties physical and chemical, degradation phenomena, release of substrates absorbed, swelling ratio, among other important aspects [13,27].

In this context, this work analyzed two types of hydrogels based on potassium and sodium polyacrylate exposed to different types of commercial waters and aqueous solutions based on NaCl, KCl, $CaCl_2 \bullet 2H_2O$, $MgCl_2 \bullet 6H_2O$, etc. A progressive degradation behavior was verified, changing the structure of the hydrogel and changing the physicochemical composition of the waters and the prepared aqueous solutions that interacted with the hydrogel. Furthermore, the hydrogel was characterized by gravimetric swelling experiments in different types of waters and solutions, and to being analyzed by FTIR before and after swelling characterization. On the other hand, the different types of waters and solutions were analyzed through physical-chemical analysis, DLS, Z-potential, UV-vis, metal content, turbidity, hardness, pH, conductivity, etc. The results obtained demonstrate the polymeric degradation of hydrogels, leading to a reduction in their respective performance in terms of swelling, retention, and useful life, and the physicochemical alteration of the different waters and prepared solutions. Finally, a methodology to verify and establish this type of polymeric degradation in hydrogels is presented.

2. Materials and Methods

2.1. Materials

Two types of commercial polyacrylate hydrogels based on potassium and sodium, synthesized by polymerization in aqueous solution, were purchased with characteristics recorded in Tables 1 and 2. The references of all analyzed samples can be consulted in Table 3. Different types of commercial waters were used for the swelling process; their physicochemical characteristics can be seen in Table 4. Moreover, sodium Chloride NaCl, Potassium Chloride KCl, and Calcium Chloride $CaCl_2 \bullet 2H_2O$ (MW: 56.11, 74.55, and 147.01, respectively) from Central Drug House and Magnesium Chloride $MgCl_2 \bullet 6H_2O$ (MW: 203.30) from Himedia were used to prepare the saline solutions. Additionally, Calcium Ca, Calcium Nitrate tetrahydrate $Ca(NO_3)_2 \bullet 4H_2O$, and Calcium Carbonate $CaCO_3$ (MW: 100.08) from Sigma-Aldrich were used to generate calcium-based solutions. All of the solutions were prepared by mechanical stirring at room temperature, keeping the indicated concentrations of Table 4. Solutions based on salts and solutions based on calcium were prepared with contents indicated in the physicochemical characterization of the types of waters, see Tables 3 and 4.

Table 1. Main characteristics of the commercial hydrogels.

	Potassium Polyacrylate (HK)	Sodium Polyacrylate (HNa)
Generic chemical formula	[-CH_2-CH(COOK)-]n	[-CH_2-CH(COONa)-]n
Purity (%)	~96	~95
Molecular weight by GPC (Mw)	~4000	~5100
Particle size (Mesh)	20–40	5–10
Crosslinker	Ethylene glycol dimethacrylate (EGDMA)	Ethylene glycol dimethacrylate (EGDMA)
Crosslinking density * (Croslinking unit per Monomer units)	950–1350	700–1100

* Relationship used for the synthesis of hydrogels.

Table 2. Characteristic Bands of hydrogels based on potassium and sodium polyacrylate.

	Description	Deformation	Wavenumber (cm^{-1})
HK	Stretching vibration of the hydroxyl group.	O-H	~3369
	Asymmetric and Symmetric stretch.	CH_2	~2935 & ~2860
	Deformation vibrations.	C-OH	~1674
	Asymmetric and symmetric stretching and another associated deformation of the group.	COO^-	~1555, ~1451, ~1404, ~1317 & ~1169
	Stretching vibrations of C-O bond and deformation vibrations of C-O-H group.	C-O & C-O-H	~1239
	Bond deformation.	C–C	1162
	Bond stretching in the carboxyl acid structure.	C=O	~1113
	Other characteristic deformations of polymeric hydrogel based on potassium.	–	~855, ~820, ~784~638 & ~616
HNa	Stretching vibration of the hydroxyl group.	O-H	~3383
	Asymmetric and Symmetric stretch.	CH_2	~2930 & ~2865
	Deformation vibrations.	C-OH	~1659
	Asymmetric and symmetric stretching and another associated deformation of the group.	COO^-	~1555, ~1451, ~1404, ~1322 & ~1162
	Stretching vibrations of C-O bond and deformation vibrations of C-O-H group.	C-O & C-O-H	−1235
	Bond deformation.	C–C	1162
	Bond stretching in the carboxyl acid structure.	C=O	~1128
	Other characteristic deformations of hydrogel based on sodium.	–	~1047, ~815, ~774 & ~621

Table 3. Samples of water and residual water from the swelling process.

References Sample	Type of Sample	References Sample	Type of Sample
HK	Hydrogel based on potassium	HNa	Hydrogel based on sodium
WD	Distilled Water	HK + WD	
WO	Osmosis Water	HK + WO	
WC	Commercial water 1	HK + WC	
WSM	Commercial water 2	HK + WSM	
WSL	Commercial water 3	HK + WSL	Hydrogel after swelling process
WSN	Water of supply net	HNa + WD	
$CaCl_2$ + HK		HNa + WO	
$MgCl$ + HK	Saline solutions with distilled	HNa + WC	
KCl + HK	water after of swelling process	HNa + WSM	
NaCl + HK		HNa + WSL	
$CaCl_2$ + HNa		WD + HK	
$MgCl$ + HNa	Saline solutions with distilled	WO + HK	
KCl + HNa	water after of swelling process	WC + HK	
NaCl + HNa		WSM+ HK	
Ca + HK	Calcium-associated solutions	WSL + HK	Wastewater after the swelling process
$Ca(NO_3)_2$ + HK	with distilled water after of	WD + HNa	
$CaCO_3$ + HK	swelling process	WO+ HNa	
Ca + HNa	Calcium-associated solutions	WC + HNa	
$Ca(NO_3)_2$ + HNa	with distilled water after of	WSM+ HNa	
$CaCO_3$ + HNa	swelling process	WSL + HNa	

Table 4. Physicochemical properties of the types of waters.

Water Reference	Turbidity	Total Dissolved Solids	Total Water Hardness $CaCO_3$	pH	Electrical Conductivity	Anions				Cations			
						Chlorides (Cl^-)	Sulfates (SO_4^{-2})	Nitrates (NO_3^-)	Nitrites (NO_2^-)	Calcium	Magnesium	Sodium	Potassium
	NTU	mg/L	mg/L	-	uS/cm	mg/L	mg/L	mg/L	mg/L	mg/L	mg/L	mg/L	mg/L
WC	0.39	266	260	7.56	478.34	35.09	177.71	8.8	0.96	84.86 *	2.69	10.49	27.1
WSM	0.49	242	65	7.62	553.67	59.88	142.55	1.3	0.28	83.14	16.79 *	13.84 *	47.01 *
WSL	0.34	59	35	6.27	54,33	29.99	12.12	1.43	0.31	1.66	3.58	3.36	7.55
WSN	1.85	1880	378.33	7.09	793.4	51.98	276.08	51.85	0.01	79.95	9.84	37.71	4.07

* The concentration used for the formulation of saline and calcium solutions.

2.2. Methods and Characterizations

To verify phenomena associated with hydrogel degradation or transformations due to exposure to different types of waters, the hydrogels and waters were characterized before and after the swelling process. Thus, hydrogels based on potassium and sodium polyacrylate were tested for swelling by the tea bag methodology employing three types of commercial waters (see Table 3). For all of the swelling tests, approximately 0.05 g of hydrogel per 100 mL of water was used; also, the swelling tests were performed in duplicate. Hydrogels were tested for swelling with distilled and osmotic water as references. Additionally, hydrogels were characterized by FTIR (Shimadzu IRTracer-100 spectrometer with Pike MIRacle single reflection horizontal ATR accessory) before and after the swelling process.

On the other hand, the characteristics of turbidity were carried out using an EPA Compliant Turbidity and Free & Total Chlorine Meter–Hanna Instruments S.L. HI-93414-02

(SMEWW-APHA-AWWA-WEF Part 2130B.22nd Ed. methodology); total water hardness was determined using a Sartorius SE2 Ultra-micro balance (SMEWW-APHA-AWWA-WEF Part 2340-C.23nd Ed. methodology); total metals were carried out using a PerkinElmer Optima 4300 DV ICP-OES Spectrometer (EPA method 200.7); total dissolved solids were determined using a Sartorius SE2 Ultra-micro balance (SMEWW-APHA-AWWA-WEF Part 2540C B.23 nd Ed. methodology); electrical conductivity was determined using a HI-9033 Heavy Duty Waterproof Portable Conductivity Meter; pH was determined using a HI-2020-01 edge® Multiparameter pH Meter; Dynamic Light Scattering (DLS) by number-weighted and Zeta-potential by Phase Analysis Light Scattering and Frequency Analysis modes tests were performed on a NICOMP Nano Z3000 System; and UV-vis test performed on a Libra S22, Biochrom, Ltd., Cambridge, England. All of these tests were performed before and after the swelling process.

3. Results and Discussion

During the swelling test, several changes and transitions were visually observed in the hydrogels when submitted to commercial waters (WC and WSM) compared to distilled and/or osmotic waters. When commercial hydrogels (HK and HNa) are exposed to commercial waters (WC and WSM), there are changes such as a lower swelling, an inflection point showing a smaller maximum swelling compared to distilled water, and later, a loss of partial or total water retention capacity. On the other hand, when removing the bags (filter) that contain the commercial hydrogels from the commercial water baths, it is observed that the wastewater presents a significant change in turbidity. These visually appreciable changes are presented in-depth below using different characterization methods.

3.1. Swelling Test in Commercial Waters

Figure 1 shows the kinetic swelling curves of commercial hydrogels based on potassium and sodium exposed to different types of waters. It is observed that, in general terms, potassium-based hydrogels perform better in terms of swelling than sodium-based hydrogels, regardless of the solution used for the test. In the same Figure, it is observed that distilled water allows for better performances concerning swelling, ~475 and ~320 g/g for HK and HNa, respectively, followed by the curves related to swelling with WSL in both types of hydrogels, around ~440 and ~250 g/g for HK and RNa, respectively; note that hydrogels exposed to the two types of water previously described (WD and WSL) reach equilibrium swelling. This difference in swelling between the two types of hydrogels may be associated with factors such as variations in polymer synthesis formulations, changes in the hydrogel particle size, as well as the electronegativity of the ions that make up the polar groups of the polymer chain of each hydrogel. Moreover, hydrogels exposed to the reference waters WC, WSM, and WSN present a phenomenon that significantly limits the swelling, transforming the kinetic curve, reaching inflection points that later decrease until to lose their character as superabsorbent materials without reaching an equilibrium swelling. These repeated trends suggest that WC, WSM, and WSN referenced waters contain substrates that limit swelling dramatically, distinct from WD and WSL referenced waters.

After these swelling tests, the hydrogels were dried on their respective filters at 40 °C until reaching a constant mass to be retested for swelling for a second time. In this new swelling test, it was observed that both types of hydrogels exposed to WD, and WSL water reached kinetic curves similar to the curves of the first swelling; however, hydrogels exposed to the reference waters WC, WSM, and WSN did not have responses that indicated swelling. This phenomenon could indicate structural changes in potassium and sodium hydrogels due to chemical agents present in the water, such as salts or other types of ions. In addition, the fact that it does not respond to second swelling attempts is one of the main indications of irreversible structural changes in the polymer, reducing the functionality of the hydrogels.

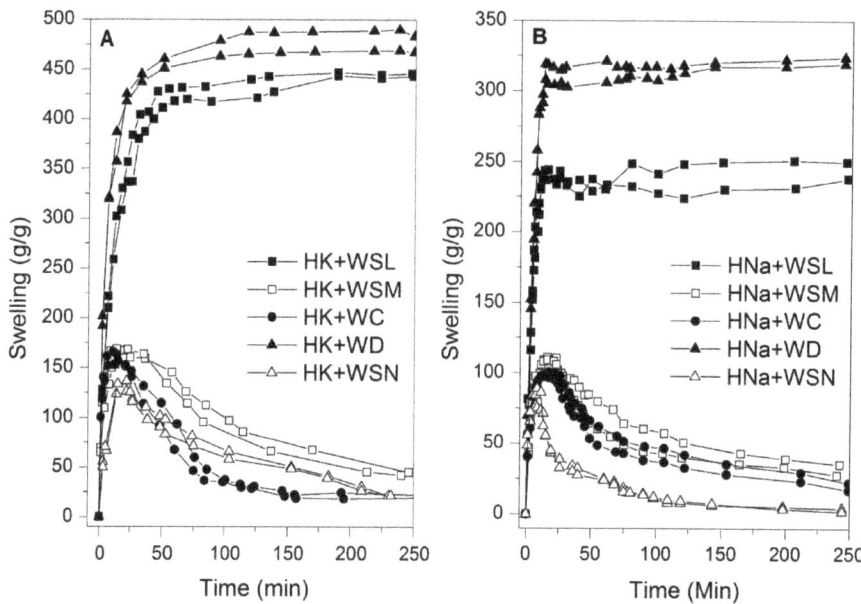

Figure 1. Swelling kinetics of hydrogels in types of waters. (**A**) HK and (**B**) HNa.

3.2. Swelling Test in Saline Solutions

In order to verify the swelling performance of hydrogels in different salts present in the different types of waters, different solutions were prepared in distilled water, maintaining the concentrations indicated in Table 4 related to calcium, magnesium, potassium, and sodium so that it was possible to determine the agent or agents that can generate the phenomenon associated with the partial or total reduction in the swelling capacity. In Figure 2, the kinetic curves of swelling of potassium and sodium-based hydrogels underexposure of the prepared solutions are shown. It should be noted that hydrogels show susceptibility depending on the type of solution, showing better swelling with potassium and sodium solutions, and lower performance with magnesium and calcium solutions. Lower swelling performances with magnesium and calcium solutions are visible, and it is observed that all of the curves reach equilibrium swelling except for the solutions prepared based on $CaCl_2$, which present a swelling peak and later drop significantly after less than two hours of testing; this phenomenon was previously reported by other authors [28–33]. Furthermore, hydrogels show swellings that gradually decrease depending on the solution and show a kinetic curve with a function close to the WC, WSM, and WSN curves when exposed to $CaCl_2$ (See Figure 2). Note that in the case of the sodium-based hydrogel, there is less susceptibility with slight variations with the $MgCl_2$, KCl, and NaCl solutions but with a swelling considered as in equilibrium, which contrasts with a significant drop in the swelling curve with the $CaCl_2$ solution.

3.3. Swelling Test in Calcium-Associated Solutions

Previous swelling results indicate that calcium-associated ions have the ability to limit and reduce the swelling capacity in potassium and sodium-based hydrogels over an exposure time. In this context, solutions were prepared based on Ca, $CaCO_3$, and $Ca(NO_3)_2$ in distilled water, respecting the indicated concentration of calcium in Table 4. Figure 3 shows the kinetic swelling curves of commercial hydrogels based on potassium and sodium in calcium-associated solutions. In said curves, the same phenomenon described in Figures 1 and 2 is observed in which the swelling reaches limited swelling peaks and a progressive drop in the solution's retention capacity, in this context, it is observed that Ca,

CaCO$_3$, and Ca(NO$_3$)$_2$ cause inflection points and subsequently generate reductions in the retention capacity, proving to be more aggressive for the solution with Ca(NO$_3$)$_2$ reducing the superabsorbent capacity of hydrogels without reaching an equilibrium swelling.

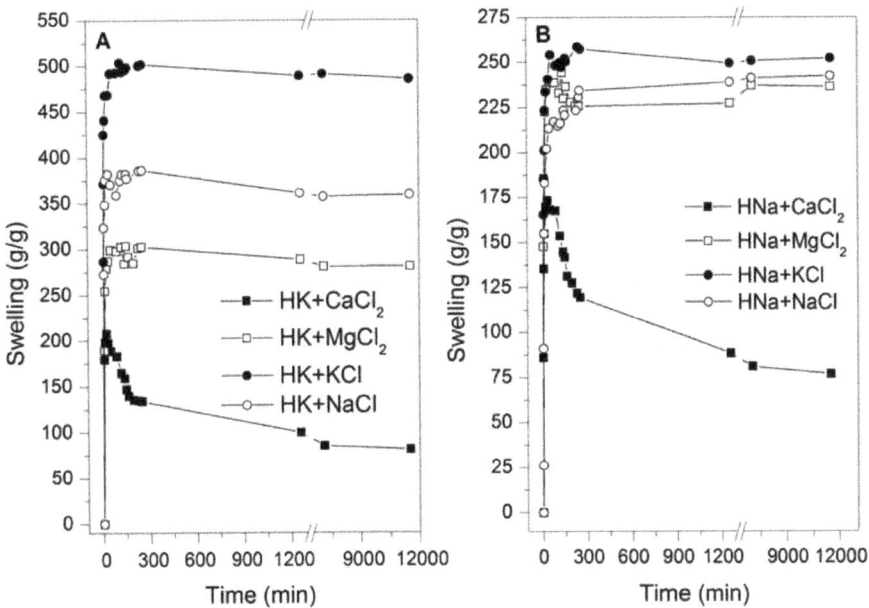

Figure 2. Swelling kinetics of hydrogels in saline solutions. (**A**) HK and (**B**) HNa.

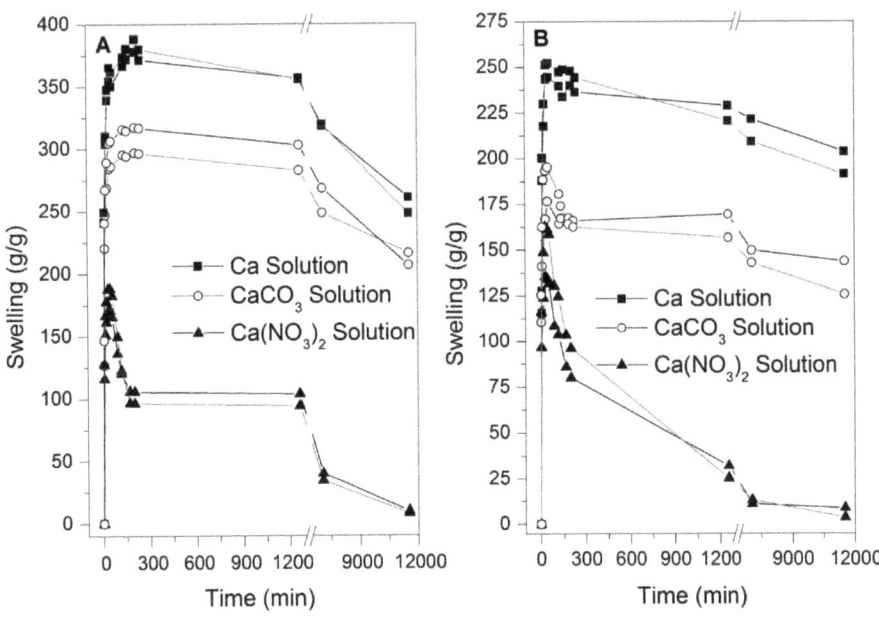

Figure 3. Swelling kinetics of hydrogels in calcium-associated solutions. (**A**) HK and (**B**) HNa.

The swelling curves presented in Figure 3 associated with calcium-based solutions, were formulated to contain the same amount of calcium in the water, presented variations in swelling between the hydrogels studied and the types of solutions, this behavior is due to the degree of ionization of these compounds in water, which in the case of $Ca(NO_3)_2$ presents complete ionization, allowing the ions of Ca^{2+} and NO_3^{-1} to completely penetrate the structure of the hydrogels generating the degradation of the hydrogel by the ion exchange and crosslinking breakage, thus the swelling is reduced to zero in both hydrogels. In the case of $CaCO_3$ and Ca, the ionization is partial and less significant compared to $Ca(NO_3)_2$, a condition that limits the number of ions generated in the solution and consequently their penetration into the polymeric networks of the hydrogel, therefore the swelling values are higher compared to the solution based on $Ca(NO_3)_2$ at all times of the test. The swelling tests associated with Figures 2 and 3 suggest that the most reactive ions for the hydrogel network are Ca^{2+} and Mg^{2+}, which significantly decrease the performance in terms of swelling and may reduce the entire swelling capacity in the case of Ca^{2+} ions. In this sense, cations such as Ca^{2+} are highly reactive, destabilizing the side groups (COO^-) of the polymer chain; hence, this ionic alteration changes the polarity and hydrophilicity of the two types of hydrogels.

The negative effects produced by solutions in the presence of ions such as Ca^{2+} start immediately after the direct interaction between the polymeric and aqueous phases; this can be observed at the beginning of the kinetic curves that present a lower swelling/time rate. Degradation is also reflected in some kinetic curves of swelling that show points of infection or maximum swelling peaks that do not reach swellings comparable to those achieved with WD. Additionally, the drop in swelling without reaching equilibrium swelling shows a degradation of the polymer, which, added to the lack of response in second swelling cycles, ratifies structural changes in the hydrogels.

3.4. FTIR Analysis

Figure 4 shows the FTIR spectra of commercial hydrogels before and after the swelling process. In the case of hydrogels without exposure to the swelling process, the characteristic bands indicated in Table 2 were identified. When comparing the spectra before and after the swelling process, insignificant changes were observed when the hydrogels were exposed to WSL; however, when hydrogels exposed to WSM, WC and WSN references undergo significant changes in intensity, wavenumber and morphology in the bands of the functional groups of OH (~3370), CH2 (~2937 and ~2860), C-OH (~1674), COO- (~1555 and ~1450), CC (~1162), and other deformations in other functional groups in bands close to ~815, 784, ~616, and ~489. These modifications suggest changes in the chemical structures of hydrogels, such as ion exchange or breaking of crosslinkings, that justify the decrease in the swelling performance in the hydrogels. One of the most evident transitions after exposure to different aqueous phases is present in the band of deformation of the OH bond manifested at ~3370, indicating the transformation of and/or decrease in this type of bond in the polymer. This change can be caused by the breakage and loss of crosslinking, as well as the decrease in the hydrophilic capacity of the polymer when the polar side groups of the chain are altered, reducing the presence of the OH bond.

3.5. Physicochemical Analysis

Figure 5 shows some physical properties of the different types of waters recovered after the swelling process. Note that turbidity has significant changes, increasing after the swelling process, indicating the possible release of substrates by the hydrogels in all aqueous media studied (Figure 5A). In this same context, changes are observed concerning the amount of total dissolved solids in several water samples, especially in the types of waters that were used in the swelling processes of potassium-based hydrogels (Figure 5B). The electrical conductivity remains constant in almost all samples except for the waters where the potassium-based hydrogel swelling process was carried out, which have slight increments as well as the sodium-based hydrogel with the WSL swelling process (Figure 5C).

Regarding the water hardness, it is observed that the WC reference has high contents of $CaCO_3$; this may be one of the reasons related to the degradation of hydrogels as observed in the swelling test (items 3.1, 3.2, and 3.3). In this context, calcium cations can interact more strongly with the ionic fraction (potassium or sodium) belonging to the polymeric chain of the hydrogel, decreasing its polar character, thus decreasing the swelling and retention capacity of these superabsorbent materials, in this context, this characteristic justifies the results of the water in terms of reduction in hardness and calcium ions (Figure 5D) and on the other hand explains the increase in sodium and potassium in the water (Figure 6C,D). Moreover, the pH shows slight changes that can be explained by the exchange in ions between the water and the polymer.

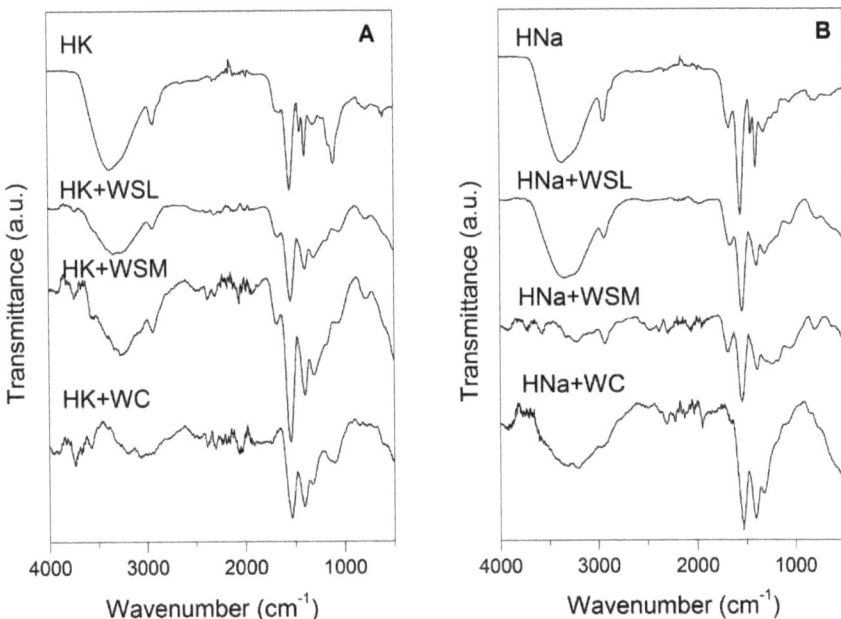

Figure 4. FTIR spectra of hydrogels before and after the swelling process. (**A**) HK and (**B**) HNa.

The contents of some metals (calcium, magnesium, sodium, and potassium) common in the types of waters are shown in Figure 6 before and after the swelling process. The results associated with the presence of these metals suggest the possibility of ionic exchange between different types of waters and hydrogels in a way that sodium and potassium contained in the hydrogel structure are exchanged by calcium and magnesium ions contained in commercial waters, previous work has shown this interchange for different types of hydrogels [15,18]. In this sense, the results of this test suggest that the calcium and magnesium cations of the aqueous phase are exchanged by the ionic fractions of sodium and potassium of the polymeric phase during the swelling process. It is noteworthy that this ion exchange phenomenon does not favor swelling according to the kinetic curves observed in Figures 1–3.

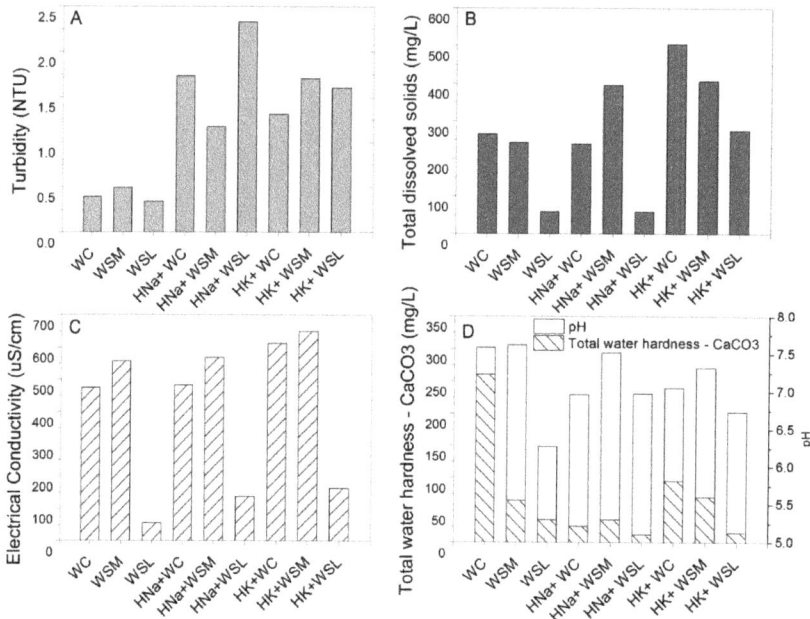

Figure 5. Physical properties comparison of the types of waters before and after the swelling process. (**A**) Turbidity. (**B**) Total dissolved solids. (**C**) Electrical conductivity. (**D**) Total water hardness—pH.

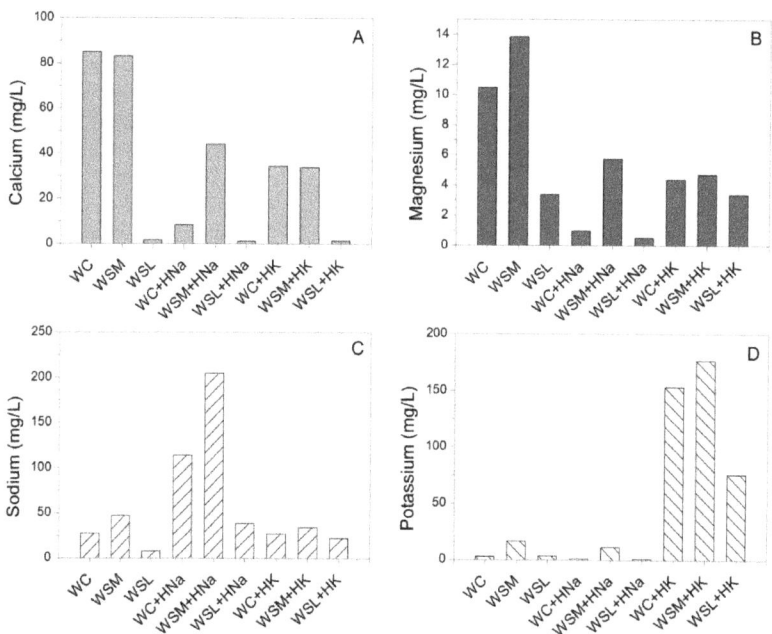

Figure 6. Cations comparison of the types of waters before and after swelling process. (**A**) Calcium, (**B**) Magnesium, (**C**) Sodium, and (**D**) Potassium.

3.6. Dynamic Light Scattering (DLS) and Zeta-Potential

The presence of particles, most of them nanometric scale, was found in the wastewaters released during the swelling process (see Figure 7), it is noteworthy that all the types of waters were examined before the swelling process and it was not possible to verify the presence of nanoparticles. Likewise, it was not possible to verify the presence of nanoparticles in the distilled water after of swelling process. In this context, the presence of these particles in wastewater from the swelling process suggests the detachment of particles from the three-dimensional structure of the hydrogel, these nanoparticles are polymeric chains of low molecular weight with fragile crosslinking bridges that are broken and consequently released into water.

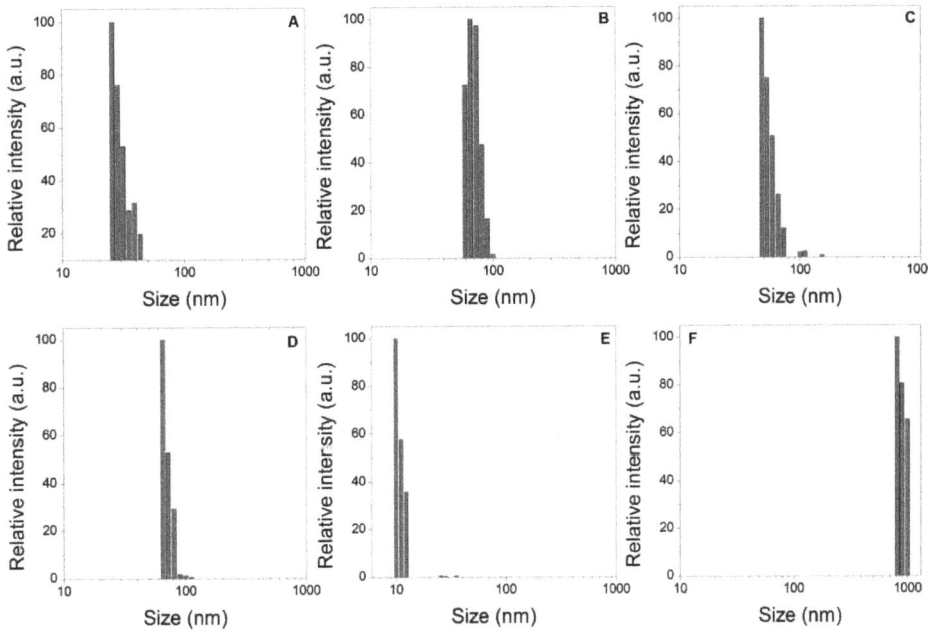

Figure 7. Dynamic light scattering particle size distribution of wastewaters: (**A**) WC + HNa. (**B**) WSM+ HNa. (**C**) WSL + HNa. (**D**) WC + HK. (**E**) WSM+ HK. (**F**) WSL + HK.

Figure 8 presents the results related to the Zeta-Potential of the types of wastewaters showing the charge present in the wastewater obtained by the Phase Analysis Light Scattering (PALS) and frequency analysis modes. In the case of commercial waters samples (WC, WSM, and WSL), relatively neutral charges are observed, close to zero; these results agree with the impossibility of detecting colloids by DLS in the same samples. On the other hand, as expected, the wastewater (WC + HNa, WSM + HNa, WSL + HNa, WC + HK, WSM + HK, and WSL + HK) show cationic charges in both Zeta-Potential techniques used, charges associated with nanoparticles that are generated by the detachment of low molecular weight polymer chains. These low molecular weight polymer chains (nanoparticles) are weakly linked to the polymer network by crosslinkings that are broken by the ionic attack of different types of water.

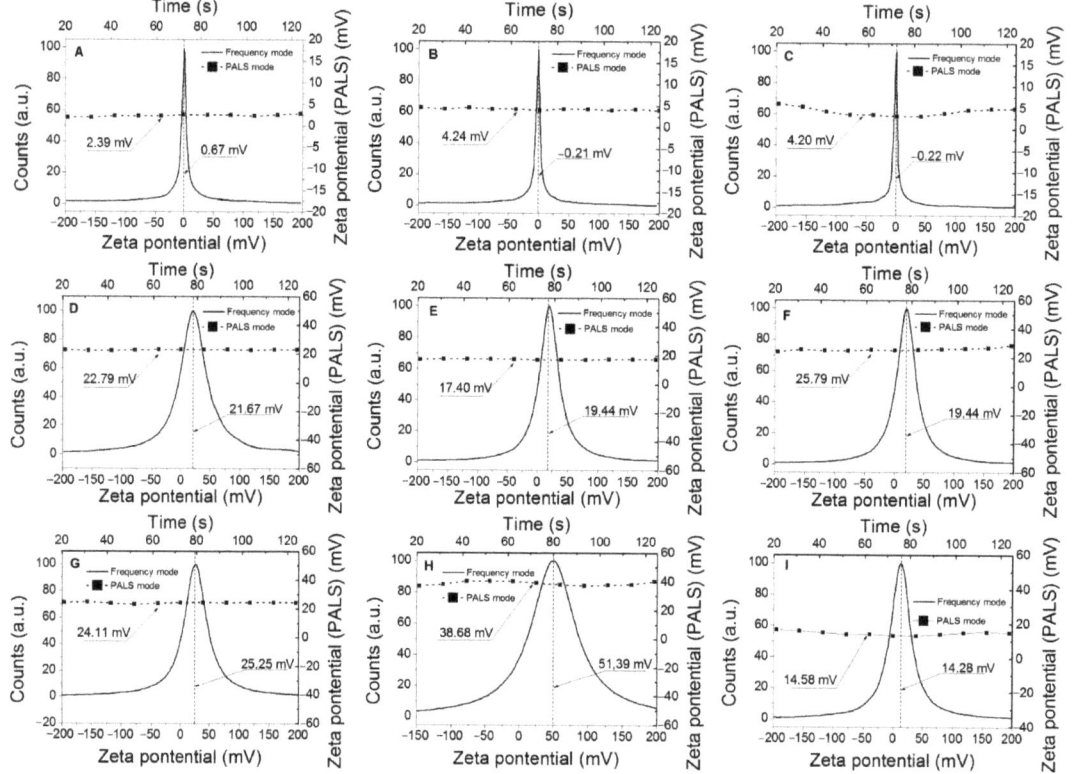

Figure 8. Zeta potential distribution of pure and wastewater: (**A**) WC. (**B**) WSM. (**C**) WSL. (**D**) WC + HNa. (**E**) WSM+ HNa. (**F**) WSL + HNa. (**G**) WC + HK. (**H**) WSM+ HK. (**I**) WSL + HK.

To verify the described phenomena, the DLS and Zeta-Potential techniques were applied to the residual solution of $Ca(NO_3)_2$, obtained after the swelling process of the two commercial hydrogels, see Figure 9. The results show the presence of suspended colloids in both residual solutions; even so, the cationic charges associated with the suspended particles can be observed, confirming the release of nanometric polymer particles from the hydrogels.

3.7. UV-vis Analysis

Figure 10 shows the UV-vis spectra of the different water samples studied. It is observed that the water samples before the swelling process (WD, WC, WSM, WSL, and WO) have detectable minimum absorbances that change significantly after the swelling process. This change in absorbance is due to the presence of released substrates (ions, salts, and/or polymeric nanoparticles) in the water.

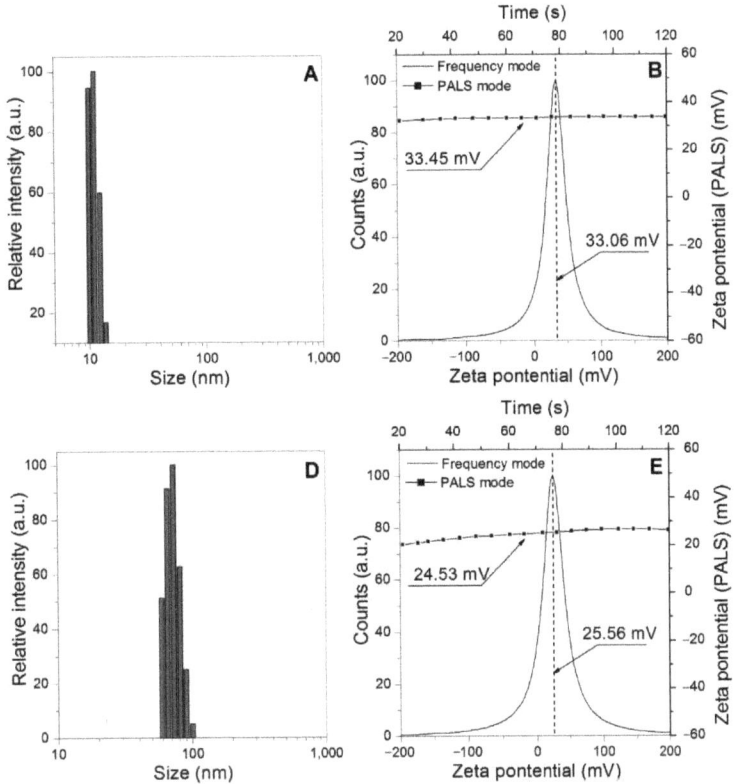

Figure 9. Dynamic light scattering particle size distribution and Zeta potential distribution of $Ca(NO_3)_2$ solution residue after of swelling process. (**A**) DLS of $Ca(NO_3)_2$+ HK. (**B**) Zeta potential of $Ca(NO_3)_2$+ HK. (**D**) DLS of $Ca(NO_3)_2$+ HNa. (**E**) Zeta potential of $Ca(NO_3)_2$+ HNa.

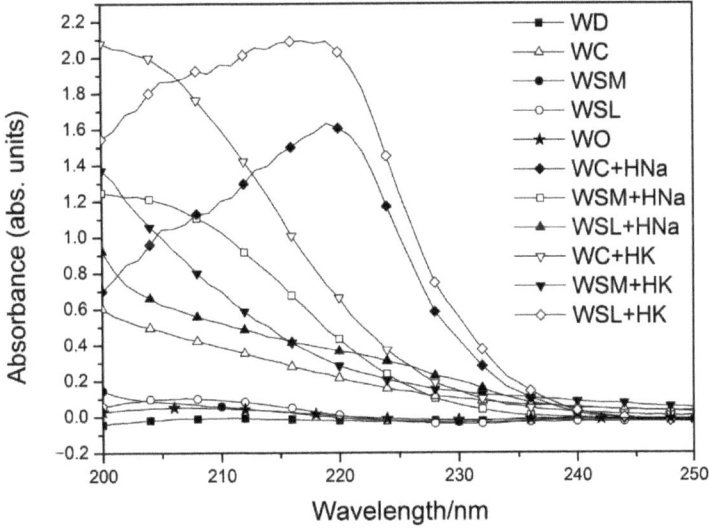

Figure 10. UV-vis spectra of pure waters and wastewaters.

The results obtained suggest several effects on the hydrogel structural network caused by the presence of some ions present in water such as calcium, in addition to a possible synergistic effect of degradation due to the joint effect of several types of ions (Ca^{2+}, Mg^{2+}, K^+, Na^+, etc.); on the other hand, physicochemical changes are observed in some types of waters. One of the effects found is the breaking or cleaving of the crosslinking bridges that connect the polymer chains, which consequently causes the release of low molecular weight polymer chains identified as nanoparticles by DLS and Zeta-potential; said nanoparticles are finally released in the solutions or wastewater after the swelling process and remain suspended after the structural collapse of the hydrogel. In addition, the ion exchange of metals present in the side polar groups of the polymeric chain of the hydrogel by other types of ions present in water and prepared solutions stand out, changing both the efficiency and functionality of the hydrogels as well as the contents of ions associated with potassium, sodium, calcium, and magnesium in commercial waters and prepared solutions, see Figure 6. In Figure 11, a schematic representation suggesting the effects observed in the interaction between solutions with different types of ions and hydrogels can be seen regarding ion exchange and crosslinking breakdown. The degradation observed in the hydrogels can be explained in the two ways suggested in Figure 11A; likewise, said degradation allows inferring the release of nanoparticles (Figure 11B).

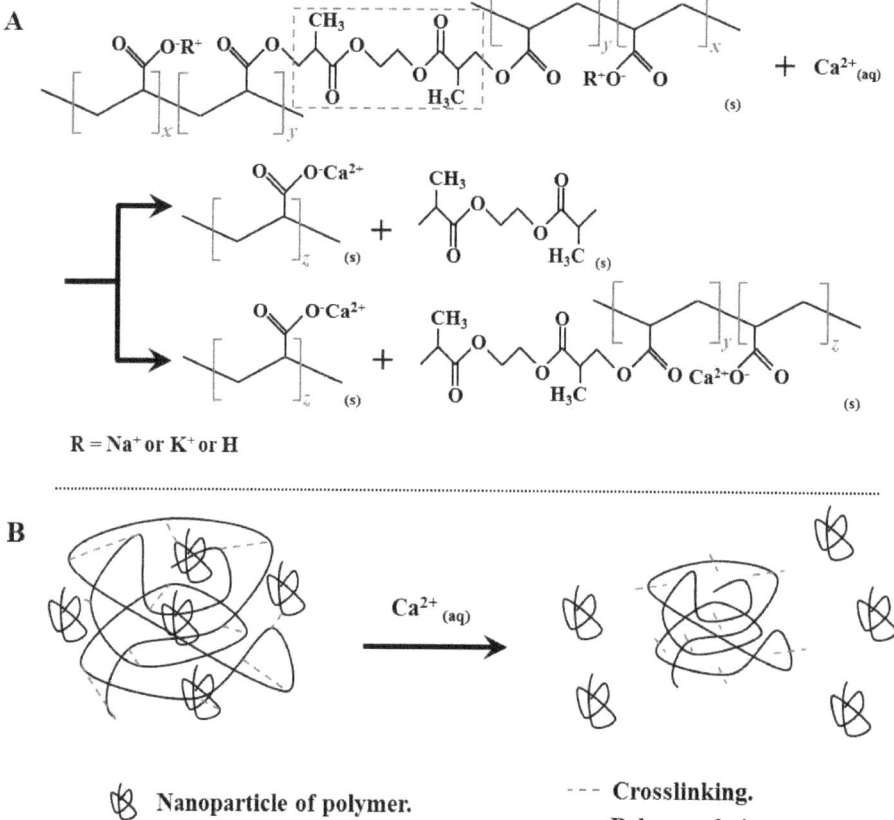

Figure 11. Schematic representation of the effects of solutions associated with calcium ions in commercial hydrogels. (A) Schematic representation of the degradation of the polymers, (B) Schematic representation of the generation of nanoparticles by crosslinking breaking.

4. Conclusions

The polymeric degradation of hydrogels based on potassium and sodium polyacrylate was studied as a result of the interaction with substrates present in absorbable fluids by this type of polymer. Likewise, this interaction between the liquid and the polymer generates transformations in both phases in physical-chemical aspects. In this context, ions present in the different types of waters or other types of solutions such as Ca^{2+}, Mg^{2+} or Na^+ can cleavage the crosslinkings and change the ionic fraction of the polymeric chain of the hydrogels, reducing the retention capacity and/or the useful life of this type of superabsorbent polymer. These changes in the three-dimensional network of the studied hydrogels lead to the structural collapse of this type of polymer, leading to the fractioning of the network and, consequently, the release of particles of low dimensional scales.

On the other hand, during the swelling process, the types of waters or chemical solutions in contact with the hydrogels are altered in several physicochemical characteristics caused by breaks of crosslinking and ionic exchange between the aqueous phase and the polymeric phase; this phenomenon has to be considered for several processes where it is essential not to generate cross-contamination in the aqueous medium surrounding the hydrogels at the time of swelling. Finally, the phenomena described in this article can be considered favorable from the point of view of ionic exchange between the studied phases; however, it can be unfavorable from a cross-contamination point of view. This is why the authors invite the development of hydrogels from renewable sources, such as biomass, that can be integrated in an environmentally friendly way.

Author Contributions: Conceptualization, D.D.P.-M.; methodology, I.R.M.-F., D.D.P.-M.; formal analysis, D.D.P.-M., Y.F.-V., I.R.M.-F.; investigation, I.R.M.-F., A.A.N.-P., D.D.P.-M.; writing—original draft preparation, D.D.P.-M.; writing—review and editing, D.D.P.-M., M.V.C.-O., A.A.N.-P., Y.F.-V., I.R.M.-F.; supervision, M.V.C.-O. All authors have read and agreed to the published version of the manuscript.

Funding: Authors are grateful to Concytec-Peru and The World Bank for the financial support of this project under the call "Mejoramiento y Ampliación de los Servicios del Sistema Nacional de Ciencia Tecnología e Innovación Tecnológica" 8682-PE, through Fondecyt Grant 017-2019 FONDECYT BM INC.INV.

Acknowledgments: Special thanks to J. Quispe and H. Loro for their valuable technical support, and to the Centro de Investigaciones Tecnológicas, Biomédicas y Medioambientales (CITBM), and the Laboratorio de Investigación en Control y Análisis de aguas, and the Laboratorio de Investigación en Nanotecnología de Alimentos of Jose Maria Arguedas National University.

Conflicts of Interest: The authors declare no conflict of interest.

References

1. Ullah, F.; Bisyrul, M.; Javed, F.; Akil, H. Classification, processing and application of hydrogels: A review. *Mater. Sci. Eng C* **2015**, *57*, 414–433. [CrossRef] [PubMed]
2. Kumar, R.; Patel, S.V.; Yadav, S.; Singh, V.; Kumar, M.; Kumar, M. Hydrogel and its effect on soil moisture status and plant growth: A review. *J. Pharm. Phytochem.* **2020**, *9*, 1746–1753. [CrossRef]
3. Yao, T.; Jia, W.; Tong, X.; Feng, Y.; Qi, Y.; Zhang, X.; Wu, J. One-step preparation of nanobeads-based polypyrrole hydrogel by a reactive-template method and their applications in adsorption and catalysis. *J. Colloid Interface Sci.* **2018**, *527*, 214–221. [CrossRef] [PubMed]
4. Liu, H.; Li, M.; Ouyang, C.; Lu, T.J.; Li, F.; Xu, F. Biofriendly, Stretchable, and Reusable Hydrogel Electronics as Wearable Force Sensors. *Small* **2018**, *14*, 1801711. [CrossRef]
5. Hina, M.; Bashir, S.; Kamran, K.; Ramesh, S.; Ramesh, K. Synthesis and characterization of self-healable poly (acrylamide) hydrogel electrolytes and their application in fabrication of aqueous supercapacitors. *Polymer* **2020**, *210*, 123020. [CrossRef]
6. Wang, H.; Liu, J.; Wang, J.; Hu, M.; Feng, Y.; Wang, P.; Wang, Y.; Nie, N.; Zhang, J.; Chen, H.; et al. Concentrated Hydrogel Electrolyte-Enabled Aqueous Rechargeable NiCo//Zn Battery Working from −20 to 50 °C. *ACS Appl. Mater. Interfaces* **2018**, *11*, 49–55. [CrossRef]
7. Kumar, N.; Mittal, H.; Alhassan, S.M.; Ray, S.S. Bionanocomposite Hydrogel for the Adsorption of Dye and Reusability of Generated Waste for the Photodegradation of Ciprofloxacin: A Demonstration of the Circularity Concept for Water Purification. *ACS Sustain. Chem. Eng.* **2018**, *6*, 17011–17025. [CrossRef]

8. Bao, Z.; Xian, C.; Yuan, Q.; Liu, G.; Wu, J. Natural Polymer-Based Hydrogels with Enhanced Mechanical Performances: Preparation, Structure, and Property. *Adv. Healthc. Mater.* **2019**, *8*, 1900670. [CrossRef]
9. Wang, Y.; Cao, H.; Wang, X. Synthesis and characterization of an injectable ε-polylysine/carboxymethyl chitosan hydrogel used in medical application. *Mater. Chem. Phys.* **2020**, *248*, 122902. [CrossRef]
10. Gradinaru, V.; Treweek, J.; Overton, K.; Deisseroth, K. Hydrogel-Tissue Chemistry: Principles and Applications. *Annu. Rev. Biophys.* **2018**, *47*, 355–376. [CrossRef]
11. Ramiah, P.; du Toit, L.C.; Choonara, Y.E.; Kondiah, P.P.D.; Pillay, V. Hydrogel-Based Bioinks for 3D Bioprinting in Tissue Regeneration. *Front. Mater.* **2020**, *7*, 76. [CrossRef]
12. Li, L.; Scheiger, J.M.; Levkin, P.A. Design and Applications of Photoresponsive Hydrogels. *Adv. Mater.* **2019**, *31*, 1807333. [CrossRef]
13. Halah, E.; López-Carrasquero, F.; Contreras, J. Applications of hydrogels in the adsorption of metallic ions. *Cienc. Ing.* **2018**, *39*, 57–70.
14. Karagöz, İ.; Yücel, G. Use of super absorbent polymers with Euonymus plants (Euonymus japonicus 'Aureomarginatus') in ornamental plant cultivation. *J Agric Sci.* **2020**, *26*, 201–211. [CrossRef]
15. Sanz-Gómez, J. Characterization and Effects of Cross-Linked Potassium Polyacrylate as Soil Amendment. Ph.D. Thesis, University of Seville, Seville, Spain, 2015. [CrossRef]
16. Maitra, J.; Shukla, V.K. Cross-linking in Hydrogels -A Review. *Am. J. Polym. Sci.* **2014**, *4*, 25–31. [CrossRef]
17. Sharma, K.; Kumar, V.; Kaith, B.S.; Kalia, S.; Swart, H.C. Conducting polymer hydrogels and their applications. In *Conducting Polymer Hybrids*, 1st ed.; Kumar, V., Kalia, S., Swart, H., Eds.; Springer: Cham, Switzerland, 2017; pp. 193–221. [CrossRef]
18. Chatzoudis, G.K.; Rigas, F. Soil salts reduce hydration of polymeric gels and affect moisture characteristics of soil. *Commun. Soil Sci. Plant. Anal.* **1999**, *30*, 2465–2474. [CrossRef]
19. Da Costa, J.P.; Nunes, A.R.; Santos, P.; Girão, A.V.; Duarte, A.; Rocha-Santos, T. Degradation of polyethylene microplastics in seawater: Insights into the environmental degradation of polymers into the environmental degradation of polymers. *J. Environ. Sci. Health Part A* **2018**, *53*, 866–875. [CrossRef]
20. Siracusa, V. Microbial Degradation of Synthetic Biopolymers Waste. *Polymers* **2019**, *11*, 1066. [CrossRef]
21. Sollehudin, I.M.; Heerwan, P.M.; Ishak, M.I.; Beams, C.; Chin, S.C.; Tong, F.S. Degradation and stability of polymer: A mini review. In Proceedings of the IOP Conference Series: Materials Science and Engineering 788, Atlanta, GA, USA, 22 April 2020; p. 012048. [CrossRef]
22. Chamas, A.; Moon, H.; Zheng, J.; Qiu, Y.; Tabassum, T.; Jang, J.H.; Abu-Omar, M.; Scott, S.L.; Suh, S. Degradation Rates of Plastics in the Environment. *ACS Sustain. Chem. Eng.* **2020**, *8*, 3494–3511. [CrossRef]
23. Li, X.; Kondo, S.; Chung, U.-I.; Sakai, T. Degradation behavior of polymer gels caused by nonspecific cleavages of network strands. *Chem. Mater.* **2014**, *26*, 5352–5357. [CrossRef]
24. Bankeeree, W.; Samathayanon, C.; Prasongsuk, S.; Lotrakul, P.; Kiatkamjornwong, S. Rapid Degradation of Superabsorbent Poly(Potassium Acrylate) and its Acrylamide Copolymer Via Thermo-Oxidation by Hydrogen Peroxide. *J. Polym. Environ.* **2021**, *29*, 3964–3976. [CrossRef]
25. Wang, Y.; Delgado-Fukushima, E.; Fu, R.X.; Doerk, G.S.; Monclare, J.K. Controlling drug absorption, release, and erosion of photopatterned protein engineered hydrogels. *Biomacromolecules* **2020**, *21*, 3608–3619. [CrossRef]
26. Zhao, Y.; Su, H.; Fang, L.; Tan, T. Superabsorbent hydrogels from poly(aspartic acid) with salt-, temperature- and pH-responsiveness properties. *Polymer* **2005**, *46*, 5368–5376. [CrossRef]
27. Tiwari, N.; Badiger, M.V. Enhanced drug release by selective cleavage of cross-links in a double-cross-linked hydrogel. *RSC Adv.* **2016**, *6*, 102453–102461. [CrossRef]
28. Ahmadian, Y.; Bakravi, A.; Hashemi, H.; Namazi, H. Synthesis of polyvinyl alcohol/CuO nanocomposite hydrogel and its application as drug delivery agent. *Polym. Bull.* **2019**, *74*, 1967–1983. [CrossRef]
29. Namazi, H.; Hasani, M.; Yadollahi, M. Antibacterial oxidized starch/ZnO nanocomposite hydrogel: Synthesis and evaluation of its swelling behaviours in various pHs and salt solutions. *Int. J. Biol. Macromol.* **2019**, *126*, 578–584. [CrossRef] [PubMed]
30. Wang, Y.; He, G.; Li, Z.; Hua, J.; Wu, M.; Gong, J.; Zhang, J.; Ban, L.-T.; Huang, L. Novel biological hydrogel: Swelling behaviors study in salt solutions with different ionic valence number. *Polymers* **2018**, *10*, 112. [CrossRef] [PubMed]
31. Gao, S.; Jiang, G.; Li, B.; Han, P. Effects of high-concentration salt solutions and pH on swelling behavior of physically and chemically cross-linked hybrid hydrophobic association hydrogels with good mechanical strength. *Soft Mater.* **2018**, *16*, 249–264. [CrossRef]
32. Wu, F.; Zhang, Y.; Liu, L.; Yao, J. Synthesis and characterization of a novel cellulose-g-poly(acrylic acid-co-acrylamide) superabsorbent composite based on flax yarn waste. *Carbohydr. Polym.* **2012**, *87*, 2519–2525. [CrossRef]
33. Deraman, N.F.; Mohamed, N.R.; Romli, A.Z. Swelling kinetics and characterization of novel superabsorbent polymer composite based on mung bean starch-filled poly(acrylic acid)-graft-waste polystyrene. *Int. J. Plast Technol.* **2019**, *23*, 188–194. [CrossRef]

Article

Partial Replacement of Municipal Incinerated Bottom Ash and PET Pellets as Fine Aggregate in Cement Mortars

Lalitsuda Phutthimethakul [1] and Nuta Supakata [2,3,*]

[1] International Program in Hazardous Substance and Environmental Management, Graduate School, Chulalongkorn University, Bangkok 10330, Thailand; adustilal@gmail.com
[2] Department of Environmental Science, Faculty of Science, Chulalongkorn University, Bangkok 10330, Thailand
[3] Research Group (STAR): Waste Utilization and Ecological Risk Assessment, The Ratchadaphiseksomphot Endowment Fund, Chulalongkorn University, Bangkok 10330, Thailand
* Correspondence: nuta.s@chula.ac.th

Abstract: The objective of this study was to examine the optimal mixing ratio of municipal incinerated bottom ash (MIBA) and PET pellets used as a partial replacement of fine aggregates in the manufacture of cement mortars. As a partial replacement for sand, 15 mortar specimens were prepared by mixing 0%, 10%, 20%, 30%, and 40% municipal incinerated bottom ash (MIBA) (A) and 0%, 10%, and 20% PET pellets (P) in 5 cm × 5 cm × 5 cm cube molds. The cement/aggregate ratio was 1:3, and the water/cement ratio was 0.5 for all specimens. The results showed that the compressive strength of cement mortars decreased when increasing the amount of MIBA and PET pellets. The mortar specimens with 10% PET pellets achieved the highest compressive strength (49.53 MPa), whereas the mortar specimens with 40% MIBA and 20% PET pellets achieved the lowest compressive strength (24.44 MPa). Based on this finding, replacing 10% and 20% sand in cement mortar with only MIBA or only PET pellets could result in compressive strengths ranging from 46.00 MPa to 49.53 MPa.

Keywords: cement mortars; municipal incinerated bottom ash; PET pellets

Citation: Phutthimethakul, L.; Supakata, N. Partial Replacement of Municipal Incinerated Bottom Ash and PET Pellets as Fine Aggregate in Cement Mortars. *Polymers* **2022**, *14*, 2597. https://doi.org/10.3390/polym14132597

Academic Editor: Fumio Narita

Received: 20 May 2022
Accepted: 23 June 2022
Published: 27 June 2022

Publisher's Note: MDPI stays neutral with regard to jurisdictional claims in published maps and institutional affiliations.

Copyright: © 2022 by the authors. Licensee MDPI, Basel, Switzerland. This article is an open access article distributed under the terms and conditions of the Creative Commons Attribution (CC BY) license (https://creativecommons.org/licenses/by/4.0/).

1. Introduction

In 2018, cities around the world generated approximately two billion tons of solid waste [1]. Improper disposal, such as landfilling and incineration, has resulted in the release of toxic elements and pollutants that contaminate air, water, and soil and endanger human health [2,3]. Waste recycling contributes to decoupling economic growth from resource use [4–6] while reducing emissions of greenhouse gases and pollutants.

The waste management problem in Si Chang Island is a long-standing problem. The main reasons are geographic problems such as rocky areas, a lack of appropriate technology, and a lack of funds. Most of the waste is treated by incineration, which generates municipal incinerated bottom ash (MIBA) that is not properly managed. Recyclable waste is also difficult to manage as it needs to be transported to recycling facilities on the shore. One type of recycled waste is PET bottles, which are imported and consumed in large quantities [7,8].

The waste composition at the Si Chang disposal site in 2020 showed food waste as the most significant component (43.46%), followed by glass waste (22.25%) and plastic waste (12.35%). Some recycling wastes were sorted before incineration, but most waste is still incinerated without complete separation. Since incinerators have a low efficiency, they can eliminate around 78% of the waste collected daily. The incineration residue consists of mixed ash (municipal incinerated bottom ash, MIBA), which is approximately 2000 kg per month and is disposed of in dumpsites with the potential to contaminate the environment [9,10]. Many studies have shown that the chemical composition of ash depends on the types of waste. They also investigated the possibilities of utilizing ash as fine and coarse aggregate in various applications such as mortar, concrete, road pavements, masonry blocks, lightweight concrete, and foamed concrete [10–12]. Therefore, this study was interested in using MIBA as a partial replacement for fine aggregates.

Data from the 2017 to 2019 waste bank projects operated by Si Chang municipality confirm that PET bottle waste is the most common type of waste compared to other plastic waste, at approximately 635 kg per year [13]. Its current management is to store it for approximately 3–6 months and send it by marine transport to a recycling facility on the mainland. However, PET bottles require more space for storage due to their low density, causing higher management costs [8]. Therefore, finding a solution for handling PET plastic bottle waste on the island is an exciting alternative.

Population growth and tourism development have increased the demand for construction. As a result, construction materials have become more expensive due to the scarcity of natural resources and rising transportation costs [14]. Numerous studies have been conducted on the compressive strength of cement mortars made from mixed materials [15–17]. Instead of dumping PET waste into the environment, it can be reused by partially replacing aggregates in concrete and mortar. Moreover, the effect of sand replacement by PET plastic waste was proven using different forms, and it was found that the shape and size can affect the concrete properties [18–20]. Saikia and de Brito [21] suggested that the replacement of sand with PET plastic waste resulted in a higher compressive strength than that of shredded PET plastic waste. However, the compressive strength was lower when the PET pellet form was used. Therefore, this research aimed to investigate the mechanical and physical properties of cement mortars using MIBA and PET bottle waste as a partial replacement for fine aggregate. The compressive strength, water absorption, and density were investigated. In addition, microstructures were examined using scanning electron microscopy. The results of this study can provide an alternative method for Si Chang Island to manage the incineration bottom ash and PET bottles as construction materials.

2. Materials and Methods
2.1. Raw Materials

The PET bottle waste used in this study was in the form of PET pellets made from recycled PET bottle waste. The PET pellets were obtained from Grand Siam Polymer Co., Ltd. The bottom ash used in this study was mixed ash consisting of fly ash and bottom ash and was obtained from an incinerator in the Si Chang municipality. Sand, MIBA, and PET pellets were sieved prior to use, and the particle size (less than 4.75 mm) was passed through a No. 4 sieve.

Sand, MIBA, and PET pellets were tested for aggregate properties, including particle size distribution, shape, surface texture, fineness modulus, water absorption, and specific gravity. Chemical composition was analyzed using an X-ray fluorescence spectrometer (Bruker model S8 Tiger). Mineral phases were identified using an X-ray diffraction instrument (Bruker AXS model S4 Pioneer, Karlsruhe, Germany). Microstructural characteristics were identified using a scanning electron microscope (Jeol JSM-6480LV). Only MIBA was analyzed for heavy metal leaching using the Toxicity Characteristic Leaching Procedure (TCLP) (USEPA, Washington, DC, USA, 1992).

2.2. Production of Mortar Specimens

The mix proportion was designed according to ASTM C109 and modified to achieve the target compressive strength at 40 MPa after 56 days of curing by immersion in water. As shown in Table 1, the binder/fine aggregate ratio was 1:3, and the water/cement ratio was 0.5 for all mixes. MIBA was used as a replacement for fine aggregate at 0%, 10%, 20%, 30%, and 40% w/v, and PET pellets were used as a replacement for fine aggregate at a replacement level of 0%, 10%, and 20% w/v. Twelve mortar specimens were prepared for each mix. Six specimens were used for compressive strength tests and another six for water absorption tests. The specimens from all mixes were tested for compressive strength and water absorption according to ASTM C39 and ASTM C64, respectively.

Table 1. Mortar formulations.

Name	Binder		Fine Aggregate		Water/Cement
	Cement	Sand (%)	MIBA (%)	PET Pellets (%)	
Control	100	100	0	0	
A10P0	100	90	10	0	
A20P0	100	80	20	0	
A30P0	100	70	30	0	
A40P0	100	60	40	0	
A0P10	100	90	0	10	
A10P10	100	80	10	10	
A20P10	100	70	20	10	0.5
A30P10	100	60	30	10	
A40P10	100	50	40	10	
A0P20	100	80	0	20	
A10P20	100	70	10	20	
A20P20	100	60	20	20	
A30P20	100	50	30	20	
A40P20	100	40	40	20	

Note: A refers to MIBA, and P is defined as PET pellets.

3. Results and Discussion

3.1. Raw Materials

3.1.1. Fine Aggregate Properties

Sand, MIBA, and PET pellets (PET) were analyzed to investigate the particle size distribution, shape, surface texture, fineness modulus, water absorption, and specific gravity.

1. Particle Size Distribution

The particle size distribution was analyzed using the sieving method to determine the average particle size of raw materials (Figure 1). The particle size distribution results indicated that the PET pellets were larger than sand and MIBA. This is due to the fixation and size consistency of PET pellets, which were retained only on sieves No. 7 (2.8 mm) and No. 16 (1.18 mm). As a result, the D_{50} values for MIBA, sand, and PET pellets were 0.3 mm, 0.5 mm, and 3 mm, respectively.

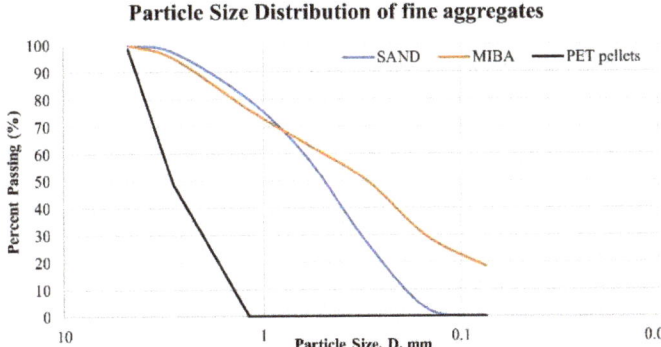

Figure 1. Particle size distribution of fine aggregates.

2. Shape and Surface Texture

Sand has different grain sizes and shapes and is rough and angular. MIBA is fine-grained, dry, and dark gray in color. In comparison, PET pellets are cylindrical with smooth surfaces and bright colors. The appearance of raw materials is shown in Figure 2.

Figure 2. General appearances of sand, MIBA, and PET pellets.

3. Fineness Modulus

Sand, MIBA, and PET pellets have a fineness modulus (F.M.) of 2.35, 1.86, and 4.53, respectively. Increasing the F.M. can affect the compressive strength of the samples and improve it [21]. The water absorption of fine aggregates was analyzed according to ASTM C128. The water absorption of sand, MIBA, and PET pellets was 1.1%, 18.5%, and 0%, respectively. The PET pellet is a nonabsorbent material with a smooth, nonporous surface [22]. Meanwhile, the high-water absorption of MIBA requires more water, reducing the actual water–cement ratio [21].

The specific gravity of fine aggregates was analyzed according to ASTM C128. The results showed that the specific gravity of sand, MIBA, and PET pellets is 2.48, 1.27, and 1.28, respectively.

4. Characteristics of Fine Aggregates

3.1.2. Chemical Composition

The chemical compositions of raw materials were analyzed using an X-ray fluorescence spectrometer (Bruker model S8 Tiger). As shown in Table 2, SiO_2 (73.6%) and CaO (31.3%) are the major components of sand and MIBA, respectively. Chlorine (Cl) is also high in MIBA (5.22%). Due to improper waste incineration without efficient waste separation, the Cl content can be inherited from PVC, chloride-contained plastics, and chloride salts in kitchen waste [23]. The chloride content in MIBA can corrode the reinforcing steel in concrete, causing the structure to collapse [24].

Table 2. Chemical composition of sand and MIBA.

Oxides	Content (wt.%)	
	Sand	MIBA
SiO_2	73.6	17.0
Al_2O_3	4.72	4.22
K_2O	2.96	3.03
Fe_2O_3	0.765	2.07

Table 2. *Cont.*

Oxides	Content (wt.%)	
	Sand	MIBA
Na_2O	0.266	3.57
TiO_2	0.125	1.19
MgO	868 PPM	2.31
CaO	868 PPM	31.3
BaO	359 PPM	729 PPM
P_2O_5	345 PPM	1.96
Rb_2O	184 PPM	60 PPM
ZrO_2	141 PPM	156 PPM
SO_3	105 PPM	3.04
MnO	87.4 PPM	875 PPM
SrO	34.7 PPM	402 PPM
PbO	27.4 PPM	300 PPM
Cl	Not detected	5.22
CoO	Not detected	Not detected
NiO	Not detected	81.6 PPM
CuO	Not detected	0.118
ZnO	Not detected	0.22
As_2O_3	Not detected	Not detected

5. Mineral Phases

The X-ray diffraction pattern of raw materials is shown in Figure 3. The major crystalline phases of sand are quartz (SiO_2) and orthoclase (KSi_3AlO_8). The major crystalline phases of MIBA are synthetic hartrurite (Ca_3SiO_5) and wollastonite ($CaSiO_3$), whereas the minor mineral phases are marcasite (FeS_2) and pyrrhotite (FeS).

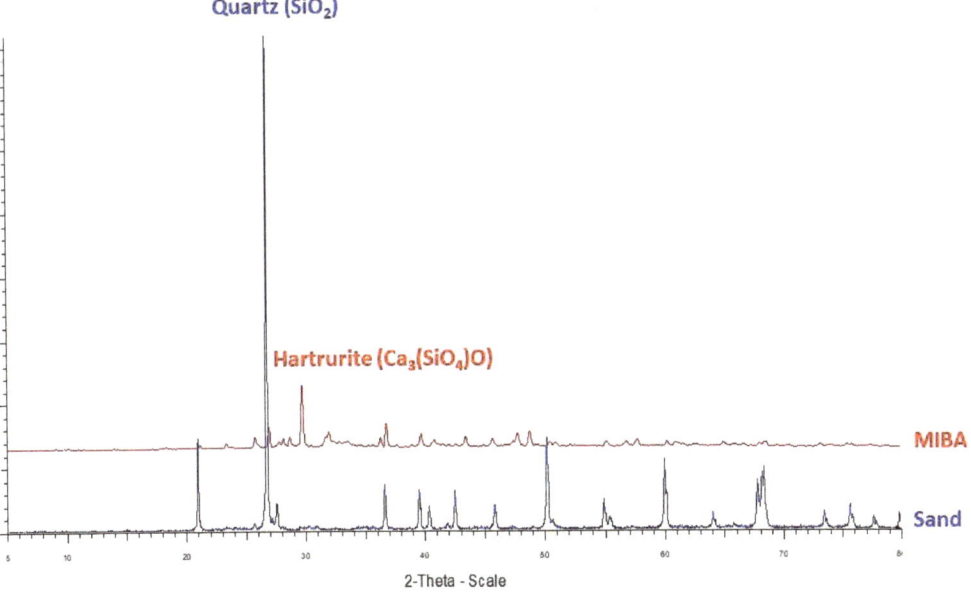

Figure 3. X-ray diffraction patterns of sand and MIBA.

6. Microstructure and Elemental Composition of Raw Materials

The microstructure and elemental composition of raw materials are shown in Figure 4. The particle morphologies were observed at 50× resolution for sand and MIBA and 30× resolution for PET pellets. The results showed that sand had a rough and angular surface, while MIBA was noticeably smaller and more porous than sand, which was confirmed at the same magnification. However, the PET pellets were much larger than sand and MIBA and had a smooth non-angular surface, and were nonporous. In addition, these products have different elemental components. For example, Si and C are the major elements of PET and sand, respectively. On the other hand, MIBA has a relatively diverse composition consisting of several elements, with CaO being the most prominent.

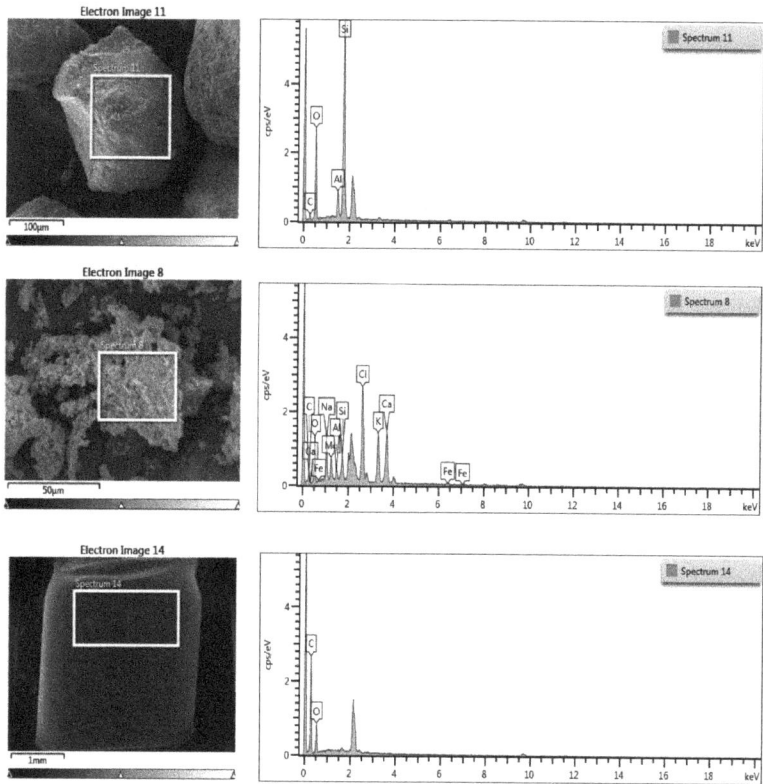

Figure 4. Microstructural characterization and chemical composition of raw materials.

7. Leaching of Heavy Metals from MIBA

Only MIBA was examined for the leaching of heavy metals, including Ba, As, Co, Cd, Fe, Cr, Mn, Cu, Se, Zn, Ni, and Pb, using the TCLP method. As shown in Table 3, the results were compared to the regulatory levels of the U.S. Code of Federal Regulations [25] and the soil quality standards of the Thai Pollution Control Department [26]. The heavy metal concentration of MIBA was within the maximum contaminant concentration for toxicity characteristics. As a result, MIBA can be classified as nonhazardous waste and used as a raw material in this study. Thus, the cement mortars made from MIBA did not require heavy metal leaching testing.

Table 3. Leaching of heavy metals from MIBA.

Heavy Metal	MIBA (mg/L)	Regulatory Level (mg/L)	Soil Quality Standards (mg/kg)
Ba	0.638 ± 0.169	100.0	-
As	Not detected	5.0	3.9
Co	Not detected	-	-
Cd	0.003 ± 0.000	1.0	37
Fe	Not detected	-	-
Cr	0.228 ± 0.028	5.0	300
Mn	Not detected	-	1800
Cu	0.188 ± 0.058	-	-
Se	0.010 ± 0.000	1.0	390
Zn	0.013 ± 0.008	-	-
Ni	Not detected	-	1600
Pb	0.011 ± 0.003	5.0	400

3.2. Mortar Specimens

Physical Properties of Mortar Specimens

1. General Appearance

As shown in Figure 5, the general appearance of mortar specimens differs slightly in those with high levels of MIBA. In addition, the mixture layers, surface roughness, and dryness can be observed compared to the control.

Figure 5. General appearance of mortar specimens.

2. Compressive Strength

Figure 6 shows the compressive strengths of mortar specimens after 56 days. The results show that the compressive strength of each group decreased as the amount of MIBA used to replace sand increased. Four mortar specimens have compressive strengths greater than 40 MPa, but less than the control (51.32 MPa), as follows: A10P0 (49.19 MPa), A20P0 (46.25 MPa), A0P10 (49.53 MPa), and A0P20 (46 MPa).

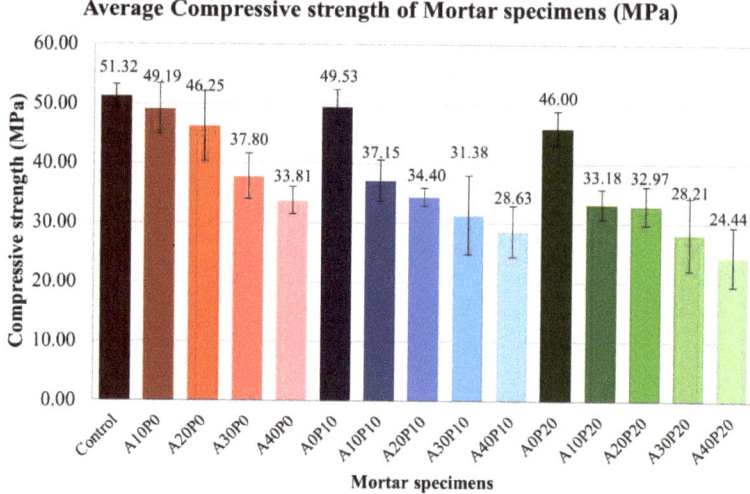

Figure 6. Average compressive strength of mortar specimens.

As shown in Table 4, data analysis two-way ANOVA for compressive strength shows no interaction between PET pellets and MIBA, F (8, 60) = 1.897, p = 0.08. This demonstrates that the combination of PET pellets and MIBA has no significant effect on specimen compressive strengths. At the same time, the effects of PET pellets and MIBA have a p-value = 0.000, indicating that the amount of PET pellets and MIBA affects the compressive strength of the specimen at a significant level of 0.01. This can describe how different replacement levels of both materials affect the compressive strength. However, as shown in Figure 7, the profile plot of estimated marginal means of compressive strength between PET pellets and MIBA shows trends in the same direction.

Table 4. Compressive strength tests between PET pellets and MIBA effects. Test of Between-Subjects Effects. Dependent Variable: Compressive strength.

Source	Type III Sum of Squares	df	Mean Square	F	Sig.	Partial Eta Squared
Corrected Model	5282.550 [a]	14	377.325	22.857	0.000	0.842
Intercept	106,129.133	1	106,129.133	6428.935	0.000	0.991
PET	1508.426	2	754.213	45.688	0.000	0.604
MIBA	3523.559	4	880.890	53.361	0.000	0.781
PET*MIBA	250.564	8	31.321	1.897	0.077	0.202
Error	990.483	60	16.508			
Total	112,402.165	75				
Corrected Total	6273.033	74				

[a] R Squared = 0.842 (Adjusted R Squared = 0.805).

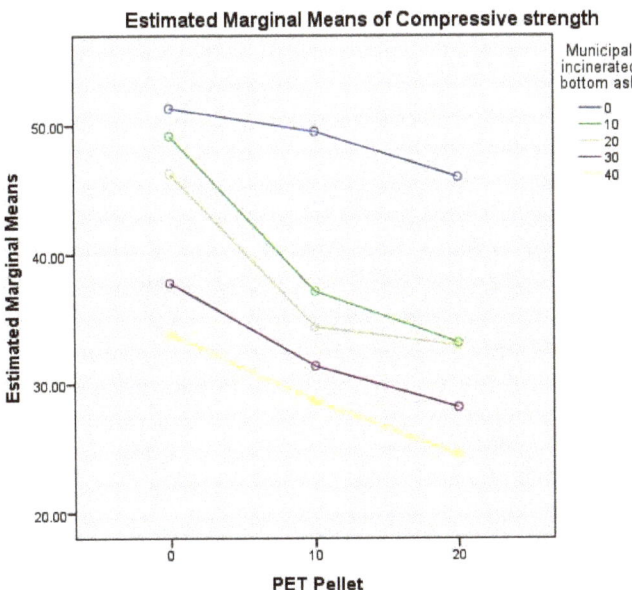

Figure 7. Estimated marginal means of compressive strength between PET pellets and MIBA.

3. Water Absorption

Figure 8 shows the water absorption of mortar specimens. The results indicate that the highest water absorption rate was A40P10 (5.5%), followed by A40P20 (5.44%), whereas the lowest was A20P20 (4.27%). This experiment shows that using 20% MIBA caused the lowest water absorption for each group. In addition, among all ratios, 20% MIBA and 20% PET pellets have the lowest water absorption. It can be concluded that replacing sand with 20% PET pellets and 20% MIBA resulted in the best aggregate arrangement with the least porous matrix and less water absorption.

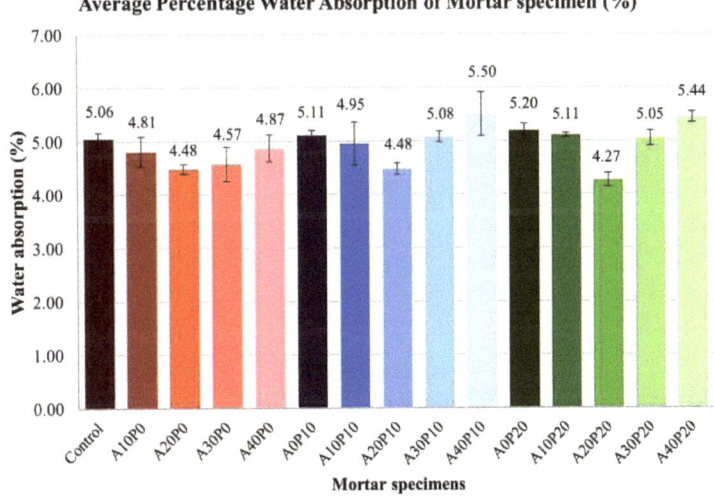

Figure 8. Average water absorption of mortar specimens.

The data analysis of water absorption (Table 5) revealed an interaction effect between PET pellets and MIBA, $F(8, 60) = 2.22$, $p = 0.004$. This demonstrates that combining different levels of PET pellets and MIBA significantly impacts the specimen's water absorption. Furthermore, the plot of estimated marginal means for different replacement levels of PET pellets and MIBA (Figure 9) shows that the use of 30% and 40% MIBA tends to differ from other scenarios.

Table 5. Water absorption tests between PET pellets and MIBA effects. Test of Between-Subjects Effects. Dependent Variable: Water absorption.

Source	Type III Sum of Squares	df	Mean Square	F	Sig.	Partial Eta Squared
Corrected Model	2413.943 [a]	14	172.425	9.764	0.000	0.695
Intercept	760,274.953	1	760,274.953	43,051.327	0.000	0.999
PET	123.614	2	61.807	3.500	0.037	0.104
MIBA	1835.554	4	458.889	25.985	0.000	0.634
PET*MIBA	454.775	8	56.847	3.219	0.004	0.300
Error	1059.584	60	17.660			
Total	763,748.480	75				
Corrected Total	3473.527	74				

[a] R Squared = 0.695 (Adjusted R Squared = 0.624).

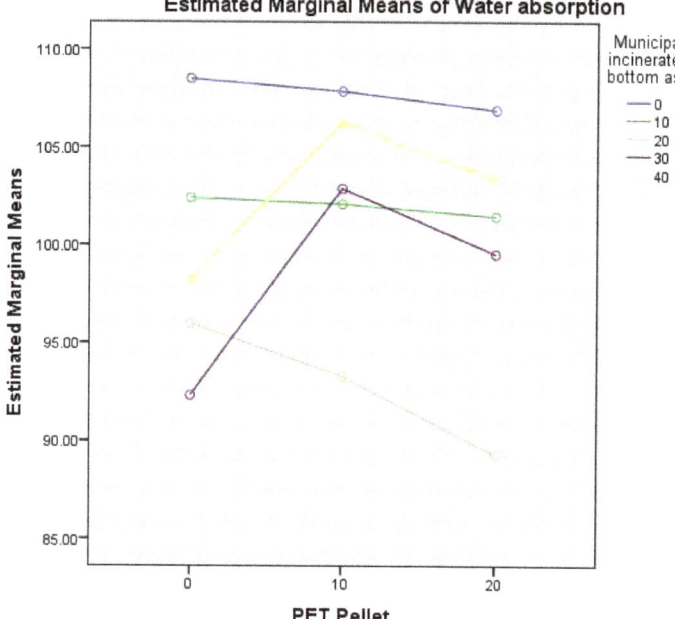

Figure 9. Estimated marginal means of water absorption between PET pellets and MIBA.

4. Density

Figure 10 shows the density of mortar specimens. The tendency of density is similar to that of compressive strength, indicating that these two parameters are correlated. The control has the highest density (2172 kg/m^3), followed by A0P10 (2158 kg/m^3) and A10P0 (2156 kg/m^3). On the other hand, A40P20 presented the lowest density (1979 kg/m^3).

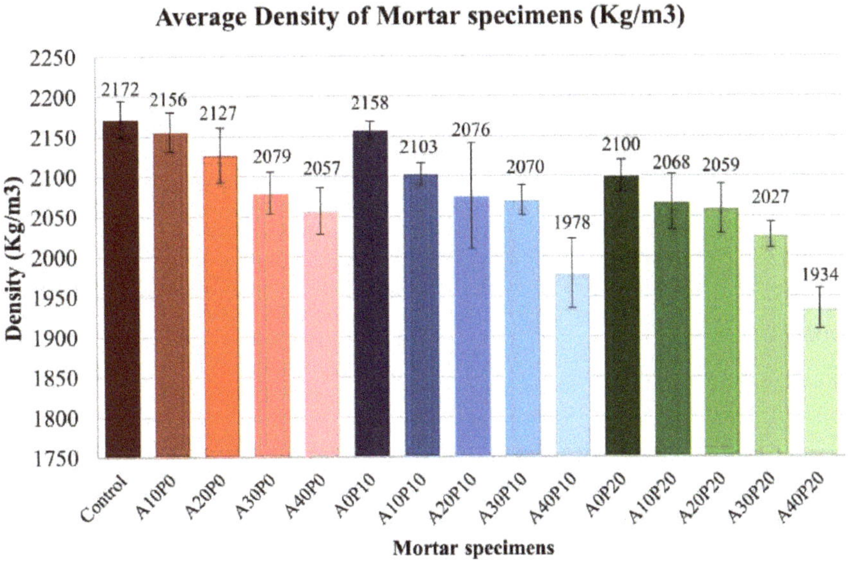

Average Density of Mortar specimens (Kg/m3)

[Bar chart showing density values (Kg/m³) for mortar specimens: Control: 2172, A10P0: 2156, A20P0: 2127, A30P0: 2079, A40P0: 2057, A0P10: 2158, A10P10: 2103, A20P10: 2076, A30P10: 2070, A40P10: 1978, A0P20: 2100, A10P20: 2068, A20P20: 2059, A30P20: 2027, A40P20: 1934]

Figure 10. The average density of mortar specimens.

The density analysis results (Table 6) revealed that the amount of PET pellets and MIBA significantly affected the specimen's density. It can be observed that there is no interaction effect when using PET pellets with MIBA, $F(8, 60) = 1.44$, $p = 0.2$. The plots of the estimated marginal mean for different replacement levels of PET and MIBA (Figure 11) show a similar trend to the compressive strength plot.

Table 6. Density tests between PET pellets and MIBA effects. Test of Between-Subjects Effects. Dependent Variable: Density.

Source	Type III Sum of Squares	df	Mean Square	F	Sig.	Partial Eta Squared
Corrected Model	294,711.711 [a]	14	21,050.837	20.365	0.000	0.826
Intercept	323,728,515.6	1	323,728,515.6	313,176.841	0.000	1.000
PET	81,567.215	2	40,783.608	39.454	0.000	0.568
MIBA	201,231.402	4	50,307.851	48.668	0.000	0.764
PET*MIBA	11,913.094	8	1489.137	1.441	0.199	0.161
Error	62,021.543	60	1033.692			
Total	324,085,248.9	75				
Corrected Total	356,733.254	74				

[a] R Squared = 0.826 (Adjusted R Squared = 0.786).

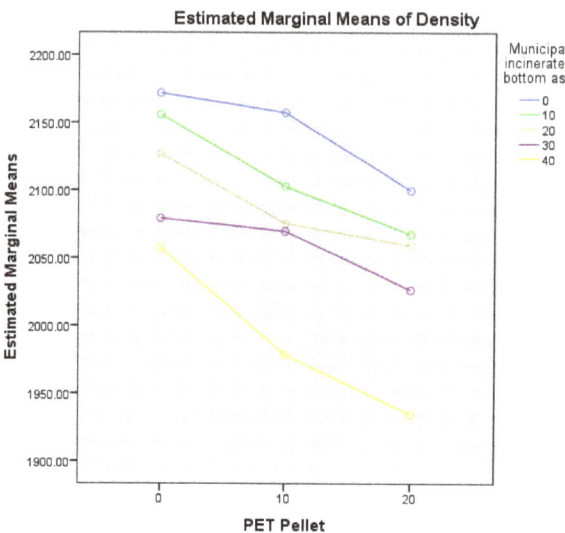

Figure 11. Estimated marginal means of density between PET pellets and MIBA.

5. Microstructure of Mortar Specimens

The microstructure of mortar specimens was identified using a scanning electron microscope (IT300) at 1500× magnification. Only mortar specimens selected from the control (no waste), A0P10 (best performance), A0P20 (highest PET pellet replacement), A40P0 (highest MIBA replacement), and A40P20 (worst performance, most uses of both wastes) were analyzed. The results revealed calcium silicate hydrate (CSH) and ettringite on the surface of mortar specimens (Figure 12). Ettringite shows needle-like crystals resulting from the hydration reaction of tricalcium aluminate (C_3A) as sulfate ions of gypsum reacted with water. On the other hand, CSH is formed by the hydration reaction of calcium silicates (C_3S and C_2S) with water. Therefore, CSH is like a gel and serves as a binder that connects the aggregates and provides strength to the specimens [27].

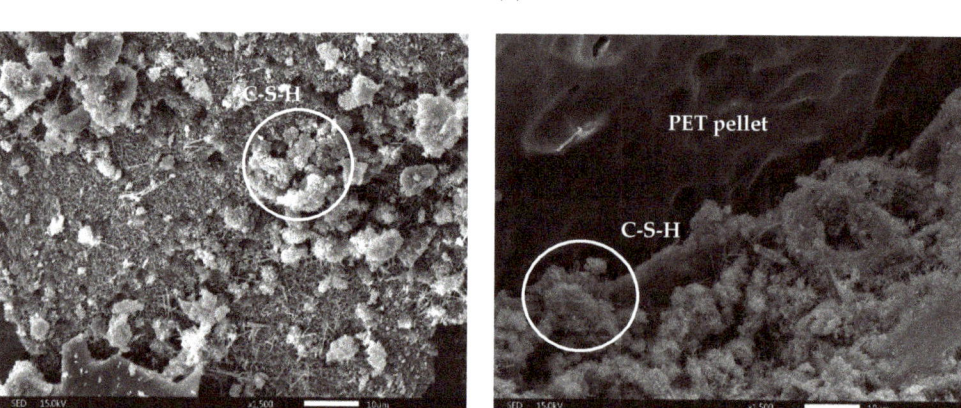

Figure 12. *Cont.*

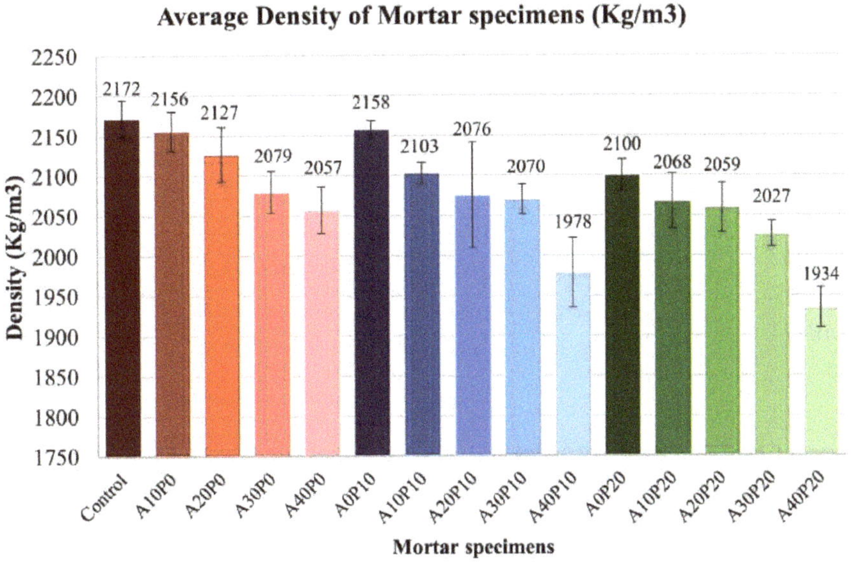

Figure 10. The average density of mortar specimens.

The density analysis results (Table 6) revealed that the amount of PET pellets and MIBA significantly affected the specimen's density. It can be observed that there is no interaction effect when using PET pellets with MIBA, $F(8, 60) = 1.44$, $p = 0.2$. The plots of the estimated marginal mean for different replacement levels of PET and MIBA (Figure 11) show a similar trend to the compressive strength plot.

Table 6. Density tests between PET pellets and MIBA effects. Test of Between-Subjects Effects. Dependent Variable: Density.

Source	Type III Sum of Squares	df	Mean Square	F	Sig.	Partial Eta Squared
Corrected Model	294,711.711 [a]	14	21,050.837	20.365	0.000	0.826
Intercept	323,728,515.6	1	323,728,515.6	313,176.841	0.000	1.000
PET	81,567.215	2	40,783.608	39.454	0.000	0.568
MIBA	201,231.402	4	50,307.851	48.668	0.000	0.764
PET*MIBA	11,913.094	8	1489.137	1.441	0.199	0.161
Error	62,021.543	60	1033.692			
Total	324,085,248.9	75				
Corrected Total	356,733.254	74				

[a] R Squared = 0.826 (Adjusted R Squared = 0.786).

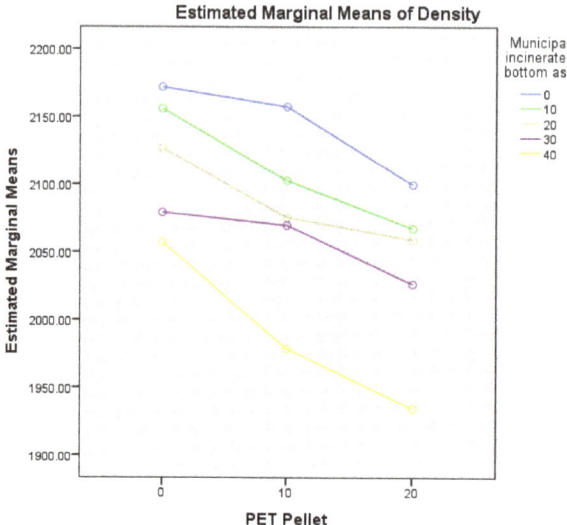

Figure 11. Estimated marginal means of density between PET pellets and MIBA.

5. Microstructure of Mortar Specimens

The microstructure of mortar specimens was identified using a scanning electron microscope (IT300) at 1500× magnification. Only mortar specimens selected from the control (no waste), A0P10 (best performance), A0P20 (highest PET pellet replacement), A40P0 (highest MIBA replacement), and A40P20 (worst performance, most uses of both wastes) were analyzed. The results revealed calcium silicate hydrate (CSH) and ettringite on the surface of mortar specimens (Figure 12). Ettringite shows needle-like crystals resulting from the hydration reaction of tricalcium aluminate (C_3A) as sulfate ions of gypsum reacted with water. On the other hand, CSH is formed by the hydration reaction of calcium silicates (C_3S and C_2S) with water. Therefore, CSH is like a gel and serves as a binder that connects the aggregates and provides strength to the specimens [27].

(**a**) Control   (**b**) A0P10

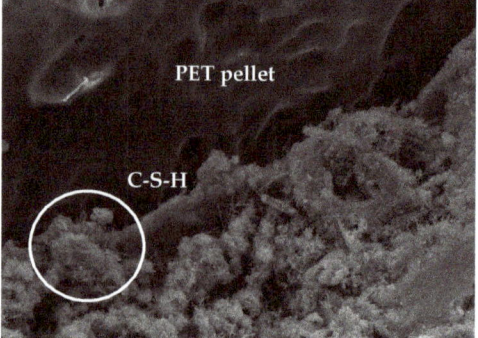

Figure 12. *Cont.*

(c) A0P20   (d) A40P0

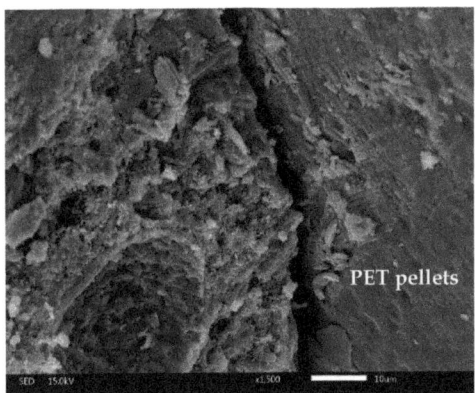 

(e) A40P20

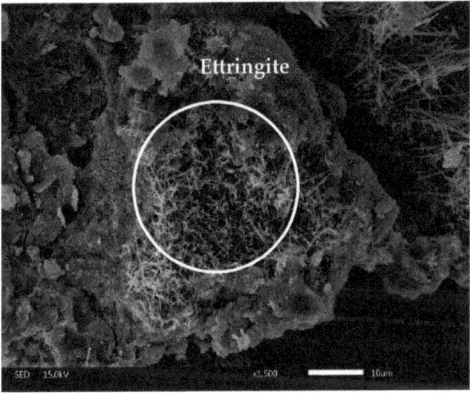

Figure 12. Microstructure of mortar specimens: control (**a**), A0P10 (**b**), A0P20 (**c**), A40P0 (**d**), and A40P20 (**e**).

The microstructures of mortar specimens with PET pellets are shown in Figure 13a,b at 50× and 3000× magnifications, respectively. The smooth surface of PET pellets was found to be poorly connected with the mixture, resulting in more free space and voids.

The physical properties of mortar specimens showed a decline in compressive strength with the increased replacement of PET pellets and MIBA. da Silva and de Brito [22] studied the replacement of sand with two types of plastic aggregates: PET pellets and PET flakes. The results showed that PET pellets with a smooth surface, low specific surface area, and no water absorption decreased the W/C ratio and improved the workability of the mortar. However, a poorly connected matrix/aggregate results in high-porosity mortars. In addition, the mortar density is reduced linearly when using more plastic aggregate due to the lower density of the plastic aggregates than sand.

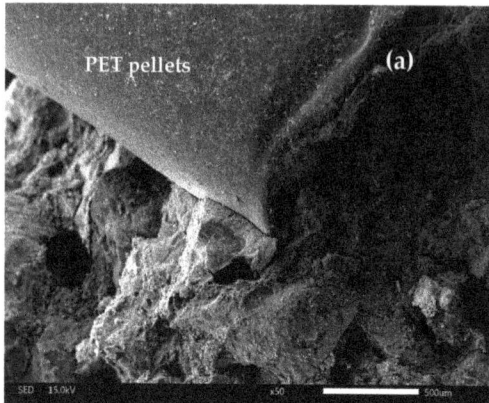 

Figure 13. Microstructure of mortar specimens with PET pellets at 50× magnification (**a**) and 3000× magnification (**b**).

At the same time, Naran and Gonzalez [28] reported that single-plastic aggregates (PA) are hydrophobic, resulting in less water absorption. The excess water was used to coat and decrease the friction between particles. Moreover, the bonding between cement and PA is not strong and causes voids, resulting in a lower compressive strength and high-water absorption.

According to the morphology of MIBA, which is irregular in shape, rough surfaces and a high porosity cause a high water absorption. In addition, MIBA has a lower density than natural aggregates, causing a lower density of the specimens when increasing the replacement of sand [29].

Kunther and Ferreiro [30] studied the influence of the Ca/Si ratio on the compressive strength of the sample and found that the high Ca/Si ratio decreased the quantity of calcium silicate hydrates. In addition, the raw material analysis showed that MIBA had higher CaO than SiO_2. Therefore, the Ca/Si ratio of mortar specimens increased when increasing the replacement levels of MIBA, decreasing the compressive strength. The Ca/Si ratio also affects the microstructure of the samples, increasing the compressive strength, reducing water absorption, and reducing microcracks [31].

A previous study of using reactive aggregate to produce concrete samples shows that the effects of the alkali–silica reaction on the compressive strength and elastic modulus of the samples can be observed after curing by immersion for 28 days, lowering the compressive strength and elastic modulus of concrete samples [32].

The relationship between density and compressive strength shows that mortar specimens with a high density had a high compressive strength. Increasing the sand replacement rate by lightweight waste results in a decrease in density. MIBA and PET pellets have a lower specific gravity than sand, with poor bonding between aggregates. Moreover, MIBA has a high CaO content. Hence, using more MIBA may increase the Ca/Si ratio. The results showed that the compressive strength decreased as the amount of CSH was reduced. On the other hand, the water absorption of mortar specimens is related to the porosity and the connection between aggregates. The results showed that using 20% MIBA with different levels of PET pellets provided the lowest water absorption for each group, and the trend was similar for each group. Therefore, it can be concluded that a 20% sand replacement by MIBA is a suitable ratio that leads to the lowest water absorption. The effect of water absorption is not directly related to compressive strength. However, as MIBA increased, the CSH formation decreased, resulting in poor aggregate connectivity, pore generation in the microstructure, and a decrease in compressive strength. However, the 10% and 20% replacement of MIBA produced different results. As a result of the diversity of MIBA sizes, substitution at 10% could result in poor alignment and the development of pores. However, at a 20% substitution, better results were obtained because the CaO content in MIBA was optimal.

4. Conclusions

This study aimed to investigate the feasibility of utilizing MIBA and PET plastic waste from Si Chang Island. Both wastes were used as partial substitutes for the fine aggregate in cement mortar production. The optimal ratio of cement mortar production was determined, and the properties of the cement mortars were examined. The following conclusions can be drawn from the results of this study:

1. The compressive strengths of mortar specimens cured for 56 days was greater than 40 MPa, and are as follows:
   - A0P10 with 10% sand replaced by PET pellets obtained a compressive strength of 49.53 MPa.
   - A10P0 with 10% sand replaced by MIBA obtained a compressive strength of 49.19 MPa.
   - A20P0 with 20% sand replaced by MIBA obtained a compressive strength of 46.25 MPa.
   - A0P20 with 20% sand replaced by PET pellets obtained a compressive strength of 46.00 MPa.

2. The properties of the mortar specimens showed that the amount of waste replaced by fine aggregate in the manufacture of cement mortar affected the reduced compressive strength and density of mortar specimens due to the poor bonding of aggregates in the mortar specimen matrix and low-density properties of the waste. However, mortar specimens with 20% sand replaced by PET pellets obtained the lowest water absorption.

The MIBA used in this study was classified as nonhazardous waste due to the amount of leaching heavy metals examined by the TCLP method not exceeding the regulatory levels of the U.S. Code of Federal Regulations and the soil quality standards of the Thai Pollution Control Department. In addition, this study showed that replacing 10% and 20% sand in cement mortar with only MIBA or only PET pellets could result in compressive strengths ranging from 46.00 MPa to 49.53 MPa. Based on these findings, the following future research topics were proposed for the alternative use of MIBA and PET bottles in building and construction work in Si Chang Island: an investigation of mechanical properties, such as tensile strength, elastic modulus, and stress–strain curves, as well as a life cycle assessment and economic feasibility.

Author Contributions: Conceptualization, N.S. and L.P.; methodology, L.P.; validation, N.S.; formal analysis, L.P.; investigation, L.P.; resources, N.S.; data curation, L.P.; writing—original draft preparation, L.P.; writing—review and editing, N.S.; visualization, N.S. and L.P.; supervision, N.S. All authors have read and agreed to the published version of the manuscript.

Funding: This research was supported by International Program in Hazardous Substance and Environmental Management and CU Graduate School Thesis Grant from Chulalongkorn University, and partially financial supported by the Ratchadaphiseksomphot Endowment Fund, Chulalongkorn University (STAR code STF 62004230).

Institutional Review Board Statement: Not applicable.

Informed Consent Statement: Not applicable.

Data Availability Statement: The data presented in this study are available on request from the corresponding author.

Acknowledgments: The authors would like to thank Koh Si Chang Subdistrict Municipality for providing information and municipal solid waste incineration bottom ash samples and for receiving PET pellet samples from Grand Siam Polymer Co., Ltd. The authors gratefully acknowledge the Department of Environmental Science in the Faculty of Science, the Department of Mining Engineering, the Department of Civil Engineering, and the Center of Laboratory Engineering in the Faculty of Engineering, and the Environmental Research Institute and the Scientific and Technological Research Equipment Centre (STREC) for providing mechanical equipment.

Conflicts of Interest: The authors declare no conflict of interest.

References

1. Word Bank. Available online: https://data.worldbank.org.cn/ (accessed on 22 May 2021).
2. Wang, H.; Wang, X.; Song, J.; Wang, S.; Liu, X. Uncovering regional energy and environmental benefits of urban waste utilization: A physical input-output analysis for a city case. *J. Clean Prod.* **2018**, *189*, 922–932. [CrossRef]
3. Nordahl, S.L.; Devkota, J.P.; Amirebrahimi, J.; Smith, S.J.; Breunig, H.M.; Preble, C.V.; Satchwell, A.J.; Jin, L.; Brown, N.J.; Kirchstetter, T.W.; et al. Life-cycle greenhouse gas emissions and human health trade-offs of organic waste management strategies. *Environ. Sci. Technol.* **2020**, *54*, 9200–9209. [CrossRef] [PubMed]
4. Corona, B.; Shen, L.; Reike, D.; Carreón, R.J.; Worrell, E. Towards sustainable development through the circular economy—A review and critical assessment on current circularity metrics. *Resour. Conserv. Recycl.* **2019**, *151*, 104498. [CrossRef]
5. Baležentis, T.; Streimikiene, D.; Zhang, T.; Liobikiene, G. The role of bioenergy in greenhouse gas emission reduction in EU countries: An environmental Kuznets curve modelling. *Resour. Conserv. Recycl.* **2019**, *142*, 225–231. [CrossRef]
6. Lanzotti, A.; Martorelli, M.; Maietta, S.; Gerbino, S.; Penta, F.; Gloria, A. A comparison between mechanical properties of specimens 3D printed with virgin and recycled PLA. *Procidia CIRP* **2019**, *79*, 143–146. [CrossRef]
7. Si Chang IslandChangSubdistrict Municipality, Recycle Waste Trading Summary. Koh Si Chang Subdistrict Municipality. 2020. Available online: https://www.kohsichang.go.th/ (accessed on 1 May 2022).
8. Interview with Phanuwat Robkob, officer of Si Chang Municipality, Chonburi province, Thailand. (interviewed on 16 August 2020).
9. Abramov, S.; He, J.; Wimmer, D.; Lemloh, M.L.; Muehe, E.M.; Gann, B.; Roehm, E.; Kirchhof, R.; Babechuk, M.G.; Schoenberg, R.; et al. Heavy metal mobility and valuable contents of processed municipal solid waste incineration residues from Southwestern Germany. *Waste Manag.* **2018**, *79*, 735–743. [CrossRef] [PubMed]
10. Lynn, C.J.; Obe, D.R.K.; Ghataora, G.S. Municipal incinerated bottom ash characteristics and potential for use as aggregate in concrete. *Constr. Build. Mater.* **2016**, *127*, 504–517. [CrossRef]
11. Lynn, C.J.; Ghataora, G.S.; Obe, D.R.K. Municipal incinerated bottom ash (MIBA) characteristics and potential for use in road pavements. *Int. J. Pavement Res. Technol.* **2017**, *10*, 185–201. [CrossRef]
12. Abdalla, A.; Mohammed, A.S. Surrogate models to predict the long-term compressive strength of cement based mortar modified with fly ash. *Arch. Comp. Methods Eng.* **2022**, 1–26. [CrossRef]
13. Koh Si Chang Subdistrict Municipality. *History of Si Chang Island*; Koh Si Chang Subdistrict Municipality: Bay of Siam, Thailand, 2017.
14. Interview with Wisit La-ongsiri, Sichangmarine Service Co.,Ltd, Chonburi province, Thailand. (interviewed on 31 August 2020).
15. Mohammed, A.; Rafiq, S.; Sihag, P.; Mahmood, W.; Ghafor, K.; Sarwar, W. ANN, M5P-tree model, and nonlinear regression approaches to predict the compression strength of cement-based mortar modified by quicklime at various water/cement ratios and curing times. *Arab. J. Geosci.* **2020**, *13*, 1216. [CrossRef]
16. Raheem, A.A.; Ikotun, B.D. Incorporation of agricultural residues as partial substitution for cement in concrete and mortar–A review. *J. Build. Eng.* **2020**, *31*, 101428. [CrossRef]
17. Mahmood, W.; Mohammed, A. Performance of ANN and M5P-tree to forecast the compressive strength of hand-mix cement-grouted sands modified with polymer using ASTM and BS standards and evaluate the outcomes using SI with OBJ assessments. *Neural Comput. Appl.* **2022**, 1–21. [CrossRef]
18. Saikia, N.; de Brito, J. Mechanical properties and abrasion behaviour of concrete containing shredded PET bottle waste as a partial substitution of natural aggregate. *Constr. Build. Mater.* **2014**, *52*, 236–244. [CrossRef]
19. Thorneycroft, J.; Orr, J.; Savoikar, P.; Ball, R.J. Performance of structural concrete with recycled plastic waste as a partial replacement for sand. *Constr. Build. Mater.* **2018**, *161*, 63–69. [CrossRef]
20. Bahij, S.; Omary, S.; Feugeas, F.; Faqiri, A. Fresh and hardened properties of concrete containing different forms of plastic waste—A review. *Waste Manag.* **2020**, *113*, 157–175. [CrossRef]
21. Silva, R.V.; de Brito, J.; Lynn, C.J.; Dhir, R.K. Environmental impacts of the use of bottom ashes from municipal solid waste incineration: A review. *Resour. Conserv. Recycl.* **2019**, *140*, 23–35. [CrossRef]
22. da Silva, A.M.; de Brito, J.; Veiga, R. Incorporation of fine plastic aggregates in rendering mortars. *Constr. Build. Mater.* **2014**, *71*, 226–236. [CrossRef]
23. Wei, M.-C.; Wey, M.; Hwang, J.; Chen, J. Stability of heavy metals in bottom ash and fly ash under various incinerating conditions. *J. Hazard. Mater.* **1998**, *57*, 145–154. [CrossRef]
24. Alam, Q.; Lazaro, A.; Schollbach, K.; Brouwers, H.J.H. Chemical speciation, distribution and leaching behavior of chlorides from municipal solid waste incineration bottom ash. *Chemosphere* **2020**, *241*, 124985. [CrossRef]

25. Environmental Protection Agency (EPA). *Identification and Listing of Hazardous Waste, Identification and Listing of Hazardous Waste*; Environmental Protection Agency (EPA): Washington, DC, USA, 2022. Available online: https://www.ecfr.gov/current/title-40/chapter-I/subchapter-I/part-261 (accessed on 1 May 2022).
26. Pollution Control Department. Announcement of the National Environment Board (BE 2547) Regarding to Determination of Soil Quality. In *Government Gazette*; Bangkok, Thailand, 2004; Volume 25. Available online: https://www.pcd.go.th/ (accessed on 1 May 2022).
27. CPAC Academy. *Concrete Technology*; The Concrete Products and Aggregate Co., Ltd.: Bangkok, Thailand, 2000; Chapter 2 Cement.
28. Naran, J.M.; Gonzalez, R.E.G.; del Rey Castillo, E.; Toma, C.L.; Almesfer, N.; van Vreden, P.; Saggi, O. Incorporating waste to develop environmentally-friendly concrete mixes. *Constr. Build. Mater.* **2022**, *314*, 125599. [CrossRef]
29. Caprai, V.; Schollbach, K.; Brouwers, H.J.H. Influence of hydrothermal treatment on the mechanical and environmental performances of mortars including MSWI bottom ash. *Waste Manag.* **2018**, *78*, 639–648. [CrossRef] [PubMed]
30. Kunther, W.; Ferreiro, S.; Skibsted, J. Influence of the Ca/Si ratio on the compressive strength of cementitious calcium–silicate–hydrate binders. *J. Mater. Chem. A* **2017**, *5*, 17401–17412. [CrossRef]
31. Hanif Khan, M.; Zhu, H.; Ali Sikandar, M.; Zamin, B.; Ahmad, M.; Muayad Sabri Sabri, M. Effects of various mineral admixtures and fibrillated polypropylene fibers on the properties of engineered cementitious composite (ECC) based mortars. *Materials* **2022**, *15*, 2880. [CrossRef] [PubMed]
32. Na, O.; Xi, Y.; Ou, E.; Saouma, V.E. The effects of alkali-silica reaction on the mechanical properties of concretes with three different types of reactive aggregate. *Struct. Concr.* **2016**, *17*, 74–83. [CrossRef]

Article

Characteristics of Microplastics and Their Affiliated PAHs in Surface Water in Ho Chi Minh City, Vietnam

Nguyen Thao Nguyen [1,2], Nguyen Thi Thanh Nhon [1,2], Ho Truong Nam Hai [1,2], Nguyen Doan Thien Chi [1,2] and To Thi Hien [1,2,*]

1. Faculty of Environment, University of Science, 227 Nguyen Van Cu Street, District 5, Ho Chi Minh City 700000, Vietnam; ngtnguyen@hcmus.edu.vn (N.T.N.); nttnhon@hcmus.edu.vn (N.T.T.N.); htnhai@hcmus.edu.vn (H.T.N.H.); ndtchi@hcmus.edu.vn (N.D.T.C.)
2. Faculty of Environment, University of Science, Vietnam National University, Ho Chi Minh City 700000, Vietnam
* Correspondence: tohien@hcmus.edu.vn; Tel.: +84-976000621

Abstract: Microplastic pollution has become a worldwide concern. However, studies on the distribution of microplastics (MPs) from inland water to the ocean and their affiliated polycyclic aromatic hydrocarbons (PAHs) are still limited in Vietnam. In this study, we investigated the distribution of MPs and PAHs associated with MPs in canals, Saigon River, and Can Gio Sea. MPs were found at all sites, with the highest average abundance of MPs being 104.17 ± 162.44 pieces/m^3 in canals, followed by 2.08 ± 2.22 pieces/m^3 in the sea, and 0.60 ± 0.38 pieces/m^3 in the river. Fragment, fiber, and granule were three common shapes, and each shape was dominant in one sampling area. White was the most common MP color at all sites. A total of 13 polymers and co-polymers were confirmed, and polyethylene, polypropylene, and ethylene-vinyl acetate were the three dominant polymers. The total concentration of MPs-affiliated PAHs ranged from 232.71 to 6448.66, from 30.94 to 8940.99, and from 432.95 to 3267.88 ng/g in Can Gio sea, canals, and Saigon River, respectively. Petrogenic sources were suggested as a major source of PAHs associated with MPs in Can Gio Sea, whereas those found in Saigon River and canals were from both petrogenic and pyrogenic sources.

Keywords: microplastics; PAHs; surface water; chemical composition; Ho Chi Minh City

Citation: Nguyen, N.T.; Nhon, N.T.T.; Hai, H.T.N.; Chi, N.D.T.; Hien, T.T. Characteristics of Microplastics and Their Affiliated PAHs in Surface Water in Ho Chi Minh City, Vietnam. *Polymers* **2022**, *14*, 2450. https://doi.org/10.3390/polym14122450

Academic Editor: Jacopo La Nasa

Received: 19 May 2022
Accepted: 9 June 2022
Published: 16 June 2022

Publisher's Note: MDPI stays neutral with regard to jurisdictional claims in published maps and institutional affiliations.

Copyright: © 2022 by the authors. Licensee MDPI, Basel, Switzerland. This article is an open access article distributed under the terms and conditions of the Creative Commons Attribution (CC BY) license (https://creativecommons.org/licenses/by/4.0/).

1. Introduction

In the 1950s, plastic was first manufactured and was considered as one of the most important inventions because its outstanding properties (durable, waterproof, easy to mold...) bring convenience to people [1]. However, plastics are non-biodegradable; they have a massive number of adverse effects on the environment and organisms. Plastic trash persisting in the environment is subjected to physical, chemical, and biological impacts, and would be fragmented into pieces smaller than 5 mm—known as microplastics (MPs) [2]. Additionally, MPs are produced directly as a primary source in cosmetics, skin care products, and resin pellets [3]. After being discharged into water bodies, MPs can cause gastrointestinal obstruction through ingestion, reduced mobility to organisms through entanglement, and enter the food chain through bioaccumulation [4–6]. In addition, with large surface areas and hydrophobic properties, MPs easily adsorb heavy metals and persistent organic pollutants (POPs). They enhance toxicity in the aquatic environment, sea creatures, and even human beings [7–9], adversely enhancing the level of ecological risk [10,11]. Polycyclic aromatic hydrocarbons (PAHs)—one among the foremost common POPs—are ubiquitous within the environment and have been well-concerned long-time ago because of their carcinogenic and mutagenic risk [12,13].

Much research has been carried out since the 2000s to understand the behavior and fate of MPs [2,14]. Initial studies on protocols for the monitoring and distribution of MPs shows that MPs were detected in many places [15–17], even in Antarctica [18]; therefore, MPs have become a global concern. Particularly, the water environment has gained the most attention, because this is often the source of more than 80% of plastic garbage [19]. However, the reported studies mainly specialize in the physical and chemical characteristics of MPs; the information of potential toxic pollutants adhered to MPs are limited, especially PAHs. PAHs in MPs may be from manufacture or adsorbed from the environment [20,21]. The study of Van [20] found high concentrations of PAHs on unexposed commercial polystyrene foam (240–1700 ng/g). PAHs are well-known because of their toxicity, carcinogenicity, and mutagenicity [22,23]. Moreover, MPs can adsorb and concentrate PAHs from the aquatic environments, then transfer and bioaccumulate through the food chain and finally affect human health [10,20,24]. After long-term exposure in the aquatic environment, concentrations of PAHs sorbed to MPs can be many times higher than those in the aqueous environment [25].

In recent years, the number of investigations on MPs in inland water has increased. MPs in canals, rivers were reported in various studies [26–28]. Inland water is the main source contributing to the MPs in the ocean; especially when pollutants adhere to MPs surface, human health and ecosystem are more seriously affected [8]. Therefore, there is a need for further studies on the distribution of MPs in the inland water environment and the ability to absorb toxic contaminants.

Vietnam is one of top countries discharging plastic wastes into the ocean, and in recent years, MPs pollution in Vietnam has been paid more attention. MPs have been detected in different media, including surface water [29–31], sediment [32,33], marine organisms [34], and even in the atmosphere [35,36]. A high abundance of MPs in the Saigon River in Ho Chi Minh City was concluded in Lahens' investigation [29]. However, the published research was only conducted locally—in one sampling area such as a river or canals, there is a lack of studies on the presence of MPs in the outflow from the inland to the receiving source (marine environment). In Vietnam, previous studies have found a remarkable level of PAHs in water bodies specifically in sediment in HCMC [37,38]. The concentration of a total of 16 PAHs in sediment in Saigon River ranged from 49 ng/g to 933 ng/g dw, with the dominance of Phe, Flt, Pyr, and BbF [36]. MPs are a chemical carrier and can penetrate the food web, therefore, PAHs in MPs can bioaccumulate in the organisms [10,24,37]. There is a lack of study on the sorption of POPs in general and PAHs specifically by MPs in Vietnam. There should be more studies on the presence of MPs in surface water, from the sources to the ocean, as well as the pollutants that adhere to MPs.

Ho Chi Minh City (HCMC)—the most populous city in Vietnam—is the leading economic, cultural, and industrial center with a population of more than 10 million people [38]. Saigon River flows through HCMC with systems of canals weaving in the urban area. They receive the direct discharge of wastewater and other anthropogenic activities, then follow the current of the Saigon River to the Can Gio Sea. Therefore, in our study, we selected three main canal systems, Saigon River, and Can Gio Sea of HCMC to investigate MPs and their affiliated PAHs. The objectives of this study are: (1) to provide information on the abundance and distribution of MPs in surface water in canals, river, and sea in HCMC; (2) to characterize physical and chemical composition of MPs; (3) to investigate the distribution of 14 PAHs affiliated to MPs and their potential sources.

2. Materials and Methods

2.1. Study Area and Sample Collection

In this study, MPs in surface water were collected in canals, Saigon River, and Can Gio sea in HCMC. A total of 45 sampling sites were set along three canals, Saigon River, and Can Gio Sea in August 2020. The details on location and sampling sites are shown in Figure 1 and Table S1.

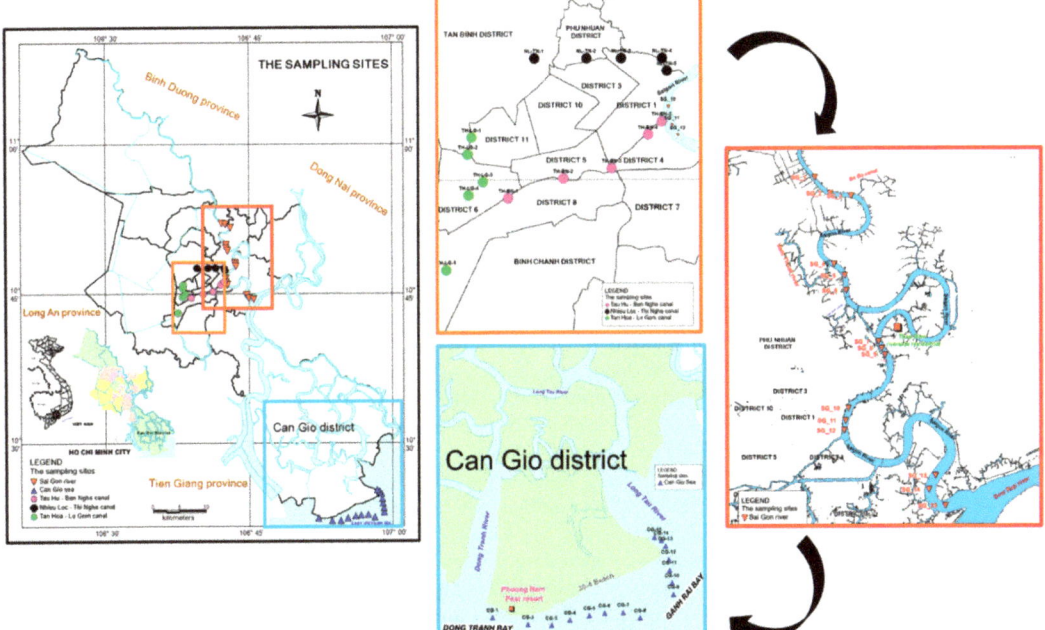

Figure 1. Sampling map of surface water MPs in HCMC, Vietnam, in August 2020.

There are three main canal systems in HCMC, these being Tan Hoa-Lo Gom (TH-LG), Nhieu Loc-Thi Nghe (NL-TN), and Tau Hu-Ben Nghe (TH-BN). These three canals belong to the network of waterways and drainage system of HCMC. TH-LG, with a length of 7.84 km, passes through 4 urban districts. NL-TN is in the city center with a length of 9.47 km, flowing through 5 districts. TH-BN is a large tributary (25.4 km in length) of Saigon River in the south of the city center, flowing through 5 urban districts. Along these canals, there are many residential areas and markets with different anthropogenic activities. Five sampling sites were chosen from the beginning to the end of each canal, evenly located along the canal, meaning a total of 15 sampling sites were chosen for three canals. At each site, 480 L of surface water (depth: 0–50 cm) was collected by using a stainless-steel bucket during the high tide of the day [39–41].

Saigon River (flows along the territory of HCMC, about 80 km) is the downstream of these canals; it is also the main waterway for ships entering and leaving Saigon harbor. After leaving the inner city, Saigon River merges with Dong Nai River and divides it into tributaries that flow into Can Gio Sea. Can Gio—a suburban district—is bordered by large estuaries, such as Soai Rap, Dong Tranh, Long Tau, and Nga Bay, transferring water through an area of muddy Sac forest and flowing into 2 bays of Dong Tranh and Ghenh Rai, and then the water flows into the sea. The trawling method was applied to collect microplastic samples from surface water on Saigon River and Can Gio Sea [42,43]. A total of 15 sampling sites of 5 evenly spaced sections of the Saigon River were chosen. The 15 sampling sites in the Can Gio Sea belong to four different communes of Can Gio and were about 500 m from the shore, and the sampling area included the estuarine zone of rivers from inland. A hydro-bios microplastic trawl with a mesh size of 330 μm was attached to the vessel's side, and samples were retained in the cod end. The sampling time was from 10 to 15 min at a speed of 3 knots. A half of the trawl net's open end was immersed under the water.

Each sample was divided into 2 parts for MPs characteristics analysis and PAHs investigation. The sample was then sieved directly through the 5 mm and 0.5 mm sieves in situ, and the upper part of the 0.5 mm sieve was rinsed and transferred directly to the brown glass bottles. They were kept at 4 °C and transported to the laboratory for further analysis. Samplers (buckets, sieves, trawl) were washed carefully by distilled water to eliminate the contaminants before the next sampling. In this study, the unit of microplastic abundance is shown as pieces per cubic meter (pieces/m^3).

2.2. Sample Preparation for Physical Characteristics and Chemical Composition Analysis of Microplastics

The microplastic extraction method is referenced from previous studies with some suitable modifications [1,27,44,45]. General principles include sieving, wet oxidation, floatation, and filtration steps. After sieving in situ, at laboratory, samples were wet oxidized with H_2O_2 and Fe(II) solution as catalyst to remove organic compounds adhered to MPs surface. Then density separation was applied to float MPs with a mixture of NaCl and $ZnCl_2$ solution (d = 1.6 g/mL). The solution was filtered through Whatman 0.45 μm filter paper to retain the supernatant. The filter paper finally was dried and observed under the microscope to determine characteristics of MPs.

In this study, polymer types of MPs were confirmed by using FTIR-ATR, an apparatus that can measure MPs from 0.5 mm in diameter. Therefore, we focus on analyzing MPs from 0.5 mm to 5 mm in diameter. The number, size, shape, and color of MPs were identified by an embedded digital microscope (MicroCapture Plus, Mustech Electronics Co., Ltd., Shenzhen, China). The number of MPs in the post-treated sample was determined by counting in three replicates. The sample was also dimensioned to know the size distribution of the MPs. MPs were also classified according to the primary colors. Shapes of MPs were classified based on their morphological features. MPs can be detected in many different shapes depending on the appearance and the formation of MPs. MPs could be fragmented due to the environmental effects (waves, UV light,...) or they were manufactured with the specific shapes to suit production purposes, and even created from synthetic fabric clothes [1]. Therefore, when observing the appearance of MPs, we classified fragments as smaller parts originating from larger pieces with irregular shapes. Fiber was defined as fibrous plastic that might come from clothes, fishing nets, etc. Another typical type that we found was granule (more three-dimensional particles); some had specific shapes (cylindrical or spherical) and white translucent color (commonly). These might have been produced as industrial resin pellets, and their appearance could change a bit depending on their state of weathering [6].

The MPs' chemical compositions were analyzed using FTIR Spectrometer (FT/IR-6600 (Jasco—Hachioji, Tokyo, Japan) wavelength 497.544–4003.5 cm^{-1}; resolution 4 cm^{-1}; 32 scans/spectrum). The device was equipped with an attenuated total reflection (ATR) single-point reflector configured with a ZnSe crystal with an incident angle of 45°, it was used to transmit the reflected infrared beam to the detector. The MPs piece with dimensions larger than 0.5 mm was placed on the ATR crystal for determination, the anvil lowered to provide good contact between MPs and the optical crystal. After each sample run, we rescanned the background with 32 scans to ensure no interference. Spectra Manager™ Suite software was used to interpret the IR spectrum of the sample. The results were compared with a standard spectrum database [43].

2.3. Analysis of PAHs in Microplastics

In this study, the procedure for PAHs determination was followed by the previous study of Tan [24]. After being immersed in H_2O_2 for 24 h to remove natural organic matter, MPs were filtered through a glass microfiber filter (Whatman GF/B). Then, the filter with MPs was placed in a desiccator at least 24 h to dry and weighed before chemical analysis. After that, PAHs associated with MPs were ultrasonically extracted in 20 mL of n-hexane three times. The extracts were combined and concentrated to 1 mL using a rotary evaporator. In the next step, the concentrated solution was then transferred to a silica gel (70–230 mesh) column. For purification, the silica gel columns were added with 5 mL hexane, 30 mL hexane, respectively, and eluted with 20 mL hexane/dichloromethane

(3/1, v/v). After elution, the extract that contains PAHs was evaporated until it was about 1 mL and dried by a gentle stream of nitrogen. A total of 1 mL of methanol was added to dissolve PAHs. Finally, PAHs were analyzed using HPLC with a fluorescence detector (Shimadzu, Japan). The 14 targeted PAHs in this study are: acenaphthalene (Ace), acenaphthylene (Acy), fluorene (Flu), phenanthrene (Phe), anthracene (Ant), fluoranthene (Flt), pyrene (Pyr), benz[a]anthracene (BaA), chrysene (Chr), benzo[b]fluoranthene (BbF), benzo[k]fluoranthene (BkF), benzo[a]pyrene (BaP), dibenz[a,h]anthracene (DahA), indeno[1,2,3-cd]pyrene (InP), and benzo[g,h,i]perylene (BghiP).

2.4. Quality Assurance and Quality Control

Throughout the whole process of sampling and analysis in the laboratory, it was that cross contamination of samples and exposure of MPs in the air was avoided. Various measures were applied, such as all investigators being equipped with cotton coats, gloves, and masks; also, only glass and metal apparatus were used during the experimental process and were washed carefully. The samples were covered by aluminum foil to avoid contamination. In addition, blank controls were also conducted using the same procedure of sample.

A mix standard of 16 PAHs (EPA 610 PAHs) was purchased from Supelco/Sigma-Aldrich (Bellefonte, PA, USA)for making standard curve and spike solutions. However, in this study, we only analyzed 14 PAHs. The blank samples were treated as the field samples to determine blank concentration of PAHs. The results were then applied to the blank correction. LOD and LOQ of the method were determined by PAHs concentration in 11 blank samples. The recovery test was conducted by analyzing unexposed commercial polyethylene (PE) pellets. A total of 1 g of PE pellets were spiked with a known amount of standard solution and left to dry in the desiccator. Other PE pellets were kept at their origin, treated as the real samples. The recoveries for PAHs ranged from 87.21 ± 5.15 to 100.48 ± 5.42.

3. Results and Discussion

3.1. Abundance and Distribution of Surface Water MPs from Canals to the Sea

MPs were detected at all sampling sites in this study with 3408 MPs pieces in total (canals—375 pieces, river—772 pieces, sea—2261 pieces), and average abundances of MPs in different sampling areas are presented in Table 1. In particular, the highest average abundance of MPs was in three canal systems (104.17 ± 162.44 pieces/m^3), followed by Can Gio sea (2.08 ± 2.22 pieces/m^3) and the lowest in the Saigon River (0.6 ± 0.38 pieces/m^3) (Figure 2). The result shows the ubiquity of MPs in aquatic environments, which leads to a potential impact on living creatures and the ecosystem.

Table 1. Mean abundance of MPs in canal systems, Saigon River, and Can Gio Sea.

Sampling Areas	MPs Abundance (Pieces/m^3)		
	Max	Min	Mean
Canals	666.67	12.50	104.17
Saigon River	1.59	0.16	0.60
Can Gio Sea	8.27	0.42	2.08

Among three canal systems, the abundance of MPs in NL-TN was the highest (165 ± 280.63 pieces/m^3), followed by TH-BN (114.17 ± 46.45 pieces/m^3), and TH-LG (33.33 ± 20.41 pieces/m^3). In TH-BN, MPs abundance in most of sampling sites was higher 100 pieces/m^3 (except TH-BN-3, 45.83 pieces/m^3). TH-BN is a large canal of HCMC with 20 km in length stretching though central districts to Saigon River. The canal connects with other canals, such as TH-LG, and intersects with Saigon River. Therefore, pollutants including MPs from other canal systems could be emitted into TH-BN. The abundance of MPs was higher at two ends of the canal. At the beginning of TH-BN, there are still makeshift houses along the bank of the canal, meanwhile, at the downstream (central district of HCMC), there are many commercial activities such as tourist yachts, restaurants on the river, and water transportation. The reception of domestic wastewater from these places led to higher abundances in these sampling sites. NL-TN pours into Saigon River at NL-TN-5. Most of sampling sites had abundances lower than 60 pieces/m^3, and the abundance of MPs tended to decrease from the upstream to the downstream of the canal (from 666.67 pieces/m^3 to 25 pieces/m^3). This can be explained: since NL-TN-1 is the starting point of the canal route, the water was stagnant and could not be circulated. Furthermore, domestic wastewater from the outlet of the sewage-collection system at the beginning of the canal might cause the increase of MPs abundance. TH-LG, located deep in the inner city, is the shortest canal among three canals, and there are mostly residential activities along the canal. This canal intersects with TH-BN at TH-LG-3, and the abundance of MPs increased at two end sampling sites of the canal (TH-LG-4 and TH-LG-5). Results of MPs in canals in this study were compared with similar sampling areas (Table 2): the MPs' abundance were lower than that of Wuliangsuhai lake, China (3120–11,250 pieces/m^3) [44] và Dongting Lake and Hong Lake, China (900–4650 pieces/m^3) [45], while being higher than that of Lake Victoria, Uganda (0.02–2.19 pieces/m^3) [46].

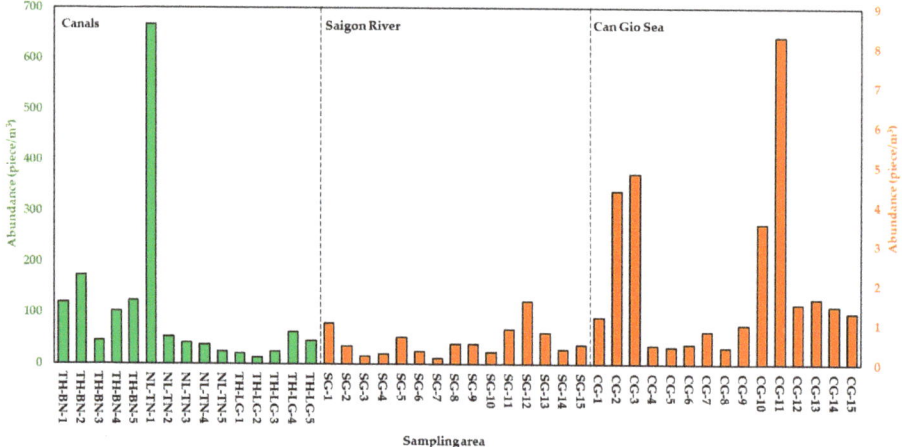

Figure 2. Abundance of MPs in canal systems, Saigon River, and Can Gio Sea in HCMC, Vietnam, in August 2020.

MPs from urban canals followed the current and were discharged into Saigon River which joins with Dong Nai River and flows into Can Gio Sea (Figure 1). Therefore, rivers are the main source for plastic discharge in the ocean [47]. The abundance of MPs in surface water of Saigon River was less than in three canal systems. The lowest abundance was at SG-7 (0.16 pieces/m^3) and highest at SG-12 (1.60 pieces/m^3). Lower abundance of MPs in Saigon River might be caused by the dilution of the higher flow of Saigon River in a larger area. In addition, some polymers have higher density (for instance: Polyethylene terephthalate, d = 1.38 g/mL) than fresh water (d = 1 g/mL); therefore, these polymers might deposit under the river bed. Furthermore, a portion of MPs may have penetrated into the food chain through aquatic animals [4,48–51]. The inconsistent MPs abundances among sampling sites in Saigon shows that, in addition to the water provided from canals, the surface water of Saigon River is also influenced by other anthropogenic activities. Saigon River is the waterway for ships entering or leaving Saigon's ports, and cargo ships could incidentally discharge wastes, which could be another potential source of MPs. Along the river, higher abundances of MPs at some sampling sites (SG-1, SG-5, SG-8) compared to adjacent sites were observed near the locations receiving flow from inner city's canals

(Figure 1). In particular, SG-12 (1.59 pieces/m^3)—the highest abundance of MPs site—not only receives the flow from TH-BN canal, but also experiences many riverside activities such as the riverside park and a harbor (one of the biggest harbors of HCMC). MPs abundance in river of this study was also compared with other research (Table 2). Compared to Cisadane River, Indonesia (13.33 to 113.33 particles/m^3), Ganges River, India (38 ± 4 MP/m^3), and Yellow River near its estuary, China (497,000–930,000 items/m^3), the abundance of MPs in this study was lower. The abundance of MPs, on the other hand, was higher than that of Beijiang River, China (0.56 ± 0.45 items/m^3). The above studies also stated that the possible origins of MPs may come from domestic wastewater, and anthropogenic activities on the river and along the river banks. Previous studies reported that MPs in road dust contributed to the accumulation of MPs in surface water through runoff (O'Brien et al., 2021; Vogelsang et al., 2018; Yukioka et al., 2020). In European countries, 50% of MPs in road dust derived from tires and road marking paint enter the environment every year, and MPs were also fragmented from plastic waste on the road such as packaging or plastic bottles under the impact of sun radiation or physical force (Monira et al., 2021). Therefore, the emission sources of MPs in canals and river in this study could be from residential activities, waterway transport, tourism along the canal, wastewater as well as road dust. The populations of urban districts are approximately 40,000 people/km^2, showing the high potential for MP pollution.

Can Gio Sea receives not only inland water from Saigon River but also from other Rivers such as Long Tau, Vam Lang, and Dong Tranh. MPs abundance in Can Gio Sea varied from 0.42 pieces/m^3 to 8.27 pieces/m^3. Compared to Saigon River, MPs abundance in Can Gio Sea is higher, but lower than that in canals. It is interesting that even though coastal water bodies (e.g., bays and estuaries) receive the inland flow of rivers, these areas are more polluted than rivers [49]. This may be explained by the emission of multiple sources and circulation patterns there (semi-enclosed basins). MPs in the surface water of Can Gio sea were collected 500 m offshore; this water body not only receives MPs from the inland currents but also from other coastal activities—aquaculture, tourism, seaports, and seafood markets [50]. In sampling sites CG-1 to CG-3 located near the estuaries of Vam Lang River and Dong Tranh River, the abundance of MPs increased from site CG-1 to site CG-3 (from 1.18 to 4.81 pieces/m^3) (Figure 2). The abundance of MPs at CG-3 was higher probably because, near this site, there is the popular seaside resort of Can Gio. In sites CG-4 to CG-9, MPs abundance was significantly lower than other sites (<1 pieces/m^3). This could be explained by the farther sampling locations from the shore compared to other sites (Figure 1). MPs abundance at CG-8 was lowest, as this site was farthest from the coast. MPs abundance at CG-10 and CG-11 (especially CG-11, with 8.27 pieces/m^3) was much higher than other adjacent sites, possibly because these two locations also intersect with the flow from Ghenh Rai Bay (Figure 1). Interestingly, the chemical composition results also show similarities in the distribution of MPs in Can Gio Sea. Particularly, nearshore sampling sites (CG-1 to CG-3) had fewer and consistent polymer types (Figure 3), while offshore sites (CG-4 to CG-9) had more diverse polymer types (Figure 3). The results show that MPs abundance in Can Gio Sea (2.08 ± 2.21 pieces/m^3) was higher than that of the North-East Atlantic (0–1.5 items/m^3) [51] and mid-North Pacific Ocean (0.51 ± 0.36 items/m^3) [52]. However, MPs abundance near estuaries in other studies were reported to be higher than this study, for example: Sebou Estuary and Atlantic Coast, Morocco (10 to 168 particles/m^3) [53] and Yangtze Estuary, China (0–259 items/m^3) [54]. The higher abundance of MPs in the seawater near the estuaries than that of rivers was reported previously, for example: Xiangxi Bay contained 0.11–68 pieces/m^3 compared to the Beijiang River of South China (0.56 ± 0.45 pieces/m^3) [24]. Compared to other studies, MPs abundance in HCMC was relatively low. However, we focused on MPs from 0.5 mm to 5 mm; therefore, the abundance of smaller MPs (<0.5 mm) might be underestimated. In the report of Peter [55], research concerning MPs with sizes from 10 μm may have higher abundance, 1000 times than that of those investigating MPs from 100 μm. Different sampling methods would lead to inconsistent results. For example, using bucket or pump as samplers, the number of detected MPs was much higher than using a manta trawl; this result was stated in the research of Felishmino [56]. This can be explained by the larger mesh size of the net (manta trawl; 300 μm) leading to the loss of small MPs and fibrous MPs. Even the above comparisons might be inaccurate due to the differences in the studies' MPs size, sampling and analysis methodologies, sampling time, etc. These research studies still contribute to the general overview of MPs pollution in surface water in the world [57]. Thus, in this study, microplastics were found in all surface water samples from canals, rivers, and oceans. The abundances of MPs varied within the same sampling area and in different sampling areas. The abundance of MPs tends to decrease gradually when transporting from canals to Saigon River and increasing in Can Gio Sea. This difference in abundance may be due to the influence of water flow and different emission sources on the water bodies and along the banks of canals, river, and the coast.

Table 2. Microplastics in surface water of previous studies.

No.	Region	Sampler	Abundance	Common MPs Size	Common MPs Shape	Common MPs Color	Common MPs Composition
1	Wuliangsuhai Lake, northern China [44]	Stainless steel bucket	3120–11,250 pieces/m^3	<2 mm (98.2%)	Fiber	No data	PS, PE
2	Dongting Lake and Hong Lake, China [45]	12 V DC Teflon pump	900–4650 pieces/m^3	<2 mm (65%)	Fiber	Transparent, blue	PP, PE
3	Lake Simcoe, Ontario, Canada [56]	Low volume grabs and manta trawls	0–700 particles/L (grab), 0.4–1.3 particles/m^3 (manta trawl)	No data	Fiber (grab) Fragment (manta trawl)	No data	PP, PE
4	Lake Victoria, Uganda [46]	Manta trawl	0.02–2.19 pieces/m^3	<1 mm (36%)	Fragment (36.7%)	White/transparent (59.1%)	PE, PP
5	Three canal systems of HCMC, Vietnam (this study)	Stainless steel bucket	104.17 ± 162.44 pieces/m^3	1.0–2.8 mm	Fiber	White, transparent, blue, green	PP, PE, EVA
6	Cisadane River, Indonesia [58]	Stainless-steel bucket	13.33 and 113.33 particles/m^3	0.5–1.0 mm	Fragment	No data	PE, PS, PP
7	Beijiang River, China [24]	Plankton net (mesh size, 0.112 mm and diameter, 20 cm)	0.56 ± 0.45 items/m^3	0.6–2 mm	Film	No data	PP, PE
8	Ganges River, India [59]	Hand-operated bilge pump	38 ± 4 MP/m^3	2.459 ± 0.209 mm average	Fiber	blue	Rayon
9	Yellow River near estuary, China [60]	Stainless steel bucket	497,000–930,000 items/m^3	<0.2 mm	Fiber	No data	PE, PS, PP
10	Snake and Lower Columbia rivers, USA [61]	Grab and net	0 to 13.7 MPs/m^3	<0.5 mm	Fiber	No data	PE, PP, PET
11	Saigon River, HCMC, Vietnam (this study)	Hydro-bios trawl	0.60 ± 0.38 pieces/m^3	2.8–5.0 mm	Granule/Pellet	White, transparent, blue, green	PP, PE
12	Yangtze Estuary, China [54]	Metal cylinder	0–259 items/m^3	<1mm (79%)	Fragment	White and transparent	PE, PP, α-cellulose
13	Sebou Estuary and Atlantic Coast, Morocco [53]	Steel sampler	10 to 168 particles/m^3	0.1–0.5 mm	Fragment	While and blue	No data
14	North-East Atlantic [51]	Manta trawl	0–1.5 items/m^3	1.00–2.79 mm	Fragment (63%)	White, transparent, and black	No data
15	Mid-North Pacific Ocean [52]	Manta trawl	0.51 ± 0.36 items/m^3	1.0–2.5 mm	Fragment (31%)	White and transparent	PP (53%)
16	Can Gio Sea, HCMC, Vietnam (this study)	Hydro-bios trawl	2.08 ± 2.22	1.0–2.8 mm	Fragment	White, transparent, blue, green	PP, PE, EVA

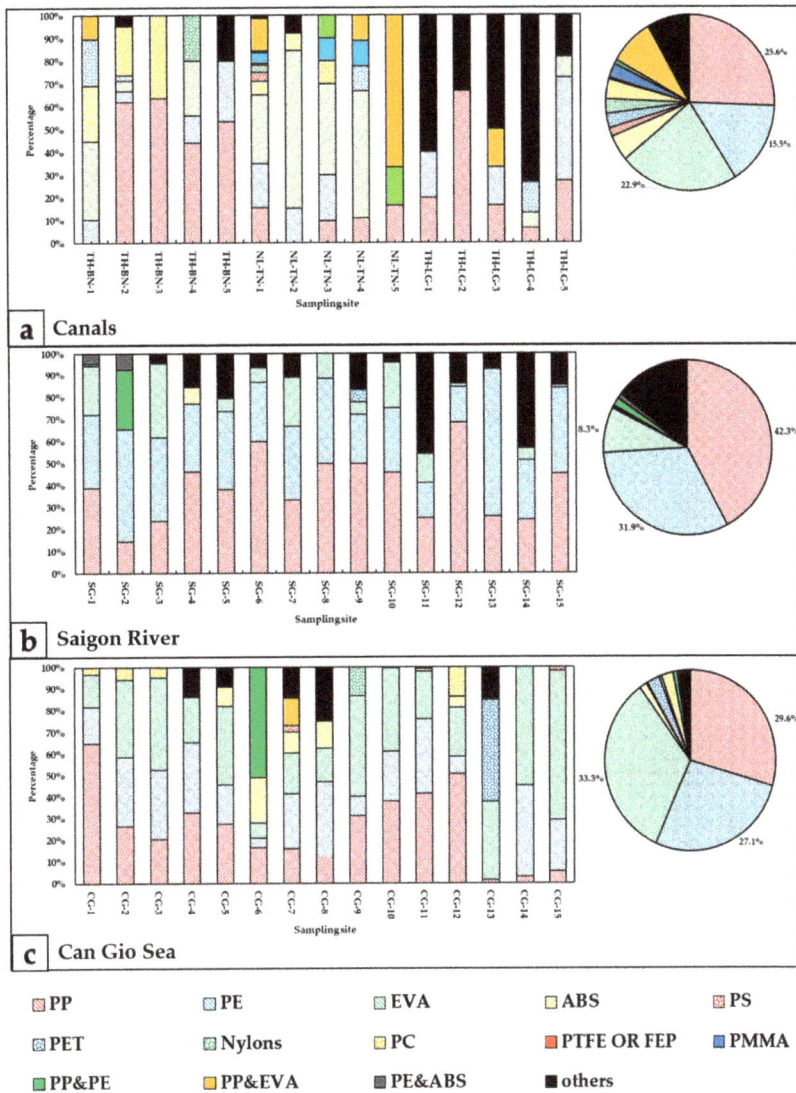

Figure 3. Chemical composition of MPs: (**a**) Canals, (**b**) Saigon River, and (**c**) Can Gio Sea, in August 2020.

3.2. Physical Characteristics of Surface Water MPs

In this study, MPs from 0.5 mm to 5 mm were analyzed for their distribution in terms of size, shape, and color. MP size was classified into groups: from 0.5 mm to 1.0 mm, from 1.0 mm to 2.8 mm, and from 2.8 mm to 5 mm (Figure 4). The size distribution varied in different sampling areas. The three canal systems and Can Gio Sea showed the highest proportion of size class of 1–2.8 mm (40.8%—canal systems; 49.1%—Can Gio Sea) while 2.8–5 mm MPs were dominant in Saigon River (42.6%). On average, MPs smaller than 2.8 mm accounted for the majority in all sampling sites (69.4%). Granule, fragment, and fiber were three common shapes found in all sampling sites (Figure 5). However, their distribution was uneven among sampling areas (Figure 6). The results show that MPs in canals were mainly fiber (37.3%) while granule accounted for the largest proportion of MPs in Saigon River (43.5%). On the other hand, Can Gio Sea data show the highest percentage of fragments (48.4%). Eight different colors of MPs were detected, some common colors were white, transparent, green, and blue (Figure 7). There was a similarity in the distribution of color among sampling areas, as white was the dominant one, and the percentage was the highest in Can Gio Sea (52.1%).

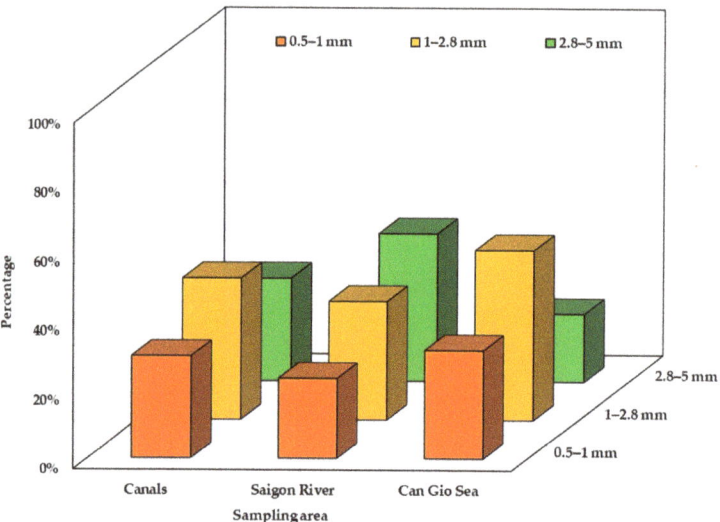

Figure 4. Size distribution of MPs in canal systems, Saigon River, and Can Gio Sea, in August 2020.

In three canal systems, MPs larger than 1 mm accounted for the majority (40.8%), canals are the primary receiver of MPs from inland emission sources plastic wastes, and MPs have not yet experienced environmental effects (sun radiation, physical impacts of water, chemical, biological impacts) so they have not suffered much fragmentation. The high percentage of fibrous MPs may relate to the wash-holding process from domestic wastewater; this is considered as the main source of microfibers in the environment [62]. Studies worldwide also give similar results with fiber being the most dominant shape of MPs [62–65]. The study by Chenxi Wu stated that fiberous MPs tend to be more abundant in more populated areas [63].

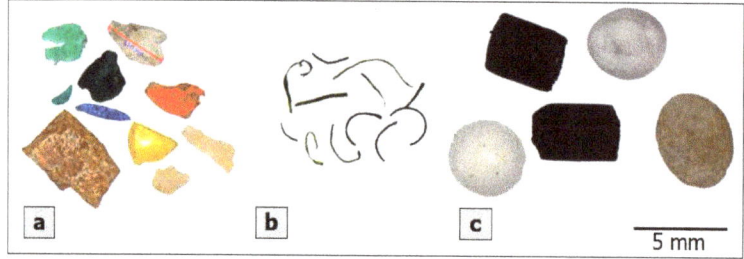

Figure 5. Typical microplastics of different shapes of this study: fragment (**a**), fiber (**b**), granule (**c**).

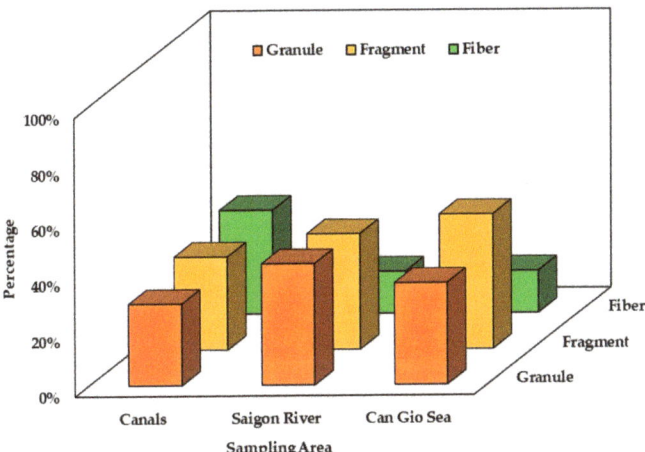

Figure 6. Shape distribution of MPs in canal systems, Saigon River, and Can Gio Sea, in August 2020.

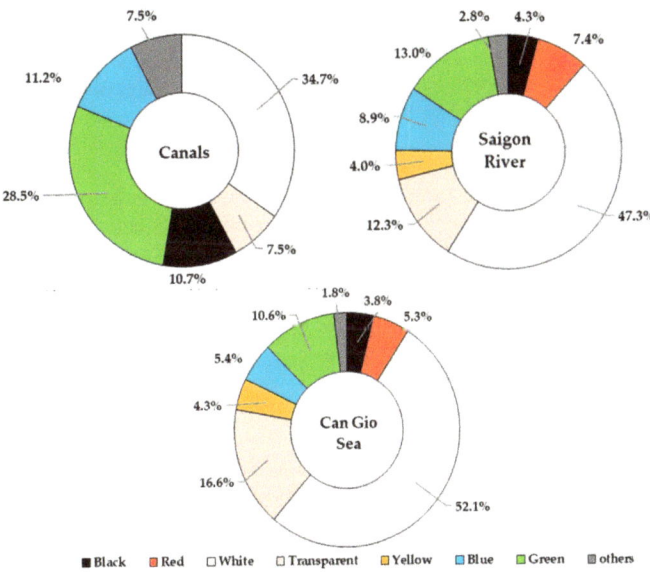

Figure 7. Color distribution of MPs in canal systems, Saigon River, and Can Gio Sea, in August 2020.

In Saigon River, the proportion of granular MPs increased, as well as the size of MPs (2.8 mm to 5 mm). This result is reasonable because we found a high number of resin pellets in this sampling area. These pellets' size were from 2.8 mm to 5 mm. HCMC is the industrial center of Vietnam with many plastic factories, resin pellets are regular imports to produce plastic items [64]. Saigon River is the waterway for ships entering and leaving Saigon ports, so cargo boats could incidentally discharge wastes, which could be another potential source of MPs. Other studies show that MPs found in rivers were fiber or fragment shape (Table 2). Concerned size of MPs were different in each study, however, the common MPs size tended to be inversely proportional to the number of MPs [47,48,63].

In Can Gio Sea, fragment was the dominant shape (48.4%). The sources of these MPs could be from agricultural activities along the coast where mulch films are utilized to conserve soil moisture [66]; aquacultural and tourism activities also contributed to the high proportion of fragmented MPs through the fragmentation of utensils and single-use plastic items. In addition, MPs in the sea experience more environmental impacts and exist for a longer period of time, making MPs more susceptible to fragmentation. Therefore, the highest proportion of fragment was rational. Compared to other published research, the most abundant MP shape found in seawater was also fragment (Table 2). This was quite reasonable because the majority of MPs in this study and other studies were Polyethylene (PE) and Polypropylene (PP). These polymers have lighter density (d = 0.9–1.0 g/mL) than seawater (d = 1.025 g/mL); therefore, they can easily float on the surface water. Mesh samplers (manta trawl, Hydro-bios trawl, ...) could cause the loss of a number of fiberous MPs, which was stated in the research of Felismino [56]. This may also be the cause of the lower percentage of fiberous MPs in Saigon River and Can Gio Sea.

In this study, 6 different colors were detected in canals, whereas we found 8 different colors in Saigon River and Can Gio Sea (Figure 7). The number of colors in each sampling area also indicate the diversity of emission sources. White was the most common color and transparent was one of major colors in all sampling areas. Transparent MPs pieces appearing to be common in surface water possibly originated from the single-use packaging plastic products (containers, bottles, cups, or bags). This finding is suitable since white and transparent are the most common color manufactured [67].

Previous studies found that fish might mistakenly intake white, yellow, or brown MPs (as they closely resemble their zooplankton prey) [68], while sea turtles commonly ingest translucent and light-colored plastics [69]. Blue MPs were reported to have been accidently consumed by a variety of marine animals such as fishes [70,71] and clams [48]. Further research into the surface water in Ho Chi Minh City is needed to verify which among the attributes of MPs aid the ingestion of marine organisms.

3.3. Chemical Composition of Surface Water MPs

All MPs in canals were analyzed for chemical composition by FTIR-ATR. Due to the large number of MPs detected in Saigon River and Can Gio Sea, MPs were classified into groups by color and shape, and representatives of each group were selected to determine their polymer (328 spectrums for Saigon River, 382 spectrums for Can Gio Sea). In total, there were 13 polymers identified in all sampling areas, including 10 homopolymers—PP (Polypropylene); PE (Polyethylene); EVA (Ethylene-vinyl acetate); ABS (Acrylonitrile-butadiene-styrene); PS (Polystyrene); PET (Polyethylene terephthalate); Nylon (all polyamides); PC (Polycarbonate); PTFE (Polytetrafluoroethylene) or FEP (Fluorinated ethylene propylene); PMMA (Poly(methyl methacrylate))—and 3 copolymers—copolymer PP and PE; copolymer PP and EVA; copolymer PE and ABS (Table S2).

In three canal systems, we detected 12 types of polymers, of which PP, PE, and EVA accounted for the highest proportion (64%). PC, PTFE, and PMMA were the minority of the total MPs (5%; 0.3%; and 2.1%, respectively). Figure 3a shows that 11 polymer types were found at NL-TN-1, where the stagnant water phenomenon occurs and is also a start of the canal. In Saigon River, 7 polymers were confirmed (PP, PE, EVA, ABS, PET, PP/PE, and PE/ABS), of which PP and PE were the two predominant polymers (74.1% in total) (Figure 3b). The percentage of PP was the highest (42.3%), PE was the second most popular polymer (31.9%), and they were detected at all sites. In addition, some polymers with relatively small percentages were found, such as ABS at SG-12, PET at SG-7, (PP/PE) and (PE/ABS) at SG-14, SG-15. There are similarities between the chemical composition of MPs in Saigon River and Can Gio Sea. The highest percentage of polymers belongs to EVA with 33.3%, then PP with 29.6%, and PE with 27.1% (Figure 3c). PS only appeared at 3/15 sites (0.2%). Other polymers (ABS, PC, PET, nylons, PP/PE, and PP/EVA) accounted for about 7.9%. PP and EVA were presented in all sampling sites, from sites CG-5 to CG-8, there were various polymers (5–7 types) while the other sites only observed 2 to 3 polymer types.

In this study, in terms of the abundance in collected samples, 3 types of plastic PE, EVA, and PP were high ubiquity due to the following reasons. On the one hand, these plastics have a buoyant density PE (0.88–0.96 g/cm^3), EVA (0.937 g/cm^3), PP (0.905 g/cm^3) while the density of river water and seawater is from 1 to 1.025 g/cm^3, which makes them tend to float on the water's surface [1]. Moreover, as a result of Chubarenko's research determining the properties and settling behaviors of MPs, we know that polyethylene fiber present in the euphotic zone from 6 to 8 months before they are sunk by the biofouling process [72]. In fact, more than half of the plastic produced globally is floating plastic, so the dominant occurrence of these polymers in the environment is understandable [73]. On the other hand, the distribution of MPs endures impacts due to wind redistribution, convergence zones, gyre entrainment, the activity of sediment microorganisms, and so on, which leads to MPs concentrating at some hot spots while another site may be insignificant [74].

Marine MPs accumulation, along with oceanographic processes (dense shelf water cascading, severe coastal storms, offshore convection, saline subduction, and the adhesion of plants and plankton) enhance the sedimentation ability of high-density plastics (ABS, PS, PET, PA, PC, PTFE or FEP, PMMA), which explains why their percentage in the collected samples was relatively low [75]. These MPs take less than 18 h to enter the marine sediments process [72]. PA's origin comes from textiles, the automotive industry, carpets, kitchenware, and sportswear; PS is used in single-use products such as containers, lids, bottles, and trays [76]. PC is principally employed in electronics, data storage, and components in phones [77]. PET is used for products such as bags, packaging, wrappers, and engineering resins [78]. PMMA is used in a wide range of fields which include vehicles, lenses for glasses, and medical and dental applications [79]. PTFE, another name for Teflon, is most commonly used as a nonstick coating for cookware, but also for the manufacture of semiconductors and medical devices, and as an inert ingredient of pesticides [80].

Interestingly, copolymers (PP/PE, PP/EVA, and PE/ABS) were found in this study. The copolymer is produced from two or more different types of monomers resulting in a more complex structure to enhance the material properties. For instance, PP/PE is a material with low temperature resistance and improved tensile and toughness [1]. Fused filament fabrication (FFF), used in 3D printers, is made from ABS. However, it has poor thermal stability due to the thermo-oxidative degradation of butadiene monomers [81]. The combination of high-density PE (for instance, HDPE) and ABS increase mechanical strength (tensile, flexure, and compressive), which significantly improves material properties. Additionally, the recycling process of plastic waste also contributes to the creation of copolymers [82].

Overall, types of polymers have quite consistent distribution in all sampling areas due to the fact that the distribution of buoyant plastics had more dominance than high-density plastic in surface water. The results of this study were compared with previous reports in the same region and around the world, which resulted in a similar tendency. PP, Nylon 6, and PE accounting for 49.06% were found in the Chao Phraya River, Thailand [83]; the percentage of PP, PE, and PS was 93.66% in the Pearl River, China [84]; in three rivers in southeastern Norway (Akerselva river, Hobøl river, and Gryta river), two types of plastic (PP and PE) were from 52.9% to 85.7%, respectively [85]. MPs in surface water and sediment was dominated by PE fragments (53–67%) followed by PP (16–30%) and PS (16–17%) in the Bay of Brest, Brittany, France [86]. These results reflect the huge production and consumption of PP and PE globally [87].

3.4. Characteristics of PAHs Associated with of Surface Water MPs from Canals to the Sea

In this section, we report the concentrations of 14 PAHs in MPs. Table S3 shows the levels of MPs-bound PAHs collected in the canals (n = 15), Saigon River (n = 15), and Can Gio Sea (n = 15). Overall, PAHs were detected in most of all MPs samples in 3 sampling areas. Total PAHs in MPs found in Can Gio Sea had the highest concentration (1398.99 ± 1612.14 ng/g), followed by that collected in canals and finally in Saigon River with a concentration of 1070.34 ± 1613.93 ng/g, and 926.68 ± 695.55 ng/g, respectively. This result suggests that there were more PAHs accumulating in MPs in Can Gio Sea. The canals carrying MPs flow into Saigon River and finally enter Can Gio Sea. During the transportation, MPs experienced exposure to PAHs, leading to the absorption and the enrichment of these compounds.

In the canals, the total concentration of 14 PAHs absorbed by MPs ranged from 30.94 to 8940.99 ng/g, with the average concentration being 1070.34 ± 1613.93 ng/g. Phe (3-ring PAH), BbF and InP (larger molecular weight PAHs compounds—5, 6 rings) had high concentration in MPs. BbF had the highest concentration of 225.6 ± 272.14 ng/g in MPs in the canals. The total concentration of PAHs in MPs in Saigon River varied from 432.95 to 3267.88 ng/g. Phe, AnT, and Pyr were dominant in MPs collected in Saigon River with values of 109.7 ± 142.98, 186.28 ± 101.35, and 174.94 ± 100 ng/g, respectively. Table S4 shows the concentration of PAHs associated with MPs, which was collected in several types of environments from previous studies. Our results were consistent with the study of Tan [24] in freshwater. In the Feilaixia reservoir, the concentration of MPs-bound PAHs with 3 and 4 rings, such as Phe and Chr, were higher than other PAHs. PAHs with 5 and 6 rings, for instance BghiP and InP, tend to have higher concentrations than the rest. Furthermore, DahA (5-ring PAH) was found at low or undetectable concentration [24]. The concentration of PAHs bound to MPs in Can Gio Sea (232.71–6448.66 ng/g) was lower than the results in the coastal area of Huanghai Sea and Bohai Sea, China [37], Zhengmingsi Beach and Dongshan Beach, China [88]. On the other hand, the PAHs concentration in this study was higher than that of Seal beach, USA, Thinh Long and Tonking Bay beaches, Vietnam [89], and the Southwest coast of Taiwan [10].

Figure 8 illustrates the proportion of PAHs by molecular weight including lower molecular weight PAHs (LMW-PAHs) (Ace, Flu, Phe, and Ant), medium molecular weight PAHs (MMW-PAHs) (Flt, Pyr, BaA, and Chr), and high molecular weight PAHs (HMW-PAHs) (BbF, BkF, BaP, DahA, BghiP, and InP). There was a difference of PAHs on MPs in the sampling areas in this study. HMW-PAHs were dominant in MPs found in canals, accounting for 63% of the total PAHs, whereas, in Saigon River, MMW-PAHs were found to have the highest concentration with a percentage of 57.5%. The concentration of LMW, MMW, and HMW- PAHs in MPs were similar in Can Gio Sea. MPs with granule shape accounted for a large number of MPs collected in Saigon River. In these granules, a pellet is an original form of commercial polymer. The adsorption capacity of pellets was smaller than other forms (foam, fiber) of MPs because of their small contact areas. The study of Van [20] found that high concentrations of PAHs were found in commercial PS foam (240–1700 ng/g) and a lower one in commercial PS pellets (12–15 ng/g). This is the reason why PAHs concentration found in MPs in rivers were smaller than in canals and sea in this study. We also analyzed concentrations of PAHs in the original PE pellet and found concentrations of total PAHs ranging from 25.85 to 104.85 ng/g. Moreover, 3 and 4 ring PAHs, for instance Flu, Phe, Flt, and Pyr were dominant in PE pellets. This could explain the dominance of MMW-PAHs in Saigon MPs samples.

We applied the ratio of PAH isomers to diagnose the sources of PAHs. The ratios of Flu/(Flu + Pyr) and Ant/(Ant + Phe) have been commonly used to examine the sources of PAHs [10,23,39]. The Ant/(Ant + Phe) ratio is <0.1, and Flu/(Flu + Pyr) ratio is <0.4 indicating that PAHs were mainly contributed by petrogenic sources. Vice versa, if these values are larger than 0.1 and 0.4, respectively, PAHs were primarily from pyrogenic sources [23,40]. Figure 9 shows the diagnostic ratios of PAHs on MPs samples collected in the three sampling areas. There were 9 of 15 MPs samples found in Can Gio Sea that had a Flu/(Flu + Pyr) ratio < 0.1 and Ant/(Ant + Phe) ratio < 0.4. This result reveals that PAHs on MPs in Can Gio Sea were mainly from petrogenic sources. Our results are consistent with many previous studies about PAHs on MPs on the sea surface water or coast in previous studies (Table S4). It is very likely that MPs on the sea water surface are exposed to floating oil leaking from ships. PAHs associated with MPs on 6 samples found in Saigon River, were also strongly affected by PAHs from petrogenic sources. Moreover, PAHs on MPs in canals and river were also from a mix of petrogenic and pyrogenic sources. The HMW-PAHs are good indicators for vehicular emission sources [88]. Particularly, BghiP and InP are considered to represent a maker for diesel vehicles [10]. Low-ring PAHs are primarily derived from petroleum sources [88]. The HMW-PAHs are good indicators for vehicular emission sources [90]. In this study, the high level of HMW-PAHs associated with MPs in canals reflects vehicle emission source of PAHs in MPs. In addition, the canals in HCMC are the main system of rainwater discharge that carries particles, road dust from the air and on the streets into the marine environment. Many studies about PAHs in road dust showed that HMW-PAHs were the most abundant PAHs [91,92]. On the other hand, MPs were also reported to have high concentration in road dust and could be possible carriers for PAHs [93]. Nevertheless, the study of Ida [94] also reported high MPs concentration on roads and in nearby stormwater, sweep sand, and wash water. Tire and road wear particles have been identified as a potential major source of MPs in road dust and stormwater and could contribute to the abundance of MPs in the aquatic environment [66,94].

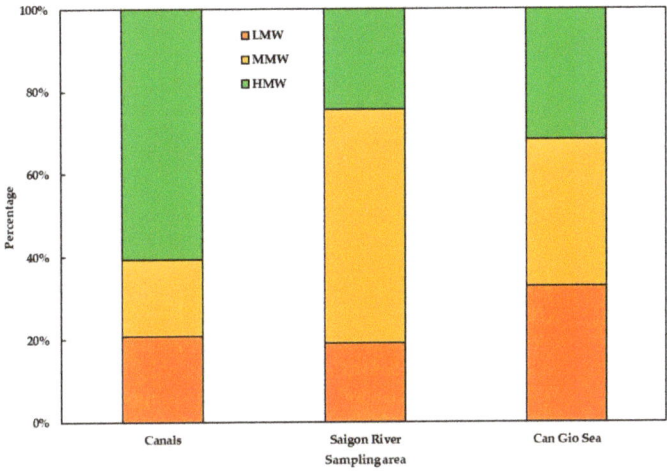

Figure 8. Percentage of PAHs bound to MPs by molecular weight in canals, Saigon River, and Can Gio Sea in August 2020.

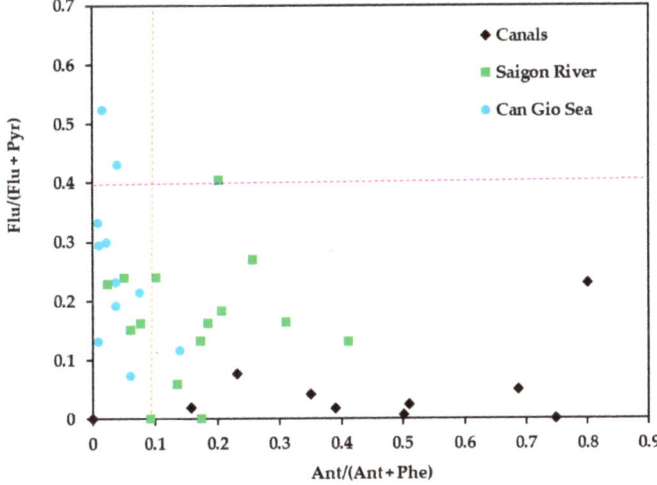

Figure 9. The bi-diagnostic ratios of Flu/(Flu + Pyr) and Ant/(Ant + Phe) in MPs found in Canals (black diamond), Saigon River (green square), and Can Gio sea (blue dot).

4. Conclusions

This study contributed to our understanding of the abundance and physical and chemical compositions of MPs, as well as provided the first information of PAHs associated with MPs in the surface water (canals, river, and sea) in Ho Chi Minh City, Vietnam. MPs were found at all sampling sites, with the highest abundance found in the urban canals (104.17 ± 162.44 pieces/m^3). The abundance of MPs decreased from canals to Saigon River and increased in Can Gio Sea. The three common shapes were fragment, fiber, and granule, and each shape was predominant in a specific sampling area. White accounted for the highest proportion of all MPs' colors (34.7%–52.1%). In this study, 13 types of polymers (including 3 co-polymers) were confirmed by FTIR-ATR, of which PE, PP, and EVA were three dominant polymers. Total PAHs in MPs found in Can Gio Sea had the highest concentration (1398.99 ± 1612.14 ng/g), followed by that collected in canals and finally in Saigon River with concentration of 1070.34 ± 1613.93 ng/g, and 926.68 ± 695.55 ng/g. PAHs on MPs in Can Gio Sea were mainly from petrogenic sources due to oil leaking from ships. Vehicle emission sources could be a significant pyrogenic source of PAHs in MPs in canals and rivers. This study provides basic data on the fate of MPs in the surface water of HCMC for future investigations. This is also the first study on PAHs associated with MPs in Vietnam.

Supplementary Materials: The following supporting information can be downloaded at: https://www.mdpi.com/article/10.3390/polym14122450/s1, Figure S1: (A) Microplastics were observed on Micro Capture Plus. (B) The FT/IR6600 FT-IR Spectrometer was used in the study. (C) Single point reflector ATR retrofit to the machine. One arrow points to the anvil, the other to the ZnSe ATR crystal. (D) Microplastics were placed on the crystal. (E) The anvil was lowered to press the microplastic onto the crystal for photo transmission into the sample; Figure S2: IR spectra of some collected MPs; Table S1: Coordinates of sampling sites: 3 canals, Saigon River, and Can Gio Sea; Table S2: Characterization of MPs chemical composition in canals, Saigon River and Can Gio Sea in 2020; Table S3: PAHs concentration affiliated with MPs (mean ± SD, min − max, unit: ng/g) in three aquatic environment in HCMC; Table S4: The characteristic of MPs-affiliated PAHs collected in different types of aquatic environment among several studies in the world.

Author Contributions: N.T.N.: data curation, formal analysis, investigation, roles/writing—original draft, visualization. N.T.T.N.: data curation, investigation, methodology, roles/writing—original draft, writing—review and editing, visualization. H.T.N.H.: data curation, formal analysis, investigation, methodology, roles/writing—original draft, visualization. N.D.T.C.: data curation, formal analysis, investigation, methodology, roles/writing—original draft, visualization. T.T.H.: conceptualization, funding acquisition, methodology, project administration, supervision, writing—review and editing. All authors have read and agreed to the published version of the manuscript.

Funding: This research is funded by Vietnam National University, Ho Chi Minh City (VNUHCM) under grant number **B2020-18-04**.

Institutional Review Board Statement: Not applicable.

Informed Consent Statement: Not applicable.

Data Availability Statement: Not applicable.

Acknowledgments: This research is funded by Vietnam National University, Ho Chi Minh City (VNUHCM) under grant number **B2020-18-04**. The authors would like to acknowledge our undergraduate students for supporting us in sampling and determination. We also thank the Air and water pollution—Human Health—Climate Change research group of Faculty of Environment, University of Science, VNUHCM for all support in this study. We are also grateful to Nguyen Huong Viet—Faculty of Electronics and Telecommunications; Faculty of Chemistry, University of Science, VNUHCM, Ho Chi Minh City, Vietnam for your support in terms of machinery and equipment.

Conflicts of Interest: The authors declare no conflict of interest. The funders had no role in the design of the study; in the collection, analyses, or interpretation of data; in the writing of the manuscript, or in the decision to publish the results.

References

1. Crawford, C.B.; Quinn, B. *Microplastic Pollutants*; Elsevier Science: Amsterdam, The Netherlands, 2016. [CrossRef]
2. Thompson, R.C.; Olsen, Y.; Mitchell, R.P.; Davis, A.; Rowland, S.J.; John, A.W.; McGonigle, D.; Russell, A.E. Lost at Sea: Where Is All the Plastic? *Science* **2004**, *304*, 838. [CrossRef] [PubMed]
3. Praveena, S.M.; Shaifuddin, S.N.M.; Akizuki, S. Exploration of microplastics from personal care and cosmetic products and its estimated emissions to marine environment: An evidence from Malaysia. *Mar. Pollut. Bull.* **2018**, *136*, 135–140. [CrossRef] [PubMed]
4. Digka, N.; Tsangaris, C.; Torre, M.; Anastasopoulou, A.; Zeri, C. Microplastics in mussels and fish from the Northern Ionian Sea. *Mar. Pollut. Bull.* **2018**, *135*, 30–40. [CrossRef] [PubMed]
5. Mao, Y.; Ai, H.; Chen, Y.; Zhang, Z.; Zeng, P.; Kang, L.; Li, W.; Gu, W.; He, Q.; Li, H. Phytoplankton response to polystyrene microplastics: Perspective from an entire growth period. *Chemosphere* **2018**, *208*, 59–68. [CrossRef]
6. Rocha-Santos, T.A.P.; Duarte, A.C. *Characterization and Analysis of Microplastics*; Elsevier: Amsterdam, The Netherlands, 2017. [CrossRef]
7. Khalid, N.; Aqeel, M.; Noman, A.; Khan, S.M.; Akhter, N. Interactions and effects of microplastics with heavy metals in aquatic and terrestrial environments. *Environ. Pollut.* **2021**, *290*, 118104. [CrossRef]
8. Sun, S.; Shi, W.; Tang, Y.; Han, Y.; Du, X.; Zhou, W.; Zhang, W.; Sun, C.; Liu, G. The toxic impacts of microplastics (MPs) and polycyclic aromatic hydrocarbons (PAHs) on haematic parameters in a marine bivalve species and their potential mechanisms of action. *Sci. Total Environ.* **2021**, *783*, 147003. [CrossRef]
9. Liu, Y.; Zhang, K.; Xu, S.; Yan, M.; Tao, D.; Chen, L.; Wei, Y.; Wu, C.; Liu, G.; Lam, P.K. Heavy metals in the 'plastisphere' of marine microplastics: Adsorption mechanisms and composite risk. *Gondwana Res.* **2021**. [CrossRef]
10. Chen, C.-F.; Ju, Y.-R.; Lim, Y.C.; Hsu, N.-H.; Lu, K.-T.; Hsieh, S.-L.; Dong, C.-D.; Chen, C.-W. Microplastics and their affiliated PAHs in the sea surface connected to the southwest coast of Taiwan. *Chemosphere* **2020**, *254*, 126818. [CrossRef]
11. Mai, L.; Bao, L.-J.; Shi, L.; Wong, C.S.; Zeng, E.Y. A review of methods for measuring microplastics in aquatic environments. *Environ. Sci. Pollut. Res.* **2018**, *25*, 11319–11332. [CrossRef]
12. Arey, J.S.; Zielj Dska, B.; Atkinson, R.; Winer, A.M. Polycyclic aromatic hydrocarbon and nitroarene concentrations in ambient air during a wintertime high-NOx episode in the Los Angeles basin. *Atmos. Environ.* **1967**, *21*, 1437–1444. [CrossRef]
13. Khalili, N.R.; Scheff, P.A.; Holsen, T.M. PAH source fingerprints for coke ovens, diesel and, gasoline engines, highway tunnels, and wood combustion emissions. *Atmos. Environ.* **1995**, *29*, 533–542. [CrossRef]
14. Fendall, L.S.; Sewell, M.A. Contributing to marine pollution by washing your face: Microplastics in facial cleansers. *Mar. Pollut. Bull.* **2009**, *58*, 1225–1228. [CrossRef] [PubMed]
15. Iwasaki, S.; Isobe, A.; Kako, S.; Uchida, K.; Tokai, T. Fate of microplastics and mesoplastics carried by surface currents and wind waves: A numerical model approach in the Sea of Japan. *Mar. Pollut. Bull.* **2017**, *121*, 85–96. [CrossRef] [PubMed]
16. Lots, F.A.E.; Behrens, P.; Vijver, M.G.; Horton, A.A.; Bosker, T. A large-scale investigation of microplastic contamination: Abundance and characteristics of microplastics in European beach sediment. *Mar. Pollut. Bull.* **2017**, *123*, 219–226. [CrossRef]
17. Zhang, J.; Wang, L.; Kannan, K. Quantitative analysis of polyethylene terephthalate and polycarbonate microplastics in sediment collected from South Korea, Japan and the USA. *Chemosphere* **2021**, *279*, 130551. [CrossRef]
18. Reed, S.; Clark, M.; Thompson, R.; Hughes, K.A. Microplastics in marine sediments near Rothera research station, Antarctica. *Mar. Pollut. Bull.* **2018**, *133*, 460–463. [CrossRef]
19. IUCN. Marine Plastics. 2018. Available online: https://www.iucn.org/resources/issues-briefs/marine-plastics (accessed on 1 December 2021).
20. Van, A.; Rochman, C.M.; Flores, E.M.; Hill, K.L.; Vargas, E.; Vargas, S.A.; Hoh, E. Persistent organic pollutants in plastic marine debris found on beaches in San Diego, California. *Chemosphere* **2012**, *86*, 258–263. [CrossRef]
21. Pittura, L.; Avio, C.G.; Giuliani, M.E.; D'Errico, G.; Keiter, S.; Cormier, B.; Gorbi, S.; Regoli, F. Microplastics as vehicles of environmental PAHs to marine organisms: Combined chemical and physical hazards to the mediterranean mussels, Mytilus galloprovincialis. *Front. Mar. Sci.* **2018**, *5*, 103. [CrossRef]
22. Li, R.J.; Kou, X.J.; Geng, H.; Dong, C.; Cai, Z.W. Pollution characteristics of ambient PM2.5-bound PAHs and NPAHs in a typical winter time period in Taiyuan. *Chin. Chem. Lett.* **2014**, *25*, 663–666. [CrossRef]
23. Tang, G.; Liu, M.; Zhou, Q.; He, H.; Chen, K.; Zhang, H.; Hu, J.; Huang, Q.; Luo, Y.; Ke, H.; et al. Microplastics and polycyclic aromatic hydrocarbons (PAHs) in Xiamen coastal areas: Implications for anthropogenic impacts. *Sci. Total Environ.* **2018**, *634*, 811–820. [CrossRef]
24. Tan, X.; Yu, X.; Cai, L.; Wang, J.; Peng, J. Microplastics and associated PAHs in surface water from the Feilaixia Reservoir in the Beijiang River, China. *Chemosphere* **2019**, *221*, 834–840. [CrossRef] [PubMed]
25. Wang, Z.; Chen, M.; Zhang, L.; Wang, K.; Yu, X.; Zheng, Z.; Zheng, R. Sorption behaviors of phenanthrene on the microplastics identified in a mariculture farm in Xiangshan Bay, southeastern China. *Sci. Total Environ.* **2018**, *628–629*, 1617–1626. [CrossRef] [PubMed]
26. Eriksen, M.; Mason, S.; Wilson, S.; Box, C.; Zellers, A.; Edwards, W.; Farley, H.; Amato, S. Microplastic pollution in the surface waters of the Laurentian Great Lakes. *Mar. Pollut. Bull.* **2013**, *77*, 177–182. [CrossRef] [PubMed]

27. Leslie, H.A.; Brandsma, S.H.; van Velzen, M.J.M.; Vethaak, A.D. Microplastics en route: Field measurements in the Dutch river delta and Amsterdam canals, wastewater treatment plants, North Sea sediments and biota. *Environ. Int.* **2017**, *101*, 133–142. [CrossRef]
28. Zhou, Q.; Tu, C.; Fu, C.; Li, Y.; Zhang, H.; Xiong, K.; Zhao, X.; Li, L.; Waniek, J.J.; Luo, Y. Characteristics and distribution of microplastics in the coastal mangrove sediments of China. *Sci. Total Environ.* **2020**, *703*, 134807. [CrossRef]
29. Lahens, L.; Strady, E.; Kieu-Le, T.C.; Dris, R.; Boukerma, K.; Rinnert, E.; Gasperi, J.; Tassin, B. Macroplastic and microplastic contamination assessment of a tropical river (Saigon River, Vietnam) transversed by a developing megacity. *Environ. Pollut.* **2018**, *236*, 661–671. [CrossRef]
30. Strady, E.; Kieu-Le, T.C.; Gasperi, J.; Tassin, B. Temporal dynamic of anthropogenic fibers in a tropical river-estuarine system. *Environ. Pollut.* **2020**, *259*, 113897. [CrossRef]
31. Strady, E.; Dang, T.H.; Dao, T.D.; Dinh, H.N.; Do, T.T.D.; Duong, T.N.; Duong, T.T.; Hoang, D.A.; Kieu-Le, T.C.; Le, T.P.Q.; et al. Baseline assessment of microplastic concentrations in marine and freshwater environments of a developing Southeast Asian country, Viet Nam. *Mar. Pollut. Bull.* **2021**, *162*, 111870. [CrossRef]
32. Hien, T.T.; Nhon, N.T.T.; Thu, V.T.M.; Quyen, D.T.T.; Nguyen, N.T. The distribution of microplastics in beach sand in Tien Giang Province and Vung Tau City, Vietnam. *J. Eng. Technol. Sci.* **2020**, *52*, 208–221. [CrossRef]
33. Nguyen, Q.A.T.; Nguyen, H.N.Y.; Strady, E.; Nguyen, Q.T.; Trinh-Dang, M.; Vo, V.M. Characteristics of microplastics in shoreline sediments from a tropical and urbanized beach (Da Nang, Vietnam). *Mar. Pollut. Bull.* **2020**, *161*, 111768. [CrossRef]
34. Mỹ, T.T.Á.; Dũng, P.T. Analytical conditions for determination of microplastics in fish. *Hue Univ. J. Sci. Nat. Sci.* **2020**, *129*, 85–92. [CrossRef]
35. Kishida, M.; Imamura, K.; Maeda, Y.; Lan, T.T.N.; Thao, N.T.P.; Pham, H.V. Distribution of persistent organic pollutants and polycyclic aromatic hydrocarbons in sediment samples from Vietnam. *J. Health Sci.* **2007**, *53*, 291–301. [CrossRef]
36. Babut, M.; Mourier, B.; Desmet, M.; Simonnet-Laprade, C.; Labadie, P.; Budzinski, H.; De Alencastro, L.F.; Tu, T.A.; Strady, E.; Gratiot, N. Where has the pollution gone? A survey of organic contaminants in Ho Chi Minh city / Saigon River (Vietnam) bed sediments. *Chemosphere* **2019**, *217*, 261–269. [CrossRef] [PubMed]
37. Mai, L.; Bao, L.J.; Shi, L.; Liu, L.Y.; Zeng, E.Y. Polycyclic aromatic hydrocarbons affiliated with microplastics in surface waters of Bohai and Huanghai Seas, China. *Environ. Pollut.* **2018**, *241*, 834–840. [CrossRef] [PubMed]
38. Statistical Yearbook of Vietnam. 2020. Available online: https://www.gso.gov.vn/wp-content/uploads/2021/07/Sach-NGTK-2020Ban-quyen.pdf (accessed on 12 April 2022).
39. Campanale, C.; Savino, I.; Pojar, I.; Massarelli, C.; Uricchio, V.F. A Practical Overview of Methodologies for Sampling and Analysis of Microplastics in Riverine Environments. *Sustainability* **2020**, *12*, 6755. [CrossRef]
40. Frias, J.P.G.L.; Lyashevska, O.; Joyce, H.; Pagter, E.; Nash, R. Floating microplastics in a coastal embayment: A multifaceted issue. *Mar. Pollut. Bull.* **2020**, *158*, 111361. [CrossRef]
41. Jiang, Y.; Zhao, Y.; Wang, X.; Yang, F.; Chen, M.; Wang, J. Characterization of microplastics in the surface seawater of the South Yellow Sea as affected by season. *Sci. Total Environ.* **2020**, *724*, 138375. [CrossRef]
42. Zhou, G.; Wang, Q.; Zhang, J.; Li, Q.; Wang, Y.; Wang, M.; Huang, X. Distribution and characteristics of microplastics in urban waters of seven cities in the Tuojiang River basin, China. *Environ. Res.* **2020**, *189*, 109893. [CrossRef]
43. Jung, M.R.; Horgen, F.D.; Orski, S.V.; Rodriguez, C.V.; Beers, K.L.; Balazs, G.H.; Jones, T.T.; Work, T.; Brignac, K.C.; Royer, S.-J.; et al. Validation of ATR FT-IR to identify polymers of plastic marine debris, including those ingested by marine organisms. *Mar. Pollut. Bull.* **2018**, *127*, 704–716. [CrossRef]
44. Mao, R.; Hu, Y.; Zhang, S.; Wu, R.; Guo, X. Microplastics in the surface water of Wuliangsuhai Lake, northern China. *Sci. Total Environ.* **2020**, *723*, 137820. [CrossRef]
45. Wang, W.; Yuan, W.; Chen, Y.; Wang, J. Microplastics in surface waters of Dongting Lake and Hong Lake, China. *Sci. Total Environ.* **2018**, *633*, 539–545. [CrossRef] [PubMed]
46. Egess, R.; Nankabirwa, A.; Ocaya, H.; Pabire, W.G. Microplastic pollution in surface water of Lake Victoria. *Sci. Total Environ.* **2020**, *741*, 140201. [CrossRef] [PubMed]
47. Cheung, P.K.; Cheung, L.T.O.; Fok, L. Seasonal variation in the abundance of marine plastic debris in the estuary of a subtropical macro-scale drainage basin in South China. *Sci. Total Environ.* **2016**, *562*, 658–665. [CrossRef] [PubMed]
48. Su, L.; Cai, H.; Kolandhasamy, P.; Wu, C.; Rochman, C.M.; Shi, H. Using the Asian clam as an indicator of microplastic pollution in freshwater ecosystems. *Environ. Pollut.* **2018**, *234*, 347–355. [CrossRef] [PubMed]
49. Hale, R.; Seeley, M.E.; la Guardia, M.J.; Mai, L.; Zeng, E.Y. A Global Perspective on Microplastics. *J. Geophys. Res. Ocean.* **2020**, *125*, e2018JC014719. [CrossRef]
50. Naidoo, T.; Glassom, D.; Smit, A.J. Plastic pollution in five urban estuaries of KwaZulu-Natal, South Africa. *Mar. Pollut. Bull.* **2015**, *101*, 473–480. [CrossRef] [PubMed]
51. Maes, T.; Van der Meulen, M.D.; Devriese, L.I.; Leslie, H.A.; Huvet, A.; Frère, L.; Robbens, J.; Vethaak, A.D. Microplastics Baseline Surveys at the Water Surface and in Sediments of the North-East Atlantic. *Front. Mar. Sci.* **2017**, *4*, 135. [CrossRef]
52. Pan, Z.; Liu, Q.; Sun, X.; Li, W.; Zou, Q.; Cai, S.; Lin, H. Widespread occurrence of microplastic pollution in open sea surface waters: Evidence from the mid-North Pacific Ocean. *Gondwana Res.* **2021**. [CrossRef]
53. Haddout, S.; Gimiliani, G.T.; Priya, K.L.; Hoguane, A.M.; Casila, J.C.C.; Ljubenkov, I. Microplastics in Surface Waters and Sediments in the Sebou Estuary and Atlantic Coast, Morocco. *Anal. Lett.* **2021**, *55*, 256–268. [CrossRef]

54. Li, Y.; Lu, Z.; Zheng, H.; Wang, J.; Chen, C. Microplastics in surface water and sediments of Chongming Island in the Yangtze Estuary, China. *Environ. Sci. Eur.* **2020**, *32*, 15. [CrossRef]
55. Harris, P.T. The fate of microplastic in marine sedimentary environments: A review and synthesis. *Mar. Pollut. Bull.* **2020**, *158*, 111398. [CrossRef] [PubMed]
56. Felismino, M.E.L.; Helm, P.A.; Rochman, C.M. Microplastic and other anthropogenic microparticles in water and sediments of Lake Simcoe. *J. Great Lakes Res.* **2021**, *47*, 180–189. [CrossRef]
57. Tanchuling, M.A.N.; Osorio, E.D. The Microplastics in Metro Manila Rivers: Characteristics, Sources, and Abatement. In *Plastics in the Aquatic Environment—Part I: Current Status and Challenges*; Stock, F., Reifferscheid, G., Brennholt, N., Kostianaia, E., Eds.; Springer International Publishing: Cham, Switzerland, 2022; pp. 405–426.
58. Sulistyowati, L.; Nurhasanah, E.R.; Cordova, M.R. The occurrence and abundance of microplastics in surface water of the midstream and downstream of the Cisadane River, Indonesia. *Chemosphere* **2022**, *291*, 133071. [CrossRef] [PubMed]
59. Napper, I.E.; Baroth, A.; Barrett, A.C.; Bhola, S.; Chowdhury, G.W.; Davies, B.F.; Duncan, E.M.; Kumar, S.; Nelms, S.E.; Niloy, N.H.; et al. The abundance and characteristics of microplastics in surface water in the transboundary Ganges River. *Environ. Pollut.* **2021**, *274*, 116348. [CrossRef] [PubMed]
60. Han, M.; Niu, X.; Tang, M.; Zhang, B.-T.; Wang, G.; Yue, W.; Kong, X.; Zhu, J. Distribution of microplastics in surface water of the lower Yellow River near estuary. *Sci. Total Environ.* **2020**, *707*, 135601. [CrossRef] [PubMed]
61. Kapp, K.J.; Yeatman, E. Microplastic hotspots in the Snake and Lower Columbia rivers: A journey from the Greater Yellowstone Ecosystem to the Pacific Ocean. *Environ. Pollut.* **2018**, *241*, 1082–1090. [CrossRef]
62. de Piñon-Colin, T.; Rodriguez-Jimenez, R.; Pastrana-Corral, M.A.; Rogel-Hernandez, E.; Wakida, F.T. Microplastics on sandy beaches of the Baja California Peninsula, Mexico. *Mar. Pollut. Bull.* **2018**, *131*, 63–71. [CrossRef]
63. Wu, C.; Zhang, K.; Xiong, X. Microplastic Pollution in Inland Waters Focusing on Asia. In *Freshwater Microplastics: Emerging Environmental Contaminants?* Wagner, M., Lambert, S., Eds.; Springer International Publishing: Cham, Switzerland, 2018; pp. 85–99.
64. Ta, V.P. Báo cáo ngành nhựa (Vietnamese) No. 8424. 2019. p. 64. Available online: http://www.fpts.com.vn/FileStore2/File/2019/09/13/FPTSPlastic_Industry_ReportAug2019_e5e64506.pdf (accessed on 18 May 2022).
65. Zhu, L.; Bai, H.; Chen, B.; Sun, X.; Qu, K.; Xia, B. Microplastic pollution in North Yellow Sea, China: Observations on occurrence, distribution and identification. *Sci. Total Environ.* **2018**, *636*, 20–29. [CrossRef]
66. Huang, Y.; Liu, Q.; Jia, W.; Yan, C.; Wang, J. Agricultural plastic mulching as a source of microplastics in the terrestrial environment. *Environ. Pollut.* **2020**, *260*, 114096. [CrossRef]
67. Derraik, J.G.B. The pollution of the marine environment by plastic debris: A review. *Mar. Pollut. Bull.* **2002**, *44*, 842–852. [CrossRef]
68. Shaw, D.G.; Day, R.H. Colour- and form-dependent loss of plastic micro-debris from the North Pacific Ocean. *Mar. Pollut. Bull.* **1994**, *28*, 39–43. [CrossRef]
69. Boerger, C.M.; Lattin, G.L.; Moore, S.L.; Moore, C.J. Plastic ingestion by planktivorous fishes in the North Pacific Central Gyre. *Mar. Pollut. Bull.* **2010**, *60*, 2275–2278. [CrossRef] [PubMed]
70. Güven, O.; Gökdağ, K.; Jovanović, B.; Kıdeyş, A.E. Microplastic litter composition of the Turkish territorial waters of the Mediterranean Sea, and its occurrence in the gastrointestinal tract of fish. *Environ. Pollut.* **2017**, *223*, 286–294. [CrossRef] [PubMed]
71. Woodall, L.C.; Sanchez-Vidal, A.; Canals, M.; Paterson, G.L.J.; Coppock, R.; Sleight, V.; Calafat, A.; Rogers, A.D.; Narayanaswamy, B.E.; Thompson, R.C. The deep sea is a major sink for microplastic debris. *Open Sci.* **2014**, *1*, 140317. [CrossRef] [PubMed]
72. Chubarenko, I.; Bagaev, A.; Zobkov, M.; Esiukova, E. On some physical and dynamical properties of microplastic particles in marine environment. *Mar. Pollut. Bull.* **2016**, *108*, 105–112. [CrossRef]
73. Kukulka, T.; Proskurowski, G.; Morét-Ferguson, S.; Meyer, D.W.; Law, K.L. The effect of wind mixing on the vertical distribution of buoyant plastic debris. *Geophys. Res. Lett.* **2012**, *39*, 88. [CrossRef]
74. Welden, N.A.C.; Lusher, A.L. Impacts of changing ocean circulation on the distribution of marine microplastic litter. *Integr. Environ. Assess. Manag.* **2017**, *13*, 483–487. [CrossRef] [PubMed]
75. Ripken, C.; Kotsifaki, D.G.; Chormaic, S.N. Analysis of small microplastics in coastal surface water samples of the subtropical island of Okinawa, Japan. *Sci. Total Environ.* **2021**, *760*, 143927. [CrossRef]
76. Maul, J.; Frushour, B.G.; Kontoff, J.R.; Eichenauer, H.; Ott, K.-H.; Schade, C. Polystyrene and Styrene Copolymers. *Ullmann's Encycl. Ind. Chem.* **2007**, *29*, 475–522. [CrossRef]
77. Serini, V. Polycarbonates. *Ullmann's Encycl. Ind. Chem.* **2000**. [CrossRef]
78. de Vos, L.; van de Voorde, B.; van Daele, L.; Dubruel, P.; van Vlierberghe, S. Poly(alkylene terephthalate)s: From current developments in synthetic strategies towards applications. *Eur. Polym. J.* **2021**, *161*, 110840. [CrossRef]
79. Vogelsang, C.; Lusher, A.L.; Dadkhah, M.E.; Sundvor, I.; Umar, M.; Ranneklev, S.B.; Eidsvoll, D.; Meland, S. *Microplastics in Road Dust—Characteristics, Pathways and Measures*; REPORT SNO. 7526-2020. 2018. p. 174. Available online: https://www.miljodirektoratet.no/globalassets/publikasjoner/M959/M959.pdf (accessed on 1 December 2021).
80. Radulovic, L.L.; Wojcinski, Z.W. PTFE (Polytetrafluoroethylene; Teflon®). In *Encyclopedia of Toxicology*, 3rd ed.; Elsevier: Amsterdam, The Netherlands, 2014; pp. 1133–1136. [CrossRef]
81. Harris, M.; Potgieter, J.; Ray, S.; Archer, R.; Arif, K.M. Preparation and characterization of thermally stable ABS/HDPE blend for fused filament fabrication. *Mater. Manuf. Process.* **2020**, *35*, 230–240. [CrossRef]

82. Eagan, J.M.; Xu, J.; Di Girolamo, R.; Thurber, C.M.; Macosko, C.W.; LaPointe, A.M.; Bates, F.S.; Coates, G.W. Combining polyethylene and polypropylene: Enhanced performance with PE/iPP multiblock polymers. *Science* **2017**, *355*, 814–816. [CrossRef] [PubMed]
83. Ta, A.; Babel, S. Microplastic pollution in surface water of the chao phraya river in ang thong area. *EnvironmentAsia* **2019**, *12*, 48–53. [CrossRef]
84. Mai, L.; You, S.-N.; He, H.; Bao, L.-J.; Liu, L.-Y.; Zeng, E.Y. Riverine Microplastic Pollution in the Pearl River Delta, China: Are Modeled Estimates Accurate? *Environ. Sci. Technol.* **2019**, *53*, 11810–11817. [CrossRef]
85. Lorenz, C.; Dolven, J.K.; Værøy, N.; Stephansen, D.; Vollertsen, J. Microplastic Pollution in Three Rivers in South Eastern Norway. 2020. Available online: https://www.miljodirektoratet.no/globalassets/publikasjoner/m1572/m1572.pdf (accessed on 1 December 2021).
86. Frère, L.; Paul-Pont, I.; Rinnert, E.; Petton, S.; Jaffré, J.; Bihannic, I.; Soudant, P.; Lambert, C.; Huvet, A. Influence of environmental and anthropogenic factors on the composition, concentration and spatial distribution of microplastics: A case study of the Bay of Brest (Brittany, France). *Environ. Pollut.* **2017**, *225*, 211–222. [CrossRef]
87. Kershaw, P. Sources, Fate and Effects of Microplastics in the Marine Environment: A Global Assessment. 2015. Available online: http://41.89.141.8/kmfri/handle/123456789/735 (accessed on 18 May 2022).
88. Zhang, W.; Ma, X.; Zhang, Z.; Wang, Y.; Wang, J.; Ma, D. Persistent organic pollutants carried on plastic resin pellets from two beaches in China. *Mar. Pollut. Bull.* **2015**, *99*, 28–34. [CrossRef]
89. Hirai, H.; Takada, H.; Ogata, Y.; Yamashita, R.; Mizukawa, K.; Saha, M.; Kwan, C.; Moore, C.; Gray, H.; Laursen, D.; et al. Organic micropollutants in marine plastics debris from the open ocean and remote and urban beaches. *Mar. Pollut. Bull.* **2011**, *62*, 1683–1692. [CrossRef]
90. Marr, L.C.; Kirchstetter, T.W.; Harley, R.A.; Miguel, A.H.; Hering, S.V.; Hammond, S.K. Characterization of Polycyclic Aromatic Hydrocarbons in Motor Vehicle Fuels and Exhaust Emissions. *Environ. Sci. Technol.* **1999**, *33*, 3091–3099. [CrossRef]
91. Dong, T.T.T.; Lee, B.-K. Characteristics, toxicity, and source apportionment of polycylic aromatic hydrocarbons (PAHs) in road dust of Ulsan, Korea. *Chemosphere* **2009**, *74*, 1245–1253. [CrossRef]
92. Liu, M.; Cheng, S.; Ou, D.; Hou, L.; Gao, L.; Wang, L.; Xie, Y.; Yang, Y.; Xu, S. Characterization, identification of road dust PAHs in central Shanghai areas, China. *Atmos. Environ.* **2007**, *41*, 8785–8795. [CrossRef]
93. Patchaiyappan, A.; Dowarah, K.; Zaki Ahmed, S.; Prabakaran, M.; Jayakumar, S.; Thirunavukkarasu, C.; Devipriya, S.P. Prevalence and characteristics of microplastics present in the street dust collected from Chennai metropolitan city, India. *Chemosphere* **2021**, *269*, 128757. [CrossRef] [PubMed]
94. Järlskog, I.; Strömvall, A.-M.; Magnusson, K.; Gustafsson, M.; Polukarova, M.; Galfi, H.; Aronsson, M.; Andersson-Sköld, Y. Occurrence of tire and bitumen wear microplastics on urban streets and in sweepsand and washwater. *Sci. Total Environ.* **2020**, *729*, 138950. [CrossRef] [PubMed]

Article

Effects and Impacts of Different Oxidative Digestion Treatments on Virgin and Aged Microplastic Particles

Ilaria Savino [1,2], Claudia Campanale [1,*], Pasquale Trotti [3], Carmine Massarelli [1], Giuseppe Corriero [2] and Vito Felice Uricchio [1]

[1] Italian National Council of Research, Water Research Institute, 70132 Bari, Italy; ilaria.savino@ba.irsa.cnr.it (I.S.); carmine.massarelli@ba.irsa.cnr.it (C.M.); vito.uricchio@ba.irsa.cnr.it (V.F.U.)
[2] Department of Biology, University of Bari Aldo Moro, 70121 Bari, Italy; giuseppe.corriero@uniba.it
[3] Sezione di Entomologia e Zoologia Agraria, Dipartimento di Scienze del Suolo, della Pianta e degli Alimenti, University of Bari Aldo Moro, 70121 Bari, Italy; pasquale.trotti@uniba.it
* Correspondence: claudia.campanale@ba.irsa.cnr.it

Abstract: Although several sample preparation methods for analyzing microplastics (MPs) in environmental matrices have been implemented in recent years, important uncertainties and criticalities in the approaches adopted still persist. Preliminary purification of samples, based on oxidative digestion, is an important phase to isolate microplastics from the environmental matrix; it should guarantee both efficacy and minimal damage to the particles. In this context, our study aims to evaluate Fenton's reaction digestion pre-treatment used to isolate and extract microplastics from environmental matrices. We evaluated the particle recovery efficiency and the impact of the oxidation method on the integrity of the MPs subjected to digestion considering different particles' polymeric composition, size, and morphology. For this purpose, two laboratory experiments were set up: the first one to evaluate the efficacy of various digestion protocols in the MPs extraction from a complex matrix, and the second one to assess the possible harm of different treatments, differing in temperatures and volume reagents used, on virgin and aged MPs. Morphological, physicochemical, and dimensional changes were verified by Scanning Electron Microscope (SEM) and Fourier Transformed Infrared (FTIR) spectroscopy. The findings of the first experiment showed the greatest difference in recovery rates especially for polyvinyl chloride and polyethylene terephthalate particles, indicating the role of temperature and the kind of polymer as the major factors influencing MPs extraction. In the second experiment, the SEM analysis revealed morphological and particle size alterations of various entities, in particular for the particles treated at 75 °C and with major evident alterations of aged MPs to virgin ones. In conclusion, this study highlights how several factors, including temperature and polymer, influence the integrity of the particles altering the quality of the final data.

Keywords: microplastics; oxidative digestion; Fenton's reagent; virgin; aged; weathering; SEM; FTIR

Citation: Savino, I.; Campanale, C.; Trotti, P.; Massarelli, C.; Corriero, G.; Uricchio, V.F. Effects and Impacts of Different Oxidative Digestion Treatments on Virgin and Aged Microplastic Particles. *Polymers* 2022, 14, 1958. https://doi.org/10.3390/polym14101958

Academic Editor: Jacopo La Nasa

Received: 1 April 2022
Accepted: 9 May 2022
Published: 11 May 2022

Publisher's Note: MDPI stays neutral with regard to jurisdictional claims in published maps and institutional affiliations.

Copyright: © 2022 by the authors. Licensee MDPI, Basel, Switzerland. This article is an open access article distributed under the terms and conditions of the Creative Commons Attribution (CC BY) license (https://creativecommons.org/licenses/by/4.0/).

1. Introduction

Microplastics (MPs) are "synthetic solid particle or polymeric matrix, with regular or irregular shape from 1 μm to 5 mm size, of either primary or secondary manufacturing origin" [1]. Their presence has been reported in all environmental matrices, becoming an emerging problem worldwide [2–4]. Due to their small size, high volume surface ratio, and their ability to adsorb or release pollutants [5], MPs' threats mainly concern their effects on organisms and human health [6–9]. Therefore, MPs monitoring is important to understand their presence in the environment. Microplastic studies require several methodological approaches to isolate, identify and quantify particles spread in environmental matrices [10–12].

However, when environmental matrices are rich in organic matter, a chemical digestion treatment is necessary to remove it and release particles. Organic residues have a density similar to that of polymers, and they may float together MPs during the density separation phase, hindering extraction and quantitative analysis of particles [13]. Organic digestion treatments may be based on oxidizing agents, acids, basics, or enzymes [10,14–16]. However, not all procedures remove organic matter without damaging polymers [17–19].

Applying strong acids, such as nitric acid (HNO_3), produced efficient digestion of biota but they are toxic, corrosive, and cause polymers degradation such as polystyrene, polyamide, and polyethylene [10,20,21]. Alternatively, studies have used alkaline solutions and enzymes for biota digestion, but these require much time, may damage some polymers, and are very expensive [22–24].

Oxidizing agents are increasingly used for water, soil, and sediment because the type of organic matter is more difficult to digest (leaves, woody debris, algae, etc.) [13,25,26]. However, at high concentrations and temperatures, agents such as hydrogen peroxide (H_2O_2) could destroy polyamide particles, reduce their size and alter the colour of polypropylene particles [27]. Digestion protocols should have minimal impact on the morphology, colour, and weight of MPs [25,28].

Several studies analyzed the effects of digestion treatments on MPs, testing different reaction times, temperatures, and regent volumes [19,27,29–33]. However, most methodological studies tested treatments on virgin MP rather than aged, neglecting their effect on fragile and damaged particles, more representative of reality [28–30]. Despite the recent development of biodegradable plastics, less impactful on the environment [34], many biotic and abiotic factors act on plastics and MPs, leading to changes in polymer properties through different degradation mechanisms [35–39]. Light and temperature, for example, involve free radical formation, chain scission, and subsequent reduction of molecular weight. This, together with mechanical and biotic stress, makes the polymers fragile and more susceptible to fragmentation. The formation of superficial cracks becomes, then, a site of other degradation reactions, leading to the disintegration of material [35,40].

In this context, the present study aims to assess the goodness of the most popular protocol of oxidative digestion used as a preparative step, to purify samples isolating and extracting MPs from complex environmental matrices. The method has been evaluated in terms of efficiency of extraction and recovery of MPs from the environmental matrix and, the impact and aggressiveness of the chemical digestion on the integrity of particles.

Moreover, we tested different experimental digestion conditions on virgin and aged MPs of various morphology, polymer, and size to assess if a different reaction to the chemical digestion and, an eventual alteration of items, occur based on MPs properties. We hypnotize that the rapid oxidation and the stringent exothermic reaction could destroy some polymer particles, especially the most aged ones. These particles are already fragile due to weathering caused by the time of permanence in the environment. The final objective is to advise a less impactful digestion protocol for the extraction of MPs from environmental matrices.

2. Materials and Methods

2.1. Experimental Design

Two different laboratory experiments were set up to assess the efficiency of the most used digestion protocol [14] and its impact on MPs integrity. For this purpose, the oxidative treatment, based on the Fenton reaction ($Fe^{2+} + H_2O_2 \rightarrow Fe^{3+} + OH + OH^-$) [25], was tested using different temperature ranges and reagent volumes. Moreover, to reproduce the difficulties linked to the MPs extraction from complex environmental samples, virgin and aged MPs standards of different sizes and compositions were added to unpolluted soil samples. The integrity of particles and the level of alteration before and after the different treatments were evaluated through Scanning Electron Microscopy (SEM), Fourier Transformed Infrared (FTIR) spectroscopy.

In Table 1 the experimental set-up of the two trials is shown.

Table 1. Indication of the experimental conditions followed for each trial.

	Experiment One
AIM	Evaluate the efficiency of extraction of the most commonly used chemical digestion protocol (based on Wet Peroxide Oxidation [14]) on the recovery of virgin MPs standards from a complex matrix
Particle selection	Virgin MPs
Matrix selected	Soil
Starting digestion condition	Reagents volume: 20 mL of 30% H_2O_2 solution add to 20 mL of 0.05 M iron sulphate heptahydrate ($FeSO_4 \cdot 7H_2O$) every 30' until complete sample digestion. The temperature of reaction: 75 °C.
Density separation	NaI (1.8 g cm^3)
Qualitative analysis	Stereomicroscope
	Experiment Two
AIM	Evaluate the impact of the most commonly used chemical digestion protocol on the integrity of virgin and aged MPs standards
Particle selection	Virgin and aged MPs
Matrix selected	Soil
Starting digestion condition	Reagents volume: 20 mL of 30% H_2O_2 solution add to 20 mL of 0.05 M iron sulphate heptahydrate ($FeSO_4 \cdot 7H_2O$) every 30' until complete sample digestion. The temperature of reaction: 75 °C.
Density separation	NaI (1.8 g cm^3)
Qualitative analysis	FTIR—SEM

2.2. Microplastic Standards Selection

Virgin MPs of different shapes and polymers were selected by common plastic items (Table 2). Particle colour was chosen to facilitate detection and counting during the extraction phase. They were cut and smoothed in the laboratory, and particles were passed through sieves with mesh sizes from 5 mm to 1 mm, from 1 mm to 500 µm, and from 500 µm to 100 µm, obtaining MPs of three size ranges. Even, 5 mm size PE, and PP pre-production pellets were added to evaluate the impact of the most commonly used chemical digestion protocol [14] on the integrity of MPs standards (Figure 1).

Table 2. Polymer, density, source, colour, and shape of MPs selected as standards for the experiments. (*): image reworked from source [10].

Polymers	Density (g cm^3) (*)	Source	Colour	Shape
Polystyrene (PS)	0.01–1.06	Food box	White	Fragment
Polypropylene (PP)	0.85–0.92	Disposable glass	Red	Fragment
Polyethylene (PE)	0.89–0.98	Mulching films	Black	Fragment
Polyamide (PA)	1.12–1.15	Textile	Black	Fibre
Polyvinyl chloride (PVC)	1.38–1.41	Building material	Black	Fragment
Polyethylene terephthalate (PET)	1.38–1.41	Plastics bottle	Green	Fragment

2.3. Experiment One: Evaluating the Efficiency of MPs Digestion Treatment through Recovery Tests

Digestion Treatment Conditions

Virgin MPs, 30 particles for each polymer (PE, PP, PET, PVC, PS), underwent six oxidative digestion treatments at three different temperatures (75 °C, 50 °C, and 30 °C), and reagent volumes (100 or 60 mL of H_2O_2 + 20 mL of $FeSO_4 \cdot 7H_2O$) (Table 3). The ferrous ion (Fe^{2+}) of the iron sulphate heptahydrate initiates and catalyses the reaction leading to the generation of hydroxyl and hydroperoxyl radicals, powerful oxidants that degrade organic compounds [41].

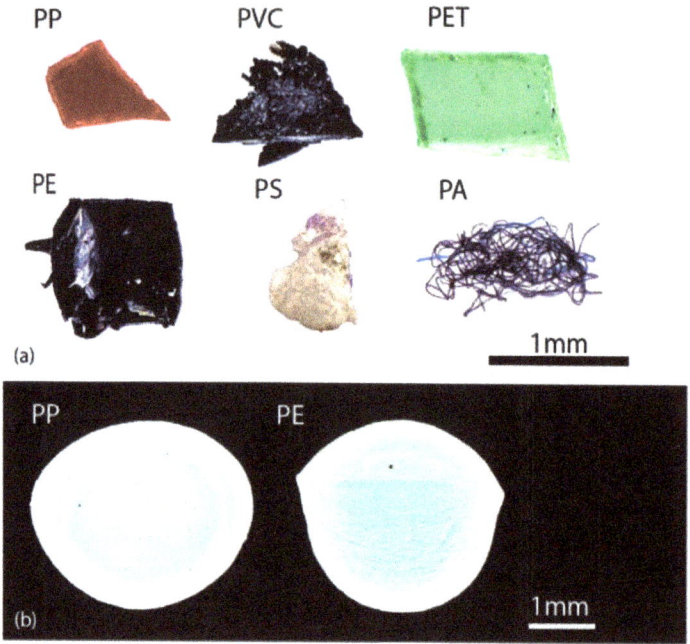

Figure 1. (**a**) Virgin MPs products by cutting common plastic items. (**b**) Pre-production pellets added in experiment two. Images produced by Carl Zeiss Tessovar Microscope.

Table 3. Summary of the different oxidative digestion conditions used in experiment one. "-" Treatment made in the absence of the matrix on particle sizes from 500 to 100 µm.

Treatment	Reagent Volumes	Temperature (°C)	Polymers	Size	Soil Matrix (g)
1	100 mL H_2O_2 + 20 mL $FeSO_4 \cdot 7H_2O$	75 °C	PE, PP, PET, PVC, PS	5–1 mm 1 mm–500 µm	50
				500–100 µm	-
2	60 mL H_2O_2 + 20 mL $FeSO_4 \cdot 7H_2O$	75 °C	PE, PP, PET, PVC, PS	5–1 mm 1 mm–500 µm	50
				500–100 µm	-
3	100 mL H_2O_2 + 20 mL $FeSO_4 \cdot 7H_2O$	50 °C	PE, PP, PET, PVC, PS	5 mm–1 mm 1 mm–500 µm	50
				500–100 µm	-
4	60 mL H_2O_2 + 20 mL $FeSO_4 \cdot 7H_2O$	50 °C	PE, PP, PET, PVC, PS	5–1 mm 1 mm–500 µm	50
				500–100 µm	-
5	100 mL H_2O_2 + 20 mL $FeSO_4 \cdot 7H_2O$	30 °C	PE, PP, PET, PVC; PS	5–1 mm 1 mm–500 µm	50
				500–100 µm	-
6	60 mL H_2O_2 + 20 mL $FeSO_4 \cdot 7H_2O$	30 °C	PE, PP, PET, PVC, PS	5–1 mm 1 mm–500 µm	50
				500–100 µm	-

Three size ranges of particles (5–1 mm; 1 mm–500 μm; 500–100 μm) were assessed for recoveries. The biggest particles (5–1 mm; 1 mm–500 μm) were added to 50 g of soil to simulate the extraction from a complex matrix while the smallest ones (500–100 μm) were added just of digestion reagents to exclude the influence of matrix on the recovery of particles and evaluate just the effect of the digestion protocol.

A solution of NaI (1.8 g cm^3) was prepared by dissolving the salt in distilled water, to extract MPs from the environmental matrix. After the digestion treatments, the solution was added to the sample, it was shaken for about 10 s and decanted for 1 h. The supernatant was filtrated by a vacuum filtration unit (Sartorius, Goettingen, Germany) using a nitrocellulose filter (Whatman nitrocellulose membrane filters diam. 47 mm, pore size 0.45 μm) and particles were observed under a stereomicroscope (Motic SMZ – 171, Hong Kong, China).

The polymer recovery rate was calculated as the number of extracted particles on the number of added particles. The final value was expressed as a percentage.

2.4. Experiment Two: Evaluating the Impact of Digestion on Virgin vs. Aged MPs Integrity

2.4.1. Ageing of Microplastics

Some virgin microplastics were artificially weathered in a climate room equipped with UVA lamps, calibrated at 340 nm, and programmed at a temperature of 22 °C, an irradiance of 12 h, and humidity at 60%, for a total of 20 days. Afterwards, samples were subjected to thermally ageing at 45 °C in an air-circulated oven, for another 20 days (Figure 2).

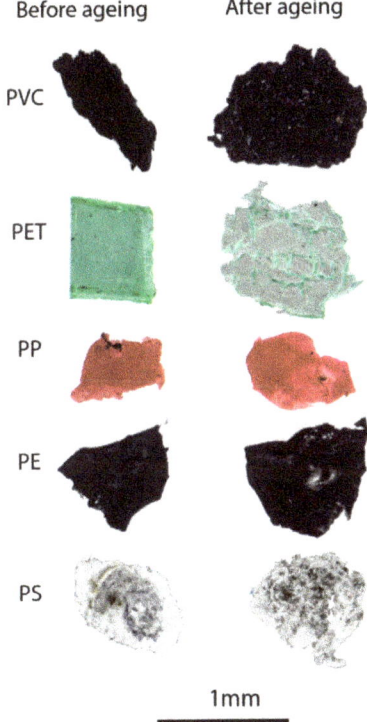

Figure 2. The morphological aspect of some polymers (PVC, PET, PP, PE, PS) before and after the ageing process. These polymers, together with PA fibre and pellets (PP, PE), were exposed to UVA (photo-oxidation) in the climatic chamber for 20 days and then at a temperature of 45 °C for a further 20 days in dry conditions.

2.4.2. Digestion Treatment Conditions

Three oxidative digestion treatments, with different experimental conditions, were tested on PVC, PE, PP, PS, PET fragment; PP and PE pellets, and PA fibre to evaluate the impact on virgin vs. aged MPs by SEM analysis. For each polymer, were selected particles of size from 5 mm to 500 µm and added to 13 g of soil, with the exception of PA to exclude possible contamination from fibre present in the environmental matrix (Table 4). After digestion, particles were separated by the matrix using a saturated NaI solution, and their integrity was observed by SEM, before and after treatments. Virgin and aged fibre were analyzed as tangles before treatments for the difficulty of obtaining single filaments and handling especially those being aged and fragile.

Table 4. Summary of different oxidative digestion conditions used in experiment two. "-" Treatment made in the absence of the matrix on particle sizes from 5 to 1 mm.

Treatment	Reagent Volumes	Temperature (°C)	Polymers	Size	Soil Matrix (g)
a	100 mL H_2O_2 + 20 mL $FeSO_4 \cdot 7H_2O$	75 °C	PE, PP, PET, PVC, PS	5–1 mm 1 mm–500 µm	13
			PA	5–1 mm	-
b	60 mL H_2O_2 + 20 mL $FeSO_4 \cdot 7H_2O$	50 °C	PE, PP, PET, PVC, PS	5–1 mm 1 mm–500 µm	13
			PA	5–1 mm	-
c	60 mL H_2O_2 + 20 mL $FeSO_4 \cdot 7H_2O$	30 °C	PE, PP, PET, PVC, PS	5–1 mm 1 mm–500 µm	13
			PA	5–1 mm	-

2.5. Fourier Transform Infrared Spectroscopy (FTIR) Acquisition

Aged particles were analyzed, before and after weathering by Fourier Transform Infrared spectroscopy using a Thermo Scientific NICOLET Summit FTIR Spectrometer (Waltham, MA, USA) equipped with an Everest ATR with a diamond Crystal plate and a DTGS KBr detector. The FTIR spectra were recorded in the region of 4000–400 cm^{-1} with 32 scans at a resolution of 4 cm^{-1}.

2.6. Scanning Electron Microscopy (SEM) Acquisition

Scanning Electron Microscopy (HITACHI TM 3000 Tabletop, Tokyo, Japan) was used to observe morphology polymers before and after oxidative digestion treatments. Particles were fixed on carbon adhesive and coated with a thin layer of gold and palladium for 2 min and 10 mA to avoid charging during electron microscopy. Larger particles, such as pellets, were measured operating at 5 kV, while for other particles it was operated at 15 kV. The size of some particles was measured before and after treatments by SEM image software (Hitachi TM 3000, ver. 02-03-02, Tokyo, Japan).

2.7. Quality Control

A cotton coat was worn during the laboratory procedures, preventing any contamination from synthetic clothing. All glass instruments were washed three times with Milli-Q water and covered with aluminium foil. The NaI solution was filtered through a nitrocellulose filter before its use. All analytical steps were performed in a laminar flow cabinet to avoid laboratory airborne contamination.

3. Results

3.1. Results of Experiment One: Evaluating the Efficiency of MPs Digestion Treatment through Recovery Tests

As an overall result, the experimental tests performed in different temperature and peroxide volume conditions showed a recovery efficiency of about 100% for most of the MP materials used (Figure 3). The major criticalities in the extraction efficiency emerged above all for PVC and PET items.

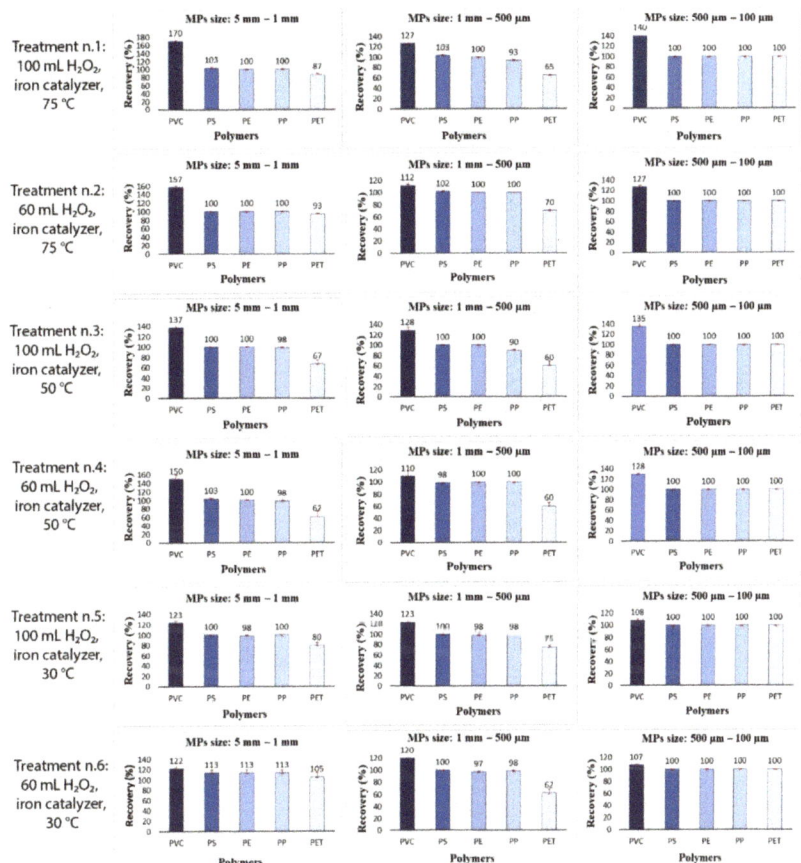

Figure 3. Recovery rates of PVC, PS, PE, PP, and PET, of three-dimensional sizes, after six different treatments varying for H_2O_2 volumes and temperatures. Values are expressed as a percentage mean value of two replicates ± the standard deviation.

The extraction of 1–5 mm PVC particles testing the treatment n. 1 (75 °C, 100 mL H_2O_2) showed the highest rate of recovery above the 100% (170 ± 1.4%) followed by treatment n. 2 (75 °C, 60 mL) with a recovery rate of 157 ± 3.5%, treatment n. 4 (50 °C, 60 mL) with 150 ± 10%, treatment n. 3 (50 °C, 100 mL) 137 ± 4%, treatment n. 5 (30 °C, 100 mL) 123 ± 4% and treatment n. 6 (30 °C, 60 mL) 122 ± 2%.

This enhancement of observed particles with respect to the initial number of added items is due to the aggressiveness of the digestion treatment on PVC, which led to its fragmentation in smaller particles observed and identified both in suspension and in the soil matrix used (Figure 4).

Figure 4. Particulars of PVC fractures (yellow boxes on the left) due to the abrupt oxidation reaction at 75 °C that led to the fragmentation of PVC in tiny particles trapped in the soil matrix (red circles on the right).

Likewise, the smallest PVC fragments (1 mm–500 µm; 500–100 µm) showed the same behaviour with all recoveries above 100%, confirming a fragmentation of this polymer. In each size range, the treatment n.6 (30 °C, 60 mL) resulted in having a lower impact on particles with recoveries close to 100% (116 ± 2%).

Many other small and tiny PVC particles lower than 100 µm were also observed especially in treatments 1 and 2, probably generated from the fragmentation of bigger particles. As shown in Figure 4 the surface of bigger PVC particles appears greatly modified with holes and cracks.

Differently from the PVC behaviour, all the treatments tested in the different ranges of temperature and reagents volume (treatments n. 1, 2, 3, 4, 5, and 6) demonstrated a good recovery efficiency for PS, PE and PP polymers in all the three size ranges evaluated (5–1 mm, 1 mm–500 µm and 500–100 m). Indeed, the recoveries obtained were 101 ± 1%, 100.3 ± 0.5%, 99.3 ± 0.9%, for PS, PE, and PP, respectively, showing good resistance to the oxidation reaction and an equally satisfying recovery efficiency from the matrix.

Otherwise, the recoveries of 5–1 mm and 1 mm–500 µm PET fragments showed the lowest recovery rates among all polymers, ranging from 24 and 93% of recovered particles. However, this particle loss is mostly attributable to an effect of the soil matrix used which made difficult the recovery of MPs trapping them in the bottom. Indeed, the tests on the smallest size range of PET MPs (500–100 µm), set up without the soil matrix, showed recovery rates equal to 100% for all the six treatments evaluated.

3.2. Results of Experiment Two: Evaluating the Impact of Digestion Treatment on Virgin and Aged MPs through Qualitative Evaluations

3.2.1. Ageing of Microplastics: FTIR Acquisition

The FTIR acquisitions of MP standards of different polymer compositions made before (black lines, Figure 5) and after the ageing of particles (red lines, Figure 5), show that new absorption peaks were formed consecutively to weathering suggesting strong differences with the pristine materials probably due to their degradation (red lines, Figure 5). In the spectra of each artificially weathered particle, is evident the presence of broad peaks in the region from 3100 to 3700 cm^{-1} (OH stretching).

The ageing process produces new bands at 3423 cm^{-1} in the IR spectra of PET, PA, and PP (pellets and fragments) (Figure 5a).

In the spectra of aged PET particles, forty days after the artificial weathering, a new peak at 1614 cm^{-1}, non-existent in the same particles before the ageing process, appears. (Figure 5c).

Regarding the PE particles, compared to the unaltered pristine MPs acquired by FTIR at the time zero before ageing, the weathered fragment spectra show the presence of new intense peaks at 3414 and 1577 cm^{-1} and others less intense at 873 and 777 cm^{-1}. Similarly, the aged PE pellet spectra show new weathering bands at 3458 and 1618 cm^{-1}.

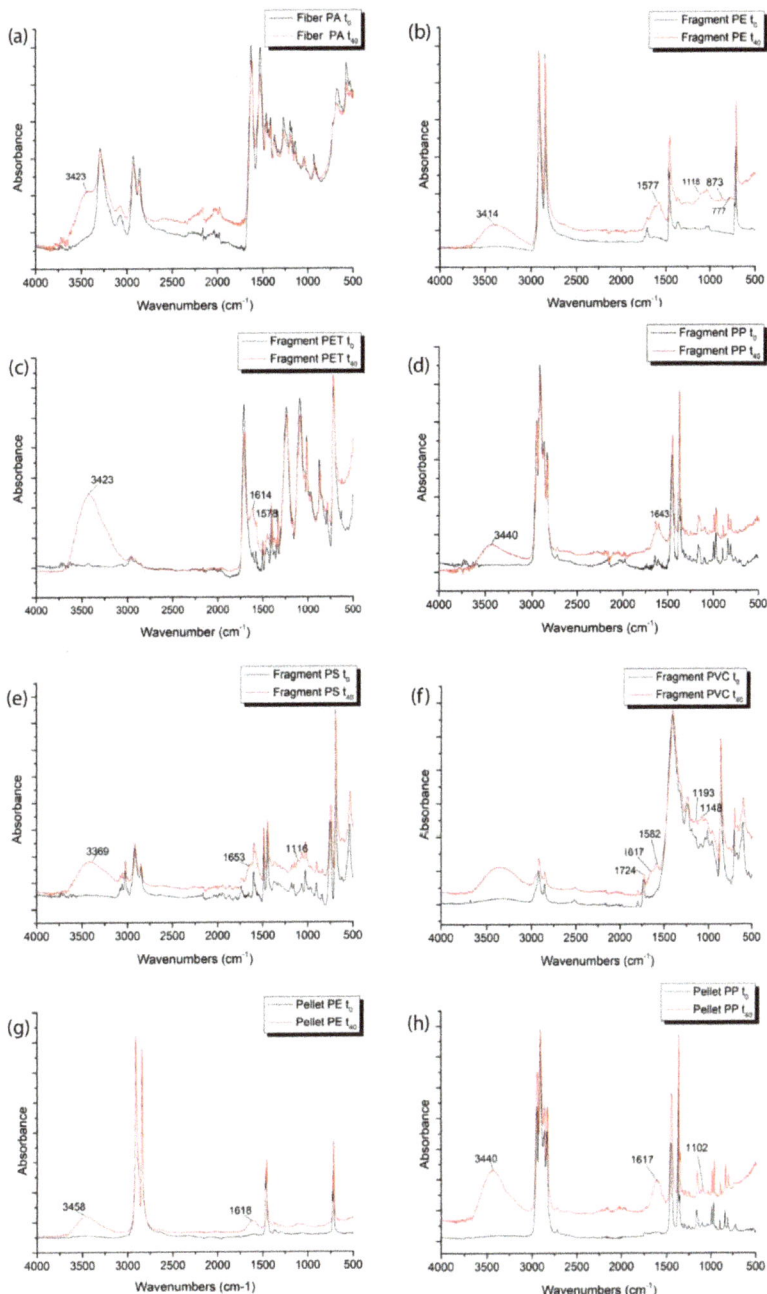

Figure 5. Comparison between the IR spectra of different polymers before (black lines, t_0) and after 40 days of ageing (red lines, t_{40}). New peaks formed after ageing are indicated in the spectra of each polymer. Absorption areas related to ageing from 3100 to 3700 cm^{-1} (hydroxyl groups) are evident in all polymers. (**a**) PA fiber; (**b**) PE fragment (**c**) PET fragment; (**d**) PP fragment; (**e**) PS fragment; (**f**) PVC fragment; (**g**) PE pellet; (**h**) PP pellet.

Otherwise, the following peaks appear in the IR spectrum of PP fragments and pellets after ageing: 3404, 3440 cm^{-1}, and 1643 cm^{-1} (OH bending) (Figure 5d–h). Moreover, the PP pellets show a new band at 1102 cm^{-1}. The IR spectra of PS show changes corresponding to the formation of new bands at 3369 cm^{-1}, 1653 cm^{-1}, and 1116 cm^{-1} (Figure 5e). Regarding the analysis of the IR spectrum of weathered PVC, a broad peak of moderate intensity can be detected in the region 3000–3500 cm^{-1} and new peaks, with respect to virgin materials, are evident at 1617 cm^{-1}, 1582 cm^{-1}, 1193 cm^{-1}, and 1148 cm^{-1} (Figure 5f).

3.2.2. Scanning Electron Microscopy (SEM) Acquisition

Scanning Electron Microscopy acquisition shows the physical effects of three different oxidative treatments on the integrity of both virgin and aged particles.

Virgin MPs appear compact and solid, with a three-dimensional structure and smooth surfaces. The treatment at 30 °C (treatment c) generates a dimensional reduction of PET MPs associated with margins corrosion (Figure 6a). An expansion of the PVC (Figure 6b), showing its surface damaged by small holes, is also visible together with the PP and PS particles abrasion (Figure S1).

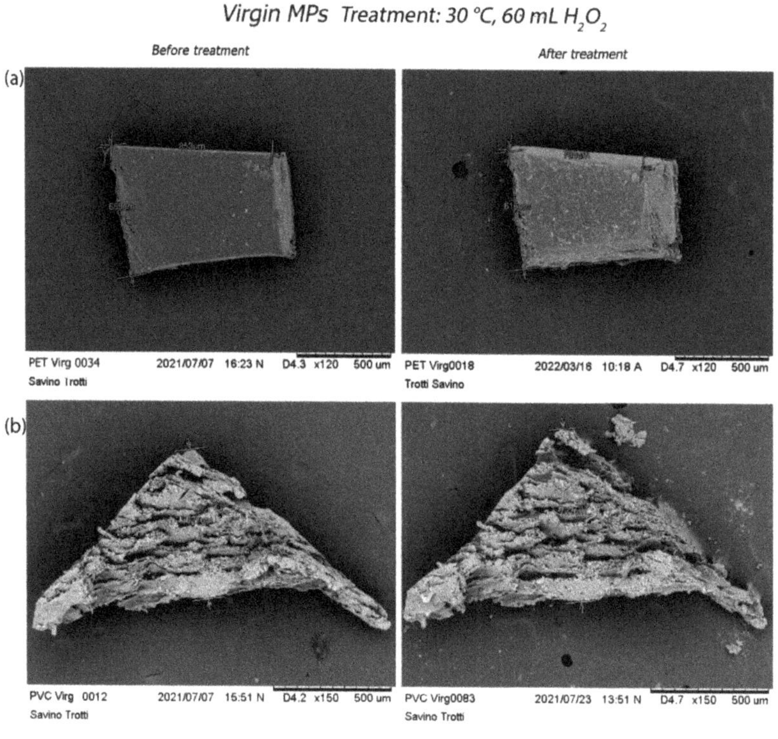

Figure 6. Alterations caused by treatment at 30 °C on virgin PET and PVC: (**a**) size reduction and corrosion of virgin PET margins, (from 853 to 708 µm); (**b**) PVC dimensional expansion from 627 µm to 722 µm.

The treatment at 50 °C (treatment b) affects PVC and PS particles by forming holes, material loss, and corrosion (Figure 7). Even in this case, a slight reduction in the size of PET and PP fragments is visible (Figure S2).

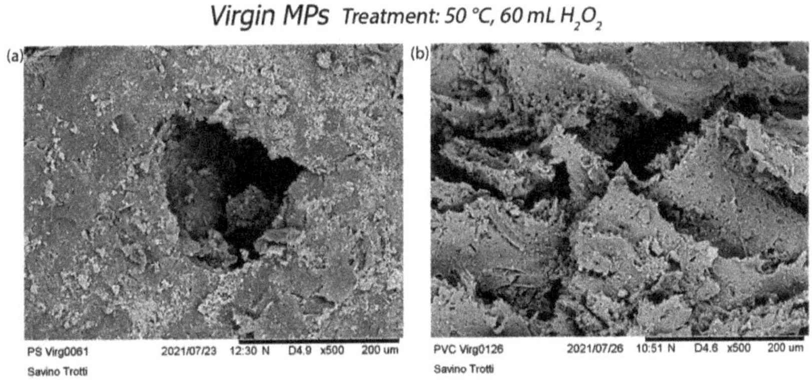

Virgin MPs Treatment: 50 °C, 60 mL H$_2$O$_2$

Figure 7. Effects of treatment at 50 °C on virgin PS and PVC: (**a**) formation of a hole on the surface of the PS; (**b**) PVC particle surface with small holes.

However, the greatest surface changes occur after treatment at 75 °C (treatment a) (Figure S3). On the one hand, PVC particles manifest wide holes (Figure 8) and a lost material of PS fragment. On the other hand, PP and PET fragments show corroded margins (Figure 9). In each treatment, virgin PE and PP pellets (Figure 10) and PE fragments highlight high resistance to oxidative digestion (Figure S4).

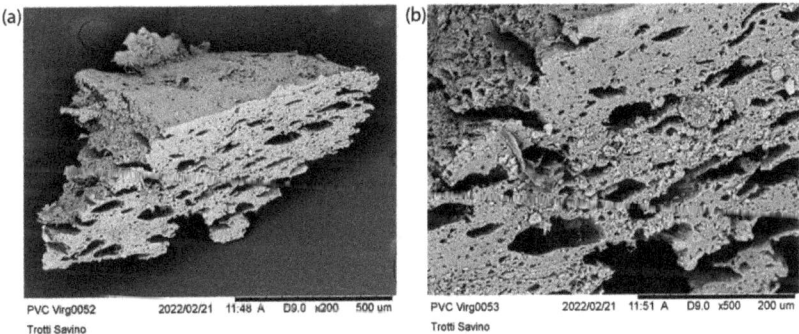

Virgin MPs Treatment: 75 °C, 100 mL H$_2$O$_2$

Figure 8. Effects of oxidative digestion treatment at 75 °C on the virgin PVC: (**a**) morphological acquisition of the entire PVC particle after treatment; (**b**) detail of large holes inside the PVC particle.

Compared to virgin standards, aged MPs appear, before the digestion treatments, flattened, brittle, with undefined shapes with some cracks on their surface.

As well as for virgin MPs, the milder and intermediate treatments do not produce strong changes in aged particles. In all treatments, a fraying of the fibre from the initial tangle is evident.

The treatment at 30 °C generates new cracks in PVC and PS fragments, and accentuates those already present, due to weathering, in PE and PET (Figure S5).

The treatment at 50 °C produces curling of PE fragments, a fraying of the fibre, and many cracks in PET ones. Even in this case, PVC particles show a surface full of small holes (Figure 11).

Figure 9. Corrosive treatment effects on PET and PP particles: comparison of size measurements, before and after treatment, emphasizes the corrosion of virgin PET (**a**) and PP (**b**).

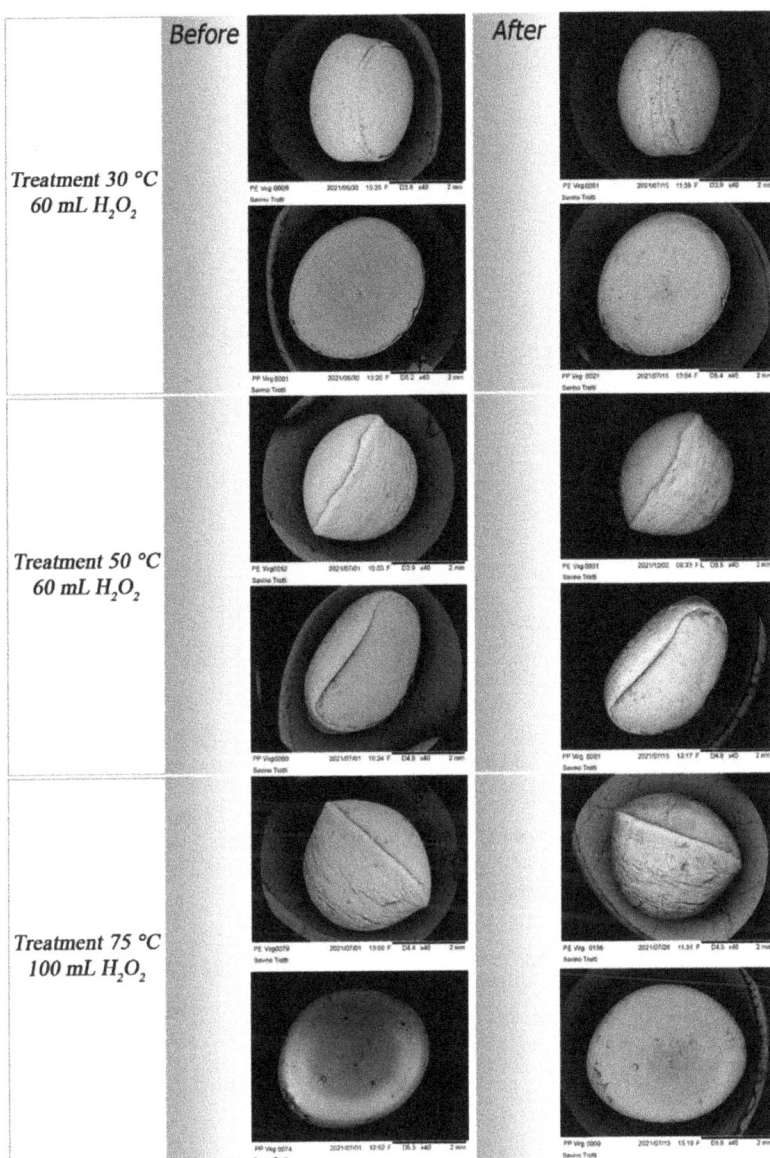

Figure 10. Morphological aspects of pellets before and after different treatments. Virgin PE and PP pellets highlight high resistance to oxidative digestion.

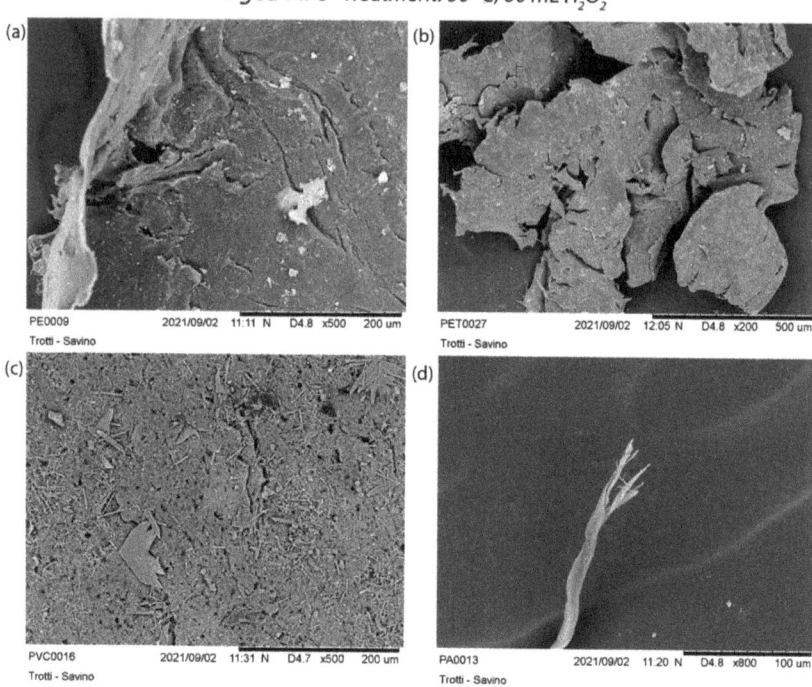

Figure 11. Focus on the effects of treatment at 50 °C on aged particles: (**a**) detail of PE curling; (**b**) more cracks in the PET; (**c**) small holes on the PVC surface; (**d**) fraying of PA fibre.

The digestion treatment at 75 °C, applied on aged MPs, causes a radical alteration of most particles (Figure S6) with evident changes such as the loss of material of PVC, PS, and PP particles, cracks expansion of PET, PE corrosion, and a fraying of PA (Figure 12). Aged pellets show several abrasions on their surface, especially after the most aggressive treatment at 75 °C (Figure 13).

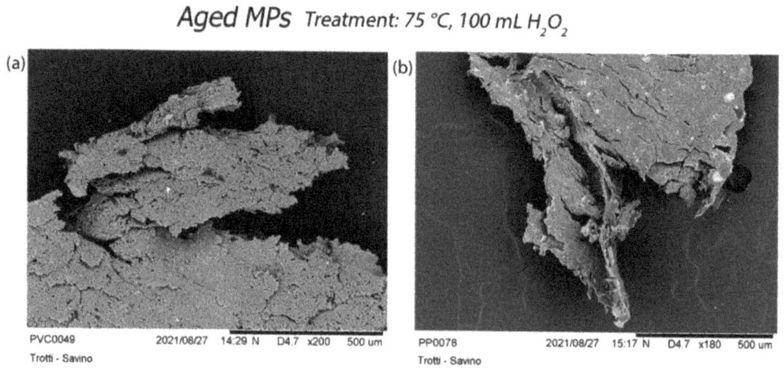

Figure 12. *Cont.*

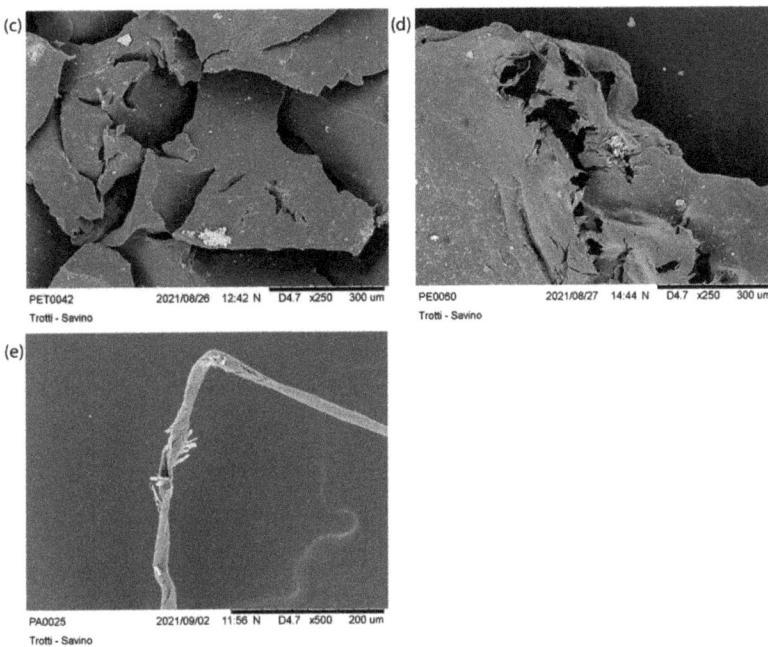

Figure 12. Focus on the effects of treatment at 75 °C on aged particles: (**a**) small holes on the PVC surface and loss of the polymer material; (**b**) corrosion and loss of PP polymer material; (**c**) formation of large cracks in PET; (**d**) corrosion and loss of PE polymer material; (**e**) breaking and fraying of the fibre.

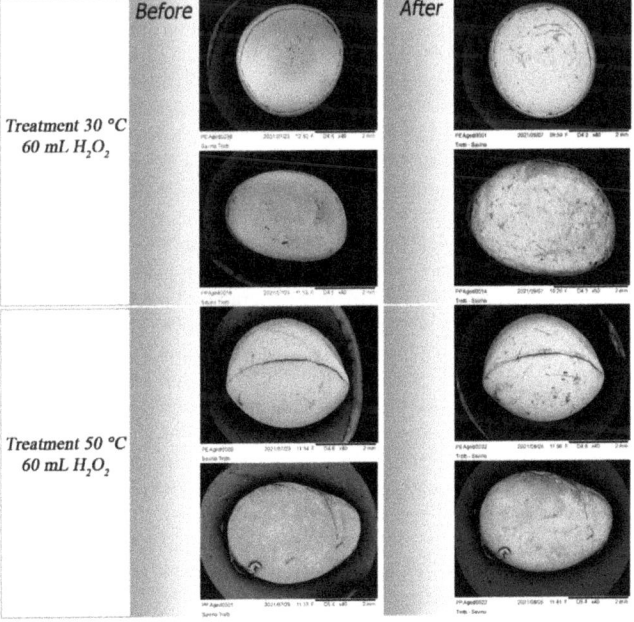

Figure 13. *Cont.*

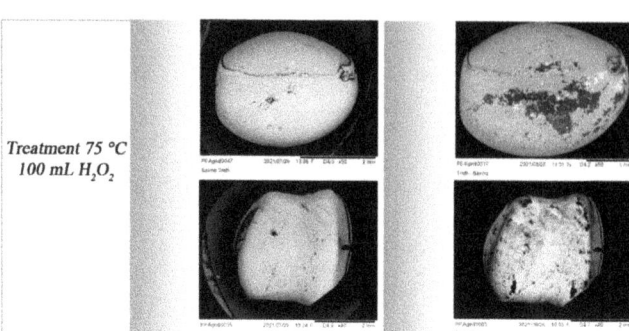

Figure 13. Morphological aspects of aged pellets before and after different treatments: abraded areas are highlighted as the temperature increases.

4. Discussion

Chemical digestion treatment is a crucial step of MPs analysis, especially when environmental samples are rich in organic matter; indeed, removing the natural debris promotes the subsequent extraction of MPs from the matrix. However, the choice of the protocol must consider its digestion effectiveness and, at the same time, the integrity of polymers, so as not to compromise the identification and quantification of particles.

In this regard in the present study, attention has been paid to the impact that one of the most common digestion protocols, based on Fenton's reaction, could have on MPs in terms of recoveries and integrity of particles.

In experiment number one, we hypothesized that the recovery of virgin MPs may be subjected to the type of sample in which the particles are dispersed. Although the density of the NaI solution, adopted for MPs extraction, is higher than high-density polymer standards (e.g., PVC and PET), it was not always possible to fully recover all particles.

Most low-density polymers were recovered almost always at 100%, instead the PET of dimensions from 5 mm to 500 microns, and the PVC of dimensions between 1 mm and 500 microns, were recovered with low efficiency because they remained inside the matrix. A similar result was also observed in a previous study [42], where small PVC fragments were trapped in the sediment after the density separation, but the protocol used does not correspond to this study. Nevertheless, the recoveries of these two polymers were also demonstrated to be low in other studies [43,44]. This behaviour may depend on the effect of the oxidation of the polymer surface, which leads to increased hydrophilicity, reducing the number of possible air bubbles and the buoyancy of the material [43,45]. Previous studies have already reported a relationship between size and recovery rate highlighting that separation of particles with diameters lower than one mm is more difficult than large MPs [27,46,47]. Electrostatic interactions between MPs and other particles depend on the morphology, composition, size, and surface charge of MPs [48]. Therefore, the presence of the matrix can interfere with the MPs extraction phase from the soil and our study suggests this with the recovery of 100% of PET, in the absence of soil.

Our results also highlight the presence of more PVC particles in the suspension and in the soil than initially added. It is appended both in the presence of the matrix and in the absence, after all treatments, especially at 75 °C using 100 mL of H_2O_2, suggesting the influence of temperature in polymer fragmentation. Indeed, an increase in the temperature of the digestion treatment caused the loss of some types of MPs also in previous studies [27,28]. The temperature is an important factor in the extraction procedure as it can affect the characteristics of a polymer based on its glass transition temperature (Tg), which is the critical temperature at which a material changes its features from "glassy material" to "rubbery material" [49].

In experiment number two, we obtained a visual demonstration of how Fenton's reagent impacts MPs. In this case, both virgin and aged particles were analyzed with the addition of PE and PP pellets, and PA fibres. Various previous studies tested digestion protocols on polymers but pellets are the most used type followed by fragments and fibres [27–29,50,51].

The greater number of small PVC particles counted could be explained by observing the formation of large pores with a slight expansion of the size of the PVC polymer. These pores, favoured by an increase in temperature, may generate a release of tiny fragments of PVC such as those found in the soil. The presence of larger cavities has been reported in [52] for some polymers after the Fenton oxidation process.

SEM analysis also shows a size reduction and margin corrosion of PET and PP, after treatment. Even in [53] particles of PE and PP show a size reduction of about 10%, after long treatment with H_2O_2 but in our case, virgin PE showed high resistance to treatment along with pellets. The corrosion of PET margins and the formation of roughness may have favoured the creation of active sites on the polymer increasing the adhesion of soil particles on the MPs surface. In addition, a particle size reduction could affect those results that include the evaluation of MPs size ranges because particles of a lower size range may not be detected because they are strongly aggravated by the treatment.

As for the virgin PS, particles were damaged at 50 °C and 75 °C as reported by [27]. We have not observed any major changes in this polymer despite the known damage caused by a temperature from 70 °C to 100 °C on the PA [29,54]. However, our data cannot be fully compared with these studies because we show a morphological alteration of the particles and not an assessment of weight.

The results observed so far suggest that the oxidative treatment generates an impact on virgin particles caused by high temperatures. However, environmental MPs are particles already altered and damaged due to biotic and abiotic degradation [35,55,56].

Indeed, the FTIR acquisitions highlighted the formation of new peaks in the polymer spectra consecutively to the ageing of virgin polymers in artificial conditions. As previously observed by other authors [57–59], the regions reflecting ageing-related changes (hydroxyl groups, peaks from 3100 to 3700 cm^{-1}, carbon double bonds, 1600 and 1680 cm^{-1} and carbonyl groups, 1690 and 1810 cm^{-1}) appeared greatly modified compared to the pristine materials in each type of MPs.

Several previous works conducted under simulated environmental conditions and on field-collected samples showed that photo and thermal oxidation together with humidity could alter the physicochemical structure of MPs leading to the introduction of oxygen into the polymer chain with the formation of carbonyl (CO) and hydroxyl (OH) functional groups [60–63].

As a matter of fact, in the spectra of our degraded MPs, exposed to UVA followed by a period of incubation at 45 °C in dry conditions, new broad hydroxyl peaks (centred at 3300–3400 cm^{-1}), appeared in the aged particles and were the most readily identified. The forced weathering, resulted, in the concomitant occurrence of oxidation reactions, chain degradation, and the formation of surface cracks and fractures. This alteration could allow an easier and deeper infiltration of water and oxygen from the atmosphere into the sample leading, in time, to an increased effect of ageing [64]. However, our weathered samples have been dried immediately before FTIR acquisition, therefore, the OH bands origin could be minimally related to the presence of adsorbed humidity from the atmosphere and more likely to the polymeric alcohols, used as a lubricant in plastic, or other by-products easily released during the degradation process [5].

Moreover, the entire region between 1550 and 1810 cm^{-1} usually referred to as the "carbonyl groups" is indicative of oxidized carbon in the plastic hydrocarbon chain. The presence of carbonyls, in almost all our weathered polymers, even if less distinctive compared to hydroxyl groups, suggests that oxygen has bonded with the hydrocarbon chain [60,65,66].

Indeed, the use of a "carbonyl index" (CI) is frequently used to measure the light-induced photo-oxidation, since it increases with increasing exposure time of plastic into the environment or progressive ageing of MPs [57,59,67].

Moreover, in our experiments, new peaks are present almost all in the fingerprint region (1500–500 cm^{-1}).

However, the spectral changes during these processes are not yet fully understood and it is complex to monitor and predict these changes.

The increased peaks around 1760 and 1690 in PVC, PP, and PE fragment spectra, indicates the presence of carbonyl groups formed by photo-oxidation in the climate chamber [57,67].

Therefore, the assessment of methodological protocols for MPs studies should consider the intrinsic physicochemical difference between virgin and aged starting materials preferring the use of weathered particles to simulate the real environmental conditions. To our knowledge, only one study used aged particles for protocol testing [18].

Compared to virgin particles, PA fibres and PE fragments, already damaged by the ageing process, showed signs of breakage and fraying. The impact of the treatment temperature was strongly evident on the aged particles because it caused the loss of polymeric material and a strong alteration of all particles. Even aged pellets showed a rough surface already after the milder treatments until to show the most evident abrasions after the digestion at 75 °C.

5. Conclusions

This study evaluated the impact of the most common oxidative digestion protocol used to extract MPs from environmental samples. We have provided a visual demonstration of particle alterations by SEM analysis and we find that this technique can help in the evaluation of protocols. Wet peroxide oxidation is an effective organic digestion method for different environmental matrices but our results showed morphological changes in polymers not observed in other studies. Both virgin and aged MPs were damaged, especially after treatment at 75 °C. Several factors must be considered in the assessment of experimental conditions. Furthermore, methods recommending temperatures below 50 °C should be preferred at the expense of longer digestion times. We also suggest evaluating the type of matrix, particle size, and shape and exploring a broad type of polymer in methodological analysis, to ensure a comprehensive assessment of the impact on MPs. Moreover, the weathering of particles in simulated environmental conditions, showed great alterations of FTIR spectra influencing the correct polymer characterization of real environmental samples and making difficult the interpretation of spectra. The degree of degradation is probably connected to the time of exposure of particles to environmental weather conditions.

Nonetheless, the evaluation of methodological approaches on aged MPs rather than on pristine materials is essential to ensure a more realistic vision of obtained results and a better quality of the final data.

Supplementary Materials: The following supporting information can be downloaded at: https://www.mdpi.com/article/10.3390/polym14101958/s1, Figure S1. Detail on the effects of treatment at 30 °C on polymers: (a) slight damage to the PS; (b) formation of small holes on the surface of the PVC; (c) corrosion of PET margins; (d) abrasion and corrosion of PP margins. Figure S2. Slight size alteration caused by treatment at 50 °C on virgin PP (a) and PET (b). Figure S3. Overview of virgin polymer morphology before and after treatment at 75 °C: loss of polymer material in PS; formation of large holes in the inside of PVC; corrosion of PP and PET margins, high resistance to PE and PA treatment. Figure S4. Visual demonstration of the high resistance of virgin PE after treatment at 30 °C (a), treatment at 50 °C (b), and 75 °C (c). Figure S5. Focus on damage caused by treatment at 30 °C on aged MPs: formation of cracks and breaks on PVC (a); PS (b); PET (c); and PE (d). Figure S6. Overview of aged polymer morphology before and after treatment at 75 °C: loss of material of PVC, PS, and PP particles, cracks expansion of PET, PE corrosion, and a fraying of PA.

Author Contributions: Conceptualization, C.C. and I.S.; methodology, C.C., I.S. and P.T.; formal analysis, I.S. and P.T.; investigation, I.S.; resources, V.F.U. and C.M.; data curation, I.S. and C.C.; writing—original draft preparation, I.S. and C.C.; writing—review and editing, C.C.; visualization,

I.S. and P.T.; supervision, C.C. and G.C.; project administration, C.M.; funding acquisition, C.M. and V.F.U. All authors have read and agreed to the published version of the manuscript.

Funding: This research received no external funding.

Institutional Review Board Statement: Not applicable.

Informed Consent Statement: Not applicable.

Data Availability Statement: Not applicable.

Acknowledgments: The authors gratefully acknowledge Domenico Bellifemine for his precious support in technical improvements in laboratory equipment and Francesco Porcelli for his support and availability in SEM acquisitions.

Conflicts of Interest: The authors declare no conflict of interest.

References

1. Frias, J.P.G.L.; Nash, R. Microplastics: Finding a consensus on the definition. *Mar. Pollut. Bull.* **2019**, *138*, 145–147. [CrossRef] [PubMed]
2. Gasperi, J.; Wright, S.L.; Dris, R.; Collard, F.; Mandin, C.; Guerrouache, M.; Langlois, V.; Kelly, F.J.; Tassin, B. Microplastics in air: Are we breathing it in? *Curr. Opin. Environ. Sci. Health* **2018**, *1*, 1–5. [CrossRef]
3. Kanhai, L.D.K.; Gårdfeldt, K.; Lyashevska, O.; Hassellöv, M.; Thompson, R.C.; O'Connor, I. Microplastics in sub-surface waters of the Arctic Central Basin. *Mar. Pollut. Bull.* **2018**, *130*, 8–18. [CrossRef] [PubMed]
4. Campanale, C.; Galafassi, S.; Savino, I.; Massarelli, C.; Ancona, V.; Volta, P.; Uricchio, V.F. Microplastics pollution in the terrestrial environments: Poorly known diffuse sources and implications for plants. *Sci. Total Environ.* **2022**, *805*, 150431. [CrossRef]
5. Campanale, C.; Dierkes, G.; Massarelli, C.; Bagnuolo, G.; Uricchio, V.F. A Relevant Screening of Organic Contaminants Present on Freshwater and Pre-Production Microplastics. *Toxics* **2020**, *8*, 100. [CrossRef]
6. Batel, A.; Borchert, F.; Reinwald, H.; Erdinger, L.; Braunbeck, T. Microplastic accumulation patterns and transfer of benzo[a]pyrene to adult zebrafish (*Danio rerio*) gills and zebrafish embryos. *Environ. Pollut.* **2018**, *235*, 918–930. [CrossRef]
7. Barboza, L.G.A.; Lopes, C.; Oliveira, P.; Bessa, F.; Otero, V.; Henriques, B.; Raimundo, J.; Caetano, M.; Vale, C.; Guilhermino, L. Microplastics in wild fish from North East Atlantic Ocean and its potential for causing neurotoxic effects, lipid oxidative damage, and human health risks associated with ingestion exposure. *Sci. Total Environ.* **2020**, *717*, 134625. [CrossRef]
8. Campanale, C.; Massarelli, C.; Savino, I.; Locaputo, V.; Uricchio, V.F. A Detailed Review Study on Potential Effects of Microplastics and Additives of Concern on Human Health. *Int. J. Environ. Res. Public Health* **2020**, *17*, 1212. [CrossRef]
9. Cormier, B.; Gambardella, C.; Tato, T.; Perdriat, Q.; Costa, E.; Veclin, C.; Le Bihanic, F.; Grassl, B.; Dubocq, F.; Kärrman, A.; et al. Chemicals sorbed to environmental microplastics are toxic to early life stages of aquatic organisms. *Ecotoxicol. Environ. Saf.* **2021**, *208*, 111665. [CrossRef]
10. Campanale, C.; Savino, I.; Pojar, I.; Massarelli, C.; Uricchio, V.F. A Practical Overview of Methodologies for Sampling and Analysis of Microplastics in Riverine Environments. *Sustainability* **2020**, *12*, 6755. [CrossRef]
11. Massarelli, C.; Campanale, C.; Uricchio, V.F. A handy open-source application based on computer vision and machine learning algorithms to count and classify microplastics. *Water* **2021**, *13*, 2104. [CrossRef]
12. Lusher, A.L.; Munno, K.; Hermabessiere, L.; Carr, S. Isolation and Extraction of Microplastics from Environmental Samples: An Evaluation of Practical Approaches and Recommendations for Further Harmonization. *Appl. Spectrosc.* **2020**, *74*, 1049–1065. [CrossRef] [PubMed]
13. Li, Q.; Wu, J.; Zhao, X.; Gu, X.; Ji, R. Separation and identification of microplastics from soil and sewage sludge. *Environ. Pollut.* **2019**, *254*, 113076. [CrossRef]
14. Masura, J.; Baker, J.; Foster, G.; Arthur, C. Laboratory Methods for the Analysis of Microplastics in the Marine Environment: Recommendations for Quantifying Synthetic Particles in Waters and Sediments. NOAA Technical Memorandum NOS-OR&R-48. 2015. Available online: https://repository.library.noaa.gov/view/noaa/10296 (accessed on 1 January 2022).
15. von Friesen, L.W.; Granberg, M.E.; Hassellöv, M.; Gabrielsen, G.W.; Magnusson, K. An efficient and gentle enzymatic digestion protocol for the extraction of microplastics from bivalve tissue. *Mar. Pollut. Bull.* **2019**, *142*, 129–134. [CrossRef] [PubMed]
16. Bretas Alvim, C.; Bes-Piá, M.A.; Mendoza-Roca, J.A. Separation and identification of microplastics from primary and secondary effluents and activated sludge from wastewater treatment plants. *Chem. Eng. J.* **2020**, *402*, 126293. [CrossRef]
17. Thiele, C.J.; Hudson, M.D.; Russell, A.E. Evaluation of existing methods to extract microplastics from bivalve tissue: Adapted KOH digestion protocol improves filtration at single-digit pore size. *Mar. Pollut. Bull.* **2019**, *142*, 384–393. [CrossRef]
18. Prata, J.C.; da Costa, J.P.; Girão, A.V.; Lopes, I.; Duarte, A.C.; Rocha-Santos, T. Identifying a quick and efficient method of removing organic matter without damaging microplastic samples. *Sci. Total Environ.* **2019**, *686*, 131–139. [CrossRef]
19. Pfohl, P.; Roth, C.; Meyer, L.; Heinemeyer, U.; Gruendling, T.; Lang, C.; Nestle, N.; Hofmann, T.; Wohlleben, W.; Jessl, S. Microplastic extraction protocols can impact the polymer structure. *Microplast. Nanoplast.* **2021**, *1*, 8. [CrossRef]
20. Naidoo, T.; Goordiyal, K.; Glassom, D. Are Nitric Acid (HNO3) Digestions Efficient in Isolating Microplastics from Juvenile Fish? *Water Air Soil Pollut.* **2017**, *228*, 470. [CrossRef]

21. Stock, F.; Kochleus, C.; Bänsch-Baltruschat, B.; Brennholt, N.; Reifferscheid, G. Sampling techniques and preparation methods for microplastic analyses in the aquatic environment—A review. *TrAC Trends Anal. Chem.* **2019**, *113*, 84–92. [CrossRef]
22. Löder, M.G.J.; Imhof, H.K.; Ladehoff, M.; Löschel, L.A.; Lorenz, C.; Mintenig, S.; Piehl, S.; Primpke, S.; Schrank, I.; Laforsch, C.; et al. Enzymatic Purification of Microplastics in Environmental Samples. *Environ. Sci. Technol.* **2017**, *51*, 14283–14292. [CrossRef] [PubMed]
23. Kühn, S.; van Werven, B.; van Oyen, A.; Meijboom, A.; Bravo Rebolledo, E.L.; van Franeker, J.A. The use of potassium hydroxide (KOH) solution as a suitable approach to isolate plastics ingested by marine organisms. *Mar. Pollut. Bull.* **2017**, *115*, 86–90. [CrossRef] [PubMed]
24. Mbachu, O.; Jenkins, G.; Pratt, C.; Kaparaju, P. Enzymatic purification of microplastics in soil. *MethodsX* **2021**, *8*, 101254. [CrossRef] [PubMed]
25. Tagg, A.S.; Harrison, J.P.; Ju-Nam, Y.; Sapp, M.; Bradley, E.L.; Sinclair, C.J.; Ojeda, J.J. Fenton's reagent for the rapid and efficient isolation of microplastics from wastewater. *Chem. Commun.* **2017**, *53*, 372–375. [CrossRef]
26. Zobkov, M.; Belkina, N.; Kovalevski, V.; Zobkova, M.; Efremova, T.; Galakhina, N. Microplastic abundance and accumulation behavior in Lake Onego sediments: A journey from the river mouth to pelagic waters of the large boreal lake. *J. Environ. Chem. Eng.* **2020**, *8*, 104367. [CrossRef]
27. Hurley, R.R.; Lusher, A.L.; Olsen, M.; Nizzetto, L. Validation of a Method for Extracting Microplastics from Complex, Organic-Rich, Environmental Matrices. *Environ. Sci. Technol.* **2018**, *52*, 7409–7417. [CrossRef]
28. Munno, K.; Helm, P.A.; Jackson, D.A.; Rochman, C.; Sims, A. Impacts of temperature and selected chemical digestion methods on microplastic particles. *Environ. Toxicol. Chem.* **2018**, *37*, 91–98. [CrossRef]
29. Pfeiffer, F.; Fischer, E.K. Various Digestion Protocols Within Microplastic Sample Processing—Evaluating the Resistance of Different Synthetic Polymers and the Efficiency of Biogenic Organic Matter Destruction. *Front. Environ. Sci.* **2020**, *8*, 263. [CrossRef]
30. Enders, K.; Lenz, R.; Beer, S.; Stedmon, C.A. Extraction of microplastic from biota: Recommended acidic digestion destroys common plastic polymers. *ICES J. Mar. Sci.* **2017**, *74*, 326–331. [CrossRef]
31. Gulizia, A.M.; Brodie, E.; Daumuller, R.; Bloom, S.B.; Corbett, T.; Santana, M.M.F.; Motti, C.A.; Vamvounis, G. Evaluating the Effect of Chemical Digestion Treatments on Polystyrene Microplastics: Recommended Updates to Chemical Digestion Protocols. *Macromol. Chem. Phys.* **2022**, 2100485. [CrossRef]
32. Alfonso, M.B.; Takashima, K.; Yamaguchi, S.; Tanaka, M.; Isobe, A. Microplastics on plankton samples: Multiple digestion techniques assessment based on weight, size, and FTIR spectroscopy analyses. *Mar. Pollut. Bull.* **2021**, *173*, 113027. [CrossRef] [PubMed]
33. Kallenbach, E.M.F.; Hurley, R.R.; Lusher, A.; Friberg, N. Chitinase digestion for the analysis of microplastics in chitinaceous organisms using the terrestrial isopod Oniscus asellus L. as a model organism. *Sci. Total Environ.* **2021**, *786*, 147455. [CrossRef] [PubMed]
34. Dordevic, D.; Necasova, L.; Antonic, B.; Jancikova, S.; Tremlová, B. Plastic cutlery alternative: Case study with biodegradable spoons. *Foods* **2021**, *10*, 1612. [CrossRef] [PubMed]
35. Zhang, K.; Hamidian, A.H.; Tubić, A.; Zhang, Y.; Fang, J.K.H.; Wu, C.; Lam, P.K.S. Understanding plastic degradation and microplastic formation in the environment: A review. *Environ. Pollut.* **2021**, *274*, 116554. [CrossRef]
36. Lang, K.; Bhattacharya, S.; Ning, Z.; Sánchez-Leija, R.J.; Bramson, M.T.K.; Centore, R.; Corr, D.T.; Linhardt, R.J.; Gross, R.A. Enzymatic Polymerization of Poly(glycerol-1,8-octanediol-sebacate): Versatile Poly(glycerol sebacate) Analogues that Form Monocomponent Biodegradable Fiber Scaffolds. *Biomacromolecules* **2020**, *21*, 3197–3206. [CrossRef]
37. Chamas, A.; Moon, H.; Zheng, J.; Qiu, Y.; Tabassum, T.; Jang, J.H.; Abu-Omar, M.; Scott, S.L.; Suh, S. Degradation Rates of Plastics in the Environment. *ACS Sustain. Chem. Eng.* **2020**, *8*, 3494–3511. [CrossRef]
38. Lear, G.; Maday, S.D.M.; Gambarini, V.; Northcott, G.; Abbel, R.; Kingsbury, J.M.; Weaver, L.; Wallbank, J.A.; Pantos, O. Microbial abilities to degrade global environmental plastic polymer waste are overstated. *Environ. Res. Lett.* **2022**, *17*, 043002. [CrossRef]
39. Glaser, J.A. Biological Degradation of Polymers in the Environment. In *Plastics in the Environment*; IntechOpen: London, UK, 2019.
40. Fairbrother, A.; Hsueh, H.-C.; Kim, J.H.; Jacobs, D.; Perry, L.; Goodwin, D.; White, C.; Watson, S.; Sung, L.-P. Temperature and light intensity effects on photodegradation of high-density polyethylene. *Polym. Degrad. Stab.* **2019**, *165*, 153–160. [CrossRef]
41. Lyngsie, G.; Krumina, L.; Tunlid, A.; Persson, P. Generation of hydroxyl radicals from reactions between a dimethoxyhydroquinone and iron oxide nanoparticles. *Sci. Rep.* **2018**, *8*, 10834. [CrossRef]
42. Monteiro, S.S.; Rocha-Santos, T.; Prata, J.C.; Duarte, A.C.; Girão, A.V.; Lopes, P.; Cristovão, J.; da Costa, J.P. A straightforward method for microplastic extraction from organic-rich freshwater samples. *Sci. Total Environ.* **2022**, *815*, 152941. [CrossRef]
43. Grause, G.; Kuniyasu, Y.; Chien, M.-F.; Inoue, C. Separation of microplastic from soil by centrifugation and its application to agricultural soil. *Chemosphere* **2022**, *288*, 132654. [CrossRef] [PubMed]
44. Duong, T.T.; Le, P.T.; Nguyen, T.N.H.; Hoang, T.Q.; Ngo, H.M.; Doan, T.O.; Le, T.P.Q.; Bui, H.T.; Bui, M.H.; Trinh, V.T.; et al. Selection of a density separation solution to study microplastics in tropical riverine sediment. *Environ. Monit. Assess.* **2022**, *194*, 65. [CrossRef] [PubMed]
45. Pongstabodee, S.; Kunachitpimol, N.; Damronglerd, S. Combination of three-stage sink–float method and selective flotation technique for separation of mixed post-consumer plastic waste. *Waste Manag.* **2008**, *28*, 475–483. [CrossRef] [PubMed]

46. Cashman, M.A.; Ho, K.T.; Boving, T.B.; Russo, S.; Robinson, S.; Burgess, R.M. Comparison of microplastic isolation and extraction procedures from marine sediments. *Mar. Pollut. Bull.* **2020**, *159*, 111507. [CrossRef]
47. Nguyen, B.; Claveau-Mallet, D.; Hernandez, L.M.; Xu, E.G.; Farner, J.M.; Tufenkji, N. Separation and Analysis of Microplastics and Nanoplastics in Complex Environmental Samples. *Acc. Chem. Res.* **2019**, *52*, 858–866. [CrossRef]
48. Brewer, A.; Dror, I.; Berkowitz, B. The Mobility of Plastic Nanoparticles in Aqueous and Soil Environments: A Critical Review. *ACS ES&T Water* **2021**, *1*, 48–57. [CrossRef]
49. He, J.; Liu, W.; Huang, Y.-X. Simultaneous Determination of Glass Transition Temperatures of Several Polymers. *PLoS ONE* **2016**, *11*, e0151454. [CrossRef]
50. Catarino, A.I.; Thompson, R.; Sanderson, W.; Henry, T.B. Development and optimization of a standard method for extraction of microplastics in mussels by enzyme digestion of soft tissues. *Environ. Toxicol. Chem.* **2017**, *36*, 947–951. [CrossRef]
51. Treilles, R.; Cayla, A.; Gaspéri, J.; Strich, B.; Ausset, P.; Tassin, B. Impacts of organic matter digestion protocols on synthetic, artificial and natural raw fibers. *Sci. Total Environ.* **2020**, *748*, 141230. [CrossRef]
52. Hu, K.; Zhou, P.; Yang, Y.; Hall, T.; Nie, G.; Yao, Y.; Duan, X.; Wang, S. Degradation of Microplastics by a Thermal Fenton Reaction. *ACS ES&T Eng.* **2021**, *2*, 110–120. [CrossRef]
53. Nuelle, M.-T.; Dekiff, J.H.; Remy, D.; Fries, E. A new analytical approach for monitoring microplastics in marine sediments. *Environ. Pollut.* **2014**, *184*, 161–169. [CrossRef] [PubMed]
54. Duan, J.; Han, J.; Zhou, H.; Lau, Y.L.; An, W.; Wei, P.; Cheung, S.G.; Yang, Y.; Tam, N.F. Development of a digestion method for determining microplastic pollution in vegetal-rich clayey mangrove sediments. *Sci. Total Environ.* **2020**, *707*, 136030. [CrossRef] [PubMed]
55. Fotopoulou, K.N.; Karapanagioti, H.K. Degradation of Various Plastics in the Environment. In *Hazardous Chemicals Associated with Plastics in the Marine Environment*; Barceló, D., Kostianoy, A.G., Eds.; The Handbook of Environmental Chemistry; Springer: Cham, Switzerland; New York, NY, USA, 2017; Volume 78, pp. 71–92. [CrossRef]
56. McGivney, E.; Cederholm, L.; Barth, A.; Hakkarainen, M.; Hamacher-Barth, E.; Ogonowski, M.; Gorokhova, E. Rapid Physico-chemical Changes in Microplastic Induced by Biofilm Formation. *Front. Bioeng. Biotechnol.* **2020**, *8*, 205. [CrossRef] [PubMed]
57. Fernández-González, V.; Andrade-Garda, J.M.; López-Mahía, P.; Muniategui-Lorenzo, S. Impact of weathering on the chemical identification of microplastics from usual packaging polymers in the marine environment. *Anal. Chim. Acta* **2021**, *1142*, 179–188. [CrossRef] [PubMed]
58. Prata, J.C.; Reis, V.; Paço, A.; Martins, P.; Cruz, A.; da Costa, J.P.; Duarte, A.C.; Rocha-Santos, T. Effects of spatial and seasonal factors on the characteristics and carbonyl index of (micro)plastics in a sandy beach in Aveiro, Portugal. *Sci. Total Environ.* **2020**, *709*, 135892. [CrossRef]
59. Rodrigues, M.O.; Abrantes, N.; Gonçalves, F.J.M.; Nogueira, H.; Marques, J.C.; Gonçalves, A.M.M. Spatial and temporal distribution of microplastics in water and sediments of a freshwater system (Antuã River, Portugal). *Sci. Total Environ.* **2018**, *633*, 1549–1559. [CrossRef]
60. Brandon, J.; Goldstein, M.; Ohman, M.D. Long-term aging and degradation of microplastic particles: Comparing in situ oceanic and experimental weathering patterns. *Mar. Pollut. Bull.* **2016**, *110*, 299–308. [CrossRef]
61. Miranda, M.N.; Sampaio, M.J.; Tavares, P.B.; Silva, A.M.T.; Pereira, M.F.R. Aging assessment of microplastics (LDPE, PET and uPVC) under urban environment stressors. *Sci. Total Environ.* **2021**, *796*, 148914. [CrossRef]
62. Xu, J.L.; Thomas, K.V.; Luo, Z.; Gowen, A.A. FTIR and Raman imaging for microplastics analysis: State of the art, challenges and prospects. *TrAC Trends Anal. Chem.* **2019**, *119*, 115629. [CrossRef]
63. Liu, P.; Qian, L.; Wang, H.; Zhan, X.; Lu, K.; Gu, C.; Gao, S. New Insights into the Aging Behavior of Microplastics Accelerated by Advanced Oxidation Processes. *Environ. Sci. Technol.* **2019**, *53*, 3579–3588. [CrossRef]
64. Gulmine, J.V.; Akcelrud, L. FTIR characterization of aged XLPE. *Polym. Test.* **2006**, *25*, 932–942. [CrossRef]
65. Lacoste, J.; Carlsson, D.J. Gamma-, photo-, and thermally-initiated oxidation of linear low density polyethylene: A quantitative comparison of oxidation products. *J. Polym. Sci. Part A Polym. Chem.* **1992**, *30*, 493–500. [CrossRef]
66. Socrates, G. *Infrared and Raman Characteristic Group Frequencies: Tables and Charts—George Socrates—Google Libri*, 3rd ed.; Chichester, E., Ed.; John Wiley & Sons, Ltd.: Hoboken, NJ, USA, 2004.
67. Veerasingam, S.; Ranjani, M.; Venkatachalapathy, R.; Bagaev, A.; Mukhanov, V.; Litvinyuk, D.; Mugilarasan, M.; Gurumoorthi, K.; Guganathan, L.; Aboobacker, V.M.; et al. Contributions of Fourier transform infrared spectroscopy in microplastic pollution research: A review. *Crit. Rev. Environ. Sci. Technol.* **2020**, *51*, 2681–2743. [CrossRef]

Article

Yellowing, Weathering and Degradation of Marine Pellets and Their Influence on the Adsorption of Chemical Pollutants

Bárbara Abaroa-Pérez [1], Sara Ortiz-Montosa [2], José Joaquín Hernández-Brito [3] and Daura Vega-Moreno [2],*

[1] Marine Litter Observatory of Reserva de la Biosfera de Fuerteventura, 35620 Las Palmas, CI, Spain; b.abaroa@observatoriodebasuramarina.com
[2] Chemistry Department, Universidad de Las Palmas de Gran Canaria (ULPGC), 35017 Las Palmas, CI, Spain; sara.ortiz101@alu.ulpgc.es
[3] Plataforma Oceánica de Canarias (PLOCAN), 35214 Telde, CI, Spain; joaquin.brito@plocan.eu
* Correspondence: daura.vega@ulpgc.es; Tel.: +34-928454429

Abstract: Marine microplastics (MPs) are exposed to environmental factors, which produce aging, weathering, surface cracking, yellowing, fragmentation and degradation, thereby changing the structure and behavior of the plastic. This degradation also has an influence on the adsorption of persistent organic pollutants over the microplastic surface, leading to increased concentration with aging. The degradation state affects the microplastic color over time; this is called yellowing, which can be quantified using the Yellowness Index (YI). Weathering and surface cracking is also related with the microplastic yellowing, which can be identified by Fourier transform infrared spectroscopy (FTIR). In this study, the degradation state of marine microplastic polyethylene pellets with different aging stages is evaluated and quantified with YI determination and the analysis of FTIR spectra. A color palette, which relates to the microplastic color and YI, was developed to obtain a visual percentage of this index. The relation with the adsorption rate of persistent organic pollutant over the microplastic surface was also determined.

Keywords: microplastic pellets; weathering; degradation; Yellowness Index; Fourier transform infrared spectroscopy; persistent organic pollutants

Citation: Abaroa-Pérez, B.; Ortiz-Montosa, S.; Hernández-Brito, J.J.; Vega-Moreno, D. Yellowing, Weathering and Degradation of Marine Pellets and Their Influence on the Adsorption of Chemical Pollutants. *Polymers* 2022, 14, 1305. https://doi.org/10.3390/polym14071305

Academic Editor: Antonio Pizzi

Received: 28 January 2022
Accepted: 18 March 2022
Published: 24 March 2022

Publisher's Note: MDPI stays neutral with regard to jurisdictional claims in published maps and institutional affiliations.

Copyright: © 2022 by the authors. Licensee MDPI, Basel, Switzerland. This article is an open access article distributed under the terms and conditions of the Creative Commons Attribution (CC BY) license (https://creativecommons.org/licenses/by/4.0/).

1. Introduction

Marine microplastics (MPs) are plastic fragments which are smaller than 5 mm [1,2] and can be present in the marine environment for long periods of time. Most of them have been in the ocean for between 5 and 25 years [3] and could have been transported thousands of kilometers in respect to the areas where they were originally dumped [4–6]. There are essentially two types: primary MPs, which were already manufactured with this size range, and secondary MPs, which have been formed from the fragmentation of larger plastics [7].

Plastics are derived from petrochemical products. Their high durability and plasticity, low cost and ease of manufacture make them widely used in everyday life. An estimated 4.9 billion tons of plastic waste has been dumped into the environment so far [8], making this product the most abundant component of marine litter [9].

Although their degradation is slow, they undergo aging, weathering and fragmentation processes [2]; factors such as ultraviolet irradiation, exposure to oxygen, salinity, physical and chemical properties of the soil, grains of sand and the presence of microorganisms influence the degree of aging of MP [10], with the weathering being more intense under marine conditions [10]. These processes progressively reduce their size while increasing their relative abundance, due to the division and fragmentation of the macroplastic and mesoplastic into smaller-sized MPs [11]. The longer they remain in the ocean, the more fragmentation the plastic undergoes [12], and the greater the change in its chemical composition [10,13–15].

The physical–chemical processes that plastic waste undergoes in the marine environment leads to a progressive degradation of the chemical structure of the polymer [13,15,16], causing a modification in the original characteristics and properties of the plastic [12,14,17] and a contamination of the seawater surrounding the marine debris [18].

1.1. Microplastic Photo-Oxidation Processes

The weathering of plastic caused by photo-oxidative processes causes changes in its chemical composition and in its physical–chemical properties, such as absorbance and reflection of light [15], which causes degradation on the surface [13,14] and changes in tonality [19].

The degradation and photo-oxidation suffered by MPs depends, among other factors, on the type and composition of the original plastic [13,20], the crystallinity, polymer type and particle size [21], as well as environmental factors, for which it has a high variability, such as the properties of the medium (salinity, pH and ionic strength), the formation of biofilms or the presence of other competing compounds (that is, organic matter) and marine conditions, where the greatest degradation suffers due to the number of influencing parameters [22].

The degradation-state identification is based on their physical characteristics, determined by microscopy [23], and on their chemical characterization, determined through methodologies such as Fourier transform infrared spectroscopy (FTIR) [13,20,24], Raman spectroscopy [25–27] or thermal analysis as pyrolysis gas chromatography–mass spectrometry (Py-GC–MS) [23,28,29]. With these methodologies, the increment of hydroxyl and carbonyl groups on different MP pellets can be determined [12,15].

The photo-oxidation of MP modifies its original color and turns it yellow; this is known as the yellowing process [15]. Although the characterization of the color of the MP is usually included in the studies carried out and is an indicator of its aging [19], it is usually based solely on the objectivity of the researcher as it is often carried out by visual identification [30].

This limits the knowledge about the states of degradation, fragmentation, or the estimation of the time in which the MP has been at the marine environment prior to its sampling [25], which is known to possibly be several decades [3]. To circumvent this limitation, various authors suggest using the Yellowness Index (YI) as an objective and quantifiable measurement parameter of the state of yellowing and degradation of microplastic [19].

1.2. Yellowness Index (YI)

The YI value (%) is based on the yellowing of the sample and increases according to MP degradation [19]. This value can be measured objectively using a colorimeter [1]. However, not all routine laboratories that analyze marine MP have this type of equipment, and there is also no single standardized protocol for its determination, causing many authors to make a visual evaluation of the color without quantifying the state of degradation of the sample.

Visually, the yellowing suffered by plastic is associated with burns, soiling and general degradation of the product [31]. According to the ASTM D 1925–70 method or the standardized method E313-15e1, the YI is a mathematical expression that allows the colorimetry in solids to be quantified ([32], pp. 1–6).

1.3. Sorption of Chemical Pollutants in Marine MP

Persistent chemical pollutants (POPs) are present in many regions of the ocean due to their high persistence and low rates of degradation [33]. This is due to the presence of aromatic rings or aliphatic halogen compounds, which also give them low solubility in water and affinity to lipophilic compounds [34].

Various studies demonstrate the ability of MP to adsorb and accumulate POPs on its surface [22,35–37]. MP acts as a transporter and accumulator of these pollutants over their surface through the ocean [38,39].

This capacity varies depending on the state of fragmentation and degradation of MP [40,41], and varies according to the plastic composition [18,22].

Age, the degree of erosion of the plastic and the chemical properties of the pollutant [21] influence the sorption of pollutants; a greater accumulation of POPs in MPs that presents greater aging and YI [30,36,40] has been found, possibly due to increased surface porosity and roughness [36], which increases the surface/volume ratio of MP due to surface cracking [22,42].

Previous work has shown that the aging process increased the surface negative charges of MPs; this increased the electrostatic attraction or repulsion, which improved the adsorption property [17,22].

In addition, although the hydrophobic interactions are abundant [17] with the degradation, the oxidation of the surface increases, which favors the adsorption of hydrophilic analytes [36,42,43]. However, there remain few studies that address the relationship between the degradation of plastics and their rate of adsorption of pollutants.

The arrival of a relevant quantity of microplastic pellets on the coast of the Canary Islands is common [44–46]. In this study, various marine MP pellet samples were collected in the region of the Canary Islands (Spain) during 2019. Their specific composition was evaluated by FTIR, and each YI value was quantitatively determined. In addition, an attempt was made to relate the adsorption rates of the POPs on these samples according to their state of aging, weathering and degradation, calculated according to their YI value and FTIR spectrum, to create a value scale that allows the quick identification of YI in the laboratory and determination of its possible impact.

2. Materials and Methods

2.1. Marine MP Samples

For this study, only primary MP, marine pellets and virgin high-density polyethylene (HDPE) microplastic pellets (CheMondis®, Köln, Germany) were used.

Pellet samples were collected on sandy beaches in the region of the Canary Islands (Gran Canaria and Fuerteventura islands, Spain) in 2019. The pellets were selected and collected directly from the sand with gloves and tweezers. Given the objectives of this study, the amount of marine MP present on the beach at the time of sampling was not counted. Other kinds of plastic and secondary MP, such as fragments or mesoplastic, were discarded. To preserve the possible presence of POPs, pellet samples were wrapped in aluminum foil and stored at a temperature of −80 °C until their subsequent analysis [30].

2.2. Equipment

To confirm the composition of each pellet, a FTIR spectrometer equipped with an Attenuated Total Reflection module (ATR-FTIR), model Cary 630 (Agilent Technologies®, Santa Clara, CA, USA), was used.

The determination of the color and the identification of the YI value of each pellet was carried out using a colorimeter, video spectrocomparator VSC5000 (Foster Freeman®, Evesham, UK).

To evaluate the adsorption rate of pollutants in pellets based on their degradation state, POPs were analyzed using GC-MS equipment (Agilent Technologies®, Santa Clara, CA, USA), model 7820A and 5977B MSD, with an HP-5MS Ultra Inert 19091S-433UI column, following a methodology for multiresidual analysis [47]. An orbital shaker was used for pellet conditioning (Supelco®, Darmstadt, Germany).

2.3. Yellowness INDEX (YI) Determination

The YI of marine MP was quantified using coordinates in a CIE three-dimensional color-space diagram (Figure 1) [1]. This system allows the brightness or luminosity of a color to be quantified (determined by Y, also called L); the color varies from black to white, with values from 0 to 100. In addition, the colorimeter records the chromatic components, represented in the CIE through the variables a and b, varying from the range of green (-a) to red (+a) and from blue (-b) to yellow (+b) [1,48]. These chromatic components can also be represented by the x and y values, determined by the cut of the Cartesian coordinates in the representation of the CIE diagram (Figure 1).

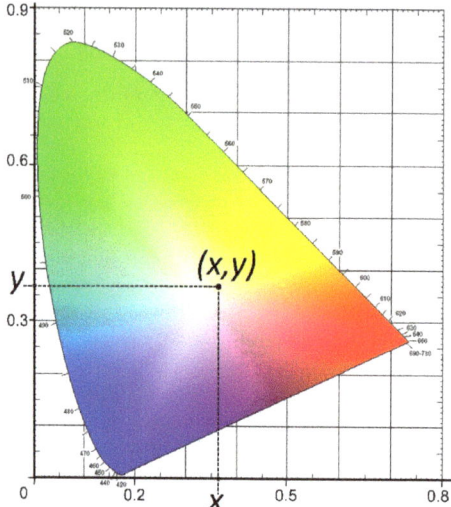

Figure 1. Representation of a CIE three–dimensional color–space diagram.

The YI, measured in percentage (%), was determined for the marine microplastic following the ASTM E131 method [31] and ([32], pp. 4–9):

$$YI\ (\%) = 100\frac{C_X X - C_Z Z}{Y} \quad (1)$$

where C_X and C_Z are constants of values 1.2769 and 1.0592, respectively.

The X and Z values are calculated using the following equations:

$$X = \tfrac{Y}{y} x$$
$$Z = \tfrac{Y}{y}(1 - x - y)$$

The values (x, y, Y) are given by the colorimeter through the CIE diagram (Figure 1). The parameters X, Z and YI are determined through the indicated equations.

2.4. Microplastic Degradation Identified by ATR-FTIR Analysis

The factors that most affect the degradation rate of the pellets are, in addition to the exposure time, environmental conditions such as UV light, salinity and oxygen exposure [13]. This photo-oxidative degradation generates variations in the chemical structure of the MP, which increases as the time the marine MPs spend in the ocean increase [15,16].

Polyethylene (PE), together with polypropylene (PP), are the most widely used hydrocarbon polymers [12]. It is a raw material, which is cheap and easy to process in a multitude of products. It has a linear chemical formula: $H(CH_2CH_2)_n H$; however, when affected by physical–chemical processes such as photo-oxidation in seawater, oxygen atoms are added to the hydrocarbon chain [12].

The PE backbones are built exclusively from single C-C bonds that do not readily undergo hydrolysis and that resist photo-oxidative degradation due to the lack of UV-visible chromophores [15]. However, weathering that occurs in PE during aging can act as chromophores [15].

Oxidation causes hydroxyl and carbonyl groups to form in the chain (see Figure 2) [12,13,15], which can be identified by FTIR. Therefore, this level of degradation can be evaluated using this technique.

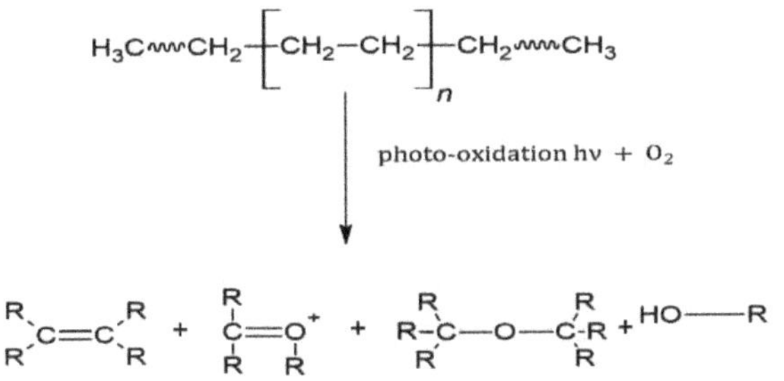

$$H_3C\text{\textasciitilde}CH_2-[CH_2-CH_2]_n-CH_2\text{\textasciitilde}CH_3$$

$$\downarrow \text{photo-oxidation } h\nu + O_2$$

$$\underset{R}{\overset{R}{>}}C=C\underset{R}{\overset{R}{<}} \quad + \quad \underset{R}{\overset{R}{>}}\overset{+}{C}=O \quad + \quad \underset{R}{\overset{R}{>}}C-O-C\underset{R}{\overset{R}{<}} \quad + \quad HO-R$$

Figure 2. Polyethylene photo-oxidation degradation process (with the carbonyl groups formation).

The Fourier transform infrared spectrometer was fitted with a golden-gate single-reflection ATR (Attenuated Total Reflection) system, a top plate attached to an optical beam having multiple optical paths. The instrument was operated in double-sided, forward–backward mode. FTIR spectra were collected over the wavenumber range of 5100–600 cm^{-1} at a resolution of 8 cm^{-1} and at a rate of 16 scans per sample.

2.5. Determination of Concentration of POPs on Marine MP

A laboratory trial with spiked samples was run to assess the POP adsorption rate according to the kind of plastic involved and its degradation state, simulating the adsorption conditions of chemical pollutants in the ocean.

The concentration of several POPs, which accumulated over different microplastic pellets, was determined. Specifically, 15 organochlorinated pesticides (OCPs), 8 polychlorinated biphenyls (PCBs) and 6 polycyclic aromatic hydrocarbons (PAHs) were analyzed, following an optimized methodology used in previous studies [49]. For POP extraction a solid–liquid–liquid microextraction (μSLLE) was applied, using 10 mL of methanol as extractant with 1 g of MP pellet in an ultrasound bath for 4 min, with the later addition of 100 μL n-hexane based on internal standards. The n-hexane extract was analyzed using GC–MS, following the analytical method described in Table S1 (Supplemental Material).

POP concentrations were determined at spiked samples for some HDPE pellets at different states of physical and chemical degradation [50]. Seawater spiked with 2 ng·L^{-1} of each of the POPs analyzed was brought into contact with 1 g of each kind of clean MP pellet (any trace of POPs was removed before the experiment) for 24 h, simulating the movement of the sea in an orbital shaker. The variability on POP concentration obtained with commercial virgin, artificial degraded and environmental degraded MP pellets (with different YI values) was studied through the same spiked conditions. Surface/volume of commercial pellets was modified by friction and sun exposure. Moreover, the POP adsorption rate was evaluated according to different levels of the YI (low YI: 25–35%; medium YI: 50–60% and high YI: 70–80%) [30].

3. Results and Discussion

3.1. ATR-FTIR Analysis

All the sampled pellets were analyzed by ATR-FTIR (between 200–220 pellets), selecting an adequate number of samples for each range of yellowing (at least three pellets per color), initially ordered from a visual point of view (see Figure 3).

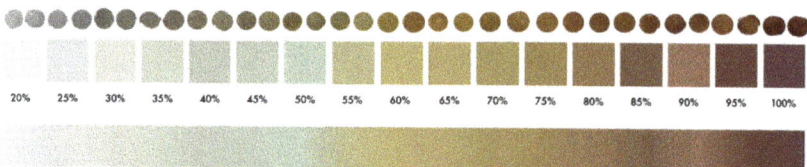

Figure 3. Color palette for Yellowness Index values (%) compared polyethylene pellets classified according to their photo-oxidation degradation process (from a visual point of view).

Among the marine pellets collected, 120 pellets identified as polyethylene (PE) compounds by ATR-FTIR analysis were selected. Other types of pellets, with a different composition to that indicated, were discarded, thus allowing us to obtain comparable samples.

The samples with more intense yellowing (oranges and browns) obtained lower values of correspondence with the PE reference FTIR spectrum when compared with the data of the spectrum library (values in percentage); nonetheless, they had a correspondence of higher than 70% [10].

In general, spectral indices are not adequate to quantify the aging time of plastics because their behaviors are not linear [10], the chemical properties of MPs change in various ways after aging depending on environmental conditions [22]; however, there is a relationship between the YI and the formation of new carbonyl groups [51].

3.2. Yellowness Index (YI) Determination for Microplastic Samples

Following the methodology previously described (Section 2.3), YI values were determined for the 120 PE pellets selected. Results obtained were compared with the color of these samples, by developing a color-grading system based on the 120 samples with values between 20–100% of YI.

Pellets with similar colors obtained similar values of YI (%). This allows for the creation of a reference color palette where the YI value (measured in percentage) is related to the color of the MP pellet (Figure 3, downloadable at https://bit.ly/microplasticYI, accessed on 22 February 2022). This makes it possible to roughly obtain the numerical value of YI of a pellet (for PE composition), without requiring specific analytical equipment, simply comparing the color of the sample with the given color palette.

3.3. Polyethylene Photo-Oxidation Process Compared with Its YI Value

The YI value gives a numerical value of the degradation state of a MP pellet, but this also can be evaluated by ATR-FTIR analysis [13].

The characteristic peaks of PE are given by two peaks at wavenumbers 2918 and 2851 cm^{-1} (Figure 4), the peaks at 1468 cm^{-1} (bending deformation) and 1373 cm^{-1} (CH_3 symmetric deformation), and another peak at 718 cm^{-1} (rocking deformation) corresponding to the native bonds present at the polymer [12].

Higher values of YI (%) indicate high degradation of the PE microplastic samples due to the photo-oxidation process suffered by the plastic, generating carbonyl groups that increase the absorbance signal in the range between 1000 and 1400 cm^{-1} of the FTIR spectrum, as previously reported [12,15] (Figure 4).

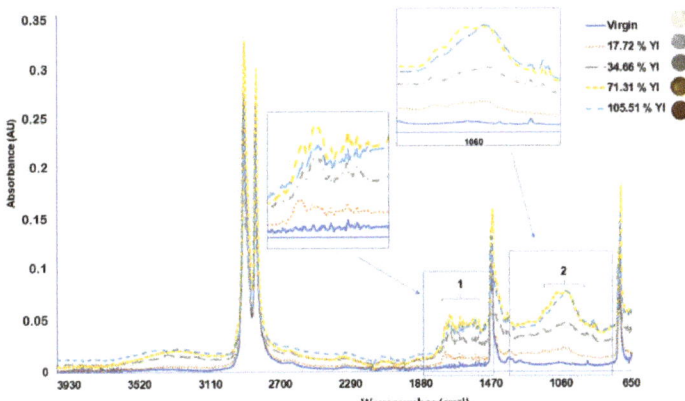

Figure 4. Polyethylene (PE) pellet FTIR-ATR spectrum with different Yellowness Index value; (**1**) PE oxidation with C=O; C=C bonds; (**2**) PE oxidation with C-O-C bonds.

3.4. Persistent Organic-Pollutant Adsorption Rate According to Microplastic Degradation Level

MP degradation and fragmentation includes two different processes between others: photo-oxidation of the plastic compound (polyethylene in this case) and the MP surface cracking, which increases the relative surface/volume on MP [52]. Although both processes are related at the marine environment, a lab trial was conducted in this study to evaluate both separately.

Seven different MP pellet samples were evaluated to assess the POPs adsorption rate depending on the degradation state of the MP: varying the relative surface/volume through physical fragmentation processes and varying the photo-oxidation degradation level using MP with different YI values.

The characteristics of each type of MP samples studied were (see Figure 5):

(a) HDPE virgin pellets;
(b) HDPE virgin pellets subjected to intense friction with metal blades of a food mixer to increase their surface/volume ratio;
(c) HDPE virgin pellets subjected to intense friction with clean beach sand to increase their surface/volume ratio;
(d) HDPE virgin pellets exposed to 1 year in seawater, in the sun under controlled conditions;
(e) Marine MP pellets collected at beaches with low YI value (25–35%);
(f) Marine MP pellets collected at beaches with medium YI value (50–60%);
(g) Marine MP pellets collected at beaches with higher YI value (70–80%).

Marine MP with different YI values collected at beaches were cleaned before spiking to ensure that they did not have additional POPs apart from those added in this experiment. For the experiment, a clean seawater without POPs (before spiked), confirmed by a seawater blank, was used.

The seven different MP pellet samples under study (1 gram each) were introduced into 100 mL of seawater spiked with the POP mixture (final concentration 2 ng·L^{-1}). They were left in an orbital shaker for 24 h at 100 rpm to simulate the conditions of ocean movement. After adsorption, desorption was carried out following a solid–liquid–liquid microextraction (μSLLE) with methanol and n-hexane [49] and analyzed by GC–MS to obtain the POP concentration determination [47].

Figure 5. Images of the analyzed pellets under a binocular magnifying glass: (**a**) virgin; (**b**) virgin with high relative surface/volume by artificial friction; (**c**) virgin with high relative surface/volume by sand; (**d**) virgin with sun degradation; (**e**) marine with low YI; (**f**) marine with medium YI; (**g**) marine with higher YI.

With results equivalent to previous studies [30,40,42,53–55], it was observed that the most degraded MP granules (with a high YI value) had a higher rate of adsorption of POPs (higher concentration of POPs under the same adsorption conditions), which is more significant in the case of OCPs in this study (Figure 6). In relation to this, the samples with the highest adsorption rate are those that have undergone a degradation process by photo-oxidation in the marine environment (Figure 6II, right). The high exposure of marine pellets to different environmental factors also causes a physical change in the microplastic surface [12–14,22,56] producing surface cracking, thus increasing the surface-to-volume ratio of the MP sample. This increase also influences the POP adsorption rate (in addition to producing the high YI value due to degradation by photo-oxidation) [17,21,36]. Figure 6I (left) shows the comparison of virgin granules and those with a high surface/volume ratio but with a low YI value (as shown in Figure 5b,c).

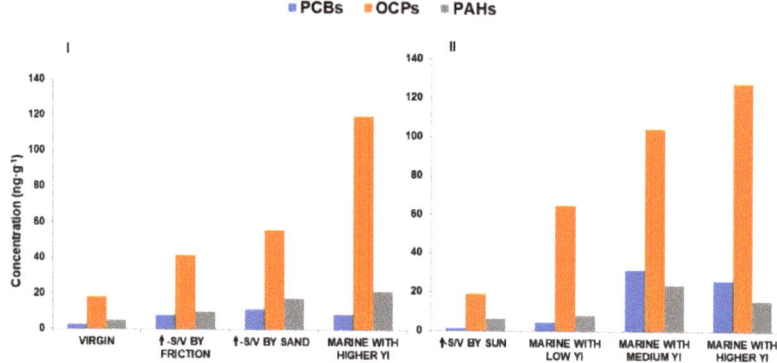

Figure 6. Total concentration of POPs (PCBs, OCPs and PAHs) adsorbed onto PE pellets obtained according to the degradation state of the MP sample. (**I**) Physical degradation and (**II**) chemical degradation.

4. Conclusions

There is a proportional relationship between the time that the MP has been exposed to the marine environment and its consequent weathering. The efficient estimation of the weathering and degradation state in marine microplastic samples is a necessity for the laboratories that analyze these types of samples.

MP yellowing is related to its degradation state by solar radiation exposure in marine environments. Although YI (%) is a parameter that allows the quantification of yellowing of a plastic, it is not widely used by the scientific community due to the necessity of specific equipment and because it is a time-consuming procedure.

Identifying a relationship between the color of a sample (visually determined), its yellowing value (YI) and the associated FTIR spectrum allows us to estimate the weathering and degradation state of an MP sample both quantitatively and qualitatively.

The proposed quantified color palette allows an approximate YI value from the MP color for PE pellets to be obtained. Moreover, samples with higher YI values presented an increase in FTIR signals in the area corresponding to the carbonyl groups, which are related to a greater degree of decomposition of PE due to the high exposure of the sample in marine environments. These results make it possible to evaluate, using FTIR analysis, the degradation state of a MP sample.

The aging, degradation, weathering, surface cracking and yellowing of marine MPs over time in the environment are all directly related to the adsorption rate of POPs onto MPs. This has been studied in PE pellets for 15 OCPs, 8 PCBs and 6 PAHs. The increment on POP adsorption rate has two different (but related) reasons: the high relative value of surface/volume due to fragmentation and MP surface cracking; and the increased age of the MPs, which have suffered processes of weathering and yellowing due to the photo-oxidation of the plastic composition. Further studies are needed to explore the effect of plastic oxidation on the adsorption of chemical pollutants in marine environments.

Supplementary Materials: The following supporting information can be downloaded at: https://www.mdpi.com/article/10.3390/polym14071305/s1, Table S1: Compounds list analyzed by GC–MS, classified as PCBs (group A), OCP (group B) or PAH (group C).

Author Contributions: Conceptualization, D.V.-M. and B.A.-P.; methodology, B.A.-P. and S.O.-M.; investigation, D.V.-M., S.O.-M. and B.A.-P.; resources, J.J.H.-B. and D.V.-M.; writing—original draft preparation, B.A.-P. and D.V.-M.; writing—review and editing, D.V.-M. and J.J.H.-B.; supervision, D.V.-M. and J.J.H.-B. All authors have read and agreed to the published version of the manuscript.

Funding: This research was funded by DeepPLAS project (Microplastics evaluation at deep water at Canary region and their chemical pollutants associated, ProID2020010030) from Canary Government with FEDER co-financing European Regional Development Fund.

Institutional Review Board Statement: Not applicable.

Informed Consent Statement: Not applicable.

Data Availability Statement: Not applicable.

Acknowledgments: Some microplastic pellet samples were provided by the Marine Litter Observatory (OBAM). This project belongs to Fuerteventura Biosphere Reserve in collaboration with the Biodiversity Foundation of the Ministry of Ecological Transition and Demographic Challenge co-financed by the PLEAMAR program of the European Maritime Fishing Fund. We would like to thank the Oceanic Platform of the Canary Islands (Plataforma Oceánica de Canarias, PLOCAN) for their support and sharing their research facilities with us.

Conflicts of Interest: The authors declare no conflict of interest. The funders had no role in the design of the study; in the collection, analyses, or interpretation of data; in the writing of the manuscript, or in the decision to publish the results.

References

1. Andrady, A.L. The plastic in microplastics: A review. *Mar. Pollut. Bull.* **2017**, *119*, 12–22. [CrossRef] [PubMed]
2. Andrady, A.L. Microplastics in the marine environment. *Mar. Pollut. Bull.* **2011**, *62*, 1596–1605. [CrossRef] [PubMed]
3. Lebreton, L.; Egger, M.; Slat, B. A global mass budget for positively buoyant macroplastic debris in the ocean. *Sci. Rep.* **2019**, *9*, 12922. [CrossRef] [PubMed]
4. Eriksen, M.; Lebreton, L.C.M.; Carson, H.S.; Thiel, M.; Moore, C.J.; Borerro, J.C.; Galgani, F.; Ryan, P.G.; Reisser, J. Plastic Pollution in the World's Oceans: More than 5 Trillion Plastic Pieces Weighing over 250,000 Tons Afloat at Sea. *PLoS ONE* **2014**, *9*, e111913. [CrossRef]
5. Kanhai, L.D.K.; Officer, R.; Lyashevska, O.; Thompson, R.C.; O'Connor, I. Microplastic abundance, distribution and composition along a latitudinal gradient in the Atlantic Ocean. *Mar. Pollut. Bull.* **2017**, *115*, 307–314. [CrossRef]
6. Nerland, I.L.; Halsband, C.; Allan, I.; Thomas, K.V. *Microplastics in Marine Environments: Occurrence, Distribution and Effects*; Norwegian Institute for Water Research: Oslo, Norway, 2014; ISBN 9788257764890.
7. Barnes, D.K.A.; Galgani, F.; Thompson, R.C.; Barlaz, M. Accumulation and fragmentation of plastic debris in global environments. *Philos. Trans. R. Soc. B Biol. Sci.* **2009**, *364*, 1985–1998. [CrossRef]
8. Geyer, R.; Jambeck, J.R.; Law, K.L. Production, use, and fate of all plastics ever made. *Sci. Adv.* **2017**, *3*, 25–29. [CrossRef]
9. Bergmann, M.; Gutow, L.; Klages, M. *Marine Anthropogenic Litter*; Springer: Cham, Switzerland, 2015; pp. 1–447. [CrossRef]
10. Fernández-González, V.; Andrade-Garda, J.M.; López-Mahía, P.; Muniategui-Lorenzo, S. Impact of weathering on the chemical identification of microplastics from usual packaging polymers in the marine environment. *Anal. Chim. Acta* **2021**, *1142*, 179–188. [CrossRef]
11. Lambert, S.; Wagner, M. Formation of microscopic particles during the degradation of different polymers. *Chemosphere* **2016**, *161*, 510–517. [CrossRef]
12. Da Costa, J.P.; Nunes, A.R.; Santos, P.S.M.; Girão, A.V.; Duarte, A.C.; Rocha-Santos, T. Degradation of polyethylene microplastics in seawater: Insights into the environmental degradation of polymers. *J. Environ. Sci. Health Part A* **2018**, *53*, 866–875. [CrossRef]
13. Cai, L.; Wang, J.; Peng, J.; Wu, Z.; Tan, X. Observation of the degradation of three types of plastic pellets exposed to UV irradiation in three different environments. *Sci. Total Environ.* **2018**, *628–629*, 740–747. [CrossRef]
14. Tang, C.C.; Chen, H.I.; Brimblecombe, P.; Lee, C.L. Morphology and chemical properties of polypropylene pellets degraded in simulated terrestrial and marine environments. *Mar. Pollut. Bull.* **2019**, *149*, 110626. [CrossRef]
15. Chamas, A.; Moon, H.; Zheng, J.; Qiu, Y.; Tabassum, T.; Jang, J.H.; Abu-Omar, M.; Scott, S.L.; Suh, S. Degradation Rates of Plastics in the Environment. *ACS Sustain. Chem. Eng.* **2020**, *8*, 3494–3511. [CrossRef]
16. Beltrán-Sanahuja, A.; Casado-Coy, N.; Simó-Cabrera, L.; Sanz-Lázaro, C. Monitoring polymer degradation under different conditions in the marine environment. *Environ. Pollut.* **2020**, *259*, 113836. [CrossRef] [PubMed]
17. Tourinho, P.S.; Ko, V.; Loureiro, S.; Gestel, C.A.M. Partitioning of chemical contaminants to microplastics: Sorption mechanisms, environmental distribution and effects on toxicity and bioaccumulation. *Environ. Pollut.* **2019**, *252*, 1246–1256. [CrossRef]
18. Rochman, C.M.; Manzano, C.; Hentschel, B.T.; Simonich, S.L.M.; Hoh, E. Polystyrene plastic: A source and sink for polycyclic aromatic hydrocarbons in the marine environment. *Environ. Sci. Technol.* **2013**, *47*, 13976–13984. [CrossRef]
19. Martí, E.; Martin, C.; Galli, M.; Echevarría, F.; Duarte, C.M.; Cózar, A. The Colors of the Ocean Plastics. *Environ. Sci. Technol.* **2020**, *54*, 6594–6601. [CrossRef]
20. Brandon, J.; Goldstein, M.; Ohman, M.D. Long-term aging and degradation of microplastic particles: Comparing in situ oceanic and experimental weathering patterns. *Mar. Pollut. Bull.* **2016**, *110*, 299–308. [CrossRef]
21. Fred-Ahmadu, O.H.; Bhagwat, G.; Oluyoye, I.; Benson, N.U.; Ayejuyo, O.O.; Palanisami, T. Interaction of chemical contaminants with microplastics: Principles and perspectives. *Sci. Total Environ.* **2020**, *706*, 135978. [CrossRef]
22. Gao, L.; Fu, D.; Zhao, J.; Wu, W.; Wang, Z.; Su, Y.; Peng, L. Microplastics aged in various environmental media exhibited strong sorption to heavy metals in seawater. *Mar. Pollut. Bull.* **2021**, *169*, 112480. [CrossRef]
23. Shim, W.J.; Hong, S.H.; Eo, S.E. Identification methods in microplastic analysis: A review. *Anal. Methods* **2017**, *9*, 1384–1391. [CrossRef]
24. Mecozzi, M.; Pietroletti, M.; Monakhova, Y.B. FTIR spectroscopy supported by statistical techniques for the structural characterization of plastic debris in the marine environment: Application to monitoring studies. *Mar. Pollut. Bull.* **2016**, *106*, 155–161. [CrossRef] [PubMed]
25. Birch, Q.T.; Potter, P.M.; Pinto, P.X.; Dionysiou, D.D.; Al-Abed, S.R. Isotope ratio mass spectrometry and spectroscopic techniques for microplastics characterization. *Talanta* **2021**, *224*, 121743. [CrossRef] [PubMed]
26. Edo, C.; Tamayo-Belda, M.; Martínez-Campos, S.; Martín-Betancor, K.; González-Pleiter, M.; Pulido-Reyes, G.; García-Ruiz, C.; Zapata, F.; Leganés, F.; Fernández-Piñas, F.; et al. Occurrence and identification of microplastics along a beach in the Biosphere Reserve of Lanzarote. *Mar. Pollut. Bull.* **2019**, *143*, 220–227. [CrossRef]
27. Enders, K.; Lenz, R.; Stedmon, C.A.; Nielsen, T.G. Abundance, size and polymer composition of marine microplastics ≥10 µm in the Atlantic Ocean and their modelled vertical distribution. *Mar. Pollut. Bull.* **2015**, *100*, 70–81. [CrossRef]
28. González-Pérez, J.A.; Jiménez-Morillo, N.T.; de la Rosa, J.M.; Almendros, G.; González-Vila, F.J. Pyrolysis-gas chromatography-isotope ratio mass spectrometry of polyethylene. *J. Chromatogr. A* **2015**, *1388*, 236–243. [CrossRef]
29. Tsuge, S.; Ohtani, H.; Watanabe, C. *Pyrolysis–GC/MS Data Book of Synthetic Polymers—Pyrograms, Thermograms and MS of Pyrolyzates*; Elsevier: Amsterdam, The Netherlands, 2011; ISBN 978-0-444-53892-5.

30. Ogata, Y.; Takada, H.; Mizukawa, K.; Hirai, H.; Iwasa, S.; Endo, S.; Mato, Y.; Saha, M.; Okuda, K.; Nakashima, A.; et al. International Pellet Watch: Global monitoring of persistent organic pollutants (POPs) in coastal waters. 1. Initial phase data on PCBs, DDTs, and HCHs. *Mar. Pollut. Bull.* **2009**, *58*, 1437–1446. [CrossRef]
31. Laboratoy, H.A. *Yellowness Indices*; Insight on Color Hunter Associates Laboratory, Inc.: Reston, VA, USA, 2008; Volume 8, pp. 5–6.
32. *ASTM Standard E313-15E01*; Standard Practice for Calculating Yellowness and Whiteness Indices from Instrumentally Measured Color Coordinates. ASTM International: West Conshohocken, PA, USA, 2015; Volume 6, pp. 1–9. [CrossRef]
33. UNEP. *Stockholm Convention on Persistent Organic Pollutants (POPs)*; United Nations Environment Programme: Nairobi, Kenya, 2009.
34. Jones, K.C.K.C.; de Voogt, P. Persistent organic pollutants (POPs): State of the science. *Environ. Pollut.* **1999**, *100*, 209–221. [CrossRef]
35. Rios, L.M.; Moore, C.; Jones, P.R. Persistent organic pollutants carried by synthetic polymers in the ocean environment. *Mar. Pollut. Bull.* **2007**, *54*, 1230–1237. [CrossRef]
36. Jiménez-Skrzypek, G.; Hernández-Sánchez, C.; Ortega-Zamora, C.; González-Sálamo, J.; González-Curbelo, M.Á.; Hernández-Borges, J. Microplastic-adsorbed organic contaminants: Analytical methods and occurrence. *TrAC Trends Anal. Chem.* **2021**, *136*, 116186. [CrossRef]
37. Lee, H.; Shim, W.J.; Kwon, J.H. Sorption capacity of plastic debris for hydrophobic organic chemicals. *Sci. Total Environ.* **2014**, *470–471*, 1545–1552. [CrossRef] [PubMed]
38. Bakir, A.; O'Connor, I.A.; Rowland, S.J.; Hendriks, A.J.; Thompson, R.C. Relative importance of microplastics as a pathway for the transfer of hydrophobic organic chemicals to marine life. *Environ. Pollut.* **2016**, *219*, 56–65. [CrossRef] [PubMed]
39. Ivar Do Sul, J.A.; Costa, M.F. The present and future of microplastic pollution in the marine environment. *Environ. Pollut.* **2014**, *185*, 352–364. [CrossRef] [PubMed]
40. Endo, S.; Takizawa, R.; Okuda, K.; Takada, H.; Chiba, K.; Kanehiro, H.; Ogi, H.; Yamashita, R.; Date, T. Concentration of polychlorinated biphenyls (PCBs) in beached resin pellets: Variability among individual particles and regional differences. *Mar. Pollut. Bull.* **2005**, *50*, 1103–1114. [CrossRef] [PubMed]
41. Frias, J.P.G.L.; Sobral, P.; Ferreira, A.M. Organic pollutants in microplastics from two beaches of the Portuguese coast. *Mar. Pollut. Bull.* **2010**, *60*, 1988–1992. [CrossRef]
42. Liu, P.; Wu, X.; Huang, H.; Wang, H.; Shi, Y.; Gao, S. Simulation of natural aging property of microplastics in Yangtze River water samples via a rooftop exposure protocol. *Sci. Total Environ.* **2021**, *785*, 147265. [CrossRef]
43. Yu, F.; Yang, C.; Zhu, Z.; Bai, X.; Ma, J. Adsorption behavior of organic pollutants and metals on micro/nanoplastics in the aquatic environment. *Sci. Total Environ.* **2019**, *694*, 133643. [CrossRef]
44. Álvarez-Hernández, C.; Cairós, C.; López-Darias, J.; Mazzetti, E.; Hernández-Sánchez, C.; González-Sálamo, J.; Hernández-Borges, J. Microplastic debris in beaches of Tenerife (Canary Islands, Spain). *Mar. Pollut. Bull.* **2019**, *146*, 26–32. [CrossRef]
45. González-Hernández, M.; Hernández Sánchez, C.; González-Sálamo, J.; Lopez-Darias, J.; Hernández-Borges, J. Monitoring of meso and microplastic debris in Playa Grande beach (Tenerife, Canary Islands, Spain) during a moon cycle. *Mar. Pollut. Bull.* **2020**, *150*, 110757. [CrossRef]
46. Hernández-Sánchez, C.; González-Sálamo, J.; Díaz-Peña, F.J.; Fraile-Nuez, E.; Hernández-Borges, J. Arenas Blancas (El Hierro island), a new hotspot of plastic debris in the Canary Islands (Spain). *Mar. Pollut. Bull.* **2021**, *169*, 112548. [CrossRef]
47. Provoost, L. *Multi-Residue Analysis of PAHs, PCBs and OCPs Using an Agilent*; Agilent Technologies, Inc.: Santa Clara, CA, USA, 2010.
48. Kirillova, N.P.; Vodyanitskii, Y.N.; Sileva, T.M. Conversion of soil color parameters from the Munsell system to the CIE-L*a*b* system. *Eurasian Soil Sci.* **2015**, *48*, 468–475. [CrossRef]
49. Abaroa-Pérez, B.; Caballero-Martel, A.; Hernández-Brito, J.J.; Vega-Moreno, D. Solid-Liquid-Liquid Microextraction (μSLLE) Method for Determining Persistent Pollutants in Microplastics. *Water Res.* **2021**, *232*, 171. [CrossRef]
50. Hidalgo-Ruz, V.; Gutow, L.; Thompson, R.C.; Thiel, M. Microplastics in the marine environment: A review of the methods used for identification and quantification. *Sci. Technol.* **2013**, *46*, 3060–3075. [CrossRef]
51. Chen, X.; Xu, M.; Yuan, L.-m.; Huang, G.; Chen, X.; Shi, W. Degradation degree analysis of environmental microplastics by micro FT-IR imaging technology. *Chemosphere* **2021**, *274*, 129779. [CrossRef]
52. Galgani, F.; Hanke, G.; Maes, T. Global Distribution, Composition and Abundance of Marine Litter. In *Marine Anthropogenic Litter*; Springer: Cham, Switzerland, 2015; pp. 1–447, ISBN 9783319165103.
53. Bakir, A.; Rowland, S.J.; Thompson, R.C. Enhanced desorption of persistent organic pollutants from microplastics under simulated physiological conditions. *Environ. Pollut.* **2014**, *185*, 16–23. [CrossRef]
54. Heskett, M.; Takada, H.; Yamashita, R.; Yuyama, M.; Ito, M.; Geok, Y.B.; Ogata, Y.; Kwan, C.; Heckhausen, A.; Taylor, H.; et al. Measurement of persistent organic pollutants (POPs) in plastic resin pellets from remote islands: Toward establishment of background concentrations for International Pellet Watch. *Mar. Pollut. Bull.* **2012**, *64*, 445–448. [CrossRef]
55. Hong, S.H.; Shim, W.J.; Hong, L. Methods of analysing chemicals associated with microplastics: A review. *Anal. Methods* **2017**, *9*, 1361–1368. [CrossRef]
56. Luo, H.; Liu, C.; He, D.; Xu, J.; Sun, J.; Li, J.; Pan, X. Environmental behaviors of microplastics in aquatic systems: A systematic review on degradation, adsorption, toxicity and biofilm under aging conditions. *J. Hazard. Mater.* **2021**, *423*, 126915. [CrossRef]

Article

Assessment of the Influence of Size and Concentration on the Ecotoxicity of Microplastics to Microalgae *Scenedesmus* sp., Bacterium *Pseudomonas putida* and Yeast *Saccharomyces cerevisiae*

Martina Miloloža [1], Kristina Bule [1], Viktorija Prevarić [1], Matija Cvetnić [1], Šime Ukić [1,*], Tomislav Bolanča [1,2] and Dajana Kučić Grgić [1,*]

[1] Faculty of Chemical Engineering and Technology, University of Zagreb, Marulićev trg 19, 10000 Zagreb, Croatia; miloloza@fkit.hr (M.M.); kbule@fkit.hr (K.B.); vprevaric@fkit.hr (V.P.); mcvetnic@fkit.hr (M.C.); tbolanca@fkit.hr (T.B.)
[2] Department for Packaging, Recycling and Environmental Protection, University North, 48000 Koprivnica, Croatia
* Correspondence: sukic@fkit.hr (Š.U.); dkucic@fkit.hr (D.K.G.); Tel.: +385-1-4597-217 (Š.U.); Fax: +385-1-4597-250 (Š.U.)

Citation: Miloloža, M.; Bule, K.; Prevarić, V.; Cvetnić, M.; Ukić, Š.; Bolanča, T.; Kučić Grgić, D. Assessment of the Influence of Size and Concentration on the Ecotoxicity of Microplastics to Microalgae *Scenedesmus* sp., Bacterium *Pseudomonas putida* and Yeast *Saccharomyces cerevisiae*. *Polymers* **2022**, *14*, 1246. https://doi.org/10.3390/polym14061246

Academic Editors: Jacopo La Nasa and Antonio Pizzi

Received: 29 January 2022
Accepted: 17 March 2022
Published: 19 March 2022

Publisher's Note: MDPI stays neutral with regard to jurisdictional claims in published maps and institutional affiliations.

Copyright: © 2022 by the authors. Licensee MDPI, Basel, Switzerland. This article is an open access article distributed under the terms and conditions of the Creative Commons Attribution (CC BY) license (https://creativecommons.org/licenses/by/4.0/).

Abstract: The harmful effects of microplastics are not yet fully revealed. This study tested harmful effects of polyethylene (PE), polypropylene (PP), polystyrene (PS), polyvinyl chloride (PVC), and polyethylene terephthalate (PET) microplastics were tested. Growth inhibition tests were conducted using three microorganisms with different characteristics: *Scenedesmus* sp., *Pseudomonas putida*, and *Saccharomyces cerevisiae*. The growth inhibition test with *Scenedesmus* sp. is relatively widely used, while the tests with *Pseudomonas putida* and *Saccharomyces cerevisiae* were, to our knowledge, applied to microplastics for the first time. The influence of concentration and size of microplastic particles, in the range of 50–1000 mg/L and 200–600 µm, was tested. Determined inhibitions on all three microorganisms confirmed the hazardous potential of the microplastics used. Modeling of the inhibition surface showed the increase in harmfulness with increasing concentration of the microplastics. Particle size showed no effect for *Scenedesmus* with PE, PP and PET, *Pseudomonas putida* with PS, and *Saccharomyces cerevisiae* with PP. In the remaining cases, higher inhibitions followed a decrease in particle size. The exception was *Scenedesmus* sp. with PS, where the lowest inhibitions were obtained at 400 µm. Finally, among the applied tests, the test with *Saccharomyces cerevisiae* proved to be the most sensitive to microplastics.

Keywords: microplastics; ecotoxicity assessment; size influence; concentration influence

1. Introduction

Plastic pollution has become a serious environmental problem [1–4]. Nevertheless, the annual world's production of plastics is continuously increasing. According to the latest report of PlasticsEurope [5], the annual plastic production for the year 2021 was 367 million metric tons, excluding the production of recycled plastics. In addition, the EU Environment Agency presented concerning projections of further production growth to over 25 billion metric tons in 2050 [6]. Looking the type of plastic polymers, polyethylene (PE), polypropylene (PP), polystyrene (PS), polyvinyl chloride (PVC) and polyethylene terephthalate (PET) account for 90% of the world production. Accordingly, these are also the most common types of plastics in the environment [7].

Plastic particles smaller than 5 mm, popularly known as microplastics (MPs), are of particular concern to scientific community [8]. MPs particles have a high hazardous potential. They can affect organisms physically (after ingestion) [9] and chemically (polymer type and chemical composition) [10], and can serve as carriers for various pollutants or as substrates for pathogenic organisms [9]. In addition, MPs particles are chemically very stable and usually non-biodegradable; therefore, they can remain in the environment for hundreds of years. Due to their small size, they are ubiquitous in practically all environments. However, their presence in the aquatic environment is of particular concern. Namely, once these contaminants enter the water, aquatic organisms feed on them, and MPs enter the food chain [11]. In addition, aquatic macrophytes have been reported to retain MPs to a significant extent, increasing thus the contamination in areas covered by aquatic grasses [12,13] and accordingly increasing the exposure of organisms living there.

The harmful nature of MPs particles has not been fully explored, but it is undeniable that there are harmful effects [7,8,14,15]. Therefore, scientists are still intensively conducting toxicity tests on various organisms. Test on the toxicity of MPs often use crustaceans *Daphnia magna* or zebrafish *Danio rerio* as test organisms [16]. Thus, ingested MPs have been reported to accumulate in body of *Daphnia magna* without significant effects on survival and reproduction for particle sizes 63–75 μm [17], while increased mortality and reduced reproductive capacity have been reported for sizes below 5 μm [18,19]. Lei et al. [20] reported that MPs caused intestinal damage in *Danio rerio*. The use of microorganisms in toxicity tests has also become popular: microorganisms are very available, it is easy to cultivate them, and their life-cycle is relatively short which provides fast results. In addition, these tests are simple, inexpensive, and provide accurate results [21]. The most commonly used microorganisms are algae and their growth inhibition is monitored as a toxicity effect [16]. A decrease in chlorophyll content and photosynthetic activity due to lower expression of photosynthetic genes, shading effect, growth inhibition, oxidative stress, physical deterioration of microalgal cells, homoaggregation and heteroaggregation have been identified as the main negative effects for microalgae [22].

Numerous factors can influence the impact of MPs on a selected organism [23]: type, shape, size and concentration of MPs, presence of additives, etc. For example, Tunali et al. [24] studied the impact of different PS concentrations on the growth and chlorophyll content of microalga *Chlorella vulgaris*; the size of PS particles was 0.5 μm. The authors reported maximal harmfulness (28.9% of growth inhibition and 21.3% decrease in chlorophyll content) at the highest concentration tested (1000 mg/L), while no effect was observed at concentrations below 50 mg/L. Zhang et al. [25] investigated the negative effect of PVC MPs on the microalga *Skeletonema costatum* during a 4-day exposure. They tested the effect of concentration at two sizes: 1 μm and 1 mm; the concentration ranges were 0–50 mg/L and 0–2000 mg/L, respectively. Concentrations of 1 μm particles had a significant effect on microalgae: higher concentrations resulted in a stronger negative effect. PVC particles of 1 mm also caused negative effect, but no significant influence of particle concentration was found. Lagarde et al. [26] tested the influence of PP and high-density PE microparticles on the freshwater microalgae *Chlamydomas reinhardtii* and confirmed that the type of MPs played a significant role in the harmfulness of plastic waste. The tested polymers acted similarly during short-term exposure: rapid colonization by *C. reinhardtii* was observed in both cases. However, a difference was reported in case of long-term exposures. In the case of PP, hetero-aggregates appeared after 20 days of contact and their size continued to increase during the experiment. In contrast, no aggregation was observed in case of high-density PE.

The aim of this work was to test the toxicity of 5 of the most-common types of MPs: PE, PP, PS, PVC, and PET. In general, hazardous substances do not have the same effects on all organisms. The effects differ not only in intensity but in mode of action as well. Therefore, in order to obtain more relevant information about the toxicity of the tested MPs, we decided to perform toxicity tests on three different microorganisms: the freshwater microalga *Scenedesmus* sp., the bacterium *Pseudomonas putida*, and the yeast *Saccharomyces cerevisiae*.

2. Materials and Methods

2.1. Design of the Experiment

In this study, five types of MPs most commonly found in the environment were used: PE, PP, PS, PVC, and PET (Figure S1). Two factors that might influence the harmfulness of MPs were tested: concentration and particle size. Five concentration levels (50, 250, 500, 750, and 1000 mg/L) and three size levels (200, 400, and 600 µm) were combined using a full factorial methodology: i.e., the experiment was designed using all possible combinations of levels of the two factors.

Three different tests for acute toxicity were performed: growth inhibition test with the microalga *Scenedesmus* sp., growth inhibition test with *Pseudomonas putida*, and yeast toxicity test with *Saccharomyces cerevisiae*. The microalga *Scenedesmus* sp. was selected because microalgae are primary producers that play an important role in the food chain [27], the bacterium *Pseudomonas putida* is commonly used as a representative of heterotrophic microorganisms in freshwaters [28], and yeasts, as eukaryotes, are considered good organisms for toxicity evaluation [29]. The selected microorganisms were easy to cultivate, they all had a short life cycle, their sensitivity to various contaminants has already been confirmed, and the cost of their toxicity tests was low.

The growth inhibition test with *Scenedesmus* sp. is relatively common. Bacterial growth inhibition tests are rarely used to determine the toxicity of MPs, and to our knowledge, this is the first study applying *Pseudomonas putida*. Further, we found no report in which *Saccharomyces cerevisiae* was used in determining the toxicity of MPs. Nomura et al. [30] applied such test to determine the toxicity of PS latex nanoparticles, which were far below the size-range tested in our study.

The standard ecotoxicity test with *Scenedesmus* sp. lasts 72 h, whereas the duration of the standard tests with *Pseudomonas putida* and *Saccharomyces cerevisiae* is 16 h. However, the tests with *Pseudomonas putida* and *Saccharomyces cerevisiae* originally refer to solutions and not suspensions, as was the case in our study. Therefore, we performed some preliminary experiments for these two tests to see if the standard methods were applicable. The preliminary experiments showed that CFU did not change within 16 h, which was not the case at 72 h. Therefore, we set an identical contact time of 72 h for all three tests.

2.2. Preparation of Microplastics

Various plastic products were used as a source of MPs: bags for PE, spoons for PP, knives for PS, packaging boxes for PVC, and bottles for PET. These products were purchased from the store. The products were first cut into smaller pieces with scissors and then ground in a cryo-mill (Retsch, Haan, Germany) with liquid nitrogen to improve the grinding. The ground particles were air-dried at room temperature for 48 h; afterwards they were sieved on stainless steel screens to obtain size classes: 100–300, 300–500, and 500–700 µm. The sieved MPs particles were stored in glass bottles. The Attenuated Total Reflectance Fourier Transform Infrared (ATR-FTIR) spectroscopic analysis (Spectrum One, Perkin Elmer, Waltham, MA, USA) was performed to verify the type of the plastics. The characteristic ATR-FTIR spectra are shown in Figure S2.

An appropriate amount of MPs particles (considering the final MPs concentrations of 50, 250, 500, 750, and 1000 mg/L) was placed in a glass flask before conducting the toxicity experiments. The flasks were filled with 70% ethanol and shaken on a rotary shaker (Unimax 1010, Heidolph, Schwabach, Germany) at 160 rpm for 10 min to sterilize the MPs particles. The sterilized particles were separated from the ethanol suspension by vacuum membrane filtration using cellulose nitrate 0.45 µm sterile filters (ReliaDisc™, Ahlstrom-Munksjö, Helsinki, Finland) and washed with sterile deionized water. Finally, the particles were quantitatively transferred into sterile flasks for the microalgae and bacteria or sterile bottles for the yeast. The transfer was performed using sterile technique.

2.3. Preparation of Test Microorganisms

Microalgae *Scenedesmus* sp. (Figure S3A) were obtained from the Ruđer Bošković Institute (Zagreb, Croatia). *Scenedesmus* sp. was activated in sterilized liquid basal medium at 25 ± 2 °C for 12/12 h light/dark cycle. Sedimentation of microalgae was prevented by aeration through a 0.45 µm sterile filter (ReliaDisc^{TM}, Ahlstrom-Munksjö, Helsinki, Finland). The number of live algal cells (shown as *Colony Forming Unit*, CFU) was determined using an optical microscope (Olympus BX50, Olympus Optical, Tokyo, Japan) with Thoma counting chamber. If necessary, the suspension was diluted to the initial number of 10^5 cells/mL.

Bacterium *Pseudomonas putida* (Figure S3B) was cultivated according to the guidelines [31]. Prior to the experiment, the culture was additionally cultivated in mineral media during 5 ± 0.5 h [32] on a rotary shaker at 160 rpm and room temperature. The optical density of the bacterial suspension was measured at 436 nm using a spectrophotometer DR/2400 (Hach, Loveland, CA, USA) and the initial value was set to 0.2 by diluting the suspension.

Yeast *Saccharomyces cerevisiae* (Figure S3C) was cultivated on yeast medium agar (3 g/L of yeast extract, 3 g/L of malt extract, 5 g/L of peptone, 10 g/L of glucose, and 15 g/L of agar; pH value was 7.0 ± 0.2) for 10–12 h at 30 ± 0.1 °C. The yeast suspension was prepared in sterile deionized water, and the initial optical density was adjusted to an absorbance value of 3.0, which was measured at 550 nm using a spectrophotometer DR/2400 (Hach, Loveland, CA, USA).

2.4. Toxicity Tests

The algal growth inhibition test with the freshwater microalga *Scenedesmus* sp. was performed according to OECD guidelines [33,34]. The test was performed in 250 mL sterile Erlenmeyer flasks on a rotary shaker at 160 rpm and 23 ± 2 °C; the working volume was 100 mL. The working flasks contained suspension of algae, basal medium, and MPs particles. The test required a control flask, which was used for comparison. The control flask did not contain MPs particles, only the algal suspension and basal medium only. The initial concentration of dissolved oxygen was 8.65 ± 0.24 mg/L, pH was 8.08 ± 0.17, and the initial CFU value was 4.2×10^5 cells/mL.

ISO guideline 10712 [31] was applied to perform the bacterial growth inhibition test with *Pseudomonas putida*. The experiments were performed in 100 mL sterile Erlenmeyer flasks on a rotary shaker at 160 rpm and 23 ± 2 °C. The working volume was 25 mL. The working flasks contained suspension of bacterium, a mineral medium (composition according to the guideline), and MPs particles. Control flasks were prepared analogously to the previously described algal test. Initial experimental conditions included dissolved oxygen at a concentration of 8.08 ± 0.22 mg/L, pH of 7.04 ± 0.90 and 5.4×10^6 cells/mL. The CFU values were determined each day during the experiments, for both the algal and bacterial tests. The value of the third-day was used to calculate the growth inhibition (INH) according to Equation (1).

$$INH = \frac{\log CFU_{CONTROL\ FLASK} - \log CFU_{WORKING\ FLASK}}{\log CFU_{CONTROL\ FLASK}} \cdot 100\% \quad (1)$$

Yeast toxicity test was based on the inhibition of saccharose fermentation by *Saccharomyces cerevisiae* [35]. The experiments were performed in hermetically sealed sterile glass bottles (working volume of 30 mL) at 28.0 ± 0.1 °C. The working bottles contained 0.6 mL of yeast suspension, 5 mL of liquid medium (composition according to Hrenovic et al. [36]), and MPs particles. Control bottles were used as well. CO_2 gas is formed during the saccharose fermentation which increases the pressure in the bottle. Therefore, a syringe was inserted through the bottle cap to collect the fluid (liquid and/or produced CO_2 gas) pressed from the bottle. The pressed volume was equal to the volume of CO_2 produced. The volume was collected daily and the three-day cumulative value was used to calculate the inhibition (INH) according to Equation (2).

$$INH = \frac{V_{CONTROL} - V_{SAMPLE}}{V_{CONTROL}} \cdot 100\% \qquad (2)$$

2.5. Response Surface Modeling

Modeling of the response (i.e., inhibition) surface was applied to define the influence of concentration (γ) and size (x) of MPs particles on the growth inhibition (INH) of each test organism. For this purpose, the size intervals of MPs were replaced by corresponding average values: 200, 400, and 600 µm. Three regression models of different complexities were applied to describe the response surface. The models were presented by Equations (3)–(5).

$$INH = a_0 + a_1\gamma + a_2 x \qquad (3)$$

$$INH = a_0 + a_1\gamma + a_2 x + a_3 \gamma^2 + a_4 x^2 \qquad (4)$$

$$INH = a_0 + a_1\gamma + a_2 x + a_3 \gamma^2 + a_4 x^2 + a_5 \gamma x \qquad (5)$$

The letter a used in these models stands for models coefficients. MODEL I (Equation (3)) contains linear contributions of the concentration and the particle size, while MODEL II (Equation (4)) is actually MODEL I extended by two quadratic terms. It is expected that first- or second-order polynomials should be adequate for description of dependent variable in most cases involving two independent variables [37]. In order to enclose eventual joint activity of concentration and particle size, MODEL III (Equation (5)) with the interaction term ($\gamma \cdot x$) was also used.

Calculations and analyses were performed using MATLAB R2010b software (MathWorks®, Natick, MA, USA).

3. Results and Discussion

Table 1 shows the experimentally determined inhibition values for predefined MPs sizes and concentrations. To statistically determine whether the concentration or the size had a significant influence on MPs harmfulness within the experimental range, the inhibition data were fitted by one linear (Equation (3)) and two polynomial regression models (Equations (4) and (5)). Statistical analysis was performed for each model applied and the results were presented in Tables 1–3. Calculations were done with 95% confidence, i.e., the significance level was 0.05. The applied regression models were compared based on related R^2_{adj} values, and the best model for each combination of applied test organism and MPs type was selected. R^2_{adj} was given primacy in front of R^2 because it is considered superior in cases where models with different number of terms need to be compared [38]. The best models were projected at Figures 1–3. Based on the best models, conclusions were drawn about the influence of MPs size and concentration.

3.1. Inhibition of Scenedesmus sp.

Experimentally determined values of growth inhibition of *Scenedesmus* sp. showed a general trend for all five plastics studied: higher concentrations resulted in more intensive inhibitions (Figure 1). No trend was evident for size variations, except perhaps for PS (Figure 1C). However, the highest inhibition values for all five plastics (Figure 1 and Table 1) were found at the lowest particle size (200 µm) and the highest concentration (1000 mg/L), suggesting that the size has an impact on the algal growth, at least at higher MPs concentrations. To check these observations, we fitted the inhibition data with three regression models (Equations (3)–(5)) and performed statistical analysis. The results are presented in Table 2. All applied models proved to be significant in describing the of variability of the dependent variable (i.e., inhibition of algal growth). Namely, high F-values and p-values below the predefined significance of 0.05 were obtained for all applied models (Table 2).

Table 1. Experimentally determined inhibition values (INH) for polyethylene (PE), polypropylene (PP), polystyrene (PS), polyvinyl chloride (PVC), and polyethylene terephthalate (PET) microplastics.

Size/ μm	Conc./ mg/L	Scenedesmus sp.					Pseudomonas putida					Saccharomyces cerevisiae				
		PE	PP	PS	PVC	PET	PE	PP	PS	PVC	PET	PE	PP	PS	PVC	PET
		INH/%					INH/%					INH/%				
200	50	3.61	3.03	4.56	2.13	4.11	1.87	2.57	3.62	4.49	2.75	49.63	42.86	50.00	63.24	35.24
	250	7.24	4.03	6.27	5.16	4.51	5.04	4.97	8.70	7.35	4.88	65.63	61.86	64.29	66.67	52.13
	500	8.08	6.55	9.26	8.43	5.89	6.53	5.65	11.60	8.56	5.31	80.31	76.29	69.05	73.33	68.09
	750	11.41	6.74	11.34	10.50	5.89	12.19	6.31	12.90	9.00	6.48	98.44	73.20	95.24	83.09	88.10
	1000	12.99	9.13	11.34	12.10	7.63	12.35	8.48	14.74	11.89	7.42	98.76	80.95	100.00	100.00	91.30
400	50	3.61	2.98	2.49	3.10	2.43	1.35	2.33	2.90	2.83	1.73	39.06	25.77	41.62	20.59	20.83
	250	4.93	4.62	3.27	4.13	3.47	5.37	4.57	4.87	4.35	4.99	58.44	54.76	52.38	32.35	43.48
	500	6.97	5.18	6.54	6.54	4.51	4.66	10.39	5.37	5.52	5.52	73.44	74.23	66.67	35.90	58.33
	750	7.81	6.55	6.94	10.23	5.89	6.67	5.23	11.60	5.83	5.30	79.06	75.26	88.10	42.86	68.64
	1000	11.15	8.22	9.26	10.5	5.89	8.00	7.17	12.49	7.37	5.72	90.63	76.19	88.59	76.92	76.19
600	50	2.73	2.70	0.72	2.24	2.43	1.54	1.93	3.10	2.33	1.02	7.810	15.46	9.520	16.67	7.450
	250	4.36	5.18	4.14	3.10	3.47	4.52	3.10	5.85	3.77	1.45	29.06	24.74	26.19	30.95	23.85
	500	5.75	6.74	6.27	4.13	4.51	5.03	3.65	8.10	5.52	1.52	40.48	57.73	54.76	33.33	37.23
	750	8.76	6.95	9.26	6.54	4.51	5.26	4.76	9.75	5.33	1.52	61.54	62.89	64.29	46.67	47.14
	1000	9.86	6.95	9.26	8.43	4.51	7.03	5.56	11.60	6.44	2.84	73.44	64.29	88.10	72.06	54.76

Table 2. Statistical analysis of the regression models (Equations (3)–(5)) for the microalgae *Scenedesmus* sp. The microalgae were exposed to microplastics (MPs) particles of: polyethylene (PE), polypropylene (PP), polystyrene (PS), polyvinyl chloride (PVC), and polyethylene terephthalate (PET). The analyzed factors were concentration and size of the particles.

MPs	No.	Model				Coefficients						
		R^2	R^2_{adj}	F	p	Coefficient	Value	Lower 95%	Upper 95%	p	Significant Term?	Influential Factor
PE	MODEL I	0.9535	0.9458	123.17	0.00	a_0	5.45	4.19	6.70	0.00		concentration & size
						$a_1\ (10^{-3})$	8.26	7.04	9.47	0.00	YES	
						$a_2\ (10^{-3})$	−5.93	−8.46	−3.41	0.00	YES	
	MODEL II	0.9618	0.9465	62.95	0.00	a_0	7.37	4.07	10.67	0.00		concentration
						$a_1\ (10^{-2})$	0.84	0.35	1.33	0.00	YES	
						$a_2\ (10^{-2})$	−1.76	−3.56	0.03	0.94		
						$a_3\ (10^{-6})$	−0.14	−4.64	4.35	0.17		
						$a_4\ (10^{-5})$	1.46	−0.76	3.68	0.05		
	MODEL III	0.9680	0.9502	54.40	0.00	a_0	6.50	2.93	10.07	0.00		concentration
						$a_1\ (10^{-2})$	1.01	0.45	1.57	0.00	YES	
						$a_2\ (10^{-2})$	−1.54	−3.34	0.25	0.08		
						$a_3\ (10^{-6})$	−0.14	−4.55	4.26	0.22		
						$a_4\ (10^{-5})$	1.46	−0.71	3.64	0.94		
						$a_5\ (10^{-5})$	−0.43	−1.17	0.31	0.16		
PP	MODEL I	0.8824	0.8628	45.01	0.00	a_0	3.25	2.02	4.48	0.00		concentration
						$a_1\ (10^{-3})$	5.19	3.99	6.38	0.00	YES	
						$a_2\ (10^{-3})$	−0.48	−2.96	2.00	0.68		
	MODEL II	0.9055	0.8677	23.96	0.00	a_0	3.79	0.58	6.99	0.02		concentration
						$a_1\ (10^{-2})$	0.80	0.33	1.28	0.00	YES	
						$a_2\ (10^{-2})$	−0.63	−2.37	1.11	0.44		
						$a_3\ (10^{-6})$	−2.70	−7.07	1.67	0.20		
						$a_4\ (10^{-5})$	0.72	−1.43	2.88	0.47		
	MODEL III	0.9279	0.8878	23.15	0.00	a_0	2.76	−0.55	6.06	0.09		concentration
						$a_1\ (10^{-2})$	1.00	0.48	1.52	0.00	YES	
						$a_2\ (10^{-2})$	−0.37	−2.04	1.30	0.63		
						$a_3\ (10^{-6})$	−2.70	−6.79	1.39	0.17		
						$a_4\ (10^{-5})$	0.72	−1.29	2.74	0.44		
						$a_5\ (10^{-5})$	−0.51	−1.19	0.18	0.13		

Table 2. Cont.

MPs	Model					Coefficients						
	No.	R^2	R^2_{adj}	F	p	Coefficient	Value	Lower 95%	Upper 95%	p	Significant Term?	Influential Factor
PS	MODEL I	0.8570	0.8332	35.95	0.00	a_0 $a_1\,(10^{-2})$ $a_2\,(10^{-2})$	5.12 0.80 −0.66	2.83 0.58 −1.12	7.41 1.03 −0.19	0.00 0.00 0.01	 YES YES	concentration & size
	MODEL II	0.9525	0.9335	50.11	0.00	a_0 $a_1\,(10^{-2})$ $a_2\,(10^{-2})$ $a_3\,(10^{-6})$ $a_4\,(10^{-5})$	1.10 1.20 −4.53 −3.78 4.84	0.71 0.63 −6.62 −9.01 2.26	1.48 1.77 −2.45 1.45 7.43	0.00 0.00 0.00 0.14 0.00	 YES YES YES	concentration & size
	MODEL III	0.9565	0.9323	39.54	0.00	$a_0\,(10)$ $a_1\,(10^{-2})$ $a_2\,(10^{-2})$ $a_3\,(10^{-6})$ $a_4\,(10^{-5})$ $a_5\,(10^{-5})$	1.17 1.06 −4.72 −3.78 4.85 0.36	0.74 0.37 −6.90 −9.14 2.20 −0.54	1.60 1.74 −2.53 1.58 7.49 1.26	0.00 0.01 0.00 0.14 0.00 0.39	 YES YES YES 	concentration & size
PVC	MODEL I	0.9331	0.9219	83.67	0.00	a_0 $a_1\,(10^{-2})$ $a_2\,(10^{-2})$	4.87 0.86 −0.69	3.26 0.71 −1.02	6.48 1.02 −0.37	0.00 0.00 0.00	 YES YES	concentration & size
	MODEL II	0.9435	0.9208	41.71	0.00	a_0 $a_1\,(10^{-2})$ $a_2\,(10^{-2})$ $a_3\,(10^{-6})$ $a_4\,(10^{-5})$	2.53 1.03 0.55 −1.62 −1.56	−1.76 0.40 −1.78 −7.47 −4.44	6.82 1.66 2.89 4.23 1.32	0.22 0.00 0.61 0.55 0.26	 YES 	concentration
	MODEL III	0.9705	0.9541	59.18	0.00	a_0 $a_1\,(10^{-2})$ $a_2\,(10^{-2})$ $a_3\,(10^{-6})$ $a_4\,(10^{-5})$ $a_5\,(10^{-5})$	0.57 1.42 1.04 −1.62 −1.56 −0.96	−3.09 0.84 −0.80 −6.14 −3.79 −1.72	4.23 2.00 2.89 2.90 0.67 −0.20	0.73 0.00 0.23 0.44 0.15 0.02	 YES YES	concentration & size
PET	MODEL I	0.8845	0.8653	45.95	0.00	a_0 $a_1\,(10^{-3})$ $a_2\,(10^{-3})$	4.75 3.17 −4.30	3.86 2.31 −6.09	5.63 4.03 −2.51	0.00 0.00 0.00	 YES YES	concentration & size
	MODEL II	0.9136	0.8790	26.42	0.00	a_0 $a_1\,(10^{-3})$ $a_2\,(10^{-2})$ $a_3\,(10^{-6})$ $a_4\,(10^{-5})$	5.46 5.23 −1.05 −1.96 0.77	3.23 1.93 −2.26 −5.01 −0.73	7.69 8.53 0.17 1.09 2.27	0.00 0.00 0.08 0.18 0.28	 YES 	concentration
	MODEL III	0.9335	0.8965	25.25	0.00	a_0 $a_1\,(10^{-2})$ $a_2\,(10^{-2})$ $a_3\,(10^{-6})$ $a_4\,(10^{-5})$ $a_5\,(10^{-6})$	4.75 0.66 −0.87 −1.96 0.77 −3.48	2.43 0.30 −2.03 −4.82 −0.64 −8.27	7.06 1.03 0.30 0.90 2.18 1.32	0.00 0.00 0.13 0.15 0.25 0.13	 YES 	concentration

The linear model (MODEL I), applied for PE experiments, successfully described 95.35% of the variance of the inhibition ($R^2 = 0.9535$), indicating the existence of a high influence of at least one of the examined factors on the algal growth (Table 2). And indeed, the confidence intervals of the estimated model-coefficients for both coefficients related to the independent variables, did not contain the value 0 and the corresponding p-values were below the predefined significance level. This implied that both factors studied: the concentration and the particle size have a significant influence on algal growth. The introduction of quadratic terms in MODEL II resulted in a better fit (higher R^2_{adj} values) and demonstrated the superiority of the model over MODEL I. The analysis of variance performed for MODEL II refuted the conclusion derived from MODEL I and revealed PE concentration as the only influential factor. The coefficient of the size-related quadratic-term had a p-value at the boundary of predefined significance. However, the associated confidence interval had a value 0 included, which gave additional conformation that size was not an influential factor. MODEL III, which had the best fit ($R^2_{adj} = 0.9502$), confirmed this statement. In addition, the inhibition surface described by MODEL III

(Figure 1A) clearly showed that higher PE concentrations were more harmfull for the algae. For the lowest concentration tested (50 mg/L), inhibitions of 3.61% or less were obtained (Figure 1A). Accordingly, it can be assumed that the concentration of PE has no significant effect at lower concentration levels. This is in agreement with report of Garrido et al. [39] who performed a similar experiment, but with the microalga *Isochrysis galbana*. Garrido et al. conducted the experiment using smaller PE particles (up to 22 μm) and lower concentrations (up to 25 mg/L) and found no harmful effects on the algae. The exposure time was identical to that in our study.

Table 3. Statistical analysis of the regression models (Equations (3)–(5)) for the bacteria *Pseudomonas putida*. The bacteria were exposed to microplastics (MPs) particles of: polyethylene (PE), polypropylene (PP), polystyrene (PS), polyvinyl chloride (PVC), and polyethylene terephthalate (PET). The analyzed factors were concentration and size of the particles.

MPs	Model					Coefficients						
	No.	R^2	R^2_{adj}	F	p	Coefficient	Value	Lower 95%	Upper 95%	p	Significant Term?	Influentia Factor
PE	MODEL I	0.8002	0.7669	24.03	0.00	a_0	5.01	2.32	7.69	0.00		concentration & size
						a_1 (10^{-2})	0.75	0.49	1.01	0.00	YES	
						a_2 (10^{-2})	-0.73	-1.27	-0.19	0.01	YES	
	MODEL II	0.8306	0.7628	12.26	0.00	a_0 (10)	0.65	-0.06	1.37	0.07		concentration
						a_1 (10^{-2})	1.25	0.18	2.31	0.03	YES	
						a_2 (10^{-2})	-2.11	-6.01	1.80	0.26		
						a_3 (10^{-5})	-0.47	-1.45	0.51	0.31		
						a_4 (10^{-5})	1.72	-3.11	6.55	0.45		
	MODEL III	0.9228	0.8799	21.52	0.00	a_0	3.04	-2.68	8.76	0.26		concentration & size
						a_1 (10^{-2})	1.93	1.03	2.83	0.00	YES	
						a_2 (10^{-2})	-1.23	-4.11	1.65	0.36		
						a_3 (10^{-5})	-0.47	-1.18	0.23	0.16		
						a_4 (10^{-5})	1.72	-1.77	5.21	0.23		
						a_5 (10^{-5})	-1.72	-2.90	-0.53	0.01	YES	
PP	MODEL I	0.9176	0.9039	66.83	0.00	a_0	4.25	3.28	5.21	0.00		concentration & size
						a_1 (10^{-3})	4.47	3.53	5.41	0.00	YES	
						a_2 (10^{-3})	-4.49	-6.44	-2.54	0.00	YES	
	MODEL II	0.9192	0.8869	28.44	0.00	a_0	3.84	1.06	6.62	0.01		concentration
						a_1 (10^{-3})	5.08	0.98	9.19	0.02	YES	
						a_2 (10^{-2})	-0.26	-1.77	1.25	0.71		
						a_3 (10^{-6})	-0.59	-4.38	3.20	0.74		
						a_4 (10^{-5})	-0.23	-2.10	1.63	0.78		
	MODEL III	0.9379	0.9035	27.20	0.00	a_0	2.95	0.08	5.83	0.04		concentration
						a_1 (10^{-2})	0.68	0.23	1.14	0.01	YES	
						a_2 (10^{-2})	-0.04	-1.49	1.41	0.95		
						a_3 (10^{-6})	-0.59	-4.14	2.97	0.72		
						a_4 (10^{-5})	-0.23	-1.99	1.52	0.77		
						a_5 (10^{-5})	-0.43	-1.03	0.62	0.13		
PS	MODEL I	0.9159	0.9019	65.37	0.00	a_0	6.30	4.22	8.38	0.00		concentration & size
						a_1 (10^{-2})	1.01	0.81	1.21	0.00	YES	
						a_2 (10^{-2})	-0.66	-1.08	-0.24	0.00	YES	
	MODEL II	0.9656	0.9519	70.26	0.00	a_0 (10)	0.68	0.29	1.06	0.00		concentration
						a_1 (10^{-2})	1.91	1.34	2.48	0.00	YES	
						a_2 (10^{-2})	-1.75	-3.85	0.35	0.09		
						a_3 (10^{-5})	-0.86	-1.38	-0.33	0.00	YES	
						a_4 (10^{-5})	1.36	-1.24	3.97	0.27		
	MODEL III	0.9723	0.9570	63.29	0.00	a_0	5.62	1.53	9.72	0.01		concentration
						a_1 (10^{-2})	2.13	1.48	2.77	0.00	YES	
						a_2 (10^{-2})	-1.47	-3.53	0.60	0.14		
						a_3 (10^{-5})	-0.86	-1.36	-0.35	0.00	YES	
						a_4 (10^{-5})	1.36	-1.13	3.86	0.25		
						a_5 (10^{-5})	-0.55	-1.40	0.29	0.17		

Table 3. Cont.

MPs	Model					Coefficients						
	No.	R^2	R^2_{adj}	F	p	Coefficient	Value	Lower 95%	Upper 95%	p	Significant Term?	Influential Factor
PVC	MODEL I	0.8626	0.8397	37.66	0.00	a_0 $a_1 (10^{-3})$ $a_2 (10^{-2})$	7.02 5.07 −0.90	5.30 3.40 −1.24	8.75 6.74 −0.55	0.00 0.00 0.00	 YES YES	concentration & size
	MODEL I	0.9340	0.9076	35.37	0.00	$a_0 (10)$ $a_1 (10^{-2})$ $a_2 (10^{-2})$ $a_3 (10^{-6})$ $a_4 (10^{-5})$	1.11 0.72 −3.53 −2.05 3.29	0.76 0.21 −5.42 −6.78 0.96	1.46 1.24 −1.64 2.68 5.63	0.00 0.01 0.00 0.36 0.01	YES YES YES YES	concentration & size
	MODEL III	0.9589	0.9360	41.97	0.00	$a_0 (10)$ $a_1 (10^{-2})$ $a_2 (10^{-2})$ $a_3 (10^{-6})$ $a_4 (10^{-5})$ $a_5 (10^{-5})$	0.97 1.00 −3.18 −2.05 3.29 −0.69	0.64 0.49 −4.81 −6.05 1.32 −1.36	1.29 1.51 −1.55 1.95 5.27 −0.02	0.00 0.00 0.00 0.28 0.00 0.04	YES YES YES YES YES	concentration & size
PET	MODEL I	0.8016	0.7686	24.25	0.00	a_0 $a_1 (10^{-3})$ $a_2 (10^{-2})$	6.00 3.13 −0.92	4.24 1.43 −1.28	7.75 4.84 −0.57	0.00 0.00 0.00	 YES YES	concentration & size
	MODEL II	0.8850	0.8389	19.23	0.00	a_0 $a_1 (10^{-2})$ $a_2 (10^{-2})$ $a_3 (10^{-6})$ $a_4 (10^{-5})$	1.77 0.61 1.34 −2.83 −2.83	−2.12 0.04 −0.77 −8.14 −5.45	5.66 1.19 3.46 2.47 −0.22	0.33 0.04 0.19 0.26 0.04	 YES YES	concentration & size
	MODEL III	0.9246	0.8827	22.07	0.00	a_0 $a_1 (10^{-2})$ $a_2 (10^{-2})$ $a_3 (10^{-6})$ $a_4 (10^{-5})$ $a_5 (10^{-5})$	0.26 0.91 1.72 −2.83 −2.83 −0.74	−3.46 0.32 −0.15 −7.43 −5.10 −1.51	3.97 1.49 3.59 1.76 −0.57 0.03	0.88 0.01 0.07 0.20 0.02 0.06	 YES YES 	concentration & size

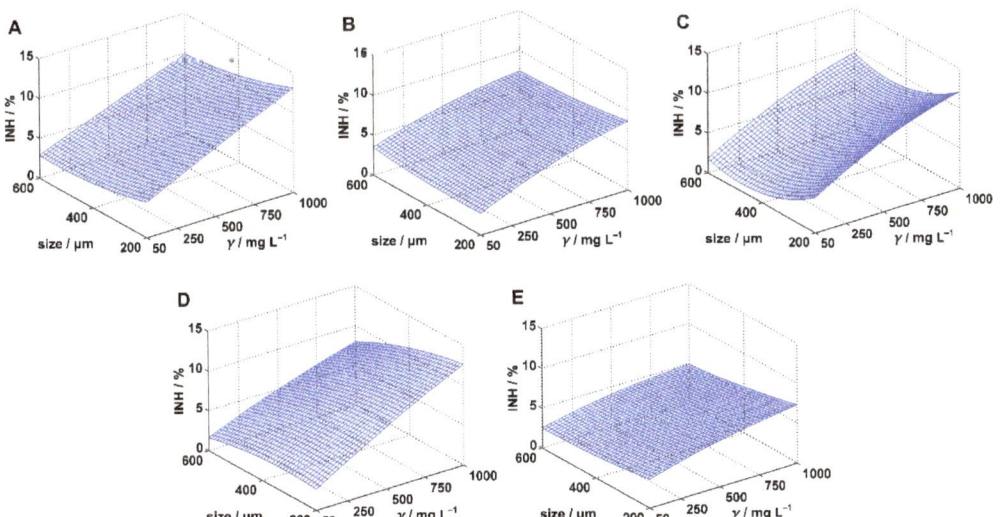

Figure 1. Inhibition surfaces estimated for *Scenedesmus* sp. by the best regression models. The cases are: (**A**) polyethylene, (**B**) polypropylene, (**C**) polystyrene, (**D**) polyvinyl chloride, and (**E**) polyethylene terephthalate.

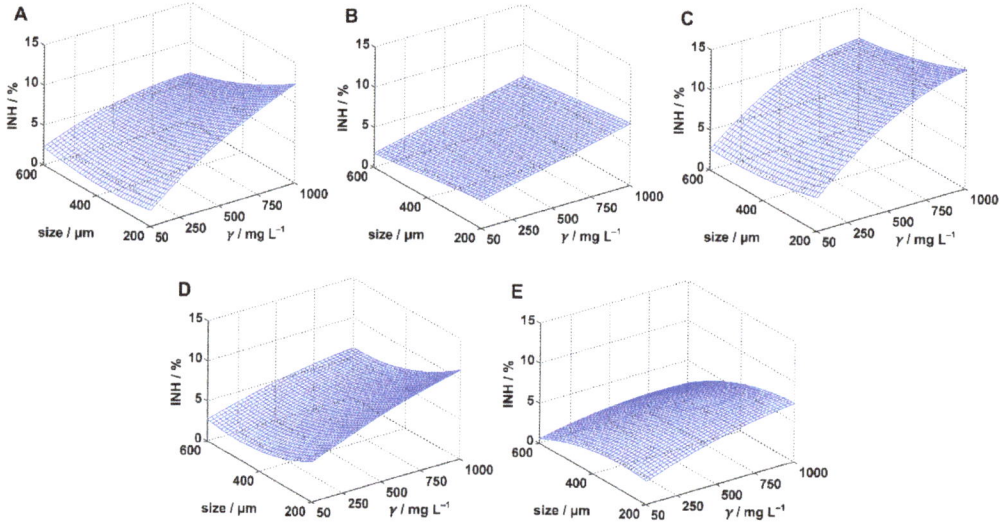

Figure 2. Inhibition surfaces estimated for *Pseudomonas putida* by the best regression models. The cases are: (**A**) polyethylene, (**B**) polypropylene, (**C**) polystyrene, (**D**) polyvinyl chloride, and (**E**) polyethylene terephthalate.

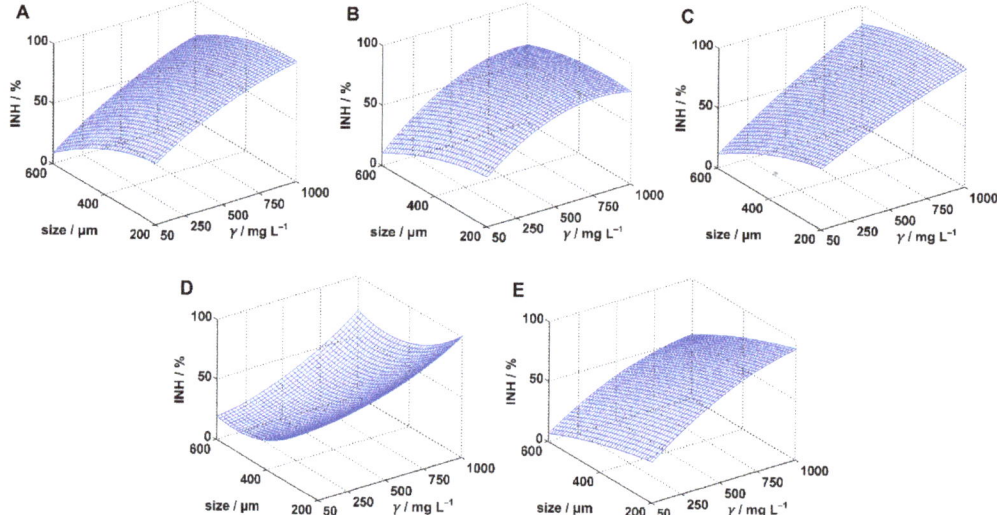

Figure 3. Inhibition surfaces estimated for *Saccharomyces cerevisiae* by the best regression models. The cases are: (**A**) polyethylene, (**B**) polypropylene, (**C**) polystyrene, (**D**) polyvinyl chloride, and (**E**) polyethylene terephthalate.

Somewhat lower values of the determination coefficients (R^2, R^2_{adj}) were observed while modeling the inhibition surface in the PP experiment (Table 2), although all three applied models were still significant. Though the best model was MODEL III, all three models gave an identical conclusion that, within the experimental range, the concentration of PP microparticles had a significant influence on the algal growth while changes in the particle size had no statistically significant effect. The influence of PP concentration is more expressed at higher concentrations (Figure 1B).

Application of MODEL I in the case of the PS experiment resulted in generally lowest percentage of the explained inhibition variance ($R^2 = 0.8570$; Table 2). However, the introduction of quadratic terms (MODEL II) greatly improved fitting of the inhibition data: the increase in R^2_{adj} was 0.1003. Both these models indicated concentration and size of PS microparticles as influential factors in *Scenedesmus* sp. inhibition. Inclusion of the interaction term in the regression model (MODEL III) did not bring additional benefit to the fitting, indicating lack of joint concentration-size activity of PS microparticles. The inhibition surface described by MODEL II is presented in Figure 1C. The surface clearly shows that higher PS concentrations are associated with higher inhibition values. For the size of PS, the surface shows the lowest inhibition for 400 μm particles. The reason why larger and smaller particles had a higher influence on the inhibition could be due to the different inhibition mechanisms. Namely, Bhattacharya et al. [40] reported that one of the reasons for inhibition of the algal growth was adsorption of the charged PS beads on the cell, which obstructed algal photosynthesis. There are two very likely types of the obstruction. The first one is a physical blockage of gas-transfer through the cell membrane (reduced uptake of CO_2) which should be more expressed for smaller particles due to more complete coverage of the cell surface. The second is a shading effect. It seems important to point out that, of the five tested MPs polymers, four were transparent while only PS had white nontransparent particles. Adsorption of such particles on the cell surface reduces the amount of light which is essential for the normal functioning of photosynthetic organisms. The shading effect should be more evident when larger particles are adsorbed. Furthermore, Besseling et al. [19] performed an experiment very similar to that described in this chapter, but for PE nanoparticles with an exact size of 0.070 μm. The test organism was the microalga *Scenedesmus obliquus*, the exposure time was the same as in our study, and the concentration interval (44–1100 mg/L) was almost identical to ours. Besseling et al. reported that increasing PS concentration resulted in increased inhibition of algal growth. Despite the fact that, based on MODEL III shape (Figure 1C), we expected higher values of inhibition in nano-size area, Besseling et al. reported much lower inhibition values. For example, at a concentration of 1000 mg/L, they achieved an inhibition of about 2.5%, whereas in our study 11.34% was obtained experimentally at the same concentration (for 200 μm-sized particles). During the experiment, Besseling et al. monitored not only the inhibition of algal growth but the level of Chlorophyll A in the algal cells as well. From the observed changes in the level of Chlorophyll A, they concluded that in the case of nano-size PS particles additional mechanism must be included in obstruction of *Scenedesmus* sp. photosynthesis, beside the previously mentioned blockage of CO_2 transfer through the cell membrane.

Interesting results were obtained in the regression analysis in the case of the PVC experiment (Table 2). Namely, MODEL I and III pointed to concentration and size of PVC microparticles as statistically influential factors, while MODEL II recognized only the concentration. Comparison of R^2_{adj} values showed the reason for this inconsistency. R^2_{adj} value of MODEL II was lower than that of MODEL I. This proved that the introduction of quadratic terms was insignificant for describing the inhibition data and made MODEL II inferior to MODEL I. In contrast, introduction of the interaction term into the regression model (MODEL III) was very beneficial for data fitting and made MODEL III the best option for the regression. According to MODEL III (Figure 1D), at the lowest concentration (50 mg/L), there is a very slight variation in inhibition values within the size-range of 200-600 μm. However, as the PVC concentration increases, the effect of the particle size variation becomes more pronounced, showing increasing harmfulness of the smaller PVC microparticles. This behavior corresponds with some similar studies [41,42] which reported higher toxicity of smaller PVC particles on algae.

MODEL I indicated a significant influence of the concentration and particle size in the case of the PET experiment. However, analysis of the other two models, which were found to be superior to MODEL I, showed that this was not the case. These two models only recognized significant influence of PET concentration. The best fitting model to describe the experimental data was MODEL III, as it gave the in best data-fit (R^2_{adj} = 0.8965). Comparison of the entire set of data presented for *Scenedesmus* sp. (Figure 1) showed that, within the experimental range, *Scenedesmus* sp. was least sensitive to PET as the exposure resulted in the lowest inhibition values. Also, the increase in inhibition with the increase in MPs concentration was least expressed in the case of PET. Recent researches [43–45] have confirmed that some microalgae can produce PET hydrolyzing enzymes called PETases and use PET as substrate. Although it is difficult to claim that this was the case in our experiment without performing a detailed analysis, especially when dealing with a 3-day exposure period, this assumption cannot be dismissed. For example, Moog et al. [43] exposed the photosynthetic microalga *Phaeodactylum tricornutum* to PET and observed a progressive increase in the concentrations of mono(2-hydroxyethyl) terephthalic acid and terephthalic acid, which are the main products of the PET hydrolysis, after only 3 days of exposure.

3.2. Inhibition of Pseudomonas putida

Exposure of the bacterium *Pseudomonas putida* to MPs particles in predefined ranges of the concentrations and particle sizes resulted in inhibition values (Figure 2) similar to those obtained for *Scenedesmus* sp. Again, higher values of maximal inhibition were determined in the PE, PS, and PVC experiments (12.35, 14.74, and 11.89%, respectively; Table 1) than in the PP and PET cases (8.48 and 7.42%, respectively). Furthermore, these maximal inhibitions were all determined at maximal MPs concentration (1000 mg/L) and the minimal size (200 μm), suggesting the potential influence of the concentration and the size of MPs particles on *Pseudomonas putida* growth. The influence of MPs concentration became even more evident after analyzing all experimental data, as an apparent trend of the inhibition increase with the increase in the concentration existed for all five applied MPs (Table 1). The influence of the particle size was not so obvious (similar to the case of *Scenedesmus* sp.), except perhaps in the cases of PVC and PET, where it appeared that smaller MPs sizes resulted in higher inhibitions. To confirm or refute all these observations, we modeled the inhibition surface (Equations (3)–(5)) and performed a statistical analysis. The results of the analysis are presented in Table 3.

Exposure of *Pseudomonas putida* to PE resulted in relatively high inhibitions (only PP resulted in higher inhibitions; Figure 2). All three regression models applied to describe the inhibition surface were statistically significant (*p*-values of the models were less than 0.05; Table 3), although some of them presented rather poor correlation (MODEL I). MODEL I and MODEL III recognized both tested factors, PE concentration and particle size, as statistically influential, while MODEL II recognized the concentration only. However, MODEL II had the lowest R^2_{adj} values, indicating its inferiority to MODEL I and especially MODEL III. Obviously, the inclusion of the quadratic terms in the case of MODEL II was not beneficial for describing the response surface. In contrast the inclusion of the interaction term (MODEL III) overcame issues of MODEL II and resulted in a huge improvement in R^2_{adj} value; consequently, MODEL III provided the best fit of the experimental data. The significance of the interaction term, i.e., of the joint concentration-size activity, was confirmed by the calculated *p*-value for the coefficient a_5 (p = 0.01) which was far below the predefined significance level. The inhibition surface plotted by MODEL III (Figure 2A) showed an interesting behavior of the inhibition. Namely, at the lowest concentration (50 mg/L), inhibition values decreased with reduction in the particle size up to 400 μm and afterwards it stagnated. However, at higher concentrations, the inhibition increased with the reduction of the particle size. Increasing of the concentration resulted in increasing of the inhibition, as observed previously.

MODEL I provided the best fit of the experimental data in the case of *Pseudomonas putida* exposure to PP microparticles (Table 3). Analysis of the model showed PP concentration and size as significant factors that affected bacterial growth. The introduction of quadratic terms (MODEL II) and the interaction term (MODEL III) reduced the suitability of the regression models to describe the inhibition surface. Graphical presentation of the best model

(Figure 2B) showed that negative influence of PP microparticles on *Pseudomonas putida* increased with higher concentrations and smaller particles sizes.

Statistically, the concentration of microparticles was the only factor that had an influence of *Pseudomonas putida* growth in the PS experiment (Table 3). MODEL I also indicated the influence of the size, but this model was inferior to MODEL II and MODEL III. MODEL III proved to be the best choice for description of the inhibition surface. Inclusion of the interaction factor in the case of MODEL III improved R^2_{adj} value for 0.0051; however, no significance of the interaction term was found (p = 0.17). The model confirmed that an increase in inhibition followed an increase in the concentration (Figure 2C). Exposure of *Pseudomonas putida* to PS microparticles resulted in the highest inhibitions (Figure 2) among the five MPs used. This is most likely related to the styrene contained in PS structure (Figure S1C). Namely, styrene is an aromatic hydrophobic compound, known to be a toxic pollutant that can cause negative effects on bacterial cells [46].

All three models applied in the case of PVC pointed to concentration and size as the influential factors. According to MODEL II, the introduction of quadratic term associated with particle size was particularly beneficial for fitting the inhibition values. This is consistent with the parabolic nature of the dependence between inhibition and the particle size seen in Figure 2D. The quadratic term associated with concentration was found to be insignificant. The best model to describe the inhibition surface was MODEL III, whose analysis showed a significance influence of the interaction term (p-value for a_5 was less than 0.05) and, accordingly, of the joint concentration-size harmful activity. Giacomucci et al. [47] reported that the bacteria *Pseudomonas citronellolis* are able to degrade PVC, and based on the reported CFU values, it is clear that biodegradation started very early, practically at the 3rd day of exposure. Therefore, it cannot be excluded that the slightly lower inhibition levels of *Pseudomonas putida* that we observed during PVC exposure were due to the onset of PVC biodegradation. However, this remains to be confirmed by future studies.

A very similar conclusion can be given for the PET experiment, as all three models pointed concentration and particle size as influential factors. Again, the inclusion of the quadratic term was significant for the size and not for the concentration. Finally, MODEL III was the best choice for describing the inhibition surface although no significant joint concentration-size activity was confirmed in this case. Based on the structure of the PET polymer, which contains a benzoic ring (Figure S1E), we expected slightly higher inhibition values than those obtained. It was known that *Pseudomonas putida* is able to metabolize ethylene glycol [48], which is one of the products of PET depolymerization [49]. Jet, although several other bacteria of the genus Pseudomonas are capable of producing PETase [50–52], to our knowledge there are no reports confirming this ability for *Pseudomonas putida*.

3.3. Inhibition of Saccharomyces cerevisiae

Toxicity experiments performed with the yeast *Saccharomyces cerevisiae* resulted in much higher inhibition values compared to experiments with *Scenedesmus* sp. and *Pseudomonas putida* (Table 1). Complete or almost complete inhibitions were obtained for four of the five MPs tested: PE, PS, PVC, and PET, indicating a high sensitivity of *Saccharomyces cerevisiae* to the presence of the selected plastic materials. These maximal values were all obtained at maximal MPs concentration (1000 mg/L) and minimal MPs size (200 μm). Therefore, it was not surprising that inhibition of *Saccharomyces cerevisiae*, considering the changes of the concentration, followed the same trend as the inhibitions of *Scenedesmus* sp. and *Pseudomonas putida*: higher MPs concentrations resulted in higher inhibitions. However, unlike the experiments performed with *Scenedesmus* sp. and *Pseudomonas putida*, there also appears to be a clear trend with respect to MPs size, with higher inhibitions obtained for smaller MPs particles. We determined whether this was true by analyzing the applied regression models (Equations (3)–(5)).

Concerning the PE experiment, all three applied models resulted in high R^2_{adj} and F-values (Table 4), confirming a good selection of regression models. For example, MODEL I, the simplest of the models used, described 94.49% of the inhibition variance; this means that only 5.51% of the variance remained unexplained by the model. MODEL III, which gave the highest R^2_{adj} value and, accordingly, best described the inhibition of the yeast growth, implied that PE concentration and size were both the influential factors. In addition, the other two models implied the same conclusion. Finally, no joint concentration-size activity was found to be significant because the p-value for the interaction term (MODEL III, coefficient a_5) was 0.09. The best model was plotted at Figure 3A. The model confirmed that concentration and size followed the trends assumed by comparing the inhibition data: higher concentrations and smaller sizes of PS microparticles favored the inhibition.

Table 4. Statistical analysis of the regression models (Equations (3)–(5)) for the yeast *Saccharomyces cerevisiae*. The yeast was exposed to microplastics (MPs) particles of: polyethylene (PE), polypropylene (PP), polystyrene (PS), polyvinyl chloride (PVC), and polyethylene terephthalate (PET). The analyzed factors were concentration and size of the particles.

MPs	Model No.	R^2	R^2_{adj}	F	p	Coefficient	Value	Lower 95%	Upper 95%	p	Significant Term?	Influential Factor
PE	MODEL I	0.9449	0.9358	102.97	0.00	a_0 (10)	6.97	5.83	8.10	0.00		concentration & size
						a_1 (10^{-2})	5.78	4.68	6.87	0.00	YES	
						a_2 (10^{-1})	−0.90	−1.13	−0.67	0.00	YES	
	MODEL II	0.9804	0.9725	124.86	0.00	a_0 (10)	3.90	1.93	5.86	0.00		concentration & size
						a_1 (10^{-1})	0.93	0.64	1.22	0.00	YES	
						a_2 (10^{-1})	0.62	−0.45	1.69	0.22		
						a_3 (10^{-5})	−3.35	−6.03	−0.67	0.02	YES	
						a_4 (10^{-4})	−1.90	−3.23	−0.58	0.01	YES	
	MODEL III	0.9860	0.9782	126.77	0.00	a_0 (10)	4.59	2.63	6.55	0.00		concentration & size
						a_1 (10^{-1})	0.79	0.48	1.10	0.00	YES	
						a_2 (10^{-1})	0.45	−0.54	1.43	0.33		
						a_3 (10^{-5})	−3.35	−5.77	−0.93	0.01	YES	
						a_4 (10^{-4})	−1.90	−3.10	−0.71	0.01	YES	
						a_5 (10^{-5})	3.42	−0.65	7.48	0.09		
PP	MODEL I	0.8100	0.7784	25.58	0.00	a_0 (10)	5.57	3.87	7.27	0.00		concentration & size
						a_1 (10^{-2})	4.72	3.07	6.37	0.00	YES	
						a_2 (10^{-2})	−5.50	−8.93	−2.07	0.00	YES	
	MODEL II	0.9387	0.9142	28.28	0.00	a_0 (10)	2.65	−0.16	5.46	0.06		concentration
						a_1 (10^{-1})	1.25	0.83	1.67	0.00	YES	
						a_2 (10^{-1})	0.49	−1.03	2.02	0.49		
						a_3 (10^{-4})	−0.74	−1.13	−0.36	0.00	YES	
						a_4 (10^{-4})	−1.30	−3.19	0.59	0.15		
	MODEL III	0.9580	0.9347	41.05	0.00	a_0 (10)	3.69	0.95	6.44	0.01		concentration
						a_1 (10^{-1})	1.05	0.61	1.48	0.00	YES	
						a_2 (10^{-1})	0.23	−1.15	1.62	0.71		
						a_3 (10^{-4})	−0.74	−1.08	−0.40	0.00	YES	
						a_4 (10^{-4})	−1.30	−2.98	0.37	0.11		
						a_5 (10^{-4})	0.51	−0.06	1.08	0.07		
PS	MODEL I	0.9334	0.9223	84.09	0.00	a_0 (10)	5.88	4.65	7.11	0.00		concentration & size
						a_1 (10^{-2})	6.32	5.13	7.52	0.00	YES	
						a_2 (10^{-2})	−6.79	−9.27	−0.04	0.00	YES	
	MODEL II	0.9468	0.9255	44.50	0.00	a_0 (10)	3.86	0.65	7.06	0.02		concentration
						a_1 (10^{-1})	0.79	0.32	1.27	0.00	YES	
						a_2 (10^{-1})	0.39	−1.36	2.13	0.63		
						a_3 (10^{-5})	−1.55	−5.92	2.82	0.45		
						a_4 (10^{-4})	−1.33	−3.49	0.82	0.20		
	MODEL III	0.9687	0.9514	55.76	0.00	a_0 (10)	5.22	2.32	8.12	0.00		concentration & size
						a_1 (10^{-2})	5.28	0.71	9.85	0.03	YES	
						a_2 (10^{-1})	0.05	−1.42	1.51	0.94		
						a_3 (10^{-5})	−1.55	−5.14	2.04	0.35		
						a_4 (10^{-4})	−1.33	−3.10	0.44	0.12		
						a_5 (10^{-4})	0.67	0.06	1.27	0.03	YES	

Table 4. Cont.

MPs	Model					Coefficients						
	No.	R^2	R^2_{adj}	F	p	Coefficient	Value	Lower 95%	Upper 95%	p	Significant Term?	Influential Factor
PVC	MODEL I	0.8314	0.8033	29.59	0.00	a_0 (10)	6.61	4.71	8.52	0.00		concentration & size
						a_1 (10^{-2})	4.74	2.89	6.58	0.00	YES	
						a_2 (10^{-1})	−0.93	−1.32	−0.55	0.00	YES	
	MODEL II	0.9672	0.9541	73.75	0.00	a_0 (10^2)	1.29	1.05	1.54	0.00		concentration & size
						a_1 (10^{-2})	0.10	−3.51	3.71	0.95		
						a_2 (10^{-1})	−4.31	−5.64	−2.98	0.00	YES	
						a_3 (10^{-5})	4.42	1.09	7.75	0.01	YES	
						a_4 (10^{-4})	4.22	2.57	5.86	0.00	YES	
	MODEL III	0.9746	0.9605	69.01	0.00	a_0 (10^2)	1.37	1.12	1.63	0.00		concentration & size
						a_1 (10^{-2})	−1.40	−5.41	2.60	0.45		
						a_2 (10^{-1})	−4.50	−5.78	−3.22	0.00	YES	
						a_3 (10^{-5})	4.42	1.28	7.56	0.01	YES	
						a_4 (10^{-4})	4.22	2.67	5.77	0.00	YES	
						a_5 (10^{-5})	3.76	−1.51	9.03	0.14		
PET	MODEL I	0.9596	0.9528	142.33	0.00	a_0 (10)	5.61	4.71	6.52	0.00		concentration & size
						a_1 ($\cdot 10^{-2}$)	5.54	4.66	6.42	0.00	YES	
						a_2 ($\cdot 10^{-1}$)	−0.82	−1.00	−0.64	0.00	YES	
	MODEL II	0.9889	0.9844	222.23	0.00	a_0 (10)	3.98	2.59	5.36	0.00		concentration
						a_1 (10^{-1})	0.98	0.78	1.19	0.00	YES	
						a_2 (10^{-2})	−2.29	−9.81	5.23	0.51		
						a_3 (10^{-5})	−4.08	−5.96	−2.19	0.00	YES	
						a_4 (10^{-4})	−0.74	−1.67	0.19	0.11		
	MODEL III	0.9944	0.9913	318.85	0.00	a_0 (10)	3.33	2.17	4.49	0.00		concentration & size
						a_1 (10^{-1})	1.11	0.93	1.29	0.00	YES	
						a_2 (10^{-2})	−0.68	−6.53	5.17	0.80		
						a_3 (10^{-5})	−4.08	−5.51	−2.64	0.00	YES	
						a_4 (10^{-4})	−0.74	−1.45	−0.03	0.04	YES	
						a_5 (10^{-5})	−3.16	−5.56	−0.75	0.02	YES	

Within the experimental range, exposure to PP microparticles did not result in complete inhibition: maximal value of 80.95% was reached for 200 μm particles at a concentration of 1000 mg/L. Application of the regression models in PP experiments (Table 4) resulted in the lowest values of the coefficients of determination (R^2, R^2_{adj}), implying a slightly worse fit of the inhibition data. Nevertheless, all three models proved their significance. MODEL I was quite inferior to MODEL II and especially MODEL III. Therefore, we ignored the idea coming from MODEL I that both factors studied had a significant influence and concluded that only the concentration is influential within the selected experimental range (MODEL II and MODEL III). Although MODEL III provided the best description of the inhibition data, no significant influence of the joint concentration-size activity was statistically confirmed (p-value for a_5 was 0.07). The inhibition surface, obtained by the best model (Figure 3B), shows that above 500 mg/L the surface loses its slope and is almost parallel to the xy-plane. Therefore, we assume that PP concentrations higher than 1000 mg/L do not additionally affect the inhibition of the yeast growth.

The quality of the fitting of the inhibition data in the PS experiment increased with the complexity of the model applied (Table 4). However, the improvement in the case of MODEL II compared with MODEL I was rather small (R^2_{adj} increase was 0.0032 only) and none of the quadratic terms included in the model proved to be statistically significant, calling into question the superiority of MODEL II. MODEL III, which had the best R^2_{adj} value, also found quadratic terms to be insignificant. Based on the analysis of MODEL III, the concentration and size of PS microparticles were found to be influential in the experimental range. Influence of the particle size was manifested through the interaction term, i.e., through the joint concentration-size activity. The plotted inhibition surface (Figure 3C) confirmed the previously mentioned observation that inhibition increases with higher PS concentrations.

Analysis of the regression models used in the PVC experiment led to the consistent conclusion that both tested factors had a significant influence on the yeast growth. The best model was MODEL III; the model indicated no significant joint concentration-size activity of PVC particles. Analysis of the inhibition surface (Figure 3D), plotted using the best regression model, showed a reduction (at high concentrations) or stagnation (at low concentrations) in yeast inhibition as particle size was decreased from 600 to 400 μm. For particle sizes below 400 μm, there was a rapid increase in the inhibition.

In the case of PET, the inclusion of quadratic terms (MODEL II) improved the fit of the inhibition data. This refers to the square of the concentration, since the analysis of the model did not confirm the significance of the square of the particle size. However, MODEL III as the best model in the PET experiment showed that both quadratic terms were significant. In addition, the analysis of the model implied significance of the interaction term. The inhibition surface described by MODEL III is shown in Figure 3E.

4. Conclusions

The acute toxicity of five MPs: PE, PP, PS, PVC and PET, was determined for the microalga *Scenedesmus* sp., the bacterium *Pseudomonas putida* and the yeast *Saccharomyces cerevisiae*. The monitored toxicity effect was the inhibition of the growth of microorganism. The influence of size and concentration of MPs on inhibition was tested.

Within the experimental range (200–600 μm and 50–1000 mg/L), the maximum value of experimentally determined inhibition for each microorganism/MPs combination was obtained at the lowest particle size and maximum concentration. These values follow further sequences: PE > PVC > PS > PP > PET for *Scenedesmus* sp.; PS > PE > PVC > PP > PET *Pseudomonas putida*; and PS = PVC > PE > PET > PP for *Saccharomyces cerevisiae*. Among the three toxicity tests used, the *Saccharomyces cerevisiae* test proved to be the most sensitive to MPs.

The concentration of MPs proved to have a significant influence on inhibition of all three organisms: higher concentrations resulted in higher inhibitions for all five MPs used. The shape of the inhibition surface for *Saccharomyces cerevisiae* exposure to PP suggests that PP concentrations above 1000 mg/L do not contribute to a further increase in inhibition.

The influence of MPs size was not statistically confirmed in all cases. These exceptions were: exposure of *Scenedesmus* sp. to PE, PP, and PET, exposure of *Pseudomonas putida* to PS, and exposure of *Saccharomyces cerevisiae* to PP. In most cases where a size effect was demonstrated, the inhibition increased as the particles became smaller. A parabolic inhibition-size dependence observed for *Scenedesmus* sp. exposed to PS implies that different inhibition mechanisms prevail at sizes above and below 400 μm.

MODEL I (the model with linear-contributions of concentration and size only) was best for describing inhibition of *Pseudomonas putida* when exposed to PP. In all other cases, MODEL II (model with quadratic terms), and especially MODEL III, were superior to MODEL I. Despite the fact that MODEL III contained the interaction term, the joint concentration-size influence on the inhibition was statistically confirmed in only 6 cases: *Scenedesmus* sp. with PVC, *Pseudomonas putida* with PE and PVC, and *Saccharomyces cerevisiae* with PE, PS and PET.

The observed inhibitions of the microorganisms confirmed the high hazardous potential of 200–600 μm MPs particles when present at concentrations of 50–1000 mg/L. The information on inhibition trends may be an indicator of possible mechanisms of harmful activity. However, additional experiments are required to reveal the true nature of the harmful effects of MPs.

Supplementary Materials: The following supporting information can be downloaded at: https://www.mdpi.com/article/10.3390/polym14061246/s1, Figure S1: Structures of the applied microplastics: (A) polyethylene, (B) polypropylene, (C) polystyrene, (D) polyvinyl chloride, and (E) polyethylene terephthalate; Figure S2: ATR-FTIR spectra of: (A) polyethylene, (B) polypropylene, (C) polystyrene, (D) polyvinyl chloride, and (E) polyethylene terephthalate; Figure S3. Microscopic image of test organisms used in the research: (A) microalgae *Scenedesmus* sp. (magnification 400×),

(B) bacteria *Pseudomonas putida* (magnification 1000×), and (C) yeast *Saccharomyces cerevisiae* (magnification 400×).

Author Contributions: Conceptualization, Š.U., D.K.G., and T.B.; Writing—Original Draft Preparation, M.M., K.B., V.P., and M.C.; Writing—Review & Editing, D.K.G., T.B., and Š.U. All authors have read and agreed to the published version of the manuscript.

Funding: The authors would like to acknowledge financial support of the Croatian Science Foundation through project entitled Advanced Water Treatment Technologies for Microplastics Removal (AdWaTMiR; IP-2019-04-9661).

Institutional Review Board Statement: Not applicable.

Informed Consent Statement: Not applicable.

Data Availability Statement: Data are contained within the article.

Conflicts of Interest: The authors declare no conflict of interest.

References

1. Calabrò, P.S.; Grosso, M. Bioplastics and waste management. *Waste Manag.* **2018**, *78*, 800–801. [CrossRef]
2. Kumar, R.; Pandit, P.; Kumar, D.; Patel, Z.; Pandya, L.; Kumar, M.; Joshi, C.; Joshi, M. Landfill microbiome harbour plastic degrading genes: A metagenomic study of solid waste dumping site of Gujarat, India. *Sci. Total Environ.* **2021**, *779*, 146184. [CrossRef] [PubMed]
3. Erceg, M.; Tutman, P.; Bojanić Varezić, D.; Bobanović, A. Characterization of Microplastics in Prapratno Beach Sediment. *Kem. Ind.* **2020**, *69*, 253–260. [CrossRef]
4. Bule, K.; Zadro, K.; Tolić, A.; Radin, E.; Miloloža, M.; Ocelić Bulatović, V.; Kučić Grgić, D. Microplastics in the Marine Environment of the Adriatic Sea. *Kem. Ind.* **2020**, *69*, 303–310. [CrossRef]
5. Plastics—the Facts 2021. Available online: https://plasticseurope.org/wp-content/uploads/2021/12/Plastics-the-Facts-2021-web-final.pdf (accessed on 26 January 2022).
6. Reichel, A.; Trier, X.; Fernandez, R.; Bakas, I.; Zeiger, B. *Plastics, the Circular Economy and Europe's Environment—A Priority for Action*; European Environmental Agency: Copenhagen, Denmark, 2021. [CrossRef]
7. Wang, C.; Zhao, J.; Xing, B. Environmental source, fate, and toxicity of microplastics. *J. Hazard. Mater.* **2020**, *407*, 124357. [CrossRef] [PubMed]
8. Xiang, Y.; Jiang, L.; Zhou, Y.; Luo, Z.; Zhi, D.; Yang, J.; Lam, S.S. Microplastics and environmental pollutants: Key interaction and toxicology in aquatic and soil environments. *J. Hazard. Mater.* **2022**, *422*, 126843. [CrossRef]
9. Duis, K.; Coors, A. Microplastics in the aquatic and terrestrial environment: Sources (with a specific focus on personal care products), fate and effects. *Environ. Sci. Eur.* **2016**, *28*, 2. [CrossRef]
10. Zimmermann, L.; Göttlich, S.; Oehlmann, J.; Wagner, M.; Völker, C. What are the drivers of microplastic toxicity? Comparing the toxicity of plastic chemicals and particles to *Daphnia magna*. *Environ. Pollut.* **2020**, *267*, 115392. [CrossRef]
11. Vivekanand, A.C.; Mohapatra, S.; Tyagi, V.K. Microplastics in aquatic environment: Challenges and perspectives. *Chemosphere* **2021**, *282*, 131151. [CrossRef]
12. Esiukova, E.E.; Lobchuk, O.I.; Volodina, A.A.; Chubarenko, I.P. Marine macrophytes retain microplastics. *Mar. Pollut. Bull.* **2021**, *171*, 112738. [CrossRef] [PubMed]
13. Yin, L.; Wen, X.; Huang, D.; Zeng, G.; Deng, R.; Liu, R.; Zhou, Z.; Tao, J.; Xiao, R.; Pan, H. Microplastics retention by reeds in freshwater environment. *Sci. Total Environ.* **2021**, *790*, 148200. [CrossRef] [PubMed]
14. Tang, Y.; Liu, Y.; Chen, Y.; Zhang, W.; Zhao, J.; He, S.; Yanga, C.; Zhang, T.; Tang, C.; Yang, Z. A review: Research progress on microplastic pollutants in aquatic environments. *Sci. Total Environ.* **2020**, *766*, 142572. [CrossRef] [PubMed]
15. Bhagat, J.; Zang, L.; Nishimura, N.; Shimada, Y. Zebrafish: An emerging model to study microplastic and nanoplastic toxicity. *Sci. Total Environ.* **2020**, *728*, 138707. [CrossRef] [PubMed]
16. Miloloža, M.; Kučić Grgić, D.; Bolanča, T.; Ukić, Š.; Cvetnić, M.; Ocelić Bulatović, V.; Dionysiou, D.D.; Kušić, H. Ecotoxicological assessment of microplastics in freshwater sources—A review. *Water* **2021**, *13*, 56. [CrossRef]
17. Canniff, P.M.; Hoang, T.C. Microplastic ingestion by *Daphnia magna* and its enhancement on algal growth. *Sci. Total Environ.* **2018**, *633*, 500–507. [CrossRef] [PubMed]
18. Pacheco, A.; Martins, A.; Guilhermino, L. Toxicological interactions induced by chronic exposure to gold nanoparticles and microplastics mixtures in *Daphnia magna*. *Sci. Total Environ.* **2018**, *628–629*, 474–483. [CrossRef] [PubMed]
19. Besseling, E.; Wang, B.; Lürling, M.; Koelmans, A.A. Nanoplastic affects growth of *S. obliquus* and reproduction of *D. magna*. *Environ. Sci. Technol.* **2014**, *48*, 12336–12343. [CrossRef]
20. Lei, L.; Wu, S.; Lu, S.; Liu, M.; Song, Y.; Fu, Z.; Shi, H.; Raley-Susman, K.M.; He, D. Microplastic particles cause intestinal damage and other adverse effects in zebrafish *Danio rerio* and nematode *Caenorhabditis elegans*. *Sci. Total Environ.* **2018**, *619–620*, 1–8. [CrossRef]

21. Jeong, J.; Choi, J. Adverse outcome pathways potentially related to hazard identification of microplastics based on toxicity mechanisms. *Chemosphere* **2019**, *231*, 249–255. [CrossRef]
22. Prata, J.C.; da Costa, J.P.; Lopes, I.; Duarte, A.C.; Rocha-Santos, T. Effects of microplastics on microalgae populations: A critical review. *Sci. Total Environ.* **2019**, *665*, 400–405. [CrossRef]
23. De Sá, L.C.; Oliveira, M.; Ribeiro, F.; Rocha, T.L.; Futter, M.N. Studies of the effects of microplastics on aquatic organisms: What do we know and where should we focus our efforts in the future? *Sci. Total Environ.* **2018**, *645*, 1029–1039. [CrossRef] [PubMed]
24. Tunali, M.; Uzoefuna, E.N.; Tunali, M.M.; Yenigün, O. Effect of microplastics and microplastic-metal combinations on growth and chlorophyll a concentration of *Chlorella vulgaris*. *Sci. Total Environ.* **2020**, *743*, 140479. [CrossRef] [PubMed]
25. Zhang, C.; Chen, X.; Wang, J.; Tan, L. Toxic effects of microplastic on marine microalgae *Skeletonema costatum*: Interactions between microplastic and algae. *Environ. Pollut.* **2016**, *220*, 1282–1288. [CrossRef] [PubMed]
26. Lagarde, F.; Olivier, O.; Zanella, M.; Daniel, P.; Hiard, S.; Caruso, A. Microplastic interactions with freshwater microalgae: Hetero-aggregation and changes in plastic density appear strongly dependent on polymer type. *Environ. Pollut.* **2016**, *215*, 331–339. [CrossRef] [PubMed]
27. Malapascua, J.R.F.; Jerez, C.G.; Sergejevová, M.; Figueroa, F.L.; Masojídek, J. Photosynthesis monitoring to optimize growth of microalgal mass cultures: Application of chlorophyll fluorescence techniques. *Aquat. Biol.* **2014**, *22*, 123–140. [CrossRef]
28. den Besten, P.J.; Munawar, M.; Suter, G. Ecotoxicological testing of marine and freshwater ecosystems: Emerging techniques, trends and strategies. *Integr. Environ. Assess. Manag.* **2007**, *3*, 305–306. [CrossRef]
29. Ribeiro, I.C.; Veríssimo, I.; Moniz, L.; Cardoso, H.; Sousa, M.J.; Soares, A.M.V.M.; Leão, C. Yeasts as a model for assessing the toxicity of the fungicides Penconazol, Cymoxanil and Dichlofluanid. *Chemosphere* **2000**, *41*, 1637–1642. [CrossRef]
30. Nomura, T.; Miyazaki, J.; Miyamoto, A.; Kuriyama, Y.; Tokumoto, H.; Konishi, Y. Exposure of the yeast *Saccharomyces cerevisiae* to functionalized polystyrene latex nanoparticles: Influence of surface charge on toxicity. *Environ. Sci. Technol.* **2013**, *47*, 3417–3423. [CrossRef] [PubMed]
31. ISO 10712; Water Quality—Pseudomonas Putida Growth Inhibition Test Pseudomonas Cell Multiplication Inhibition Test; International Organization for Standardization: Geneva, Switzerland, 1995.
32. Hafner, C. Pseudomonas putida growth inhibition test. In *Ecotoxicological Characterization of Waste*; Moser, H., Römbke, J., Eds.; Springer: New York, NY, USA, 2009; pp. 153–159. [CrossRef]
33. OECD. *Test No. 201: Alga, Growth Inhibition Test, OECD Guidelines for Testing of Chemicals, Section 2*; OECD Publishing: Paris, France, 1984.
34. OECD. *Test No. 201: Freshwater Alga and Cyanobacteria, Growth Inhibition Test, OECD Guidelines for the Testing of Chemicals, Section 2*; OECD Publishing: Paris, France, 2011. [CrossRef]
35. Stilinović, B. *Saccharomyces cerevisiae* test (YTT) as the water toxicity determination method. *Acta Bot. Croat.* **1981**, *40*, 127–131.
36. Hrenovic, J.; Stilinović, B.; Dvoracek, L. Use of prokaryotic and eukaryotic biotests to assess. *Acta Chim. Slov.* **2005**, *52*, 119–125.
37. Myers, R.H.; Montgomery, D.C.; Anderson-Cook, C.M. *Response Surface Methodology: Process and Product Optimization Using Designed Experiments*, 3rd ed.; John Wiley & Sons: Hoboken, NJ, USA, 2009.
38. Washington, S.; Karlaftis, M.G.; Mannering, F.; Anastasopoulos, P. *Statistical and Econometric Methods for Transportation Data Analysis*, 3rd ed.; CRC Press: Boca Raton, FL, USA; London, UK; New York, NY, USA, 2020.
39. Garrido, S.; Linares, M.; Campillo, J.M.; Albentosa, M. Effect of microplastics on the toxicity of chlorpyrifos to the microalgae *Isochrysis galbana*, clone t-ISO. *Ecotoxicol. Environ. Saf.* **2019**, *173*, 103–109. [CrossRef] [PubMed]
40. Bhattacharya, P.; Lin, S.; Turner, J.P.; Ke, P.C. Physical Adsorption of Charged Plastic Nanoparticles Affects Algal Photosynthesis. *J. Phys. Chem. C* **2010**, *114*, 16556–16561. [CrossRef]
41. Guo, Y.; Ma, W.; Li, J.; Liu, W.; Qi, P.; Ye, Y.; Guo, B.; Zhang, J.; Qu, C. Effects of microplastics on growth, phenanthrene stress, and lipid accumulation in a diatom. *Phaeodactylum Tricornutum*. *Environ. Pollut.* **2019**, *257*, 113628. [CrossRef] [PubMed]
42. Zhang, W.; Zhang, S.; Wang, J.; Wang, Y.; Mu, J.; Wang, P.; Lin, X.; Ma, D. Microplastic pollution in the surface waters of the Bohai Sea. *China Environ. Pollut.* **2017**, *231*, 541–548. [CrossRef] [PubMed]
43. Moog, D.; Schmitt, J.; Senger, J.; Zarzycki, J.; Rexer, K.H.; Linne, U.; Erb, T.; Maier, U.G. Using a marine microalga as a chassis for polyethylene terephthalate (PET) degradation. *Microb. Cell. Fact.* **2019**, *18*, 171. [CrossRef] [PubMed]
44. Kim, J.W.; Park, S.B.; Tran, Q.G.; Cho, D.H.; Choi, D.Y.; Lee, Y.J.; Kim, H.S. Functional expression of polyethylene terephthalate-degrading enzyme (PETase) in green microalgae. *Microb. Cell. Fact.* **2020**, *19*, 97. [CrossRef] [PubMed]
45. Falah, W.; Chen, F.; Zeb, B.S.; Hayat, M.T.; Mahmood, Q.; Ebadi, A.; Toughani, M.; Li, E. Polyethylene Terephthalate Degradation by Microalga Chlorella vulgaris along with Pretreatment. *Mater. Plast.* **2020**, *57*, 260–270. [CrossRef]
46. Sikkema, J.; Poolman, B.; Konings, W.N.; de Bont, J.A.M. Effects of the membrane action of tetralin on the functional and structural properties of artificial and bacterial membranes. *J. Bacteriol.* **1992**, *174*, 2986–2992. [CrossRef] [PubMed]
47. Giacomucci, L.; Raddadi, N.; Soccio, M.; Lotti, N.; Fava, F. Polyvinyl chloride biodegradation by *Pseudomonas citronellolis* and *Bacillus flexus*. *New Biotechnol.* **2019**, *52*, 35–41. [CrossRef]
48. Li, W.J.; Jayakody, L.N.; Franden, M.A.; Wehrmann, M.; Daun, T.; Hauer, B.; Blank, L.M.; Beckham, G.T.; Klebensberger, J.; Wierckx, N. Laboratory evolution reveals the metabolic and regulatory basis of ethylene glycol metabolism by *Pseudomonas putida* KT2440. *Environ. Microbiol.* **2019**, *21*, 3669–3682. [CrossRef]
49. Taniguchi, I.; Yoshida, S.; Hiraga, K.; Miyamoto, K.; Kimura, Y.; Oda, K. Biodegradation of PET: Current status and application aspects. *Acs Catal.* **2019**, *9*, 4089–4105. [CrossRef]

50. Roberts, C.; Edwards, S.; Vague, M.; León-Zayas, R.; Scheffer, H.; Chan, G.; Swartz, N.A.; Melliesa, J.L. Environmental consortium containing *Pseudomonas* and *Bacillus* species synergistically degrades polyethylene terephthalate plastic. *Appl. Environ. Sci.* **2020**, *5*, e01151-20. [CrossRef] [PubMed]
51. Bollinger, A.; Thies, S.; Knieps-Grünhagen, E.; Gertzen, C.; Kobus, S.; Höppner, A.; Ferrer, M.; Gohlke, H.; Smits, S.H.J.; Jaeger, K.E. A Novel Polyester Hydrolase from the Marine Bacterium *Pseudomonas aestusnigri*—Structural and Functional Insights. *Front. Microbiol.* **2020**, *11*, 114. [CrossRef] [PubMed]
52. Joo, S.; Cho, I.J.; Seo, H.; Son, H.F.; Sagong, H.Y.; Shin, T.J.; Choi, S.Y.; Lee, S.Y.; Kim, K.J. Structural insight into molecular mechanism of poly(ethylene terephthalate) degradation. *Nat. Commun.* **2018**, *9*, 382. [CrossRef] [PubMed]

Article

Detecting Micro- and Nanoplastics Released from Food Packaging: Challenges and Analytical Strategies

Claudia Cella, Rita La Spina, Dora Mehn, Francesco Fumagalli, Giacomo Ceccone, Andrea Valsesia and Douglas Gilliland *

European Commission, Joint Research Centre (JRC), Ispra, Italy; claudia.cella@ec.europa.eu (C.C.); rita.la-spina@ec.europa.eu (R.L.S.); dora.mehn@ec.europa.eu (D.M.); francesco-sirio.fumagalli@ec.europa.eu (F.F.); giacomo.ceccone@ec.europa.eu (G.C.); andrea.valsesia@ec.europa.eu (A.V.)
* Correspondence: douglas.gilliland@ec.europa.eu

Abstract: Micro- and nanoplastic (pMP and pNP, respectively) release is an emerging issue since these particles constitute a ubiquitous and growing pollutant, which not only threatens the environment but may have potential consequences for human health. In particular, there is concern about the release of secondary pMP and pNP from the degradation of plastic consumer products. The phenomenon is well-documented in relation to plastic waste in the environment but, more recently, reports of pMP generated even during the normal use of plastic food contact materials, such as water bottles, tea bags, and containers, have been published. So far, a validated and harmonized strategy to tackle the issue is not available. In this study, we demonstrate that plastic breakdown to pMP and pNP can occur during the normal use of polyethylene (PE) rice cooking bags and ice-cube bags as well as of nylon teabags. A multi-instrumental approach based on Raman microscopy, X-ray photoelectron spectroscopy (XPS), scanning electron microscopy (SEM), and particular attention on the importance of sample preparation were applied to evaluate the chemical nature of the released material and their morphology. In addition, a simple method based on Fourier transform infrared (FT-IR) spectroscopy is proposed for pNP mass quantification, resulting in the release of 1.13 ± 0.07 mg of nylon 6 from each teabag. However, temperature was shown to have a strong impact on the morphology and aggregation status of the released materials, posing to scientists and legislators a challenging question: are they micro- or nanoplastics or something else altogether?

Keywords: microplastics; nanoplastics; microplastic detection and identification; microplastic quantification; food packaging; particle release; plastic consumption

Citation: Cella, C.; La Spina, R.; Mehn, D.; Fumagalli, F.; Ceccone, G.; Valsesia, A.; Gilliland, D. Detecting Micro- and Nanoplastics Released from Food Packaging: Challenges and Analytical Strategies. *Polymers* **2022**, *14*, 1238. https://doi.org/10.3390/polym14061238

Academic Editor: Jacopo La Nasa

Received: 14 February 2022
Accepted: 16 March 2022
Published: 18 March 2022

Publisher's Note: MDPI stays neutral with regard to jurisdictional claims in published maps and institutional affiliations.

Copyright: © 2022 by the authors. Licensee MDPI, Basel, Switzerland. This article is an open access article distributed under the terms and conditions of the Creative Commons Attribution (CC BY) license (https:// creativecommons.org/licenses/by/ 4.0/).

1. Introduction

The spread of microplastic pollution across our planet may be likened to the sword of Damocles: a threat which grows while knowledge gaps—due to inadequate analytical tools—continue leave us uncertain of when and how the negative effects may be manifest [1–3].

It is now well established that particulate microplastics (pMP) are present in detectable concentrations in the oceans, surface waters, soil, and atmosphere, and researchers suspect that they might harm the ecosystem [4–8]. In contrast, the real-world occurrence of particulate nanoplastics (pNP, i.e., typical size of less than 1 μm), even in simple matrixes, remains largely unknown due to the lack of effective, harmonized, and commonly available analytical methods tailored to the nano-range [9–14].

The precautionary principle is promoting a number of actions across the globe to tackle pMP release in the environment [15–17]. On the one hand, actions to improve our knowledge on biodegradable plastics and increment their use are promoted from both scientific and policy points of view, especially in the framework of the circular economy and end-of-life management routes [18–20]. Additionally, in 2016, the European Food Safety

Authority (EFSA) highlighted the need to monitor the occurrence of pMP in food products and the development of validated methods for their detection and quantification [21]. This call has been reiterated in the recent literature [22–25]. pMP in food are typically thought to originate from contaminated environmental sources, as demonstrated for seafood, such mussels and krill [23,26] and fish [26,27]. Beverages, such as beer, honey, and drinking water were also reported to contain microplastics [24,25,27–31] with various suspected origins. In particular, for bottled water, many of the identified pMP were found to be chemically identical to the food packaging material [28,29,31–34]. Emerging studies additionally reported the release of plastic particulates from a variety of food contact materials and especially food packaging [35–42].

The use of plastics in food packaging has undoubtable had an enormous positive impact on human life by helping in food preservation, while its low density and weight provide benefits for transport and logistics. The polymers in food packaging materials are generally considered to be chemically inert, and the legal controls of material migration into food products consider mainly low molecular weight compounds (<1 kD), such as unreacted or partially unreacted monomers, processing aids, and additives. For such controls, standardized methods have been developed in order to simulate typical food contact scenarios and components [43,44]. However, most of the methods focusing on the release of additives and oligomers are chromatography- and/or mass spectrometry-based and include a filtration step in the analysis process. This filtration, even unintentionally as in the case of HPLC column guards, serves mainly as instrument protection but essentially prevents the detection of particulate plastic material in food contact experiments [45,46].

From various publications reporting the release of pMP and pNP from food packaging, it appears that there is still an absence of any common or systematic analytical strategies, and authors have explored different approaches for particle detection and characterization. In some cases, chemically specific methods, such as Fourier transform infrared (FT-IR) microscopy and pyrolysis gas chromatography/mass spectrometry, have been used, while others report non-specific methods, such as scanning electron microscopy (SEM), nanoparticle tracking analysis, fluorescent microscopy, or quartz crystal microbalance [35–42]. Recently, a call for more knowledge and experimental data generated under realistic scenarios was conducted, with a particular focus on the assessment of the amount of pMP and pNP [14], and the additional request to improve study comparisons by developing and validating certified reference materials that can mimic pMP and pNP in the environment [14,47].

In this report, we will present the results of an investigation of pMP and pNP release from polyethylene (PE) rice cooking bags, ice cube bags, and nylon teabags, examples of everyday food packaging subjected very different use conditions. Our working scheme for sample treatment, based on a multi-instrumental approach, took advantage of the existing literature about pMP and pNP in the environment [48]. In particular, we will discuss critical preparation steps to minimize particle agglomeration, thus improving the detection of individual particles (nanometer and micron-sized). We especially focus on spectroscopic techniques, such as Raman microscopy (μRaman), as key methods for the identification of plastics and other materials coming from the packaging, such as dyes or food residues. Moreover, in case of higher quantities of polymer release, we demonstrate how classical FT-IR spectroscopy using potassium bromide (KBr) pellets can be a simple and useful tool for plastic particle mass quantification. Finally, the same approach was explored as a fast and cost-effective tool to determine the structural nature of the released material: oligomers vs. pMP/pNP.

2. Materials and Methods

2.1. Measures to Avoid Contamination

Ultrapure water (MilliQ, Millipore, filtered 0.2 μm) was used in all the experiments. In order to reduce sample contamination coming from the laboratory environment, plastic labware was avoided wherever possible. All glassware and metallic items were cleaned with Hellmanex® III detergent (Sigma Aldrich, St. Louis, MI, USA) 2% v/v, then rinsed with MilliQ water. Cotton lab coats were used. In order to estimate the contribution of airborne contamination, procedural blanks were performed as described.

2.2. Sample Preparation

Ice cube bags. Commercially available ice cube bags were filled up with 300 mL of MilliQ water and placed in a freezer (-20 °C) for 24 h. The ice cubes were then manually released from the bag and transferred into a clean glass beaker. After thawing at room temperature, the water was filtered through an Anodisc filter with a 0.1 µm pore size (25 mm with a polypropylene ring, Sigma Aldrich). As a negative control, ice cube bags were filled with the same amount of water, stored at room temperature for 24 h, and filtered as described above. Filters were placed on a glass support and analyzed by µRaman.

Rice bags. Commercially available cooking bags containing rice were opened with clean metallic scissors. After the removal of rice, the bag was placed in 250 mL of boiling water in a 500 mL glass beaker. Following the instructions for preparation time indicated on the packaging, the bag was kept under the liquid surface at 100 °C for 10 min with a glass rod. The bag was then removed from the beaker and the remaining liquid was filtered through an Anodisc filter with a 0.1 µm pore size. A blank control experiment was performed in parallel by boiling the same amount of MilliQ water for the same time and filtering the liquid on the same type of filter. Filters were placed on a glass support and analyzed by µRaman.

Teabags. Commercially available empty plastic teabags were purchased from an online platform. The attached strings with paper labels were removed before proceeding. Each teabag was steeped at 95 °C for 10 min in 10 mL of MilliQ water. The teabags were removed and squeezed with the help of a metallic tweezer. The obtained leachate was freeze dried (Martin Christ, Model Alpha 1–4 LD plus) after a pre-cooling period of 2 h at -20 °C and a freezing step in liquid nitrogen for some minutes. Residues were resuspended in 1 mL of fresh MilliQ water and immediately used for further analysis. Negative controls were prepared with the same procedure without the heating step. For µRaman mapping, 1 µL of the resuspended leachate was pipetted on a clean silicon wafer and allowed to dry at room temperature. Additionally, µRaman measurements were performed on the solid residues obtained by filtering the teabag leachate through Si_3N_4 filters (pore size 0.45 µm, Aquamarijn, The Netherlands). The filtrate that passed through the pores was collected and spotted on a Si-PTFE-PDDA chip using a microspotter (Sci Flexarrayer, Scienion) (drop volume 300–360 pL, 1000 drops spotted). The resulting spots (approx. 0.8 mm in diameter) were viewed by the optical microscope and then characterized by µRaman.

2.3. Identification Methods

Raman spectroscopy. A WiTec alpha 300 Raman microscope equipped with a 532 nm laser was used to analyze the original food packaging and the released particles. The spectra of the original materials were collected using either 10× or 100× objectives by averaging at least 10 spectra. Spectral recognition was performed with the help of the OpenSpecy database [49,50] and by comparison with a home-built polymer database (UVIR Manager, ACD labs, Toronto, ON, Canada). Pigment identification (in case of the ice cube bags) was conducted using the open-source database of the Pigments checker of Cultural Heritage Science Open Source [51] and the UVIR manager software. For the rice cooking bag samples, Raman mapping of the rice and polymeric particles on Anodisc filters was performed over a 150 µm × 150 µm area at a 2 µm resolution (step size) using a 10× objective and a 2 s integration time. In the case of ice cube bags, a Raman analysis on the Anodisc filters was carried out over a 31 µm × 31 µm area at a 1 µm resolution (step size) with a 10 s integration time and a 100× objective. The same parameters were used also for the teabag leachate deposited onto a silicon substrate. The WiTec Project software was additionally used to generate 2D chemical maps based on the base component analysis or peak signal intensity.

X-ray photoelectron spectroscopy (XPS). XPS measurements were performed with an Axis Ultra spectrometer (Kratos, Manchester, UK) using a Kα Al monochromatic source (hν = 1486.6 eV) operating at 150 W and an X-ray spot size of 400 µm × 700 µm in the hybrid mode. The residual pressure of the analysis chamber during the analysis was less than 8×10^{-9} Torr. For each sample, both survey spectra (0–1150 eV, pass energy 80 eV) and high-resolution spectra (pass energy at 40 eV) were recorded. The surface charge was compensated by a magnetic charge compensation system, and the energy scale was calibrated by setting the C 1s hydrocarbon peak to 285.00 eV in binding energy [52,53]. The data were processed using vision2 software (Kratos Analytical, Manchester, UK), and the analysis of the XPS peaks was carried out using a commercial software package (CasaXPS v2.3.18PR1, Casa Software, Ltd., Teignmouth, UK). Peak fitting was performed with no preliminary smoothing. Symmetric Gaussian-Lorentzian (70% Gaussian and 30% Lorentzian) product functions were used to approximate the line shapes of the fitting components after a Shirley-type background subtraction. In order to produce relatively wide, uniformly covered surfaces suitable for XPS analysis, samples were diluted in MilliQ water, drop-cast, and vacuum-dried in triplicate on clean Teflon substrates.

2.4. Quantification Method

Application of Beers law to Fourier Transform Infrared (FT-IR) spectroscopy. In order to quantify the plastic material released from the food contact containers during use, we adopted a general approach using Beer's law and FT-IR spectral analysis as described in detail in Figure S1.

For this study, specific calibration series were created with nylon 6 (nominal range 5–50 µm, Sigma Aldrich) or PE powder (Mw 4000 Da, Sigma Aldrich) for teabag leachate and rice bag samples, respectively. Due to their micron size, the nylon 6 particles were mixed with KBr powder in a mass ratio of 1:100 and micronized by cryo-milling in a SPEX 6875 Freezer/Mill. Known amounts of the plastic-containing KBr powder were weighed and mixed with additional KBr before being mechanically pressed to create 133.7 ± 3.9 mg pellets containing 0.64 ± 0.01/0.09 ± 0.01 mg of nylon 6 plus a plastic-free KBr pellet for use in background estimation (n = 4, values are presented as means ± standard deviations). In the case of polyethylene, calibration points were registered in the 2.00–0.13 mg PE/pellet range plus a plastic-free KBr pellet. Polyethylene residues were quantified based on the peak at about 720 cm^{-1} (CH_2, rocking), while for nylon 6, the region between 1490 and 1810 cm^{-1} was selected. In all cases, the FT-IR spectra were collected from 4000 to 600 cm^{-1} with a resolution of 4 cm^{-1} (Tensor27, Bruker, Bremen, Germany).

For the preparation of the pellets of residues from the heat-treated rice bag samples, 5 mL aliquots of the cooking liquid were pipetted into glass test tubes, and 200 mg of KBr was added to each tube. Each sample was then lyophilized and carefully mixed before 130 mg of the resulting powder was used to produce KBr pellets.

For the teabag samples, 400 mg of KBr was added to the leachates, and the salt was allowed to completely dissolve before freeze drying (n = 6). Multiplication with the appropriate dilution factor allowed the total amount of nylon 6 in the original teabag leachate to be determined. A procedural blank was obtained by performing the complete procedure without steeping any teabags.

Additionally, teabag leachates used immediately after preparation ("hot-filtered") or after cooling at room temperature ("cold-filtered") were filtered with different pore sizes, namely, Anodisc filter 0.02 µm, Anodisc filter 0.2 µm, silver filter 0.45 µm, and polycarbonate track edge filter 2.0 µm (Figure S2). All the filters were 25 mm in diameter and were purchased from Sigma Aldrich. The liquid that passed through the filter into the receiving flask was mixed with 300 mg of KBr and freeze dried. The resulting powder was used to create a pellet as previously described. All the data are presented as means ± standard deviations.

2.5. Statistical Analysis

The effect size was estimated in our study by means of Cohen's d value. Cohen's d expresses the size of the difference between two observable variables independent from the sample size. The definition [54,55] used in this work is represented in Equation (1):

$$\text{Cohen's } d = \frac{\text{mean}_1 - \text{mean}_2}{\sqrt{\frac{SD_1^2 + SD_2^2}{2}}} \quad (1)$$

where mean_1 and mean_2 refer to the two variable means, while SD_1 and SD_2 represent their respective standard deviations. In accordance with Sawilowsky [56], $d > 0.8$ was considered as the boundary for a "large" effect size, implicating both (i) meaningfulness of the observed difference between the measured means and (ii) sufficient statistical power of the sample to support a subsequent significance test. To estimate the statistical significance of the difference between the two measurement groups, we performed a two-sample t-test (Origin, Version 2019b, OriginLab Corporation, Northampton, MA, USA).

2.6. Morpholigical Study

Scanning Electron Microscopy (SEM). SEM images were obtained by a FEI-NovaNanolab 600I microscope by detecting the secondary electrons at a 5 kV acceleration voltage with different apertures. Two different approaches were used to prepare the teabag leachate samples for direct imaging by SEM. In the first method, a drop of the sample was spotted and allowed to air dry on a clean silicon substrate previously rinsed with ethanol. In the second method, 10 μL of phosphate-buffered saline solution (Sigma Aldrich) was added to 90 μL of leachate sample and then spotted onto a silicon surface whose hydrophobicity and charge had been modified (Si-PTFE-PDDA) as described in the literature [57,58]. Briefly, a polytetrafluoroethylene (PTFE) coating was plasma-deposited to generate a hydrophobic surface. The deposition was performed using pure octofluorocyclobutane (C_4F_8) as the gas precursor at a pressure of 3.5 Pa (27 mTorr), applying a power of 142 W for 5 min. The plasma-modified substrate was incubated for 2 min in poly(diallyldimethylammonium chloride) (PDDA, Sigma Aldrich, Milan, Italy) 2% v/v before rinsing with MilliQ water and drying under nitrogen flow. Spotting was conducted by pipetting 1 μL of solution in triplicate. In order to allow the time for adsorption of the particulates from the liquid onto the surface, the samples were incubated in a closed box overnight at 4 °C at a relative humidity >85%. Then, they were gently rinsed with MilliQ water to remove any unbound particles and/or any soluble residues.

3. Results and Discussion

The identification and quantification of the micro- and nanoplastics released from food contact materials and food packaging is still considered a challenge despite the increasing number of literature reports on the topic [36–42].

To assess if and how food packaging can release pMP and pNP when subjected to their normal, everyday use conditions, we focused on a selection of three food packaging types: PE rice cooking bags, ice cube bags, and nylon teabags. These food containers were chosen partially on the basis of their widespread use but also because their normal handlings expose them to conditions well beyond normal ambient temperature. Moreover, we want to concretely answer to the need of protocol harmonization in the field by adopting a multi-instrumental approach as already validated in the environmental field [48].

The first step that was considered in development of our testing strategy was to ensure that the particles released were produced under conditions that mimic their intended use. In general, the scenarios of the everyday use of food packaging include the application of direct heating, microwave irradiation, freeze-thaw cycling, or steeping directly in boiling water. In this study, given the intended use of the selected containers, water was selected as the most appropriate food simulant, as indicated by the Council Directive 85/572/EEC [43]. The food packaging was treated with water under hot conditions (boiling water) in the case of nylon teabags and rice cooking bags or with a freeze-thaw process in the case of PE ice bags. In contrast to some other cases reported in the literature [21,59], this removed the need to eliminate the organic contamination that could hinder correct particle counting and identification.

In all investigated cases, it was necessary to concentrate and collect the released particles on a surface suitable for further analysis. Freeze-drying was found to be a good method for concentrating small volumes of liquid as in the case of teabag leachates. For later analytical investigations, it was possible to redisperse the lyophilized powder in an appropriate medium or to use the powder directly as described later in the text. To our knowledge, few other authors have explored solvent evaporation [48,60] or freeze-drying as methods to concentrate pMP and pNP for analytical studies. The main drawbacks of such concentration steps include the risk of aggregation and a possible loss of particles, either by adhesion to the vessel walls or by flying away due to repulsion by electrostatic charges, which are especially relevant in the case of the few particles present in a large amount of liquid. In our case, the freeze-drying process was further explored as a preparation step for the quantification of released materials, as will be explained later in the text. When the volume was higher than 100 mL, as in the case of PE ice bag and PE rice bag sample preparation, filtration was preferred as the concentration method. Alumina filters with a pore size of 0.1 µm were selected because of their suitability for use with different spectroscopic techniques, e.g., as a support for µRaman while their low absorption of IR light above 1500 cm^{-1} makes them acceptable for use with µFTIR in transmission mode.

Particles collected on supports (filter surface or silicon substrate) were then analyzed by µRaman to assess their chemical nature and establish if they were actually released from the food packaging (Figure 1). Spectral matches with reference materials in both the open-source [49,50] and home-built libraries were higher than 0.95 in all the tested cases. Additional comparisons were made with the spectra of the original packaging. Figure 1 shows three selected examples with optical images of particles and spectral identification.

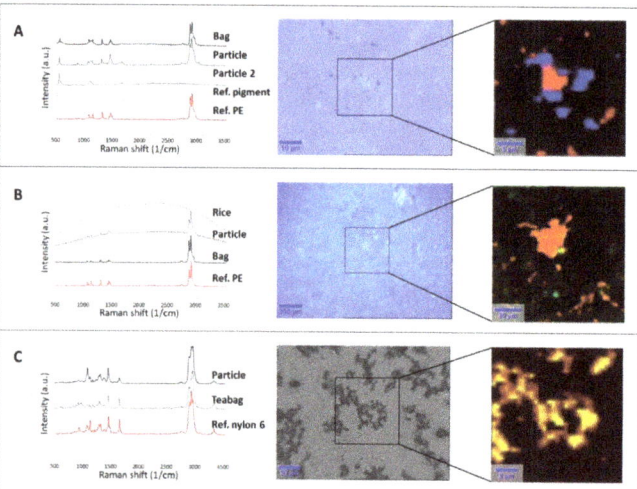

Figure 1. Raman spectral identification (left), optical image (middle), and chemical heat maps (right) of the collected particles released from food packaging materials: (**A**) ice bag, (**B**) rice cooking bag, (**C**) teabag. The rectangular area in the optical images shows the area analyzed by 2D mapping and the subsequent base component analysis or intensity peak heat map. Chemical maps of the filtrate of

the ice bag and cooked rice bag on Anodisc filters were generated by base component analysis. For the ice bag (**A**), the red and blue colors identify different particle types (PE and ultramarine blue pigment, respectively). For the rice bag (**B**), the orange and red color scale represents the PE signal intensity, while the green dots indicate rice residues. The chemical heat map of the teabag filtrate on a silicon support (**C**) was generated from the intensity of the signal between 1611 and 1688 cm^{-1} (amide I, C=O stretching). In the color scale here, yellow represents the strongest signal coming from the polymer with respect to dark brown (the lowest signal).

Figure 1A (middle panel) refers to the particles released from the water frozen in ice cube bags. The image illustrates the presence of at least two populations of particulates with different colors (white and dark blue in the middle optical image). Raman analysis of the white particles provided spectra where most of the peaks were characteristic of PE, which was the raw material of the plastic bags. For the full assignment of the additional peaks, it was necessary to also consider the analysis of the dark blue particles, attributable to the ultramarine blue pigment generally used to dye food packaging [61,62]. In this case, the characteristic spectral peaks of the pigment at 548, 1096, and 1647 cm^{-1} [63] were also present in the spectra of the white particulates. The analysis of different spots and the chemical mapping of a selected area (Figure 1A, right panel) on the filters confirmed that the released material was composed of PE particles mixed with the pigment (red area in Figure 1A right panel) and individual particles of the ultramarine blue pigment (blue area in Figure 1A right panel). This suggests that both components are likely to be present when the ice cubes are prepared and, consequently, could be ingested. As a control, the same analysis was carried out using water samples which contacted the plastic only at room temperature. In this case, only a few particles were visible on the filter, and they were mainly recognized as PE. The difference of the control and the frozen sample suggests that the release of PE and ultramarine blue pigment was mainly due to mechanical stress, either during the freezing process or when removing of the ice cubes from the bags.

As in the case of freezer bag, the water used to boil the rice cooking bags was filtered on Anodisc filters, leaving a clearly visible layer of particulate on the filter surface (Figure 1B, middle panel). Spectra attributable to PE (the material of the selected rice bag) were detected at different locations on the filter. However, some of the measured spectra showed a strong fluorescence, similar to that observed in the analysis of the rice itself. Fluorescence due to the presence of organic compounds when performing Raman investigation at short wavelength excitation is a well-known phenomenon in the literature [64,65]. Although the rice had been removed from the packaging before cooking, a base component analysis of the spectral map revealed that spectra fully or partially attributable to rice were randomly present (Figure 1B, right panel). This co-localization of spectra suggests that hydrophobic PE particles released from the bag will probably be adsorbed on the rice during typical cooking process. In the blank sample, no particles attributable to rice or PE were found on the filter. The chemical recognition of the particles is a fundamental step before proceeding with any further investigation, as already pointed out in the recent literature [14,23,48]. As presented here, not only pMP and pNP can be present in a sample but also food residues, pigments, and additives associated with the plastic packaging.

In the case of the material released from the teabags after steeping in water, the freeze-dried leachate was resuspended in MilliQ water and spotted on a silicon support. As can be seen in the middle panel of Figure 1C, only one component was detected in the sample, namely, nylon 6 (the material of the original teabags). In some cases, fluorescence was observed, probably due to a weak auto-fluorescence of the original material (data not shown). Figure 1C (right panel) illustrates the chemical heat map created on the basis of the intensity signal of the peak between 1611 and 1688 cm^{-1}, which is characteristic of the amide group. Since this chemical group is present only in the polyamide (nylon) and polyurethane families, its presence reasonably indicates that all the detected particles were associated with the nylon released by the original teabag due to the heating process.

Additionally, the chemical nature of the released particles in different particle size ranges was investigated. To conduct this analysis, teabag leachate was filtered through a Si_3N_4 filter (0.45 µm pore size) to produce two distinct particle size fractions, namely, below and above the filter cut-off. The particulates larger than 0.45 µm retained on the filter surface (residue) were directly analyzed with µRaman, while the filtrate liquid containing the smaller particles was microspotted on a silicon surface to concentrate the solids into a small uniform spot prior to the Raman analysis. In both cases, the spectra matched that of the native teabag material (Figure S3). A microspotter is an effective tool to deposit small quantities of dispersed solids into very small, well-defined areas for microanalysis. Moreover, by concentrating many particulates into one highly localized and well-defined position, a spectrum can be obtained that is the sum of the signals coming from many particulates. In this way, µRaman characterization becomes feasible, even when the individual particle size is expected to be below the instrument's resolution and sensitivity [66]. As far we know, there are no descriptions in literature of the use of this tool for pMP and pNP investigation.

The sub-micron particulate fraction (pNP fraction) obtained after filtration was further analyzed by XPS spectroscopy and compared with the surface chemistry of the pristine teabag. The spectra and quantification of the elemental surface composition for both the pNP fraction and the teabag surface are shown in Figures S4 and S5 and Table S1. In both cases, the dominant peaks detected in the spectra were assigned to C1s, O1s, and N1s. This set is compatible with the Raman spectral assignment for nylon. In the pNP fraction spectrum, several minor low-intensity peaks were also observed, and they were assigned to trace contaminants (Na 1s, Ca 1s, Si 2p, B 1s, P 2p, and S 2p), either resulting from the water dispersion, sample preparation, or airborne contamination. In both analyzed samples, the measured stoichiometric elemental ratio N/C is compatible with the theoretical value expected from the nylon composition (Figure S6). The O/C elemental ratio measured on the teabag net is also compatible with nylon assignment, but, in the case of pNP, a much larger value was obtained. It is likely that the large specific surface area of the nano-fraction filtered material was readily oxidized when exposed to air and MilliQ water during sample preparation, a phenomenon already described in the literature [48]. In addition, Nylon hydrolysis could increase the O/C ratio by forming –COOH groups, which would probably be concentrated at the surface of the particles, making them more hydrophilic. The surface oxidation occurring in the pNP fraction was also confirmed by the high-resolution scans performed over the C1s peaks (Figure S7). The slight underestimation of the measured N/C values in the two samples and the slight overestimation of the O/C ratio in the case of the teabag net, as compared to the theoretical values, indicated a small but detectable amount of organic contamination on the material's surface. Even if XPS alone is not able to unambiguously identify the polymer type or to completely exclude the presence of other components, the overall characterization of the sub-micron fraction of the teabag leachate indicated that it is mostly represented by nylon 6.

In order to quantify the plastic material released from the food packaging during use, we adopted the approach described in the Materials and Methods Section of using Beer's law and FT-IR spectral analysis.

PE residues were quantified based on the peak at around 720 cm^{-1} (CH_2 rocking), while for nylon 6, the region between 1490 and 1810 cm^{-1} was selected because it is specific for amide peaks (amide I C=O stretching and amide II C–N stretching and N–H bending) and it is less affected by the overlay with the OH peak in the region around 3200–3300 cm^{-1} (NH stretching). The calibration curve for nylon 6 in the selected mass range (0.65–0.00 mg nylon 6/pellet) showed good linearity and a Pearson correlation coefficient of 0.9982 (Figure S8). From the teabags, a release of 1.13 ± 0.07 mg of nylon 6 (n = 6) was found, as can be seen in Table S2. This corresponds to the release of 5.98 ± 0.81 mg/g of the original teabag. The value in terms of mg of nylon 6 obtained with the procedural blank was less than the 3% of the nylon 6 found in the teabag leachates. This also includes the potential contamination deriving from the condensation of atmospheric humidity during the freeze-drying procedure.

The same approach was also used for the materials released by the rice bags, although, in this case, the amount of PE residue was below the detection limit in the KBr pellets (Figure S9). Similarly, for the particulate released from the ice bags, this approach was not applicable due to the low quantity of material released.

Commonly available methods for pMP quantification as described in the literature include thermal degradation combined with gas chromatography–mass spectrometry (GC-MS) and additional methods that are extensively described elsewhere [67–69]. Different from these methods, the approach presented here can directly evaluate the mass concentration of the released plastics. The method is a simple, acceptably sensitive, and inexpensive option for analyzing the residues of small particulates, which are difficult to quantify on a particle number basis by µRaman and µFT-IR. However, a series of limitations can be found. To avoid signal saturation, the plastic particle size should be sufficiently small (i.e., below few microns) to let the IR beam pass through the sample. Moreover, the method is not as sensitive as the thermal degradation methods combined with GC-MS since for two samples we were not able to see any spectrum. In addition, the presence of more than one polymer or other contaminants can complicate the spectral evaluation and compromise the calibration curve linearity. All of these aspects need to be taken into account, eventually with a case-by-case approach, when exporting these methods to more complex samples.

Our findings were partially in agreement with what expressed by Busse at al. [70] in terms of mass of the nylon 6 released from teabags. However, the conclusions of the mentioned paper were that the material released from teabags was not consistent with true released particles (both pMP or pNP) but, rather, partially soluble oligomers.

We, therefore, hypothesized that what we observed during the µRaman investigation (Figure 1C) resulted from the precipitation of such oligomers due to the cooling process. Accordingly, we filtered the teabag leachates with different pore size membranes, either when hot immediately after boiling ("hot-filtered" leachate) or after cooling at room temperature ("cold-filtered" leachate). The results are summarized in Table 1.

Table 1. Quantities of nylon 6 (mg) from teabag leachates after filtration with different pore sizes. The mass (mg) of nylon 6 quantified with the FT-IR method using KBr pellets on teabag leachates after filtration with different pore size filters immediately after teabag removal ("hot-filtered" leachate) or after cooling at room temperature ("cold-filtered" leachate). The values refer to the amount of nylon 6 found in the liquid portion recovered in the receiving flask (n = 3, data are presented as means ± standard deviations). In the last columns, the Cohen's d and the p values for the two-sample t-test are reported.

Filter Pore Size µm	mg of Nylon 6 in Hot-Filtered Leachate	mg of Nylon 6 in Cold-Filtered Leachate	Cohen's d	p Value (Two Sample t-Test)
0.02	1.13 ± 0.05	0.71 ± 0.10	5.3	0.003
0.2	0.60 ± 0.10	0.84 ± 0.14	1.9	0.080
0.4	0.91 ± 0.14	0.93 ± 0.03	0.2	NA [1]
2.0	0.92 ± 0.08	1.03 ± 0.10	1.2	0.271

[1] NA = not appropriate because Cohen's d < 0.8, see description in the Materials and Methods Section.

Under "hot-filtered" conditions, it appears that the material present in the leachates passed through the filters independently from the cut off (except for the 0.2 µm filter). In particular, the mass of nylon 6 found after passing through the smallest pore size filter was not different from the overall quantity found when no filtration was applied (1.13 ± 0.07 mg of nylon 6, Cohen's d 0.1). In contrast, if the leachate was cooled to room temperature before filtration, a different trend was noted. With a 2.0 µm cut off, almost all the material passed through the filter, while an increase in material retention was observed for smaller pore size filters. With a 0.02 µm cut off filter, approximately 65% of the overall material mass passed through the filter, meaning that about 35% of the total mass belonged to >20 nm particles. Table 1 reports the Cohen's *d* values obtained by comparing the measured masses retrieved after filtration at the different cut-offs in the cold- or hot-filtered conditions. As can be seen, the highest difference in recovered mass was found at the smallest cut-offs (0.02 µm) that retained more materials in the cold-filtered conditions (Cohen's d value 5.3, *p* value 0.003). On the contrary, with larger pore sized filters (0.4 µm and above) such variations were reduced, probably because the differences in filtered mass between the hot and cold conditions were not sufficient to determine a statistically relevant effect, as expected.

Analytical ultracentrifuge (AUC) investigations confirmed the presence of submicron particles suspended in the leachate after cooling (see Figure S10). These findings suggested that the nylon 6 released from the teabags during the heating process may actually be attributable to oligomers that are partially soluble when heated, as suggested by Busse et al. [70]. However, such oligomers, if allowed to cool before filtration, could precipitate from solution and aggregate or crystalize into submicron particles, which would be trapped during cold filtering. As an additional support for this finding, leachates after cooling were re-heated at 95 °C and immediately filtered with the smallest pore size filter (0.02 µm). The nylon 6 quantified in these samples was 1.01 ± 0.04 mg, meaning that the formed aggregates were able to dissolve again because of the heating process.

An additional observation on the aggregation status of the released material and its importance in the quantification of pMP and pNP comes from the analysis of SEM images. SEM is one of the most used techniques for the investigation and counting of plastics, especially in the nano scale [48,69] but is critically dependent on appropriate sample preparation. If the sample preparation technique is not adequate, there is a high risk of obtaining misleading information. The first sample preparation approach was based on leachate spotted and dried directly on a polished silicon substrate that was previously cleaned with ethanol. From the resulting SEM image, a variety of particle sizes were detected (Figure 2A), including micron size particles and probable agglomerates, similar to what was already presented in the middle panel of Figure 1C.

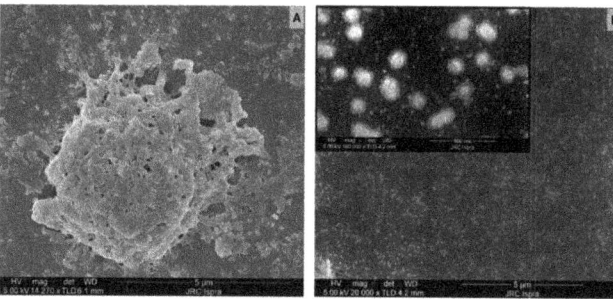

Figure 2. SEM images of teabag leachates prepared on different silicon substrates: (**A**) untreated Si (**B**) hydrophobically modified Si-PTFE-PDDA supports. Scale bar: 5 µm (500 nm for the insert in B).

To achieve a more homogeneous distribution on the surface, we took advantage of modified Si-PTFE-PDDA supports, which have been designed to have a surface with a controllably increased surface hydrophobicity. In these conditions, the particles bind to the surface by electrostatic forces, and their absorption is driven by diffusion. Unless particles aggregation occurs in the solution, no particle aggregation occurs on the surface [58]. As can be seen in Figure 2B, no microparticles were detected. This suggests that the micron-sized particles observed during the Raman investigation (Figure 1C) were actually an artifact of the sample preparation. In addition, the morphology of the micron-sized particles (Figure 2A) appears more similar to that of an aggregate than a single compact particle. If this is the case, particle counting on the basis of SEM images could introduce errors and huge variability with respect to the real number of particles present in the sample.

Together with the findings of the temperature effects on oligomer aggregation, it is clear that proper quantification or counting of pMP, and especially pNP, in diluted samples is still a challenge. The simple spotting of sample leachate on a silicon surface can help the visualization of particles and, eventually, their identification as demonstrated with the Raman investigation (Figure 1), but this is not sufficient to properly quantify the number of plastics released during sample preparation.

4. Conclusions

In conclusion, here we present a study on the release of pMP and pNP from different types of food packaging, i.e., PE rice of cooking bags and ice cube bags and nylon teabags. We demonstrated that the real use of the selected plastic packaging in everyday life can produce pMP and pNP detachment and consequent contamination of food for human consumption. The adaptation of multi-instrumental approaches already described in the environmental field was found to be fundamental for a complete characterization and a global understanding of what happens during the process. In particular, sample preparation was found to be a key point for the study of individual particles and to minimize the aggregation process. The introduction of the heating and cooling steps during the investigation revealed how this seems to affect the agglomeration status of the released oligomers, possibly inducing the formation of thermally unstable pNP. This finding poses challenging issues for legislators: should we consider keeping the heat-released material after cooling as being "oligomeric" or should we take into account the possible formation of pNP? The different ways in which these materials can be classified may potentially have different consequences when subjected to evaluation under different types of legislative controls.

Supplementary Materials: The following supporting information can be downloaded at: https://www.mdpi.com/article/10.3390/polym14061238/s1, Figure S1: General scheme of the overall procedure for applying Beers law to Fourier transform infrared (FT-IR) spectroscopy; Figure S2: General scheme of sample preparation for suspensions after filtration with different pore sizes immediately after teabag removal or after cooling completely at room temperature; Figure S3: Chemical recognition by Raman microscopy on particles after filtration with Si_3N_4 filters; Figure S4: XPS spectra from the surface of the teabag net wire (reference) and the dropcast sample form the filtered nano-fraction from the teabag leachate; Figure S5: Elemental quantitative XPS analysis from the leachate nano-fraction sample and the teabag and substrate reference; Figure S6: XPS atomic ratios (O/C and C/N) for the leachate nanofraction and the teabag reference; Figure S7: XPS high-resolution spectra of the C1s peaks for the leachate nanofraction and the teabag reference; Figure S8: Nylon 6 spectra for calibration curve building; Figure S9: PE spectra for calibration curve building; Figure S10: Multimodal size distribution of the teabag leachate, as obtained from an analytical ultracentrifuge investigation; Table S1: Elemental quantitative XPS analysis from the leachate nano-fraction sample, and the teabag and substrate reference. Table S2: Quantities of nylon 6 (mg) realized from each single teabag after boiling. References [71–73] are cited in the supplementary materials.

Author Contributions: Conceptualization, C.C., R.L.S., D.M. and D.G.; investigation, C.C., R.L.S., D.M., F.F. and A.V.; writing—original draft preparation, C.C., R.L.S., D.M. and D.G.; writing—review and editing, C.C., R.L.S., D.M., F.F., G.C., A.V. and D.G.; visualization, C.C., R.L.S. and D.M.; supervision, G.C. and D.G. All authors have read and agreed to the published version of the manuscript.

Funding: This research received no external funding.

Institutional Review Board Statement: Not applicable.

Informed Consent Statement: Not applicable.

Data Availability Statement: Not applicable.

Conflicts of Interest: The authors declare no conflict of interest. The scientific output expressed does not imply a policy position of the European Commission. Neither the European Commission nor any person acting on behalf of the Commission is responsible for the use that might be made of this publication. For information on the methodology and quality underlying the data used in this publication for which the source is neither Eurostat nor other Commission services, users should contact the referenced source.

References

1. World Health Organization. *Microplastics in Drinking-Water*; World Health Organization: Geneva, Switzerland, 2019.
2. O'Neill, S.M.; Lawler, J. Knowledge gaps on micro and nanoplastics and human health: A critical review. *Case Stud. Chem. Environ. Eng.* **2021**, *3*, 100091. [CrossRef]
3. Burns, E.E.; Boxall, A.B.A. Microplastics in the aquatic environment: Evidence for or against adverse impacts and major knowledge gaps. *Environ. Toxicol. Chem.* **2018**, *37*, 2776–2796. [CrossRef] [PubMed]
4. Jiang, J.Q. Occurrence of microplastics and its pollution in the environment: A review. *Sustain. Prod. Consum.* **2018**, *13*, 16–23. [CrossRef]
5. Ivleva, N.P.; Wiesheu, A.C.; Niessner, R. Microplastic in Aquatic Ecosystems. *Angew. Chem. Int. Ed.* **2017**, *56*, 1720–1739. [CrossRef]
6. Boyle, K.; Örmeci, B. Microplastics and nanoplastics in the freshwater and terrestrial environment: A review. *Water* **2020**, *12*, 2633. [CrossRef]
7. de Souza Machado, A.A.; Kloas, W.; Zarfl, C.; Hempel, S.; Rillig, M.C. Microplastics as an emerging threat to terrestrial ecosystems. *Glob. Chang. Biol.* **2018**, *24*, 1405–1416. [CrossRef]
8. da Costa, J.P. Micro- and nanoplastics in the environment: Research and policymaking. *Curr. Opin. Environ. Sci. Health* **2018**, *1*, 12–16. [CrossRef]
9. Pico, Y.; Alfarhan, A.; Barcelo, D. Nano- and microplastic analysis: Focus on their occurrence in freshwater ecosystems and remediation technologies. *TrAC Trends Anal. Chem.* **2019**, *113*, 409–425. [CrossRef]
10. Ter Halle, A.; Jeanneau, L.; Martignac, M.; Jardé, E.; Pedrono, B.; Brach, L.; Gigault, J. Nanoplastic in the North Atlantic Subtropical Gyre. *Environ. Sci. Technol.* **2017**, *51*, 13689–13697. [CrossRef]
11. Materić, D.; Kasper-Giebl, A.; Kau, D.; Anten, M.; Greilinger, M.; Ludewig, E.; Van Sebille, E.; Röckmann, T.; Holzinger, R. Micro-and Nanoplastics in Alpine Snow: A New Method for Chemical Identification and (Semi)Quantification in the Nanogram Range. *Environ. Sci. Technol.* **2020**, *54*, 2353–2359. [CrossRef]
12. Wahl, A.; Le Juge, C.; Davranche, M.; El Hadri, H.; Grassl, B.; Reynaud, S.; Gigault, J. Nanoplastic occurrence in a soil amended with plastic debris. *Chemosphere* **2021**, *262*, 127784. [CrossRef] [PubMed]
13. Valsesia, A.; Quarato, M.; Ponti, J.; Fumagalli, F.; Gilliland, D.; Colpo, P. Combining microcavity size selection with Raman microscopy for the characterization of Nanoplastics in complex matrices. *Sci. Rep.* **2021**, *11*, 362. [CrossRef] [PubMed]
14. Alexy, P.; Anklam, E.; Emans, T.; Furfari, A.; Galgani, F.; Hanke, G.; Koelmans, A.; Pant, R.; Saveyn, H.; Sokull Kluettgen, B. Managing the analytical challenges related to micro- and nanoplastics in the environment and food: Filling the knowledge gaps. *Food Addit. Contam. Part A Chem. Anal. Control. Expo. Risk Assess.* **2020**, *37*, 1–10. [CrossRef] [PubMed]
15. ECHA. *Annex XV Restriction Report 2019*; European Chemicals Agency: Helsinki, Finland, 2019.
16. ECHA. *Committee for Risk Assessment (RAC) Committee for Socio-Economic Analysis (SEAC) Background Document*; European Chemicals Agency: Helsinki, Finland, 2020; Volume 1.
17. Kentin, E.; Kaarto, H. An EU ban on microplastics in cosmetic products and the right to regulate. *Rev. Eur. Comp. Int. Environ. Law* **2018**, *27*, 254–266. [CrossRef]
18. Hann, S.; Scholes, R.; Molteno, S.; Hilton, M.; Favoino, E.; Geest Jakobsen, L. *Relevance of Biodegradable and Compostable Consumer Plastic Products and Packaging in a Circular Economy*; Publications Office of the European Union: Luxembourg, 2020.
19. Hann, S.; Fletcher, E.; Molteno, S.; Sherrington, C.; Elliott, L.; Kong, M.; Koite, A.; Sastre, S.; Martinez, V. *Relevance of Conventional and Biodegradable Plastics in Agriculture*; Publications Office of the European Union: Luxembourg, 2021.
20. Dordevic, D.; Necasova, L.; Antonic, B.; Jancikova, S. Plastic Cutlery Alternative: Case Study with Biodegradable Spoons. *Food* **2021**, *10*, 1612. [CrossRef]
21. EFSA Panel on Contaminants in the Food Chain Statement on the presence of microplastics and nanoplastics in food, with particular focus on seafood. *EFSA J.* **2016**, *14*, 4501–4530. [CrossRef]
22. Hantoro, I.; Löhr, A.J.; Van Belleghem, F.G.A.J.; Widianarko, B.; Ragas, A.M.J. Microplastics in coastal areas and seafood: Implications for food safety. *Food Addit. Contam.-Part A Chem. Anal. Control Expo. Risk Assess.* **2019**, *36*, 674–711. [CrossRef]
23. Toussaint, B.; Raffael, B.; Angers-Loustau, A.; Gilliland, D.; Kestens, V.; Petrillo, M.; Rio-Echevarria, I.M.; Van den Eede, G. Review of micro- and nanoplastic contamination in the food chain. *Food Addit. Contam. Part A Chem. Anal. Control Expo. Risk Assess.* **2019**, *36*, 639–673. [CrossRef]
24. Van Raamsdonk, L.W.D.; Van Der Zande, M.; Koelmans, A.A.; Ron, L.A.P.; Hoogenboom, R.J.B.P.; Groot, M.J.; Peijnenburg, A.A.C.M.; Weesepoel, Y.J.A. Current Insights into Monitoring, Bioaccumulation and Potential Health Effects of Microplastics Present in the Food Chain. *Foods* **2020**, *9*, 72. [CrossRef]
25. Shruti, V.C.; Pérez-Guevara, F.; Elizalde-Martínez, I.; Kutralam-Muniasamy, G. Toward a unified framework for investigating micro(nano)plastics in packaged beverages intended for human consumption. *Environ. Pollut.* **2021**, *268*, 115811. [CrossRef]

26. Llorca, M.; Álvarez-Muñoz, D.; Ábalos, M.; Rodríguez-Mozaz, S.; Santos, L.H.M.L.M.; León, V.M.; Campillo, J.A.; Martínez-Gómez, C.; Abad, E.; Farré, M. Microplastics in Mediterranean coastal area: Toxicity and impact for the environment and human health. *Trends Environ. Anal. Chem.* **2020**, *27*, e00090. [CrossRef]
27. Karbalaei, S.; Hanachi, P.; Walker, T.R.; Cole, M. Occurrence, sources, human health impacts and mitigation of microplastic pollution. *Environ. Sci. Pollut. Res.* **2018**, *25*, 36046–36063. [CrossRef] [PubMed]
28. Mason, S.A.; Welch, V.G.; Neratko, J. Synthetic Polymer Contamination in Bottled Water. *Front. Chem.* **2018**, *6*, 407. [CrossRef] [PubMed]
29. Oßmann, B.E.; Sarau, G.; Holtmannspötter, H.; Pischetsrieder, M.; Christiansen, S.H.; Dicke, W. Small-sized microplastics and pigmented particles in bottled mineral water. *Water Res.* **2018**, *141*, 307–316. [CrossRef]
30. Schymanski, D.; Goldbeck, C.; Humpf, H.U.; Fürst, P. Analysis of microplastics in water by micro-Raman spectroscopy: Release of plastic particles from different packaging into mineral water. *Water Res.* **2018**, *129*, 154–162. [CrossRef]
31. Zhang, Q.; Xu, E.G.; Li, J.; Chen, Q.; Ma, L.; Zeng, E.Y.; Shi, H. A Review of Microplastics in Table Salt, Drinking Water, and Air: Direct Human Exposure. *Environ. Sci. Technol.* **2020**, *54*, 3740–3751. [CrossRef]
32. Winkler, A.; Santo, N.; Ortenzi, M.A.; Bolzoni, E.; Bacchetta, R.; Tremolada, P. Does mechanical stress cause microplastic release from plastic water bottles? *Water Res.* **2019**, *166*, 115082. [CrossRef]
33. Danopoulos, E.; Twiddy, M.; Rotchell, J.M. Microplastic contamination of drinking water: A systematic review. *PLoS ONE* **2020**, *15*, e0236838. [CrossRef]
34. Mortensen, N.P.; Fennell, T.R.; Johnson, L.M. Unintended human ingestion of nanoplastics and small microplastics through drinking water, beverages, and food sources. *NanoImpact* **2021**, *21*, 100302. [CrossRef]
35. Kedzierski, M.; Lechat, B.; Sire, O.; Le Maguer, G.; Le Tilly, V.; Bruzaud, S. Microplastic contamination of packaged meat: Occurrence and associated risks. *Food Packag. Shelf Life* **2020**, *24*, 100489. [CrossRef]
36. Fadare, O.O.; Wan, B.; Guo, L.H.; Zhao, L. Microplastics from consumer plastic food containers: Are we consuming it? *Chemosphere* **2020**, *253*, 126787. [CrossRef]
37. Du, F.; Cai, H.; Zhang, Q.; Chen, Q.; Shi, H. Microplastics in take-out food containers. *J. Hazard. Mater.* **2020**, *399*, 122969. [CrossRef] [PubMed]
38. Hernandez, L.M.; Xu, E.G.; Larsson, H.C.E.; Tahara, R.; Maisuria, V.B.; Tufenkji, N. Plastic Teabags Release Billions of Microparticles and Nanoparticles into Tea. *Environ. Sci. Technol.* **2019**, *53*, 12300–12310. [CrossRef] [PubMed]
39. Li, D.; Shi, Y.; Yang, L.; Xiao, L.; Kehoe, D.K.; Gun, Y.K.; Boland, J.J.; Wang, J.J. Microplastic release from the degradation of polypropylene feeding bottles during infant formula preparation. *Nat. Food* **2020**, *1*, 746–754. [CrossRef]
40. Dessì, C.; Okoffo, E.D.; O'Brien, J.W.; Gallen, M.; Samanipour, S.; Kaserzon, S.; Rauert, C.; Wang, X.; Thomas, K.V. Plastics contamination of store-bought rice. *J. Hazard. Mater.* **2021**, *416*, 125778. [CrossRef]
41. Ranjan, V.P.; Joseph, A.; Goel, S. Microplastics and other harmful substances released from disposable paper cups into hot water. *J. Hazard. Mater.* **2021**, *404*, 124118. [CrossRef]
42. Sobhani, Z.; Lei, Y.; Tang, Y.; Wu, L.; Zhang, X.; Naidu, R.; Megharaj, M.; Fang, C. Microplastics generated when opening plastic packaging. *Sci. Rep.* **2020**, *10*, 4841. [CrossRef]
43. European Union. Council directive of 19 December 1985 laying down the list of simulants to be used for testing migration of constituents of plastic materials and articles intended to come into contact with foodstuffs. *Off. J. Eur. Communities* **1985**, *372*, 14–21.
44. Jakubowska, N.; Beldi, G.; Robouch, P.; Hoekstra, E. *Testing Conditions for Kitchenware Articles in Contact with Foodstuffs: Plastics and Metals*; Technical Report JRC121622; European Commission: Ispra, Italy, 2020.
45. Reed, C.R.; Loscombe, G.D. The Use of Microparticulate Guard Columns in Reverse-Phase High-Performance Liquid Chromatography. *Chromatographia* **1982**, *15*, 15–17. [CrossRef]
46. Majors, R. Current Trends in HPLC Column Technology. *LCGC N. Am.* **2012**, *30*, 20–34.
47. Seghers, J.; Stefaniak, E.A.; La Spina, R.; Cella, C.; Mehn, D.; Gilliland, D.; Held, A.; Jacobsson, U.; Emteborg, H. Preparation of a reference material for microplastics in water—evaluation of homogeneity. *Anal. Bioanal. Chem.* **2022**, *414*, 385–397. [CrossRef] [PubMed]
48. Schwaferts, C.; Niessner, R.; Elsner, M.; Ivleva, N.P. Methods for the analysis of submicrometer- and nanoplastic particles in the environment. *TrAC Trends Anal. Chem.* **2019**, *112*, 52–65. [CrossRef]
49. Cowger, W.; Gray, A.; Hapich, H.; Rochman, C.; Lynch, J.; Primpke, S.; Munno, K.; De Frond, H.O.H. Open Specy. Available online: www.openspecy.org (accessed on 17 March 2022).
50. Cowger, W.; Steinmetz, Z.; Gray, A.; Munno, K.; Lynch, J.; Hapich, H.; Primpke, S.; De Frond, H.; Rochman, C.; Herodotou, O. Microplastic Spectral Classification Needs an Open Source Community: Open Specy to the Rescue! *Anal. Chem.* **2021**, *93*, 7543–7548. [CrossRef] [PubMed]
51. Pigments Checker, v.5. Available online: https://chsopensource.org/pigments-checker/ (accessed on 17 March 2022).
52. Shard, A.G. Practical guides for X-ray photoelectron spectroscopy: Quantitative XPS. *J. Vac. Sci. Technol. A* **2020**, *38*, 041201. [CrossRef]
53. Baer, D.R.; Artyushkova, K.; Cohen, H.; Easton, C.D.; Engelhard, M.; Gengenbach, T.R.; Greczynski, G.; Mack, P.; Morgan, D.J.; Roberts, A. XPS guide: Charge neutralization and binding energy referencing for insulating samples. *J. Vac. Sci. Technol. A* **2020**, *38*, 031204. [CrossRef]

54. Cohen, J. *Statistical Power Analysis for the Behavioral Sciences*, 2nd ed.; Lawrence Erlbaum Associate: Mahwah, NJ, USA, 1988; ISBN 0-8058-0283-5.
55. Lenth, R.V. Some Practical Guidelines for Effective Sample Size Determination. *Am. Stat.* **2001**, *55*, 187–193. [CrossRef]
56. Sawilowsky, S.S. New Effect Size Rules of Thumb. *J. Mod. Appl. Stat. Methods* **2009**, *8*, 26. [CrossRef]
57. Valsesia, A.; Desmet, C.; Ojea-Jiménez, I.; Oddo, A.; Capomaccio, R.; Rossi, F.; Colpo, P. Direct quantification of nanoparticle surface hydrophobicity. *Commun. Chem.* **2018**, *1*, 53. [CrossRef]
58. Desmet, C.; Valsesia, A.; Oddo, A.; Ceccone, G.; Spampinato, V.; Rossi, F.; Colpo, P. Characterisation of nanomaterial hydrophobicity using engineered surfaces. *J. Nanopart. Res.* **2017**, *19*, 117. [CrossRef]
59. Diaz-Basantes, M.F.; Conesa, J.A.; Fullana, A. Microplastics in honey, beer, milk and refreshments in Ecuador as emerging contaminants. *Sustainability* **2020**, *12*, 5514. [CrossRef]
60. Magrì, D.; Sánchez-Moreno, P.; Caputo, G.; Gatto, F.; Veronesi, M.; Bardi, G.; Catelani, T.; Guarnieri, D.; Athanassiou, A.; Pompa, P.P.; et al. Laser ablation as a versatile tool to mimic polyethylene terephthalate nanoplastic pollutants: Characterization and toxicology assessment. *ACS Nano* **2018**, *12*, 7690–7700. [CrossRef] [PubMed]
61. Electronic Code of Federal Regulations PART 178—Indirect Food Additives: Adjuvants, Production Aids, and Sanitizers. Available online: https://www.ecfr.gov/cgi-bin/text-idx?SID=a1ef4942d858446924d35877fa61effc&mc=true&node=pt21.3.178&rgn=div5 (accessed on 17 March 2022).
62. European Union. Commission Regulation (EU) No 10/2011 of 14 January 2011 on plastic materials and articles intended to come into contact with food (Text with EEA relevance). *Off. J. Eur. Union* **2011**, *45*, 1–89.
63. Osticioli, I.; Mendes, N.F.C.; Nevin, A.; Gil, F.P.S.C.; Becucci, M.; Castellucci, E. Analysis of natural and artificial ultramarine blue pigments using laser induced breakdown and pulsed Raman spectroscopy, statistical analysis and light microscopy. *Spectrochim. Acta Part A Mol. Biomol. Spectrosc.* **2009**, *73*, 525–531. [CrossRef] [PubMed]
64. Wei, D.; Chen, S.; Liu, Q. Review of Fluorescence Suppression Techniques in Raman Spectroscopy. *Appl. Spectrosc. Rev.* **2015**, *50*, 387–406. [CrossRef]
65. Yakubovskaya, E.; Zaliznyak, T.; Martínez Martínez, J.; Taylor, G.T. Tear Down the Fluorescent Curtain: A New Fluorescence Suppression Method for Raman Microspectroscopic Analyses. *Sci. Rep.* **2019**, *9*, 15785. [CrossRef] [PubMed]
66. Valsesia, A.; Parot, J.; Ponti, J.; Mehn, D.; Marino, R.; Melillo, D.; Muramoto, S.; Verkouteren, M.; Hackley, V.A.; Colpo, P. Detection, counting and characterization of nanoplastics in marine bioindicators: A proof of principle study. *Microplast. Nanoplast.* **2021**, *1*, 5. [CrossRef]
67. Ivleva, N.P. Chemical Analysis of Microplastics and Nanoplastics: Challenges, Advanced Methods, and Perspectives. *Chem. Rev.* **2021**, *121*, 11886–11936. [CrossRef]
68. Li, J.; Liu, H.; Paul Chen, J. Microplastics in freshwater systems: A review on occurrence, environmental effects, and methods for microplastics detection. *Water Res.* **2018**, *137*, 362–374. [CrossRef]
69. Shim, W.J.; Hong, S.H.; Eo, S.E. Identification methods in microplastic analysis: A review. *Anal. Methods* **2017**, *9*, 1384–1391. [CrossRef]
70. Busse, K.; Ebner, I.; Humpf, H.U.; Ivleva, N.; Kaeppler, A.; Oßmann, B.E.; Schymanski, D. Comment on "plastic Teabags Release Billions of Microparticles and Nanoparticles into Tea". *Environ. Sci. Technol.* **2020**, *54*, 14134–14135. [CrossRef]
71. Schuck, P.; Perugini, M.A.; Gonzales, N.R.; Hewlett, G.J.; Schubert, D. Size-distribution analysis of proteins by analytical ultracentrifugation: Strategies and application to model systems. *Biophys. J.* **2002**, *82*, 1096–1111. [CrossRef]
72. Mehn, D.; Capomaccio, R.; Gioria, S.; Gilliland, D.; Calzolai, L. Analytical ultracentrifugation for measuring drug distribution of doxorubicin loaded liposomes in human serum. *J. Nanopart. Res.* **2020**, *22*, 158. [CrossRef]
73. Mehn, D.; Iavicoli, P.; Cabaleiro, N.; Borgos, S.E.; Caputo, F.; Geiss, O.; Calzolai, L.; Rossi, F.; Gilliland, D. Analytical ultracentrifugation for analysis of doxorubicin loaded liposomes. *Int. J. Pharm.* **2017**, *523*, 320–326. [CrossRef] [PubMed]

Article

An In Situ Experiment to Evaluate the Aging and Degradation Phenomena Induced by Marine Environment Conditions on Commercial Plastic Granules

Cristina De Monte [1], Marina Locritani [2,*], Silvia Merlino [3], Lucia Ricci [1], Agnese Pistolesi [1,4] and Simona Bronco [1]

1. Istituto per i Processi Chimico-Fisici, Sede di Pisa, del Consiglio Nazionale delle Ricerche, (IPCF-CNR), Via G. Moruzzi 1, 56124 Pisa, Italy; cristina.demonte@pi.ipcf.cnr.it (C.D.M.); lucia.ricci@pi.ipcf.cnr.it (L.R.); a.pistolesi@studenti.unipi.it (A.P.); simona.bronco@pi.ipcf.cnr.it (S.B.)
2. Istituto Nazionale di Geofisica e Vulcanologia, Via di Vigna Murata 605, 00143 Roma, Italy
3. Istituto di Scienze Marine, Sede di Lerici, del Consiglio Nazionale delle Ricerche (ISMAR-CNR), 19032 Lerici, Italy; silvia.merlino@sp.ismar.cnr.it
4. Dipartimento di Chimica e Chimica Industriale, Università di Pisa, Via G. Moruzzi 13, 56124 Pisa, Italy
* Correspondence: marina.locritani@ingv.it

Abstract: In this paper, we present two novel experimental setups specifically designed to perform in situ long-term monitoring of the aging behaviour of commercial plastic granules (HDPE, PP, PLA and PBAT). The results of the first six months of a three year monitoring campaign are presented. The two experimental setups consist of: (i) special cages positioned close to the sea floor at a depth of about 10 m, and (ii) a box containing sand exposed to atmospheric agents to simulate the surface of a beach. Starting from March 2020, plastic granules were put into the cages and plunged in seawater and in a sandboxe. Chemical spectroscopic and thermal analyses (GPC, SEM, FTIR-ATR, DSC, TGA) were performed on the granules before and after exposure to natural elements for six months, in order to identify the physical-chemical modifications occurring in marine environmental conditions (both in seawater and in sandy coastal conditions). Changes in colour, surface morphology, chemical composition, thermal properties, molecular weight and polydispersity, showed the different influences of the environmental conditions. Photooxidative reaction pathways were prevalent in the sandbox. Abrasive phenomena acted specially in the sea environment. PLA and PBAT did not show significant degradation after six months, making the possible reduction of marine pollution due to this process negligible.

Keywords: multi-parametric platform; bioplastics; polymer degradation; marine environment; microplastics; spectroscopy; resin pellets

Citation: De Monte, C.; Locritani, M.; Merlino, S.; Ricci, L.; Pistolesi, A.; Bronco, S. An In Situ Experiment to Evaluate the Aging and Degradation Phenomena Induced by Marine Environment Conditions on Commercial Plastic Granules. *Polymers* **2022**, *14*, 1111. https://doi.org/10.3390/polym14061111

Academic Editor: Jacopo La Nasa

Received: 9 February 2022
Accepted: 3 March 2022
Published: 10 March 2022

Publisher's Note: MDPI stays neutral with regard to jurisdictional claims in published maps and institutional affiliations.

Copyright: © 2022 by the authors. Licensee MDPI, Basel, Switzerland. This article is an open access article distributed under the terms and conditions of the Creative Commons Attribution (CC BY) license (https://creativecommons.org/licenses/by/4.0/).

1. Introduction

The United Nations Environment Programme (UNEP) defines "marine litter" as "any persistent, manufactured or processed solid material discarded, disposed of, or abandoned in the marine and coastal environment", including items made of metals, wood, glass, polymeric materials and some other materials [1]. The amount of marine litter has increased enormously since the middle of the twentieth century, when the awareness of the problem first started to grow, so much so that its size is tentatively considered as one of the factors defining the time boundaries of a new geological era. Many studies confirm that plastic is the main component of marine litter [2–4]. However, this quantity can be roughly estimated only and existing reports lead to different results [5].

In 2020, Napper and Thompson established that more than 75% of "marine litters" are plastics [6]. Polymers by their nature do not chemically degrade significantly in the marine environment and, transported by currents, end up accumulating and being deposited on beaches, in the sea (surface, seabed and water column) and even in Arctic Sea ice, depending on the specific density of the materials [7–10]. For decades, plastic debris in the oceans, coastlines and sea floor has been of great concern as a conspicuous form of pollution [11–18].

In 2021, Ali et al. analysing many studies in recent years, related the abundance of use of various plastics to the abundance of plastic waste dispersed in the environment (air, soil, water) and the consequent negative impact on the environment, animals (in terms of health and food chain), humans (inhalation, ingestion, food chain) and human health [19].

The exposure of macro and meso-plastics (>5 mm) to mechanical, chemical, and biological degradation can lead to their breakdown into smaller pieces at the scale of microplastics, defined as plastic particles with a size of 1 µm to 5 mm [20], or even nanoparticles <1 µm [6,13,21–26].

Photooxidation, in particular, leads to the weakening and fracturing of polymers, giving rise to so called "second generation microplastics". There are also "first generation microplastics" ([5,11,22,27]) that are directly introduced into the marine environment in the form in which they were industrially produced (e.g., microbeads, resin pellets or pre-production granules) [28].

Micro and nanoplastic can be composed of different plastic polymers, such as, among others: polyethylene terephthalate (PET); high-density polyethylene (HDPE); polyvinyl chloride (PVC), which can also vary as rigid PVC, flexible PVC, and polyvinyl acetate (PVAc); low-density polyethylene (LDPE); polypropylene (PP); and polystyrene (PS). Moreover, such plastic is found in a wide variety of structures and shapes, being classified principally as fragments, pellets, filaments, films, foam plastic, granules, and styrofoam [16,29].

The small size of the fragments leads to greater complexity in their recovery and identification, and challenges in the accuracy of the analyses that need to be performed on them [30].

The presence of microplastics in the coastal and marine environment has been widely reported worldwide [18,29], see also [20,31–43].

In addition to physical damage (e.g., entanglement, suffocation, hypoxia, anoxia [15,39,44–52]), several studies have reported toxicological damage caused by the presence of contaminants in the environment (e.g., pesticides, phthalates, PCBs and bisphenol A), adsorbed and released by the plastic debris ingested by the biota and their possible movement along the food chain, with consequences that are still not completely understood [24,53–70]. In addition, several plastic items are composed of hazardous monomers and additives (e.g., plasticizer, stabilizer, flame retardants, antioxidants, etc. ([67,71]), and/or they adsorb chemical pollutants from aquatic environments [53,72–77].

To date, detrimental effects of microplastics have been reported in fishes [78], mussels [79], meiofauna [80], shrimps [81], crabs [82], and seabirds [83].

In recent years, experimental studies have also focused on the mechanisms of object (especially plastic) degradation [84], and on the connections between degradation and pollutant absorption/desorption mechanisms. The topic is extremely important, because, in addition to its scientific interest, a better understanding of these phenomena could contribute to the development of mitigation actions. It has been stressed by several authors that there are several factors in the sea that lead to a reduction in the speed of degradation. These include lower solar radiation exposure, lower temperatures and changes in salinity [7,22]. Fouling by micro-organisms, limiting the exposure of the surface to UV light, blocks surface photodegradation [85], and reduces the buoyancy of the objects [84], causing a further reduction in exposure to direct sunlight. An understanding of the degradation mechanisms of abandoned plastics in the environment, and in particular in the marine environment, is complicated by variability of the plastics in question, the complexity and extent of the environmental matrix and the boundary conditions. Other issues that cannot be neglected are the mobility and determination of the exact residence times of the materials to be studied.

For example, polyethylene resin pellets kept for eight weeks in the dark, in saltwater under mechanical stirring, showed more morphological and structural changes than under UV exposure [86].

The properties and characteristics of the various polymeric species are related to the mechanisms of chemical and physical degradation (e.g., thermal degradation, photo degradation, hydrolysis) and biological degradation, and the climatic and environmental factors that can affect them, although there is currently no process capable of degrading waste plastic [19].

1.1. Bioplastic Case

Several solutions have been proposed in an effort to curb the problem of the increasing presence of plastics in the sea. These include replacing traditional plastics with so-called "bio-plastics" or "compostable plastics". In part, this has already been implemented in the European Union as a result of EU Directive 2019/904. The goal of the directive was to reduce the consumption of standard single-use plastic items, but, in fact, resulted in the replacement of traditional plastic disposables with biodegradable/compostable plastic disposables. This kind of action does not solve the problem, if it is not accompanied by other equally important actions, such as improving waste management (e.g., sorting, chemical recycling, pyrolysis), preventing waste reaching the sea (e.g., barriers in rivers, improving the transport system and landfills, etc.), and environmental education (e.g., recognizing different types of materials, raising awareness of reuse before recycling). In fact, with implementation of the law, the type of material arriving in the sea changed, with little effect on its quantity. The assumption behind the decision was that biodegradable materials in the sea degrade much faster than traditional materials. However, is this really the case? Some recent studies, focusing on the behaviour of different biodegradable polymers in the environment, cast serious doubt on this.

In 2010, O'Brine and Thompspon [87] compared marine degradation (in Devon, UK) of different plastics, including compostable Mater Bi, two oxo-degradable polyethylenes, and traditional PE, partially from recycled plastics, and observed discharge of 100% of the mass of Mater Bi in 24 weeks, with minimal change observed in the other plastics after 40 weeks. They note that fragmentation of Mater-Bi caused its dispersion in free seawater with no debris being recovered for analysis.

In 2017, Bagheri et al. [88] carried out a study on biodegradable polyesters, including PLGA, PCL, PLA, PHB, Ecoflex and the non-biodegradable polyester polymer PET, maintaining them in artificial seawater and freshwater under controlled conditions for one year. They observed 100% loss only for PLGA, while the other polymers remained intact over 400 days, and concluded that so-called biodegradable polymers do not degrade in water under natural conditions.

In 2020, Kliem et al. [10] analysed the biological degradation of certain polymers under specific conditions of temperature and pH that simulate natural environments, such as soil, fresh water and seawater. They showed that PLA, a biodegradable polymer currently used in industry, does not decompose significantly in seawater or fresh water, but only after industrial composting, with temperature together with battery activity being the most important factors affecting and enabling degradation. Experimental investigations involving biodegradability comparison of chitosan, PBAT (Ecoflex) and HDPE, were carried out in saltwater at the Aquarius underwater laboratory in 2020 [89], with good biodegradation observed for chitosan, 16% of biodegradation for PBAT, and only 7.8 % for HDPE, with FTIR used to evidence structural changes.

Numerous degradation experiments and studies have been performed under controlled conditions and in simulated environments, with only a small number conducted in seawater; however, even in a laboratory environment, it is possible to evaluate photodegradation and thermal-degradation.

1.2. Resin Pellets Case

Among first generation microplastics, in the millimetre range (from 1 to 5 mm) "resin pellets" represent a significant proportion. Pellets are granules of different polymer types, used to produce macroplastic objects, through melting followed by extrusion and moulding. Since there is still no strict regulation for the adoption of measures to prevent the possible loss of these millimetre-sized plastics during their transport, storage and processing, the pellets are easily dispersed in the environment, and are now present in many regions, including the polar regions [90]. Recent studies have shown that their percentage content ranges from 3% [91] to about 30% of all microplastics surveyed on beaches [40,92]. The occurrence of hydrophobic contaminants in marine plastic pellets has been widely reported, and their measurement has established a global association between pollutants and plastic pellets discarded at sea [72–74,77,93–101].

Laboratory experiments, theoretical modelling and several field experiments have been performed to investigate how the physical and chemical properties of each type of polymer (e.g., surface area, diffusivity, and crystallinity) influence the sorption of chemicals by resin pellets [53,72,95,98,102,103]. The results of these studies confirm that the sorption patterns are dependent on the properties of the particular compound considered (e.g., hydrophobicity, molecular weight (MW)) and on the polymer type, but also on the environmental conditions, such as weathering and water salinity. In many studies, polyethylene (PE) has been shown to present a larger surface area than polypropylene (PP) and to have an affinity for a wide range of organic contaminants varying in hydrophobicity [72,104,105].

PE, together with PP, is produced in very large quantities in west European countries and represents a significant proportion of plastic debris in the environment [75]. This evidence led to the decision to use PE pellets as passive samplers of POP pollution in the International Pellet Watch project [94,97].

Predictive modelling for low-density polyethylene (LDPE) acting as a passive sampling device suggests that the time to reach saturation could vary from days to months, depending on the chemical and environmental conditions considered [97], but also on the sorption/desorption models used, and on the weathering state of the sample [53,95,102].

Most of these studies were carried out in the laboratory, i.e., with the application of precise doses of contaminants in the case of adsorption experiments, or the establishment of controlled conditions of temperature and pressure for degradation experiments. For example, Da Costa et al. [87] studied morphological and thermic variation (using FTIR-ATR analysis, RAMAN spectra, TGA analysis, SEM images) in PE pellets maintained in artificial seawater for eight weeks under controlled temperature, light exposure and movement conditions, and observed quantifiable chemical and physical impacts on the structural and morphological characteristics of granules. They recommended this type of approach to investigate the degradation and behaviour of microplastics.

Such controlled experiments, on the other hand, overlook many factors dependent on environmental and unpredictable weather conditions. Moreover, in laboratory experiments the use of organic-free water prevents a proper assessment of the role of dissolved organic material in the marine environment, for example, by favouring the transfer of chemicals into the pellets, extending the equilibrium time scale for pollutants sorption [103], or influencing weathering mechanisms at the pellet surface.

With respect to the few experiments conducted in free seawater, some of them measured sorption of POPs onto previously uncontaminated plastic pre-production pellets ([73,104]), others focused on degradation of the polymeric matrix [106], while still others focused of the colonization by microorganisms of traditional-plastic and bio-plastic pre-production pellets [107–109]. The importance of weathering is emphasized in all these studies. Rochmann et al. [103] focused on the importance of weathering influencing the diffusion behaviour of chemical compounds within the pellets in marine environments. Weathering, leading to an increase of the surface area and pore size of the polymer, promotes an even greater diffusivity in PE pellets, allowing them to sorb a higher quantity of substance. In 2016, Brandon et al. [110] focused only on pellet degradation processes, considering pellets weathered in dry or seawater simulated environment, both in darkness and sunlight conditions. Their results showed non-linear changes in chemical bonds with exposure time to selected environmental conditions.

The long-term effect of atmospheric agents on the mechanisms of degradation of plastics (and, in particular, bioplastics and compostable plastics) in the marine environment, and the connection between degradation and the uptake of chemicals, appears to be a still evolving field of study.

Based on the background described above, our experiment sought to investigate the behaviour of plastic items in the marine environment. These items were both artefacts and pellets of commonly used plastic materials. They were made up of traditional and biodegradable polymers contained in special structures built for the purpose and placed in seawater, or on a simulated beach, and left to age for a minimum of three years. At intervals of a few months, the materials were sampled, and the necessary measurements were taken, to verify any intervening structural and morphological changes. The experiment is still in progress after twenty four months from the positioning of the granules in both environments.

In the Materials and Methods section, we describe the setup of the experiment, the preparation of the cages and samples, the sampling of a portion of the granules of each material after six months, and the types of analyses performed on the samples. The remainder of the granules was left for further sampling in the following months. In the Results section, we report the data obtained for all materials introduced into both the cages and the sandbox, as well as the first results obtained after six months from deployment. In the Discussion section, we comment on these results in the light of what is currently available in the literature. In the Conclusion section, we summarize what has been described above, and describe what our future objectives will be with respect to continuation of the analyses of the material obtained to date, and to continuation of the experiment itself.

2. Materials and Methods

2.1. Materials

Two types of standard polymer pellets (high density polyethylene—HDPE, Auser Polimeri, Coreglia Antelminelli (LU), Italy, and polypropylene—PP, PoliEko, Celje, Slovenia) were used, together with two types of biodegradable polymer pellets (polylactic acid—PLA, Ingeo 2002D®, NatureWorks, Plymouth, MN, USA, and polybutylene adipate-co-terephthalate—PBAT, Ecoflex® F Blend C1200, BASF, Ludwigshafen, Germany). They were used as received.

2.2. Experimental Set Up

The experiment was carried out in the Bay of Santa Teresa, a small bay inside the tourist and commercial Gulf of La Spezia, Italy (Figure 1).

The site hosts precious marine and terrestrial ecosystems and lies in very close proximity to the headquarters of the research institutes of the National Research Council (CNR), the National Institute of Geophysics and Volcanology (INGV) and the National Agency for New Technologies, Energy and Sustainable Economic Development (ENEA), which has made the bay a natural laboratory in which to undertake research and evaluate technology. Within this framework, the "Smart Bay Santa Teresa" (https://smartbaysteresa.com/, accessed on 7 February 2022) initiative was inaugurated, a collaboration platform that hosts different scientific projects and an underwater observatory called LabMARE.

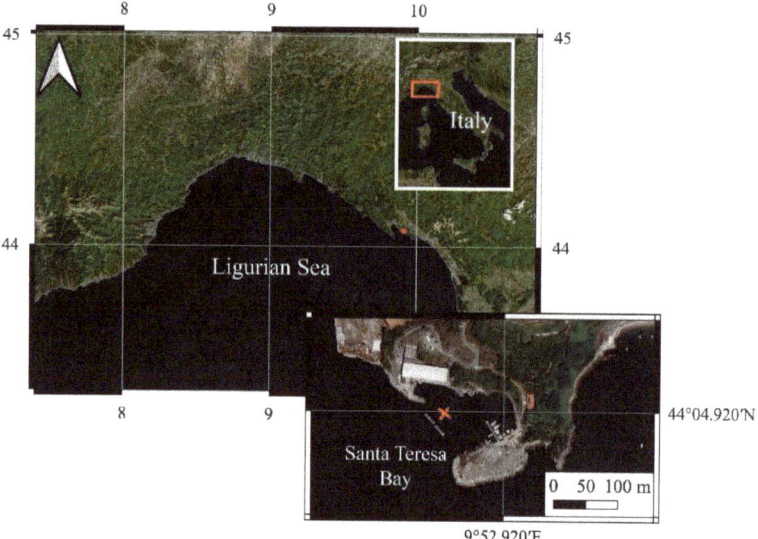

Figure 1. Geographic localization of LabMARE station.

This underwater multi-parametric platform was developed by the Ligurian Cluster of the Marine Technologies (DLTM), in collaboration with the INGV, CNR, ENEA, the Hydrographic Institute of the Navy with the support of the Municipality of Lerici and the Cooperative Mitilicoltori Associati, as part of the LabMARE project funded by the Liguria Region.

The main aim of the underwater observatory is experimentation with new technologies and environmental status monitoring. LabMARE is equipped with sensors for monitoring environmental parameters (temperature and salinity) and special cages for studying the degradation of plastics in the marine environment (Figure 2).

The underwater observatory was deployed on 3 March 2020 at 44°4′55.08″ N –9°52′50.46″ E at 10 m depth, about 60 m from the coast (Figure 1) and provides real time temperature data through the underwater cable connected to the land station and via transmission of data by internet connection.

The two special cylindric cages (about 40 cm × 30 cm, Figure 2) installed on the structure of the underwater observatory were self-produced by the researchers in stainless steel 318, each one containing three "baskets" (about 15 cm × 10 cm, Figure 2 in stainless steel wire mesh (AISI 316, mesh 0.24 mm)), electropolished and with anodic passivation. The cages were closed with plastic ties and anchored to the underwater observatory with large plastic ties. Different types of plastic objects, both macro and micro, made of both standard polymers and biopolymers, were placed inside the cages (Table 1). The choice to use metal, avoiding any plastic material to build the cage and baskets was made to prevent any possible contamination, such as possible release of additives from the plastic material (Plastic Additive Standards Guide. AccuStandard, 2018 [106]) or interference with the pellets in the process of absorbing chemicals from the seawater.

Despite the presence of sacrificial anodes placed at one end of each of the two cages, as well as at the end of each of the inner baskets, during August and September 2020, the metal of the cages was affected by corrosion due to a galvanic currents effect. This rapid corrosion was probably due to the large presence of cation metals in the port area. On November 2020, the cages and baskets were changed with new ones constructed by the researchers in collaboration with an enterprise expert in the industrial and naval field (Vamp s.nc., La Spezia): two large cages in stainless steel 316 (size: 70 × 30 cm), 6 small cages in stainless steel 316 (size: 20 × 15 cm) and 8 baskets in stainless steel 316 with meshed net of 1 mm of two different sizes (size: 19 × 14 cm and 14 × 14 cm). Each was equipped with sacrificial anodes to avoid galvanic corrosion. Cages were closed with plastic ties and anchored to the underwater observatory with large plastic ties.

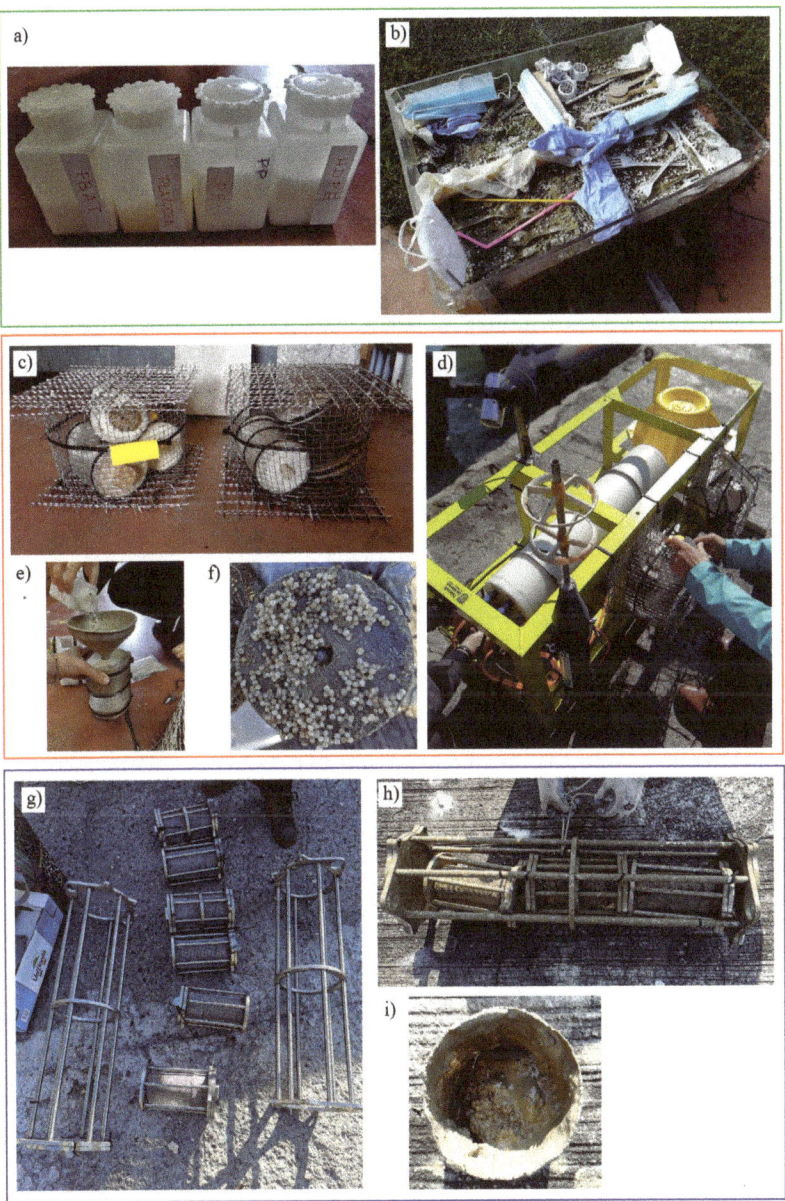

Figure 2. The figure shows: (**a**) containers with raw pellets, (**b**) the sandbox, (**c**) the old, and (**g**) the new version of the cages, (**d**) the underwater observatory before the deployment with the cages installed, (**e**) the insertion of pellets in basket, (**f**,**i**) the pellets in the basket during sampling, (**h**) the content of the cages after first sampling.

Table 1. DSC programs for the polymers. Rate of the scans: 10 °C/min.

	First Heating (from/to)	Hold (min)	Cooling (from/to)	Hold (min)	Second Heating (from/to)	Hold (min)
HDPE/PP	20–220 °C	2	220/−80 °C	5	−80/220	2
PLA	20–180 °C	1	180/−80 °C	1	−80/180	1
PBAT	20–200 °C	1	200/−80 °C	1	−80/200	1

A sandbox containing sand taken from the Gulf of La Spezia (Le Grazie Bay) was set up near the dock where the cages were located, and the same materials introduced into the cages were placed in it, in order to carry out a comparative study of the degradation and absorption of chemicals at sea and on a "simulated beach". Due to the COVID-19 health emergency, and the consequent total lockdown that prevented free circulation in Italy from February to June 2020, the sandbox was promptly transported to an area near the home (44°7′51.04″ N; 9°57′28.88″ E, in Sarzana) of one of the authors, where it was possible to check that it was not tampered with or damaged.

2.3. Experimental Design

The sampling campaign started on 3 March 2020, corresponding to point zero contextually of the first deployment of the LabMARE station and the cages, for all inserted objects, both in the sea cages and in the simulated beach.

The experiment was designed to last three years in total. The objects placed in the cages were of different type, size and material: polymer pellets (see Section 2.1), coffee capsules, single use tableware, personal protective equipment and cigarettes.

In order to study the evolution of ageing of the objects in the different environments, samples were taken on the same day from the cages in the sea and from the sandbox every few months. The sampling dates for the first two years of the experiment were as follows: September 2020, November 2020, March 2021, June 2021, October 2021 and the next sampling is scheduled for the first week of March 2022. All the samples taken in the sea and in the sandbox were washed with ultra-pure water and dried in air before being analysed.

In this investigation, comparison between the characteristics and properties of the raw materials (four types of pellets) and those placed in the two different environments for six months (sampled on September 2020) is discussed. This is the period during which a biodegradable material should be able to biodegrade, according to the UNI EN ISO 14855-2:2018, although the legislation considers composting conditions.

The sampling of material contained in the cages was performed with the support of scubadivers of Dipartimento Polizia di Stat—Centro Nautico Sommozzatori of La Spezia (Italy).

2.4. Instruments and Analysis Methods

The marine station LabMARE was equipped with the sensor SBE37SM (supplied by Sea-Bird Scientific) just before the lockdown due to SARS-CoV-2 to record data every 10 min from March 2020. All the data collected were evaluated by calculation of monthly averaged values from the average daily temperatures with relative standard deviations and the maximum and minimum temperatures for each month.

A hygrometer thermometer wireless Bluetooth sensor supplied by ORIA was put in a sandbox located at the CNR research area of Pisa to record the temperature data every 10 min from March 2021. All the data collected were evaluated by calculation of monthly averaged values from the average daily temperatures with relative standard deviations and the maximum and minimum temperatures for each month.

Attenuated total reflectance (ATR) spectra were registered using a Fourier Transform—InfraRed Jasco 6200 (Jasco, Tokyo, Japan) equipped with a PIKE MIRacle (Madison, WT,

USA) accessory. Each sample underwent 64 scans from 4000 to 650 cm^{-1}, after the collection of background data.

Thermogravimetric analysis (TGA) was performed with an SII TG/DTA 7200 EXSTAR Seiko analyser (Seiko, Chiba, Japa), under heating from 30 to 700 °C, at 10 °C/min rate. Air was fluxed at 200 mL/min during all measurements. The sample amount used for TGA was 5–10 mg. The first derivative (DTG) for each curve was recorded for the different samples and analysed. T_{onset} (temperature corresponding to starting degradation) and residue amount at 700 °C were determined from each TGA profile. T_{max} (temperature at which the maximum rate of mass loss occurs) was calculated from each DTG curve.

Differential scanning calorimetry (DSC) analysis was carried out in order to assess any effects induced by the degradation processes triggered during the months of immersion and in the sandbox relating to plasticisation phenomena or changes in crystallinity. DSC analysis was performed using a Seiko SII ExtarDSC7020 calorimeter (Seiko, Chiba, Japan) with a different thermal programme for each type of polymer (Table 1). Each measurement was carried out on approximately 5–10 mg of materials. The software associated with the instrument was used to determine the values of glass transition temperature (T_g), thermal capacity variation associated with the glass transition (ΔC_p), melting temperature (T_m) with relative enthalpy of fusion (ΔH_m), and cold crystallization temperature (T_{cc}), with relative enthalpy (ΔH_{cc}) estimated from the second heating scan. The degree of crystallinity ($\chi\%$) was determined using the following Equations (1) and (2) [107]:

$$\chi\% = \frac{\Delta H_m}{\Delta H_{0m}} \times 100 \quad (1)$$

$$\chi\% = \frac{\Delta H_m + \Delta H_{cc}}{\Delta H_{0m}} \times 100 \quad (2)$$

with ΔH_m = mean enthalpy of fusion of the sample, ΔH_{0m} = enthalpy of fusion of material 100% crystalline and ΔH_{cc} enthalpy of cold crystallization.

GPC measurements were carried out on PLA and PBAT polymers using an HPLC Agilent 1260 Infinity (Agilent, Santa Clara, CA, United States) equipped with a three-way valve as the injection system and a styrene-divinylbenzene resin as the stationary phase, at an operating pressure of 80 bar, with CHCl$_3$ as eluent at 0.3 mL/min and a refractometer as a detector. Solutions were prepared by dissolving the materials in CHCl$_3$ (HPLC grade) with a concentration of 3 mg/mL and filtered two times with an Agilent filter with a porosity of 2 µm.

SEM images were recorded with an FEI Quanta 450 ESEM FEG scanning electron microscope (Thermo Fisher, Waltham, MA, USA) at CISUP Laboratories (Centro per la Integrazione della Strumentazione) at University of Pisa. The specimens were analysed on the external surface without any specific manipulations.

3. Results

As stated in Section 2, the results obtained from the above-mentioned analyses on standard (HDPE and PP) and biodegradable (PLA and PBAT) polymeric granules during the first six months of the study were reported. The choice of six months was not accidental. The idea was to follow modification due to aging and eventual degradation occurring to the granules in a marine environment (direct immersion in sea or deposition on the surface of the sand in the sandbox) in the first six months after positioning, the period during which a biodegradable material should be able to biodegrade. The results obtained on each kind of "bio" material were compared with the corresponding results for standard polymer granules.

3.1. Environmental Analyses

Analysis of the temperatures reached in the sea and in the sandbox in the period under review (March 2020–August 2020) was considered important in order to provide data on whether this variable could affect the characteristics of the aged samples. Table 2 shows the result of the analysis of the data collected at the LabMARE station.

Table 2. Environmental parameters measured in LabMARE.

	Average Temperaures (°C)	Standard Deviation * (°C)	Standard Deviation ** (°C)	Tmin (°C)	Tmax (°C)
March 2020	13.70	0.077	0.32	12.93	14.32
April 2020	15.07	0.055	0.78	13.69	17.33
May 2020	18.20	0.064	0.72	16.93	20.16
June 2020	20.88	0.072	0.85	18.97	23.19
July 2020	21.25	0.11	0.59	19.75	23.35
August 2020	23.68	0.20	2.20	22.12	25.24

* Calculated on daily averages; ** Calculated on all measures.

Unfortunately, it was not possible to place a similar sensor in the sandbox near Sarzana (La Spezia). The following were considered in this case. The temperature data were collected in the period from March 2021 to August 2021 using the sensor in the sandbox at CNR and are reported here in Table 3 according to the same analysis carried out for the data in Table 2. In a similar way, the temperatures of the air recorded in Pisa in the period from March 2021 to August 2021 by the San Giusto Weather Station were compared with the data recorded in Sarzana in the period from March 2020 to August 2020 as measured by the Sarzana Luni Weather Station and reported on the website https://www.ilmeteo.it/portale/archivio-meteo, accessed on 7 February 2022 and in Table 4. No significant differences were observed from one year to the other in the air temperatures; the data recorded in the sandbox in Pisa in 2021 were considered consistent with those of the Sarzana sandbox in 2020 in the same months.

Table 3. Environmental parameters measured in a sandbox placed in Pisa.

	Average Temperaures (°C)	Standard Deviation * (°C)	Standard Deviation ** (°C)	Tmin (°C)	Tmax (°C)
March 2021	12.88	2.94	7.15	1.95	36.16
April 2021	14.98	2.53	6.45	0.123	36.21
May 2021	18.64	2.43	6.05	8.28	42.08
June 2021	25.64	2.15	7.48	11.79	45.60
July 2021	27.69	1.99	6.82	17.84	48.69
August 2021	29.06	2.09	7.19	18.81	51.63

* Calculated on daily averages; ** Calculated on all measures.

Given this premise, the analysis of the results indicated that in both the sandbox and the sea similar average temperatures were reached. The substantial difference was the thermal excursion during the day with a maximum of 51 °C in the sandbox and 25 °C in the sea and the corresponding standard deviations. As a consequence, the materials in the sandbox were more thermally stressed than the same materials in the sea.

Table 4. Comparison of the air temperature data of Pisa and Sarzana.

Sarzana	Average Temperaures (°C)	Standard Deviation (°C)	T_{min} (°C)	T_{max} (°C)	Pisa	Average Temperaures (°C)	Standard Deviation (°C)	T_{min} (°C)	T_{max} (°C)
March 2020	11.09	2.22	2	20	March 2021	9.65	1.85	−1	22
April 2020	14.48	2.58	2	21	April 2021	11.70	2.15	−2	21
May 2020	19.35	1.98	12	25	May 2021	15.39	1.67	5	24
June 2020	20.67	2.67	14	30	June 2021	21.93	1.86	10	32
July 2020	24.65	1.53	18	30	July 2021	24.23	1.89	16	33
August 2020	25.19	1.86	18	32	August 2021	24.74	1.97	15	36

3.2. Six-Months-Aged Standard Polymers

HDPE pellets left in the two different environments (sand and seawater) were analysed using the different techniques described above. The photos collected of the starting granules and of the sampled granules of HDPE after six months in both environments (HDPE_6SW for seawater and HDPE_6S for sand) are compared in Figure 3a.

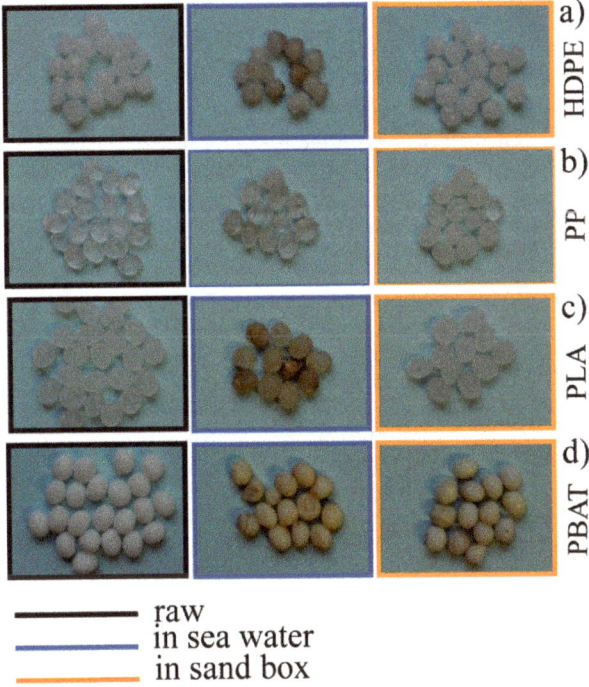

Figure 3. Digital photos of the raw pellets (in black), 6-months-aged in seawater (in blue) and 6-months-aged in sandbox (in orange) (**a–d**).

A colour deviation towards yellow/amber was apparent. For all kinds of granules considered in this experiment, this was much more pronounced in the sea samples. In detail, the effect of an aggressive attack from the environment was evident from the comparison of SEM images (Figure 4a). Modification of the surface morphology taking place correlated with the different action of the sea compared to sand, with formation of an evident porous surface in samples kept in the sea.

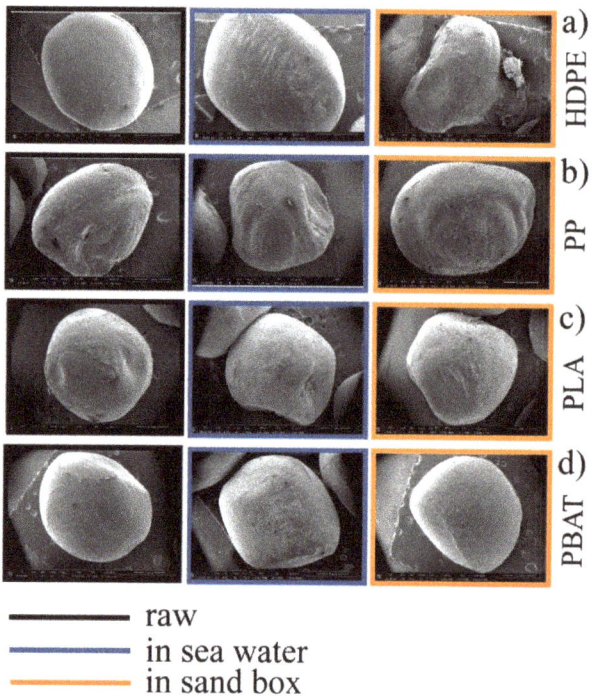

Figure 4. SEM images of the series (a–d): raw granules in black; 6-months-aged in seawater in blue; 6-months-aged in sandbox in orange.

These effects were correlated with chemical modification of the polymer chains on the surface of the granules and/or eventual chemical absorption by ATR analyses. New absorption peaks appeared in the region of OH stretching (at about 3400 cm^{-1}), between 1740–1550 cm^{-1} due to the stretching mode of carboxyl groups and double bonds in both HDPE_6SW and HDPE_6S (Figure 5a). A large and intense peak between 900–1200 cm^{-1} (centred at about 1010 cm^{-1}) was also observed after six months in both environments in the stretching Si-O region.

As far as thermal properties are concerned, TGA and DSC analyses were carried out on the HDPE series. The thermal profiles TGA and DTG are compared in Figure 6a,b and the characteristic temperatures calculated from the thermal curves are summarized in Table 5. A decrease in the thermal stability of the polymer, associated with a decrease in the degradation starting temperature (T_{onset}) was observed for both the sea and sand samples after six months. T_{max} decreased only for seawater samples. A slight decrease in the melting temperature for both the seawater and sand samples was observed, as well as in the crystallinity percentage results from the analysis of the second heating scan in the DSC thermogram (Figure 7a and Table 5). The effect was more apparent for HDPE_6SW, as in the TGA results.

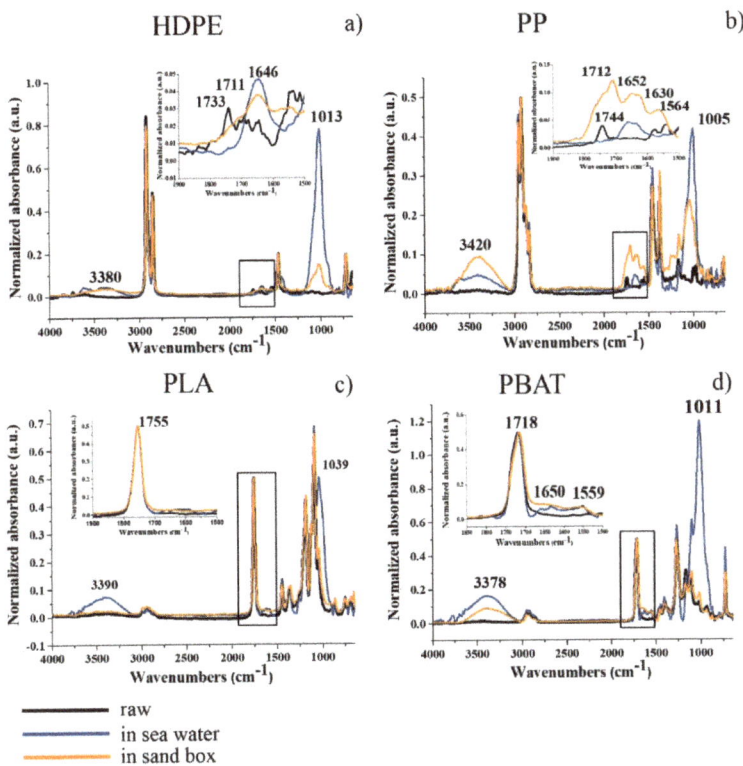

Figure 5. ATR spectra of (**a–d**). raw granules in black; six-months-aged in seawater in blue; six-months-aged in sandbox in orange. The inset shows the enlargement of the spectra highlighted from the square.

Table 5. Significant temperatures from TGA and DSC curves of HDPE samples.

Sample	TGA Results			DSC Results		
	T_{onset} (°C)	T_{max} (°C)	Residue at 700 °C (%)	T_m (°C)	ΔH_m (J/g)	$\chi\%$ *
HDPE	255.5	464.3	0.8	135.1	228	77.8
HDPE_6SW	249.1	444.7	0.0	130.9	183	62.5
HDPE_6S	242.1	465.0	0.0	132.8	207	70.6

* $\Delta H^0_m = 293$ J/g [108]; $\chi\% = \frac{\Delta H_m}{\Delta H_{0m}} \times 100$.

The PP granules (Figure 3b) showed a light colour deviation towards yellow which was more pronounced in the seawater sample (PP_6SW) than in the sand sample (PP_6S), as for HDPE. The appearance of some cracks on the surface of the granules was evident in the SEM images of PP_6S (Figure 4b). Formation of absorption bands in the carbonyl and double bond region was observed in the ATR spectra (Figure 5b) and the disappearance of the peak at 1744 cm^{-1} was noticeable. Absorption peaks appeared in the region of OH stretching (centred at about 3420 cm^{-1}), both for the sample PP-6SW and PP-6S. As for the HDPE samples, a large and intense peak appearing between 900–1200 cm^{-1} (centred at about 1005 cm^{-1}) also formed after six months in both the materials in the stretching Si-O region.

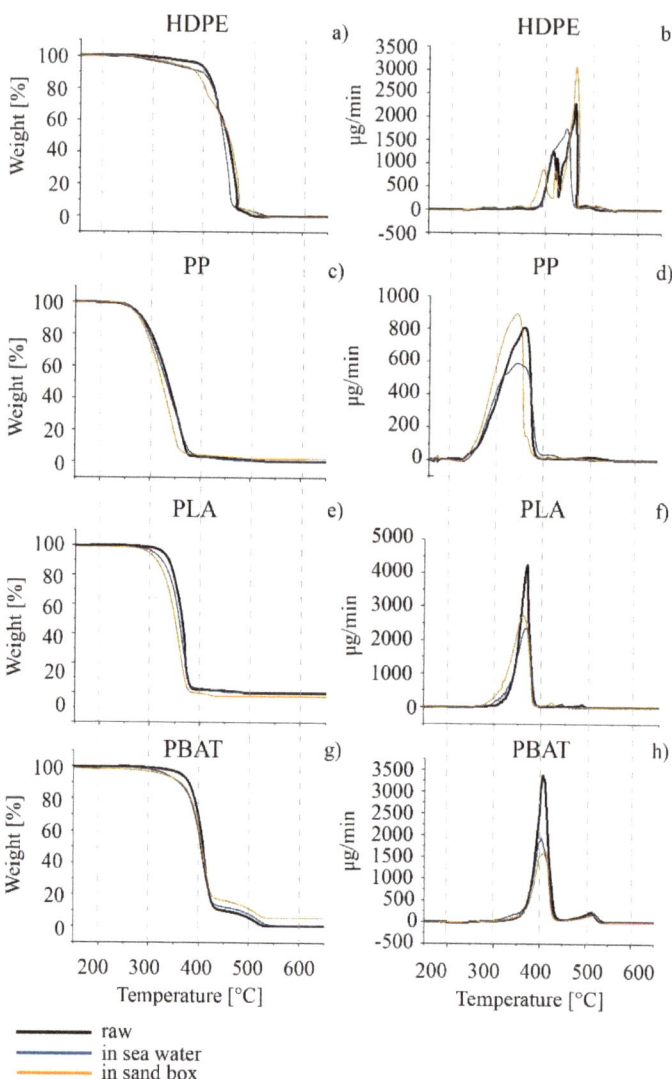

Figure 6. TGA thermograms on the left and the relative DTG on the right of the series (**a–h**): raw granules in black; 6-months-aged in seawater in blue; 6-months-aged in sandbox in orange. The scale on the abscissa is the same for all the curves.

A decrease in T_{onset} and T_{max} with time in both seawater and in sand was observed by comparison of the TGA thermal analyses. An increase in the TGA residue was measured for the PP_6S sample with respect to the other samples (Figure 6c,d and Table 6). A splitting of the T_m peak of PP_6S with respect to PP with the presence of a shoulder and a decrease in the crystallinity was measured in the DSC profiles (Figure 7b and Table 6).

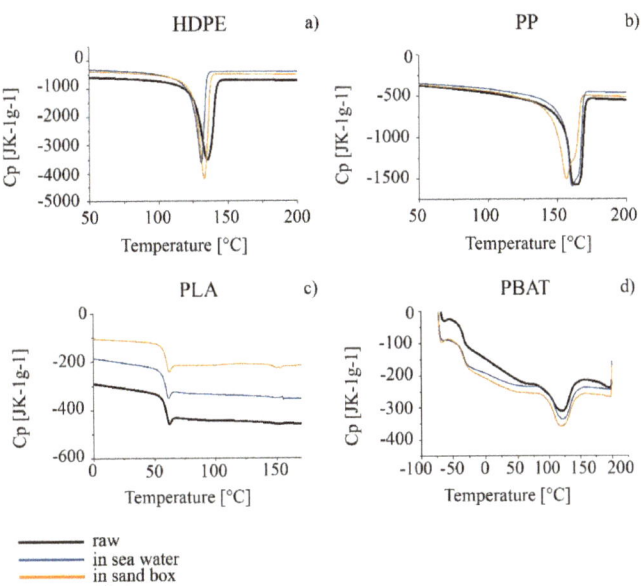

Figure 7. DSC thermograms of the series (**a–d**): raw granules in black; 6-months-aged in seawater in blue; 6-months-aged in sandbox in orange.

Table 6. TGA and DSC results of PP samples.

Sample	TGA Results			DSC Results		
	T_{onset} (°C)	T_{max} (°C)	Residue at 700 °C (%)	T_m (°C)	ΔH_m (J/g)	χ% *
PP	261.5	355.7	0.3	162.5	108.0	52.2
PP_6SW	261.4	340.6	0.9	160.6	102.0	49.3
PP_6S	255.0	339.4	2.2	156.3 (162.4)	98.6	47.6

* $\Delta H^0{}_m$ = 207 J/g [108]; $\chi\% = \frac{\Delta H_m}{\Delta H_{0m}} \times 100$.

3.3. Six-Months-Aged Biodegradable Polymers

PLA granules showed a colour deviation towards yellow especially for the sea samples (PLA_6SW) compared to the sand samples (PLA_6S), as previously described for standard polymers (Figure 3c). Noticeable changes in morphology after six months were not observed in both degradative environments (SEM images in Figure 4c). These results were confirmed by ATR analysis. Only a small difference in the spectrum of seawater samples was observed due to the presence of a large peak centred at about 3400 cm^{-1} in the region of -OH stretching (Figure 5c). A broad band between 1700 and 1550 cm^{-1} was detectable in this spectrum with an absorption band also at 1039 cm^{-1}.

The results from the ATR and SEM images were confirmed by TGA and DSC analyses (Figures 6e,f and 7c, and Table 7). In the DSC profiles, similarly, the samples showed only minimal variations in the characteristic T_g, T_m and crystallinity degree, statistically not different from those of the PLA.

Table 7. TGA and DSC results of PLA samples.

Sample	TGA Results			DSC Results						
	T_{onset} (°C)	T_{max} (°C)	Residue at 700 °C (%)	T_g (°C)	ΔC_p (J/g°C)	T_{cc} (°C)	ΔH_{cc} (J/g)	T_m (°C)	ΔH_m (J/g)	χ% *
PLA	323.1	367.7	0.0	58.3	0.55	122.7	−0.19	150.9	0.21	0.022
PLA_6SW	320.2	365.4	0.6	57.5	0.57	-	-	147.9	0.16	0.17
PLA_6S	315.0	359.2	0.0	58.5	0.51	117.6	−1.89	147.5 (152.4)	2.04	0.16

* $\Delta H^0_m = 93$ J/g [109]; $\chi\% = \frac{\Delta H_m + \Delta H_{cc}}{\Delta H_{0m}} \times 100$.

The molecular weight of the samples and the polydispersity index are given in Table 6. In all cases there was an invariance in polydispersity, even though an improvement in $\overline{Mn}$ and $\overline{Mw}$ values was measured for PLA_6S.

In the case of the PBAT polymer, the effect of the environment on the colour of the pellets was more marked. The pellets developed an amber colour both in the sea (PBAT_6SW) and sand (PBAT_6S) (Figure 3d). The surface of the PBAT granules showed grooves that ran through the granule (Figure 4d). The appearance of a large band in the -OH stretching region (3380 cm^{-1}) was evident from ATR analyses (Figure 5d). The presence of additional peaks in the region 1640–1550 cm^{-1} of the double C=C bond was observed in both aged materials, but was more pronounced in the PBAT_6S samples. An intensive peak was also present at 1011 cm^{-1} in the seawater sample.

Regarding the TGA results, the DTG curves in all cases showed two main peaks: the first can be considered the principal degradative step and was used to calculate T_{max}, while the second, at higher temperatures, represents a subsequent degradation step which occurred with definitely lower speed. The degradation (Figure 6g,h and Table 8) of the polymer started at a lower temperature for the materials aged in both seawater and in sand. A higher percentage of residue was measured of PBAT_6S. Some effects were also evident in the GPC results, especially for PBAT_6S, where the polydispersity increased from 2.2 of starting material to 2.8 with a decrease in $\overline{Mn}$ and $\overline{Mw}$ (Table 9). The DSC results (Figure 7d and Table 8), however, did not undergo significant change [89].

Table 8. TGA and DSC results of PBAT samples.

Sample	TGA Results			DSC Results				
	T_{onset} (°C)	T_{max} (°C)	Residue at 700 °C (%)	T_g (°C)	ΔC_p (J/g°C)	T_m (°C)	ΔH_m (J/g)	χ% *
PBAT	369.9	407.1	0.5	−35.4	0.40	120.6	18.4	16.1
PBAT_6SW	359.3	403.6	0.6	−35.0	0.41	122.1	18.2	16.0
PBAT_6S	357.2	412.6	5.6	−36.0	0.44	119.5	18.7	16.4

* $\Delta H^0_m = 114$ J/g [109]; $\chi\% = \frac{\Delta H_m}{\Delta H_{0m}} \times 100$.

Table 9. GPC results of PLA and PBAT samples.

Sample	$\overline{Mn}$ (KDa)	$\overline{Mw}$ (KDa)	PDI
PLA	84.6	146.3	1.7
PLA_6SW	82.2	145.2	1.8
PLA_6S	88.9	149.7	1.7
PBAT	21.5	47.3	2.2
PBAT_6SW	19.8	45.2	2.3
PBAT_6S	12.6	35.8	2.8

4. Discussion

The polymers chosen for this investigation can be classified, in relation to their chemical structure, into two large families: the C–C backbone and C–O backbone polymers, as reported in Figure 8. HDPE and PP belong to the C–C backbone polymers while PLA and PBAT belong to the C–O backbone polymers.

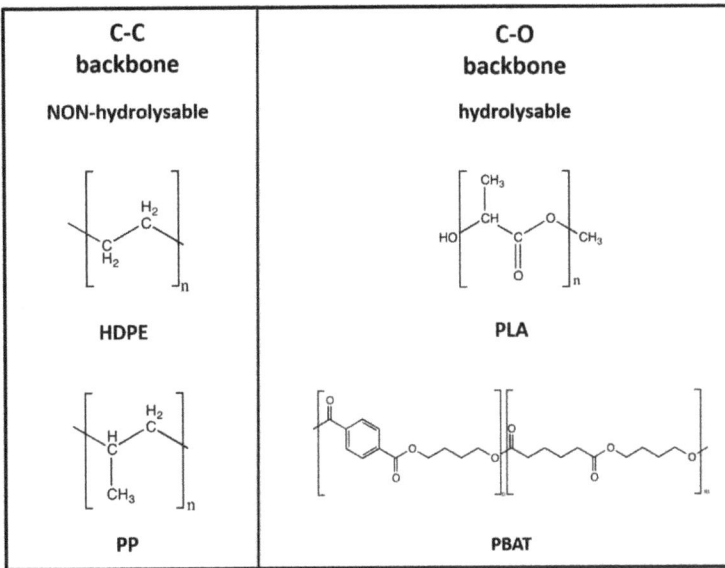

Figure 8. Classification of HDPE, PP, PLA and PBAT based on chemical backbone structure.

Overall, this classification translates into a differentiation of properties and chemical behaviour from the point of view of the possibility of being prone to hydrolysis for C–O backbone polymers (PLA and PBAT) or being non-hydrolysable for C–C backbone polymers (HDPE or PP).

HPDE and PP are traditional non-biodegradable plastics and are mainly fossil-fuel based and partly bio-based, as in bio-PE, bio-PP, and bio-PET. However, PLA and PBAT are biodegradable bioplastics, bio-based in the case of PLA (such as PHA or PBS), and fossil-fuel based in the case of PBAT [111].

The effects of the two environments investigated (seawater and sandbox) on the selected materials in the form of granules during the six months are illustrated well by the photos reported in Figure 3. The change in colour can be attributed to several factors, as reported by Andrady et al. [22] and Ali et al. [19], and increased with time of exposure, as observed by Brandon [110]. The colouring (yellowing) of the granules can be mainly related to the start of the degradation process (i.e., photo-oxidation, hydrolysis) or the absorption of pollutants by the surrounding environment [73]. Overall, photo-oxidation and hydrolysis mechanisms lead to the scission of chains with decrease in molecular weight and formation of double bonds C=C and C=O, or the formation of OH groups. Endo et al. [73] hypothesized that the colour change of the granule surface may be due to the facilitated absorption of additives due to the formation of a porous surface with aging. The more pronounced colour deviation towards yellow/amber, found in our study for all sea samples compared to sand samples, seems to indicate that in the first environment the processes causing yellowing are enhanced (whatever they are).

Anyway, noticeable changes in morphology were already evident after six months from the comparison of SEM images (Figure 4) for all the polymers tested in both sea and in sand, with the exception of PLA. The different effects induced by sea or sand conditions on the change in the surface morphology was evident. The formation of a porous surface, or, at least, areas with cavities due to the effect of abrasive phenomena [112], was confirmed in HDPE_6SW and PP_6SW, but particularly in PBAT_6SW. The formation of grooves that ran through the granule observed on the surface of PP_6S granules and PBAT_6S granules can be correlated with the greater exposure of the granule to solar radiation and to the higher temperature reached in the sand environment, which increases with seasonality very differently from sea-water.

The ATR analyses helped to better understand which processes occurred in both granule environments.

The absorption at about 3400 cm^{-1} highlighted in many spectra could be attributed to OH-stretching and was assumed to be the result of a hydrolysis reaction in the polymer chains with an increase in hydroxyl groups.

The peak at about 1010–1030 cm^{-1}, shown in the spectra can, alternatively, can be attributed to contamination by sand absorbed on the granule surface (stretching Si-O).

With respect to the standard polymers, the HDPE and PP samples, the appearance of the bands due to the stretching mode of double C=C bond or C=O, that occurred more intensely in PP_6S than HDPE_6S, can be attributed to a more extensive photodegradation process, occurring preferentially in sand compared to seawater. The presence of the methyl sidechain of PP allows β-scission reactions to occur more easily with formation of unsaturated bonds and decrease in molecular weight [113]. The disappearance of the peak at 1744 cm-1 from raw PP can be attributed to additives present in the pristine polymer.

The results appeared to be different for both biodegradable polymers PLA and PBAT. Both PLA_6SW and PLA_6S did not show new peaks, with the exception of that around 3500 cm^{-1}. The invariance of the ATR spectra was confirmed by GPC results.

The value of molecular weight remained quite similar, even when the measurements were carried out not by sampling of the surface but on the entire solubilised granule. The percentage of degraded material was low compared to the bulk and did not affect the average molecular weight value [10,88].

Weak peaks appeared in the ATR spectra of the PBAT samples between 1550 and 1640 cm^{-1} which could be associated with C=C stretching, attributable to Norrish II mechanisms [19], and were more intense in the sand environment than in the sea environment.

Materials is not transferred from the surface of the granules to the surface of the crystals after the recording of the ATR spectra and this result indicates the low extent of contamination of the surface of the granules after six months

GPC analysis confirmed these results. An evident decrease was observed in both samples after six months with an increase in polydispersity, especially for the sample in the sandbox.

From the results of the TGA thermal analyses, a decrease in thermal stability was observed for all samples in the sea and in the sandbox, with the greatest degradation occurring for the sandbox samples. These results confirm the higher photodegradation extent occurring for the samples in the sandbox than for those in the sea. The increase in residue percentage for PP_6S (as for PBAT_6S) can be attributed to photo-oxidised samples as reported by Contact-Rodrigo [114].

The DSC results confirmed the findings from the TGA analyses. The splitting of the melting peak in the PP_6S sample and the decrease in the crystallinity can be hypothesized to reflect a possible decrease in the length of the macromolecular chains which could affect the crystallization capacity of the system and therefore give rise to less ordered crystalline forms [114]. On the other hand, Brandon associated increase in brittleness of the material with progression in the environmental exposure time and observed greater action precisely on the PP [110].

The average temperature evolution and temperature values reached during the entire experimental period could explain the limited molecular weight variation for PLA compared to that of PBAT. In the six months of the experiment, the maximum recorded temperature did not exceed 25.24 °C for the sea and 51.63 °C for the sandbox, with a maximum monthly average that reached a maximum only in August 2020 with a value of 23.68 °C (seawater) and 29.06 °C (sandbox). The temperatures recorded remained very far from both the Tg of PLA and Tm temperatures of both polymers.

Unfortunately, while several experiments have been carried out on the biodegradation of PLA (almost always in film form) following the standardized procedures required by international standards (ASTM, UNI EN, ISO) [115–118], there are few studies regarding the biodegradation of PLA in the marine environment that show little evidence of microbial degradation observed in such environments, temperature conditions and type of microbial species present [119,120].

Although PBAT showed good biodegradation in soil, even at low temperatures [121], PLA showed good biodegradation in compost at 58 °C, at 60% relative humidity and with maintenance of aerobic conditions through oxygen insufflations, while, in contrast, in free soil, its degradation was not significant [118]. Furthermore, in a 100-day experiment to investigate the influence of temperature and type of soil substrates variation on the biodegradation of pure PLA, Gil Castel et al. [116] concluded that chain scission occurring as a function of time and temperature was the main effect influencing the degradation processes of PLA.

5. Conclusions

The results of the present study were obtained from the exposure of different types of pre-production pellets to natural sunlight and seawater/beach conditions for six months. At the end of this experiment, we will be able to collect data from a three-year in situ weathering study of standard and bioplastic pellets in the marine environment. Compared to previous studies, based on experiments that accelerated pellet weathering in laboratory or simulated environmental condition, this long-term field experiment represents a more realistic assessment of the aging processes that these particular widespread microplastics experience in the marine environment. As far as we know, this is the first such kind of experiment conducted in the Mediterranean Sea.

In this study, two metal cages containing commercial and commonly used industrial filament plastic granules were placed on a multi-parameter platform at a depth of 10 m in seawater; this was flanked by a sandbox with the same granules placed on the sand surface (simulated beach). The behaviour during aging in the marine environment of both traditional plastics, such as HDPE and PP, and bioplastics, such as PLA and PBAT, was compared.

After six months of experiment (from March to September 2020) the materials immersed in the sea result had been subjected to less thermal stress than the corresponding materials in the sandbox, because of the lower solar radiation and reduced thermal excursion. Prolonged exposure to intense solar radiation and the temperature reached in the sand produced much more visible effects than those produced on samples placed in the depths of the sea in terms of photooxidation by Norrish I and II degradation reactions. On the other hand, the formation of a porous surface and cavities on all six-months-aged seawater samples (except for PLA) was indicative of abrasive phenomena that could induce a greater increase in the absorption of chemicals dissolved in seawater that could, in turn, affect the observed change in colour.

Concerning the biopolymer pellets, the percentage by weight of lost material was low compared to the raw material and did not affect the gravimetric weight value. The molecular weight of PLA remained practically unchanged both in sea and in sand; PBAT, in contrast, showed a decrease of about 40% of the initial molecular weight and less than 10% for the sample in sea. The environmental conditions of the sandbox significantly modified the thermal properties of the materials. A decrease in thermal stability was observed, with the greatest degradation for sandbox samples in all cases and in both environments.

In the two substrates of the experiment (sandbox and sea), the degree of degradation of PLA and PBAT remained low when compared to what was detected for the biodegradation process under composting conditions.

Changes in the chemical-physical properties of the materials suggest material aging and the beginning of a degradation process but with an evolution time that is still long and to be established.

The experiment is still running, and a fuller evaluation of the aging and degradation process will be discussed in the future.

Author Contributions: Conceptualization, M.L., S.M., S.B., C.D.M. and L.R.; methodology, M.L., S.M., S.B., C.D.M. and L.R; software, L.R.; validation, S.B., C.D.M. and L.R; formal analysis, C.D.M., L.R. and A.P.; investigation, M.L. and S.M.; resources, M.L., S.M. and S.B.; data curation, L.R. and M.L.; writing—original draft preparation, C.D.M. and S.M.; writing—review and editing, M.L., S.M., S.B., C.D.M. and L.R., funding acquisition, M.L. and S.M. All authors have read and agreed to the published version of the manuscript.

Funding: The chemical analysis in this research was funded by the Italian Ministry of Health, grant number IZS PLV 16/18 RC, the production of the cages was funded by INGV (Roma2-section) and the paper pubblication was funded by an INGV project (Ricerca libera 9999.521—RL2019-Marina Locritani).

Data Availability Statement: The data that support the findings of this study are available from the authors upon reasonable request.

Acknowledgments: The authors thank the Italian Istituto Zooprofilattico Sperimentale del Piemonte, Liguria e Valle d'Aosta that, through the project "Chemistry and genomics: a synergistic strategy for the detection of contaminants associated with microplastics in food", enabled some of the materials useful for the setup of the experiment to be obtained. Other thanks are to DLTM, that made available the LabMARE coastal underwater platform to host the cages, to Andrea Bordone and Giancarlo Raiteri (ENEA) for the calibration of the CTD probe and the correction of temperature data of the seawater, and to Dipartimento Polizia di Stato—Centro Nautico Sommozzatori of La Spezia (Italy) that give their precious support to sampling of the plastic material contained in the cages. The authors also thank the Win on Waste group, at Area della Ricerca CNR of Pisa (WoW, https://wow.area.pi.cnr.it/, accessed on 7 February 2022) for their helpful discussions on environmental safety and the circular economy. The authors thank Luca Pardi (IPCF-CNR) for revision of the English language of the manuscript.

Conflicts of Interest: The authors declare no conflict of interest. The funders had no role in the design of the study; in the collection, analyses, or interpretation of data; in the writing of the manuscript, or in the decision to publish the results.

References

1. Bergmann, M.; Gutow, L.; Klages, M. *Marine Anthropogenic Litter*; Springer: Berlin/Heidelberg, Germany, 2015; pp. 1–447.
2. Crutzen, P.J. The *"Anthropocene"*; Journal De Physique. IV: JP.; Springer: Berlin/Heidelberg, Germany, 2002; pp. Pr10/1–Pr10/5.
3. Thompson, R.C.; Swan, S.H.; Moore, C.J.; Vom Saal, F.S. Our plastic age. *Philos. Trans. R. Soc. B Biol. Sci.* **2009**, *364*, 1973–1976. [CrossRef] [PubMed]
4. Porta, R. Anthropocene, the plastic age and future perspectives. *FEBS Open Bio* **2021**, *11*, 948–953. [CrossRef] [PubMed]
5. Derraik, J.G. The pollution of the marine environment by plastic debris: A review. *Mar. Pollut. Bull.* **2002**, *44*, 842–852. [CrossRef]
6. Napper, I.E.; Thompson, R.C. Plastic Debris in the Marine Environment: History and Future Challenges. *Glob. Chall.* **2020**, *4*, 1900081. [CrossRef]
7. Andrady, A.L. Persistence of Plastic Litter in the Oceans. In *Marine Anthropogenic Litter*; Springer International Publishing: Cham, Switzerland, 2015; pp. 57–72.
8. Bhuyan, S.M.; Venkatramanan, S.; Selvam, S.; Szabo, S.; Maruf Hossain, M.; Rashed-Un-Nabi, M.; Paramasivam, C.R.; Jonathan, M.P.; Islam, S.M. Plastics in marine ecosystem: A review of their sources and pollution conduits. *Reg. Stud. Mar. Sci.* **2021**, *41*, 101539. [CrossRef]
9. Barnes, D.K.A.; Galgani, F.; Thompson, R.C.; Barlaz, M. Accumulation and fragmentation of plastic debris in global environments. *Philos. Trans. R Soc. Lond. B Biol. Sci.* **2009**, *364*, 1985–1998. [CrossRef] [PubMed]
10. Kliem, S.; Kreutzbruck, M.; Bonten, C. Review on the Biological Degradation of Polymers in Various Environments. *Materials* **2020**, *13*, 4586. [CrossRef] [PubMed]
11. Carpenter, E.J.; Smith, K.L., Jr. Plastics on the Sargasso sea surface. *Science* **1972**, *175*, 1240–1241. [CrossRef] [PubMed]
12. Maes, T.; Barry, J.; Leslie, H.A.; Vethaak, A.D.; Nicolaus, E.E.M.; Law, R.J.; Lyons, B.P.; Martinez, R.; Harley, B.; Thain, J.E. Below the surface: Twenty-five years of seafloor litter monitoring in coastal seas of North West Europe (1992–2017). *Sci. Total Environ.* **2018**, *630*, 790–798. [CrossRef] [PubMed]

13. Cozar Cabañas, A.; Sanz-Martín, M.; Martí, E.; González-Gordillo, J.I.; Ubeda, B.; Gálvez, J.Á.; Irigoien, X.; Duarte, C.M. Concentrations of floating plastic debris in the Mediterranean Sea measured during MedSeA-2013 cruise. *PLoS ONE* **1972**, *4062*.
14. Cózara, A.; Echevarría, F.; González-Gordillo, J.I.; Irigoien, X.; Úbeda, B.; Hernández-León, S.; Palma, Á.T.; Navarro, S.; García-de-Lomas, J.; Ruiz, A.; et al. Plastic debris in the open ocean. *Proc. Natl. Acad. Sci. USA* **2014**, *111*, 10239–10244. [CrossRef] [PubMed]
15. Senga Green, D.; Boots, B.; Blockley, D.J.; Rocha, C.; Thompson, R. Impacts of Discarded Plastic Bags on Marine Assemblages and Ecosystem Functioning. *Environ. Sci. Technol.* **2015**, *49*, 5380–5389. [CrossRef] [PubMed]
16. Hidalgo-Ruz, V.; Honorato-Zimmer, D.; Gatta-Rosemary, M.; Nuñez, P.; Hinojosa, I.A.; Thiel, M. Spatio-temporal variation of anthropogenic marine debris on Chilean beaches. *Mar. Pollut. Bull.* **2018**, *126*, 516–524. [CrossRef] [PubMed]
17. Moore, C.J. Synthetic polymers in the marine environment: A rapidly increasing, long-term threat. *Environ. Res.* **2008**, *108*, 131–139. [CrossRef] [PubMed]
18. Hidalgo-Ruz, V.; Thiel, M. Distribution and abundance of small plastic debris on beaches in the SE Pacific (Chile): A study supported by a citizen science project. *Mar. Environ. Res.* **2013**, *87–88*, 12–18. [CrossRef] [PubMed]
19. Ali, S.S.; Elsamahy, T.; Koutra, E.; Kornaros, M.; El-Sheekh, M.; Abdelkarim, E.A.; Zhu, D.; Sun, J. Degradation of conventional plastic wastes in the environment: A review on current status of knowledge and future perspectives of disposal. *Sci. Total Environ.* **2021**, *771*, 144719. [CrossRef] [PubMed]
20. Suaria, G.; Avio, C.G.; Mineo, A.; Lattin, G.L.; Magaldi, M.G.; Belmonte, G.; Moore, C.J.; Regoli, F.; Aliani, S. The Mediterranean Plastic Soup: Synthetic polymers in Mediterranean surface waters. *Sci. Rep.* **2016**, *6*, 37551. [CrossRef] [PubMed]
21. Andrady, A.L. Microplastics in the marine environment. *Mar. Pollut. Bull.* **2011**, *62*, 1596–1605. [CrossRef] [PubMed]
22. Andrady, A.L. The plastic in microplastics: A review. *Mar. Pollut. Bull.* **2017**, *119*, 12–22. [CrossRef] [PubMed]
23. Galgani, F.; Claro, F.; Depledge, M.; Fossi, C. Monitoring the impact of litter in large vertebrates in the Mediterranean Sea within the European Marine Strategy Framework Directive (MSFD): Constraints, specificities and recommendations. *Mar. Environ. Res.* **2014**, *100*, 3–9. [CrossRef]
24. Koelmans, A.A.; Bakir, A.; Burton, G.A.; Janssen, C.R. Microplastic as a Vector for Chemicals in the Aquatic Environment: Critical Review and Model-Supported Reinterpretation of Empirical Studies. *Environ. Sci. Technol.* **2016**, *50*, 3315–3326. [CrossRef] [PubMed]
25. Koelmans, A.A.; Besseling, E.; Shim, W.J. Nanoplastics in the Aquatic Environment. Critical Review. In *Marine Anthropogenic Litter*; Springer International Publishing: Cham, Switzerland, 2015; pp. 325–340.
26. Frias, J.P.G.L.; Nash, R. Microplastics: Finding a consensus on the definition. *Mar. Pollut. Bull.* **2019**, *138*, 145–147. [CrossRef] [PubMed]
27. Hartmann, N.B.; Hüffer, T.; Thompson, R.C.; Hassellöv, M.; Verschoor, A.; Daugaard, A.E.; Rist, S.; Karlsson, T.; Brennholt, N.; Cole, M.; et al. Are We Speaking the Same Language? Recommendations for a Definition and Categorization Framework for Plastic Debris. *Environ. Sci. Technol.* **2019**, *53*, 1039–1047. [CrossRef] [PubMed]
28. Cole, M.; Lindeque, P.; Halsband, C.; Galloway, T.S. Microplastics as contaminants in the marine environment. A review. *Mar. Pollut. Bull.* **2011**, *62*, 2588–2597. [CrossRef] [PubMed]
29. Rangel-Buitrago, N.; Arroyo-Olarte, H.; Trilleras, J.; Arana, V.A.; Mantilla-Barbosa, E.; Gracia, C.A.; Mendoza, A.V.; Neal, W.J.; Williams, A.T.; Micallef, A. Microplastics pollution on Colombian Central Caribbean beaches. *Mar. Pollut. Bull.* **2021**, *170*, 112685. [CrossRef] [PubMed]
30. Jung, S.; Cho, S.-H.; Kim, K.-H.; Kwon, E.E. Progress in quantitative analysis of microplastics in the environment: A review. *Chem. Eng. J.* **2021**, *422*, 130154. [CrossRef]
31. Liebezeit, G.; Dubaish, F. Microplastics in Beaches of the East Frisian Islands Spiekeroog and Kachelotplate. *Bull. Environ. Contam. Toxicol.* **2012**, *89*, 213–217. [CrossRef] [PubMed]
32. Van Cauwenberghe, L.; Devriese, L.; Galgani, F.S.; Robbens, J.; Janssen, C.R. Microplastics in sediments: A review of techniques, occurrence and effects. *Mar. Environ. Res.* **2015**, *111*, 5–17. [CrossRef] [PubMed]
33. Peng, L.; Fu, D.; Qi, H.; Lan, C.Q.; Yu, H.; Ge, C. Micro- and nano-plastics in marine environment: Source, distribution and threats—A review. *Sci. Total Environ.* **2020**, *698*, 134254. [CrossRef] [PubMed]
34. Wang, Z.; Su, B.; Xu, X.; Di, D.; Huang, H.; Mei, K.; Dahlgren, R.A.; Zhang, M.; Shang, X. Preferential accumulation of small (<300 mm) microplastics in the sediments of a coastal plain river network in eastern China. *Water Res.* **2018**, *144*, 393–401. [CrossRef] [PubMed]
35. Frère, L.; Paul-Pont, I.; Rinnert, E.; Petton, S.; Jaffré, J.; Bihannic, I.; Soudant, P.; Lambert, C.; Huvet, A. Influence of environmental and anthropogenic factors on the composition, concentration and spatial distribution of microplastics: A case study of the Bay of Brest (Brittany, France). *Environ. Pollut.* **2017**, *225*, 211–222. [CrossRef] [PubMed]
36. Schmidt, L.K.; Bochow, M.; Imhof, H.K.; Oswald, S.E. Multi-temporal surveys for microplastic particles enabled by a novel and fast application of SWIR imaging spectroscopy—Study of an urban watercourse traversing the city of Berlin, Germany. *Environ. Pollut.* **2018**, *239*, 579–589. [CrossRef] [PubMed]
37. Zobkov, M.B.; Esiukova, E.E.; Zyubin, A.Y.; Samusev, I.G. Microplastic content variation in water column: The observations employing a novel sampling tool in stratified Baltic Sea. *Mar. Pollut. Bull.* **2019**, *138*, 193–205. [CrossRef] [PubMed]
38. De-la-Torre, G.E.; Dioses-Salinas, D.C.; Castro, J.M.; Antay, R.; Fernándeza, N.Y.; Espinoza-Morriberónb, D.; Saldaña-Serrano, M. Abundance and distribution of microplastics on sandy beaches of Lima, Peru. *Mar. Pollut. Bull.* **2020**, *151*, 110877. [CrossRef] [PubMed]

39. Giovacchini, A.; Merlino, S.; Locritani, M.; Stroobant, M. Spatial distribution of marine litter along italian coastal areas in the Pelagos sanctuary (Ligurian Sea—NW Mediterranean Sea): A focus on natural and urban beaches. *Mar. Pollut. Bull.* **2018**, *130*, 140–152. [CrossRef] [PubMed]
40. Merlino, S.; Locritani, M.; Bernardi, G.; Como, C.; Legnaioli, S.; Palleschi, V.; Abbate, M. Spatial and Temporal Distribution of Chemically Characterized Microplastics within the Protected Area of Pelagos Sanctuary (NW Mediterranean Sea): Focus on Natural and Urban Beaches. *Water* **2020**, *12*, 3389. [CrossRef]
41. Zhang, W.; Zhang, S.; Wang, J.; Wang, Y.; Mu, J.; Wang, P.; Lin, X.; Ma, D. Microplastic pollution in the surface waters of the Bohai Sea, China. *Environ. Pollut.* **2017**, *231*, 541–548. [CrossRef] [PubMed]
42. Brach, L.; Deixonne, P.; Bernard, M.F.; Durand, E.; Desjean, M.C.; Perez, E.; van Sebille, E.; ter Halle, A. Anticyclonic eddies increase accumulation of microplastic in the North Atlantic subtropical gyre. *Mar. Pollut. Bull.* **2018**, *126*, 191–196. [CrossRef] [PubMed]
43. Massetti, L.; Rangel-Buitrago, N.; Pietrelli, L.; Merlino, S. Litter impacts on marine birds: The Mediterranean Northern gannet as case study. *Mar. Pollut. Bull.* **2021**, *171*, 112779. [CrossRef]
44. van Franeker, J.A.; Blaize, C.; Danielsen, J.; Fairclough, K.; Gollan, J.; Guse, N.; Hansen, P.-L.; Heubeck, M.; Jensen, J.-K.; Le Guillou, G.; et al. Monitoring plastic ingestion by the northern fulmar Fulmarus glacialis in the North Sea. *Mar. Environ. Res.* **2011**, *159*, 2609–2615. [CrossRef] [PubMed]
45. Mordecai, G.; Tyler, P.A.; Masson, D.G.; Huvenne, V.A.I. Litter in submarine canyons off the west coast of Portugal. *Deep. Sea Res. Part II Top. Stud. Oceanogr.* **2011**, *58*, 2489–2496. [CrossRef]
46. Knowlton, A.R.; Robbins, J.; Landry, S.; McKenna, H.A.; Kraus, S.D.; Werner, T.B. Effects of fishing rope strength on the severity of large whale entanglements. *Conserv. Biol.* **2016**, *30*, 318–328. [CrossRef] [PubMed]
47. Campani, T.; Baini, M.; Giannetti, M.; Cancelli, F.; Mancusi, C.; Serena, F.; Marsili, L.; Casini, S.; Fossi, M.C. Presence of plastic debris in loggerhead turtle stranded along the Tuscany coasts of the Pelagos Sanctuary for Mediterranean Marine Mammals (Italy). *Mar. Pollut. Bull.* **2013**, *74*, 225–230. [CrossRef] [PubMed]
48. de Stephanis, R.; Giménez, J.; Carpinelli, E.; Gutierrez-Exposito, C.; Cañadas, A. As main meal for sperm whales: Plastics debris. *Mar. Pollut. Bull.* **2013**, *69*, 206–214. [CrossRef] [PubMed]
49. Waluda, C.M.; Staniland, I.J. Entanglement of Antarctic fur seals at Bird Island, South Georgia. *Mar. Pollut. Bull.* **2013**, *74*, 244–252. [CrossRef] [PubMed]
50. Kühn, S.; Bravo Rebolledo, E.L.; van Franeker, J.A. Deleterious Effects of Litter on Marine Life. In *Marine Anthropogenic Litter*; Springer International Publishing: Cham, Switzerland, 2015; pp. 75–116.
51. Rodríguez, B.; Beneharo, J.; Rodríguez, A.; Arcos, J.M. Incidence of entanglements with marine debris by northern gannets (*Morus bassanus*) in the non-breeding grounds. *Mar. Pollut. Bull.* **2013**, *75*, 259–263. [CrossRef]
52. Tetu, S.G.; Sarker, I.; Schrameyer, V.; Pickford, R.; Elbourne, L.D.H.; Moore, L.R.; Paulsen, I.T. Plastic leachates impair growth and oxygen production in Prochlorococcus, the ocean's most abundant photosynthetic bacteria. *Commun. Biol.* **2019**, *2*, 184. [CrossRef]
53. Teuten, E.L.; Saquing, J.M.; Knappe, D.R.U.; Barlaz, M.A.; Jonsson, S.; Bjorn, A.; Rowland, S.J.; Thompson, R.C.; Galloway, T.S.; Yamashita, R.; et al. Transport and release of chemicals from plastics to the environment and to wildlife. *Philos. Trans. R. A Soc. Lond. B Biol. Sci.* **2009**, *364*, 2027–2045. [CrossRef]
54. Browne, M.A.; Dissanayake, A.; Galloway, T.S.; Lowe, D.M.; Thompson, R.C. Ingested Microscopic Plastic Translocates to the Circulatory System of the Mussel, *Mytilus edulis* (L.). *Environ. Sci. Technol.* **2008**, *42*, 5026–5031. [CrossRef]
55. Oehlmann, J.; Schulte-Oehlmann, U.; Kloas, W.; Jagnytsch, O.; Lutz, I.; Kusk, K.O.; Wollenberger, L.; Santos, E.M.; Paull, G.C.; Van Look, K.J.; et al. A critical analysis of the biological impacts of plasticizers on wildlife. *Philos. Trans. R. Soc. Lond. B Biol. Sci.* **2009**, *364*, 2047–2062. [CrossRef] [PubMed]
56. Koch, H.M.; Calafat, A.M. Human body burdens of chemicals used in plastic manufacture. *Philos. Trans. R. Soc. Lond. B Biol. Sci.* **2009**, *364*, 2063–2078. [CrossRef] [PubMed]
57. Meeker, J.D.; Sathyanarayana, S.; Swan, S.H. Phthalates and other additives in plastics: Human exposure and associated health outcomes. *Philos. Trans. R. Soc. Lond. Ser. B Biol. Sci.* **2009**, *364*, 2097–2113. [CrossRef] [PubMed]
58. Fossi, M.; Coppola, D.; Baini, M.; Giannetti, M.; Guerranti, C.; Marsili, L.; Panti, C.; de Sabata, E.; Clò, S. Large filter feeding marine organisms as indicators of microplastic in the pelagic environment: The case studies of the Mediterranean basking shark (*Cetorhinus maximus*) and fin whale (*Balaenoptera physalus*). *Mar. Environ. Res.* **2014**, *100*, 17–24. [CrossRef] [PubMed]
59. Fossi, M.; Panti, C.; Guerranti, C.; Coppola, D.; Giannetti, M.; Marsili, L.; Minutoli, R. Are baleen whales exposed to the threat of microplastics? A case study of the Mediterranean fin whale (*Balaenoptera physalus*). *Mar. Pollut. Bull.* **2012**, *64*, 2374–2379. [CrossRef] [PubMed]
60. Fossi, M.C.; Panti, C.; Baini, M.; Lavers, J.L. A Review of Plastic-Associated Pressures: Cetaceans of the Mediterranean Sea and Eastern Australian Shearwaters as Case Studies. *Front. Mar. Sci.* **2018**, *5*, 173. [CrossRef]
61. Rochman, C.M.; Hoh, E.; Kurobe, T.; Teh, S.J. Ingested plastic transfers hazardous chemicals to fish and induces hepatic stress. *Sci. Rep.* **2013**, *3*, 3263. [CrossRef] [PubMed]
62. Tanaka, K.; Takada, H.; Yamashita, R.; Mizukawa, K.; Fukuwaka, M.A.; Watanuki, Y. Accumulation of plastic-derived chemicals in tissues of seabirds ingesting marine plastics. *Mar. Pollut. Bull.* **2013**, *69*, 219–222. [CrossRef] [PubMed]
63. Cole, M.; Lindeque, P.; Fileman, E.; Halsband, C.; Galloway, T.S. The impact of polystyrene microplastics on feeding, function and fecundity in the marine copepod Calanus helgolandicus. *Environ. Sci. Technol.* **2015**, *49*, 1130–1137. [CrossRef] [PubMed]

64. Avio, C.G.; Gorbi, S.; Milan, M.; Benedetti, M.; Fattorini, D.; d'Errico, G.; Pauletto, M.; Bargelloni, L.; Regoli, F. Pollutants bioavailability and toxicological risk from microplastics to marine mussels. *Environ. Pollut.* **1987**, *198*, 211–222. [CrossRef] [PubMed]
65. Karami, A.; Romano, N.; Galloway, T.; Hamzah, H. Virgin microplastics cause toxicity and modulate the impacts of phenanthrene on biomarker responses in African catfish (*Clarias gariepinus*). *Environ. Res.* **2016**, *151*, 58–70. [CrossRef]
66. Wright, S.L.; Thompson, R.C.; Galloway, T.S. The physical impacts of microplastics on marine organisms: A review. *Environ. Pollut.* **2013**, *178*, 483–492. [CrossRef] [PubMed]
67. Provencher, J.F.; Liboiron, M.; Borrelle, S.B.; Bond, A.L.; Rochman, C.; Lavers, J.L.; Avery-Gomm, S.; Yamashita, R.; Ryan, P.G.; Lusher, A.L.; et al. A Horizon Scan of research priorities to inform policies aimed at reducing the harm of plastic pollution to biota. *Sci. Total Environ.* **2020**, *733*, 139381. [CrossRef] [PubMed]
68. Campanale, C.; Massarelli, C.; Savino, I.; Locaputo, V.; Uricchio, V.F. A Detailed Review Study on Potential Effects of Microplastics and Additives of Concern on Human Health. *Int. J. Environ. Res. Public Health* **2020**, *17*, 1212. [CrossRef] [PubMed]
69. Graham, E.R.; Thompson, J.T. Deposit- and suspension-feeding sea cucumbers (Echinodermata) ingest plastic fragments. *J. Exp. Mar. Biol. Ecol.* **2009**, *368*, 22–29. [CrossRef]
70. Talsness, C.E.; Andrade, A.J.; Kuriyama, S.N.; Taylor, J.A.; vom Saal, F.S. Components of plastic: Experimental studies in animals and relevance for human health. *Philos. Trans. R. Soc. Lond. B Biol. Sci.* **2009**, *364*, 2079–2096. [CrossRef] [PubMed]
71. Hansen, E.; Nilsson, N.H.; Lithner, D.; Lassen, C. *Hazardous Substances in Plastic Materials*; COWI: Vejle, Denmark, 2013.
72. Mato, Y.; Isobe, T.; Takada, H.; Kanehiro, H.; Ohtake, C.; Kaminuma, T. Plastic Resin Pellets as a Transport Medium for Toxic Chemicals in the Marine Environment. *Environ. Sci. Technol.* **2001**, *35*, 318–324. [CrossRef] [PubMed]
73. Endo, S.; Takizawa, R.; Okuda, K.; Takada, H.; Chiba, K.; Kanehiro, H.; Ogi, H.; Yamashita, R.; Date, T. Concentration of polychlorinated biphenyls (PCBs) in beached resin pellets: Variability among individual particles and regional differences. *Mar. Pollut. Bull.* **2005**, *50*, 1103–1114. [CrossRef] [PubMed]
74. Rios, L.M.; Moore, C.; Jones, P.R. Persistent organic pollutants carried by synthetic polymers in the ocean environment. *Mar. Pollut. Bull.* **2007**, *54*, 1230–1237. [CrossRef] [PubMed]
75. Teuten, E.L.; Rowland, S.J.; Galloway, T.S.; Thompson, R.C. Potential for Plastics to Transport Hydrophobic Contaminants. *Environ. Sci. Technol.* **2007**, *41*, 7759–7764. [CrossRef] [PubMed]
76. Lithner, D.; Larsson, A.; Dave, G. Environmental and health hazard ranking and assessment of plastic polymers based on chemical composition. *Sci. Total Environ.* **2011**, *409*, 3309–3324. [CrossRef] [PubMed]
77. Hirai, H.; Takada, H.; Ogata, Y.; Yamashita, R.; Mizukawa, K.; Saha, M.; Kwan, C.; Moore, C.; Gray, H.; Laursen, D.; et al. Organic micropollutants in marine plastics debris from the open ocean and remote and urban beaches. *Mar. Pollut. Bull.* **2011**, *62*, 1683–1692. [CrossRef] [PubMed]
78. Bellas, J.; Martínez-Armental, J.; Martínez-Cámara, A.; Besada, V.; Martínez-Gómez, C. Ingestion of microplastics by demersal fish from the Spanish Atlantic and Mediterranean coasts. *Mar. Pollut. Bull.* **2016**, *109*, 55–60. [CrossRef] [PubMed]
79. Li, J.; Qu, X.; Su, L.; Zhang, W.; Yang, D.; Kolandhasamy, P.; Li, D.; Shi, H. Microplastics in mussels along the coastal waters of China. *Environ. Pollut.* **2016**, *214*, 177–184. [CrossRef] [PubMed]
80. Gusmão, F.; Di Domenico, M.; Amaral, A.C.Z.; Martínez, A.; Gonzalez, B.C.; Worsaae, K.; Ivar do Sul, J.A.; da Cunha Lana, P. In situ ingestion of microfibres by meiofauna from sandy beaches. *Environ. Pollut.* **2016**, *216*, 584–590. [CrossRef] [PubMed]
81. Costa, L.L.; Arueira, V.F.; da Costa, M.F.; Di Beneditto, A.P.M.; Zalmon, I.R. Can the Atlantic ghost crab be a potential biomonitor of microplastic pollution of sandy beaches sediment? *Mar. Pollut. Bull.* **2019**, *145*, 5–13. [CrossRef] [PubMed]
82. Wójcik-Fudalewska, D.; Normant-Saremba, M.; Anastácio, P. Occurrence of plastic debris in the stomach of the invasive crab Eriocheir sinensis. *Mar. Pollut. Bull.* **2016**, *113*, 306–311. [CrossRef] [PubMed]
83. Carlin, J.; Craig, C.; Little, S.; Donnelly, M.; Fox, D.; Zhai, L.; Walters, L. Microplastic accumulation in the gastrointestinal tracts in birds of prey in central Florida, USA. *Environ. Pollut.* **2020**, *264*, 114633. [CrossRef] [PubMed]
84. Fazey, F.M.C.; Ryan, P.G. Biofouling on buoyant marine plastics: An experimental study into the effect of size on surface longevity. *Environ. Pollut.* **2016**, *210*, 354–360. [CrossRef] [PubMed]
85. Lobelle, D.; Cunliffe, M. Early microbial biofilm formation on marine plastic debris. *Mar. Pollut. Bull.* **2011**, *62*, 197–200. [CrossRef]
86. Da Costa, J.P.; Nunes, A.R.; Santos, P.S.M.; Girão, A.V.; Duarte, A.C.; Rocha-Santos, T. Degradation of polyethylene microplastics in seawater: Insights into the environmental degradation of polymers. *J. Environ. Sci. Health A* **2018**, *53*, 866–875. [CrossRef]
87. O'Brine, T.; Thompson, R.C. Degradation of plastic carrier bags in the marine environment. *Mar. Pollut. Bull.* **2010**, *60*, 2279–2283. [CrossRef] [PubMed]
88. Bagheri, A.R.; Laforsch, C.; Greiner, A.; Agarwal, S. Fate of So-Called Biodegradable Polymers in Seawater and Freshwater. *Glob. Chall.* **2017**, *1*, 1700048. [CrossRef] [PubMed]
89. Bonilla, J.; Paiano, R.B.; Lourenço, R.V.; Bittante, A.; Sobral, P.J.A. Biodegradability in aquatic system of thin materials based on chitosan, PBAT and HDPE polymers: Respirometric and physical-chemical analysis. *Int. J. Biol. Macromol.* **2020**, *164*, 1399–1412. [CrossRef] [PubMed]
90. Carpenter, E.J.; Anderson, S.J.; Harvey, G.R.; Miklas, H.P.; Peck, B.B. Polystyrene spherules in coastal waters. *Science* **1972**, *178*, 749–750. [CrossRef] [PubMed]
91. Abelouah, M.R.; Ben-Haddad, M.; Alla, A.A.; Rangel-Buitrago, N. Marine litter in the central Atlantic coast of Morocco. *Ocean. Coast. Manag.* **2021**, *214*, 105940. [CrossRef]

92. Gündogdu, S.; Çevik, C. Mediterranean dirty edge: High level of meso and macroplastics pollution on the Turkish coast. *Environ. Pollut.* **2019**, *255*, 113351. [CrossRef] [PubMed]
93. Camacho, M.; Herrera, A.; Gómez, M.; Acosta-Dacal, A.; Martínez, I.; Henríquez-Hernández, L.A.; Luzardo, O.P. Organic pollutants in marine plastic debris from Canary Islands beaches. *Sci. Total Environ.* **2019**, *662*, 22–31. [CrossRef] [PubMed]
94. Ogata, Y.; Takada, H.; Mizukawa, K.; Hirai, H.; Iwasa, S.; Endo, S.; Mato, Y.; Saha, M.; Okuda, K.; Nakashima, A.; et al. International Pellet Watch: Global monitoring of persistent organic pollutants (POPs) in coastal waters. 1. Initial phase data on PCBs, DDTs, and HCHs. *Mar. Pollut. Bull.* **2009**, *58*, 1437–1446. [CrossRef]
95. Karapanagioti, H.K.; Klontza, I. Testing phenanthrene distribution properties of virgin plastic pellets and plastic eroded pellets found on Lesvos island beaches (Greece). *Mar. Environ. Res.* **2008**, *65*, 283–290. [CrossRef]
96. Heskett, M.; Takada, H.; Yamashita, R.; Yuyama, M.; Ito, M.; Geok, Y.B.; Ogata, Y.; Kwan, C.; Heckhausen, A.; Taylor, H.; et al. Measurement of persistent organic pollutants (POPs) in plastic resin pellets from remote islands: Toward establishment of background concentrations for International Pellet Watch. *Mar. Pollut. Bull.* **2012**, *64*, 445–448. [CrossRef]
97. Lohmann, R. Critical Review of Low-Density Polyethylene Partitioning and Diffusion Coefficients for Trace Organic Contaminants and Implications for Its Use As a Passive Sampler. *Environ. Sci. Technol.* **2012**, *46*, 606–618. [CrossRef] [PubMed]
98. Karapanagioti, H.K.; Endo, S.; Ogata, Y.; Takada, H. Diffuse pollution by persistent organic pollutants as measured in plastic pellets sampled from various beaches in Greece. *Mar. Pollut. Bull.* **2011**, *62*, 312–317. [CrossRef] [PubMed]
99. Karapanagioti, H.; Ogata, Y.; Takada, H. Eroded plastic pellets as monitoring tools for polycyclic aromatic hydrocarbons (PAH):Laboratory and field studies. *Glob. Nest J.* **2010**, *12*, 327–334.
100. Karkanorachaki, K.; Kiparissis, S.; Kalogerakis, G.C.; Yiantzi, E.; Psillakis, E.; Kalogerakis, N. Plastic pellets, meso- and microplastics on the coastline of Northern Crete: Distribution and organic pollution. *Mar. Pollut. Bull.* **2018**, *133*, 578–589. [CrossRef] [PubMed]
101. Taniguchi, S.; Colabuono, F.I.; Dias, P.S.; Oliveira, R.; Fisner, M.; Turra, A.; Izar, G.M.; Abessa, D.M.; Saha, M.; Hosoda, J.; et al. Spatial variability in persistent organic pollutants and polycyclic aromatic hydrocarbons found in beach-stranded pellets along the coast of the state of São Paulo, southeastern Brazil. *Mar. Pollut. Bull.* **2016**, *106*, 87–94. [CrossRef] [PubMed]
102. Endo, S.; Yuyama, M.; Takada, H. Desorption kinetics of hydrophobic organic contaminants from marine plastic pellets. *Mar. Pollut. Bull.* **2013**, *74*, 125–131. [CrossRef] [PubMed]
103. Rochman, C.M.; Hoh, E.; Hentschel, B.T.; Kaye, S. Long-Term Field Measurement of Sorption of Organic Contaminants to Five Types of Plastic Pellets: Implications for Plastic Marine Debris. *Environ. Sci. Technol.* **2013**, *47*, 1646–1654. [CrossRef] [PubMed]
104. Müller, J.F.; Manomanii, K.; Mortimer, M.R.; McLachlan, M.S. Partitioning of polycyclic aromatic hydrocarbons in the polyethylene/water system. *Fresenius J. Anal. Chem.* **2001**, *371*, 816–822. [CrossRef] [PubMed]
105. Smedes, F.; Geertsma, R.W.; van der Zande, T.; Booij, K. Polymer-water partition coefficients of hydrophobic compounds for passive sampling: Application of cosolvent models for validation. *Environ. Sci. Technol.* **2009**, *43*, 7047–7054. [CrossRef] [PubMed]
106. Accustandard Accustandard Plastic Additive Guide. Available online: https://www.accustandard.com/plastic-additive-catalog-2nd-edition (accessed on 7 December 2021).
107. Quero, E.; Müller, A.J.; Signori, F.; Coltelli, M.-B.; Bronco, S. Isothermal Cold-Crystallization of PLA/PBAT Blends With and Without the Addition of Acetyl Tributyl Citrate. *Macromol. Chem. Phys.* **2012**, *213*, 36–48. [CrossRef]
108. Rachtanapun, P.; Selke, S.E.M.; Matuana, L.M. Effect of the high-density polyethylene melt index on the microcellular foaming of high-density polyethylene/polypropylene blends. *J. Appl. Polym. Sci.* **2004**, *93*, 364–371. [CrossRef]
109. Xiao, H.; Lu, W.; Yeh, J.-T. Crystallization behavior of fully biodegradable poly(lactic acid)/poly(butylene adipate-co-terephthalate) blends. *J. Appl. Polym. Sci.* **2009**, *112*, 3754–3763. [CrossRef]
110. Brandon, J.; Goldstein, M.; Ohman, M.D. Long-term aging and degradation of microplastic particles: Comparing in situ oceanic and experimental weathering patterns. *Mar. Pollut. Bull.* **2016**, *110*, 299–308. [CrossRef] [PubMed]
111. European Bioplastics. Available online: https://www.european-bioplastics.org/bioplastics (accessed on 9 December 2021).
112. Corcoran, P.; Biesinger, M.; Grifi, M. Plastics and beaches: A degrading relationship. *Mar. Pollut. Bull.* **2009**, *58*, 80–84. [CrossRef] [PubMed]
113. Svoboda, P.; Trivedi, K.; Stoklasa, K.; Svobodova, D.; Ougizawa, T. Study of crystallization behaviour of electron beam irradiated polypropylene and high-density polyethylene. *R. Soc. Open Sci.* **2021**, *8*, 202250. [CrossRef] [PubMed]
114. Contat-Rodrigo, L. Thermal characterization of the oxo-degradation of polypropylene containing a pro-oxidant/pro-degradant additive. *Polym. Degrad. Stab.* **2013**, *98*, 2117–2124. [CrossRef]
115. Itävaara, M.; Karjomaa, S.; Selin, J.-F. Biodegradation of polylactide in aerobic and anaerobic thermophilic conditions. *Chemosphere* **2002**, *46*, 879–885. [CrossRef]
116. Gil-Castell, O.; Andres-Puche, R.; Dominguez, E.; Verdejo, R.; Monreal, L.; Ribes-Greus, A. Influence of substrate and temperature on the biodegradation of polyester-based materials: Polylactide and poly(3-hydroxybutyrate-co-3-hydroxyhexanoate) as model cases. *Polym. Degrad. Stab.* **2020**, *180*, 109288. [CrossRef]
117. Kalita, N.K.; Bhasney, S.M.; Mudenur, C.; Kalamdhad, A.; Katiyar, V. End-of-life evaluation and biodegradation of Poly(lactic acid) (PLA)/Polycaprolactone (PCL)/Microcrystalline cellulose (MCC) polyblends under composting conditions. *Chemosphere* **2020**, *247*, 125875. [CrossRef] [PubMed]

118. Kalita, N.K.; Hazarika, D.; Kalamdhad, A.; Katiyar, V. Biodegradation of biopolymeric composites and blends under different environmental conditions: Approach towards end-of-life panacea for crop sustainability. *Bioresour. Technol. Rep.* **2021**, *15*, 100705. [CrossRef]
119. Karamanlioglu, M.; Preziosi, R.; Robson, G.D. Abiotic and biotic environmental degradation of the bioplastic polymer poly(lactic acid): A review. *Polym. Degrad. Stab.* **2017**, *137*, 122–130. [CrossRef]
120. Tsuji, H.; Suzuyoshi, K. Environmental degradation of biodegradable polyesters 1. Poly(ε-caprolactone), poly[(R)-3-hydroxybutyrate], and poly(L-lactide) films in controlled static seawater. *Polym. Degrad. Stab.* **2002**, *75*, 347–355. [CrossRef]
121. Palsikowski, P.; Kuchnier, C.; Pinheiro, I.; Morales, A. Biodegradation in Soil of PLA/PBAT Blends Compatibilized with Chain Extender. *J. Polym. Environ.* **2018**, *26*, 1–12. [CrossRef]

Review

A Mini-Review of Strategies for Quantifying Anthropogenic Activities in Microplastic Studies in Aquatic Environments

Chun-Ting Lin, Ming-Chih Chiu * and Mei-Hwa Kuo *

Department of Entomology, National Chung Hsing University, Taichung 40227, Taiwan; b03612016@g.ntu.edu.tw
* Correspondence: mingchih.chiu@gmail.com (M.-C.C.); mhkuo@dragon.nchu.edu.tw (M.-H.K.)

Abstract: Microplastic pollution is no longer neglected worldwide, as recent studies have unveiled its potential harm to ecosystems and, even worse, to human health. Numerous studies have documented the ubiquity of microplastics, reflecting the necessity of formulating corresponding policies to mitigate the accumulation of microplastics in natural environments. Although anthropogenic activities are generally acknowledged as the primary source of microplastics, a robust approach to identify sources of microplastics is needed to provide scientific suggestions for practical policymaking. This review elucidates recent microplastic studies on various approaches for quantifying or reflecting the degree to which anthropogenic activities contribute to microplastic pollution. Population density (i.e., often used to quantify anthropogenic activities) was not always significantly correlated with microplastic abundance. Furthermore, this review argues that considering potential sources near sample sites as characteristics that may serve to predict the spatial distribution of microplastics in aquatic environments is equivocal. In this vein, a watershed-scale measure that uses land-cover datasets to calculate different percentages of land use in the watershed margins delineated by using Geographic Information System (GIS) software is discussed and suggested. Progress in strategies for quantifying anthropogenic activities is important for guiding future microplastic research and developing effective management policies to prevent microplastic contamination in aquatic ecosystems.

Keywords: anthropogenic activities; microplastics; quantification; freshwater; marine

Citation: Lin, C.-T.; Chiu, M.-C.; Kuo, M.-H. A Mini-Review of Strategies for Quantifying Anthropogenic Activities in Microplastic Studies in Aquatic Environments. *Polymers* **2022**, *14*, 198. https://doi.org/10.3390/polym14010198

Academic Editor: Jacopo La Nasa

Received: 22 December 2021
Accepted: 30 December 2021
Published: 4 January 2022

Publisher's Note: MDPI stays neutral with regard to jurisdictional claims in published maps and institutional affiliations.

Copyright: © 2022 by the authors. Licensee MDPI, Basel, Switzerland. This article is an open access article distributed under the terms and conditions of the Creative Commons Attribution (CC BY) license (https://creativecommons.org/licenses/by/4.0/).

1. Introduction

The term "microplastic", which refers to tiny debris of plastics normally defined to be smaller than 5 mm [1], was not widely used until 2004 [2]. Approximately 10% of municipal waste globally comprises plastics [3]. The vast use of plastic in human life has resulted in the ubiquity of microplastics in the environment, as they can be degraded into small, persistent, and therefore easy-to-transport plastic debris [4]. For example, microplastics have been detected in a variety of environments, such as beaches, bays, estuaries [5], ocean surfaces [6], deep-sea sediments [7], rivers [8], lakes [9], raindrop [10], the Alps and the Arctic [11], and polar waters [12]. Microplastics have also been documented in biota, including riverine macroinvertebrates [13], marine fish [14], and birds [15]. The accumulation of microplastic pollution is considered an environmental hazard that has attracted global concern. Generally, microplastics originating from terrestrial environments are either retained in freshwater systems or eventually enter the ocean [16]. In this context, this review mainly focuses on aquatic environments.

Many studies have discussed the impact of microplastics on organisms [17], and these impacts can be categorized into two types: physical and chemical [18]. Physical impacts can then be categorized as being either direct or indirect. The direct impacts of microplastics have been observed in numerous studies [17], as ecotoxicological assessments of microplastic pollution are frequently conducted on different species in the laboratory. Generally, the detrimental consequence of microplastic ingestion results from the blockage of the digestive system, which reduces nutrition intake, inhibits food assimilation [19], and causes inflammation [20], resulting in the reduction of growth, reproduction, fitness, mortality, emergence delay, and immune-system weakening [18,21]. Furthermore, indirect impacts of microplastic pollution on organisms also occur. These detrimental effects are not caused by the ingestion of microplastics per se but include the alteration of gut microbiota [22], induction of microbiota dysbiosis [23], ecosystem functioning change [24], behavioral change [25], and locomotion interruption [26]. Plastisphere, a term denoting the microbiome of microplastics, has raised global concern because its community structures are distinct from the natural environment. *Vibrio*, a genus of bacteria, is represented in the plastisphere of the North and Baltic Sea [27], and can have harmful effects on the human body [28].

Chemical impacts are caused by chemical additive consumptions, which are added to plastics during their production, and organic pollutants, which tend to attach to microplastics because of their large surface area to volume ratio [29]. These chemical substances can be easily exposed [30], especially under ultraviolet radiation and extreme heat [31,32]. For example, plasticizers added to plastic products for flexibility and malleability enhancement are not stable and can leach into the environment [33]. Additives such as bisphenol A (BPA), polybrominated diphenyl ethers (PBDEs), and phthalates are also known as endocrine-disrupting compounds (EDCs) and are harmful to the endocrine system [31]; these directly (reception of plasticizers by hormone receptors on microbes [34]), and indirectly (interruption of host hormone signaling) influence gut microbes, as gut microbes are mediated by hormones secreted by their hosts [29].

In summary, microplastics or substances attached to them can induce immediate and chronic mechanical and chemical disruptions in organisms. Preventing microplastic pollution of natural habitats is necessary to overcome these problems. Therefore, identifying microplastic sources is imperative to mitigating this damage. However, most review articles mainly focus on the risks of microplastics to organisms; the methodological progress of microplastic extraction and identification; and the comparison of microplastic occurrence, size, shape, type, color, and abundance between publications [35–37]. Discussions on how these reports attributed microplastic pollution to various anthropogenic factors have been limited. This review aims to elucidate the current advancements in the strategies used to analyze the relationship between anthropogenic activities and microplastic pollution.

2. Microplastics and Anthropogenic Activities

The major sources of microplastics are anthropogenic activities, such as human manufacturing and plastic-product usage. Humans are a major source of microplastics. The increasing world population size is a possible reason for the increasing plastic waste [2,38,39], owing to the short lifetime that these plastics are actually in use [40]. In 2019, while the world population reached 7700 million [41], the enormous demand for plastic drove the world plastic production up to 370 million metric tons a year [42], which has attracted attention as the growing rate of plastic recycling is overtaken by the growing rate of plastic production. Although the recycling rate of plastic waste from 2006 to 2018 has doubled, 25% of plastic waste is still sent to landfills [42]. Furthermore, since the COVID-19 pandemic happened, the relationship between anthropogenic activities and microplastics has become clearer. The plastic demand decreased tremendously during the pandemic in Europe in 2020, due to the quarantine, indicating less human activity, and therefore less plastic production [42]. However, the subsequent lifting of the lockdown restriction implies a resumption of plastic demand, and thus the microplastic problem remains to be solved. In this section, we introduce global publications (n = 34) that have linked microplastic abundance to potential anthropogenic factors (Figure 1), with Europe, India, and China being the top three most studied regions. Indeed, studies on the relationship between human activities and microplastic pollution in many densely populated areas are still in the developing stage, and providing in-depth focus on the link between these variables is necessary for studies that examine microplastic pollution as a function of spatial factors.

Therefore, this review aims to not only amplify the importance of defining the relationships between variables, but also to explain why a better measure than population density for quantifying anthropogenic activities is needed and why statistical analysis is essential.

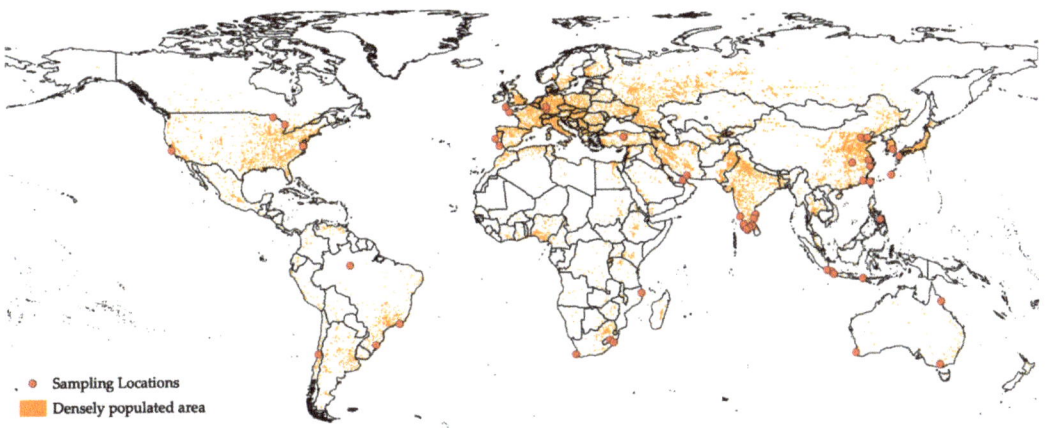

Figure 1. Distribution of the sampling sites of studies that linked microplastic pollution to anthropogenic activities. Densely populated area data were retrieved from Natural Earth (http://www.naturalearthdata.com (accessed on 20 December 2021) [43]. The Antarctica region was excluded as it is an area with limited human activities. Coordinate reference system: WGS 84, EPSG: 4326.

2.1. Population Density

Numerous studies have shown that areas with intensive anthropogenic activities tend to have higher microplastic pollution levels [17,44–50]. Previous reports related to aquatic environments ($n = 34$) are listed in Table 1, showing that 64.7% of studies sampled microplastics from water surface/column, 61.8% sampled microplastics from sediments, and only 29.4% sampled microplastics from organisms. Above all, only 50% of studies have conducted statistical analyses to investigate the relationship between anthropogenic factors and microplastic abundance, while 45, 50 and 50% of studies made statistical conclusions regarding the relationship between the two in water surfaces/columns, sediments, and organisms, respectively. Such paucity underlines the pressing need to conduct more statistics-based research in this field, and only half of the studies addressing the relationship between microplastic and anthropogenic activities is insufficient to formulate reliable microplastic control policies. Indeed, because of the heterogeneity of anthropogenic activities, previous studies have usually treated anthropogenic activities as a point source of microplastics. Those studies reflected the degree to which anthropogenic activities were responsible, mainly with regards to population density and proximity to city centers, wastewater treatment plants (WWTPs), harbors, and highly urbanized areas [5,9,44,45,51].

Browne et al. [45], for example, investigated microplastic pollution in sediments sampled from 18 sandy beaches worldwide, with microplastic abundance ranging from 2 to 31 particles in 250 mL sediment, suggesting that population density is significantly positively correlated with level of microplastic pollution ($p < 0.05$, $r^2 = 0.34$). However, it was difficult to compare this study with other sediment-focused microplastic studies on the coastline, as most relevant studies used weight/area rather than volume as the sampling unit. Yonkos et al. [52] supported this conclusion, demonstrating that variation in microplastic abundance on sampling dates (5534 to 297,927 particles km^{-2}) at the water surface of a bay was significantly correlated with population density ($p < 0.05$, $r^2 = 0.33$). In addition, Tang et al. [53] also suggested that, when their observations (514 particles m^{-3} on average) were integrated with other studies that took place in coastal areas of China, microplastic abundance at the water surface was significantly correlated with population size ($p < 0.05$, $r^2 = 0.99$) and urbanization rate ($p < 0.05$, $r^2 = 0.98$). Compared with not only a bay in South Korea, where the abundance at the water surface was 770 particles m^{-3} on average [47], but also other reports in China (see Reference [53]), the abundance observed by Tang et al. [53] was lower. This was possibly due to (1) different sampling methodologies, (2) different degrees of population density in sampling sites, and (3) samples being collected during the rainy season. More importantly, microplastic abundance in urban areas was not significantly different from that in rural areas with low population density (ANOVA, $p > 0.05$) in a bay in South Korea [47]. Furthermore, Wang et al. [51] found that, in China, distance from Wuhan City Center was significantly negatively correlated with microplastic abundance ($p < 0.05$, $r^2 = 0.90$), indicating a close relationship between human activities and microplastic pollution. Similarly, microplastic abundance in sediment (11 to 234.6 particles kg^{-1}) in heavily polluted areas in Taihu Lake, based on the index of eutrophication that generally reflects the degree of anthropogenic activities, was significantly higher than it was in clean areas (ANOVA, $p < 0.05$) [9].

In contrast, many studies provided no evidence of a relationship between population density and microplastics, as population density was not significantly associated with microplastic concentration [38,50,54]. For example, no significant relationship was found between the local municipal population and the level of microplastic abundance in water ($p > 0.05$) and sediment ($p > 0.05$) in the South African coastline, although some harbors had significantly higher microplastic loads (up to 1200 particles m^{-3}) in the water column (ANOVA, $p < 0.05$) [38]. Furthermore, Townsend et al. [50] investigated microplastic abundance in the wetlands in Australia (2 to 147 particles kg^{-1}), suggesting that neither population size ($p > 0.05$) nor population density ($p > 0.05$) was significantly correlated with microplastic abundance. Klein et al. [55] analyzed microplastics in river-shore sediments in Germany (228 to 3763 particles kg^{-1}), suggesting that population density was not significantly correlated with microplastic abundance ($p > 0.05$), and similarly microplastic abundance did not vary as a function of proximity to industrial areas or wastewater treatment plants. This disparity indicates that neither population density, a measure to quantify anthropogenic activities as a point source of microplastics, nor the characteristics of sample sites and their surroundings can fully explain the spatial variability of microplastics, with the latter measure being common in previous reports (see next section).

Table 1. Sampling condition, quantitative data and quantification strategies of anthropogenic activities in microplastic (MP) studies in aquatic environments. Note: dw, dry weight; ww, wet weight.

Environment	Sample Type and Average MP Concentration for Sampling Sites			Organism	Statistical Analysis	Anthropogenic Factors	Conclusion	Reference
	Water Surface	Water Column	Sediment					
Bay	0.24 ± 0.35 MP m^{-3} (excluding fibers, mean $\pm$ SD)	-	0.97 ± 2.08 MP kg^{-1} (excluding fibers, dw, mean $\pm$ SD)	-	-	1. Commercial port 2. Military base 3. Wastewater treatment plant 4. Shellfish farming 5. Marina	MP abundance at water surface was higher in sites next to anthropogenic factors	[5]
Bay	2.2 ± 1.4 MP L^{-1}	1.6 to 6.9 MP L^{-1}	31.1 to 256.3 MP kg^{-1} (dw)	-	-	1. Vessel activity 2. Close to coastline	MP abundance at water surface and in columns was higher in sites next to anthropogenic factors	[56]
Bay	7.62 MP m^{-3}	-	Beach: 166.50 MP kg^{-1} bay sediment: 20.74 166.5 MP kg^{-1}	-	-	1. Aquaculture 2. Fishing activity 3. recreational activities 4. Marine sports activities 5. Bars and restaurants 6. Proximity to rivers and channels 7. Urban drainage 8. Boat marina 9. Proximity to roads and waterway transport	MP abundance might be related to adjacent potential human activities, greater river inflow, and lower hydrodynamics	[57]
Bay	For each sampling date: 5534 to 297,927 MP km^{-2} or 2.7 to 245.7 g km^{-2}	-	-	-	V	1. Land use (proportion of urban/suburban area, agricultural area in catchments) 2. Population density	MP abundance was significantly correlated with population density and the proportion of urban/suburban development in the catchment	[52]
Bay	-	-	In Lumpung: 72.64 ± 25.28 MP kg^{-1} (mean $\pm$ SD) In Sumbawa: 44.19 ± 12.40 MP kg^{-1} (mean $\pm$ SD)	Sandfish: In Lampung: 3.21 ± 1.07 MP fish^{-1} (mean $\pm$ SD) or 126.34 ± 51.99 MP kg^{-1} (mean $\pm$ SD) In Sumbawa: 1.39 ± 0.86 MP fish^{-1} (mean $\pm$ SD) or 69.69 ± 52.22 MP kg^{-1} (mean $\pm$ SD)	V	1. Populated area	MP abundance in sediment and sandfish was significantly higher in Lumpung (populated area) than in Sumbawa (semi-enclosed ecosystem).	[58]

Table 1. *Cont.*

Environment	Sample Type and Average MP Concentration for Sampling Sites				Statistical Analysis	Anthropogenic Factors	Conclusion	Reference
	Water Surface	Water Column	Sediment	Organism				
Bay/Coastline	0.77 ± 0.88 MP L^{-1} (mean $\pm$ SD)	-	0.94 ± 0.69 MP g^{-1} (ww, mean $\pm$ SD)	Mussel: 1.43 ± 1.45 MP g^{-1} (ww, mean $\pm$ SD) Oyster: 1.13 ± 0.84 MP g^{-1} (ww, mean $\pm$ SD) Polychaete: 0.71 ± 1.00 MP g^{-1} (ww, mean $\pm$ SD)	V	1. Close to urban areas 2. Close to aquafarm areas	MP abundance in sediment was significantly higher in urban areas than in rural areas	[47]
Bay/Coastline/ Estuary	514.3 ± 520.0 MP m^{-3} (mean $\pm$ SD)	-	-	76 to 333 MP kg^{-1} (mean $\pm$ SD)	V	1. Total population 2. Urbanization rate 3. Farmland	MP abundance at water surface was significantly correlated with total population and urbanization rate	[53]
River	-	-	Summer: 6.3 ± 4.3 MP kg^{-1} (dw, mean $\pm$ SD) Winter: 160.1 ± 139.5 MP kg^{-1} (dw, mean $\pm$ SD)	*Chironomus* spp.: Summer: 0.37 ± 0.44 MP mg^{-1} (ww, mean $\pm$ SD) Winter: 1.12 ± 1.19 MP mg^{-1} (ww, mean $\pm$ SD)	-	1. Close to populated areas 2. Close to wastewater treatment plants	MP abundance was higher in sites next to anthropogenic factors	[59]
River	892,777 MP km^{-2}	-	-	-	-	1. Close to populated areas 2. Close to wastewater treatment plants	MP abundance was higher near populated areas and at the side of riverbanks wherein wastewater treatment plant effluents are entering	[60]
River	-	-	0.063–5 mm: 417 to 8178 MP kg^{-1} (dw) 0.063–1 mm: 0 to 5725 MP kg^{-1} (dw)	-	-	1. City	Location with highest MP concentration might be related to hydraulic conditions and proximity to the city	[61]
River	Reference area: 6.8 MP L^{-1} Textile industrial area: 13.3 MP L^{-1}	-	16.7 to 1323.3 MP kg^{-1} (dw)	-	V	1. Close to textile industrial area	MP abundance was significantly higher in the industrial area than in the reference area	[46]

Table 1. Cont.

Environment	Sample Type and Average MP Concentration for Sampling Sites				Statistical Analysis	Anthropogenic Factors	Conclusion	Reference
	Water Surface	Water Column	Sediment	Organism				
River	9.2 ± 2.2 MP L^{-1} (mean $\pm$ SD)	Intermediate: 8.4 ± 1.7 MP L^{-1} (mean $\pm$ SD) Bottom: 14.2 ± 5.6 MP L^{-1} (mean $\pm$ SD)	4328 ± 2037 MP kg^{-1} (dw, mean $\pm$ SD)	-	$\checkmark$	1. Population density of suburban area 2. Population density of urban area 3. Population density of industrial area	MP abundance in water columns was significantly correlated with population density in suburban and urban areas	[48]
River	-	-	Rhine river: 21.8 to 932 mg kg^{-1} or 228 to 3763 MP kg^{-1} Main river: 43.5 to 459 mg kg^{-1} or 786 to 1368 MP kg^{-1}	-	$\checkmark$	1. Close to industrial area 2. Population density	No significant correlation between MP masses and population density was found, and MP abundance did not increase downstream of the industrial area	[55]
River	-	-	-	Chironomidae larvae: 0.28 to 2.07 MP mg^{-1}	$\checkmark$	1. Land use (proportion of industrial area and residential area in the catchment)	The proportion of industrial areas in catchment contributes more to MP concentration in midge larvae than the proportion of residential areas	[49]
River	-	5.85 ± 3.28 MP L^{-1} (mean $\pm$ SD)	3.03 ± 1.59 MP 100 g^{-1} (dw, mean $\pm$ SD)	-	$\checkmark$	1. Industrial area 2. Slum area	MP abundance in sediment was significantly higher in sites located around industrial and slum areas	[62]
River/Coastline	8.48 to 9.37 MP m^{-3}	-	-	*Aplocheilus* sp.: 1.97 MP fish^{-1}	-	1. Tourism 2. Port 3. Industrial operation	MP abundance was higher in sites located around anthropogenic factors	[63]
River/Lake	-	1660.0 to 8925 MP m^{-3}	-	-	$\checkmark$	1. Distance from the urban left	MP abundance correlated significantly negatively with distance from the city left	[51]
Lake	$43{,}157 \pm 115{,}519$ MP km^{-2} (mean $\pm$ SD)	-	-	-	-	1. Close to populated areas 2. Close to shoreline	MP abundance was higher near populated areas and areas near the shoreline	[64]
Lake	11.9 to 61.2 MP m^{-3}	-	-	-	-	1. Population density 2. Domestic sewage	MP abundance was higher in sites located around populated area	[65]

Table 1. Cont.

Environment	Sample Type and Average MP Concentration for Sampling Sites				Statistical Analysis	Anthropogenic Factors	Conclusion	Reference
	Water Surface	Water Column	Sediment	Organism				
Lake	3.4 to 25.8 MP L^{-1}	-	11.0 to 234.6 MP kg^{-1} (dw)	Plankton: 0.01 × 10^6 to 6.8 × 10^6 MP km^{-2} Asian clams: August: 1.3 to 12.5 MP g^{-1} (ww) November: 0.2 to 9.6 MP g^{-1} (ww)	√	1. Close to populated areas 2. Index of eutrophication	MP abundance in sediment was significantly higher near areas with more human activity than areas with less human activity, according to the index of eutrophication	[9]
Lake	0.05 to 32 MP m^{-3}	-	-	-	√	1. Land use (proportion of industrial area, agricultural area (total, crops, pasture, and hay) and impervious area) 2. Population density 3. Wastewater treatment plant effluent contribution	MP abundance was significantly correlated with the proportion of urban area, agricultural area (total and crops), and impervious area in catchments; MP abundance was significantly correlated with population density	[66]
Coastline	-	-	High tide line: 439 ± 172 to 119 ± 72 MP kg^{-1} (dw, mean ± SD) Low tide line: 179 ± 68 to 33 ± 30 MP kg^{-1} (dw, mean ± SD)	-	-	1. Metropolitan city	MP abundance was highest in the location near the metropolitan city	[67]
Coastline	-	24 ± 9 to 96 ± 57 MP L^{-1} (mean ± SD)	55 ± 21 to 259 ± 88 MP kg^{-1} (mean ± SD)	-	-	1. Tourism 2. Shipping 3. Fishing 4. Aquaculture	MP abundance was higher in sites located around anthropogenic factors	[68]
Coastline	-	3.1 ± 2.3 to 23.7 ± 4.2 MP L^{-1} (mean ± SD)	-	0.11 ± 0.06 to 3.64 ± 1.7 MP fish^{-1} (mean ± SD) or 0.0002 ± 0.0001 to 0.2 ± 0.03 MP g^{-1} gut weight (mean ± SD)	-	1. Sewage effluent 2. Proximity to anthropogenic activities	MP abundance was higher in shore areas (adjacent to sewage effluent) and in epipelagic fish (adjacent to urban runoff)	[69]
Coastline	-	1.25 ± 0.88 MP m^{-3} (mean ± SD)	40.7 ± 33.2 MP m^{-2} (mean ± SD)	Fishes (not specified)	-	1. Population density 2. Industrial activities 3. Tourism 4. Sewage effluent 5. fishing	MP abundance in water and sediment was high due to proximity of urban regions, river runoff, fisheries and tourism	[70]

Table 1. Cont.

Environment	Sample Type and Average MP Concentration for Sampling Sites				Statistical Analysis	Anthropogenic Factors	Conclusion	Reference
	Water Surface	Water Column	Sediment	Organism				
Coastline	-	-	-	Zooplankton: 0.002 to 0.036 MP m^{-3}	-	1. Close to populated areas 2. Close to industrial facilities 3. Close to port facilities	MP abundance was higher near populated areas and areas close to industrial and port facilities	[71]
Coastline	-	-	43 MP 50 g^{-1} (dw, only include fragments and fibers)	-	-	1. Tourism 2. Harbor 3. Residential area	MP abundance was high in beaches with associated anthropogenic activity	[72]
Coastline	-	-	2 to 31 MP 250 mL^{-1}	-	√	1. Population density	MP abundance was significantly correlated with population density	[45]
Coastline	-	(Not specified)	86.67 ± 48.68 to 754.7 ± 393 MP m^{-2} (depth 5 cm, mean ± SD)	-	√	1. Close to harbors 2. Population density	No significant correlation between population density and MP abundance in water column and sediment was found	[38]
Coastline	-	-	High tide line: 1323 ± 1228 mg m^{-2} (mean ± SD) Low tide line: 180 ± 261 mg m^{-2} (mean ± SD) Overall: 46.6 ± 37.2 MP m^{-2} (mean ± SD)	Important fish species: 0.1 MP fish^{-1}	√	1. Tourism 2. Fishing 3. River mouth 4. Urban activities	MP abundance in beaches was insignificantly correlated with the distance of the beach from the nearest river mouth	[73]
Coastline	Proportion of MP in collected particles: 13.3 to 25.0%	-	-	-	√	1. Population density	Significantly greater proportions of MP particles were found in areas with higher population density	[74]
Pond	233 MP m^{-3}	-	-	-	-	1. Populated area	MP abundance was low in the studied area (near protected areas) compared to reference study sites (near populated areas)	[75]

Table 1. Cont.

Environment	Sample Type and Average MP Concentration for Sampling Sites				Statistical Analysis	Anthropogenic Factors	Conclusion	Reference
	Water Surface	Water Column	Sediment	Organism				
Strait	-	-	2 to 1258 MP kg^{-1} (dw)	-	-	1. The relative level of industrialization (manufacturing, oil refineries, and industrial sewage) and urbanization	MP abundance was higher near areas with elevated levels of industrialization and urbanization	[44]
Wetland	-	-	2 to 147 MP kg^{-1} (dw)	-	V	1. Land use (proportion of and absolute commercial area, industrial area, and residential area) 2. Dwelling density 3. Population density 4. Population size 5. Road/rail 6. Urban growth	MP abundance was significantly less in catchments with more open space (undeveloped catchments) The proportion of road/rail areas, commercial areas, industrial areas, and residential areas in catchments was not significantly associated with MP abundance Population density, population size, dwelling density, urban growth, and catchment size were not significantly associated with MP abundance	[50]

2.2. Importance of Statistical Analysis

It is very common to relate the effects of human activities to microplastic abundance without clear statistical analyses [5,55,56]. Previous studies tended to attribute the elevated microplastic abundance to the surrounding possible point source of microplastics, probably because it is straightforward and intuitive to infer the relationship between anthropogenic factors and microplastic abundance by associating the spatial distribution of microplastic abundance with general characterization around sample locations.

For example, although Klein et al. [55] suggested that, as mentioned above, it was difficult to visualize the relationship between microplastic abundance and proximity to industrial areas or wastewater treatment plants on a map; sample sites that were close to nature reserves had low microplastic abundance, which probably could be explained by the fewer human activities in nature reserves. In contrast, areas exhibiting high microplastic abundance on the water surface probably resulted from the proximity to marinas, military, and commercial harbors, as well as effluent from wastewater treatment plants that process sewage from more than 134,377 people [5]. In addition, we must acknowledge that those areas are located in the most densely urbanized area in the monitored region (Bay of Brest, France) [5].

Furthermore, sample sites located on the cruise route had higher microplastic abundance, supporting the inference that vessel activities produce microplastic pollution [56,76]. In addition, since certain sample sites located downstream of wastewater treatment plants showed high microplastic abundance, especially at the right river bank, and that the outlets of wastewater treatment plants entered the Rhine River from the right river bank, it can be inferred that the elevated microplastic concentration on the river surface probably resulted from the outlets of the wastewater treatment plant [60]. Additionally, consistently high microplastic abundance on the surface of Lake Erie of the Laurentian Great Lakes might be due to anthropogenic activities, as Lake Erie was the most populated lake in the monitored region [64].

In summary, reports regarding anthropogenic activities and microplastics in the field can generally be presented in two ways, depending on whether the discussion is based on statistical analyses. If yes, there were usually two kinds of mathematical results: microplastic abundance in densely urbanized areas was significantly different from that in less developed areas (reference area) [9,46,47], and there was a correlation between population density and microplastic abundance in sample sites [52,53,55]. If not, the discussion was usually made by visual inspection of anthropogenic factors surrounding the sample sites, and this can be problematic.

Microplastic distribution and abundance in monitored regions do not always depend on surrounding anthropogenic activities (e.g., location of WWTPs and harbors). According to Klein et al. [55], the four sample sites with the highest microplastic abundance, regardless of count (particles kg^{-1}) or mass (mg kg^{-1}), were also categorized as the four most populated sites in the research area; therefore, there is a trend indicating that population density can explain the high level of microplastic pollution at these sites. However, statistical analysis revealed no significant correlation between microplastic abundance and population density when all sample sites were considered. This highlights the potential scale-dependent effect on the results and the necessity of conducting appropriate statistical analyses to account for this. Linking these variables based on visual inspection of spatial distribution may lead to problematic conclusions. Therefore, in order to apply statistical analysis and produce practical results, two questions remain to be answered: (1) How can anthropogenic factors be quantified? (2) Are there other quantification strategies more appropriate than population density?

2.3. Urban Attributes

Quantifying the level of human activity by population density is simple. As mentioned above, previous studies revealed that the correlation between human activities and the level of microplastic pollution is mostly significant; on the other hand, remote and/or less developed areas showed significantly lower microplastic abundance than urbanized areas. These results, however, oversimplified anthropogenic activities and thus cannot help governments construct effective policies for controlling microplastic pollution. In other words, the anthropogenic activities that contribute microplastics to the environment predominantly remain unknown, leading to a difficult situation in which controlling microplastics from the source is the most effective way to reduce microplastics [77]. While it is no secret that human activities are the biggest source of microplastics, we still have no clue what the exact source is. There is an urgent need for detailed information on human activities.

Therefore, in addition to population density, recent studies have used other urban attributes to quantify different human activities, i.e., different land uses within the catchment of the sample location (Figure 2). Figure 2 visualizes the quantification strategy of different anthropogenic factors: delineation of the catchment margins of sample sites and calculation of the percentages of different upstream land covers (e.g., industrial area, residential area, and agricultural area) in the watershed. These percentages of land cover were used to reflect the magnitude of different anthropogenic activities. For example, Yonkos et al. [52] extrapolated not only the population density in catchments of sample locations from the 2010 US census data, but also the percentages of urban (industrial), suburban (residential), agricultural, and forested areas in catchments of sample sites from the 2006 National Land Cover Database. The study estimated the correlation between different land covers and microplastic abundance and concluded that the microplastic abundance on the water surface was significantly associated with population density, percentage of urban (industrial) area, and percentage of total developed (industrial and residential) areas.

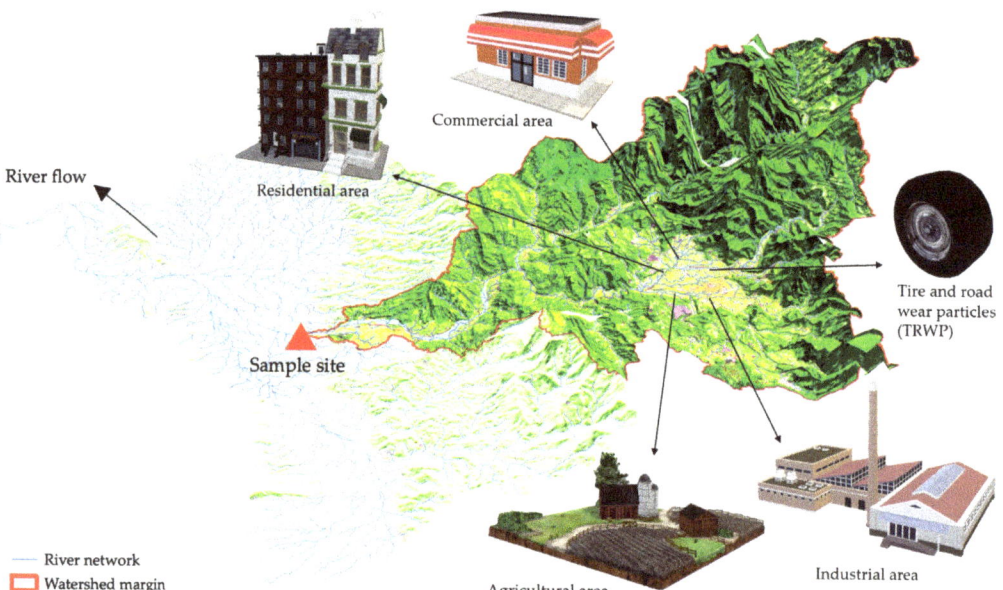

Figure 2. Watershed margin delineation of sample sites and different upstream land covers. Three-dimensional objects were retrieved from Microsoft® Office PowerPoint®.

Correspondingly, Baldwin et al. [66] analyzed the correlation between the microplastic abundance on the lake surface and different watershed characteristics, including percentages of impervious areas (e.g., roads, parking lots, and buildings), urban area, agricultural areas (total, crops, pasture, and hay), and forested area in catchments of sample sites. Land-cover datasets were retrieved from the National Land Cover Database (NLCD) Land Cover and Percent Developed Imperviousness datasets, and watershed margins were derived from the US Geological Survey Watershed Boundary Dataset (USGC-WBD). The results suggested that microplastic abundance was positively correlated with the percentage of urban area and percentage of impervious areas, and negatively correlated with the percentage of agricultural area (total and crops) [66], in line with the results of Yonkos et al. [52]. Although agricultural activities were not addressed too much in the study by Baldwin et al. [66], Yonkos et al. [52] found a similar but insignificant trend in the percentage of agricultural area, which was also dominated by crop agriculture; it was negatively associated with microplastic abundance. This can probably be explained by the lower development and lesser amount of human activities in areas with high agricultural activity.

A similar approach for the quantification of human activities was conducted in the urban wetlands of Melbourne, Australia. Briefly, the study calculated the catchment margins of sample sites with certain digital elevation models, using ArcGIS 10.3 [50]. Detailed land-use data were retrieved from the 2011 Australian Population Census, including the percentages of the following: commercial area, industrial area, undeveloped area, road/rail, residential area, percentage of rural area, and semi-rural area. Additionally, urban growth and dwelling density were also included in the analyses to reflect the different magnitudes of anthropogenic activities.

However, in contrast with the study by Yonkos et al. [52], Townsend et al. [50] indicated that only the percentage of undeveloped areas within the catchment was significantly negatively correlated with microplastic abundance in sediment. No significant correlation between the percentage of industrial area and microplastic abundance in the catchment was observed. More interestingly, if not using the percentage of land use, no significant correlation was found between microplastic abundance and the absolute area of different land use, presumably resulting from the effect of catchment size. Lin et al. [49] supported this result, as the percentage of the industrial area model was better than its logarithmic model. To elaborate further, Lin et al. [49] constructed several general linear mixed-effect models and conducted model selection to investigate the effects of different land uses (industrial and residential) on microplastic concentration found in chironomid larvae. The results showed that the percentage of industrial area in catchments contributes more to microplastic concentration, a finding that is in line with previous studies that showed that the percentage of industrial (urban) areas in catchments was a potential predictor of microplastic pollution [52,66].

3. Future Directions and Conclusions

To make progress in mitigating environmental microplastics, the source of microplastics needs to be identified. While anthropogenic activities are the most well-known source of microplastics, one should keep in mind that quantifying parameters, such as population density, might not be detailed enough to offer practical suggestions on formulating policies for microplastic pollution management. In fact, due to the environmental risks of microplastics, government agencies and environmental protection organizations have actively advocated policies and regulations to protect aquatic organisms from the detrimental effects of microplastics [78,79]. Regulation that bans plastic/microplastic production and consumption has been articulated globally in the past few years [80]. For instance, since 2003, the government of South Africa has charged for the use of thick plastic bags, and plastic-bag use has been decreased by 90%. Since 2007, Kenya has banned the manufacture and import of thin plastic bags; however, the ban was not enforced. Since 2008, Rwanda has become the first plastic-free country by banning non-biodegradable plastics. In the same year in China, while the Beijing Olympics were in full swing, plastic bags under a certain thickness were banned, and citizens were charged for the use of others, leading to effective results in mitigating damage caused by plastic bags. In 2013, Pucón was the first city in Chili to ban plastic bags. In 2014, California, USA, banned plastic shopping bags and plastic bottles, and France implemented a tax on non-biodegradable plastic bags. In

2015, an amendment bill in the United States against personal-care products containing microbeads was passed. Similar acts have been adopted in Canada [81] and Taiwan [49]. In addition, plastic regulation in India, the most populated country and the largest plastic consumers in the world, was not successful, due to poor enforcement and pressure from the rapidly growing plastic industry until 2016, when plastic bags with the thickness below 50 µm were banned [82].

It is recommended that policies focus on secondary microplastics, such as microfibers, because the major source of microfibers is the washing of clothes [45]. Policies should be developed to improve domestic wastewater treatment processes to filter out microfibers, as waste management policies directly influence microplastic abundance in the environment [83]. Furthermore, policies for regular monitoring of microplastic abundance in various ecosystems have been suggested [82]. In addition to reducing plastics from consumption or import, and capturing microplastics before they contaminate the natural environment, more studies are required to further discuss the heterogeneity of human activities and how to effectively control microplastic pollution. Recently, a few studies have attempted to further discuss different kinds of human activities based on land-use-survey data. They have delineated watershed margins, presuming that microplastics released from sources are mainly transported by rain runoff to rivers or lakes (freshwater systems), and incorporated a land-cover dataset to represent distinctive human activities. These studies have shown that industrial/urban areas within catchments are potential sources of microplastics. However, future challenges in this context will be to answer the following questions: (1) Can the resolution of land-cover data increase so that we can identify the exact industry producing the largest amount of microplastics in industrial areas, and can the resolution of microplastic properties increase (with better microplastic identification efficiency) so that we can identify a specific "marker" substance of microplastics that may be representative of a certain type of industry? (2) Because watershed margin delineation has a huge influence on the results, is the watershed margin reliable if river channels are artificially manipulated in urban river systems, and should the sewer system be considered? (3) Are there better approaches to quantifying anthropogenic activities that can help the government build related policies?

In conclusion, these challenges can be tackled by the construction of accessible and reliable land-cover surveillance data from government agencies, the development of better microplastic-substance-identification techniques and protocols, and the available data on artificial river channels and sewer systems in urban areas. Although the relationship between anthropogenic activities and microplastics is yet to be comprehensively studied, this review has observed positive and consistent progress on this issue, showing that these problems are expected to be addressed in the near future. This review integrates the global literature on environmental microplastics and anthropogenic activities, highlighting the necessity of summarizing results with statistical analyses. It is imperative to quantify anthropogenic factors by using indices other than population density, which creates equivocal results whose use by policymakers is difficult. At present, this review suggests that using watershed-scale attributes derived from land-use datasets might produce a more in-depth scientific basis for government authorities and environmental protection organizations and institutions to articulate efficient policies to reduce microplastic pollution.

Author Contributions: Conceptualization, C.-T.L.; writing—review and editing of the manuscript, C.-T.L. and M.-C.C.; supervision, project administration, and funding acquisition, M.-H.K. All authors have read and agreed to the published version of the manuscript.

Funding: This research received no external funding.

Institutional Review Board Statement: Not applicable.

Informed Consent Statement: Not applicable.

Conflicts of Interest: The authors declare no conflict of interest.

References

1. Arthur, C.; Baker, J.; Bamford, H. *Proceedings of the International Research Workshop on the Occurrence, Effects, and Fate of Microplastic Marine Debris, 9–11 September 2008*; University of Washington Tacoma: Tacoma, WA, USA, 2009; 530p.
2. Thompson, R.C.; Olsen, Y.; Mitchell, R.P.; Davis, A.; Rowland, S.J.; John, A.W.G.; McGonigle, D.; Russel, A.E. Lost at Sea: Where Is All the Plastic? *Science* **2004**, *304*, 838. [CrossRef] [PubMed]
3. Barnes, D.K.A.; Galgani, F.; Thompson, R.C.; Barlaz, M. Accumulation and fragmentation of plastic debris in global environments. *Philos. Trans. R. Soc. B Biol. Sci.* **2009**, *364*, 1985–1998. [CrossRef]
4. Singh, B.; Sharma, N. Mechanistic implications of plastic degradation. *Polym. Degrad. Stab.* **2008**, *93*, 561–584. [CrossRef]
5. Frère, L.; Paul-Pont, I.; Rinnert, E.; Petton, S.; Jaffré, J.; Bihannic, I.; Soudant, P.; Lambert, C.; Huvet, A. Influence of environmental and anthropogenic factors on the composition, concentration and spatial distribution of microplastics: A case study of the Bay of Brest (Brittany, France). *Environ. Pollut.* **2017**, *225*, 211–222. [CrossRef]
6. Eriksen, M.; Lebreton, L.C.M.; Carson, H.S.; Thiel, M.; Moore, C.J.; Borerro, J.C.; Galgani, F.; Ryan, P.G.; Reisser, J. Plastic Pollution in the World's Oceans: More than 5 Trillion Plastic Pieces Weighing over 250,000 Tons Afloat at Sea. *PLoS ONE* **2014**, *9*, e111913. [CrossRef] [PubMed]
7. Van Cauwenberghe, L.; Vanreusel, A.; Mees, J.; Janssen, C. Microplastic pollution in deep-sea sediments. *Environ. Pollut.* **2013**, *182*, 495–499. [CrossRef]
8. Li, J.; Liu, H.; Chen, J.P. Microplastics in freshwater systems: A review on occurrence, environmental effects, and methods for microplastics detection. *Water Res.* **2018**, *137*, 362–374. [CrossRef]
9. Su, L.; Xue, Y.; Li, L.; Yang, D.; Kolandhasamy, P.; Li, D.; Shi, H. Microplastics in Taihu Lake, China. *Environ. Pollut.* **2016**, *216*, 711–719. [CrossRef]
10. Wetherbee, G.A.; Baldwin, A.K.; Ranville, J.F. *It Is Raining Plastic*; U.S. Geological Survey: Reston, VA, USA, 2019. [CrossRef]
11. Bergmann, M.; Mützel, S.; Primpke, S.; Tekman, M.B.; Trachsel, J.; Gerdts, G. White and wonderful? Microplastics prevail in snow from the Alps to the Arctic. *Sci. Adv.* **2019**, *5*, eaax1157. [CrossRef]
12. Bergmann, M.; Sandhop, N.; Schewe, I.; D'Hert, D. Observations of floating anthropogenic litter in the Barents Sea and Fram Strait, Arctic. *Polar Biol.* **2015**, *39*, 553–560. [CrossRef]
13. Windsor, F.M.; Tilley, R.M.; Tyler, C.R.; Ormerod, S.J. Microplastic ingestion by riverine macroinvertebrates. *Sci. Total Environ.* **2018**, *646*, 68–74. [CrossRef]
14. Karami, A.; Golieskardi, A.; Choo, C.K.; Larat, V.; Karbalaei, S.; Salamatinia, B. Microplastic and mesoplastic contamination in canned sardines and sprats. *Sci. Total Environ.* **2018**, *612*, 1380–1386. [CrossRef]
15. Carlin, J.; Craig, C.; Little, S.; Donnelly, M.; Fox, D.; Zhai, L.; Walters, L. Microplastic accumulation in the gastrointestinal tracts in birds of prey in central Florida, USA. *Environ. Pollut.* **2020**, *264*, 114633. [CrossRef]
16. Free, C.M.; Jensen, O.P.; Mason, S.A.; Eriksen, M.; Williamson, N.J.; Boldgiv, B. High-levels of microplastic pollution in a large, remote, mountain lake. *Mar. Pollut. Bull.* **2014**, *85*, 156–163. [CrossRef]
17. Wong, J.K.H.; Lee, K.K.; Tang, K.H.D.; Yap, P.-S. Microplastics in the freshwater and terrestrial environments: Prevalence, fates, impacts and sustainable solutions. *Sci. Total Environ.* **2020**, *719*, 137512. [CrossRef]
18. Wright, S.L.; Thompson, R.C.; Galloway, T.S. The physical impacts of microplastics on marine organisms: A review. *Environ. Pollut.* **2013**, *178*, 483–492. [CrossRef] [PubMed]
19. Straub, S.; Hirsch, P.E.; Burkhardt-Holm, P. Biodegradable and Petroleum-Based Microplastics Do Not Differ in Their Ingestion and Excretion but in Their Biological Effects in a Freshwater Invertebrate Gammarus fossarum. *Int. J. Environ. Res. Public Health* **2017**, *14*, 774. [CrossRef] [PubMed]
20. Lu, Y.; Zhang, Y.; Deng, Y.; Jiang, W.; Zhao, Y.; Geng, J.; Ding, L.; Ren, H.-Q. Uptake and Accumulation of Polystyrene Microplastics in Zebrafish (Danio rerio) and Toxic Effects in Liver. *Environ. Sci. Technol.* **2016**, *50*, 4054–4060. [CrossRef] [PubMed]
21. Silva, C.J.; Beleza, S.; Campos, D.; Soares, A.M.; Silva, A.L.P.; Pestana, J.L.; Gravato, C. Immune response triggered by the ingestion of polyethylene microplastics in the dipteran larvae Chironomus riparius. *J. Hazard. Mater.* **2021**, *414*, 125401. [CrossRef]
22. Wang, K.; Li, J.; Zhao, L.; Mu, X.; Wang, C.; Wang, M.; Xue, X.; Qi, S.; Wu, L. Gut microbiota protects honey bees (Apis mellifera L.) against polystyrene microplastics exposure risks. *J. Hazard. Mater.* **2020**, *402*, 123828. [CrossRef]
23. Jin, Y.; Xia, J.; Pan, Z.; Yang, J.; Wang, W.; Fu, Z. Polystyrene microplastics induce microbiota dysbiosis and inflammation in the gut of adult zebrafish. *Environ. Pollut.* **2018**, *235*, 322–329. [CrossRef] [PubMed]
24. Huang, Y.; Li, W.; Gao, J.; Wang, F.; Yang, W.; Han, L.; Lin, D.; Min, B.; Zhi, Y.; Grieger, K.; et al. Effect of microplastics on ecosystem functioning: Microbial nitrogen removal mediated by benthic invertebrates. *Sci. Total Environ.* **2020**, *754*, 142133. [CrossRef] [PubMed]
25. Hansen, B.; Hansen, P.J.; Nielsen, T.G. Effects of large nongrazable particles on clearance and swimming behaviour of zooplankton. *J. Exp. Mar. Biol. Ecol.* **1991**, *152*, 257–269. [CrossRef]
26. Cole, M.; Lindeque, P.; Fileman, E.; Halsband, C.; Goodhead, R.; Moger, J.; Galloway, T.S. Microplastic Ingestion by Zooplankton. *Environ. Sci. Technol.* **2013**, *47*, 6646–6655. [CrossRef] [PubMed]
27. Kirstein, I.V.; Kirmizi, S.; Wichels, A.; Garin-Fernandez, A.; Erler, R.; Löder, M.; Gerdts, G. Dangerous hitchhikers? Evidence for potentially pathogenic Vibrio spp. on microplastic particles. *Mar. Environ. Res.* **2016**, *120*, 1–8. [CrossRef]
28. Oliver, J.D. Wound infections caused by Vibrio vulnificus and other marine bacteria. *Epidemiol. Infect.* **2005**, *133*, 383–391. [CrossRef]

29. Fackelmann, G.; Sommer, S. Microplastics and the gut microbiome: How chronically exposed species may suffer from gut dysbiosis. *Mar. Pollut. Bull.* **2019**, *143*, 193–203. [CrossRef] [PubMed]
30. Teuten, E.L.; Saquing, J.M.; Knappe, D.; Barlaz, M.A.; Jonsson, S.; Björn, A.; Rowland, S.J.; Thompson, R.; Galloway, T.S.; Yamashita, R.; et al. Transport and release of chemicals from plastics to the environment and to wildlife. *Philos. Trans. R. Soc. B Biol. Sci.* **2009**, *364*, 2027–2045. [CrossRef]
31. Talsness, C.E.; Andrade, A.J.M.; Kuriyama, S.N.; Taylor, J.A.; vom Saal, F.S. Components of plastic: Experimental studies in animals and relevance for human health. *Philos. Trans. R. Soc. B Biol. Sci.* **2009**, *364*, 2079–2096. [CrossRef] [PubMed]
32. Andrady, A.L. Microplastics in the marine environment. *Mar. Pollut. Bull.* **2011**, *62*, 1596–1605. [CrossRef]
33. Oehlmann, J.; Schulte-Oehlmann, U.; Kloas, W.; Jagnytsch, O.; Lutz, I.; Kusk, K.O.; Wollenberger, L.; Santos, E.; Paull, G.C.; Van Look, K.J.W.; et al. A critical analysis of the biological impacts of plasticizers on wildlife. *Philos. Trans. R. Soc. B Biol. Sci.* **2009**, *364*, 2047–2062. [CrossRef] [PubMed]
34. Neuman, H.; Debelius, J.; Knight, R.; Koren, O. Microbial endocrinology: The interplay between the microbiota and the endocrine system. *FEMS Microbiol. Rev.* **2015**, *39*, 509–521. [CrossRef] [PubMed]
35. Eerkes-Medrano, D.; Thompson, R.C.; Aldridge, D.C. Microplastics in freshwater systems: A review of the emerging threats, identification of knowledge gaps and prioritisation of research needs. *Water Res.* **2015**, *75*, 63–82. [CrossRef]
36. Hidalgo-Ruz, V.; Gutow, L.; Thompson, R.C.; Thiel, M. Microplastics in the Marine Environment: A Review of the Methods Used for Identification and Quantification. *Environ. Sci. Technol.* **2012**, *46*, 3060–3075. [CrossRef] [PubMed]
37. Yang, L.; Zhang, Y.; Kang, S.; Wang, Z.; Wu, C. Microplastics in freshwater sediment: A review on methods, occurrence, and sources. *Sci. Total Environ.* **2020**, *754*, 141948. [CrossRef]
38. Nel, H.A.; Hean, J.W.; Noundou, X.S.; Froneman, W. Do microplastic loads reflect the population demographics along the southern African coastline? *Mar. Pollut. Bull.* **2017**, *115*, 115–119. [CrossRef]
39. Jambeck, J.R.; Geyer, R.; Wilcox, C.; Siegler, T.R.; Perryman, M.; Andrady, A.; Narayan, R.; Law, K.L. Plastic waste inputs from land into the ocean. *Science* **2015**, *347*, 768–771. [CrossRef] [PubMed]
40. Geyer, R.; Jambeck, J.R.; Law, K.L. Production, use, and fate of all plastics ever made. *Sci. Adv.* **2017**, *3*, e1700782. [CrossRef]
41. World Population Prospects. *World Population Prospects 2019: Highlights*; ST/ESA/SER.A/423; UN: New York, NY, USA, 2019.
42. Plastics Europe. *Plastics—The Facts. An Analysis of European Plastics Production, Demand and Waste Data*; Plastics Europe: Brussels, Belgium, 2020.
43. Patterson, T.; Kelso, N.V. *World Urban Areas, LandScan, 1:10 Million*; North American Cartographic Information Society: Milwaukee, WI, USA, 2012.
44. Naji, A.; Esmaili, Z.; Khan, F.R. Plastic debris and microplastics along the beaches of the Strait of Hormuz, Persian Gulf. *Mar. Pollut. Bull.* **2017**, *114*, 1057–1062. [CrossRef] [PubMed]
45. Browne, M.A.; Crump, P.; Niven, S.J.; Teuten, E.; Tonkin, A.; Galloway, T.; Thompson, R. Accumulation of Microplastic on Shorelines Woldwide: Sources and Sinks. *Environ. Sci. Technol.* **2011**, *45*, 9175–9179. [CrossRef] [PubMed]
46. Deng, H.; Wei, R.; Luo, W.; Hu, L.; Li, B.; Di, Y.; Shi, H. Microplastic pollution in water and sediment in a textile industrial area. *Environ. Pollut.* **2020**, *258*, 113658. [CrossRef] [PubMed]
47. Jang, M.; Shim, W.J.; Cho, Y.; Han, G.M.; Song, Y.K.; Hong, S.H. A close relationship between microplastic contamination and coastal area use pattern. *Water Res.* **2019**, *171*, 115400. [CrossRef]
48. Liu, Y.; You, J.; Li, Y.; Zhang, J.; He, Y.; Breider, F.; Tao, S.; Liu, W. Insights into the horizontal and vertical profiles of microplastics in a river emptying into the sea affected by intensive anthropogenic activities in Northern China. *Sci. Total Environ.* **2021**, *779*, 146589. [CrossRef] [PubMed]
49. Lin, C.-T.; Chiu, M.-C.; Kuo, M.-H. Effects of anthropogenic activities on microplastics in deposit-feeders (Diptera: Chironomidae) in an urban river of Taiwan. *Sci. Rep.* **2021**, *11*, 1–8. [CrossRef]
50. Townsend, K.R.; Lu, H.-C.; Sharley, D.J.; Pettigrove, V. Associations between microplastic pollution and land use in urban wetland sediments. *Environ. Sci. Pollut. Res.* **2019**, *26*, 22551–22561. [CrossRef] [PubMed]
51. Wang, W.; Ndungu, A.W.; Li, Z.; Wang, J. Microplastics pollution in inland freshwaters of China: A case study in urban surface waters of Wuhan, China. *Sci. Total Environ.* **2017**, *575*, 1369–1374. [CrossRef]
52. Yonkos, L.T.; Friedel, E.A.; Perez-Reyes, A.C.; Ghosal, S.; Arthur, C.D. Microplastics in Four Estuarine Rivers in the Chesapeake Bay, U.S.A. *Environ. Sci. Technol.* **2014**, *48*, 14195–14202. [CrossRef] [PubMed]
53. Tang, G.; Liu, M.; Zhou, Q.; He, H.; Chen, K.; Zhang, H.; Hu, J.; Huang, Q.; Luo, Y.; Ke, H.; et al. Microplastics and polycyclic aromatic hydrocarbons (PAHs) in Xiamen coastal areas: Implications for anthropogenic impacts. *Sci. Total Environ.* **2018**, *634*, 811–820. [CrossRef] [PubMed]
54. Feng, S.; Lu, H.; Liu, Y. The occurrence of microplastics in farmland and grassland soils in the Qinghai-Tibet plateau: Different land use and mulching time in facility agriculture. *Environ. Pollut.* **2021**, *279*, 116939. [CrossRef] [PubMed]
55. Klein, S.; Worch, E.; Knepper, T.P. Occurrence and Spatial Distribution of Microplastics in River Shore Sediments of the Rhine-Main Area in Germany. *Environ. Sci. Technol.* **2015**, *49*, 6070–6076. [CrossRef] [PubMed]
56. Dai, Z.; Zhang, H.; Zhou, Q.; Tian, Y.; Chen, T.; Tu, C.; Fu, C.; Luo, Y. Occurrence of microplastics in the water column and sediment in an inland sea affected by intensive anthropogenic activities. *Environ. Pollut.* **2018**, *242*, 1557–1565. [CrossRef]

57. Castro, R.O.; da Silva, M.L.; Marques, M.R.; de Araújo, F.V. Spatio-temporal evaluation of macro, meso and microplastics in surface waters, bottom and beach sediments of two embayments in Niterói, RJ, Brazil. *Mar. Pollut. Bull.* **2020**, *160*, 111537. [CrossRef]
58. Riani, E.; Cordova, M.R. Microplastic ingestion by the sandfish Holothuria scabra in Lampung and Sumbawa, Indonesia. *Mar. Pollut. Bull.* **2021**, 113134. [CrossRef] [PubMed]
59. Nel, H.A.; Dalu, T.; Wasserman, R.J. Sinks and sources: Assessing microplastic abundance in river sediment and deposit feeders in an Austral temperate urban river system. *Sci. Total Environ.* **2018**, *612*, 950–956. [CrossRef] [PubMed]
60. Mani, T.; Hauk, A.; Walter, U.; Burkhardt-Holm, P. Microplastics profile along the Rhine River. *Sci. Rep.* **2016**, *5*, 17988. [CrossRef] [PubMed]
61. Gerolin, C.R.; Pupim, F.N.; Sawakuchi, A.O.; Grohmann, C.H.; Labuto, G.; Semensatto, D. Microplastics in sediments from Amazon rivers, Brazil. *Sci. Total Environ.* **2020**, *749*, 141604. [CrossRef] [PubMed]
62. Alam, F.C.; Sembiring, E.; Muntalif, B.S.; Suendo, V. Microplastic distribution in surface water and sediment river around slum and industrial area (case study: Ciwalengke River, Majalaya district, Indonesia). *Chemosphere* **2019**, *224*, 637–645. [CrossRef]
63. Cordova, M.R.; Riani, E.; Shiomoto, A. Microplastics ingestion by blue panchax fish (*Aplocheilus sp.*) from Ciliwung Estuary, Jakarta, Indonesia. *Mar. Pollut. Bull.* **2020**, *161*, 111763. [CrossRef] [PubMed]
64. Eriksen, M.; Mason, S.; Wilson, S.; Box, C.; Zellers, A.; Edwards, W.; Farley, H.; Amato, S. Microplastic pollution in the surface waters of the Laurentian Great Lakes. *Mar. Pollut. Bull.* **2013**, *77*, 177–182. [CrossRef] [PubMed]
65. Bertoldi, C.; Lara, L.Z.; Mizushima, F.A.D.L.; Martins, F.C.; Battisti, M.A.; Hinrichs, R.; Fernandes, A.N. First evidence of microplastic contamination in the freshwater of Lake Guaíba, Porto Alegre, Brazil. *Sci. Total Environ.* **2020**, *759*, 143503. [CrossRef] [PubMed]
66. Baldwin, A.; Corsi, S.R.; Mason, S.A. Plastic Debris in 29 Great Lakes Tributaries: Relations to Watershed Attributes and Hydrology. *Environ. Sci. Technol.* **2016**, *50*, 10377–10385. [CrossRef] [PubMed]
67. Sathish, N.; Jeyasanta, K.I.; Patterson, J. Abundance, characteristics and surface degradation features of microplastics in beach sediments of five coastal areas in Tamil Nadu, India. *Mar. Pollut. Bull.* **2019**, *142*, 112–118. [CrossRef] [PubMed]
68. Jeyasanta, K.I.; Patterson, J.; Grimsditch, G.; Edward, J.P. Occurrence and characteristics of microplastics in the coral reef, sea grass and near shore habitats of Rameswaram Island, India. *Mar. Pollut. Bull.* **2020**, *160*, 111674. [CrossRef] [PubMed]
69. Sathish, M.N.; Jeyasanta, I.; Patterson, J. Occurrence of microplastics in epipelagic and mesopelagic fishes from Tuticorin, Southeast coast of India. *Sci. Total Environ.* **2020**, *720*, 137614. [CrossRef] [PubMed]
70. Robin, R.; Karthik, R.; Purvaja, R.; Ganguly, D.; Anandavelu, I.; Mugilarasan, M.; Ramesh, R. Holistic assessment of microplastics in various coastal environmental matrices, southwest coast of India. *Sci. Total Environ.* **2019**, *703*, 134947. [CrossRef] [PubMed]
71. Frias, J.P.G.L.; Otero, V.; Sobral, P. Evidence of microplastics in samples of zooplankton from Portuguese coastal waters. *Mar. Environ. Res.* **2014**, *95*, 89–95. [CrossRef] [PubMed]
72. Sundar, S.; Chokkalingam, L.; Roy, P.D.; Usha, T. Estimation of microplastics in sediments at the southernmost coast of India (Kanyakumari). *Environ. Sci. Pollut. Res.* **2020**, *28*, 18495–18500. [CrossRef]
73. Karthik, R.; Robin, R.; Purvaja, R.; Ganguly, D.; Anandavelu, I.; Raghuraman, R.; Hariharan, G.; Ramakrishna, A.; Ramesh, R. Microplastics along the beaches of southeast coast of India. *Sci. Total Environ.* **2018**, *645*, 1388–1399. [CrossRef] [PubMed]
74. Ripken, C.; Kotsifaki, D.G.; Nic Chormaic, S. Analysis of small microplastics in coastal surface water samples of the subtropical island of Okinawa, Japan. *Sci. Total Environ.* **2020**, *760*, 143927. [CrossRef]
75. Erdogan, S. Microplastic pollution in freshwater ecosystems: A case study from Turkey. *Ege J. Fish. Aquat. Sci.* **2020**, *37*, 213–221. [CrossRef]
76. Lusher, A.L.; Tirelli, V.; O'Connor, I.; Officer, R. Microplastics in Arctic polar waters: The first reported values of particles in surface and sub-surface samples. *Sci. Rep.* **2015**, *5*, 14947. [CrossRef] [PubMed]
77. Sadri, S.S.; Thompson, R. On the quantity and composition of floating plastic debris entering and leaving the Tamar Estuary, Southwest England. *Mar. Pollut. Bull.* **2014**, *81*, 55–60. [CrossRef] [PubMed]
78. Guerranti, C.; Martellini, T.; Perra, G.; Scopetani, C.; Cincinelli, A. Microplastics in cosmetics: Environmental issues and needs for global bans. *Environ. Toxicol. Pharmacol.* **2019**, *68*, 75–79. [CrossRef] [PubMed]
79. Kumar, R.; Sharma, P.; Manna, C.; Jain, M. Abundance, interaction, ingestion, ecological concerns, and mitigation policies of microplastic pollution in riverine ecosystem: A review. *Sci. Total Environ.* **2021**, *782*, 146695. [CrossRef]
80. Larsen, J.; Venkova, S. The Downfall of the Plastic Bag: A Global Picture 2014. Available online: http://www.earth-policy.org/plan_b_updates/2014/update123 (accessed on 20 December 2021).
81. Jiang, J.-Q. Occurrence of microplastics and its pollution in the environment: A review. *Sustain. Prod. Consum.* **2018**, *13*, 16–23. [CrossRef]
82. Laskar, N.; Kumar, U. Plastics and microplastics: A threat to environment. *Environ. Technol. Innov.* **2019**, *14*, 100352. [CrossRef]
83. Li, L.; Geng, S.; Wu, C.; Song, K.; Sun, F.; Visvanathan, C.; Xie, F.; Wang, Q. Microplastics contamination in different trophic state lakes along the middle and lower reaches of Yangtze River Basin. *Environ. Pollut.* **2019**, *254*, 112951. [CrossRef] [PubMed]

Review

Microplastic Contamination in Soils: A Review from Geotechnical Engineering View

Mehmet Murat Monkul *[ ] and Hakkı O. Özhan [ ]

Department of Civil Engineering, Yeditepe University, İstanbul 34755, Turkey; hakki.ozhan@yeditepe.edu.tr
* Correspondence: murat.monkul@yeditepe.edu.tr or monkul@gmail.com

Abstract: Microplastic contamination is a growing threat to marine and freshwater ecosystems, agricultural production, groundwater, plant growth and even human and animal health. Disintegration of plastic products due to mainly biochemical or physical activities leads to the formation and existence of microplastics in significant amounts, not only in marine and freshwater environments but also in soils. There are several valuable studies on microplastics in soils, which have typically focused on environmental, chemical, agricultural and health aspects. However, there is also a need for the geotechnical engineering perspective on microplastic contamination in soils. In this review paper, first, degradation, existence and persistence of microplastics in soils are assessed by considering various studies. Then, the potential role of solid waste disposal facilities as a source for microplastics is discussed by considering their geotechnical design and addressing the risk for the migration of microplastics from landfills to soils and other environments. Even though landfills are considered as one of the main geotechnical structures that contribute to the formation of considerably high amounts of microplastics and their contamination in soils, some other geotechnical engineering applications (i.e., soil improvement with tirechips, forming engineering fills with dredged sediments, soil improvement with synthetic polymer-based fibers, polystyrene based lightweight fill applications), as potential local source for microplastics, are also mentioned. Finally, the importance of geotechnical engineering as a mitigation tool for microplastics is emphasized and several important research topics involving geotechnical engineering are suggested.

Keywords: microplastics; soil; polymers; geotechnics; landfills; geosynthetics; GCL; clay liner; hydraulic conductivity; plastics

1. Introduction

Plastic products are being produced in increasingly vast amounts in a global scale. It is estimated that about 400 million tons of plastic production is made annually, and this amount is expected to more than double by 2050 [1]. Similar numbers were also reported by two recent publications: first one (359 million tons) supported by the European Parliament's Policy Department for Citizens' Rights and Constitutional Affairs [2], and the second one (368 million tons) supported by Association of Plastic Manufacturers in Europe [3]. These numbers indicate an alarming potential of global contamination for different ecosystems (marine, fresh water, soil, arctic) by plastics and their residues.

For the past three decades, many countries including US, Japan, EU, Mexico etc. have exported their plastic waste to China and surrounding countries, which partially prevented the plastic debris going to solid waste disposal areas or incineration at those countries. However, in the last decade, China initiated restrictions on plastic waste importing policies, and finally banned the import of nonindustrial plastic waste in January 2018 [2,4]. Immediately afterwards, other Southeast Asian countries such as Malaysia rose noticeably as global plastic waste importers, but this was temporary, because such countries adopted restrictions on plastic waste importing policies as well. According to Zhao et al. [5], the center of gravity on global plastic waste trade is still evolving. For instance, the EU transformed from being a significant plastic waste exporter to being both a significant importer and exporter. Meanwhile, Turkey has become one of the growing plastic waste importers taking attention for the last few years. It suddenly entered the top 10 global plastic waste importers list in 2017, which was 10th in the list in 2017 with 26.19×10^4 tons imported plastic waste and became 7th in 2018 with 43.69×10^4 tons imported plastic waste [5]. In the same year (2018), Turkey became the second largest global importer of British plastic waste with 8×10^4 tons [6] as shown in Figure 1, which increased to 21×10^4 tons in 2021 [7]. However, it was dramatic to see that some of those imported plastic wastes were illegally dumped on fields and some burned, instead of being properly recycled [8]. In 2021, Turkey has initiated restrictions on plastic waste importing policies. Note that those reported random dump sites are not even Municipal Solid Waste (MSW) facilities, the engineering design of which would also be assessed considering microplastic contamination later in this paper.

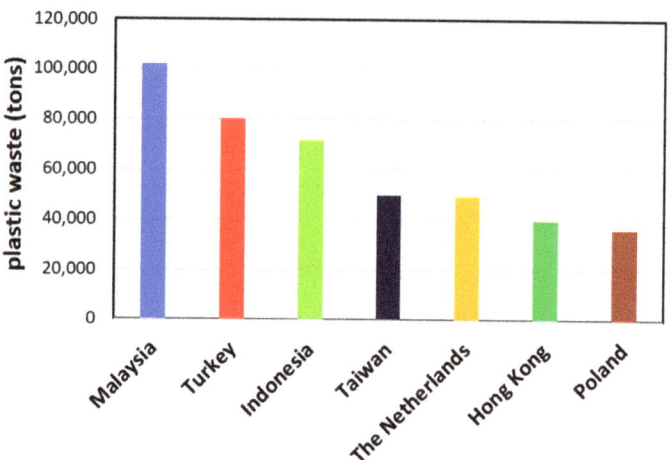

Figure 1. Top seven countries where the UK exported its plastic waste in 2018 (data from [6]).

Consequently, it is a big question not only for the major plastic waste importing countries, but also for all other developed/developing countries (which have vast amounts of plastic usage and debris) whether non-recycled plastic waste dumped illegally on fields or legally on MSW landfills significantly contaminate the soil environment by different means including macro, micro and nano plastics. For instance, the US Environmental Protection Agency reported that 35.68 million tons of plastic waste was generated in the US in 2018, and 26.97 million tons was accumulated in MSW landfills [9]. This indicates a very high percentage (i.e., 75.6%) of plastic waste dumped in MSW landfills, potentially threatening the soil and other environments in different ways, including microplastic contamination.

The term 'Microplastic' was used for the first time in 2004 by Dr. Richard Thompson, a British marine ecologist, to refer to small plastic debris [1,10]. Today, upper particle size limit (i.e., $\leq$5 mm or 5000 μm) for classifying microplastics is widely used by researchers and scientists, however additional terminology (i.e., nanoplastics, mesoplastics etc.) and corresponding size ranges to define those terminology had been introduced. In general, microplastic size range is considered between 0.1 μm and 5000 μm, while plastic particles smaller than 0.1 μm are considered as nanoplastics in literature [11,12], even though different size ranges are also proposed both for nano and microplastics [2,13]. Considering the mass production of plastics accelerating rapidly (i.e., 10-fold increase by 2025 according to Pinto Da Costa et al. [2]), it is not surprising that UN Environmental Programme (UNEP) already announced microplastic contamination in marine environment among the top 10 emerging issues in 2014 [11,14,15].

Hence, microplastic contamination is already an increasingly ubiquitous problem which requires the cooperation of different disciplines (fields). However, it seems that there is a delay of global awareness between different disciplines regarding the microplastic contamination in different environments. For instance, Figure 2 gives the top 10 disciplines and the distribution of published studies on microplastics and soils in the Web of Science Core Collection database between January 2016 and October 2021 (some studies for 2022 were also available in the database but not included in Figure 2). Accordingly, the top 10 disciplines working on microplastics and soils for the past 6 years are Environmental Sciences, Environmental Engineering, Multidisciplinary Sciences, Water Resources, Toxicology, Chemical Engineering, Chemistry, Soil Science, and Green Sustainable Science Technology. Moreover, Figure 2 shows that there are only two Engineering disciplines (i.e., Chemical and Environmental) which show sufficient awareness (conducts considerable research) about microplastics and soils.

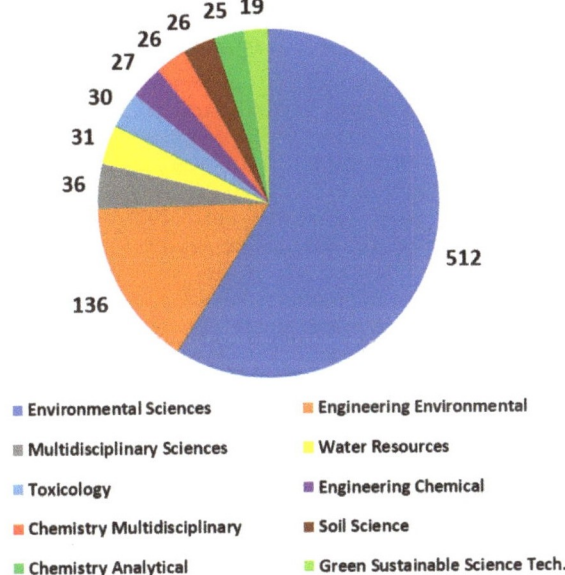

Figure 2. The distribution of the number of published studies on microplastics and soils in the Web of Science Core Collection database between January 2016 and October 2021 based on the top 10 disciplines.

Authors of this study are Civil Engineers, and could definitely state that the current global awareness of the Civil Engineering community on microplastics is extremely limited, which is one of the main motivations of this paper. Geotechnical Engineering is one of the sub-branches of Civil Engineering, which deals with soils and their engineering behavior. Soils and their engineering behavior, (including but not limited to the concepts involving permeability, compressibility, shear strength etc.) are very important for many civil engineering projects such as foundation design of structures, design of tunnels, highways, bridges, ports etc. Nevertheless, geotechnical engineering (or geotechnics) can also act as an enhancing discipline to improve our understanding on microplastic contamination in soils from a different perspective.

The second section of this paper addresses the formation and existence of microplastics in soils, including the main degradation mechanisms influencing their presence, their concentrations in various soil environments (i.e., garden soil, landfill soil, agricultural soil, marine soil etc.) in different countries. Meanwhile, a geotechnical perspective is also expressed by the authors together with some key parameters (e.g., void ratio, hydraulic conductivity etc.) that could influence migration and spatial distribution of microplastics in soils. The persistence of microplastics in soils have also been considered through relatively long term field studies in literature, investigating the persistence of different polymers in real soil conditions. In the third section of the paper, geotechnical design of MSW landfills is discussed and their potential role as a source for microplastics and soil contamination are assessed. In the fourth section of the paper, some geotechnical applications, which are beneficial/smart from an engineering perspective, are reminded as additional potential microplastic sources in the long term. These geotechnical applications mentioned in the fourth part, are ground improvement with polymer-based fibers, ground improvement with waste tire-chips, lightweight polystyrene foam applications, reclamation fills with microplastic contaminated dredged soils. In the last (fifth) section of the paper, authors explained some future research ideas regarding how to benefit Geotechnical Engineering as a mitigation tool for reducing the microplastic contamination in soils and gave concluding remarks.

2. Formation and Existence of Microplastics in Soils

Even though microplastic contamination affects various ecosystems including marine [16] and freshwater environments [17], soils [18] etc., terrestrial soils are perhaps the relatively less studied one among the three. Nizzetto et al. [19], Zhang et al. [20], Zhu et al. [21], Da Costa et al. [22], Möller et al. [23] and Wang et al. [24] have all explicitly mentioned that research on microplastics in soil environment is still relatively limited in number and content compared to other environments. Authors could list several reasons for such an observation including but not limited to: (1) relative delays for the global awareness between different disciplines on microplastic contamination (e.g., Environmental Sciences vs. Civil Engineering; Chemistry vs. Geology etc.), which was mentioned previously (Figure 2); (2) lack of multidisciplinary research on microplastics in soils between soil related disciplines and others (e.g., bwtn. Environmental Sciences, Chemistry and Geotechnical Engineering); (3) historical/gradual development of microplastic awareness, which originally initiated for the marine environment [10].

Nevertheless, research and relevant studies on microplastics in soils also gained a momentum in the last few years as well, including the analytical methods for microplastic sampling and determination in soils [22,23,25]; their migration, ecological and environmental risks [5,18,26]; their transfer and accumulation in agricultural soils [27–29]; their effects on soil quality and function [30,31].

The mentioned studies on microplastic-soil interaction understandably focused mostly on agricultural, environmental and health aspects, where there is little-to-no engineering perspective involved. For example, De Souza MacHado et al. [30] conducted an experimental study and observed that microplastics influence the bulk density and water holding capacity of soils. They interpreted the results from an agronomical perspective, since there is a good correlation between soil bulk density and rootability. More specifically, a reduction in soil bulk density would normally imply better rootability, due to increased soil porosity helping root growth [30,32]. However, as also questioned by De Souza MacHado et al. [30], such a decrease in bulk density due to microplastics could be misleading for rootability. Since it is unclear how much the soil porosity actually increased in their experimental study considering that the densest microplastic polymer used in their study (polyester) is still considerably lighter (~1370 kg/m^3) than their control soil (~1439 kg/m^3). More explicitly stated, the observed decrease in bulk density of soil due to microplastic addition in the experiments could be more affected by addition of a lighter material than the soil itself, instead of increasing soil porosity. From a typical geotechnical engineering perspective, bulk density and porosity intuitively remind shear strength and settlement/compressibility characteristics of a soil [33,34]. Both shear strength and compressibility are two very important concepts related with many aspects in civil/geotechnical engineering, including slope stability [35], foundation design of engineering structures [36], dredging and reclamation works [37], landfill design [38] etc. Though void ratio (e) (which is the ratio of the volume of voids to the volume of solids in a soil matrix) is a more popular parameter than the porosity (n) (which is the ratio of the volume of voids to the total volume in a soil matrix) in geotechnical research and practice. Nevertheless, changes in soil density due to microplastic contamination is not expected by the authors to influence its overall engineering behavior in terms of settlement or shear strength, unless extreme levels of contamination exist in the soil.

2.1. Degradation Processes

Several different degradation processes exist for microplastic formation, which are typically known to have a slow rate depending on the ambient conditions. The main mechanism of degradation could involve chemical, physical, biological or a combination of those processes. More specifically, polymeric degradations playing role on microplastic formation can be listed as thermal, mechanochemical, ozone-induced, biological, photo-oxidative, and catalytic types of processes [39,40], and interestingly, these processes seem to influence their toxicity as well [26,41]. O'Kelly et al. [15] stated that ozone-induced degradation, biodegradation and photo-oxidative degradation are the main degradation types contributing to breaking down the "macro" and "meso" plastics and consequently forming microplastics in soils.

Singh and Sharma [40] discussed the detailed mechanisms of the above-mentioned degradation processes. Accordingly, photo-oxidative degradation is a decomposition process which occurs by the absorption of the UV and visible lights by the polymers. Then, degradation and oxidation reactions start, which are determined by the groups attached to the polymer and/or the impurities in it.

Ozone-induced degradation is caused by the ozone in the air, even though it is present in very small concentrations. Nevertheless, it is enough to accelerate the aging of the polymeric materials. This phenomenon is followed by the intensive formation of oxygen-containing compounds, as well as changing the molecular weight and the mechanical and electrical properties of the materials. Ozone-induced degradation occurs by the attack of the ozone molecule to the unsaturation in unsaturated polymers where the reaction propagates in three main steps; firstly, the formation of ozone olefin adduct, secondly the decomposition of the primary ozonide to carbonyl compounds and a carbonyl oxide, and thirdly the fate of the carbonyl oxide [40].

Biodegradation of polymers is typically caused by the available microorganisms and enzymes in soils. Examples to such microorganisms could be different fungi and/or bacteria, while different enzymes (e.g., polyurethanase, lipase etc.), could contribute to the biodegradation process [40,42,43]. For example, Crabbe et al. [44] studied four fungus types isolated from soil (Curvularia sene galensis, Fusarium solani, Aureobasidium pullulans and Cladosporium sp), and concluded that they could biodegrade ester-based polyurethane. Yoshida et al. [45] discovered a novel bacterium (Ideonella sakaiensis), which could use PET (polyethylene terephthalate) as an energy and carbon source for living (i.e., it efficiently biodegrades PET). In a recent study, Feng et al. [46] investigated the biodegradation mechanism of PET with the molecular dynamics and quantum mechanics/molecular mechanics approaches. Their study focused on the two enzyme system (IsPETase and IsMHETase) in the mentioned novel bacterium (Ideonella sakaiensis). Orhan et al. [47] studied biodegradation of plastic compost bags under controlled soil conditions. In their study, Orhan et al. [47] investigated the response of supposedly degradable and non-degradable low (LDPE) and high-density polyethylene (HDPE) in soil mixed with 50% (w/w) mature municipal solid waste compost supplied from municipal refuse and mentioned that the rate of polymer biodegradation is affected by environmental factors such as moisture, temperature and biological activity. Similar environmental factors were also mentioned by Kliem et al. [42] for the biodegradation of different polymers. Moreover, there is another aspect of biodegradation in terms of terminology. As Kyrikou and Briassoulis [48] discussed in detail, a biodegradable polymer is expected to leave no harmful substances in the environment, which means that it should be entirely converted to carbon dioxide, water, mineral and biomass, without any ecotoxicity. Kyrikou and Briassoulis [48] also explained that many polymers that are considered as "biodegradable" are indeed degradable depending on environmental conditions and better be called "hydro-degradable", "photo-degradable" or "oxo-degradable".

It is also worth to mention that photo-oxidative, and ozone-induced degradations may also promote the biodegradation potential of polymers due to breaking of polymer chain and increasing the surface area for colonization of microorganisms and/or by decreasing molecular weight [15,40,49,50].

2.2. Existence of Microplastics in Soils

The existence of microplastics in soil environment is a globally emerging issue, which should alert not only the scientific community, but also the public and policy makers. Several studies reported microplastic concentrations available in different soils worldwide.

Huerta Lwanga et al. [51] investigated surficial samples (0–20 cm) from 10 home garden soils in Mexico from sites selected from similar ethnic and economic demographics, and reported microplastic concentration of 870 ± 1900 ptcl./kg (particles per kilogram of soil). Zhang and Liu [29] investigated the distribution of plastic particles over four agricultural sites involving cropped soils and another site at riparian forest buffer zone in China. All soil samples were taken from surficial layers (0–10 cm), and an average of 18,760 ptcl./kg were reported, in which 95% of the particles were in the microplastic range based on their classification (i.e., 0.05–1 mm). Zhang and Liu [29] also noted that microplastic concentrations were higher for the agricultural soils in their study compared to the forest buffer zone soil.

Crossman et al. [28] studied microplastic concentrations in soils between 0–15 cm depth at three agricultural sites in Ontario, Canada. Accordingly, average microplastic concentrations for the three sites vary significantly (i.e., 18 ptcl./kg ±22.2%; 187 ptcl./kg ± 53.1%; 541 ptcl./kg ± 56.4%, respectively). van den Berg et al. [52] inspected soils from 16 agricultural sites in Spain, and found that average microplastic concentration is 930 ± 740 ptcl./kg for light density microplastics (i.e., $\rho < 1$ g/cm^3), while it is 1100 ± 570 ptcl./kg for high density microplastics (i.e., $\rho > 1$ g/cm^3) based on surficial samples (0–30 cm).

In a recent study, Dahl et al. [53] investigated the contamination of seagrass soils (i.e., marine soils) at three sites along the Spanish Mediterranean coast, and concluded that microplastic contamination was negligible until 1975s, then increased dramatically. The samples were taken from shallow depths from the soil surface (0–15 cm) and the concentrations change (between 68 and 3819 ptcl./kg) depending on the site. However, Dahl et al. [53] stated that there is a strong relationship between the intensification of the agricultural industry at a particular region and the microplastic concentrations in the soils.

As expected, the studies mentioned above reveal that concentrations and existence of microplastics in soils vary with country (e.g., Mexico, China, Canada, Spain), location and characteristics of the site (i.e., garden soil, landfill soil, agricultural soil, marine soil etc.), regional industrial practices (e.g., agricultural applications) and possibly from many other interlinked sub-factors such as the level and quality of the waste solid/water treatment plants, regional population intensity, economic level and usage of plastic involving products etc.

Another observation from the studies reviewed above is that they all focused on the microplastic existence within the first 20 or 30 cm below the ground surface. This is understandable, considering that soil life is more active in those depths and typical rooting and ploughing depth do not exceed 30 cm [52]. However, this depth range (0–30 cm) is very shallow from a geotechnical engineering perspective. Hence, more studies are needed to quantify the existence of microplastics at greater depths from the ground surface. Permeability or hydraulic conductivity (k) is one of the key parameters in geotechnical engineering (also in geology and hydrogeology), which indicates the ability of water and other fluids to flow through the voids between the soil grains [54]. Hydraulic conductivity is a function of different factors including but not limited to density (ρ_{soil}) and void ratio (e) of the soil, viscosity of the pore fluid (e.g., clean or contaminated water), type of the soil (e.g., clay, silt, sand), effective pore size between the soil grains etc. [54,55]. Nevertheless, authors think that hydraulic conductivity (k) could be among the key parameters for spatial distribution of microplastics in soils both in horizontal and vertical directions, therefore deserves attention during future research on microplastics in soils.

Several studies are available in literature about the migration of microplastics in soils. O'Connor et al. [56] conducted an experimental study about the vertical migration of microplastics in sands from wetting-drying cycles perspective. They investigated the mobility of five different microplastics having different sizes and densities, which consist of polypropylene (PP) and polyethylene (PE) particles. They found that maximum penetration depth of microplastics through sand almost linearly increases with the number of wetting-drying cycles, and as the microplastic size becomes smaller, its mobility in the soil also increases. O'Connor et al. [56] also mentioned that microplastic concentration at the surface and volume of infiltration liquid had only negligible or weak effects on depth of migration. Moreover, they forecasted the long-term penetration depths based on weather data of 347 Chinese cities and their experimental model. Accordingly, they estimated an average penetration depth of 5.24 m in the long term ($\approx$100 years). The study of O'Connor et al. [56] is certainly very valuable in terms of the influence of wetting-drying cycles on microplastic penetration in sands. However, from geotechnical engineering view, sands are only one of the several soil types (e.g., clays, silts, gravels and their mixtures) in engineering classification, hence the penetration depth of microplastics can be expected to change with various other factors explained before (e.g., soil type, void ratio, hydraulic conductivity etc.).

De Souza MacHado et al. [30] also conducted an experimental study on loamy sand, where they measured the hydraulic conductivity (though in a different way compared to the ASTM standards used in Civil Engineering). They stated that the existence of microplastics in sand did not significantly change the hydraulic conductivity of the soil. From geotechnical point of view, sandy soils typically have quite high hydraulic conductivities ranging from 1 to 10^{-2} cm/sec for clean coarse sands, and from 10^{-2} to 10^{-4} cm/sec for clayey sands [54,57]. In fact, De Souza MacHado et al. [30] also acknowledged this aspect and wrote that the hydraulic conductivity of the sandy soil that is used in their experimental program could be high enough to be unaffected by the microplastic concentrations and adopted k measurements in their study.

Wu et al. [58] ran column experiments to investigate the vertical migration response of polystyrene nanoplastics in three natural soils from China with contrasting physicochemical properties (e.g., salt composition, ionic strength, zeta potentials etc.). They reported that soil mineralogy and pH influence the migration of nanoplastics in soil medium, where the migration of nanoplastics was also reported to be sensitive to ionic strength and cation type.

Consequently, there have been several valuable studies which investigated the existence and migration of microplastics in soils from different aspects. From geotechnical perspective, authors want to emphasize/remind the importance of following factors/research gaps for future studies: (1) distribution and migration of microplastics in non-shallow soil depths (i.e., $\geq$30 cm); (2) influence of hydraulic conductivity (k) for spatial distribution and migration of microplastics in soils both in horizontal and vertical directions; (3) consideration of the influence of different soil types used in engineering classification (e.g., clay, silt, sand, gravel and their mixtures) on microplastic distribution in soils. All the three aspects listed above deserve further systematic multi-disciplinary research involving geotechnical engineering.

2.3. Persistence of Microplastics in Soils

It is known that microplastics are quite persistent in soils (i.e., their degradation process takes a very long time). Molecular structures of some of the commonly encountered polymers in soils are shown in Figure 3, which would also be addressed in different parts of this paper based on different studies.

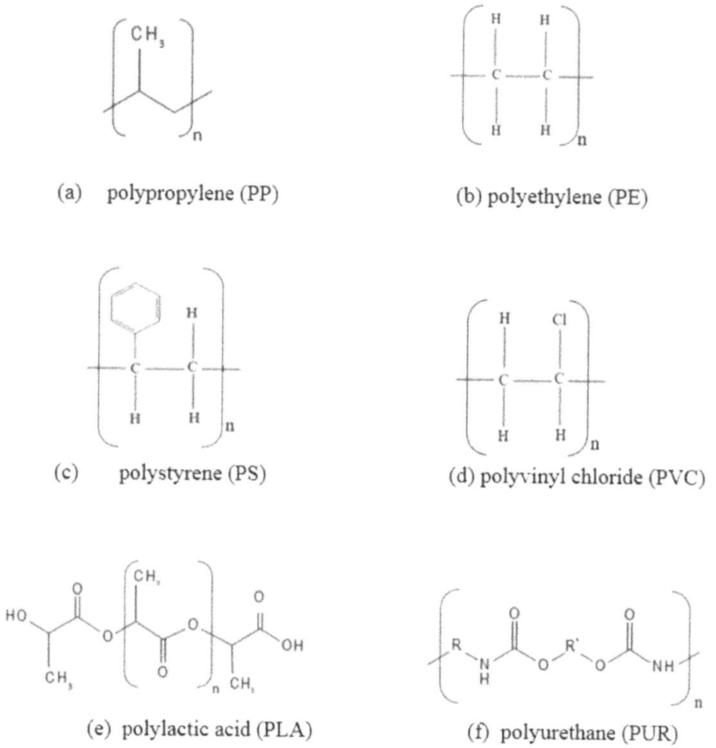

Figure 3. Molecular structures of some of the commonly encountered polymers in soils (**a**) polypropylene (PP), (**b**) polyethylene (PE), (**c**) polystyrene (PS), (**d**) polyvinyl chloride (PVC), (**e**) polylactic acid (PLA), (**f**) polyurethane (PUR).

Cooper and Corcoran [59] investigated plastic particles from 5 beach soils at Kauai Island, Hawaii. They warned that microplastics formed by disintegration of macroplastics remain in the environment almost indefinitely, which cannot be tolerated simply by using more rapidly-degrading polymer types, especially with the accelerating trend of plastic usage. A parallel observation was also made by Krueger et al. [60], who mentioned that the present-day synthetic polymers are quite persistent against biodegradation (some having degradation period of decades or even centuries), and this, in turn cannot counteract with the overwhelming pollution with plastics. Krueger et al. [60] also compiled several laboratory studies about the biodegradation of synthetic polymers in different environments, including soil, marine conditions, and mentioned that most plastics are quite recalcitrant (with low reaction rates) even under optimized laboratory conditions. More importantly, Krueger et al. [60] claimed that published laboratory studies could be strongly biased to successes obtained under optimized laboratory conditions, which have limited transferability to real environments.

Fortunately, there are also few relatively "long" term field studies in literature investigating the persistence of different polymers in real soil conditions. Briassoulis et al. [61]

buried low-density polyethylene (LDPE) mulching films in an agricultural soil (after being used for one watermelon cultivation) to simulate and observe the long-term degradation behavior in field conditions. In their experimental study, Briassoulis et al. [61] buried pro-oxidant added mulching films for 8.5-year period in soil. Pro-oxidants are special additives, which involve mainly carbonyl groups and metals blended with different ingredients (e.g., cobalt acetylacetonate, magnesium stearate etc.) which accelerate the breakdown of polyethylene [61,62]. Briassoulis et al. [61] reported that after 8.5 years staying in soil, buried low-density polyethylene mulching films were recovered almost intact with no disintegration observed, which implies the persistency of even macroplastics (i.e., PE in that study) in soils. Study of Albertsson and Karlsson [63] also gave similar results, who investigated the degradation behavior of LDPE film in laboratory conditions for a 10-year cultivation period with soil.

Otake et al. [64] examined the biodegradation of different polymers, when buried in soil for about 32 years. They determined different synthetic polymer types in a Japanese garden soil from 10 and 50 cm depths from the ground surface, including LDPE, polystyrene (PS), polyvinyl chloride (PVC), and urea formaldehyde (UF) resin buried in soil between 32 and 37 years. They reported that for PVC, PS, and UF resin, no biodegradation was observed after over 32 years, however LDPE samples have shown signs of degradation. Otake et al. [64] also mentioned that the signs of degradation were more visible for samples collected from shallow depths (~10 cm) than the ones collected from relatively deeper levels (~50 cm), possibly because of having more aerobic activity at shallow depths. Higher persistence of microplastics in deeper soil layers is due to smaller microbial population available, compared to the shallow depths, which would reduce their degradation potential. This observation makes the research topic of "distribution and migration of microplastics in non-shallow soil depths" mentioned by the authors in previous Section 2.2 also important from persistence point of view as well.

Tabone et al. [65] investigated the sustainability metrics (which include atom economy, mass from renewable sources, biodegradability, percent recycled, distance of furthest feedstock, price, life cycle, health hazards, and life cycle energy use) of 12 polymers. Accordingly, "biodegradable" polymers (e.g., polylactic acid (PLA), polyhydroxyalkanoate (PHA)) are listed on top of the green design rankings of their study, however these polymers also exhibit relatively large environmental impacts from production [65]. Since PLA (Figure 3e) is being increasingly used in short shelf-life products (i.e., in compostable food-packaging films, bags etc.), wastes involving PLA increase in the environment [66]. Karamanlioglu and Robson [66] investigated the degradation behavior of PLA in commercial packaging buried in soil and compost for a temperature range (i.e., btwn., 25 °C and 55 °C). Accordingly, no change in tensile strength or molecular weight was observed after 1 year at relatively low temperatures (i.e., 25 °C and 37 °C), which implies a problem for PLA persistence in soils. However, they observed that at elevated temperatures (i.e., 50 °C) microbes enhanced the biodegradation process of PLA, which indicates the importance of soil temperature on biodegradation.

3. Solid Waste Disposal Facilities and Their Potential Role as a Source for Microplastics

3.1. Solid Waste Disposal Facilities and Their Geotechnical Design

Rapid population growth and industrialization are the two key factors that contribute to environmental pollution. Furthermore, production diversity and changes in the consumption habits result in an increase in the amount of waste materials. These waste products are mainly composed of municipal solid wastes, which are the waste materials that human beings use and throw away every day. Although the solid wastes are partially recycled or burnt, it is not possible to eliminate all the particles of these wastes [67,68]. Therefore, it is aimed to store the municipal solid wastes (MSW) in specific waste disposal facilities that are called MSW landfills or simply landfills in order to minimize their harmful effects on the environment by isolating them from the subsoil environment and the groundwater.

In order to catch and evacuate the surface flow, a drainage system has to be installed and the landfill has to be built in accordance with the natural environmental conditions by growing plants on top of it or making a social area for the public [69]. Separation and classification of the solid wastes is a crucial step for providing some of the wastes to be recycled and reducing the use of the natural resources and as a result, for preventing environmental pollution. Considering that the separated solid wastes also have harmful effects on the environment and human health, these wastes have to be stored in the landfills that are designed for various purposes. Determination of the engineering properties of the soil profile and the groundwater conditions, measurement of the piezometric heads in the aquifers, the hydraulic conductivity of the soil, and characterization of the geochemical conditions play important roles in the selection of the location of a landfill [70].

The base and the sides of a landfill have to be covered with a barrier or liner material in order to control/prevent soil and water contamination. For this purpose, compacted clay liners (CCLs) or geosynthetics are generally preferred to be used as the barrier materials in landfills [71]. The main aim for placing the liner material in a landfill is to prevent or control the permeation of the leachate through the barrier to the subsoil and groundwater. Compacted clay liners are composed of natural clay deposits and the hydraulic conductivity of the compacted clay depends on the clay mineralogy, void ratio and water content of the clay during compaction and the method of compaction. CCLs are typically compacted at water contents greater than the optimum water content (w_{opt}) obtained from Proctor Compaction Tests in order to achieve smaller permeability for the liners. As a result, the hydraulic conductivity of the CCL is correlated with the void ratio. The CCL that is selected to be used in a landfill should satisfy several geotechnical criteria. For instance, the hydraulic conductivity of CCL, $k \leq 10^{-9}$ m/s; the dry weight percentage of the fine soil particles that pass through 0.075 mm sieve (No., 200) $\geq$ 50%; the plasticity index $\geq$ 7–10%; the dry weight percentage of the soil particles that remain on 4.75 mm sieve (No. 4) $\leq$ 20% [72,73].

Geomembranes and geotextiles are the two major geosynthetics that are used for various purposes in landfills. The geotextiles are flexible textile materials with synthetic fibers and they typically provide filtration in a landfill. The geomembranes are thin sheets of impervious plastic materials. A geomembrane layer is typically placed over the CCL to provide imperviousness against leachates [72]. A geomembrane is the additional lining layer over a CCL that is used for enhancing the barrier capacity of the clay liner in a waste disposal facility. However, the geomembrane can easily be damaged or punctured by a sharp particle that can be found in a solid waste. Moreover, installation damages on geomembranes are not uncommon if the construction quality assurance and quality control are not strictly applied [74,75]. If such damages occur, leachate involving microplastics can seep into the underlying CCL.

A geosynthetic clay liner (GCL) is a lining and barrier material that consists of a thin bentonite layer sandwiched between two geotextiles [72]. In recent years, the design engineers have preferred to use geosynthetic clay liners (GCLs) as an alternative to the CCLs as the barrier material in the waste disposal facilities [76]. Moreover, the GCLs have some advantages over the CCLs such as having lower hydraulic conductivity (<10^{-10} m/s), lower thickness (4–10 mm), lower cost, less labor work and faster installation [72,77,78]. The most critical component of a GCL that determines the hydraulic performance of the barrier material is the bentonite layer [79–81]. A waste disposal area in Kütahya, Turkey before and after the placement of a barrier system that was composed of a geomembrane-laminated GCL is shown in Figure 4.

According to their manufacturing process, the GCLs can be classified into three groups. The adhesive-bonded GCL is composed of a bentonite layer attached to the upper and lower geotextiles with a water-soluble adhesive without any reinforcement. The needle-punched GCL is manufactured by punching the needle-like fiber particles from the upper geotextile through the bentonite layer to the lower geotextile. Due to the reinforcement provided by the needle-punching process, the migration of the bentonite from the GCL is mostly prevented in this type. The stitch-bonded GCL is another reinforced GCL type. For the stitch-bonded GCL, the upper and lower geotextiles are stitched together with parallel oriented yarns by keeping the bentonite layer inside the GCL [71,72,75]. The cross-sectional views of the three different GCL types mentioned are shown in Figure 5.

Figure 4. Waste disposal area in Kütahya, Turkey: (**a**) without a barrier material; (**b**) with the barrier material that consisted of a geomembrane-laminated GCL.

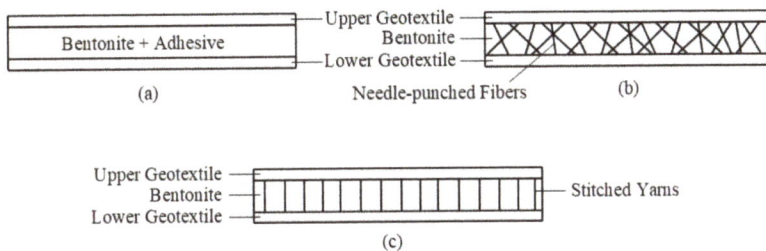

Figure 5. Cross-sectional views of GCL types: (**a**) Adhesive-bonded GCL; (**b**) Needle-punched GCL; (**c**) Stitch-bonded GCL.

In a typical landfill liner system (shown in Figure 6), first, the GCL which is used as the lining and barrier material is placed over the subsoil (natural soil). Then, a geomembrane layer is preferred to be used between the drainage layer and the GCL in order to protect the lining material against possible sharp gravel or solid waste particles that might puncture or tear the GCL as well as acting as an impermeable interface between the drainage layer and GCL. The gravel layer with a minimum thickness of 30 cm that is placed over the geomembrane, acts as the drainage layer of the landfill.

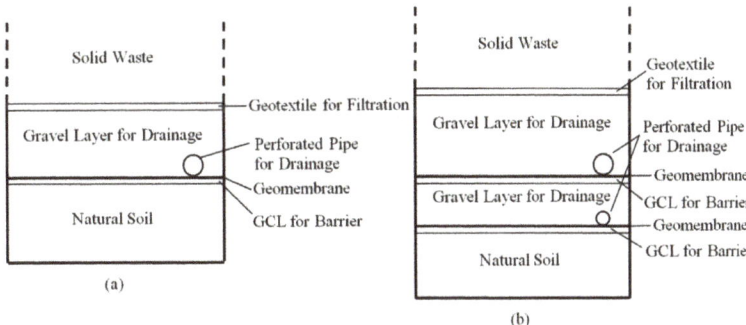

Figure 6. Cross-sectional view of landfill liner systems: (**a**) Single-liner system; (**b**) Double-liner system (Upper layer for leachate collection and lower layer for leak detection).

The efficiency of the operation of a landfill depends on the installation of a proper drainage system that collects and removes the leachate from the landfill properly [70]. The selection of the barrier materials and the transmission pipes for the leachate drainage are the crucial steps for the design of the landfill drainage systems. In order to collect the leachate, the drainage system has to be designed with an inclination of 0.5–1% [71,82,83]. The main reason for this inclination is to provide the collection of the leachate by creating the needed hydraulic head on the barrier material. The capacity of the drainage system plays the critical role for determining the exact length and inclination of the drainage pipes. The perforated pipes, which are installed in the leachate collection system, are used for removing the leachate from the landfill. As a result, the leachate is sent to the surface of the landfill for wastewater treatment [67,72,73]. The geotextile that is placed over the gravel drainage layer serves as a filtration layer as can be seen in Figure 6. The geotextile is used for preventing the clogging of the drainage layer with the particles of the solid waste. Finally, the solid waste is dumped into the landfill and can be stored in a single-liner system as shown in Figure 6a.

The geotechnical design of a landfill determines the performance and the functionality of the liner system. The construction and demolition debris are typically buried in landfills that have single-liner systems. However, single liners are not preferred to be used to store municipal solid wastes. A double-liner system is composed of two single liners as shown in Figure 6b. In a double-liner system, the upper liner is used for collecting the leachate whereas the lower liner serves as a leak-detection system. The gravel layer in the leachate-collection system has a height of at least 30 cm while the height of the gravel layer in the leak-detection system is half of the height of the upper gravel layer that functions for drainage [72].

Double-liner systems are generally used in both municipal solid waste and hazardous waste landfills [69,72]. Although a double-liner system has more barriers and lining components than a single-liner system as shown in Figure 6, there is no guarantee that the leachate will completely remain above the natural subsoil environment. Thus, there is still the risk for the contamination of the soil and the groundwater, which is also a potential threat for different levels of microplastic migration into the surrounding soil environment. In fact, the levels of microplastic contamination in soils at the vicinity of different landfills could be an important research topic, to be further studied.

3.2. The Risk of Microplastic Migration from Landfills to Soil and Other Environments

Plastics cover a significant amount of the solid wastes and 79% of the plastic wastes are either stored in landfills or released to the natural environment [84]. 21–42% of the global plastic wastes that are not recycled and not burnt, are stored in landfills [85]. Approximately 12,000 million tons of plastic wastes are estimated to be stored in landfills or released to the environment in 2050 [84]. Isolating the bottom and sides of a landfill with a barrier system could prevent or at least limit not only the leakage of the leachate from the solid wastes but also the migration of the microplastics to the subsoil and the groundwater.

The plastic wastes that are stored in the landfills may undergo both physical and biochemical changes due to many factors including biodegradation (explained in Section 2.1), drastic changes in temperature (temperature may rise up to 60–90 °C) and acidity (pH fluctuations between 4.5 and 9), CH_4 or CO_2 generation, high salinity or high pressure and compaction [86–89]. As a result of these changes, the plastic wastes can be fragmented into smaller pieces that are classified as microplastics. The soil and the groundwater can be contaminated due to the possible transfer and migration of these microplastics from the landfill to the subsoil environment. The lack of sufficient liner systems for abandoned waste dumps or defects/liner damages in landfills at different countries might cause environmental problems due to the discharge of the leachate that carries out the microplastics generated from the plastic wastes into the soil.

Approximately 52 million tons of municipal waste materials were stored in landfills of the EU countries in 2018 [90]. According to Xu et al. [90], this amount corresponds to 10.4 million m^3 of leachate formation and 3.03 billion in microplastic release to the environment. Considering that these values were only obtained in the EU countries, the amount of the microplastic release for the whole world would be much higher. In another study, the average leachate formation from 1000 kg of solid waste that was deposited in landfills in 30 different cities of China was measured as 1.3–3.2 m^3 [91].

He et al. [85] also detected microplastics in all of the 12 leachate samples taken from six different landfills in China. Polypropylene (Figure 3a) and polyethylene (Figure 3b) were found to be the two major polymer types among the microplastic particles in their study. Approximately 77.5% of the microplastics had a particle size between 0.1–1 mm and the concentration of the microplastics in the landfill leachates was within the range of 0.42–24.58 ptcl./L (particles per liter). Meanwhile, the concentration of the microplastics in the leachate of landfills from Northern Europe was determined only up to 4.51 ptcl./L in the study of Van Praagh et al. [92]. This difference in concentrations could be due to the strict environmental regulations, lower population and environmental awareness of the Northern European countries when compared with those of China. Similar to He et al. [85], Su et al. [93] also reported that the most widely encountered polymer types of the microplastics in a landfill leachate from China were determined as polyethylene and polypropylene, whereas the study of Van Praagh et al. [92] has shown that polyurethane (Figure 3f) and polyethylene (Figure 3b) were the two dominant polymers in landfill leachates in the North European countries. The particle size of the microplastics from the landfill leachates in the North European countries was within the range of 0.05–5 mm. Furthermore, the dominant shape of the microplastics in several leachate samples was angular with sharp edges [85,94].

Landfill leachate might also contain hazardous substances such as Bisphenol A (BPA). Due to hydrolysis, diffusion and physicochemical processes, BPA can easily be released from plastic materials to the environment [95,96]. As the plastics are decomposed into microplastics and the microplastics are carried with the leachate flow, the BPA can also be mixed with the soil and the groundwater. The concentrations of BPA and microplastics were measured almost the same in a study conducted by Narevski et al. [94] in which leachate samples, that were taken from three landfills in Southeastern Europe, were analyzed. The results indicated that the concentration of the microplastics was within the range of 0.64–2.16 mg/L whereas the BPA concentration was found to be 0.70–2.72 mg/L. Due to their high absorption capacity, the microplastics may easily absorb toxic organic substances [97] and increase the heavy metal concentration in soils [98]. Soils involving these microplastics, that are carried with the leachates seeping into the soil in the vicinity of landfills, are sometimes used in agricultural practices and the toxic substances that are absorbed by the microplastics might quickly deteriorate soil health and hinder plant growth.

There are also relatively limited amount of studies that indicate the occurrence of microplastics in soil layers and groundwater beneath landfills. For instance, the wrinkles in the geomembrane liner systems could cause the leachate with microplastics to migrate through the barrier material to the subsoil [99]. Manikanda et al. [100] detected microplastics in the groundwater from the subsoil beneath municipal solid waste landfills in South India. The microplastic concentrations in the groundwater were measured as 2–80 ptcl./L and the main polymer types of the microplastics were classified as polypropylene (Figure 3a) and polystyrene (Figure 3c). Microplastics were also observed in several soil samples from a sanitary landfill in Bangladesh. The polymer types of the microplastics were expressed as polyethylene and cellulose acetate. The particle diameter of the microplastics detected in the soil beneath the landfill was within the range of 0.001–2 mm [101].

Both leachate and soil samples were taken from several landfills in Thailand [102]. The soil samples were collected from the upper layer of the subsoil beneath the landfill bottom with a depth of 10–20 cm by a hand auger. It is interesting that the microplastic concentration was found to be higher in soil than in leachate. According to Puthcharoen and Leungprasert [102], the soil samples involve microplastics approximately 1500 ptcl./kg while there were only 20 ptcl./kg microplastics in the leachate samples. The main polymer types of the microplastics obtained from both the soil and the leachate samples were polyethylene, polypropylene and polyethylene terephthalate. The dominant microplastic patterns were granules and films for both the soil and the leachate samples. The other microplastic patterns observed in the soil and leachate samples were spheres, irregulars and fibers. The granule, sphere and irregular-shaped microplastics were reported to be mainly decomposed from plastic food storage containers and water bottles. On the other hand, plastic packages and bags were the main source of the film-shaped microplastics. Fiber-shaped microplastics were mainly formed by the decomposition of synthetic fibers from synthetic clothes [103]. In another study, the solid waste samples taken from different landfills in Shangai, China were examined and according to the results, the older landfills contained a wider range of microplastics in terms of shape. However, the contribution of the landfill age to the microplastic shape in leachates was not as critical as that of in solid wastes [93].

The main agents that contribute to the horizontal transportation of the microplastics are surface run-off, wind erosion, flooding and animal scavenging [15,19]. Lighter microplastics can be much easily transported by wind from the surface of the soils to other lands or water resources, while the denser microplastics are more likely to vertically migrate into soil [104,105].

Some studies indicate that the microplastics can be also transferred from landfills to some water resources or other lands. For example, microplastics were detected in a river that was close to a landfill located around a coastal area [106]. The investigations showed that the landfill leachate had carried out the microplastics to the river. According to some studies, the landfills that are built on coastal areas or beaches might collapse due to erosion. Global warming that leads to increases in sea levels is the main reason for such erosion triggered failures. As a result, the collapsed landfills might cause the scattering of the microplastics to the sea and contamination of the water [107,108]. Due to the atmospheric transport of the lighter microplastics from landfills to either other terrestrial regions or any sea, lake or river horizontally, there is the risk for the human beings to inhale the microplastics as well [109]. When inhaled, microplastics may reach the lungs and stomach, and threaten human respiratory and digestive systems [110,111].

As discussed here, the microplastic transfer from landfills to soils, water resources and atmosphere has been evaluated in different studies. According to these studies, the leachates are the main agents that carry microplastics from landfills to the soil deposits vertically. Although the particle size of the microplastics that were collected from both leachates in landfills and soil samples beneath landfill bases varied in a wide range, the main polymer types of the microplastics were found to be polypropylene and polyethylene as shown in Figure 3. Furthermore, the primary sources for the formation of the microplastics in the landfills were plastic food storage containers, plastic packages and plastic bottles. The microplastics could also be carried from landfills to water resources such as rivers, lakes or seas, to terrestrial regions/soils far away from the landfills or to the atmosphere by mainly wind and surface run-off. The microplastics in leachates or soil beneath the landfills might also contain hazardous or toxic substances. As a result, microplastics can be easily transferred to the soil from a landfill without a barrier material or with a barrier material that is damaged or not properly designed.

4. Some Geotechnical Engineering Applications as Potential Source for Microplastics

Landfills are considered by the authors as one of the significant sources of microplastic contamination in soils, which deserves further attention. However, there are also other geotechnical engineering applications, where polymer-based materials (e.g., tire chips) or already contaminated soils (e.g., dredged soils) are used. Such applications inherently have the potential to generate microplastics as well, even though their scale of contamination is expected to be smaller than the ones due to landfills. Some of these applications are explained below:

4.1. Soil Improvement with Tire Chips

Tire chips are the shredded waste tires obtained from the abrasion of road vehicle and airplane tires. Tire wastes are one of the most significant microplastic sources for the migration of the microplastics to the soil. As a result of the disintegration of the shredded waste tires, microplastics can enter the soil easily. The global emission of microplastics due to the abrasion of road vehicle tires is approximately 0.81 kg/year per capita [112]. Moreover, the most abundant source of annual microplastic mass emission per capita for the EU countries was found to be the tire wear with 1784.8 $g.cap^{-1} a^{-1}$ [113]. Microplastics formed by the disintegration of the waste tires were also observed in various landfill leachate samples [114].

On the other hand, waste tires are also used as geomaterials for enhancing the shear strength of soils in civil engineering projects. In order to reduce the negative impacts of the shredded waste tires on the environment and seeking for economical engineering solutions, soil deposits mixed with tire chips are used in various geotechnical applications including road embankments, pavement subgrades and backfills of retaining structures [115–118].

Singh and Sonthwal [117] and Solanki et al. [118] added tire chips to clayey soils and the tire chip addition in these studies led to an increase in the shear strength of the clay. Tire chips were also mixed with sandy soils in various studies. Daud [115] reported that the shear strength of a sandy deposit was increased by tire chip addition with an observed increase in the internal friction angle of the soil. Al-Neami [119] conducted California bearing ratio (CBR) tests on tire chip-sand mixtures and the results indicated that tire chip addition with a content of 8% by dry weight caused 1.6 times higher load bearing capacity than the sand without any tire chip addition. CBR and unconfined compression tests were performed on tire chip-cement-soil mixtures, and according to the results, both the shear strength and the bearing capacity of the soils increased due to tire chip addition [117,120].

4.2. Dredge Sediments from Water Resources Used as Filling Materials

Dredge sediments are typically carried from the bottom of rivers, lakes, harbors or any other water resources to the target sites and dumped to fill lowlands. The dredge sediments are mainly used for urban soil reconstruction, shoreline stabilization, coastal land reclamation, upland placement for agriculture, as fertilizers in farmlands or as covers for landfills and mine tailings dams [37,121–123]. Due to the fact that microplastics can be easily dispersed in the aquatic environment, the dredge sediments might contain significant amounts of microplastics [121].

More than 200 million m^3 of dredge sediments have been collected annually from the bottom of almost 400 ports in the USA and mainly used for urban soil reconstruction and upland placement for agriculture [123]. In another study, the dredge sediment that was mainly composed of fine sand was collected from a riverbank in Dhaka, Bangladesh and used as filling material for land reclamation [124]. The total sediment amount that has been dredged annually in China is more than 5 billion m^3 [125]. For instance, 136 and 116 million m^3 of dredge sediments were collected from the riverbanks in Zhejiang Province, China for reconstruction facilities in 2016 and 2017, respectively [125,126]. Moreover, Ji et al. [121] mentioned that only 3 out of 10 million m^3 of the collected dredge sediments from the riverbanks in Yueqing, China was transported and dumped into a landfill to serve as the cover soil in a coastal reclamation project. The remaining 7 million m^3 of the dredge sediment was placed in storage piles to be used as agricultural fertilizers in farmlands. The microplastic content in the soil around the storage piles was found to be higher in the dry season due to the wind dispersion while the river surrounded by the storage piles contained higher microplastics in the wet season due to the surface run-off [121].

4.3. Soil Improvement with Synthetic Polymer-Based Fibers

Polymers have been used in fiber form for various geotechnical engineering applications in recent years, especially for ground improvement. Polymer-based fibers are typically added to the soil in order to enhance the strength parameters of the soil. These fibers serve as the coating agents that are able to fill the voids among the soil grains as well as act as reinforcements in soil fabric. Furthermore, these fibers can be mixed with other binding materials such as lime or cement in the soil for increasing the shear and compressive strengths of the soil [127]. In various studies, synthetic polymer-based fibers were added to clayey or sandy soils or soil-cement mixtures, and geotechnical laboratory tests such as unconfined compression, direct shear and triaxial compression tests were performed on these mixtures [128–131]. The test results indicated that fiber addition up to a limiting content led to an increase in both shear and compressive strengths of the soil, which indicates improved geotechnical properties for foundation design.

For example, Ayeldeen and Kitazume [132] added a synthetic polymer-based fiber (polypropylene) with a content of up to 1% by dry weight to a clayey soil that was mixed with cement. The cement content in the clay was 15% by dry weight. The fiber content of 0.5% caused the shear strength of the clay-cement mixture to increase by almost 240%. However, further increase in the fiber content resulted in a decrease in the shear strength. The synthetic polymer-based fibers can also be used as the secondary additives to increase the ductility of stabilized soils that are mainly used as barriers in landfills or subgrade soils beneath roadways [133]. Yılmaz and Sevencan [134] added polypropylene fiber to a soil mixed with fly ash and the fiber addition led to an increase in the unconfined compressive strength. In another study, polypropylene fibers and waste ashes were added to a sulfate-rich expansive soil. Due to the fiber addition, both the swell index and the shrinkage strain of the expansive soil increased [135]. Öncü and Bilsel [133] added a polymeric fiber with a content of 2% by dry weight to an expansive soil-sand mixture. According to the results of the laboratory tests, compressive and split tensile strength values of the soil mixture increased.

However, polymer-based fibers that are added to soils for reinforcement can easily be fragmented into microplastics [96,136]. In literature, polypropylene is reported to be one of the most abundant microplastic polymer types in both wastewater and soil environments [85,93,100,102,137]. Hence, one should remember that there is a considerable risk of microplastic contamination when soils are stabilized by the addition of synthetic polymer-based fibers for various civil/geotechnical engineering projects.

4.4. Geotechnical Applications with Expanded Polystyrene as a Lightweight Fill Material
4.4.1. Polystyrene Based Microplastics in Soils

Polystyrene (PS) is the polymer type that is typically used for the manufacture of hard plastics such as food packages, food containers, beverage cups, plates and laboratory wares etc. [138,139]. However, once polystyrene-based plastic wastes are released to the environment, they can be broken into smaller pieces easily and carried to the atmosphere, water resources and terrestrial lands that are farther away from the location where the wastes are originally deposited or alternatively can penetrate into the soil vertically [100,140]. As mentioned in Section 2.3 of this paper, PS is also a very persistent polymer (Figure 3c) in soil environment.

A good example about the distribution of PS based microplastics can be seen in a case study from Belarus by Kukharchyk and Chernyuk [140], where an industrial plant deposited its plastic wastes into an area nearby a river. Soil and groundwater samples were collected from the floodplain of the river. High amounts of microplastics that were composed of polystyrene were detected in both the soil and the groundwater samples. The microplastic amount in the soil was measured as 1700 ptcl./kg while the groundwater samples contained 16,700 ptcl./kg microplastics [140]. Moreover, polystyrene particles smaller than 1 mm were found not only at the surface of the soil but also at a depth of 10–15 cm below the surface. Note that microplastic contamination at depths greater than 15 cm is unknown.

On the other hand, microplastics with a particle size of 0.001 mm (1 μm) that are formed by the disintegration of polystyrene-based plastic wastes, can also be considered as a serious threat for human beings if inhaled. Once the polystyrene based microplastic particles are inhaled or the foods that contain these particles are eaten, the polystyrene microplastics may destabilize red blood cells and eventually could lead to the disease named as hemolysis [138].

4.4.2. Geotechnical Applications with Polystyrene Based Lightweight Fills

Expanded polystyrene is one of the most preferred lightweight fill materials that is used for geotechnical applications including embankment constructions for roads and bridges, slope stabilization and backfills of retaining walls [141–144]. In road and bridge constructions over soft soils, lightweight fill embankments with expanded polystyrene are used in order to reduce the bearing pressures on the soft soil [145,146]. Expanded polystyrene as a lightweight fill material is also preferred to be used to protect culverts and buried pipelines by decreasing the soil pressures acting on these structures [137,147].

In another study, prefabricated vertical expanded polystyrene drains were used for reducing the swelling capability of an expansive soil beneath the foundations of a couple of buildings in India [148]. After a while, expanded polystyrene drains stabilized the expansive soil by decreasing the rate of consolidation and settlements. Note that the mentioned drains in the study of Daigavane et al. [148] are not lightweight fills, but drainage systems (i.e., they are completely different geotechnical applications involving PS based materials).

Although the usage of the expanded polystyrene as a lightweight fill material in various geotechnical applications has several important benefits from engineering point of view, the possible propagation of the microplastics to the soil may lead to at least local soil and groundwater pollution in the long term.

It should be reminded that the geotechnical applications mentioned in Section 4 of this review paper (i.e., soil improvement with tire chips, forming engineering fills with dredged sediments, soil improvement with synthetic polymer-based fibers, polystyrene based lightweight fill applications) are all smart engineering solutions and beneficial for different civil engineering projects. Moreover, the level/weight of microplastic contamination in soils due to such applications is probably relatively small compared to the contamination levels caused by massive amounts of global plastic waste generated systematically. Therefore, the goal of authors is not to criticize or discourage those applications, but rather make the engineers be aware of the potential risks for local microplastic contamination. More importantly, authors suggest that future multidisciplinary research on microplastics should consider assessing and quantifying such potential risks.

5. Benefitting Geotechnical Engineering as a Mitigation Tool for Microplastics, Future Research and Concluding Remarks

In this review paper, microplastic contamination in soils is assessed from a geotechnical engineering perspective. It was emphasized that there is a delay of global awareness between different disciplines regarding the microplastic contamination in soils. A basic statistical analysis on the published literature for the last six years revealed that there are only two engineering disciplines (i.e., Chemical and Environmental Engineering) among the top 10 disciplines conducting research about microplastics and soils (Figure 2). Meanwhile, as civil/geotechnical engineers, authors of this paper could definitely conclude that the global awareness of the Civil Engineering community on microplastics is extremely limited, which is one of the main motivations for this review.

Today, it is well known that microplastic contamination in soils is a global problem that needs to be seriously considered. In the second part of this review, microplastic-soil relationship is addressed from degradation, existence and persistence aspects based on various studies in literature. Meanwhile, it was observed that the mentioned studies focused mostly on agricultural, environmental and health aspects, which are all very important. However, it would be quite beneficial to adopt the geotechnical engineering perspective in order to better understand the soil-microplastic interaction. For instance, influence of several relevant key parameters in geotechnical engineering, such as void ratio and hydraulic conductivity, are reminded for future multidisciplinary research on the migration and spatial distribution of microplastics in soils. It was also observed that the relevant literature on the microplastic existence in soils typically focused for the first 20 cm below the ground surface (with one exception for 30 cm). However, this depth range (0–30 cm) is very shallow from a geotechnical engineering perspective. Hence, more studies are needed to quantify the existence of microplastics at greater depths from the ground surface. In addition, consideration of the influence of different soil types used in engineering classification (e.g., clay, silt, sand, gravel and their mixtures) on microplastic migration and distribution in soils is needed in future multi-disciplinary research studies, the effects of which are currently unknown.

In the third part of this review, the geotechnical design of MSW landfills is addressed, including the compacted clay liners (CCL) and geosynthetic clay liners (GCL). Moreover, components of single and double-liner systems and their functions are explained (Figure 6). The limited amount of studies in literature clearly reveal that leachates from various landfills involve different concentrations of microplastics, which is highly expected considering the vast amount of plastic waste dumped in landfills globally. However, the number of studies on microplastic contamination in soils beneath or in the vicinity of landfills are even more limited (authors could find extremely few studies on the subject). Authors think that this subject is very important and deserves further research with case studies, because the lack of sufficient liner systems for abandoned waste dumps or defects/liner damages in landfills might also significantly contribute to the microplastic contamination in soils.

Furthermore, it would also be an interesting and critical research topic to investigate the performance of GCLs for filtering/holding the microplastics, which is currently unknown. It is possible that GCLs could act as a mitigation tool to reduce the microplastic contamination in soils. In fact, the authors of this paper have been working on a research project proposal on the topic. It is also ironic to remind that GCLs themselves also have polymer-based components (e.g., geotextiles), which could generate microplastics in the long term. However, authors think that their pros will be more than their cons both as a possible mitigation tool for microplastic contamination in soils and as an innovative engineering tool for various projects including landfills. Though, the mentioned hypothesis of authors should be evaluated in future studies. Moreover, the overall microplastic filtering/holding capacity of landfill liners working as a system (i.e., geotextile, drainage layer, geomembrane (with defect simulation), GCL working together as a system) should also be evaluated in future research.

In the last (fourth) part of this review, some geotechnical engineering applications as potential additional sources for microplastics in soils are expressed. These are soil improvement with tire chips, forming engineering fills with microplastic contaminated dredged sediments, soil improvement with synthetic polymer-based fibers, polystyrene based lightweight fill applications. Note that, there could be more geotechnical engineering applications which are not considered in this review. It is important to mention that those applications are all smart engineering solutions and beneficial for different civil engineering projects. Moreover, the weight of microplastic contamination in soils due to such applications is expected to be relatively very small compared to the weight of contamination levels caused by massive amounts of global plastic waste generated systematically. However, Civil/Geotechnical Engineers should be aware of the potential risks for additional microplastic contamination due to such applications. This could also be a novel multidisciplinary research topic, for which no studies are currently available in the literature.

Author Contributions: Conceptualization, M.M.M.; methodology, M.M.M. and H.O.Ö.; writing—original draft preparation, M.M.M. and H.O.Ö.; writing—review and editing, M.M.M. and H.O.Ö. All authors have read and agreed to the published version of the manuscript.

Funding: The APC of this paper was funded by Yeditepe University. This funding is greatly appreciated by authors.

Acknowledgments: Authors want to thank Chief Senior Researcher Bahar Özmen-Monkul, (TÜBİTAK Marmara Research Center) for her help in drawing Figure 3 and useful discussions about degradation of polymers. Authors also want to thank Ali Hakan Ören, (Dokuz Eylül University) for useful discussions about the landfill liner systems.

Conflicts of Interest: The authors declare no conflict of interest.

References

1. Lim, X.Z. Microplastics Are Everywhere—But Are They Harmful? *Nature* **2021**, *593*, 22–25. [CrossRef] [PubMed]
2. Da Costa, J.P.; Rocha Santos, T.; Duarte, A. The Environmental Impacts of Plastics and Micro-Plastics Use, Waste and Pollution: EU and National Measures. *Eur. Union* **2020**, 76.
3. PlasticsEurope—Association of Plastics Manufactures Plastics—The Facts 2020. *PlasticsEurope* **2020**, 1–64. Available online: https://plasticseurope.org/wp-content/uploads/2021/09/Plastics_the_facts-WEB-2020_versionJun2021_final.pdf (accessed on 8 September 2021).
4. Brooks, A.L.; Wang, S.; Jambeck, J.R. The Chinese Import Ban and Its Impact on Global Plastic Waste Trade. *Sci. Adv.* **2018**, *4*, 52. [CrossRef] [PubMed]
5. Zhao, C.; Liu, M.; Du, H.; Gong, Y. The Evolutionary Trend and Impact of Global Plastic Waste Trade Network. *Sustainability* **2021**, *13*, 3662. [CrossRef]
6. Fuhr, L.; Franklin, M. *The PLASTIC ATLAS 2019*; Heinrich Böll Foundation: Berlin, Germany, 2019; ISBN 9783869282114.
7. Ataş, N.T. Plastik Atık İthalatına Karşı Zafere Giden Yolculuk. Available online: https://www.greenpeace.org/turkey/blog/plastik-atik-ithalatina-karsi-zafere-giden-yolculuk/ (accessed on 15 November 2021).
8. Snowdon, K. UK Plastic Waste Being Dumped and Burned in Turkey, Says Greenpeace. Available online: https://www.bbc.com/news/uk-57139474 (accessed on 15 November 2021).
9. EPA Advancing Sustainable Materials Management. *United States Environmental Protection Agency*; Office of Resource Conservation and Recovery: Washington, WA, USA, 2020; p. 184.
10. Thompson, R.C.; Olsen, Y.; Mitchell, R.P.; Davis, A.; Rowland, S.J.; John, A.W.G.; Mcgonigle, D.; Russell, A.E. Lost at Sea: Where Is All the Plastic? *Science* **2004**, *304*, 838. [CrossRef] [PubMed]
11. Liu, Y.; Shao, H.; Liu, J.; Cao, R.; Shang, E.; Liu, S.; Li, Y. Transport and Transformation of Microplastics and Nanoplastics in the Soil Environment: A Critical Review. *Soil Use Manag.* **2021**, *37*, 224–242. [CrossRef]
12. Ng, E.L.; Huerta Lwanga, E.; Eldridge, S.M.; Johnston, P.; Hu, H.W.; Geissen, V.; Chen, D. An Overview of Microplastic and Nanoplastic Pollution in Agroecosystems. *Sci. Total. Environ.* **2018**, *627*, 1377–1388. [CrossRef]
13. Gigault, J.; Ter Halle, A.; Baudrimont, M.; Pascal, P.Y.; Gauffre, F.; Phi, T.L.; el Hadri, H.; Grassl, B.; Reynaud, S. Current Opinion: What Is a Nanoplastic? *Environ. Pollut.* **2018**, *235*, 1030–1034. [CrossRef] [PubMed]
14. UNEP (UN Environmental Programme). *UNEP Year Book 2014: Emerging Issues in Our Global Environment*; UNEP, Division of Early Warning and Assessment: Nairobi, Kenya, 2014; p. 71.
15. O'Kelly, B.C.; El-Zein, A.; Liu, X.; Patel, A.; Fei, X.; Sharma, S.; Mohammad, A.; Goli, V.S.N.S.; Wang, J.J.; Li, D.; et al. Microplastics in Soils: An Environmental Geotechnics Perspective. *Environ. Geotech.* **2021**, *2000179*, 1–30. [CrossRef]
16. Jahnke, A.; Arp, H.P.H.; Escher, B.I.; Gewert, B.; Gorokhova, E.; Kühnel, D.; Ogonowski, M.; Potthoff, A.; Rummel, C.; Schmitt-Jansen, M.; et al. Reducing Uncertainty and Confronting Ignorance about the Possible Impacts of Weathering Plastic in the Marine Environment. *Environ. Sci. Technol. Lett.* **2017**, *4*, 85–90. [CrossRef]
17. Połeć, M.; Aleksander-Kwaterczak, U.; Wątor, K.; Kmiecik, E. The Occurrence of Microplastics in Freshwater Systems—Preliminary Results from Krakow (Poland). *Geol. Geophys. Environ.* **2018**, *44*, 391. [CrossRef]
18. Zhou, Y.; Wang, J.; Zou, M.; Jia, Z.; Zhou, S.; Li, Y. Microplastics in Soils: A Review of Methods, Occurrence, Fate, Transport, Ecological and Environmental Risks. *Sci. Total Environ.* **2020**, *748*, 141368. [CrossRef]
19. Nizzetto, L.; Bussi, G.; Futter, M.N.; Butterfield, D.; Whitehead, P.G. A Theoretical Assessment of Microplastic Transport in River Catchments and Their Retention by Soils and River Sediments. *Environ. Sci. Process. Impacts* **2016**, *18*, 1050–1059. [CrossRef] [PubMed]
20. Zhang, S.; Yang, X.; Gertsen, H.; Peters, P.; Salánki, T.; Geissen, V. A Simple Method for the Extraction and Identification of Light Density Microplastics from Soil. *Sci. Total Environ.* **2018**, *616–617*, 1056–1065. [CrossRef]
21. Zhu, F.; Zhu, C.; Wang, C.; Gu, C. Occurrence and Ecological Impacts of Microplastics in Soil Systems: A Review. *Bull. Environ. Contam. Toxicol.* **2019**, *102*, 741–749. [CrossRef] [PubMed]
22. da Costa, J.P.; Paço, A.; Santos, P.S.M.; Duarte, A.C.; Rocha-Santos, T. Microplastics in Soils: Assessment, Analytics and Risks. *Environ. Chem.* **2019**, *16*, 18–30. [CrossRef]
23. Möller, J.N.; Löder, M.G.J.; Laforsch, C. Finding Microplastics in Soils: A Review of Analytical Methods. *Environ. Sci. Technol.* **2020**, *54*, 2078–2090. [CrossRef] [PubMed]
24. Wang, W.; Ge, J.; Yu, X.; Li, H. Environmental Fate and Impacts of Microplastics in Soil Ecosystems: Progress and Perspective. *Sci. Total Environ.* **2020**, *708*, 134841. [CrossRef] [PubMed]
25. Li, W.; Wufuer, R.; Duo, J.; Wang, S.; Luo, Y.; Zhang, D.; Pan, X. Microplastics in Agricultural Soils: Extraction and Characterization after Different Periods of Polythene Film Mulching in an Arid Region. *Sci. Total Environ.* **2020**, *749*, 141420. [CrossRef] [PubMed]

26. Guo, J.J.; Huang, X.P.; Xiang, L.; Wang, Y.Z.; Li, Y.W.; Li, H.; Cai, Q.Y.; Mo, C.H.; Wong, M.H. Source, Migration and Toxicology of Microplastics in Soil. *Environ. Int.* **2020**, *137*, 105263. [CrossRef] [PubMed]
27. Corradini, F.; Meza, P.; Eguiluz, R.; Casado, F.; Huerta-Lwanga, E.; Geissen, V. Evidence of Microplastic Accumulation in Agricultural Soils from Sewage Sludge Disposal. *Sci. Total Environ.* **2019**, *671*, 411–420. [CrossRef] [PubMed]
28. Crossman, J.; Hurley, R.R.; Futter, M.; Nizzetto, L. Transfer and Transport of Microplastics from Biosolids to Agricultural Soils and the Wider Environment. *Sci. Total Environ.* **2020**, *724*, 138334. [CrossRef]
29. Zhang, G.S.; Liu, Y.F. The Distribution of Microplastics in Soil Aggregate Fractions in Southwestern China. *Sci. Total Environ.* **2018**, *642*, 12–20. [CrossRef]
30. de Souza Machado, A.A.; Lau, C.W.; Till, J.; Kloas, W.; Lehmann, A.; Becker, R.; Rillig, M.C. Impacts of Microplastics on the Soil Biophysical Environment. *Environ. Sci. Technol.* **2018**, *52*, 9656–9665. [CrossRef] [PubMed]
31. Steinmetz, Z.; Wollmann, C.; Schaefer, M.; Buchmann, C.; David, J.; Tröger, J.; Muñoz, K.; Frör, O.; Schaumann, G.E. Plastic Mulching in Agriculture. Trading Short-Term Agronomic Benefits for Long-Term Soil Degradation? *Sci. Total Environ.* **2016**, *550*, 690–705. [CrossRef]
32. Dexter, A.R. Soil Physical Quality: Part, I. Theory, Effects of Soil Texture, Density, and Organic Matter, and Effects on Root Growth. *Geoderma* **2004**, *120*, 201–214. [CrossRef]
33. Monkul, M.M.; Aydın, N.G.; Demirhan, B.; Şahin, M. Undrained Shear Strength and Monotonic Behavior of Different Nonplastic Silts: Sand-Like or Clay-Like? *Geotech. Test. J.* **2020**, *43*, 20180147. [CrossRef]
34. Monkul, M.M.; Ozden, G. Compressional Behavior of Clayey Sand and Transition Fines Content. *Eng. Geol.* **2007**, *89*, 195–205. [CrossRef]
35. Duncan, J.M.; Wright, S.G. *Soil Strength Slope Stab*, 1st ed.; Wiley: Hoboken, NJ, USA, 2005; ISBN 9780471691631.
36. Coduto, D.P. *Foundation Design: Principles and Practices*, 2nd ed. Pearson: London, UK, 2001.
37. Van't Hoff, J.; van der Kolff, A.N. *Hydraulic Fill Manual: For Dredging and Reclamation Works*; CRC Press/Balkema Taylor & Francis Group: London, UK, 2012.
38. Qian, G.; Koerner, R.M.; Gray, D. *Geotechnical Aspects of Landfill Design and Construction*; Pearson: London, UK, 2001.
39. Lambert, S.; Wagner, M. Formation of Microscopic Particles during the Degradation of Different Polymers. *Chemosphere* **2016**, *161*, 510–517. [CrossRef] [PubMed]
40. Singh, B.; Sharma, N. Mechanistic Implications of Plastic Degradation. *Polym. Degrad. Stab.* **2008**, *93*, 561–584. [CrossRef]
41. Zou, W.; Xia, M.; Jiang, K.; Cao, Z.; Zhang, X.; Hu, X. Photo-Oxidative Degradation Mitigated the Developmental Toxicity of Polyamide Microplastics to Zebrafish Larvae by Modulating Macrophage-Triggered Proinflammatory Responses and Apoptosis. *Environ. Sci. Technol.* **2020**, *54*, 13888–13898. [CrossRef] [PubMed]
42. Kliem, S.; Kreutzbruck, M.; Bonten, C. Review on the Biological Degradation of Polymers in Various Environments. *Materials* **2020**, *13*, 4586. [CrossRef]
43. Zheng, Y.; Yanful, E.K.; Bassi, A.S. A Review of Plastic Waste Biodegradation. *Crit. Rev. Biotechnol.* **2005**, *25*, 243–250. [CrossRef] [PubMed]
44. Crabbe, J.R.; Campbell, J.R.; Thompson, L.; Walz, S.L.; Schultz, W.W. Biodegradation of a Colloidal Ester-Based Polyurethane by Soil Fungi. *Int. Biodeterior. Biodegrad.* **1994**, *33*, 103–113. [CrossRef]
45. Yoshida, S.; Hiraga, K.; Takehana, T.; Taniguchi, I.; Yamaji, H.; Maeda, Y.; Toyohara, K.; Miyamoto, K.; Kimura, Y.; Oda, K. A Bacterium That Degrades and Assimilates Poly(Ethylene Terephthalate). *Science* **2016**, *351*, 1196–1199. [CrossRef] [PubMed]
46. Feng, S.; Yue, Y.; Zheng, M.; Li, Y.; Zhang, Q.; Wang, W. IsPETase- AndIsMHETase-Catalyzed Cascade Degradation Mechanism toward Polyethylene Terephthalate. *ACS Sustain. Chem. Eng.* **2021**, *9*, 9823–9832. [CrossRef]
47. Orhan, Y.; Hrenović, J.; Büyükgüngör, H. Biodegradation of Plastic Compost Bags under Controlled Soil Conditions. *Acta Chim. Slov.* **2004**, *51*, 579–588.
48. Kyrikou, I.; Briassoulis, D. Biodegradation of Agricultural Plastic Films: A Critical Review. *J. Polym. Environ.* **2007**, *15*, 125–150. [CrossRef]
49. Palmisano, A.C.; Pettigrew, C.A. Biodegradability of Plastics. *BioScience* **1992**, *42*, 680–685. [CrossRef]
50. Tokiwa, Y.; Calabia, B.P.; Ugwu, C.U.; Aiba, S. Biodegradability of Plastics. *Int. J. Mol. Sci.* **2009**, *10*, 3722–3742. [CrossRef]
51. Lwanga, E.H.; Vega, J.M.; Quej, V.K.; de los Angeles Chi, J.; Del Cid, L.S.; Chi, C.; Segura, G.E.; Gertsen, H.; Salánki, T.; van der Ploeg, M.; et al. Field Evidence for Transfer of Plastic Debris along a Terrestrial Food Chain. *Sci. Rep.* **2017**, *7*, 1–7. [CrossRef]
52. van den Berg, P.; Huerta-Lwanga, E.; Corradini, F.; Geissen, V. Sewage Sludge Application as a Vehicle for Microplastics in Eastern Spanish Agricultural Soils. *Environ. Pollut.* **2020**, *261*, 114198. [CrossRef] [PubMed]
53. Dahl, M.; Bergman, S.; Björk, M.; Diaz-Almela, E.; Granberg, M.; Gullström, M.; Leiva-Dueñas, C.; Magnusson, K.; Marco-Méndez, C.; Piñeiro-Juncal, N.; et al. A Temporal Record of Microplastic Pollution in Mediterranean Seagrass Soils. *Environ. Pollut.* **2021**, *273*, 116451. [CrossRef]
54. Coduto, D.P.; Yeung, M.R.; Kitch, W.A. *Geotechnical Engineering: Principles and Practices*, 2nd ed.; Pearson: London, UK, 2010; ISBN 0131354256.
55. Holtz, R.D.; Kovacs, W.D.; Sheahan, T.C. *An Introduction to Geotechnical Engineering*, 2nd ed.; Pearson: London, UK, 2010.
56. O'Connor, D.; Pan, S.; Shen, Z.; Song, Y.; Jin, Y.; Wu, W.M.; Hou, D. Microplastics Undergo Accelerated Vertical Migration in Sand Soil Due to Small Size and Wet-Dry Cycles. *Environ. Pollut.* **2019**, *249*, 527–534. [CrossRef] [PubMed]
57. Terzaghi, K.; Peck, R.B.; Mesri, G. *Soil Mechanics in Engineering Practice*, 3rd ed.; John Wiley & Sons: Hoboken, NJ, USA, 1996.

58. Wu, X.; Lyu, X.; Li, Z.; Gao, B.; Zeng, X.; Wu, J.; Sun, Y. Transport of Polystyrene Nanoplastics in Natural Soils: Effect of Soil Properties, Ionic Strength and Cation Type. *Sci. Total Environ.* **2020**, *707*, 136065. [CrossRef] [PubMed]
59. Cooper, D.A.; Corcoran, P.L. Effects of Mechanical and Chemical Processes on the Degradation of Plastic Beach Debris on the Island of Kauai, Hawaii. *Mar. Pollut. Bull.* **2010**, *60*, 650–654. [CrossRef]
60. Krueger, M.C.; Harms, H.; Schlosser, D. Prospects for Microbiological Solutions to Environmental Pollution with Plastics. *Appl. Microbiol. Biotechnol.* **2015**, *99*, 8857–8874. [CrossRef] [PubMed]
61. Briassoulis, D.; Babou, E.; Hiskakis, M.; Kyrikou, I. Analysis of Long-Term Degradation Behaviour of Polyethylene Mulching Films with pro-Oxidants under Real Cultivation and Soil Burial Conditions. *Environ. Sci. Pollut. Res.* **2015**, *22*, 2584–2598. [CrossRef]
62. Reddy, M.M.; Gupta, R.K.; Gupta, R.K.; Bhattacharya, S.N.; Parthasarathy, R. Abiotic Oxidation Studies of Oxo-Biodegradable Polyethylene. *J. Polym. Environ.* **2008**, *16*, 27–34. [CrossRef]
63. Albertsson, A.C.; Karlsson, S. The Three Stages in Degradation of Polymers—Polyethylene as a Model Substance. *J. Appl. Polym. Sci.* **1988**, *35*, 1289–1302. [CrossRef]
64. Otake, Y.; Kobayashi, T.; Asabe, H.; Murakami, N.; Ono, K. Biodegradation of Low-density Polyethylene, Polystyrene, Polyvinyl Chloride, and Urea Formaldehyde Resin Buried under Soil for over 32 Years. *J. Appl. Polym. Sci.* **1995**, *56*, 1789–1796. [CrossRef]
65. Tabone, M.D.; Cregg, J.J.; Beckman, E.J.; Landis, A.E. Sustainability Metrics: Life Cycle Assessment and Green Design in Polymers. *Environ. Sci. Technol.* **2010**, *45*, 8264–8269. [CrossRef] [PubMed]
66. Karamanlioglu, M.; Robson, G.D. The Influence of Biotic and Abiotic Factors on the Rate of Degradation of Poly(Lactic) Acid (PLA) Coupons Buried in Compost and Soil. *Polym. Degrad. Stab.* **2013**, *98*, 2063–2071. [CrossRef]
67. Akbulut, S. Katı atık depo alanlarının geoteknik tasarımı. *Mühendislik Bilimleri Derg.* **2003**, *9*, 223–230.
68. Ghosh, S.K.; Agamuthu, P. Plastics in municipal solid waste: What, where, how and when? *Waste Manag. Res.* **2019**, *37*, 1061–1062. [CrossRef] [PubMed]
69. Hughes, K.L.; Christy, A.D.; Heimlich, J.E. *Landfill Types and Liner Systems. Extension Fact Sheet*; The Ohio State University: Columbus, OH, USA, 2005.
70. Daniel, D.E.; Wu, Y. Compacted Clay Liners and Covers for Arid Sites. *J. Geotech. Eng.* **1993**, *119*, 223–237. [CrossRef]
71. Qian, X.; Gray, D.H.; Koerner, R.M. Estimation of Maximum Liquid Head over Landfill Barriers. *J. Geotech. Geoenviron. Eng.* **2004**, *130*, 488–497. [CrossRef]
72. Koerner, R.M. *Designing with Geosynthetics*; Pearson: London, UK, 2005.
73. Sharma, D.H.; Lewis, P.S. *Waste Containment Systems, Waste Stabilization and Landfills, Design and Evaluation, Waste Characterization and Solid-Waste Interaction*; John Wiley & Sons: Hoboken, NJ, USA, 1994.
74. Koerner, G.R. Assessing Potential Geomembrane Damage from Direct Construction Equipment Contact. *Geo Front. Congr.* **2005**, *1*, 1–3. [CrossRef]
75. Ben Othmen, A.; Bouassida, M. Detecting defects in geomembranes of landfill liner systems: Durable electrical method. *Int. J. Geotech. Eng.* **2013**, *7*, 130–135. [CrossRef]
76. Bouazza, A. Geosynthetic clay liners. *Geotext. Geomembr.* **2002**, *20*, 3–17. [CrossRef]
77. Benson, C.H.; Meer, S.R. Relative abundance of monovalent and divalent cations and the impact of desiccation on geosynthetic clay liners. *J. Geotech. Geoenviron. Eng.* **2009**, *135*, 349–358. [CrossRef]
78. Güler, E.; Özhan, H.O.; Karaoglu, S. Hydraulic performance of anionic polymer-treated bentonite-granular soil mixtures. *Appl. Clay Sci.* **2018**, *157*, 139–147. [CrossRef]
79. Liu, Y.; Gates, W.P.; Bouazza, A. Acid induced degradation of the bentonite component used in geosynthetic clay liners. *Geotext. Geomembr.* **2013**, *36*, 71–80. [CrossRef]
80. Özhan, H.O.; Güler, E. Use of Perforated Base Pedestal to Simulate the Gravel Subbase in Evaluating the Internal Erosion of Geosynthetic Clay Liners. *Geotech. Test. J.* **2013**, *36*, 418–428. [CrossRef]
81. Özhan, H.O.; Güler, E. Factors affecting failure by internal erosion of geosynthetic clay liners used in fresh water reservoirs. *Environ. Eng. Geosci.* **2016**, *22*, 157–169. [CrossRef]
82. Giroud, J.; Soderman, K. Criterion for Acceptable Bentonite Loss From a GCL Incorporated Into a Liner System. *Geosynth. Int.* **2000**, *7*, 529–581. [CrossRef]
83. Orsini, C.; Rowe, R.K. Testing procedure and results for the study of internal erosion of geosynthetic clay liners. In Proceedings of the Geosynthetics, Portland, OR, USA, 12–14 February 2001; pp. 189–201.
84. Geyer, R.; Jambeck, J.R.; Law, K.L. Production, use, and fate of all plastics ever made. *Sci. Adv.* **2017**, *3*, 25–29. [CrossRef] [PubMed]
85. He, P.; Chen, L.; Shao, L.; Zhang, H.; Lü, F. Municipal solid waste (MSW) landfill: A source of microplastics? -Evidence of microplastics in landfill leachate. *Water Res.* **2019**, *159*, 38–45. [CrossRef] [PubMed]
86. Hanson, J.L.; Yesiller, N.; Kendall, L.A. Integrated temperature and gas analysis at a municipal solid wast landfill. In Proceedings of the 16th International Conference on Soil Mechanics and Geotechnical Engineering: Geotechnology in Harmony with the Global Environment, Osaka, Japan, 12 September 2005.
87. Mahon, A.M.; O'Connell, B.; Healy, M.; O'Connor, I.; Officer, R.; Nash, R.; Morrison, L. Microplastics in Sewage Sludge: Effects of Treatment. *Environ. Sci. Technol.* **2017**, *51*, 810–818. [CrossRef] [PubMed]

88. Sundt, P.; Schulze, P.E.; Syversen, F. Sources of microplastic pollution to the marine environment. *Mepex Rep. Nor. Environ. Agency* **2014**, 49–50.
89. Zettler, E.; Mincer, T.; Amaral-Zettler, L.A. Life in the "Plastisphere": Microbial Communities on Plastic Marine Debris. *Environ. Sci. Technol.* **2013**, *47*, 7137–7146. [CrossRef]
90. Xu, Z.; Sui, Q.; Li, A.; Sun, M.; Zhang, L.; Lyu, S.; Zhao, W. How to detect small microplastics (20–100 µm) in freshwater, municipal wastewaters and landfill leachates? A trial from sampling to identification. *Sci. Total Environ.* **2020**, *733*, 139218. [CrossRef] [PubMed]
91. Yang, N.; Damgaard, A.; Kjeldsen, P.; Shao, L.-M.; He, P.-J. Quantification of regional leachate variance from municipal solid waste landfills in China. *Waste Manag.* **2015**, *46*, 362–372. [CrossRef] [PubMed]
92. Van Praagh, M.; Hartman, C.; Brandmyr, E. Brandmyr, E. Microplastics in Landfill leachates in the Nordic Countries. Tema.Nord. 2019. Nordic Council of Ministers, Copenhagen, TN2018-557. Available online: https://norden.diva-portal.org/smash/get/diva2:1277395/FULLTEXT01.pdf (accessed on 5 October 2021).
93. Su, Y.; Zhang, Z.; Wu, D.; Zhan, L.; Shi, H.; Xie, B. Occurrence of microplastics in landfill systems and their fate with landfill age. *Water Res.* **2019**, *164*, 114968. [CrossRef] [PubMed]
94. Narevski, A.C.; Novaković, M.I.; Petrović, M.Z.; Mihajlović, I.J.; Maoduš, N.B.; Vujić, G.V. Occurrence of bisphenol A and microplastics in landfill leachate: Lessons from South East Europe. *Environ. Sci. Pollut. Res.* **2021**, *28*, 42196–42203. [CrossRef] [PubMed]
95. Ficociello, G.; Gerardi, V.; Uccelletti, D.; Setini, A. Molecular and cellular responses to short exposure to bisphenols A, F, and S and eluates of microplastics in C. elegans. *Environ. Sci. Pollut. Res.* **2021**, *28*, 805–818. [CrossRef] [PubMed]
96. Liu, X.; Shi, H.; Xie, B.; Dionysiou, D.D.; Zhao, Y. Microplastics as Both a Sink and a Source of Bisphenol A in the Marine Environment. *Environ. Sci. Technol.* **2019**, *53*, 10188–10196. [CrossRef] [PubMed]
97. Beckingham, B.; Ghosh, U. Differential bioavailability of polychlorinated biphenyls associated with environmental particles: Microplastic in comparison to wood, coal and biochar. *Environ. Pollut.* **2017**, *220*, 150–158. [CrossRef] [PubMed]
98. Hodson, M.E.; Duffus-Hodson, C.A.; Clark, A.; Prendergast-Miller, M.T.; Thorpe, K.L. Plastic bag derived-microplastics as a vector for metal exposure in terrestrial invertebrates. *Environ. Sci. Technol.* **2017**, *51*, 4714–4721. [CrossRef] [PubMed]
99. Foose, G.J.; Benson, C.H.; Edil, T.B. Predicting Leakage through Composite Landfill Liners. *J. Geotech. Geoenviron. Eng.* **2001**, *127*, 510–520. [CrossRef]
100. Manikanda, B.K.; Usha, N.; Vaikunth, R.; Praveen, K.R.; Ruthra, R.; Srinivasalu, S. Spatial Distribution of Microplastic Concentration around Landfill Sites and its Potential Risk on Groundwater. *Chemosphere* **2021**, *277*, 130263. [CrossRef]
101. Afrin, S.; Uddin, K.; Rahman, M. Microplastics contamination in the soil from Urban Landfill site, Dhaka, Bangladesh. *Heliyon* **2020**, *6*, e05572. [CrossRef] [PubMed]
102. Puthcharoen, A.; Leungprasert, S. Determination of microplastics in soil and leachate from the landfills. *Thai Environ. Eng. J.* **2019**, *33*, 39–46.
103. Browne, M.A.; Underwood, A.J.; Chapman, M.G.; Williams, R.; Thompson, R.; Van Franeker, J.A. Linking effects of anthropogenic debris to ecological impacts. *Proc. R. Soc. B Biol. Sci.* **2015**, *282*, 20142929. [CrossRef]
104. Horton, A.A.; Walton, A.; Spurgeon, D.J.; Lahive, E.; Svendsen, C. Microplastics in freshwater and terrestrial environments: Evaluating the current understanding to identify the knowledge gaps and future research priorities. *Sci. Total Environ.* **2017**, *586*, 127–141. [CrossRef] [PubMed]
105. Zylstra, E. Accumulation of wind-dispersed trash in desert environments. *J. Arid. Environ.* **2013**, *89*, 13–15. [CrossRef]
106. Kilponen, J. Microplastics and harmful substances in urban runoffs and landfill leachates: Possible emission sources to marine environment. In *Faculty of Technology, Volume Environmental Technology*; Lahti University of Applied Sciences: Lahti, Finland, 2016.
107. Hurley, R.R.; Lusher, A.L.; Olsen, M.; Nizzetto, L. Validation of a Method for Extracting Microplastics from Complex, Organic-Rich, Environmental Matrices. *Environ. Sci. Technol.* **2018**, *52*, 7409–7417. [CrossRef] [PubMed]
108. Kazour, M.; Terki, S.; Rabhi, K.; Jemaa, S.; Khalaf, G.; Amara, R. Sources of microplastics pollution in the marine environment: Importance of wastewater treatment plant and coastal landfill. *Mar. Pollut. Bull.* **2019**, *146*, 608–618. [CrossRef]
109. Rezaei, M.; Riksen, M.J.; Sirjani, E.; Sameni, A.; Geissen, V. Wind erosion as a driver for transport of light density microplastics. *Sci. Total Environ.* **2019**, *669*, 273–281. [CrossRef] [PubMed]
110. Prata, J.C.; da Costa, J.P.; Lopes, I.; Duarte, A.C.; Rocha-Santos, T. Environmental exposure to microplastics: An overview on possible human health effects. *Sci. Total Environ.* **2020**, *702*, 134455. [CrossRef] [PubMed]
111. Wright, S.L.; Kelly, F.J. Plastic and Human Health: A Micro Issue? *Environ. Sci. Technol.* **2017**, *51*, 6634–6647. [CrossRef] [PubMed]
112. Kole, P.J.; Löhr, A.J.; Van Belleghem, F.G.A.J.; Ragas, A.M.J. Wear and Tear of Tyres: A Stealthy Source of Microplastics in the Environment. *Int. J. Environ. Res. Public Health* **2017**, *14*, 1265. [CrossRef] [PubMed]
113. Meixner, K.; Kubiczek, M.; Fritz, I. Microplastic in soil—Current status in Europe with special focus on method tests with Austrian samples. *AIMS Environ. Sci.* **2020**, *7*, 174–191. [CrossRef]
114. PlasticsEurope-Association of Plastics Manufacturers. Plastics–the Facts 2019. *PlasticsEurope* **2019**, 1–42. Available online: https://plasticseurope.org/wp-content/uploads/2021/10/2019-Plastics-the-facts.pdf (accessed on 27 September 2021).
115. Daud, K.A. Soil improvement using waste tire chips. *Int. J. Civ. Eng. Technol.* **2018**, *9*, 1338–1345.
116. Promputthangkoona, P.; Karnchanachetaneea, B. Geomaterial prepared from waste tyres, soil and cement. *Procedia Soc. Behav. Sci.* **2013**, *91*, 421–428. [CrossRef]

117. Singh, J.; Sonthwal, V.K. Improvement of engineering properties of clayey soil using shredded rubber tyre. *Int. J. Theor. Appl. Sci.* **2017**, *9*, 1–6.
118. Solanki, D.; Dave, M.; Purohit, D.G.M. Stabilization of clay soil mixed with rubber tyre chips for design in road construction. *Int. J. Eng. Sci. Invent.* **2017**, *6*, 88–91.
119. Al-Neami, M.A. Stabilization of sandy soil using recycle waste tire chips. *Int. J.* **2018**, *15*, 175–180. [CrossRef]
120. Reddy, V.R.; Reddy, I.S.; Prasad, D.S.V. Improvement of soil characteristics using shredded rubber. *IOSR J. Mech. Civ. Eng.* **2016**, *13*, 44–48.
121. Ji, X.; Ma, Y.; Zeng, G.; Xu, X.; Mei, K.; Wang, Z.; Chen, Z.; Dahlgren, R.; Zhang, M.; Shang, X. Transport and fate of microplastics from riverine sediment dredge piles: Implications for disposal. *J. Hazard. Mater.* **2021**, *404*, 124132. [CrossRef]
122. Monkul, M.M.; Yükselen-Aksoy, Y. Sustainable usage of dredge materials at engineering fills and investigation of their performance with different additives. In *Geotechnical Special Publication No 273: Sustainable Waste Management and Remediation*; ASCE: Reston, VI, USA, 2016; pp. 471–480. ISBN 978-0-7844-8016-8.
123. US Army Corps of Engineers (USACE). Dredging Quality Management. 2014. Available online: http://dqm.usace.army.mil/, (accessed on 19 October 2021).
124. Islam, M.; Nasrin, M.; Khan, A. Foundation alternatives in dredge fill soils overlaying organic clay. *Lowl. Technol. Int.* **2013**, *15*, 1–14. [CrossRef]
125. Zhejiang Provincial Bureau of Statistics. Power Conversion Structure Optimization and Quality Improvement-Zhejiang's Economic Performance in 2017. Available online: http://tjj.zj.gov.cn/art/2018/11/23/art_1562012_25740568.html (accessed on 19 October 2021).
126. Zhejiang Provincial Bureau of Statistics. Zhejiang's Economy is Operating Steadily in 2016. Available online: http://tjj.zj.gov.cn/art/2018/11/23/art_1562012_25740525.html (accessed on 19 October 2021).
127. Tingle, J.S.; Newman, J.K.; Larson, S.L.; Weiss, C.A.; Rushing, J.F. Stabilization Mechanisms of Nontraditional Additives. *Transp. Res. Rec. J. Transp. Res. Board* **2007**, *1989*, 59–67. [CrossRef]
128. Abdel Hadi, N.A.R. Utilization of polymer fibers and crushed limestone sand for stabilization of expansive clays in Amman area. *Int. J. Geotech. Eng.* **2016**, *10*, 428–434. [CrossRef]
129. Chen, M.; Shen, S.-L.; Arulrajah, A.; Wu, H.-N.; Hou, D.-W.; Xu, Y.-S. Laboratory evaluation on the effectiveness of polypropylene fibers on the strength of fiber-reinforced and cement-stabilized Shanghai soft clay. *Geotext. Geomembr.* **2015**, *43*, 515–523. [CrossRef]
130. Hamidi, A.; Hooresfand, M. Effect of fiber reinforcement on triaxial shear behavior of cement treated sand. *Geotext. Geomembr.* **2013**, *36*, 1–9. [CrossRef]
131. Tang, C.; Shi, B.; Gao, W.; Chen, F.; Cai, Y. Strength and mechanical behavior of short polypropylene fiber reinforced and cement stabilized clayey soil. *Geotext. Geomembr.* **2007**, *25*, 194–202. [CrossRef]
132. Ayeldeen, M.; Kitazume, M. Using fiber and liquid polymer to improve the behaviour of cement-stabilized soft clay. *Geotext. Geomembranes* **2017**, *45*, 592–602. [CrossRef]
133. Öncü, Ş.; Bilsel, H. Influence of Polymeric Fiber Reinforcement on Strength Properties of Sand-stabilized Expansive Soil. *Polym. Technol. Eng.* **2017**, *56*, 391–399. [CrossRef]
134. Yılmaz, Y.; Sevencan, Ü. Investigation of some geotechnical properties of polypropylene fiber and fly ash amended Ankara clay. In Proceedings of the 13th National Conference on Soil Mechanics and Foundation Engineering, Istanbul, Turkey, 5–10 January 1994; pp. 133–142.
135. Punthutaecha, K.; Puppala, A.J.; Vanapalli, S.K.; Inyang, H. Volume Change Behaviors of Expansive Soils Stabilized with Recycled Ashes and Fibers. *J. Mater. Civ. Eng.* **2006**, *18*, 295–306. [CrossRef]
136. Wei, L.; Chai, S.X.; Zhang, H.Y.; Shi, Q. Mechanical properties of soil reinforced with both lime and four kinds of fiber. *Constr. Build. Mater.* **2018**, *172*, 300–308. [CrossRef]
137. Ahmed, M.R.; Meguid, M.; Whalen, J.; Eng, P. Laboratory Measurement of the load reduction on buried structures overlain by EPS geofoam. In Proceedings of the 66th Canadian Geotechnical Conference, Montreal, QC, Canada, 29 September–3 October 2013.
138. Hwang, J.; Choi, D.; Han, S.; Jung, S.Y.; Choi, J.; Hong, J. Potential toxicity of polystyrene microplastic particles. *Sci. Rep.* **2020**, *10*, 7391. [CrossRef] [PubMed]
139. Schellenberg, J. *Syndiotactic Polystyrene: Synthesis, Characterization, Processing, and Applications*; John Wiley & Sons: Hoboken, NJ, USA, 2009.
140. Kukharchyk, T.; Chernyuk, V. Microplastic of polystyrene in soil and water: Fluxes study from industrial site. In Proceedings of the EGU General Assembly 2020, Vienna, Austria, 4–8 May 2020. Online.
141. Abbasimaedeh, P.; Ghanbari, A.; O'Kelly, B.; Tavanafar, M.; Irdmoosa, K. Geomechanical Behaviour of Uncemented Expanded Polystyrene (EPS) Beads–Clayey Soil Mixtures as Lightweight Fill. *Geotechnics* **2021**, *1*, 38–58. [CrossRef]
142. Özer, A.T.; Akay, O. Interface Shear Strength Characteristics of Interlocked EPS-Block Geofoam. *J. Mater. Civ. Eng.* **2016**, *28*, 04015156. [CrossRef]
143. Stark, T.D.; Arellano, D.; Horvath, J.S.; Leshchinsky, D. *Geofoam Applications in the Design and Construction of Highway Embankments*; Transportation Research Board: Washington, WA, USA, 2004.
144. Tsuchida, T.; Egashira, K. *The Lightweight Treated Soil Method-New Geomaterials for Soft Ground Engineering in Coastal Areas*; Balkema: Liden, The Netherlands, 2004.

233

145. Arellano, D.; Stark, T.D.; Horvath, J.S.; Leshchinsky, D. Guidelines for geofoam applications in slope stability projects. In *Research Results Digest 380*; National Cooperative Highway Research Program, The National Academies Press: Washington, WA, USA, 2013.
146. Gan, C.H.; Tan, S.M. Some construction experiences on soft soil using light weight materials. In Proceedings of the 2nd International Conference on Advances in Soft Soil Engineering and Technology, Putrajaya, Malaysia, 2–4 July 2003.
147. Moorsel, D.; Kilpeläinen, T.; Meuwissen, E.; Neirinckx, L.; Tepper, H.; Thompsett, D.; Zipp, K. *EPS White Book, EUMEPS Background Information on Standardization of EPS*; EUMEPS Construction: Maaseik, Belgium, 2014.
148. Daigavane, P.B.; Dawande, G.M.; Gulhane, S.W.; Chaudhari, N.D. Improvement of B. C. Soil Using Geofoam Prefabricated Vertical Drain. In Proceedings of the Indian Geotechnical Conference GEOtrendz, Mumbai, India, 16–18 December 2010.

Article

Extraction and Identification of a Wide Range of Microplastic Polymers in Soil and Compost

Franja Prosenc [1,*], Pia Leban [2], Urška Šunta [1] and Mojca Bavcon Kralj [2]

1. Research Institute, Faculty of Health Sciences, University of Ljubljana, 1000 Ljubljana, Slovenia; urska.sunta@zf.uni-lj.si
2. Department for Sanitary Engineering, Faculty of Health Sciences, University of Ljubljana, 1000 Ljubljana, Slovenia; pia.leban@ijs.si (P.L.); mojca.kralj@zf.uni-lj.si (M.B.K.)
* Correspondence: franja.prosenc@zf.uni-lj.si

Abstract: Microplastic pollution is globally widespread; however, the presence of microplastics in soil systems is poorly understood, due to the complexity of soils and a lack of standardised extraction methods. Two commonly used extraction methods were optimised and compared for the extraction of low-density (polyethylene (PE)) and high-density microplastics (polyethylene (PET)), olive-oil-based extraction, and density separation with zinc chloride (ZnCl2). Comparable recoveries in a low-organic-matter matrix (soil; most >98%) were observed, but in a high-organic-matter matrix (compost), density separation yielded higher recoveries (98 ± 4% vs. 80 ± 11%). Density separation was further tested for the extraction of five microplastic polymers spiked at different concentrations. Recoveries were >93% for both soil and compost, with no differences between matrices and individual polymers. Reduction in levels of organic matter in compost was tested before and after extraction, as well as combined. Double oxidation (Fenton's reagent and 1 M NaOH) exhibited the highest reduction in organic matter. Extracted microplastic polymers were further identified via headspace solid-phase microextraction–gas chromatography–mass spectrometry (HS-SPME–GC-MS). This method has shown the potential for descriptive quantification of microplastic polymers. A linear relationship between the number of particles and the signal response was demonstrated for PET, polystyrene (PS), polyvinyl chloride (PVC), and PE ($R^2 > 0.98$ in alluvial soil, and $R^2 > 0.80$ in compost). The extraction and identification methods were demonstrated on an environmental sample of municipal biowaste compost, with the recovery of 36 ± 9 microplastic particles per 10 g of compost, and the detection of PS and PP.

Keywords: microplastic extraction; oil extraction; density separation; GC–MS; mass spectrometry identification; plastic polymers; polyethylene terephthalate; polyethylene; terrestrial

Citation: Prosenc, F.; Leban, P.; Šunta, U.; Bavcon Kralj, M. Extraction and Identification of a Wide Range of Microplastic Polymers in Soil and Compost. *Polymers* **2021**, *13*, 4069. https://doi.org/10.3390/polym13234069

Academic Editor: Jacopo La Nasa

Received: 28 October 2021
Accepted: 19 November 2021
Published: 23 November 2021

Publisher's Note: MDPI stays neutral with regard to jurisdictional claims in published maps and institutional affiliations.

Copyright: © 2021 by the authors. Licensee MDPI, Basel, Switzerland. This article is an open access article distributed under the terms and conditions of the Creative Commons Attribution (CC BY) license (https://creativecommons.org/licenses/by/4.0/).

1. Introduction

Microplastic (MP) pollution is widespread across all ecosystems, and has been widely studied in marine systems, whereas other environmental compartments—such as soil—have only started to emerge as a field of research in the past few years [1,2]. Soil is a very versatile and complex matrix with a broad range of organic matter content (from 0.02% in desert soils to up to 100% in bog soils) [2]. Specifically, the high organic matter content of soils hampers the extraction of MPs due their having similar densities to most common plastic polymers and, as such, poses a challenge in the determination of MPs in soils. Recently, standards for the collection and preparation of water samples for the identification and quantification of MP particles and fibres have been developed (ASTM D8332-20 and D8333-20) [3,4]. However, to date, there are no such standards for soil and soil-like samples, even though various methods for the detection of MPs in soil have been proposed.

MPs enter soil via the addition of soil amendments—such as biowaste compost and waste sewage sludge/biosolids—the use of agricultural plastics, irrigation with (reclaimed) wastewater, flooding, and atmospheric deposition, as well as littering and street runoff [2,5–7]. Studies reporting on primary sources of MPs in soil consider tire wear, fibres from synthetic clothing, artificial turf, and agricultural plastics (plastic mulch films, greenhouses, fruit protection foams, etc.) as notable sources [8–10]. Agricultural soils are particularly prone to accumulating MPs [11]. A study quantifying MPs in soil reported between 7100 and 42,960 plastic particles per kilogram of soil in southwestern China, 95% of which were <1 mm [12]. A study in northwestern China found 40 ± 126 light-density polyethylene (LDPE) MPs per kilogram of soil in the top 10 cm of agricultural soil, and 100 ± 141 LDPE MPs per kilogram of soil in the 10–30 cm layer [13]. In east China, a study found 40.2 ± 15.6 MPs per kilogram of unamended agricultural soil, while in soil amended with sludge, between 68.6 ± 21.5 and 149.2 ± 52.5 MPs per kilogram of soil were found [14]. In Germany, 0.34 ± 0.36 MPs were found per kilogram of dry weight of soil that had never been fertilised with MP-containing fertilisers [15]. Organic waste from households that is composted, or anaerobically digested with further composting of digestate, is usually applied to soil as a fertiliser. This is a common practice to return nutrients, trace elements, and humus to the soil; however, most municipal organic waste is contaminated with plastics—either with non-biodegradable plastic bags and/or other plastic items [16]. A study by Weithmann et al. reported between 20 and 24 MP particles per kilogram of dry weight biowaste compost, and between 14 and 146 MPs per kilogram of dry weight biowaste digestate [17].

MPs in soil have been shown to have deleterious effects on soil organisms; however, most studies were carried out at (currently) environmentally irrelevant concentrations. These high concentrations, however, represent future, presumably higher levels, as plastic production, use, and consequent pollution increase [18]. MPs have been shown to alter the microbial activity in soil and sediments, and to increase root and shoot mass [19,20], affect the mortality and growth rate of earthworms (*Lumbricus terrestris*) [21], affect the reproduction of nematodes (*Caenorhabditis elegans*) [22], and alter the immune response in crustaceans (*Porcellio scaber*) [23]. In addition to the biotic effects of MPs, they have also been shown to influence soils' biophysical properties, such as pH, content and size distribution of water-stable aggregates, water-holding capacity, soil bulk density, and saturated hydraulic conductivity [19,24–26].

To precisely evaluate the risks of MP pollution in agroecosystems and terrestrial environments, reliable methods for the identification and quantification of MPs are needed. Extraction of MP particles from these complex matrices is a paramount step in their analysis. To date, various methods for the extraction of MPs have been developed, including density separation, oil extraction, electrostatic separation, magnetic-field separation, solvent extraction, and circular separation [27–31]. Density separation using various saturated saline solutions—such as sodium iodide (NaI), sodium chloride (NaCl), sodium bromide ($NaBr_2$), calcium chloride ($CaCl_2$), zinc bromide ($ZnBr_2$), and zinc chloride ($ZnCl_2$)—is the most frequently applied extraction method for solid samples. In this method, MP particles are separated from the solid matrix based on the difference between the density of MPs and the density of the saturated saline solution. Various technical solutions have been used to separate MP-containing supernatants without disturbing the settled sediment. Most commonly, the supernatant is decanted onto a filter after a settling period, then withdrawn with a pipette or sucked with the aid of a pump [13,32]. Imhof et al. [33] devised the "Munich Plastic Sediment Separator"—a custom-made stainless steel apparatus that separates the settled sediment from the MP-containing supernatant with a valve. Konechnaya et al. [34] separated MP-containing supernatant from various sediments by slowly overflowing the samples with saline solution and collecting the MP-containing overflow. Recently, Grause et al. [35] proposed using a centrifuge to speed up the separation step and decanting the supernatant with MPs onto a filter. Many of these methods, however, were not tested on soils or organic-rich matrices, such as compost.

Oil-based separation has recently been developed, with promising applications. In this method, MPs are extracted based on the oleophilic properties of most plastic polymers, so that the MPs, when in contact with oil, move to the oil layer independent of their density. Water and a small volume of oil are added to the solid sample, stirred, and left to settle, and then the top oil layer, containing MPs, is separated from the rest of the sample. In several studies, a separation funnel was used to drain the sediment and aqueous fraction from the sample and, in this way, retain the MP-containing oil layer [27,36,37]. Several types of oil—such as canola, castor, and olive oils—were used for the separation of MPs from sediments, soil, and sludge. It has been reported, however, that separation funnels could easily become obstructed when extracting solid samples [36]. The same issue was encountered in our preliminary experiments. Scopetani et al. [38], therefore, designed a system to freeze the samples after the settling period and cut off the frozen oil layer.

After extraction, MPs can be identified by employing advanced instrumentation, using non-destructive (A) and destructive methods (B): (A) scanning electron microscopy (SEM) combined with spectroscopic techniques, such as micro-Fourier-transform infrared (µFTIR) and Raman spectroscopies; (B) thermogravimetric analysis (TGA), thermal desorption gas chromatography coupled with mass spectrometry (TED-GC–MS), and pyrolysis employing gas chromatography coupled with mass spectrometry (Py-GC–MS) [39–43]. Even though the well-established methods can offer enhanced identification of polymers, the field strives towards developing low-cost and simple techniques to facilitate analysis in most laboratories with basic analytical equipment. For this reason, solid-phase microextraction (SPME)—as a simple, inexpensive, and easy-to-handle technique—combined with gas chromatography coupled with mass spectrometry (GC–MS) analysis was used to develop an efficient method for the identification of MPs in environmental matrices, termed headspace solid-phase microextraction–gas chromatography–mass spectrometry (HS-SPME–GC–MS) [44,45].

This study aimed to determine the most suitable method for the extraction of various MP polymers from soil and compost, based on recovery, ease of handling, the potential for operator error, and time efficiency. Two methods—an olive-oil-based method, and density separation with saturated $ZnCl_2$ solution, never compared before—were evaluated for the extraction of MPs. Moreover, the effects of different vessels on MP recovery with an oil-based method were evaluated for the first time. Additionally, a simpler identification method based on HS-SPME–GC–MS is presented in this study as a viable alternative to traditionally used identification methods, and shows the potential for descriptive quantification of MPs. This method allowed for the simultaneous identification of various polymers in a mixture, and showed good linearity by increasing the number of spiked MPs in real matrices (both alluvial soil and compost), which is a significant advantage over the traditionally used methods.

2. Materials and Methods

2.1. Preparation of Microplastics

Two types of MP polymers were used in the development and optimisation of the extraction methods—polyethylene terephthalate (PET), as a polymer with high density (ρ = 1.33–1.48 g cm^{-3}), and low-density polyethylene (LDPE), as a polymer with low density (ρ = 0.91–0.94 g cm^{-3}) (Figure 1a). PET particles were prepared by cutting a plastic bottle (Radenska, Radenci, Slovenia) into <5 mm pieces. LDPE particles were prepared by melting plastic pellets (Tera Tolmin, Ltd., Tolmin, Slovenia), and then grating the melted plastic mass using a metal cheese grater and sieving the particles with a 2 mm sieve (Retsch, Dusseldorf, Germany).

After the method optimisation, the method was tested on five MP polymers—namely, polypropylene (PP), polyethylene (PE), polystyrene (PS), polyvinyl chloride (PVC), and PET (Table 1, Figure 1b). The same particles were then identified via HS-SPME–GC–MS. The MP particles were prepared with cryo-milling according to an adapted protocol [44]. In short, 5–10 g of plastic material (pellets or small pieces) was added to a stainless-steel grinding jar, together with a stainless steel 25 mm grinding ball. The jar was submerged in liquid nitrogen for 6 min, after which the grinding was carried out in a MillMix 20 ball mill (Tehtnica, Železniki, Slovenia) at between 25 and 30 Hz for 1.5 min, depending on the polymer. For PET and PE, the procedure of cooling and grinding was repeated more than once; the fine fraction was then sieved through 2, 1, and 0.5 mm sieves (Retsch, Germany) to obtain a fine fraction of 1–2 mm.

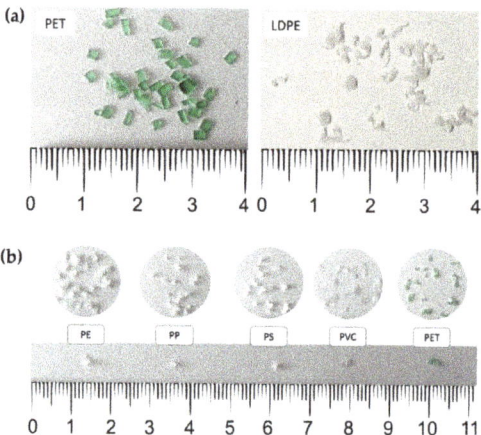

Figure 1. Microplastic particles prepared for (**a**) extraction method optimisation and (**b**) extraction of a wide range of microplastic polymers (PP: polypropylene; PE: polyethylene; PS: polystyrene; PVC: polyvinyl chloride; PET: polyethylene terephthalate).

Table 1. Plastic polymers used, their source, and their properties.

Polymer Type (1–2 mm)	Source	Density (g cm^{-3})	Melting Temperature (°C)
PP	Pellets (Golias Ltd., Ljubljana, Slovenia)	0.85–0.88	179
PE	Bottle cap (Radenska)	0.91–0.96	108–141
PS	Pellets (Golias Ltd.)	1.04–1.10	242–276
PVC	Blister pack	1.38–1.40	220–305
PET	Plastic bottle (Radenska)	1.33–1.48	264

PP: polypropylene; PE: polyethylene; PS: polystyrene; PVC: polyvinyl chloride; PET: polyethylene terephthalate.

2.2. Optimisation of Microplastic Extraction

Two methods for the extraction of MPs from soil were adapted from the literature and optimised: olive-oil-based extraction, and density separation with saturated ZnCl$_2$ solution [34,38]. Both methods were tested on two types of MP polymers: LDPE, representing MPs with low density; and PET, representing MPs with high density. In both methods, 10 and 20 MP particles of each composition were spiked into 10 g of alluvial soil or biowaste compost. Alluvial soil used in this experiment was collected on the banks of the Sava River basin (15°31′47.65″ E, 45°55′27.21″ N), in the immediate vicinity of agricultural fields, at a depth of 0–10 cm. The compost was from a biowaste composting plant that collects and processes separately collected organic waste (JP VOKA Snaga Ltd., Ljubljana, Slovenia), with a reported organic matter content of 24–44%.

2.2.1. Oil-Based Method

This method was carried out as described by Scopetani et al. [38], with some modifications, as shown in Figure 2a. As a vessel for the sample, three technical configurations were used: a polytetrafluoroethylene (PTFE or Teflon) cylinder (Dastaflon Ltd., Medvode, Slovenia), a glass cylinder (Promal, Logatec, Slovenia), and a modified plastic syringe with the hub removed to create an open, flat aperture (Soft-Ject®, Henke-Sass, Wolf GmbH, Tuttlingen, Germany). Ten grams of alluvial soil was spiked with either 10 or 20 LDPE or

PET MPs added to each sample vessel and closed on one end with a cap. Then, 30 mL of dH_2O was mixed with 3 mL of olive oil. Afterwards, the vessels were capped on the other end and shaken to homogenise the sample. Samples were left to stand for 2 h to sediment, and then frozen overnight at $-18\,°C$. After freezing, the frozen samples were pushed out of the vessels with a piston. The frozen oil layer containing MPs was removed, melted at room temperature, and filtered through a 47 mm GF/C filter (Whatman, Maidstone, UK). The MPs and remaining debris were rinsed with hexane (CARLO ERBA Reagents S.A.S., Val-de-Reuil, France) and dH_2O to remove any remaining oil. In the case of compost, the procedure was similar to that for alluvial soil, with slight modifications; the method was optimised using 10 PET MPs in a plastic syringe, and the oxidation step was added after extraction to reduce organic matter content. Oxidation was achieved by submerging the GF/C filter with the sample remaining on the filter in 60 mL of Fenton's reagent for 2 h under constant stirring [46]. The oxidised sample was again filtered through the GF/C filter. Extraction of MPs in each treatment (vessel, MP type, and concentration) was repeated six times. Recovery of MPs was calculated using the following equation:

$$R\,(\%) = \frac{N_{extracted}}{N_{spiked}} \times 100 \tag{1}$$

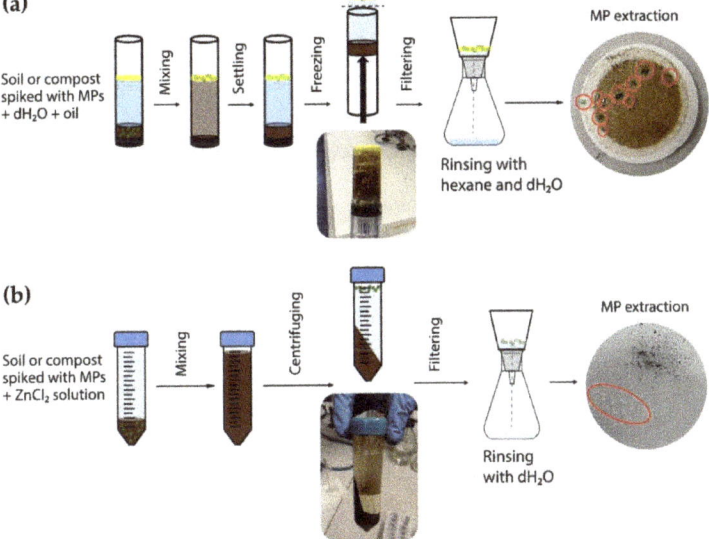

Figure 2. Schematic representation of microplastic extraction methods: (**a**) oil-based method, and (**b**) density separation.

2.2.2. Density Separation

This method was carried out as described by Konechnaya et al. [34], with some modifications (Figure 2b). Saturated $ZnCl_2$ solution with a density of 1.6 g cm^{-3} was prepared by dissolving 1 kg of $ZnCl_2$ (anhydrous RE, CARLO ERBA Reagents S.A.S., Val-de-Reuil, France) in 751.4 mL of dH_2O. The solution was adjusted to a pH of 3 with 5 M KOH, and then filtered through a 47 mm GF/C filter (Whatman, Maidstone, UK). Then, 10 g of alluvial soil was spiked with either 10 or 20 LDPE or PET MPs, and added to a 50 mL centrifuge tube. $ZnCl_2$ solution was added to the 50 mL mark in the tube, and the sample was shaken for 30 s. The separation was carried out by centrifuging the sample at 9000 rpm for 15 min (Hettich Universal 320 centrifuge, Andreas Hettich GmbH & Co. KG, Tuttlingen, Germany). The supernatant, together with floated MP particles, was filtered through a

GF/C filter. The filtrate—used ZnCl₂ solution—was reused up to 20 times, since its density did not change. In the case of compost, the procedure was similar to that for alluvial soil, with slight modifications; the method was optimised using 10 PET MPs, and an oxidation step was added after extraction to reduce organic matter content. Oxidation was achieved by submerging the GF/C filter with the sample remaining on the filter in 60 mL of Fenton's reagent for 2 h under constant stirring [46]. The oxidised sample was again filtered through the GF/C filter. Extraction of MPs in each treatment (MP type and concentration; matrix) was repeated six times. Recovery of MPs was calculated using Equation (1).

2.3. Extraction of a Wide Range of Microplastic Polymers

The more optimal of the two extraction methods (i.e., better recovery, easier to handle, less time consuming) was tested on MPs of five different polymer compositions—namely, PE, PP, PS, PVC, and PET (Table 1)—spiked into alluvial soil and compost. MPs were spiked into 10 g of matrix in different concentrations—namely, 10 MPs (2 particles of each polymer composition), 25 MPs (5 particles of each composition), and 50 MPs (10 particles of each composition). The extraction procedure was carried out as described in Section 2.2.2. Treatment of compost samples included double oxidation—one before extraction, and another after extraction. Before extraction, the whole sample (10 g of compost with MPs) was added to 60 mL of Fenton's reagent and left for 2 h under constant stirring. Afterwards, the sample was filtered through the GF/C filter, and the sample remaining on the filter was oxidised again, this time using 50 mL of 1 M NaOH under constant stirring overnight at 50 °C. After the oxidation, the sample was filtered through a 100 µm stainless steel mesh (Fipis, Ribnica, Slovenia). The experiment was conducted in four replicates. Recoveries were calculated using Equation (1), and identification of cumulative extracted MPs was done using the HS-SPME–GC–MS method, as described in Section 2.4.

2.4. Identification of Microplastic Polymers Using Headspace Solid-Phase Microextraction with GC–MS

MP polymers from alluvial soil and compost samples were identified via the HS-SPME–GC–MS method described by Šunta et al. [44], with some modifications. The HS-SPME–GC–MS method is based on the adsorption of volatile compounds emitted during melting of plastic particles onto SPME fibre, followed by GC–MS analysis and detection of characteristic fragment ions for individual polymers (PET—m/z 163; PS—m/z 104; PVC—m/z 91; PP—m/z 142; and PE—m/z 85). Due to observed difficulties in the melting of PET in polymer mixtures, the method was optimised in the steps of sample preparation and thermal decomposition. In the sample preparation step, PET MPs were separated from other MPs and analysed separately. In the thermal decomposition step, headspace vials were placed in a sand bath with a temperature of 260 °C (probe thermometer, Amarell GmbH & Co. KG, Kreuzwertheim, Germany) for 15 min and 3 min for melting of PET MPs and other MP polymers, respectively. With this improved protocol, identification of MP polymers extracted from spiked samples of alluvial soil and compost was carried out in cumulative samples composed of three replicates. A linear relationship between the number of MP particles and the signal response, with the area under the chromatographic peak being the measure, was determined using the coefficient of determination (R^2).

3. Results and Discussion

3.1. Comparison of Microplastic Extraction Methods

The two most widely reported methods for the extraction of MPs from soil and soil-like matrices—oil-based extraction, and density separation—were optimised for MP polymers of high (PET) and low density (LDPE) and compared. As per the recommendations of Scopetani et al. [38], olive oil was used for the separation of oleophilic MPs from the matrix in the oil-based method. Olive oil supposedly has the strongest affinity for a wide range of plastic polymers [38]. Different solutions to separate MPs from the matrix with the oil-based method are reported in the literature—from a custom-made PTFE cylinder with a piston, to separation funnels [27,37,38]. Separation funnels were tested, and were found to be impractical for the extraction of MPs from soil and compost, due to frequent clogging. This problem was also encountered by other researchers [36]. Three vessels were tested for the oil-based method: a PTFE cylinder, similar to the one used by Scopetani et al. [38]; a glass cylinder; and a modified plastic syringe. The glass cylinder was used as a cheaper alternative to the PTFE cylinder (i.e., EUR 15 vs. EUR 48 per piece), while the plastic syringe is a cheap and widely available laboratory consumable. Recoveries obtained in alluvial soil for all three vessels were high (>97%), and comparable for all spiked MPs (10 and 20 PET MPs, as well as 10 and 20 LDPE MPs), Table 2. The glass cylinder, however, was found to break frequently—not in the stages of freezing the sample, but rather when pushing the sample out after the freezing process. In compost, the recovery obtained for 10 PET MPs in a modified syringe was lower, at only 80%. It should be noted that in compost, only marginal conditions were tested—that is, using the lowest quantity of spiked MPs (10 MPs per 10 g of matrix) as well as the MPs with the highest density (PET), which are usually more challenging to separate. The recoveries achieved in this study for high-density MPs in soil were comparable to those obtained by Scopetani et al. [38], who used the extraction system most similar to ours (98% vs. 95%, respectively); for low-density MPs, however, the recoveries achieved were higher in this study (98% vs. 90%). In compost, Scopetani et al. [38] observed lower recovery of low-density MPs (PE and polyurethane (PU)) (80%), although the recovery of medium- and high-density MPs did not seem to be affected by the matrix. In this study, only high-density MPs (PET) were tested in the compost matrix, and their recovery was lower as compared to recovery in soil (80% vs. 98%, respectively). Other studies reporting on the use of the oil-based method used different solutions for extraction systems, e.g., separation funnels. Mani et al. [27] used separation funnels and castor oil, and achieved an average recovery of 99% of four polymers spiked into fluvial suspended surface solids, marine suspended surface solids, marine beach sediments, and agricultural soil; the matrix did not significantly influence MP recovery; however, the tested matrices were low in organic matter content and, therefore, less challenging. Crichton et al. [37] mixed samples with water and oil in an Erlenmeyer flask, and transferred the liquid fraction into a separation funnel during the sedimentation period. They achieved an average recovery of 96% of five polymers spiked into sediment beach samples.

Table 2. Comparison of recoveries of PET and LDPE microplastic particles, achieved in two extraction methods: oil-based extraction in three different vessels, and density separation. $n = 6$; results are presented as mean ± standard deviation (SD).

Type of Soil	No. and Type of MPs	Oil-Based Method (%)			Density Separation (%)
		PTFE Cylinder	Glass Cylinder	Plastic Syringe	
Alluvial soil	10 PET	96.7 ± 8.2	100.0 ± 0.0	96.7 ± 5.2	93.3 ± 5.2
	20 PET	97.5 ± 2.7	97.5 ± 2.7	97.5 ± 4.2	99.2 ± 2.0
	10 LDPE	98.3 ± 4.1	98.3 ± 4.1	98.3 ± 4.1	98.3 ± 7.5
	20 LDPE	96.7 ± 8.2	96.7 ± 4.1	97.5 ± 2.7	100.0 ± 0.0
Compost	10 PET	N.A.[1]	N.A.[1]	80.0 ± 11.0	98.3 ± 4.1

[1] Not applicable.

The oil-based method did not prove to be straightforward in this experiment. MPs were often found to be stuck to the inner walls of vessels, as well as to the surface of the frozen samples. This was due to the formation of a thin layer of oil on the walls of the vessels, which the MPs were attracted to. To combat this, the inner walls were rinsed with hexane; after the sample was removed, the surface of the frozen sample was scraped, and these fractions were further processed to isolate MPs. This method was also time-consuming (2 h for settling, overnight for freezing, and up to 45 min for melting the cut-off sample containing MPs for further processing, e.g., filtration).

The second method tested was density separation with saturated $ZnCl_2$ solution. This method followed previously published protocols, with further optimisation [32–34]. Sedimentation (usually overnight) was sped up with centrifugation (15 min). This significantly improved time efficiency, and improved decanting of the supernatant due to more compact sediment. Recoveries obtained in alluvial soil and compost were high (>93%, most >98%, respectively), and were comparable for all spiked MPs (different compositions and quantities) (Table 2). Additionally, the $ZnCl_2$ solution was reused up to 20 times without losing the desired density, which significantly reduced the cost and the environmental footprint of the method. Density separation achieved better recoveries in compost as compared to the oil-based method (98% vs. 80%), with an oxidation step after filtration in both methods (see Sections 2.2.1 and 2.2.2). For this method, oxidation with Fenton's reagent before extraction was also tested, and the recovery obtained was 100% ($n = 6$). In the study by Konechnaya et al. [34], the authors used $ZnCl_2$ solution and separated MPs from sandy matrices by overflow, and the obtained recoveries were comparable to those obtained in this study in alluvial soil (between 94 and 104% vs. 98%, respectively). Very recently, Grause et al. [35] used a similar approach to the one used in this study—density separation sped up via centrifugation using $CaCl_2$ solution to extract MPs from agricultural soil. They optimised the centrifugation protocol on PET MPs spiked into soil, and obtained 95% recoveries, which is consistent with the results of our study; their method, however, was not tested on samples rich in organic matter; likewise, $CaCl_2$ solution with a density of 1.4 g cm^{-3} could be unsuitable for the extraction of some MP polymers, e.g., some PET, chlorinated polyvinyl chloride (CPVC), polyvinylidene chloride (PVDC), etc. [35].

In addition to better recoveries in compost, the significant advantage of the density separation method over the oil-based method was in better time efficiency (approximately 30 min to obtain MPs on the filter vs. approximately 1 day, respectively), as well as in the ease of handling. The potential for scaling up—e.g., analysing multiple-kilogram samples—could be feasible with the use of a large-volume centrifuge. For this reason, density separation was further used in testing the method on a wide range of plastic polymers at various concentrations, as described in Section 3.2.

3.2. Extraction of a Wide Range of Microplastic Polymers

The extraction method that proved to be more optimal—density separation—was used for the extraction of MP particles of different polymer compositions (Table 1) spiked into alluvial soil and compost, at various concentrations. The five chosen MP polymers (PE, PP, PET, PVC, and PS) were chosen based on their frequent occurrence in the environment [15,17,47]. Cumulative recoveries were high (>93%) for all spiked MP polymers, especially at spiking quantities of 25 and 50 MPs (>98%) per 10 g of matrix (Figure 3c). Moreover, there were no significant differences in the recoveries of individual MP polymers from soil (Figure 3a) and compost (Figure 3b), nor were there any significant differences in recoveries from the two matrices. Blank samples of non-spiked alluvial soil were treated in the same way as the spiked samples to check for potential cross-contamination from aerial deposition, clothing, and/or lab equipment. Blank alluvial soil resulted in zero recovered MPs ($n = 3$). Blank compost samples were also treated in the same way; however, compost originating from the industrially composted organic fraction of household waste inherently contains plastic contamination from plastic bags and other sources [17]. A total of 32 ± 2 native MPs ($n = 3$) were recovered from the compost blank sample; however, their authenticity was not additionally checked with identification methods or the hot needle test. Native MPs did not hinder the determination of recovery of spiked MPs, since the spiked MPs were easily visually distinguishable. This method was additionally tested by inexperienced operators—primary school pupils, 14 years of age, with no prior laboratory experience; the recovery achieved ranged from 75% to 95%, which shows that the method is simple for handling.

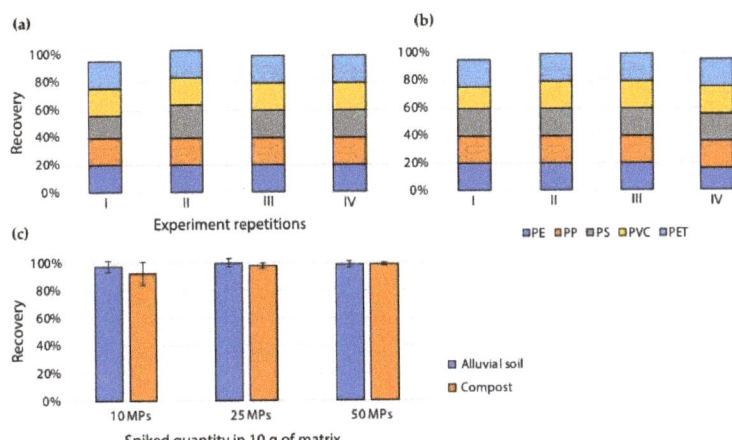

Figure 3. Recovery of MPs (polymer types: PE, PP, PET, PVC, and PS) spiked into alluvial soil and compost, extracted via the density separation method. Recovery of individual polymer types from (**a**) alluvial soil and (**b**) compost, spiked at a concentration of 5 MPs of each polymer type. (**c**) Cumulative recovery of all MPs, spiked at different concentrations. $n = 4$; results are presented as mean ± SD.

There are numerous reports on using density separation to extract various MPs from river and sea sediments; however, fewer studies have dealt with extraction from agricultural soil and organic-rich matrices, such as compost, which are considerably more difficult to process. Hurley et al. [46] tested a density separation protocol with NaI solution and an additional oxidation step on soil and sludge samples spiked with PE microbeads and PET fibres. Recoveries obtained in the reported study ranged between 92% and 100% for PE microbeads, and between 79% and 86% for PET fibres. The recoveries of PE microbeads resembled the recoveries obtained in this study (>92% vs. >93%, respectively); however, PET fibres were recovered to a lesser extent than the PET particles used in this study (>79% vs. 100%, respectively). Hurley et al. also tested the effect of an additional oxidation step on recoveries, and found that oxidation with Fenton's reagent after MP extraction led to lower recoveries than oxidation before MP extraction, but this result was not statistically significant. No such effect was observed in this study; instead, we observed a difference in the quantity of organic matter removed (Figure 4). Liu et al. [31] devised a circulation separation device and tested various saline solutions to extract a wide range of MP polymers (polyamide (PA), polycarbonate (PC), PP, acrylonitrile butadiene styrene (ABS), PE, PS, poly(methyl methacrylate) (PMMA), polyoxymethylene (POM), PET, and PVC) from soil; with the use of NaBr and $CaCl_2$, they achieved recoveries from 95% to 100%, similar to the recoveries reported herein. Similarly, Li et al. [48] constructed a circulation system for the extraction of MPs from soil via density separation using NaBr solution; the system was tested on common MP polymers (LDPE, PS, PP, and PVC) as well as biodegradable MPs (polybutylene succinate (PBS), poly(adipic acid), butylene terephthalate (PBAT), and polylactic acid (PLA)); recovery rates ranged from 92% to 99.6%, resembling the recoveries reported herein. Grause et al. [35], using density separation with centrifugation and $CaCl_2$ solution, achieved recoveries between 97% and 98% for LDPE, PP, PS, and flexible PVC MPs, comparable to the recoveries obtained in this study; PET had lower recoveries as compared to the other tested MPs, at around 95%, whereas in this study no differences in recoveries of the five tested MP polymers were observed.

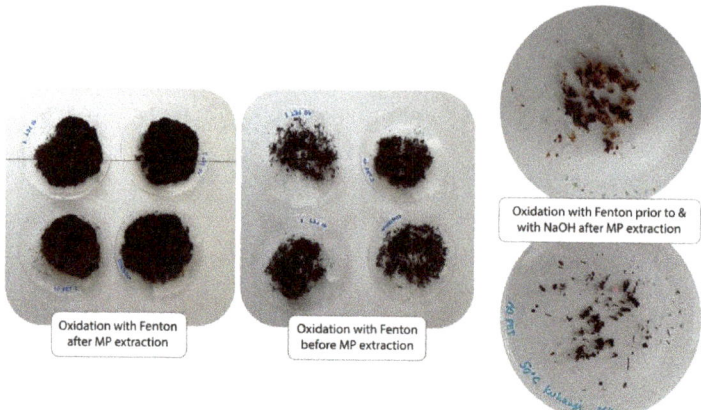

Figure 4. Reduction in organic content in compost samples with oxidation at various stages of the MP extraction procedure. The initial amount of all samples was 10 g of compost.

An oxidation step was employed for all compost samples in order to reduce the organic matter content and facilitate the identification of MP polymers. Oxidation was implemented at various stages of the MP extraction process to compare the efficiency of organic matter reduction. Oxidation with Fenton's reagent after MP extraction reduced the organic matter content to a lesser extent compared to oxidation with Fenton's reagent before MP extraction, as shown in Figure 4. Additionally, double oxidation was trialled—with Fenton's reagent before extraction, and 1 M NaOH after extraction. Double oxidation reduced organic matter content to the largest extent (Figure 4), and was therefore used in processing all compost samples used in the recovery experiments. Hurley et al. [46] tested the impacts of different oxidation protocols on MP integrity; Fenton's reagent proved to reduce organic matter to the highest degree (106%) without damaging the MPs; alkaline digestion with 1 M NaOH, on the other hand, provided a lesser reduction in organic matter (68%), but it also did not cause significant changes in the mass or size of MP particles [46].

3.3. Identification of Microplastic Polymers in Compost and Alluvial Soil

Identification of MPs extracted from spiked alluvial soil and compost samples using density separation with $ZnCl_2$ solution was possible via the HS-SPME–GC–MS method. As described in Section 2.4, the proposed method by Šunta et al. [44] was optimised for improved identification of PET. To improve the melting of PET, a thermal decomposition step was carried out in a sand bath, and PET particles were analysed separately from other MP polymers. PS, PVC, and PP in alluvial soil and compost samples were identified with previously proposed characteristic compounds for the identification of MPs after their thermal decomposition: styrene and dimer, chlorooctane, and 4,6-dimethyl 2-heptanone, respectively [39,41,44,49]. In the chromatogram of analysed MP mixtures (PS, PVC, PP, and PE), pentadecane—a characteristic compound proposed for the identification of PE—was not observed; however, the series of higher alkanes (from tetradecane, C_{14}, up to eicosane, C_{20}) was observed, and the presence of PE was confirmed. PET extracted from compost samples was identified using the previously proposed characteristic compound—dimethyl terephthalate (DMTP) [41,44,49]—while DMTP from analysed PET particles extracted from alluvial soil was not determined, and ethyl methyl terephthalate (signal 1a, Figure 5) at a retention time of 16.994 min was used instead. Ethyl methyl terephthalate was chosen in the case of alluvial soil samples because it is one of the terephthalate derivatives that are produced upon thermal decomposition of PET, according to Yakovenko et al. [50].

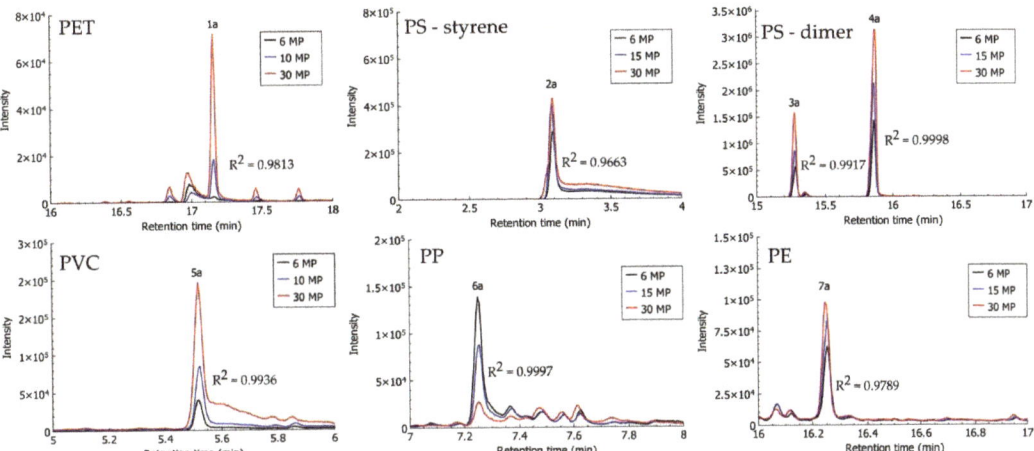

Figure 5. Headspace solid-phase microextraction–gas chromatography–mass spectrometry (HS-SPME–GC–MS) chromatograms of identification compounds from MP particles extracted from alluvial soil samples, spiked with three concentrations (2 MPs, 5 MPs, and 10 MPs) of each polymer type (PET: dimethyl terephthalate (1a); PS: styrene (2a) and dimer trans(cis)-1,2-diphenylcyclobutane (3a and 4a); PVC: chlorooctane (5a); PP: 4,6-dimethyl 2-heptanone (6a); and PE: eicosane (7a)).

After optimisation of the HS-SPME–GC–MS method for the identification and analysis of spiked samples of soil with different MP contents, the potential of this method for the descriptive quantification of MPs was observed. In a detailed analysis of chromatographic peaks of characteristic compounds for the identification of MPs, a relationship between the number of analysed particles and the signal response (determined area under the chromatographic peak) was observed (Figures 5 and 6). Cumulative samples of three replicates were analysed (6, 15, and 30 MPs).

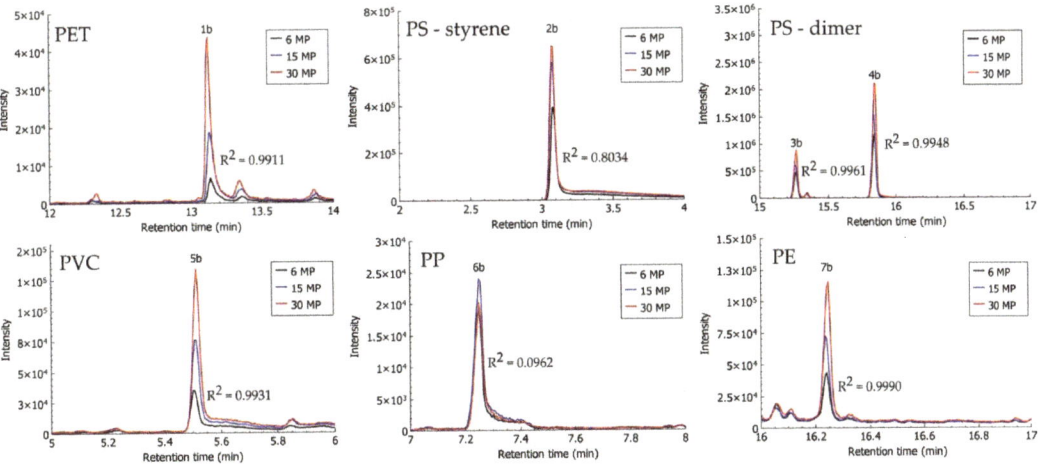

Figure 6. Headspace solid-phase microextraction–gas chromatography–mass spectrometry (HS-SPME–GC–MS) chromatograms of identification compounds from MP particles extracted from compost samples, spiked with three concentrations (2 MPs, 5 MPs, and 10 MPs) of each polymer type (PET: ethyl methyl terephthalate (1b); PS: styrene (2b) and dimer trans(cis)-1,2-diphenylcyclobutane (3b and 4b); PVC: chlorooctane (5b); PP: 4,6-dimethyl 2-heptanone (6b); and PE: eicosane (7b)).

In the case of alluvial soil samples, a linear relationship between the number of particles and the signal response with $R^2 > 0.97$ was observed for ethyl methyl terephthalate (PET), styrene and the dimer—trans(cis)-1,2-diphenylcyclobutane (PS), chlorooctane (PVC), and eicosane (PE) (Figure 5). For 4,6-dimethyl 2-heptanone (PP), the linear coefficient was very high ($R^2 = 0.9997$); nevertheless, the signal response (signal 6a in Figure 5)—or area under the chromatographic peak—was not increasing with the number of MPs, as would be expected, but decreasing.

A high linear relationship ($R^2 > 0.99$) between the signals of characteristic compounds of individual polymer types of MPs extracted from compost samples on the one hand, and the number of MPs on the other, was also observed in the case of DMTP (PET), trans(cis)-1,2-diphenylcyclobutane (PS), chlorooctane (PVC), and eicosane (PE) (Figure 6). There was a low linear relationship observed for styrene (PS) with $R^2 = 0.8034$, and none for 4,6-dimethyl 2-heptanone (PP).

Observed linearity (above $R^2 > 0.98$ in alluvial soil, and above $R^2 > 0.80$ in compost) in a controlled laboratory environment with established particle size, number, average particle mass, and polymer type demonstrated that the HS-SPME–GC–MS method could be used for tentative descriptive quantification of PET, PS, PVC, and PE in analysed matrices. However, it is a prerequisite that samples are prepared for the identification of MPs using pretreatment methods for the elimination of organic matter and extraction of MP polymers from the matrix. Quantitative measurements with the SPME headspace extraction technique depend on the type of SPME coating, the partitioning coefficient between the fibre coating and the analytes, the chemical properties of the analytes, the complexity of the sample, and temperature, while particle size, mass, and chemical structure (additives) also need to be considered in the identification of MPs [51,52].

The nonlinearity of 4,6-dimethyl 2-heptanone (PP) could be attributed to the possible interactions in the headspace vial between volatile substances emitted from MP polymers during the thermal decomposition step of the identification method. As the best signal response can be observed at the lowest number of analysed PP particles, it is also possible that competition for the active adsorption sites on the SPME fibre occurred between 4,6-dimethyl 2-heptanone and other volatile substances, as was also observed by Dutra et al. [53] and Demets et al. [54]. Therefore, future studies on the quantification of MPs in environmental matrices using HS-SPME–GC–MS would be of interest.

3.4. Applicability of the Developed Methods to Environmental Samples

Density separation was applied to real compost samples from municipal organic waste to extract natively present MPs. The procedure was carried out as described in Section 2.3. From 10 g of compost, 36 ± 9 MPs were extracted, with an average mass of 6.3 ± 1.8 mg ($n = 3$). Extracted particles were placed in vials and identified via HS-SPME–GC–MS as described in Section 2.4. The presence of two polymers was determined: PS and PP, the latter being tentatively identified using methylated higher alkanes in the absence of the characteristic compound for PP, i.e., 4,6-dimethyl 2-heptanone. It should be noted, however, that the identification method has currently only been developed for the five aforementioned plastic polymers (PET, PVC, PE, PS, and PE), while the determination of characteristic compounds for other plastics is underway. Hence, the detection of other potentially present MP polymers was not possible.

Biowaste compost is frequently contaminated with MPs, originating from incompletely removed plastic bags, plastic packaging of food, incompletely degraded biodegradable plastics, and other sources [16]. A study quantifying MPs in organic fertilisers found between 20 and 24 MPs (>1 mm) per kg of compost [17], which is 150–180 times less than the amount isolated in this study; the employed method for MP extraction, however, was different (wet sieving), which could, in part, explain the differences in MP numbers. Other studies reported up to 2800 MPs per kg of compost, which more closely aligns with the abundance reported herein (2900–4600 MPs per kg) [55].

The strengths of the methods reported in this study include high MP recoveries from soil and compost via the density separation method, as well as the methods being easy to handle and widely accessible. The identification method has the advantage of simultaneous multi-polymer analysis and the use of common analytical equipment. There are, however, some drawbacks: Neither method was optimised for large sample processing (e.g., >1 kg). Additionally, the HS-SPME–GC–MS method is not yet developed for the identification of other plastic polymers, e.g., PA and biodegradable plastic polymers. Furthermore, the presence of organic matter in the MP isolate hinders polymer identification, as the limit of detection (LOD) rises with increasing organic matter content; therefore, efforts to significantly reduce organic matter should be made prior to sample analysis.

4. Conclusions

Two methods for the extraction of MP particles were tested and optimised: oil-based extraction, and density separation. Both methods achieved high recoveries of spiked MP polymers of low and high density in alluvial soil; however, in compost, density separation with $ZnCl_2$ solution proved to have better recoveries. In addition, considering other factors—such as time efficiency, ease of handling, and the potential for scaling up—density separation was shown to be the better option. Additionally, oxidation before and after extraction was necessary in order to considerably reduce the organic matter content of compost samples. Identification of MPs via the HS-SPME–GC–MS method was carried out successfully via the detection of characteristic compounds for each polymer type. The potential to use this method for descriptive quantification was proven by a linear relationship between the number of particles and the signal response on chromatograms in the case of PET, PS, PVC, and PE. Density separation was employed to isolate natively present MPs from municipal biowaste compost (36 ± 9 MPs per 10 g). The presence of PS and PP was confirmed with HS-SPME–GC–MS.

This study contributes towards the understanding of the suitability of available extraction methods for soil and compost samples, and additionally presents the possibility of using commonly available analytical equipment for MP identification.

Author Contributions: Conceptualisation, F.P. and M.B.K.; methodology, U.Š.; formal analysis, P.L.; investigation, P.L.; resources, U.Š. and M.B.K.; data curation, P.L.; writing—original draft preparation, F.P., P.L., U.Š., and M.B.K.; writing—review and editing, F.P., U.Š., and M.B.K.; visualisation, F.P., P.L., and U.Š.; supervision, F.P. and M.B.K.; project administration, F.P.; funding acquisition, F.P. and M.B.K. All authors have read and agreed to the published version of the manuscript.

Funding: This research was funded by the Slovenian Research Agency, grant number Z2-2643 (Agri-iMPact: Microplastic Prevalence and Impact in Agricultural Fields), grant number SP-0510/21 (Postgraduate Research Funding Programme: Young Researchers), and research core funding number P3-0388 (Mechanisms of Health Maintenance). The APC was funded by the Slovenian Research Agency, grant number Z2-2643.

Institutional Review Board Statement: Not applicable.

Informed Consent Statement: Not applicable.

Data Availability Statement: The data presented in this study are available in the Zenodo digital repository at https://doi.org/10.5281/zenodo.5607477 (accessed on 21 November 2021).

Acknowledgments: The authors acknowledge the Research Nature Institute, Ljubljana, Slovenia for analytical support.

Conflicts of Interest: The authors declare no conflict of interest. The funders had no role in the design of the study, in the collection, analyses, or interpretation of data, in the writing of the manuscript, or in the decision to publish the results.

References

1. Möller, J.N.; Löder, M.G.J.; Laforsch, C. Finding Microplastics in Soils: A Review of Analytical Methods. *Environ. Sci. Technol.* **2020**, *54*, 2078–2090. [CrossRef] [PubMed]
2. Bläsing, M.; Amelung, W. Plastics in soil: Analytical methods and possible sources. *Sci. Total Environ.* **2018**, *612*, 422–435. [CrossRef] [PubMed]
3. ASTM D8332—20 Standard Practice for Collection of Water Samples with High, Medium, or Low Suspended Solids for Identification and Quantification of Microplastic Particles and Fibers; ASTM International: West Conshohocken, PA, USA, 2020.
4. ASTM D8333—20 Standard Practice for Preparation of Water Samples with High, Medium, or Low Suspended Solids for Identification and Quantification of Microplastic Particles and Fibers Using Raman Spectroscopy, IR Spectroscopy, or Pyrolysis-GC/MS; ASTM International: West Conshohocken, PA, USA, 2020.
5. de Souza Machado, A.A.; Lau, C.W.; Till, J.; Kloas, W.; Lehmann, A.; Becker, R.; Rillig, M.C. Impacts of Microplastics on the Soil Biophysical Environment. *Environ. Sci. Technol.* **2018**, *52*, 9656–9665. [CrossRef]
6. Brahney, J.; Hallerud, M.; Heim, E.; Hahnenberger, M.; Sukumaran, S. Plastic rain in protected areas of the United States. *Science* **2020**, *368*, 1257–1260. [CrossRef]
7. Rillig, M.C. Microplastic in terrestrial ecosystems and the soil? *Environ. Sci. Technol.* **2012**, *46*, 6453–6454. [CrossRef] [PubMed]
8. Jan Kole, P.; Löhr, A.J.; Van Belleghem, F.G.A.J.; Ragas, A.M.J. Wear and tear of tyres: A stealthy source of microplastics in the environment. *Int. J. Environ. Res. Public Health* **2017**, *14*, 1265. [CrossRef]
9. Zhou, Y.; Wang, J.; Zou, M.; Jia, Z.; Zhou, S.; Li, Y. Microplastics in soils: A review of methods, occurrence, fate, transport, ecological and environmental risks. *Sci. Total Environ.* **2020**, *748*, 141368. [CrossRef]
10. Fakour, H.; Lo, S.L.; Yoashi, N.T.; Massao, A.M.; Lema, N.N.; Mkhontfo, F.B.; Jomalema, P.C.; Jumanne, N.S.; Mbuya, B.H.; Mtweve, J.T.; et al. Quantification and Analysis of Microplastics in Farmland Soils: Characterization, Sources, and Pathways. *Agriculture* **2021**, *11*, 330. [CrossRef]
11. Nizzetto, L.; Langaas, S.; Futter, M. Pollution: Do microplastics spill on to farm soils? *Nature* **2016**, *537*, 488. [CrossRef] [PubMed]
12. Zhang, G.S.; Liu, Y.F. The distribution of microplastics in soil aggregate fractions in southwestern China. *Sci. Total Environ.* **2018**, *642*, 12–20. [CrossRef] [PubMed]
13. Zhang, S.; Yang, X.; Gertsen, H.; Peters, P.; Salánki, T.; Geissen, V. A simple method for the extraction and identification of light density microplastics from soil. *Sci. Total Environ.* **2018**, *616–617*, 1056–1065. [CrossRef] [PubMed]
14. Yang, J.; Li, L.; Li, R.; Xu, L.; Shen, Y.; Li, S.; Tu, C.; Wu, L.; Christie, P.; Luo, Y. Microplastics in an agricultural soil following repeated application of three types of sewage sludge: A field study. *Environ. Pollut.* **2021**, *289*, 117943. [CrossRef]
15. Piehl, S.; Leibner, A.; Löder, M.G.J.; Dris, R.; Bogner, C.; Laforsch, C. Identification and quantification of macro- and microplastics on an agricultural farmland. *Sci. Rep.* **2018**, *8*, 17950. [CrossRef]
16. Stubenrauch, J.; Ekardt, F. Plastic Pollution in Soils: Governance Approaches to Foster Soil Health and Closed Nutrient Cycles. *Environments* **2020**, *7*, 38. [CrossRef]
17. Weithmann, N.; Möller, J.N.; Löder, M.G.J.; Piehl, S.; Laforsch, C.; Freitag, R. Organic fertilizer as a vehicle for the entry of microplastic into the environment. *Sci. Adv.* **2018**, *4*, eaap8060. [CrossRef] [PubMed]
18. Rillig, M.C.; Leifheit, E.; Lehmann, J. Microplastic effects on carbon cycling processes in soils. *PLOS Biol.* **2021**, *19*, e3001130. [CrossRef] [PubMed]
19. Lozano, Y.M.; Lehnert, T.; Linck, L.T.; Lehmann, A.; Rillig, M.C. Microplastic Shape, Polymer Type, and Concentration Affect Soil Properties and Plant Biomass. *Front. Plant Sci.* **2021**, *12*, 169. [CrossRef] [PubMed]
20. Seeley, M.E.; Song, B.; Passie, R.; Hale, R.C. Microplastics affect sedimentary microbial communities and nitrogen cycling. *Nat. Commun.* **2020**, *11*, 2372. [CrossRef]
21. Huerta Lwanga, E.; Gertsen, H.; Gooren, H.; Peters, P.; Salánki, T.; van der Ploeg, M.; Besseling, E.; Koelmans, A.A.; Geissen, V. Microplastics in the Terrestrial Ecosystem: Implications for *Lumbricus terrestris* (Oligochaeta, Lumbricidae). *Environ. Sci. Technol.* **2016**, *50*, 2685–2691. [CrossRef]
22. Lei, Y.; Wu, S.; Lu, S.; Liu, M.; Song, Y.; Fu, Z.; Shi, H.; Raley-Susman, K.M.; He, D. Microplastic particles cause intestinal damage and other adverse effects in zebrafish Danio rerio and nematode Caenorhabditis elegans. *Sci. Total Environ.* **2018**, *619–620*, 1–8. [CrossRef]
23. Kokalj, A.J.; Dolar, A.; Titova, J.; Visnapuu, M.; Škrlep, L.; Drobne, D.; Vija, H.; Kisand, V.; Heinlaan, M. Long Term Exposure to Virgin and Recycled LDPE Microplastics Induced Minor Effects in the Freshwater and Terrestrial Crustaceans Daphnia magna and Porcellio scaber. *Polymers* **2021**, *13*, 771. [CrossRef] [PubMed]
24. de Souza Machado, A.A.; Lau, C.W.; Kloas, W.; Bergmann, J.; Bachelier, J.B.; Faltin, E.; Becker, R.; Görlich, A.S.; Rillig, M.C. Microplastics Can Change Soil Properties and Affect Plant Performance. *Environ. Sci. Technol.* **2019**, *53*, 6044–6052. [CrossRef] [PubMed]
25. Zhang, G.S.; Zhang, F.X.; Li, X.T. Effects of polyester microfibers on soil physical properties: Perception from a field and a pot experiment. *Sci. Total Environ.* **2019**, *670*, 1–7. [CrossRef]
26. Boots, B.; Russell, C.W.; Green, D.S. Effects of Microplastics in Soil Ecosystems: Above and Below Ground. *Environ. Sci. Technol.* **2019**, *53*, 11496–11506. [CrossRef] [PubMed]
27. Mani, T.; Frehland, S.; Kalberer, A.; Burkhardt-Holm, P. Using castor oil to separate microplastics from four different environmental matrices. *Anal. Methods* **2019**, *11*, 1788–1794. [CrossRef]

28. Felsing, S.; Kochleus, C.; Buchinger, S.; Brennholt, N.; Stock, F.; Reifferscheid, G. A new approach in separating microplastics from environmental samples based on their electrostatic behavior. *Environ. Pollut.* **2018**, *234*, 20–28. [CrossRef] [PubMed]
29. Rhein, F.; Scholl, F.; Nirschl, H. Magnetic seeded filtration for the separation of fine polymer particles from dilute suspensions: Microplastics. *Chem. Eng. Sci.* **2019**, *207*, 1278–1287. [CrossRef]
30. La Nasa, J.; Biale, G.; Mattonai, M.; Modugno, F. Microwave-assisted solvent extraction and double-shot analytical pyrolysis for the quali-quantitation of plasticizers and microplastics in beach sand samples. *J. Hazard. Mater.* **2021**, *401*, 123287. [CrossRef]
31. Liu, M.; Song, Y.; Lu, S.; Qiu, R.; Hu, J.; Li, X.; Bigalke, M.; Shi, H.; He, D. A method for extracting soil microplastics through circulation of sodium bromide solutions. *Sci. Total Environ.* **2019**, *691*, 341–347. [CrossRef]
32. Korez, Š.; Gutow, L.; Saborowski, R. Microplastics at the strandlines of Slovenian beaches. *Mar. Pollut. Bull.* **2019**, *145*, 334–342. [CrossRef]
33. Imhof, H.K.; Schmid, J.; Niessner, R.; Ivleva, N.P.; Laforsch, C. A novel, highly efficient method for the separation and quantification of plastic particles in sediments of aquatic environments. *Limnol. Oceanogr. Methods* **2012**, *10*, 524–537. [CrossRef]
34. Konechnaya, O.; Lüchtrath, S.; Dsikowitzky, L.; Schwarzbauer, J. Optimized microplastic analysis based on size fractionation, density separation and μ-FTIR. *Water Sci. Technol.* **2020**, *81*, 834–844. [CrossRef]
35. Grause, G.; Kuniyasu, Y.; Chien, M.-F.; Inoue, C. Separation of microplastic from soil by centrifugation and its application to agricultural soil. *Chemosphere* **2021**, 132654. [CrossRef] [PubMed]
36. Lares, M.; Ncibi, M.C.; Sillanpää, M.; Sillanpää, M. Intercomparison study on commonly used methods to determine microplastics in wastewater and sludge samples. *Environ. Sci. Pollut. Res.* **2019**, *26*, 12109–12122. [CrossRef]
37. Crichton, E.M.; Noël, M.; Gies, E.A.; Ross, P.S. A novel, density-independent and FTIR-compatible approach for the rapid extraction of microplastics from aquatic sediments. *Anal. Methods* **2017**, *9*, 1419–1428. [CrossRef]
38. Scopetani, C.; Chelazzi, D.; Mikola, J.; Leiniö, V.; Heikkinen, R.; Cincinelli, A.; Pellinen, J. Olive oil-based method for the extraction, quantification and identification of microplastics in soil and compost samples. *Sci. Total Environ.* **2020**, *733*, 139338. [CrossRef] [PubMed]
39. Dümichen, E.; Eisentraut, P.; Bannick, C.G.; Barthel, A.K.; Senz, R.; Braun, U. Fast identification of microplastics in complex environmental samples by a thermal degradation method. *Chemosphere* **2017**, *174*, 572–584. [CrossRef]
40. Dümichen, E.; Eisentraut, P.; Celina, M.; Braun, U. Automated thermal extraction-desorption gas chromatography mass spectrometry: A multifunctional tool for comprehensive characterization of polymers and their degradation products. *J. Chromatogr. A* **2019**, *1592*, 133–142. [CrossRef] [PubMed]
41. Krauskopf, L.M.; Hemmerich, H.; Dsikowitzky, L.; Schwarzbauer, J. Critical aspects on off-line pyrolysis-based quantification of microplastic in environmental samples. *J. Anal. Appl. Pyrolysis* **2020**, *152*, 104830. [CrossRef]
42. Peñalver, R.; Arroyo-Manzanares, N.; López-García, I.; Hernández-Córdoba, M. An overview of microplastics characterization by thermal analysis. *Chemosphere* **2020**, *242*, 125170. [CrossRef]
43. Steinmetz, Z.; Kintzi, A.; Muñoz, K.; Schaumann, G.E. A simple method for the selective quantification of polyethylene, polypropylene, and polystyrene plastic debris in soil by pyrolysis-gas chromatography/mass spectrometry. *J. Anal. Appl. Pyrolysis* **2020**, *147*, 104803. [CrossRef]
44. Šunta, U.; Trebše, P.; Kralj, M.B. Simply Applicable Method for Microplastics Determination in Environmental Samples. *Molecules* **2021**, *26*, 1840. [CrossRef] [PubMed]
45. Boyaci, E.; Rodríguez-Lafuente, Á.; Gorynski, K.; Mirnaghi, F.; Souza-Silva, É.A.; Hein, D.; Pawliszyn, J. Sample preparation with solid phase microextraction and exhaustive extraction approaches: Comparison for challenging cases. *Anal. Chim. Acta* **2015**, *873*, 14–30. [CrossRef] [PubMed]
46. Hurley, R.R.; Lusher, A.L.; Olsen, M.; Nizzetto, L. Validation of a Method for Extracting Microplastics from Complex, Organic-Rich, Environmental Matrices - Supporting Information. *Environ. Sci. Technol.* **2018**, *52*, 7409–7417. [CrossRef] [PubMed]
47. Erni-Cassola, G.; Zadjelovic, V.; Gibson, M.I.; Christie-Oleza, J.A. Distribution of plastic polymer types in the marine environment; A meta-analysis. *J. Hazard. Mater.* **2019**, *369*, 691–698. [CrossRef] [PubMed]
48. Li, C.; Cui, Q.; Zhang, M.; Vogt, R.D.; Lu, X. A commonly available and easily assembled device for extraction of bio/non-degradable microplastics from soil by flotation in NaBr solution. *Sci. Total Environ.* **2021**, *759*, 143482. [CrossRef] [PubMed]
49. Fischer, M.; Scholz-Böttcher, B.M. Simultaneous Trace Identification and Quantification of Common Types of Microplastics in Environmental Samples by Pyrolysis-Gas Chromatography–Mass Spectrometry. *Environ. Sci. Technol.* **2017**, *51*, 5052–5060. [CrossRef] [PubMed]
50. Yakovenko, N.; Carvalho, A.; ter Halle, A. Emerging use thermo-analytical method coupled with mass spectrometry for the quantification of micro(nano)plastics in environmental samples. *TrAC Trends Anal. Chem.* **2020**, *131*, 115979. [CrossRef]
51. Hakkarainen, M.; Albertsson, A.-C.; Karlsson1, S. Solid phase microextraction (SPME) as an effective means to isolate degradation products in polymers. *J. Environ. Polym. Degrad.* **1997**, *5*, 67–73. [CrossRef]
52. Shirey, R.E. SPME Commercial Devices and Fibre Coatings. In *Handbook of Solid Phase Microextraction*; Elsevier: Amsterdam, The Netherlands, 2012; pp. 99–133. [CrossRef]
53. Dutra, C.; Pezo, D.; de Alvarenga Freire, M.T.; Nerín, C.; Reyes, F.G.R. Determination of volatile organic compounds in recycled polyethylene terephthalate and high-density polyethylene by headspace solid phase microextraction gas chromatography mass spectrometry to evaluate the efficiency of recycling processes. *J. Chromatogr. A* **2011**, *1218*, 1319–1330. [CrossRef] [PubMed]

54. Demets, R.; Roosen, M.; Vandermeersch, L.; Ragaert, K.; Walgraeve, C.; De Meester, S. Development and application of an analytical method to quantify odour removal in plastic waste recycling processes. *Resour. Conserv. Recycl.* **2020**, *161*, 104907. [CrossRef]
55. Vithanage, M.; Ramanayaka, S.; Hasinthara, S.; Navaratne, A. Compost as a carrier for microplastics and plastic-bound toxic metals into agroecosystems. *Curr. Opin. Environ. Sci. Heal.* **2021**, *24*, 100297. [CrossRef]

Article

Analysis of Microplastics Released from Plain Woven Classified by Yarn Types during Washing and Drying

Sola Choi [1,2], Miyeon Kwon [1], Myung-Ja Park [2,*] and Juhea Kim [1,*]

1. Department of Material & Component Convergence R&D, Korea Institute of Industrial Technology, Ansan 15588, Korea; solacs@kitech.re.kr (S.C.); mykwon@kitech.re.kr (M.K.)
2. Human Tech Convergence Program, Department of Clothing and Textiles, Hanyang University, Seoul 04763, Korea
* Correspondence: mjapark@hanyang.ac.kr (M.-J.P.); juheakim@kitech.re.kr (J.K.); Tel.: +82-2-2220-1192 (M.-J.P.); +82-31-9082-6221 (J.K.)

Abstract: Microplastics reach the aquatic environment through wastewater. Larger debris is removed in sewage treatment plants, but filters are not explicitly designed to retain sewage sludge's microplastic or terrestrial soils. Therefore, the effective quantification of filtration system to mitigate microplastics is needed. To mitigate microplastics, various devices have been designed, and the removal efficiency of devices was compared. However, this study focused on identifying different fabrics that shed fewer microplastics. Therefore, in this study, fabric-specific analyses of microplastics of three different fabrics during washing and drying processes were studied. Also, the change in the generation of microplastics for each washing process of standard washing was investigated. The amount of microplastics released according to the washing process was analyzed, and the collected microplastics' weight, length, and diameter were measured and recorded. According to the different types of yarn, the amount of microplastic fibers produced during washing and drying varied. As the washing processes proceed, the amount of microplastics gradually decreased. The minimum length (>40 μm) of micro-plastics generated were in plain-woven fabric. These results will be helpful to mitigate microplastics in the production of textiles and in selecting built-in filters, and focusing on the strict control of other parameters will be useful for the development of textile-based filters, such as washing bags.

Keywords: microplastics; microplastic fiber; washing textile; drying textile; polyester yarn types

Citation: Choi, S.; Kwon, M.; Park, M.-J.; Kim, J. Analysis of Microplastics Released from Plain Woven Classified by Yarn Types during Washing and Drying. *Polymers* **2021**, *13*, 2988. https://doi.org/10.3390/polym13172988

Academic Editor: Jacopo La Nasa

Received: 20 July 2021
Accepted: 1 September 2021
Published: 3 September 2021

Publisher's Note: MDPI stays neutral with regard to jurisdictional claims in published maps and institutional affiliations.

Copyright: © 2021 by the authors. Licensee MDPI, Basel, Switzerland. This article is an open access article distributed under the terms and conditions of the Creative Commons Attribution (CC BY) license (https://creativecommons.org/licenses/by/4.0/).

1. Introduction

The use and production of plastics have rapidly increased globally since the 1960s; the amount of production rose from 2 million tons in 1950 to 380 million tons in 2015 [1]. Plastic waste is fragmented through various processes, and these fragments have become a cause of marine pollution. In general, microplastic refers to a material composed of small or fine solid particles made of synthetic polymers smaller than 5 mm [2]. On the basis of their origin, microplastics are further defined as "primary microplastics" if they are intentionally produced either for direct use or as precursors to other products, and "secondary microplastics" if they are formed in the environment from the breakdown of larger plastic material [3,4].

Microplastics are used in various consumer, professional, agricultural, and industrial products such as cosmetics, detergents, fabric softeners, maintenance products, paints, coatings, inks, chemicals, construction, pharmaceuticals, and food supplements [2]. They reach the aquatic environment through the wastewater, which is a large conduit [5]. Larger debris is removed in sewage treatment plants, but filters are not explicitly designed to retain microplastic and terrestrial soils that contain microplastic fibers [6]. Therefore, the effective quantification of a filtration system to mitigate microplastics is needed for preventing plastic waste from entering large bodies of water.

The production of 65 million tons of synthetic fiber in 2016 suggests that the massive production of synthetic textiles is the driving cause of microplastic fibers in the environment [6–8]. According to the 2020 report from the European Chemicals Agency [2], fiber-shaped microplastics are 3 nm or more, but less than 15 mm in length, with a length to diameter ratio exceeding 3. Natural polymers and biodegradable polymers are excluded [2]. A significant amount of fibers (260–320 × 103 particles/m^3) was found in the Seine River [9], and after examining microplastics of less than 1 mm in the discharged from a sewage treatment plant located in Jinhae-gu, Changwon-si, in 2013, it was found that 26% of the fibers met the criteria for microplastics (867 ± 470 particles/m^3) [10]. When fiber-shaped microplastics are absorbed into the body of microorganisms, there is a risk of eating disorders caused by the wrapping of the intestines, thereby hampering growth and reproduction [11]. When PET microfibers of 62–1400 μm in length were exposed to Daphnia magna at a concentration of 12.5–100 mg/L for 48 h, the lethality due to ingestion increased [6]. Therefore, it is necessary to change the perception of collection, reduce the production of microplastics that pose a risk to the environment and human health, and analyze the microplastics emitted from fibers and derive collection methods.

A study that directly estimates the release of microplastics from laundry was first attempted by Browne et al. in 2011 [6]. Starting with Folkö's research in 2015, various studies on microplastics were produced during laundry focused on microplastics during the laundry [12]. The fabrics almost used were polyester-based [12,13]. In recent studies, fabrics were clearly classified by chemical composition, yarn types, and fabric construction [14–18].

The experiment was conducted on an experimental Launder-O-meter, a laboratory instrument used for conducting accelerated laundering, which uses a small sample size. The Launder-O-meter differs in physical force and movement of fabric compared to that of a large load laundry machine used for home laundry [19–21]. Zambrano et al. (2019) used a Launder-O-meter and a large load laundry machine to compare the amount of microplastics released. The Launder-O-meter created approximately 40 times more microplastics than the large load laundry machine. This result was caused by the mechanical action of the stainless-steel balls [20]. Therefore, the use of a large load laundry machine, used for home laundry, was recommended for studies. It is also necessary to measure the microplastics produced by various parameters, such as the individual washing processes, including washing, rinsing, and drying.

Various devices have been designed to divert and capture released microfibers to mitigate the release of microplastics in wastewater effluent. Several studies previously compared the removal efficiency of drum washing machines based on weight [21,22]. However, this study focused on identifying different fabrics that shed fewer microplastics. Focusing on fabrics while strictly controlling other parameters will be useful for developing textile-based filters, such as washing bags. The direction of twist, the number of twists, the thickness and length of the yarn affect the properties of yarn and textiles [23]. Thus, the primary aim of our study was to investigate the release of microfibers according to different yarn types. In addition, other textile parameters, such as chemical composition and fabric construction, were controlled. Three different yarn types with the same chemical composition and fabric construction were selected. Several processes were developed. A secondary aim was to detect microplastics that were reattached to the fabric during the drying process. Finally, to observe the change in the generation of microplastics for each washing process, the entire wastewater effluent was collected separately after each washing process and filtered.

2. Experimental
2.1. Materials

Three different yarn-type samples with the same chemical composition and weaving method were selected as specimens. Three different fabrics with two types of filament yarns (hard twist yarn, non-twist yarn) and one spun yarn of plain woven polyester were used. The filament hard twist yarn was purchased from Hwan Tex (Korea), the filament non-twist yarn from Yoomyung (Korea), and the spun yarn from Basic (Korea). Cutting edges of the specimen were overlocked using white polyester yarn to prevent thread loosening from the sample. The fabric density of the specimen was measured according to ISO 7211-2, and the weight (ISO 3801), thickness (ISO 4603), and size of the specimens are shown (Table 1). Since the physical forces during washing vary depending on the fabric weight, the fabric weight of the fabric to be tested was set at 500 g [24]. After preparing each specimen, the experiment was performed using five pieces, each weighing 100 g. Since it was challenging to find information about the raw resin of the polyester used in commercially available fabrics, to confirm the fabric composition and construction Fourier Transform Infrared (FTIR, Vertex 80v, Bruker) was used to determine the fabric composition, and a scanning electron microscope (SU8010, HITACHI) was used to confirm the fabric structure and yarn shape.

Table 1. Characteristics of Specimens.

Specimen		Density		Weight (g/m^2)	Thickness (mm)	Size (m × m per 100 g)
Yarn Type	Code	Warp (e.p.i [1])	Weft (p.p.i [2])			
Filament Hard Twist	F(HT)	123.04	88.90	115.0	0.27	1.40 × 0.65
Filament Non Twist	F(NT)	120.09	108.37	85.6	0.18	1.47 × 0.80
Spun	Spun	102.62	84.33	84.4	0.22	1.44 × 0.81

[1]: Ends per inch is the number of warp threads per inch of woven fabric [2]: Picks per inch is the number of warp threads per inch of woven fabric.

2.2. Washing and Drying

The drum-type washing machine (F9WK, LG electronics, Seoul, Korea) with a capacity of 9 kg was used, and washing was carried out through a standard course without a dummy load. One cycle of the standard washing course is comprised of one laundering and three rinsing processes. The temperature was 40 ± 2 °C of the laundering process, and 20 ± 2 °C in each of the rinsing processes. Washing was performed for a total of 1 h and 20 min, with 40 min for laundering, 10 min for rinsing 1, 10 min for rinsing 2, and 20 min for rinsing 3. Washing water from each cycle was separately collected to determine the amount of microplastics generated during each process. Tap water was used for the washing water and no detergent was added. After the experiment was completed, the washing machine was run empty three times to eliminate microplastic residue or other contaminants in the washing machine. The same experiment was repeated three times.

The dryer used was a dual inverter heat pump type drum dryer (RH9WGN, LG Elec-tronics, Seoul, Korea) with a capacity of 9 kg. Washed fabrics were dried in a standard drying course, with a drying temperature of about 60 °C for 1 h and 40 min. After drying, the drying machine was run three times with no fabric to ensure that no residue was left and the experiment was repeated three times.

2.3. Filtration

The wastewater discharged during each cycle was collected in four 20 L containers and was filtered separately. The washing effluents were filtered by means of a peristaltic pump connected with Teflon tubes; throughout quantitative filter papers consisting of cellulose fibers, with a pore size of 5 μm and a diameter of 185 mm (Grade 30, Hyundai micro, Seoul, Korea) were utilized. The filter pore size smaller than the fiber diameter was chosen. The filter paper was placed on a Buchner funnel and with a 5 L filtering flask. The washing water stored in a 20 L container was filtered using a manual liquid pump, and after filtering all water, the container was rinsed with 1 L of tap water to re-move all remaining fibers. The 1 L of tap water was also filtered. The used filtering flask, Buchner funnel, and manual transfer pump were washed with 6 L of tap water before each of the subsequent experiments.

2.4. Analysis of Microplastics

Methodologies in previous studies have assessed a select area of filter paper and scaled the count for all fibers from the specimen. However, manual counting of fibers leaves considerable potential for counting errors since fibrous forms, especially microplastics during washing, are intertwined across a 3-dimensional spaghetti-like structure [25]. Tiffin et al. (2021) discussed a homogenous distribution of fibers across the entire filter area. Thus, this study quantified microplastic released by weight. The collected microplastics were compared by weight, length, and diameter measurements. The microplastics' weight was measured using a precision balance after drying with filter paper at a temperature of $26 \pm 2\ °C$ and relative humidity of 20% for 24 h. The microplastics generated after drying were measured by comparing the built-in filter's weight inside the dryer before and after drying. The filter paper's weight before and after filtering the washing water was calculated according to the ppm equation and com-pared. The mass of collected materials per the mass of the textile, ppm (mg/kg) is calculated by the Equation (1):

$$\text{ppm (mg/kg)} = (M_m \times 1000)/M_{kg} \qquad (1)$$

where

ppm is the mass of the collected microplastics per mass of textile in mg/kg
M_m is the mass of the collected microplastics during washing and drying in mg
M_{kg} is the mass of test specimens in kg

The length identification of the three samples was carried out using a digital microscope (magnification of $40\times$). The captured fibers were spread evenly on the filter paper, and the fiber lengths and diameters were analyzed using the Image J program (NIH, National Institutes of Health, Bethesda, Maryland, MD, USA).

3. Results and Discussion

3.1. Fourier Transform Infrared and Scanning Electron Microscopy of Specimens

Figure 1 shows the FTIR spectrum of the three different fabrics classified by yarn type. All three fabrics exhibited almost the same peak as polyester resin. It can be observed that the unsaturated polyester had a weak band at $2967\ cm^{-1}$, which can be attributed to the C–H elongation [26]. The polyester showed important characteristic absorption in the $1713\ cm^{-1}$ band, which represents the carbonyl group, C=O. The bands close to $740\ cm^{-1}$ represent the elongation of the aromatic nucleus C=C. This occurred due to the presence of the unsaturated double bond (C=C) in the polyester and refers to the vinyl group present in the styrene monomer. The bands close to 1260 and $1117\ cm^{-1}$ occurred due to stretching vibrations C–O–C connected to the aliphatic and aromatic groups, respectively. Through these spectral results, the three samples were confirmed to be polyester, and it was confirmed that the fabric components were chemically similar through FTIR.

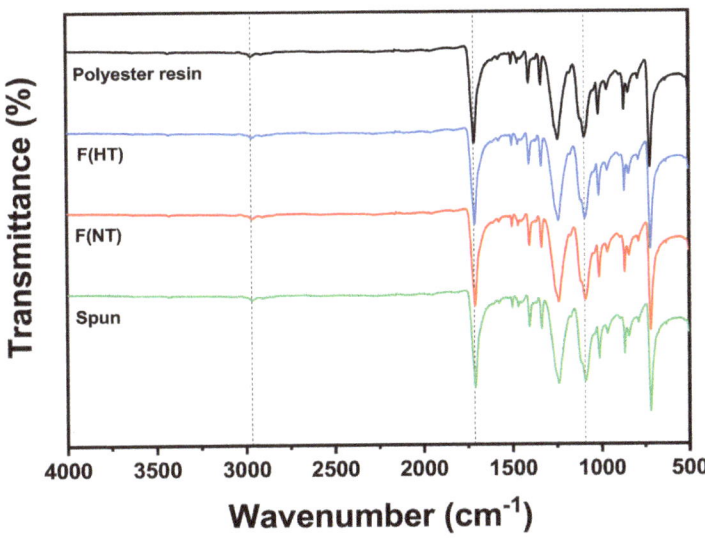

Figure 1. FTIR spectroscopy of polyester resin and three different fabrics classified by yarn types.

Also, Figure 2 shows the scanning electron microscopy image of the three different fabrics classified by yarn type. As a result of a scanning electron microscopy, it was confirmed that the types of yarn were different from hard twist filament yarn (Figure 2a), non-twist filament yarn (Figure 2b), and spun yarn (Figure 2c), but all three fabrics were the same as the plain woven. Figure 2a is a hard-twisted filament yarn in which a bundle of fibers is tightly twisted in one direction, compared to Figure 2b is a yarn in which a bundle of fibers is arranged fiber length direction without twisting. Figure 2c is a spun yarn that is composed of a short bundle of fibers that is twisted in one direction more loosely than hard twist yarn. Through these scanning electron microscopy results, the three samples confirmed the same as plain woven and difference in the type of yarn.

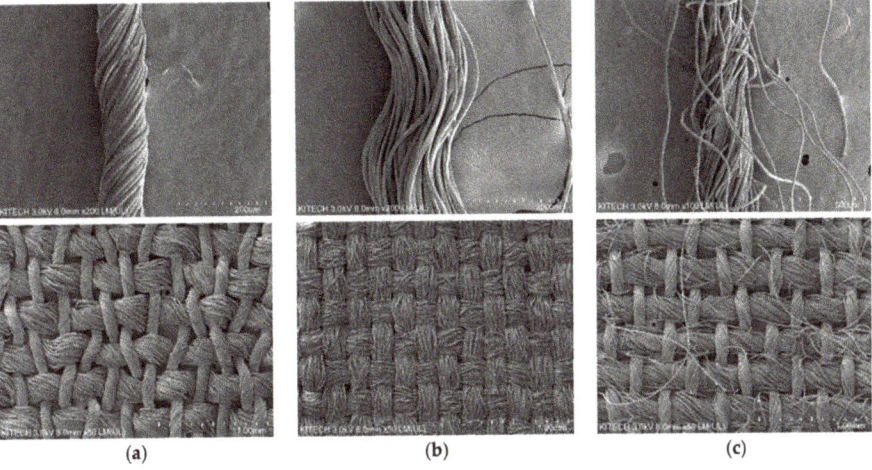

Figure 2. Scanning electron microscope images for identifying yarn type and fabric construction. (**a**) Plain woven with hard twist filament yarn; (**b**) Plain woven with non-twist filament yarn; (**c**) Plain woven with spun yarn.

At the results of FTIR and SEM, it was confirmed that the three fabrics had similar properties physically and chemically and were only different in yarn type.

3.2. Mass of Microplastics Released from Textiles According to Yarn Types during Washing

The results comparing the release of microplastics during washing of plain-woven from hard twist filament yarn, non-twist filament yarn, and spun yarns are shown in Figure 3. The release of microplastics was 51.6 (±6.9) ppm for the hard twist filament yarn, 88.7 (±24.7) ppm for the non-twist filament yarn, and 107.7 (±14.5) ppm for the spun yarn.

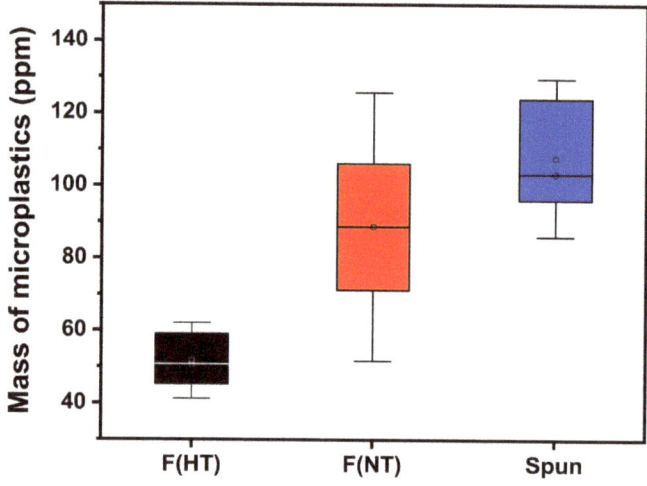

Figure 3. Amount of microplastics according to yarn types during washing.

The spun yarn showed a higher release of microplastics than the non-twist filament yarn and the hard twist filament yarn. This is because the spun yarn has a higher degree of freedom due to the shorter fiber length than the filament yarn, and the fibers were more easily released from the fabric, resulting in the generation of more microplastics. As a result of comparing the amount of microplastics released by the degree of twist in the filament yarn, non-twist yarn released more microplastics than high-twist yarn. It is considered that the non-twist yarn is more likely to generate microplastics since the friction between the fibers is decreased, resulting in increased fiber freedom.

A similar tendency was observed in Almroth et al. (2018), in which polyester knit with different types of multifilament yarns or spun yarns were used. This research used the counting method inferred from the length measurement results for analyzing the amount of microplastics released [15]. Comparing the overall quantities of microplastics released during washing to those reported by De Falco of 244–296 mg/kg of PES-knit fabric, the release for the present study was lower [17]. Considering the fabric construction, unlike the woven fabric used in this study, knitted fabrics were used in the De Falco et al. (2018). Therefore, it can be expected that fabric construction affected the release of microplastics [4].

As a result of the amount of microplastics produced by the three fabrics with different yarn types, the higher the degree of freedom of the fibers, the higher the amount of microplastics produced. Therefore, to mitigate the release of microplastics, the use of filament yarns, rather than spun yarns, should be applied in synthetic fibers, and a finishing method, such as more twisting or increasing the density of the yarn, is required.

3.3. Release of Microplastics as Washing Process

One cycle of the standard washing course comprises one laundering process and three rinsing processes, and the release of microplastics in each process is shown in Figure 4.

Microplastics released in all three fabrics showed the same tendency in the washing process; most microplastics were generated in the laundering process, and the microplastics tended to decrease as the washing proceeded.

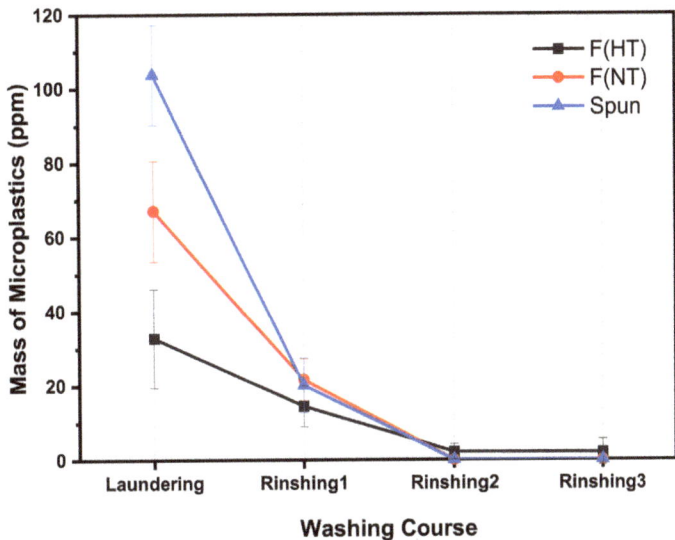

Figure 4. Amount of microplastics generated according to the laundering process.

Washing was performed for a total of 1 h and 20 min, and the time is longer in the laundering processes (40 min for laundering) than the rinsing processes (10 min for rinsing 1 and 2, and 20 min for rinsing 3). Since the time to receive physical force (washing water, falling off the fabric, and movement of the fabric) is longer in the laundering process than in the rinsing processes, it is understandable that the release of microplastics was released higher in the laundering process. Additionally, since the temperature is higher during laundering (40 °C), the fabrics' binding forces may be weakened due to the heat energy. Therefore, as the washing process proceeds, the release of microplastics gradually decreases since the time the fabric receives physical force is shortened, and the temperature is lower. In short, the physical forces differ according to each washing process considering temperature and washing time.

3.4. Release of Microplastics during Drying

Microplastics released during drying were on the order of 59.9 (±20.1) ppm for spun yarn, 29.9 (±13.9) ppm for non-twist filament yarn, 20.6 ppm for hard twist filament yarn, and showed the same tendency as that during washing (Figure 5). As a result of comparing the microplastics released during washing and drying, the amount of microplastics during drying decreased to 66% for spun yarn, 60% for hard twist filament yarn, and 44% for non-twist filament yarn. The amount of microplastics released during the drying process showed the same tendency as the washing process in the order of spun yarn, non-twist filament yarn, and hard twist yarn.

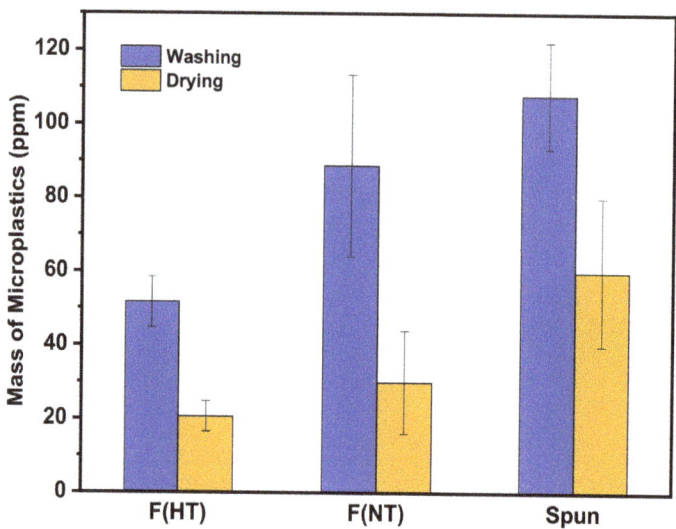

Figure 5. Comparison of microplastics during washing and drying.

As shown in washing, the same correlation was observed during drying between the degree of freedom of fibers and the release of microplastics depending on the type of yarn. The reason that amount of microplastics during the drying process was reduced compared to washing is as follows: The principal difference between washing and drying processes is the difference in the physical force of water that affects fabrics during washing. Moisture seeping into the fabric during washing makes fabric lose structure, so fibers easily drop out of the fabric. Furthermore, during washing, the fabric receives the physical force of the washing water, the falling of the fabric, and the fabric movement, but only the physical forces of movement and falling of the fabric were present during the drying process. To reduce the production of microplastics, it may be effective to increase the fabric weight to minimize the physical force imparted.

The drying results suggest two things: First, the drying results showed the same tendency as the washing results in the order of spun yarn, non-twist filament yarn, and hard-twist filament yarn, thereby proving our hypothesis again that fabric properties can affect the amount of microfibers. Second, the detachment of microplastics by the drying process is thought to be caused by damage to the fabric due to abrasion during washing. In this case, new microplastics are possibly generated every time the fabric is washed. The results of Napper and Thompson (2016), showed that garments had an initial peak of microplastics shedding in the first 1–4 washes and showed new microplastics are possible to be generated every time the fabric is washed [13]. Although microplastic generation continued to appear after drying, the drying results can be used as an alternative to solve wastewater effluent, and marine pollution as drying does not cause wastewater.

3.5. Length of Microplastic Fibers

The length distribution of fibers released by yarn types showed various length distributions below 1000 μm and a tendency to peak at 1500 μm (Figure 6). As a result of the fiber length distribution from 0 μm to 5000 μm are as follows: the peak points were at 200–300 μm and 1500 μm for hard twist yarn, 100–300 μm and 1500 μm for non-twist filament yarn, and 300–400 μm and 1500 μm for spun yarn. A similar high peak point appeared at 1500 μm in all three samples.

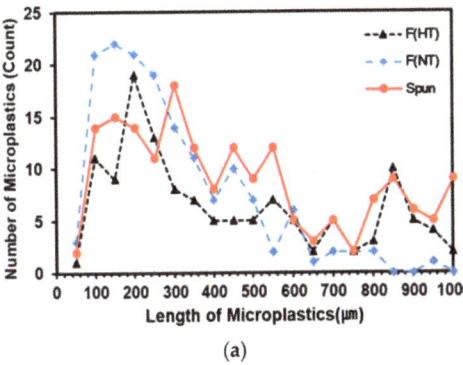 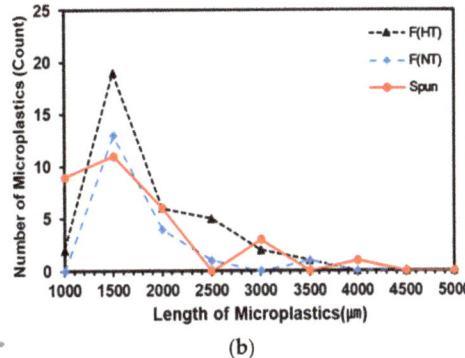

Figure 6. Frequency distribution diagram of microplastics length. (**a**) Length of microplastics from 0 μm to 1000 μm (interval: 100 μm) (**b**) Length of microplastics from 1000 μm to 5000 μm (interval: 500 μm).

Unlike hard and non-twist filament yarns that showed peaks at 100–300 μm and 1500 μm, slight max peaks appeared at 300–400 μm and 900–1000 μm in spun yarn. Based on the characteristics of the spun yarn, which has lengths of 20–46 mm and is shorter than filament yarn, it is considered that the twists of the spun yarn are easily separated due to swelling and physical force. In contrast, it is believed that the relatively long filament yarns were fragmented after being separated from the fabric.

The diameter used in this study was found to be as small as 10.78 μm, and a smaller filter pore size was used (Table 2). The result of the length analysis of released fibers showed that the minimum lengths were 40.49 μm (FHT), 45.81 μm (FNT), and 44.75 μm (Spun). These are considerably more than 40 μm of microplastics generated in plain-woven fabric. The minimum length (>40 μm) of microplastics is generated in plain woven. Therefore, these results demonstrate that selecting a built-in filter without too small filter pore sizes can mitigate microplastics.

Table 2. Descriptive Statistics of Fiber Length and Diameter of Microplastics Released.

Specimen	Length (μm)						Diameter (μm)		
	n	M	SD	Md	SE	Min.	Max.	n	M	SD
F(HT)	161	648.97	591.54	470.03	46.62	40.49	3108.03	10	14.67	2.20
F(NT)	170	402.40	458.49	240.94	35.16	45.81	3377.19	10	10.78	1.23
Spun	199	572.30	544.88	412.17	38.63	44.75	3679.33	10	12.53	1.94

4. Conclusions

This study proposes a method to reduce microplastics released from textiles during washing and drying. Focusing on fabrics construction, with strict control of other parameters, will be useful for developing textile-based filters, such as washing bags. The higher the degree of freedom of fibers, the higher the amount of microplastics released, so a finishing method, such as more twisting or increasing the density of the yarn, is required. The results of the washing process can lead to a method for consumers to reduce the amount of microplastics produced in the home. A washing method that shortens the washing time and lowers the temperature is required to reduce physical force. The difference between washing and drying reconfirmed the effect of physical force, and tumble drying is proposed as another alternative for microplastics mitigation. Lastly, fiber length results are expected to contribute as basic data for developing optimal filter pore sizes in washing machine filtration devices to collect microplastics efficiently. To minimize some of the avoidable environmental challenges we currently face, further investigations are needed to develop technology [22]. At the design stage of methods to minimize unintended environmental consequences, we believe that our results are one source for developing effective mitigation tools.

Author Contributions: Methodology, Data curation, Investigation, Writing—Original draft, S.C.; Conceptualization, Methodology, Investigation, Reviewing and Editing M.K.; Writing—Reviewing and Editing, Supervision, M.-J.P.; Conceptualization, Supervision, Writing—Original draft preparation, J.K. All authors have read and agreed to the published version of the manuscript.

Funding: This work was supported by the National Research Council of Science & Technology (NST) grant by the Korean government (MSTI) (No. CAP-20-02-KITOX).

Institutional Review Board Statement: Not applicable.

Informed Consent Statement: Not applicable.

Data Availability Statement: Data sharing not applicable.

Acknowledgments: This work was supported by the National Research Council of Science & Technology (NST) grant by the Korean government (MSTI) (No. CAP-20-02-KITOX).

Conflicts of Interest: The authors declare no conflict of interest.

References

1. Geuer, R.; Jambeck, J.R.; Law, K.R. Production, use, and fate of all plastics ever made. *Sci. Adv.* **2017**, *3*, e1700782.
2. European Chemicals Agency. Available online: https://echa.europa.eu/documents/10162/05bd96e3-b969-0a7c-c6d0-441182893720 (accessed on 3 November 2020).
3. Kershaw, P.J. Sources, fate and effects of microplastics in the marine environment: A global assessment. In *Reports and Studies GESAMP, Joint Group of Experts on the Scientific Aspects of Marine Environmental Protection*; Maritime Organization: London, UK, 2015; Volume 90, pp. 1–96.
4. De Falco, F.; Cocca, M.; Avella, M.; Thompson, R.C. Microfiber release to water, via laundering, and to air, via everyday use: A comparison between polyester clothing with differing textile parameters. *Environ. Sci. Technol.* **2020**, *54*, 3288–3296.
5. McIlwraith, H.K.; Lin, J.; Erdle, L.M.; Mallos, N.; Diamond, M.L.; Rochman, C.M. Capturing microfibers–marketed technologies reduce microfiber emissions from washing machines. *Mar. Pollut. Bull.* **2019**, *139*, 40–45.
6. Browne, M.A.; Crump, P.; Niven, S.J.; Teuten, E.; Tonkin, A.; Galloway, T.; Thompson, R. Accumulation of microplastic on shorelines woldwide: Sources and sinks. *Environ. Sci. Technol.* **2011**, *45*, 9175–9184.
7. Nonwovens Industry. Available online: https://www.nonwovens-industry.com/contents/view_online-exclusives/2017-05-23/the-fiber-year-reports-on-2016-world-fiber-market (accessed on 3 November 2020).
8. Boucher, J.; Friot, D. *Primary Microplastics in the Oceans: A Global Evaluation of Sources*; Lucn: Gland, Switzerland, 2017; pp. 227–229.
9. Dris, R.; Gasperi, J.; Rocher, V.; Saad, M.; Renault, N.; Tassin, B. Microplastic contamination in an urban area: A case study in Greater Paris. *Environ. Chem.* **2015**, *12*, 592–599.
10. Korea Institute of Ocean Science & Technology. Available online: http://wwwsciwatch.kiost.ac.kr/kor.dohandle/2020.kiost/36946 (accessed on 4 November 2020).
11. Jemec, A.; Horvat, P.; Kunej, U.; Bele, M.; Kržan, A. Uptake and effects of microplastic textile fibers on freshwater crustacean Daphnia magna. *Environ. Pollut.* **2016**, *219*, 201–209.
12. Folkö, A. *Quantification and Characterization of Fibers Emitted from Common Synthetic Materials during Washinglaundering*; Universitet och Käppalaförbundet: Stockholms, Switzerland, 2015.
13. Napper, I.E.; Thompson, R.C. Release of synthetic microplastic plastic fibres from domestic washinglaundering machines: Effects of fabric type and washinglaundering conditions. *Mar. Pollut. Bull.* **2016**, *112*, 39–45.
14. Yang, L.; Qiao, F.; Lei, K.; Li, H.; Kang, Y.; Cui, S.; An, L. Microfiber release from different fabrics during washinglaundering. *Environ. Pollut.* **2019**, *249*, 136–143.
15. Almroth, B.M.C.; Ström, L.Å.; Roslund, S.; Petersson, H.; Johansson, M.; Persson, N.K. Quantifying shedding of synthetic fibers from textiles; a source of microplastics released into the environment. *Envrionmental Sci. Pollut.* **2018**, *25*, 1191–1199.
16. Cai, Y.; Yang, T.; Mitrano, D.M.; Heuberger, M.; Hufenus, R.; Nowack, B. Systematic study of microplastic fiber release from 12 different polyester textiles during washinglaundering. *Environ. Science Technol.* **2020**, *54*, 4847–4855.
17. De Falco, F.; Gullo, M.P.; Gentile, G.; Di Pace, E.; Cocca, M.; Gelabert, L.; Brouta-Agnésa, M.; Rovira, A.; Escudero, R.; Avella, M.; et al. Evaluation of microplastic release caused by textile washing processes of synthetic fabrics. *Environ. Pollut.* **2018**, *236*, 916–925.
18. De Falco, F.; Di Pace, E.; Cocca, M.; Avella, M. The contribution of washing processes of synthetic clothes to microplastic pollution. *Sci. Rep.* **2019**, *9*, 1–11.
19. Hernandez, E.; Nowack, B.; Mitrano, D.M. Polyester textiles as a source of microplastics from households: A mechanistic study to understand microfiber release during washing. *Environ. Sci. Technol.* **2017**, *51*, 7036–7046.

20. Zambrano, M.C.; Pawlak, J.J.; Daystar, J.; Ankeny, M.; Cheng, J.J.; Venditti, R.A. Microfibers generated from the laundering of cotton, rayon and polyester based fabrics and their aquatic biodegradation. *Mar. Pollut. Bull.* **2019**, *142*, 394–407.
21. Jönsson, C.; Levenstam Arturin, O.; Hanning, A.C.; Landin, R.; Holmström, E.; Roos, S. Microplastics shedding from textiles—Developing analytical method for measurement of shed material representing release during domestic washing. *Sustainability* **2018**, *10*, 2457.
22. Napper, I.E.; Barrett, A.C.; Thompson, R.C. The efficiency of devices intended to reduce microfibre release during clothes washing. *Sci. Total Environ.* **2020**, *738*, 140412.
23. Kim, S.R. *Clothing and Materials*, 3rd ed.; Gyomoonsa: Gyoenggi, Korea, 2000; pp. 240–246.
24. Kelly, M.R.; Lant, N.J.; Kurr, M.; Burgess, J.G. Importance of water-volume on the release of microplastic fibers from laundry. *Environ. Sci. Technol.* **2019**, *53*, 11735–11744.
25. Tiffin, L.; Hazlehurst, A.; Sumner, M.; Taylor, M. Reliable quantification of microplastic release from the domestic laundry of textile fabrics. *J. Text. Inst.* **2021**, 1–9. [CrossRef]
26. Maradini, G.D.S.; Oliveira, M.P.; Carreira, L.G.; Guimarães, D.; Profeti, D.; Dias Júnior, A.F.; Boschetti, W.T.N.; Oliveira, B.F.D.; Pereira, A.C.; Monteiro, S.N. Impact and Tensile Properties of Polyester Nanocomposites Reinforced with Conifer Fiber Cellulose Nanocrystal: A Previous Study Extension. *Polymers* **2021**, *13*, 1878.

Article

A Systematic Study on the Degradation Products Generated from Artificially Aged Microplastics

Greta Biale [1], Jacopo La Nasa [1,2,*], Marco Mattonai [1], Andrea Corti [1], Virginia Vinciguerra [1], Valter Castelvetro [1,3] and Francesca Modugno [1,3]

- [1] Department of Chemistry and Industrial Chemistry, University of Pisa, 56124 Pisa, Italy; greta.biale@gmail.com (G.B.); marco.mattonai@dcci.unipi.it (M.M.); andrea.corti@unipi.it (A.C.); virgi-vinci@hotmail.it (V.V.); valter.castelvetro@unipi.it (V.C.); francesca.modugno@unipi.it (F.M.)
- [2] National Interuniversity Consortium of Materials Science and Technology, 50121 Florence, Italy
- [3] CISUP—Center for the Integration of Scientific Instruments of the University of Pisa, University of Pisa, 56124 Pisa, Italy
- * Correspondence: jacopo.lanasa@for.unipi.it; Tel.: +39-050-221-9258

Abstract: Most of the analytical studies focused on microplastics (MPs) are based on the detection and identification of the polymers constituting the particles. On the other hand, plastic debris in the environment undergoes chemical and physical degradation processes leading not only to mechanical but also to molecular fragmentation quickly resulting in the formation of leachable, soluble and/or volatile degradation products that are released in the environment. We performed the analysis of reference MPs–polymer micropowders obtained by grinding a set of five polymer types down to final size in the 857–509 µm range, namely high- and low-density polyethylene, polystyrene (PS), polypropylene (PP), and polyethylene terephthalate (PET). The reference MPs were artificially aged in a solar-box to investigate their degradation processes by characterizing the aged (photo-oxidized) MPs and their low molecular weight and/or highly oxidized fraction. For this purpose, the artificially aged MPs were subjected to extraction in polar organic solvents, targeting selective recovery of the low molecular weight fractions generated during the artificial aging. Analysis of the extractable fractions and of the residues was carried out by a multi technique approach combining evolved gas analysis–mass spectrometry (EGA–MS), pyrolysis–gas chromatography–mass spectrometry (Py–GC–MS), and size exclusion chromatography (SEC). The results provided information on the degradation products formed during accelerated aging. Up to 18 wt% of extractable, low molecular weight fraction was recovered from the photo-aged MPs, depending on the polymer type. The photo-degradation products of polyolefins (PE and PP) included a wide range of long chain alcohols, aldehydes, ketones, carboxylic acids, and hydroxy acids, as detected in the soluble fractions of aged samples. SEC analyses also showed a marked decrease in the average molecular weight of PP polymer chains, whereas cross-linking was observed in the case of PS. The most abundant low molecular weight photo-degradation products of PS were benzoic acid and 1,4-benzenedicarboxylic acid, while PET had the highest stability towards aging, as indicated by the modest generation of low molecular weight species.

Keywords: microplastics; polymer degradation; artificial ageing; polyolefins; polystyrene; polyethylene terephthalate

Citation: Biale, G.; La Nasa, J.; Mattonai, M.; Corti, A.; Vinciguerra, V.; Castelvetro, V.; Modugno, F. A Systematic Study on the Degradation Products Generated from Artificially Aged Microplastics. *Polymers* **2021**, *13*, 1997. https://doi.org/10.3390/polym13121997

Academic Editor: Francesca Lionetto

Received: 23 April 2021
Accepted: 15 June 2021
Published: 18 June 2021

Publisher's Note: MDPI stays neutral with regard to jurisdictional claims in published maps and institutional affiliations.

Copyright: © 2021 by the authors. Licensee MDPI, Basel, Switzerland. This article is an open access article distributed under the terms and conditions of the Creative Commons Attribution (CC BY) license (https://creativecommons.org/licenses/by/4.0/).

1. Introduction

Due to the steadily increasing production of plastic materials since the second half of the 20th century, the mismanagement of a significant fraction of plastic waste has resulted in massive plastic pollution becoming an environmental threat worldwide with potential risks for biota and also for human health [1,2]. Therefore, the assessment of the amount, distribution and nature of the plastic debris in ecosystems, and especially in the marine environment, is the focus of intense multidisciplinary research [3,4]. Currently, about 5–13 million tons of plastic waste are estimated to enter the ocean every year [5]. A generally accepted picture based on an increasing number of environmental studies suggests that the largest fraction of it consists of microplastics (MPs) [6]. While there is currently no scientific or regulatory agreement on the definition of MPs size range, the commonly adopted 5 mm as the upper limit [7,8] is being questioned as it relates to the ingestion by fish rather than to the physical and chemical properties. In fact, the latter are mainly related to the specific surface area and its rapid increase (roughly by a half of the third power) with the reduction of particle size; on the other hand, the specific surface area is likely to be more strictly related to the rate and extent of polymer degradation under environmental conditions, and thus with the interaction of MPs with the environment. A scientifically more appropriate size range of 1–1000 μm for MPs has recently been proposed [3,9]. The relevance of the size range of the MPs is exemplified by the observation that the higher the specific surface area, the faster the chemical degradation, mainly due to photo-oxidation processes, leading to the formation of leachable, soluble and/or volatile degradation products that need to be considered as an emerging source of environmental pollution deriving from MPs [10–13]. The most common analytical techniques used for the analysis of MPs are micro Fourier transform infrared (μ-FTIR), and micro-Raman spectroscopy [14,15], both providing information that does not allow to clearly highlight the differences between the surface and bulk composition and to pinpoint the presence and relative amount of highly degraded fractions and of degradation products that are likely to affect the most the chemical behavior of MPs. On the other hand, in the last few years thermo-analytical approaches have emerged as powerful tools for studying MPs [16–18], also in routine analysis of contaminated environmental samples [19]. Among them, analytical pyrolysis [20–22] allows the sensitive and accurate characterization of polymers and polymer mixtures through their pyrolytic profiles. The systematic study presented here is based on a combination of analytical pyrolysis approaches and mass spectrometry detection to evaluate the effects of simulated environmental photo-oxidative degradation processes in five different reference MPs. We describe the parallel characterization of the bulk chemical modifications, and of the low-molecular-weight, leachable or soluble organic fraction resulting from the polymer degradation. Investigating the correlation between the modifications of the different polymer types under simulated environmental aging, and the nature and amount of the low molecular weight species resulting from the polymer degradation, is fundamental in order to assess the possible role of such low molecular weight species as emerging environmental pollutants as they may leach out of the MPs, being thus potentially harmful to living organisms. From the most recent literature, a need emerges for studies about the potential harmfulness of the oxidation products (low molecular weight and oxidized oligomeric species) that may leach out from plastic debris dispersed in the environment. The formation of such species is generally neglected in the studies on MPs even if these compounds may pose even higher risks for the environment and the biota than the MP particles themselves, risks that are far from being understood and assessed.

In this context, the aim of this study was the thorough characterization of aged (photo-oxidized) MPs and of their low molecular weight and/or highly oxidized fraction as evaluated by their enhanced solubility in polar organic solvents. Five reference polymers were selected among those most commonly found in the form of MPs polluting the environment, namely: polypropylene (PP), polystyrene (PS), polyethylene terephthalate (PET), low-density polyethylene (LDPE) and high-density polyethylene (HDPE). The virgin polymers in the form of micropowders (average particle size in the 857–509 μm range) were aged in a solar-box for four weeks. Samples were periodically collected and analyzed by means of evolved gas analysis–mass spectrometry (EGA–MS) and the results were compared with those obtained for the unaged polymers in order to study the thermal degradation behavior specific for each polymer during photo-aging. The unaged and the polymer samples aged 4 weeks were also characterized by means of pyrolysis–gas chromatography–mass spectrometry (Py–GC–MS) to detect and identify the produced alteration and degradation products. Finally, all samples were subjected to solvent extraction in either methanol (for PS) or refluxing dichloromethane (for PP, PET, LDPE and HDPE), that are non-solvents for the virgin polymers, to selectively extract the low molecular weight fractions generated by chain scissions as a result of extensive degradation. The solvent extracts were then analyzed by size exclusion chromatography (SEC) and Py–GC–MS; for the latter analyses hexamethyldisilazane (HMDS) derivatization was performed to allow detection of high polarity and low-volatility species such as those resulting from photo-oxidative processes entailing oxygen pickup through free radical reactions. The insoluble polymer fractions were also analyzed by means of Py–GC–MS in an attempt to evaluate the chemical modifications preliminary to the most extensive degradation processes, and compare them with those of the most highly degraded, solvent extractable fractions.

2. Materials and Methods

2.1. Chemicals

Dichloromethane (DCM, high-performance liquid chromatography (HPLC) grade, Sigma-Aldrich, St. Louis, MO, USA) and methanol (MeOH, HPLC grade, Sigma-Aldrich) were used as solvents in the extractions. Hexamethyldisilazane (HMDS $\geq$ 99%, Sigma-Aldrich) was used as derivatizing agent for the in situ thermally assisted silylation of the pyrolysis products bearing carboxylic and hydroxyl groups in the Py–GC–MS analysis of the polymer extracts.

2.2. Reference Polymers

Micronized polypropylene (PP), polystyrene (PS), polyethylene terephthalate (PET), low-density polyethylene (LDPE) and high-density polyethylene (HDPE), with average particle size in the 857–509 μm range depending on the polymer as reported elsewhere [23,24], were kind gifts from Poliplast S.p.A (Casnigo, Italy).

2.3. Artificial Aging and Extraction

PP, PS, PET, LDPE and HDPE micropowders were artificially aged for four weeks using a solar-box system (CO.FO.ME.GRA. Srl, Milan, Italy) equipped with a Xenon-arc lamp and outdoor filter. The conditions for the aging were: temperature 40 °C, irradiance 750 W/m^2, relative humidity around 60%. Aliquots (ca. 200 mg) of each polymer were collected before (0w) and after 1 (1w), 2 (2w), 3 (3w), and 4 (4w) weeks of artificial aging and stored in sealed glass vials at -20 °C until analysis [23]. About 150 mg of each unaged and aged polymer sample was extracted for 6 h with 30 mL either MeOH (for PS) or DCM (all other polymers) in a Soxhlet apparatus, collecting both the residues and the extractable fractions for the subsequent characterizations. MeOH and DCM, that act as non-solvents for the virgin polymers, were chosen to selectively extract the degraded, low molecular weight fractions. The extracted fractions were dried in a rotary evaporator until constant weight and then stored in glass vials before the Py-GC-MS and SEC analyses. Procedural blanks (DCM and MeOH) were also prepared and analyzed along with the polymer extracts. The overall procedure is schematically summarized in Figure 1.

2.4. Analytical Methods and Instrumentation

2.4.1. Evolved Gas Analysis–Mass Spectrometry (EGA–MS)

The EGA-MS analyses of unaged and artificially aged bulk polymers were performed with an EGA/PY-3030D micro-furnace pyrolyzer (Frontier Laboratories Ltd., Koriyama, Japan) coupled to a 6890 gas chromatograph and a 5973 mass spectrometric detector (Agilent Technologies, Santa Clara, CA, USA). The experimental conditions were the following: temperature ramp for the furnace from 50 °C to 700 °C at 10 °C/min; interface between the pyrolysis furnace and the GC-MS system set at a temperature 100 °C higher than that of the furnace but limited to a maximum of 300 °C; GC injector operated in split mode (20:1 ratio) at 280 °C [25,26]. The evolved pyrolysis products were directly sent to the mass spectrometer using a UADTM-2.5N deactivated stainless-steel capillary tube (3 m × 0.15 mm, Frontier Laboratories Ltd., Japan) held at 300 °C, and using helium (1 mL/min) as the carrier gas. The temperature of the transfer line to the mass spectrometer was 280 °C. The mass spectrometer was operated in electronic impact (EI) positive mode (70 eV, m/z range 15–700). The temperatures of the ion source and quadrupole analyzer were 230 °C and 150 °C, respectively. Each 100–250 µg sample was directly weighed in the deactivated stainless-steel pyrolysis cup with an XS3DU microanalytical scale (Mettler-Toledo, Columbus, OH, USA) with seven digits and a precision of 1 µg.

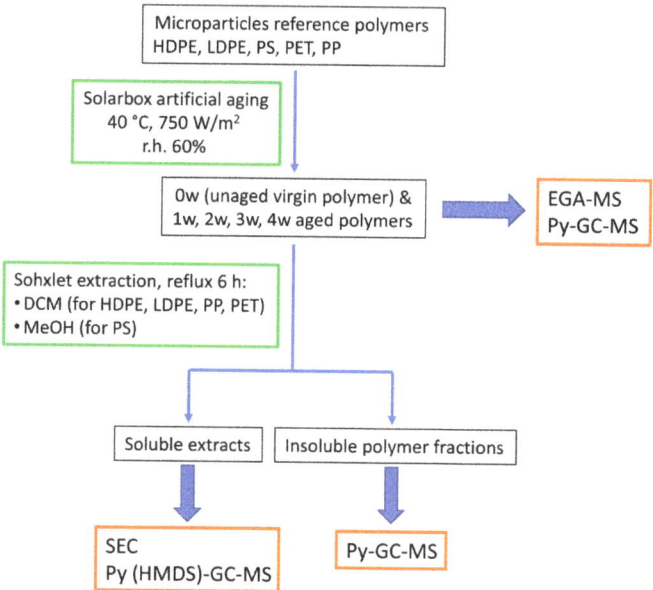

Figure 1. Flowsheet of the overall procedure for the investigation of the polymer degradation products generated upon accelerated photo-oxidative aging.

2.4.2. Pyrolysis–Gas Chromatography–Mass Spectrometry (Py–GC–MS)

Analyses of unaged and artificially aged bulk polymers, their extracts, and the corresponding extraction residues (insoluble fractions) were performed using a multi-shot pyrolyzer EGA/PY-3030D (Frontier Lab.) coupled with a 6890N gas chromatography system with a split/splitless injection port and combined with a 5973 mass selective single quadrupole mass spectrometer (Agilent Technologies). The samples (50–100 µg) were placed in deactivated stainless-steel sample cups. Pyrolysis conditions were optimized as follows: pyrolysis temperatures were selected based on the samples analyzed [25–27]; interface 280 °C; GC injector temperature 280 °C; GC injection operated in split mode with an optimized 10:1 split ratio. The chromatographic separation of the pyrolysis products was performed on a fused silica capillary column HP-5MS (5% diphenyl–95% dimethyl-polysiloxane, 30 m × 0.25 mm internal diameter., 0.25 µm film thickness, J&W Scientific, Agilent Technologies) preceded by a deactivated fused silica pre-column (2 m × 0.32 mm i.d.). The chromatographic conditions were: 40 °C for 5 min, 10 °C/min to 310 °C for 20 min, carrier gas (He, 99.9995%) flow 1.2 mL/min. MS parameters: electron impact ionization (EI, 70 eV) in positive mode; ion source temperature 230 °C; scan range 35–700 m/z; interface temperature 280 °C. Perfluorotributylamine (PFTBA) was used for mass spectrometer tuning. MSD ChemStation (Agilent Technologies) software was used for data analysis and peak assignment was based on mass spectra libraries (NIST 8, score higher than 80%) and literature data [20]. For the analysis of the polymer samples and the polymer residues (after extraction), the pyrolysis furnace was set at 600 °C and samples ranging from 50 to 100 µg were directly weighed in the deactivated stainless-steel pyrolysis cup with an XS3DU microanalytical scale (Mettler-Toledo, USA) with seven digits and a precision of 1 µg. For the analysis of the extracted fraction of the reference polymer samples, 1 mL of DCM was added to each vial containing the polymer dried extracts; regarding PS dried extract, 1 mL of methanol was used, since it is the solvent used for its extraction. Then, different volumes of the extracts (20–340 µL) were directly dried in the pyrolysis cup and then weighed with an XS3DU microanalytical scale in order to obtain about 100 µg of sample. 4 µL of HMDS were added in the pyrolysis cup as derivatizing agent in order to detect polar and low-volatility compounds. Pyrolysis temperature was set at 550 °C. HMDS was used in the cleaning pre-treatment of the Py–GC–MS system.

2.4.3. Size-Exclusion Chromatography (SEC)

Size-exclusion chromatography (SEC) analysis of the extracts of unaged and aged polymers was performed with a Jasco (Jasco Europe srl, Cremella, LC, Italy) instrument comprising a PU-2080 Plus four-channel pump with degasser, two in series PL gel MIXED-E Mesopore (Polymer Laboratories, Church Stretton, UK) columns placed in a Jasco CO-2063 column oven thermostated at 30 °C, a Jasco RI 2031 Plus refractive index detector, and a Jasco UV-2077 Plus multi-channel ultraviolet (UV) 120 spectrometer; the ChromNav Jasco software was used for data acquisition and analysis. The eluent was trichloromethane (HPLC grade Sigma-Aldrich) at 1 mL/min flow rate.

3. Results and Discussion

The micropowders of all reference polymers were artificially aged 4 weeks under conditions roughly corresponding to a 6-month exposure at the latitude of the Tuscany region, central Italy. The pristine (unaged virgin polymers) and the irradiated samples collected after each subsequent 1-week period of artificial aging were analyzed by means of EGA–MS and Py–GC–MS to gain information about the extent of the photo-oxidative degradation and the type of chemical damage induced by the photo-aging. Pristine and artificially aged polymers were also extracted with solvents suitable for separating the low molecular weight photo-oxidized fragments from the insoluble bulk polymer. Finally, the two fractions obtained upon solvent extraction of each sample (organic extract containing the soluble degradation products, and residue containing the insoluble polymers) were characterized by Py–GC–MS and, in the case of the extracts containing the soluble fraction produced upon aging, also by SEC. The extraction yields and their variation during artificial aging are reported in Figure 2. The extraction yields increased steadily throughout the investigated aging time for LDPE, PP, and PS. The higher sensitivity of these polymers towards aging can be related to the presence of tertiary and benzyl carbon atoms, which are more prone to be attacked by free radical species directly or indirectly generated by photo-irradiation, and therefore more susceptible to undergo C–C bond cleavage as a result of secondary processes (e.g., β-cleavage of oxy-radicals generated upon decomposition of peroxy-radicals, the latter resulting from oxygen pickup by the primary radicals produced by H-abstraction). However, while the extractable fraction increases linearly with the irradiation time for the three polyolefins (i.e., including HDPE), for PS a tendency resembling an exponential growth or indicating some initial inhibition effect is observed. This result can be interpreted as an effect of possible radical scavenging associated with the presence of the monosubstituted phenyl ring and/or with the higher glass transition temperature (Tg) of PS: a lower diffusional mobility of free radical species and thus an induction time associated with slower increase of their concentration is typically associated with free radical transfer and oxygen pickup.

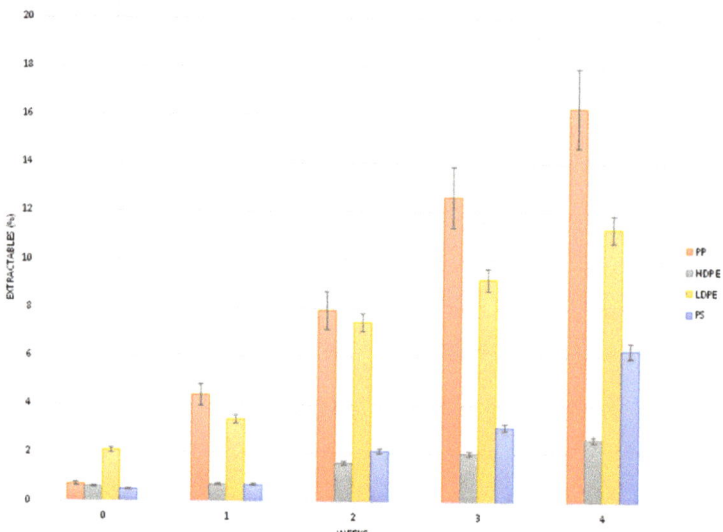

Figure 2. Extractable fractions (w%) for each reference polymer sample upon artificial aging (10% error bars).

3.1. EGA–MS and Py–GC–MS Analysis of Polymers during Artificial Aging

EGA analyses were performed on the 0w (unaged), 1w, 3w, and 4w polymer samples to evaluate changes in the thermal degradation temperature profiles. The 0w and 4w samples, representing the initial and final situation, were also analyzed by Py–GC–MS. The results are reported and discussed in the following paragraphs, separately for each polymer type.

3.1.1. Polypropylene

The EGA profiles for virgin and aged PP are shown in Figure 3, each curve being the average of five replicates.

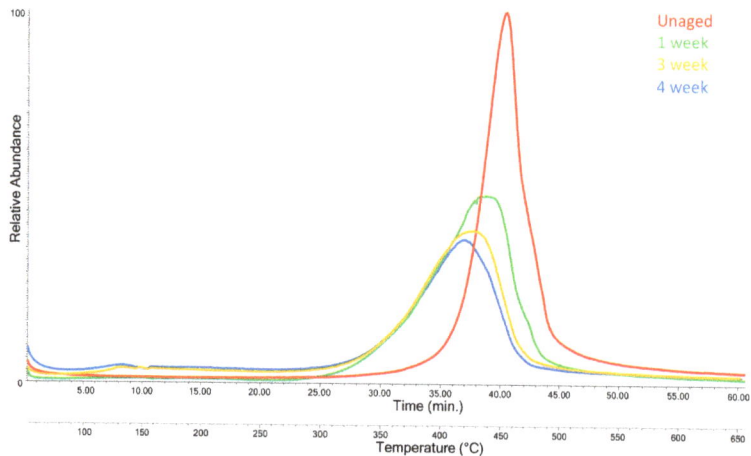

Figure 3. Evolved gas analysis (EGA) profiles of polypropylene (PP) after different artificial aging times: PP-0w (red), PP-1w (green), PP-3w (yellow), and PP-4w (blue) accelerated aging.

Based on a statistical evaluation of the five replicated analyses (*t*-test, 95%), a difference higher than 5.2 °C in the thermal degradation temperature, T_D, taken as the temperature corresponding to the maximum of the EGA peak, can be considered as significant. The EGA profiles show a change in the thermal degradation profile of PP upon aging. In particular, photodegradation induces a progressive decrease of T_D and a broadening of the thermal degradation peak (Table 1), with a T_D drop of 35 °C from T_D = 453 °C for the unaged PP-0w down to 418 °C for PP-4w. An even more consistent drop is observed for the onset temperature of the thermal degradation peak: from 421 °C for the unaged PP-0w down to 350 °C for PP-4w.

Table 1. Degradation temperature ranges and peak maxima (T_D max) obtained from the evolved gas analysis–mass spectrometry (EGA–MS) profiles of the PP samples.

Polypropylene	Degradation Temperature Range (°C) [a]	T_D Max (°C)
PP-0w	421–480	453
PP-1w	350–484	437
PP-3w	350–463	424
PP-4w	350–458	418

[a] onset and offset temperatures determined at the intersection of the baseline with the tangent at the inflection point of the upward and downward slope of the EGA peak, respectively.

Each average mass spectrum is obtained by averaging the EGA–MS mass spectra in the temperature range corresponding to the peak. The most abundant ion fragments in the average EGA mass spectrum of PP-0w (see Figure S1 in the Supplementary Materials; spectra collected in the temperature range from 421 °C to 480 °C) are m/z 43, 69, 83, 97, 111, 125, 153, corresponding to PP oligomers with different chain lengths (Table S1 in Supplementary Materials) [20].

The progressive shift of TD max towards lower temperatures at increasing aging time is in agreement with the known strong susceptibility of PP to chain scission reactions as a result of photo-induced degradation processes, being PP reactivity higher than that of both HDPE and LDPE, due to the higher concentration of tertiary C–H bonds [10,28,29]. This behavior is in full agreement with the evolution of the EGA profiles during artificial aging as observed in our experiments, featuring the progressive T_D lowering and peak broadening discussed above, as a result of the growing concentration of chain scission products and of reactive oxidized and unsaturated groups, less thermally stable than saturated hydrocarbon structures. Such structural changes did not induce any significant change in the EGA–MS mass spectra of the polymer thermal degradation products.

The Py–GC–MS chromatogram of PP-0w is shown in Figure 4, while its peak assignments are listed in Table S1 in Supplementary Materials. The most abundant pyrolysis products are 2,4-dimethyl-1-heptene (n° 6), 2,4,6-trimethyl-1-nonene (n° 12, 13), 2,4,6,8-tetramethyl-1-undecene (n° 16, 17, 18) along with other polypropylene oligomers of increasing chain length (n° 19–44), in agreement with the well-known pyrolytic behavior of PP involving random C–C scissions followed by intramolecular H transfer yielding alkanes and 1-alkenes [20,30]. The pyrolysis profiles of PP-0w and PP-4w are very similar. In particular, no pyrolysis products indicative of the occurrence of photoaging-related polymer degradation could be detected, even if the EGA profile did highlight the occurrence and progress of polymer degradation processes during the aging experiment. The lack of detectable molecular fragments that could be associated with the increasing fraction of oxidized products resulting from the photo-oxidative degradation may be explained by their low concentration in the bulk polymer, and by the poor effectiveness of the GC–MS separation and detection section in a conventional Py–GC–MS apparatus when oxidation products are involved, unless in situ thermally assisted derivatization of the carboxylic/hydroxyl functions is applied prior to the analysis

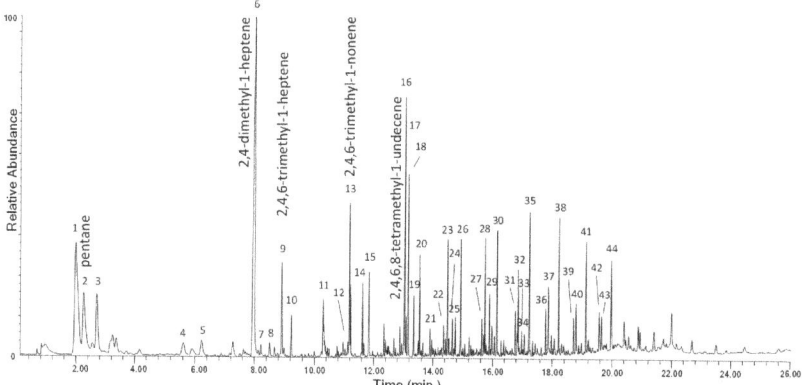

Figure 4. Chromatogram obtained in the pyrolysis–gas chromatography–mass spectrometry (Py–GC–MS) analysis of the unaged PP (PP-0w). Peak identification is reported in Table S1 in Supplementary Materials.

3.1.2. Polystyrene (PS)

All the EGA–MS curves feature a single peak in the 350–450 °C range (Figure S2 in Supplementary Materials), with T_D max at about 406 °C. A slight decrease of the onset temperature ($\Delta T = 8$ °C) observed in the EGA profile of the PS-3w and PS-4w samples can be associated to the formation of photo-oxidized products, although in a comparatively smaller amount with respect to the case of PP as seen before. In the average mass spectrum of PS (350–450 °C, Figure S3 in Supplementary Materials), the most abundant ions are fragments with m/z 51, 65, 78, 91, 104, 117, 207, corresponding to the ions in the mass spectra of the well-known pyrolysis products of the polymer (toluene, styrene, styrene dimer and styrene trimer, Table S2 in Supplementary Materials) [20]. By comparing the mass spectra of the four samples, no significant differences in the relative abundance of the main ions are detected. In the Py–GC–MS chromatogram of PS-0w (Figure 5 and Table S2 in Supplementary Materials) the most abundant pyrolysis products are styrene (n° 2), α-methylstyrene (n° 5), 3-butene-1,3-diyldibenzene (styrene dimer, n° 13) and 5-hexene-1,3,5-triyltribenzene (styrene trimer, n° 17). The observed decrease in the onset temperature is in agreement with the typical thermal degradation processes of PS, mainly characterized by depolymerization pathways.

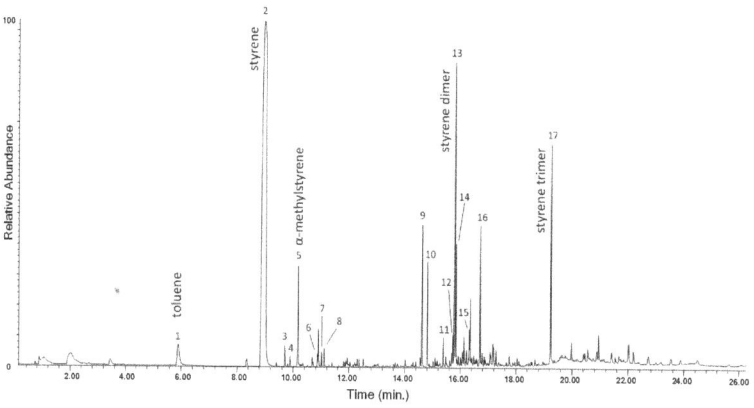

Figure 5. Chromatogram obtained in the Py–GC–MS analysis of the unaged polystyrene (PS-0w). Peak identification is reported in Table S2 in Supplementary Materials.

The pyrolysis profile of PS-4w (Figure S4 in Supplementary Materials) and PS-0w are very similar. This lack of differences between the two pyrolysis profiles can be explained by the tendency of PS to produce hydroquinonic structures as a result of photo-oxidation, which are known for their antioxidant and free radical scavenging properties and are thus likely to generate a chemically altered surface layer protecting the bulk polymer from further photo-oxidative degradation [28,31].

The yellowing observed in the PS-4w sample is a direct consequence of such surface-limited formation of oxidized and possibly conjugated aromatic structures, which do not significantly affect the EGA–MS curves and the pyrolysis profiles of the aged PS because degradation only involves a small fraction of the overall polymer mass.

3.1.3. Polyethylene Terephthalate (PET)

The EGA profile of the unaged PET-0w sample (Figure S5 in Supplementary Materials) shows a peak from 381 °C to 463 °C with a maximum (T_D) at 410 °C. The EGA profiles of the artificially aged samples do not show any significant variation in the T_D or in the relative abundance of the main ions in the average mass spectrum (381–463 °C, Figure S6 in Supplementary Materials). The latter are fragments with m/z 44, 77, 105, 122, 149, 297 which correspond to the ions in the mass spectra of the thermal degradation of the polymer: vinyl benzoate, benzoic acid, divinyl terephthalate, and 2-(benzoyloxy) ethyl vinyl terephthalate. As in the case of PS, a slight decrease of the onset temperature ($\Delta T = 7$ °C) is only observed in the EGA profile of the PET-4w sample, that can be related to the formation of photo-oxidation products; an extended artificial aging time would be necessary to better investigate the degradation behavior.

The chromatogram obtained in the Py–GC–MS analysis of the unaged PET-0w is shown in Figure 6, while peak identification is listed in Table S3 in Supplementary Materials. The main pyrolysis products of PET are vinyl benzoate (n° 9), benzoic acid (n° 10), biphenyl (n° 11), divinyl terephthalate (n° 12) and ethan-1,2-divinyldibenzoate (n° 18). The Py–GC–MS profile of the PET-4w sample (Figure S7 in Supplementary Materials) is also in this case very similar to that of the unaged polymer, in agreement with its well-known higher photostability compared to polyolefins [24,28]. In the Py–GC–MS profile of the artificially aged polymer only a slight increase of the relative abundance of acetophenone, benzaldehyde, vinyl benzoate, dibenzofuran, and fluorenone is observed.

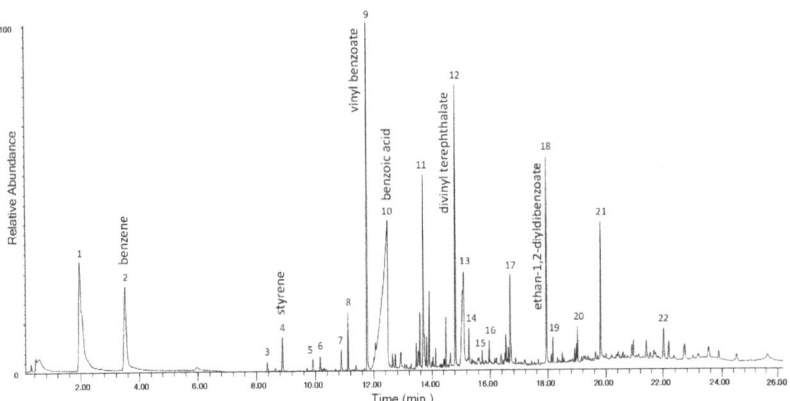

Figure 6. Chromatogram obtained in the Py–GC–MS analysis of the unaged polyethylene terephthalate (PET-0w). Peak identification is reported in Table S3 in Supplementary Materials.

3.1.4. Polyethylene (PE)

Both HDPE and LDPE were investigated, and are both discussed in this section. The EGA profile of the unaged (Figure S8 in Supplementary Materials) LDPE-0w sample shows a peak with T_D at 454 °C; artificial aging induces a slight increase of the baseline and a shift (ΔT = 7 °C) of the onset to lower temperatures, indicative of the formation of oxidation products. No significant changes are observed in the average mass spectra of the artificially aged LDPE samples (Figure S10 in Supplementary Materials). The main ions are the fragments with m/z 43, 57, 69, 83, 97, 111, 125, 139, 154 which correspond to polyethylene oligomers of different chain lengths (Table S4) [20]. The EGA curves of the unaged and aged HDPE samples (Figure S9 in Supplementary Materials) show a peak with T_D at 474 °C, 20 degrees higher than that recorder for LDPE samples. Even in this case, artificial aging induces a slight decrease of the onset temperature (ΔT = 6 °C) probably due to the presence of oxidation products at low concentration as a result of photo-oxidative degradation. The average mass spectra from the EGA profiles of the HDPE samples (Figure S11 in Supplementary Materials) are equivalent to the LDPE ones. The results obtained in the Py–GC–MS of the LDPE-0w are reported in Figure 7 and Table S4 in Supplementary Materials. The pyrogram features a series of clusters comprising three main peaks each, assigned to the diene (most likely an α,ω-diene, $C_{n:2}$), the monoalkene (most likely a 1-alkene, $C_{n:1}$), and the alkane, respectively, for any given C_n hydrocarbon in the chain lengths range C_6–C_{26}. At the right of each cluster, the peak corresponding to the C_{n-2} linear aldehyde is observed, with low relative intensity.

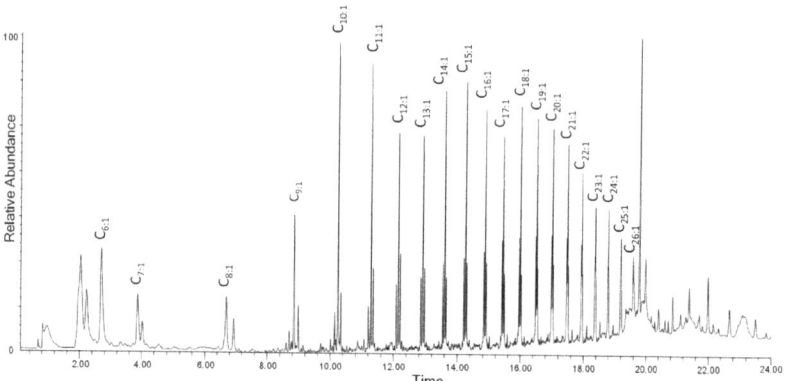

Figure 7. Chromatogram obtained in the Py–GC–MS analysis of the unaged low-density polyethylene (LDPE-0w). $C_{n:1}$ refers to 1-alkenes of a given C_n hydrocarbon. Peak identification is reported in Table S4 in Supplementary Materials.

After artificial aging, new peaks next to the triplets are observed in the pyrolytic profiles of both HDPE and LDPE; these can be associated to oxidation products and in particular to linear ketones, linear saturated alcohols, and monocarboxylic acids. Peaks corresponding to saturated aldehydes, with lengths up to C24, increase in their relative intensity with aging (Figure 8) [32]. The complete list of all the pyrolysis products detected in the chromatogram obtained in the Py–GC–MS analysis of LDPE-4w is reported in Table S5 in the Supplementary Materials. The oxidation products detected in the chromatogram can either be the products of the thermolytic cleavage of mildly oxidized high molecular weight polymer chains, or smaller oxidized fragments produced by photolytic oxidation and chain scission as a result of photo-oxidative artificial aging, or both. Even though the general features of the pyrolysis profiles of the two aged polymers are very similar, in the case of LDPE a slightly higher number of oxidized pyrolysis products were detected compared to HDPE, probably due to the different intermolecular forces in the structures of the two polymers and consequence differences in the ageing behaviors. Tables S6 and S7 report the complete list of the pyrolysis products identified in the chromatogram obtained in the Py-GC-MS analysis of the HDPE-0w and HDPE-4w samples, respectively.

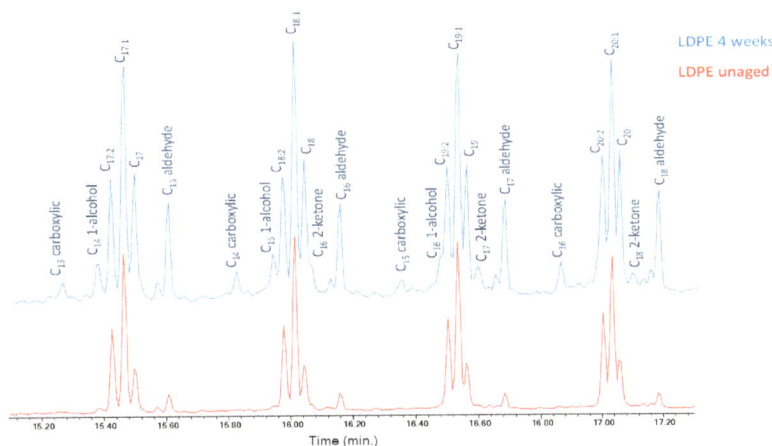

Figure 8. Chromatograms (15.20–17.20 min) obtained in the Py–GC–MS analysis of LDPE-0w (red) and LDPE-4w (blue). $C_{n:2}$ refers to α,ω-dienes, $C_{n:1}$ to 1-alkenes, and C_n to alkanes of a hydrocarbon with n carbon atoms.

3.2. Analysis of Extractable Fraction of Reference Polymers before and after Artificial Aging

The results obtained suggest that low molecular weight degradation and oxidation products are not detectable by Py–GC–MS analysis of the aged polymer, because the peaks deriving from the unaltered fraction of the polymer hinder the detection of low-intensity peaks associated to the lower fraction of altered portion of the polymer. The alteration induced on the five reference plastics by the photoaging was thus investigated also by a complementary approach: the aged samples were extracted with polar organic solvent and the composition of the extracts was investigated by SEC and Py–GC–MS, and compared with the extracts obtained from the unaged corresponding polymer. PP, PET, HDPE, and LDPE were subjected to DCM extraction, while MeOH was used for PS. Procedural blanks were also prepared for comparison. Py–GC–MS analysis of the extractable fractions allowed us to achieve an enhanced sensitivity towards the degradation products, focusing on the extractable and leachable components to gain additional information on the degradation processes occurring during the aging of MPs. Py–GC–MS was carried out with the addition of HMDS in order to detect and characterize the low-volatile and polar compounds such as aldehydes, alcohols and carboxylic acids strictly related to photo-oxidative degradation; HMDS also achieves the derivatization of PET pyrolysis products.

3.2.1. Polypropylene

The chromatogram obtained in the Py(HMDS)–GC–MS analysis of the DCM extract of the PP-4w sample is reported in Figure 9, with peak identification in Table 2. The pyrolytic profile shows oxidized products such as different chain length mono- and dicarboxylic acid in the first part of the chromatogram (14–24 min). In particular, low-molecular weight differently branched monocarboxylic acids are detected in the C_2–C_6 range, such as butanoic acid (n° 5), 2-butenoic acid (n° 6), 2-methyl-4-pentenoic acid (n° 8), 3-methyl-3-butenoic acid (n° 9), 2-hydroxypropanoic acid (n° 12), hydroxyacetic acid (n° 13) 2-hydroxy-2-propenoic acid (n° 15), 2-hydroxybutanoic acid (n° 16), 4-oxopentanoic acid (n° 17), 3-hydroxypropanoic acid (n° 18), 3-hydroxybutanoic acid (n° 19), and 3-hydroxy-3-butenoic acid (n° 20). Low molecular weight dicarboxylic acids such as butanedioic acid (n° 22) and methylbutanedioic acid (n° 23) are also observed in the first part of the profile. The second part of the chromatogram (24–35 min) is mainly characterized by different chain-length PP oligomers that are soluble in DCM.

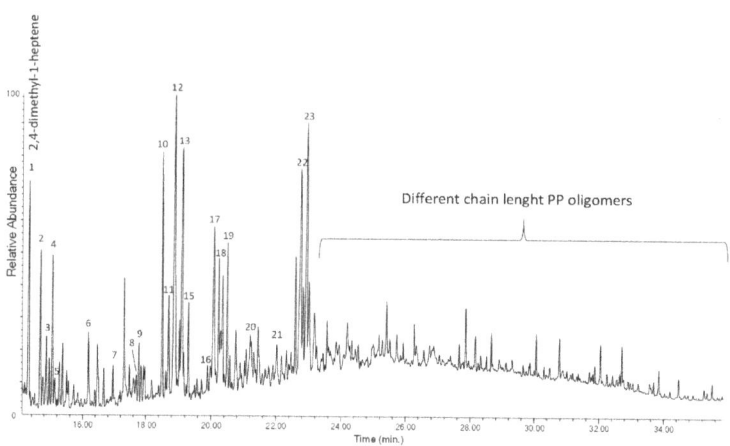

Figure 9. Chromatogram obtained in the Py(HMDS, hexamethyldisilazane)–GC–MS analysis of the dichloromethane (DCM) extract of PP-4w. Peak identification is reported in Table 2.

Table 2. Identification of peaks in the chromatogram obtained in the Py(HMDS)–GC–MS analysis of the DCM extract of PP-4w (Figure 9). Bold: most abundant species in the chromatogram; most abundant ions in the mass sppectra.

No.	t_r (min)	Peak Identification	Main Ions (m/z)
1	14.3	**2,4-dimethyl-1-heptene**	126, 83, **70**, 55
2	14.7	xylene	106, **91**
3	14.9	ethoxytriethylsilane	131, 103, **73**
4	15.05	**octamethyltrisiloxane**	**221**, 73
5	15.1	butanoic acid, trimethylsilyl ester	145, 117, **75**
6	16.2	2-butenoic acid, tert-butyldimethylsilyl ester	**143**, 99, 75, 59
7	16.9	1,2,3-trimethylbenzene	120, **105**
8	17.6	4-pentenoic acid, 2-methyl, trimethylsilyl ester	186, 171, 157, 117, **73**
9	17.8	3-butenoic acid,3-methyl, trimethylsilyl ester	172, 157, 127, 113, **73**, 54
10	18.5	**methyltris(trimethylsiloxy)silane**	295, **207**, 191, 73
11	18.7	unknown	171, 146, 133, 117, **73**
12	18.9	**propanoic acid, 2-[(trimethylsilyl)oxy]-, trimethylsilyl ester**	233, 129, 191, **147**, 133, 117, 73
13	19.1	acetic acid, [(trimethylsilyl)oxy]-, trimethylsilyl ester	205, 190, 161, **147**, 133, 117, 103
14	19.2	unknown	171, 157, 145, 129, 117, 103, **75**
15	19.3	2-propenoic acid, 2-[(trimethylsilyl)oxy]-, trimetylsilyl ester	217, **147**, 131, 73
16	19.9	butanoic acid, 2-[(trimethylsilyl)oxy]-, trimethylsilyl ester	233, 205, 190, 147, **131**, 73
17	20.1	**pentanoic acid, 4-oxo-, trimethylsilyl ester**	173, 155, 145, 131, **75**
18	20.2	**propanoic acid, 3-[(trimethylsilyl)oxy]-, trimethylsilyl ester**	219, 177, **147**, 133, 116, 73
19	20.5	**butanoic acid, 3-[(trimethylsilyl)oxy]-, trimethylsilyl ester**	223, 191, **147**, 130, 117, 73
20	21.2	3-butenoic acid,3-(trimethylsilyloxy)-,trimethylsilyl ester	231, 157, **147**, **73**
21	22.1	malic acid, O-(trimethylsilyl)-, bis(trimethylsilyl) ester	245, 233, **147**, **73**
22	22.7	**butanedioic acid, bis(trimethylsilyl) ester**	247, **147**, 129, 73
23	22.9	**butanedioic acid, methyl-, bis(trimethylsilyl) ester**	261, 217, **147**, 129, 73

The results show the presence of oxidation products that could not be detected in the pyrogram of the aged bulk sample analyzed without in situ thermally assisted derivatization with HMDS of the carboxylic/hydroxyl functions, due both to their low concentration in the bulk polymer, and to their polarity—low volatility—that prevented their GC separation as such. The carboxylic and dicarboxylic acids observed in the Py–GC–MS analysis of the extract of artificially aged PP-4w are not detected in the extract of unaged PP-0w (Figure S12 in Supplementary Materials), clearly indicating that they are the result of photo-oxidative degradation rather than of pyrolytic fragmentation. Figure 10 reports the SEC chromatograms (10–25 min) obtained for the DCM extracts of PP-0w (red), PP-1w (green), PP-3w (yellow), and PP-4w (blue). The profiles show two peaks at high retention times (about 20.4 min and 22.5 min) corresponding to low molecular weight fractions. This is expected since high molecular weight polyolefins are insoluble in DCM. No significant differences are highlighted when comparing the SEC profiles of the unaged and aged PP extracts.

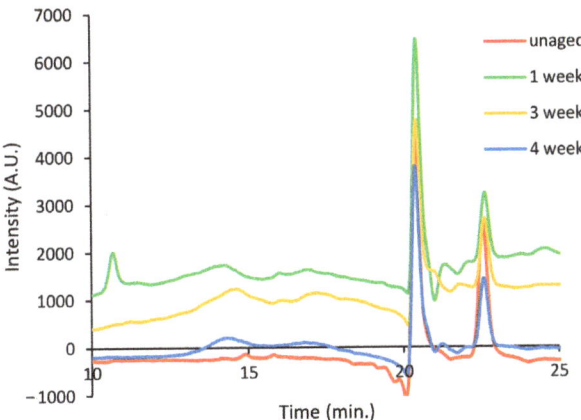

Figure 10. Size exclusion chromatography (SEC) chromatograms (10–25 min) of the DCM extracts of PP-0w (red), PP-1w (green), PP-3w (yellow), and PP-4w (blue); refractive index detector was used.

3.2.2. Polystyrene

The chromatogram obtained in the Py(HMDS)–GC–MS analysis of the MeOH extract of PS-4w is reported in Figure 11. Peak identification is in Table 3. The main pyrolysis products are the same as those observed in the Py–GC–MS chromatogram of PS-4w analyzed in "bulk" (non-subjected to extraction, Figure 5): styrene (n° 2), benzoic acid (n° 26), its dimer (n° 39), and its trimer (n° 53). Different acids and alcohols deriving from benzoic acid are detected, like 4-methylphenol (n° 20), 1-phenylethenol (n° 25), 3-methylphenol (n° 27), phenylacetic acid (n° 28), phenylpropanoic acid (n° 32), and 4-hydroxybenzoic acid (n° 36); dicarboxylic acids and other carboxylic acids are also found: 2-hydroxy-propanoic acid (n° 14), hydroxyacetic acid (n° 16), 4-hydroxy pentanoic acid (n° 17), 3-hydroxypropanoic acid (n° 19), butanedioic acid (n° 29), methylbutanedioic acid (n° 30), 1,4-benzenedicarboxylic acid (n° 42), pentadecanoic acid (n° 48), hexadecenoic acid (n° 51), and octadecanoic acid (n° 52).

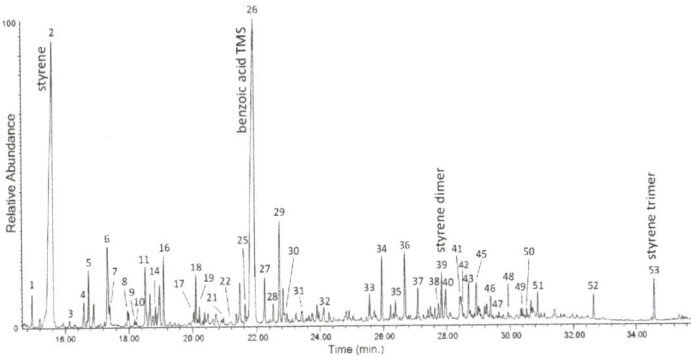

Figure 11. Chromatogram obtained in the Py(HMDS)–GC–MS analysis of the MeOH extract of PS-4w. Peak identification is reported in Table 3.

Table 3. Identification of peaks in the chromatogram obtained in the Py(HMDS)–GC–MS analysis of the MeOH extract of PS-4w (Figure 11). Bold: most abundant species in the chromatogram; most abundant ions in the mass sppectra.

No.	t_r (min)	Peak Identification	Main Ions (m/z)
1	14.9	ethylbenzene	106, **91**, 77, 65, 51
2	15.5	styrene	**104**, 89, 78, 63, 51
3	16.1	benzene, (1-methylethyl)-	120, **105**, 91, 77, 51
4	16.6	unknown	175, 146, **132**, 115, 102
5	16.7	HMDS unknown	**222**, 206, 190, 132, 74
6	17.3	HMDS unknown	**220**, 207, 188, 132, 73
7	17.4	cyclotrisiloxane, hexamethyl-	**207**, 191, 133, 96
8	17.9	cyclotetrasiloxane, octamethyl-	**281**, 265, 207, 191, 133, 73
9	18.1	benzene, 1-propenyl-	**117**, 103, 91, 77, 63, 51
10	18.2	benzene, (1-methylene-2-propenyl)-	**130**, 115, 102, 91, 77, 63, 51
11	18.5	**tetrasiloxane, decamethyl-**	295, **207**, 191, 73
12	18.6	silane, trimethylphenoxy-	166, **151**, 135, 91, 77
13	18.7	benzene, (1-methylenepropyl)-	132, **117**, 103, 91, 77, 63, 51
14	18.8	propanoic acid, 2-[(trimethylsilyl)oxy]-, trimethylsilyl ester	191, **147**, 133, 117, 73
15	18.9	acetophenone	120, **105**, 77, 51
16	19.1	acetic acid, [trimethylsilyl)oxy]-, trimethylsilyl ester	205, 177, **147**, 133, 73
17	20.0	pentanoic acid, 4-oxo, trimethylsilyl ester	173, 145, 131, **75**
18	20.1	4,6-dioxa-5-aza-2,3,7,8-tetrasilanonane-2,2,3,3,7,7,8,8-octamethyl-	294, **206**, 190, 130, 73
19	20.2	propanoic acid, 3-[(trimethylsilyl)oxy]-, trimethylsilyl ester	219, 177, **147**, 133, 116, 73
20	20.3	silane, trimethyl(4-methylphenoxy)-	180, **165**, 149, 135, 91
21	20.9	cyclopentasiloxane, decamethyl	355, 267, 251, 187, **73**
22	21.2	propanedioic acid, bis(trimethylsilyl) ester	233, 179, **147**, 73
23	21.3	pentasiloxane, dodecamethyl-	369, 353, **281**, 265, 207, 147, 43
24	21.6	unknown	281, 192, **117**, 151, 135, 115, 73
25	21.7	1-phenyl-1-(trimethylsilyloxy)ethylene	**191**, 177, 135, 103, 91, 75
26	21.9	**benzoic acid trimethylsilyl ester**	194, **179**, 135, 105, 77, 51
27	22.2	1-dimethylvinylsilyloxy-3-methylbenzene	192, **117**, 165, 151, 135, 91
28	22.5	phenylacetic acid, trimethylsilyl ester	193, 164, 91, **73**
29	22.7	**butanedioic acid, bis(trimethylsilyl) ester**	147, 172, **147**, 73
30	22.9	butanedioic acid, methyl-, bis(trimethylsilyl) ester	261, 217, **147**, 73
31	23.4	hexasiloxane, tetradecamethyl-	443, 355, 281, 267, 221, 147, **73**
32	24.1	phenylpropanoic acid, trimethylsilyl ester	222, 207, **104**, 91, 75
33	25.5	bibenzyl	182, **91**, 65
34	25.9	**1-pentene-2,4-diyldibenzen**	194, 115, **105**, 91
35	26.4	benzene, 1,1'-(1,2-dimethyl-1,2-ethanediyl)bis-	210, **105**, 91, 77
36	26.7	**benzoic acid, 4-[(trimethylsilyl)oxy]-, trimethylsilyl ester**	282, **267**, 223, 193, 73
37	27.1	benzene, 1,1'-(1,3-propanediy)bis-	196, 117, 105, **92**, 77, 65, 51
38	27.7	stilbene	**179**, 165, 152, 102, 89, 76, 51
39	27.8	3-butene-1,3-diyldibenzene (styrene dimer)	208, 130, 115, 104, **91**, 77, 65
40	28.0	unknown	**194**, 165, 152, 115, 91, 77, 51
41	28.4	1H-indene, 2-phenyl-	**192**, 165, 115, 91
42	28.5	1,4-benzenedicarboxylic acid, bis(trimethylsilyl) ester	310, **295**, 251, 221, 140, 103, 73
43	28.7	naphthalene, 1,2-dihydro-4-phenyl-	**206**, 191, 128, 115, 91
44	28.8	anthracene	**178**, 152, 89, 76
45	28.9	1,3-butadiene, 1,4-diphenyl-	**206**, 191, 178, 165, 128, 115, 91
46	29.4	naphthalene, 1-phenyl-	**204**, 101, 89
47	29.6	2,5-diphenyl-1,5-hexadiene	234, 143, **130**, 115, 104, 91, 77
48	29.9	pentadecanoic acid, trimethylsilyl ester	297, 145, 129, 117, **73**
49	30.4	fluoranthene, 1,2,3,10b-tetrahydro-	206, 190, **178**, 165, 152, 89, 76
50	30.5	naphthalene, 2-phenyl	**204**, 101, 89
51	30.9	hexadecanoic acid, trimethylsilyl ester	328, **313**, 145, 129, 117, 73
52	32.6	octadecanoic acid, trimethylsilyl ester	341, 145, 129, 117, **73**
53	34.6	**5-hexene-1,3,5-triyltribenzene (styrene trimer)**	312, 207, 194, 117, **91**, 77

None of the oxidized compounds identified in the pyrolysis profile of the MeOH extract of PS-4w are detectable in the extract of the PS-0w (Figure S13 in Supplementary Materials), indicating that the former are the result of extensive photo-oxidative degradation occurred during artificial aging. Figure 12a,b report the SEC chromatograms (10–27 min.) of the MeOH extracts of PS-0w (red), PS-1w (green), PS-2w (purple), PS-3w (yellow), PS-4w (blue), acquired at 260 nm and 340 nm, respectively.

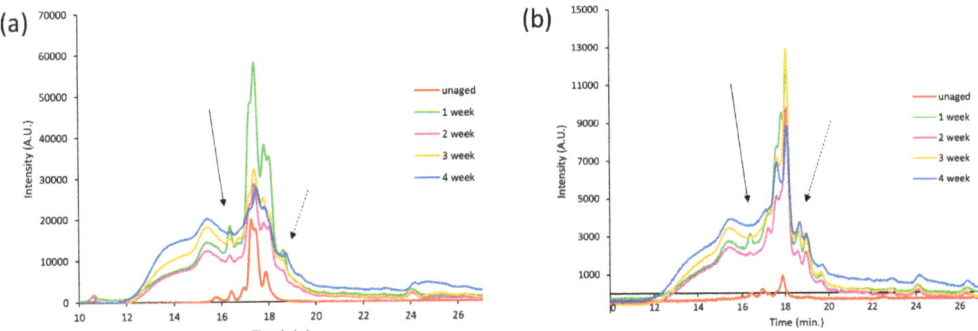

Figure 12. SEC chromatograms (10–27 min) of the MeOH extracts of PS-0w (red), PS-1w (green), PS-2w, PS-3w (yellow), and PS-4w (blue) acquired at 260 nm (**a**) and 340 nm (**b**).

The extracts of the aged PS samples show a broad structured band in the 12–20 min range, which corresponds to fractions with molecular weight ranging from 10,200 Da to values below 500 Da; whereas the profile of the extract of the PS-0w (red) appears less-intense in the 260 nm chromatogram, and it is almost non-detectable in the 340 nm chromatogram, showing a band in the 16–19 min range. Aging induces the disappearance of the peak at 16 min (black arrow) and the appearance of a new one at about 19 min (dotted black arrow) which corresponds to low-molecular weight fraction (below 500 Da).

3.2.3. Polyethylene Terephthalate

The chromatogram obtained in the Py(HMDS)–GC–MS analysis of the DCM extract of the PET-4w sample is reported in Figure 13, with peak identification in Table 4. The pyrograms of the DCM extract of PET-0w (Figure S14 in Supplementary Materials) and PET-4w samples are very similar, in agreement with the results obtained by EGA–MS and Py–GC–MS for the bulk (non-extracted) polymer, highlighting PET photo-oxidative stability. In particular, the typical pyrolysis products of the polymer are observed: benzoic acid (n° 14), vinyl benzoate (n° 22), and divinyl terephthalate (n° 24). The identified carboxylic acids and benzenedicarboxylic acids—hydroxybenzoic acid (n° 11), 3-phenyl-2-propenoic acid (n° 20) and 1,4-benezenedicarboxylic acid (n° 26)—were also present in the pyrogram of the DCM extract of the unaged sample, suggesting that they are not the result of photo-oxidative degradation but more likely they are the result of thermolytic cleavage and rearrangements occurring during pyrolysis, and that irradiation of PET microparticles did not produce a significant amount of extractable oxidation products, highlighting the stability of PET in the adopted conditions.

The PET extracts were not analyzed by means of SEC since the amount of the extractable fraction and the pyrolysis profiles of the extracts of PET-0w and PET-4w samples were nearly identical. The profile observed in the analysis of the extract of the artificially aged PET is essentially the same as that observed for the DCM-soluble fraction of the unaged polymer, and thus the contribution from polymer degradation processes can be assumed to be negligible.

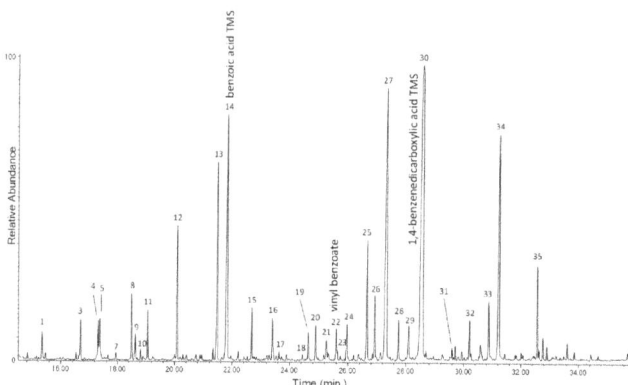

Figure 13. Chromatogram obtained in the Py(HMDS)–GC–MS analysis of the DCM extract of PET-4w. Peak identification is reported in Table 4.

Table 4. Identification of peaks in the chromatogram obtained in the Py(HMDS)–GC–MS analysis of the DCM extract of PET-4w (Figure 13). Bold: most abundant species in the chromatogram; most abundant ions in the mass sppectra.

No.	t_r (min)	Peak Identification	Main Ions (*m/z*)
1	15.3	styrene	**104**, 89, 78, 63, 51
2	16.4	unknown	175, 146, **132**, 115, 73
3	16.7	HMDS unknown	**222**, 206, 190, 132, 74
4	17.2	HMDS unknown	**220**, 204, 132, 73
5	17.3	HMDS unknown	**220**, 207, 188, 132, 73
6	17.4	unknown	223, 207, 191, **147**, 73
7	17.9	cyclotetrasiloxane, octamethyl	**281**, 265, 249, 193, 73
8	18.5	tetrasiloxane, decamethyl	295, **207**, 191, 73
9	18.6	silane, trimethylphenoxy-	166, **151**, 135, 91, 77
10	18.9	acetophenone	120, **105**, 77, 51
11	19.1	acetic acid, [trimethylsilyl)oxy]-, trimethylsilyl ester	205, 177, **147**, 133, 73
12	20.1	**2,2,3,3,7,7,8,8-octamethyl-4,6-dioxa-5-aza-2,3,7,8-tetrasilanonane**	**294**, 206, 190, 73
13	21.5	unknown	**293**, 205, 146, 130, 73
14	21.8	**benzoic acid trimethylsilyl ester**	194, **179**, 135, 105, 77, 51
15	22.7	butanedioic acid, bis(trimethylsilyl) ester	**247**, 172, **147**, 73
16	23.4	benzoic acid, 2-methyl-, trimethylsilyl ester	208, **193**, 149, 119, 91, 65
17	23.7	naphthalene, 2-ethenyl-	**154**, 128, 76
18	24.4	unknown	442, 354, 266, 206, **146**, 130, 73
19	24.5	decanoic acid, trimethylsilyl ester	**229**, 145, 129, 117, 73
20	24.9	2-propenoic acid, 3-phenyl-,trimethylsilyl ester	220, **205**, 161, 131, 103, 75
21	25.2	unknown	275, 147, 117, **73**
22	25.6	vinyl benzoate	**105**, 77, 51
23	25.9	benzoic acid, 3-[(trimethylsilyl)oxy]-, trimethylisilyl ester	282, **267**, 223, 193, 73
24	26.0	divinyl terephthalate	**175**, 147, 132, 104, 76
25	26.7	unknown	236, **221**, 177, 147, 91
26	26.9	1,4-benzenedicarboxylic acid, methyl trimethylsilyl ester	252, **237**, 221, 163, 135, 103
27	27.3	unknown	249, **221**, 205, 170, 103
28	27.7	1,4-benzenedicarboxylic acid, ethyl trimethylsilyl ester	**251**, 221, 207, 177, 149, 103, 76
29	28.1	1,3-benzenedicarboxylic acid, bis(trimethylsilyl) ester	**295**, 279, 221, 205, 140, 103, 73
30	28.5	**1,4-benzenedicarboxylic acid, bis(trimethylsilyl) ester**	310, **295**, 251, 221, 140, 103, 73
31	29.7	unknown	265, 221, **147**, 103, 73
32	30.2	unknown	**265**, 249, 175, 149, 104
33	30.9	unknown	313, **295**, 251, 221, 149, 117, 73
34	31.3	unknown	339, **221**, 140, 103, 73
35	32.6	**2,2-bis[(4-trimethylsilyloxy)phenyl]propane**	372, **357**, 207, 73

3.2.4. Polyethylene

The DCM extracts of LDPE and HDPE, 0w and 4w, were analyzed by Py(HMDS)–GC–MS. The chromatogram obtained from LDPE-4w is shown in Figure 14, while the main pyrolysis products are listed in Table S10 of the Supplementary Materials.

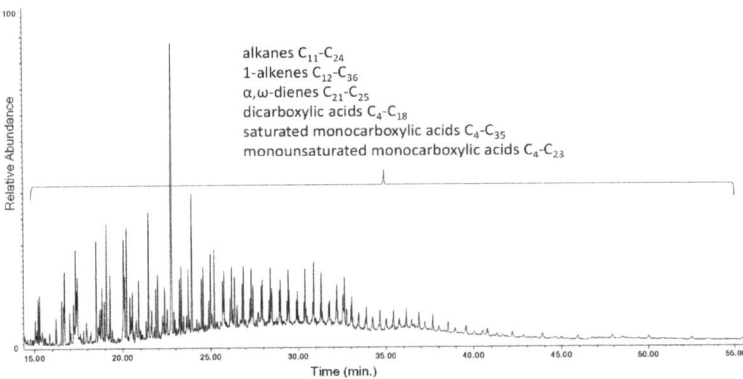

Figure 14. Chromatogram obtained in the Py(HMDS)–GC–MS analysis of the DCM extract of LDPE-4w. The list of the main pyrolysis products is reported in Table S10 in Supplementary Materials.

Saturated monocarboxylic acids, monounsaturated monocarboxylic acids, and dicarboxylic acids are detected in the Py(HMDS)–GC–MS analysis of the extract of the aged sample, together with the pyrolysis products that were observed in the analysis of the unaged bulk (non-extracted) polymer: α,ω-dienes, 1-alkenes and alkanes. These oxidized products are not observed in the Py–GC–MS profile of the extract of the LDPE-0w (in Supplementary Materials, see Figure S15 along with the list of all its main pyrolysis products in Table S11). The complete lists of the pyrolysis products detected in the Py(HMDS)–GC–MS chromatogram of the extracts of the HDPE-0w and HDPE-4w are reported in the Supplementary Materials in Table S12 and Table S13, respectively. The Py(HMDS)–GC–MS profiles of the pyrograms for the extracts of LDPE-4w and HDPE-4w show very similar general features; however, in the case of aged LDPE a higher number of oxidized pyrolysis products is observed. In particular, the analysis of the extract of the LDPE-4w highlights saturated monocarboxylic acids in the C_4–C_{35} range, monounsaturated monocarboxylic acids in the C_4–C_{23} range, and dicarboxylic acids in the C_4–C_{18} range. For the extract of the HDPE-4w sample, saturated monocarboxylic acids in the Py–GC–MS chromatograms are in the C_5–C_{20} range, monounsaturated monocarboxylic acids in the C_4–C_{18} range, and dicarboxylic acids in the C_4–C_{12} range (Table S13). Even in this case a recurrent pyrolysis pattern can be highlighted (Figure 15): along with the formation of the α,ω-diene and 1-alkene, we observe the presence of the dicarboxylic acid (C_{n-9}), the monounsaturated monocarboxylic acid (C_{n-4}) and the monocarboxylic acid (C_{n-4}) of a given hydrocarbon (C_n).

Figure 16a,b show the SEC chromatograms (10–24 min.) of the DCM extracts of HDPE-0w (red), HDPE-1w (green), HDPE-2w (purple), HDPE-3w (yellow), and HDPE-4w (blue) acquired with a RI detector and at 260 nm, respectively. The SEC chromatograms acquired with a RI detector of the HDPE samples show two peaks at around 20 and 22 min corresponding to the low molecular weight fractions that are expected, as for PP, due to the fact that the high molecular weight polyolefins are insoluble in DCM. All the extracts show two signals in the 260 nm chromatographic profiles that are not observed in the chromatograms acquired at 340 nm: the first one at about 10 min, corresponding to the one found in the RI chromatograms, and a second one at about 16 min characterized by low intensity that has no corresponding peak in the RI chromatogram, which suggests the presence of impurities. As for LDPE, the SEC-RI chromatograms are completely identical to the ones of the HDPE extracts, while the 260 nm SEC-UV chromatographic profiles do not show any peak.

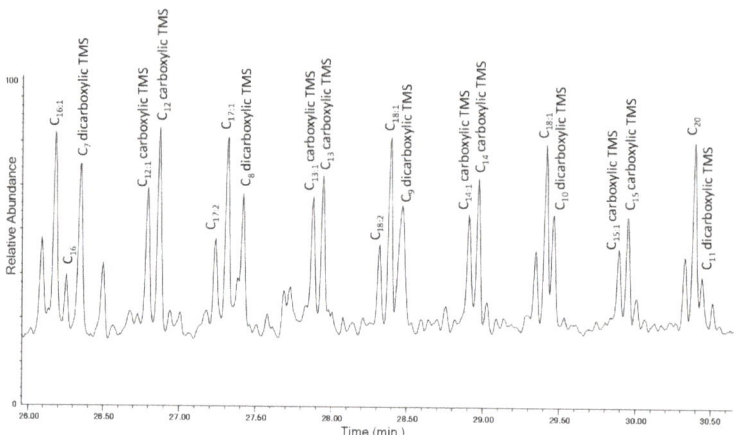

Figure 15. Chromatogram obtained (26.00–30.50 min) in the Py(HMDS)–GC–MS analysis of the DCM extract of LDPE-4w. $C_{n:2}$ refers to α,ω-dienes, $C_{n:1}$ to 1-alkenes, and C_n to alkanes of a given C_n hydrocarbon.

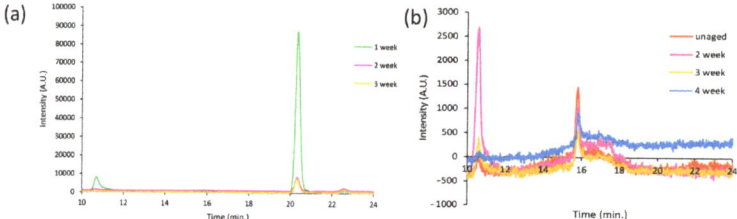

Figure 16. SEC chromatograms (10–24 min) of the DCM extracts of HDPE-0w (red), HDPE-1w (green), HDPE-2w (purple), HDPE-3w (yellow), and HDPE-4w (blue) acquired with a refractive index (RI) detector (**a**) and at 260 nm (**b**).

3.3. Analysis of Insoluble Fractions of Reference Polymers before and after Artificial Aging

The insoluble residues from the DCM (MeOH for PS) extractions were analyzed by means of Py–GC–MS to complement the analysis of the aged polymers and of the extracts, and to evaluate the effectiveness of the extraction procedure.

3.3.1. Polypropylene

The Py–GC–MS profiles of the extraction residues (DCM insoluble fraction) of PP-0w (red) and PP-4w (blue) are reported in Figure 17. The list of the pyrolysis products identified in both the chromatograms is in Table S14 in the Supplementary Materials. The two pyrolytic profiles do not show significant differences, and both feature the usual fragmentation pattern produced in the pyrolysis of PP, indicating that the DCM extraction has proven to be very effective. The extractable fraction of PP is observed to increase with aging: for the unaged PP-0w, the extractable fraction is 0.6% of the sample, while for PP-4w, it reaches 16.3%. PP oligomers with chain length up to C40 are detected in the Py–GC–MS analysis of the extracts, whereas in the pyrolytic profiles of the bulk (non-extracted) PP samples only PP oligomers up to C34 were detected. The usefulness of the extraction procedure is demonstrated by the fact that the oxidized fraction, which cannot be detected in the pyrogram of the aged bulk sample, is detected in the pyrogram of the corresponding extract, allowing the identification of higher molecular weight PP oligomers in the pyrogram of the corresponding extract residue.

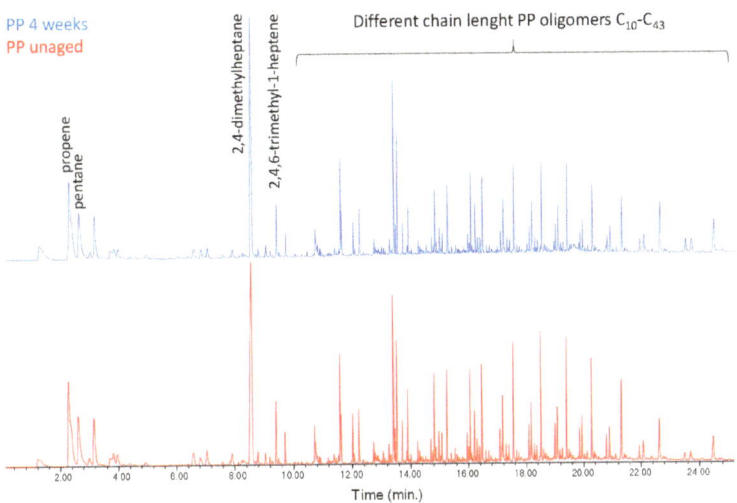

Figure 17. Chromatograms obtained in the Py–GC–MS analysis of the extraction residues of PP-0w (red) and PP-4w (blue). The complete list of the main pyrolysis products is reported in Table S14 in the Supplementary Materials.

3.3.2. Polystyrene

The Py-GC-MS profiles of the extraction residues (MeOH insoluble fraction) of PS-0w (red) and PS-4w (blue) are reported in Figure 18. The complete list of the pyrolysis products identified in both chromatograms in Table S15 in the Supplementary Materials. The pyrolytic profiles of the extraction residues of the unaged PS-0w and aged PS-4w samples are very similar, showing mainly the pyrolysis products of reference unaged PS: toluene, styrene and its dimer and trimer. By comparing the two profiles, the only difference is the lower relative abundance of the styrene dimer and styrene trimer in the profile of the aged PS-4w. After MeOH extraction, the insoluble residue of the yellowed PS-4w becomes colorless, indicating that the colored degradation products formed during artificial aging were efficiently extracted in MeOH.

3.3.3. Polyethylene Terephthalate

The Py-GC-MS profiles of the extraction residues (DCM insoluble fraction) of PET-0w and PET-4w resulted in being completely undistinguishable, as observed also for the DCM extracts. They are also very similar to the pyrograms of the corresponding bulk (non-extracted) samples featuring benzene, vinyl benzoate, benzoic acid, biphenyl, divinyl terephthalate and ethan-1,2-diyldibenzoate. As expected, PET appears to be only mildly affected by photo-oxidation under the adopted conditions. Such higher stability is the result of the absence of labile tertiary C–H bonds and of the prevalence of aromatic carbon atoms in the polymer structure. Figure S16 in the Supplementary Materials reports the chromatograms obtained in the Py–GC–MS analysis of the extraction residues of the PET-0w (red) and PET-4w (blue). The complete list of the pyrolysis products identified in both chromatograms is reported in Table S16 in the Supplementary Materials.

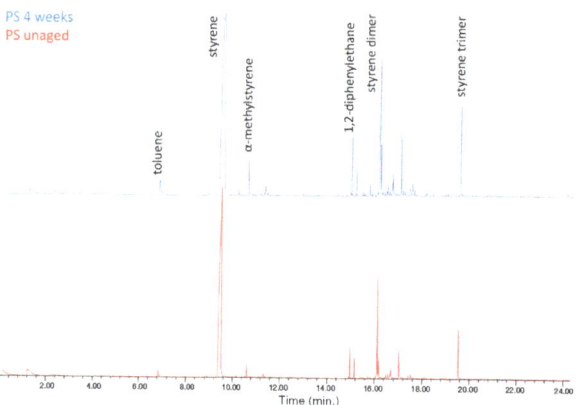

Figure 18. Chromatograms obtained in the Py–GC–MS analysis of the extraction residues of PS-0w (red) and PS-4w (blue). The complete list of the main pyrolysis products is reported in Table S15 in Supplementary Materials.

3.3.4. Polyethylene

The Py–GC–MS profiles of the extraction residues (DCM insoluble fraction) of LDPE, both unaged 0w and aged 4w, are very similar to those of the corresponding profiles for HDPE. Therefore, only LDPE is discussed here. Figure 19 reports the chromatograms obtained in the Py–GC–MS analysis of the extraction residues of the LDPE-0w (red) and LDPE-4w (blue). The complete list of the pyrolysis products identified in both the chromatograms is reported in Table S17 in the Supplementary Materials. The pyrogram of the extraction residue of LDPE-4w does not show the presence of any of the oxidized products (saturated aldehydes, monocarboxylic acids, alcohols, and ketones) that were observed in the analysis of the extract. DCM has proven to be as effective in the extraction of samples HDPE-4w and LDPE-4w as in the case of samples PP-4w and PS-4w. The extractable fraction increases with aging for both LDPE (from 2.1% for LDPE-0w to 11.3% for LDPE-4w) and HDPE (from 0.6% for HDPE-0w to 2.6% for HDPE-4w). Moreover, even in this case, in the pyrograms of the insoluble residues of both samples PE oligomers are observed to present higher molecular weight (up to C_{33}) than in the pyrograms of the corresponding bulk samples (molecular weight up to C_{26}).

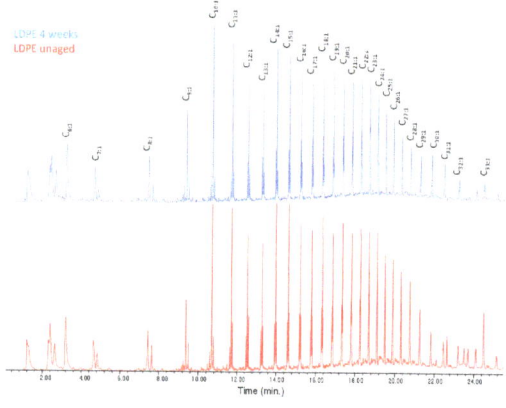

Figure 19. Chromatograms obtained in the Py–GC–MS analysis of the extraction residues of the LDPE-0w (red) LDPE-4w (blue). The complete list of the main pyrolysis products is reported in Table S17 in the Supplementary Materials.

4. Conclusions

The results of this paper demonstrate the potential of EGA–MS, Py–GC–MS, and SEC to provide a comprehensive overview of the effects of aging on the most common synthetic polymers. EGA–MS analyses allowed us to assess the changes in the thermal behavior of bulk polymers induced by photo-oxidative degradation. Py–GC–MS and Py(HMDS)–GC–MS analyses provided information on the degradation products formed during induced aging, even when the degradation products contained highly polar functional groups. The pyrolytic profiles of the insoluble extraction residues of unaged polymers were nearly identical to those of the corresponding insoluble fractions of polymers aged 4 weeks. This demonstrates the effectiveness and the necessity of a solvent extraction step to selectively investigate photo-oxidation products. Finally, SEC analyses allowed us to correlate the extent of oxidation and the reduction of molecular weight due to photo-oxidative degradation.

The stability of polyolefins towards aging was found, as expected, to be strongly related to their structure, and in particular to the number of tertiary carbon atoms, due to the higher stability of the radicals resulting from hydrogen abstraction. Thus, LDPE was more susceptible to degradation than HDPE due to the higher branching, and PP was significantly more susceptible than both types of PE. The main low molecular weight photo-degradation products of polyolefins were long chain alcohols, aldehydes, ketones, carboxylic acids, and hydroxy acids, which were observed in the soluble fractions of aged samples. EGA–MS profiles of PP showed the most marked decrease in the average molecular weight of the polymer chains as the aging time increased.

The two investigated aromatic polymers showed different behaviors. PS was significantly degraded by artificial aging. The main low molecular weight photo-degradation products were alcohols and carboxylic acid, the most abundant being benzoic acid and 1,4-benzenedicarboxylic acid. Interestingly, cross-linking was also observed as a consequence of aging, as highlighted by the SEC analyses; again, this may be the result of the higher stability of the generated free radicals that are more likely to undergo biradical coupling as their concentration in the polymeric material increases. Finally, PET showed the highest stability towards aging, as very small differences were observed comparing fresh and aged samples.

The described experimentation paves the ground for a more thorough and detailed characterization of the oxidation products leaching out from plastic debris dispersed in the environment, which may provide a better understanding of the mechanisms of interaction of MPs with the biosphere as they are likely to go well beyond the simple interaction of biota with solid polymer particles. Future studies should aim at investigating the possible connections between the pollution by microplastics and the toxicity potential of the small, oxidized species that may leach out of MPs in into the environment.

Supplementary Materials: The following are available online at https://www.mdpi.com/article/10.3390/polym13121997/s1, Figures S1–S16; Tables S1–S17.

Author Contributions: Conceptualization, A.C., V.C., Methodology and formal analysis, G.B., J.L.N., M.M., A.C., V.V.; all the authors participated in the investigation; all the authors participated in the writing—review and editing of the manuscript; visualization, J.L.N., A.C., V.V., V.C., F.M.; supervision, V.C., F.M.; project administration, V.C., F.M.; funding acquisition, V.C., F.M.; All authors have read and agreed to the published version of the manuscript.

Funding: Funding for the fellowship of V.V. came from Fondazione Cassa di Risparmio di Lucca (Bando Ricerca 2019-21, CISUP project "Micro- e nano-plastiche: metodologie di quantificazione, valutazione dell'impatto in ecosistemi marini e lacustri, strategie di remediation ambientale"). Additional funding from the University of Pisa, PRA project 2020_27 on Micro- and nanoplastics, is also gratefully acknowledged.

Institutional Review Board Statement: Not applicable.

Informed Consent Statement: Not applicable.

Data Availability Statement: The data presented in this study are available in the Supplementary Materials.

Conflicts of Interest: The authors declare no conflict of interest.

References

1. Andrady, A.L. Microplastics in the marine environment. *Mar. Pollut. Bull.* **2011**, *62*, 1596–1605. [CrossRef]
2. Betts, K. Why small plastic particles may pose a big problem in the oceans. *Environ. Sci. Technol.* **2008**, *42*, 8995. [CrossRef]
3. Browne, M.A.; Crump, P.; Niven, S.J.; Teuten, E.; Tonkin, A.; Galloway, T.; Thompson, R. Accumulation of microplastic on shorelines woldwide: Sources and sinks. *Environ. Sci. Technol.* **2011**, *45*, 9175–9179. [CrossRef]
4. Frere, L.; Paul-Pont, I.; Rinnert, E.; Petton, S.; Jaffré, J.; Bihannic, I.; Soudant, P.; Lambert, C.; Huvet, A. Influence of environmental and anthropogenic factors on the composition, concentration and spatial distribution of microplastics: A case study of the Bay of Brest (Brittany, France). *Environ. Pollut.* **2017**, *225*, 211–222. [CrossRef]
5. Jambeck, J.R.; Geyer, R.; Wilcox, C.; Siegler, T.R.; Perryman, M.; Andrady, A.; Narayan, R.; Law, K.L. Plastic waste inputs from land into the ocean. *Science* **2015**, *347*, 768. [CrossRef]
6. Shahul Hamid, F.; Bhatti, M.S.; Anuar, N.; Anuar, N.; Mohan, P.; Periathamby, A. Worldwide distribution and abundance of microplastic: How dire is the situation? *Waste Manag. Res.* **2018**, *36*, 873–897. [CrossRef]
7. Horton, A.A.; Walton, A.; Spurgeon, D.J.; Lahive, E.; Svendsen, K. Microplastics in freshwater and terrestrial environments: Evaluating the current understanding to identify the knowledge gaps and future research priorities. *Sci. Total. Environ.* **2017**, *586*, 127–141. [CrossRef]
8. Frias, J.; Nash, R. Microplastics: Finding a consensus on the definition. *Mar. Pollut. Bull.* **2019**, *138*, 145–147. [CrossRef]
9. Fendall, L.S.; Sewell, M.A. Contributing to marine pollution by washing your face: Microplastics in facial cleansers. *Mar. Pollut. Bull.* **2009**, *58*, 1225–1228. [CrossRef]
10. Gewert, B.; Plassmann, M.; Sandblom, O.; MacLeod, M. Identification of Chain Scission Products Released to Water by Plastic Exposed to Ultraviolet Light. *Environ. Sci. Technol. Lett.* **2018**, *5*, 272–276. [CrossRef]
11. Ceccarini, A.; Corti, A.; Erba, F.; Modugno, F.; La Nasa, J.; Bianchi, S.; Castelvetro, V. The Hidden Microplastics: New Insights and Figures from the Thorough Separation and Characterization of Microplastics and of Their Degradation Byproducts in Coastal Sediments. *Environ. Sci. Technol.* **2018**, *52*, 5634–5643. [CrossRef]
12. Bejgarn, S.; MacLeod, M.; Bogdal, C.; Breitholtz, M. Toxicity of leachate from weathering plastics: An exploratory screening study with Nitocra spinipes. *Chemosphere* **2015**, *132*, 114–119. [CrossRef] [PubMed]
13. La Nasa, J.; Lomonaco, T.; Manco, E.; Ceccarini, A.; Fuoco, R.; Corti, A.; Modugno, F.; Castelvetro, V.; Degano, I. Plastic breeze: Volatile organic compounds (VOCs) emitted by degrading macro- and microplastics analyzed by selected ion flow-tube mass spectrometry. *Chemosphere* **2021**, *270*, 128612. [CrossRef] [PubMed]
14. Schwaferts, C.; Niessner, R.; Elsner, M.; Ivleva, N.P. Methods for the analysis of submicrometer- and nanoplastic particles in the environment. *TrAC Trends Anal. Chem.* **2019**, *112*, 52–65. [CrossRef]
15. Zhang, S.; Wang, J.; Liu, X.; Qu, F.; Wang, X.; Wang, X.; Li, Y.; Sun, Y. Microplastics in the environment: A review of analytical methods, distribution, and biological effects. *TrAC Trends Anal. Chem.* **2019**, *111*, 62–72. [CrossRef]
16. Dümichen, E.; Eisentraut, P.; Bannick, C.G.; Barthel, A.-K.; Senz, R.; Braun, U. Fast identification of microplastics in complex environmental samples by a thermal degradation method. *Chemosphere* **2017**, *174*, 572–584. [CrossRef]
17. David, J.; Weissmannová, H.D.; Steinmetz, Z.; Kabelíková, L.; Demyan, M.S.; Šimečková, J.; Tokarski, D.; Siewert, C.; Schaumann, G.E.; Kučerík, J. Introducing a soil universal model method (SUMM) and its application for qualitative and quantitative determination of poly (ethylene), poly (styrene), poly (vinyl chloride) and poly (ethylene terephthalate) microplastics in a model soil. *Chemosphere* **2019**, *225*, 810–819. [CrossRef]
18. David, J.; Steinmetz, Z.; Kučerík, J.; Schaumann, G.E. Quantitative Analysis of Poly(ethylene terephthalate) Microplastics in Soil via Thermogravimetry–Mass Spectrometry. *Anal. Chem.* **2018**, *90*, 8793–8799. [CrossRef]
19. Primpke, S.; Fischer, M.; Lorenz, C.; Gerdts, G.; Scholz-Böttcher, B.M. Comparison of pyrolysis gas chromatography/mass spectrometry and hyperspectral FTIR imaging spectroscopy for the analysis of microplastics. *Anal. Bioanal. Chem.* **2020**, *412*, 8283–8298. [CrossRef]
20. Tsuge, S.; Ohtani, H.; Watanabe, C. *Pyrolysis-GC/MS Data Book of Synthetic Polymers: Pyrograms, Thermograms and MS of Pyrolyzates*; Elsevier: Amsterdam, The Netherlands, 2011.
21. Moldoveanu, S.C. *Analytical Pyrolysis of Synthetic Organic Polymers*; Elsevier: Amsterdam, The Netherlands, 2005.
22. La Nasa, J.; Biale, G.; Fabbri, D.; Modugno, F. A review on challenges and developments of analytical pyrolysis and other thermoanalytical techniques for the quali-quantitative determination of microplastics. *J. Anal. Appl. Pyrolysis* **2020**, *149*, 104841. [CrossRef]
23. Lomonaco, T.; Manco, E.; Corti, A.; La Nasa, J.; Ghimenti, S.; Biagini, D.; di Francesco, F.; Modugno, F.; Ceccarini, A.; Castelvetro, V.; et al. Release of harmful volatile organic compounds (VOCs) from photo-degraded plastic debris: A neglected source of environmental pollution. *J. Hazard. Mater.* **2020**, *394*, 122596. [CrossRef] [PubMed]
24. Castelvetro, V.; Corti, A.; Bianchi, S.; Ceccarini, A.; Manariti, A.; Vinciguerra, V. Quantification of poly(ethylene terephthalate) micro- and nanoparticle contaminants in marine sediments and other environmental matrices. *J. Hazard. Mater.* **2020**, *385*, 121517. [CrossRef]

25. La Nasa, J.; Biale, G.; Mattonai, M.; Modugno, F. Microwave-assisted solvent extraction and double-shot analytical pyrolysis for the quali-quantitation of plasticizers and microplastics in beach sand samples. *J. Hazard. Mater.* **2021**, *401*, 123287. [CrossRef]
26. La Nasa, J.; Biale, G.; Ferriani, B.; Trevisan, R.; Colombini, M.P.; Modugno, F. Plastics in Heritage Science: Analytical Pyrolysis Techniques Applied to Objects of Design. *Molecules* **2020**, *25*, 1705. [CrossRef] [PubMed]
27. Castelvetro, V.; Corti, A.; La Nasa, J.; Modugno, F.; Ceccarini, A.; Giannarelli, S.; Vinciguerra, V.; Bertoldo, M. Polymer Identification and Specific Analysis (PISA) of Microplastic Total Mass in Sediments of the Protected Marine Area of the Meloria Shoals. *Polymers* **2021**, *13*, 796. [CrossRef]
28. Gewert, B.; Plassmann, M.M.; MacLeod, M. Pathways for degradation of plastic polymers floating in the marine environment. *Environ. Sci. Process. Impacts* **2015**, *17*, 1513. [CrossRef]
29. Chiellini, E.; Corti, A.; D'Antone, S.; Baciu, R. Oxo-biodegradable carbon backbone polymers—Oxidative degradation of polyethylene under accelerated test conditions. *Polym. Degrad. Stab.* **2006**, *91*, 2739–2747. [CrossRef]
30. Luda, M.P.; Dall'Anese, R. On the microstructure of polypropylenes by pyrolysis GC-MS. *Polym. Degrad. Stab.* **2014**, *110*, 35–43. [CrossRef]
31. Bottino, F.A.; Cinquegrani, A.R.; Di Pasquale, G.; Leonardi, L.; Pollicino, A. Chemical modifications, mechanical properties and surface photo-oxidation of films of polystyrene (PS). *Polym. Test.* **2004**, *23*, 405–411. [CrossRef]
32. Castelvetro, V.; Corti, A.; Biale, G.; Ceccarini, A.; Degano, I.; La Nasa, J.; Lomonaco, T.; Manariti, A.; Manco, E.; Modugno, F.; et al. New methodologies for the detection, identification, and quantification of microplastics and their environmental degradation by-products. *Environ. Sci. Pollut. Res.* **2021**. [CrossRef]

Article

A Selective Ratiometric Fluorescent Probe for No-Wash Detection of PVC Microplastic

Valeria Caponetti [1,2,†], Alexandra Mavridi-Printezi [1,†], Matteo Cingolani [1], Enrico Rampazzo [1], Damiano Genovese [1], Luca Prodi [1], Daniele Fabbri [1,2] and Marco Montalti [1,2,*]

1. Department of Chemistry "Giacomo Ciamician", University of Bologna, Via Selmi 2, 40126 Bologna, Italy; valeria.caponetti2@unibo.it (V.C.); alexandra.mavridi2@unibo.it (A.M.-P.); matteo.cingolani4@unibo.it (M.C.); enrico.rampazzo@unibo.it (E.R.); damiano.genovese2@unibo.it (D.G.); luca.prodi@unibo.it (L.P.); dani.fabbri@unibo.it (D.F.)
2. Tecnopolo di Rimini, Via Dario Campana, 71, 47922 Rimini, Italy
* Correspondence: marco.montalti2@unibo.it
† These authors contribute equally.

Abstract: Microplastics (MP) are micrometric plastic particles present in drinking water, food and the environment that constitute an emerging pollutant and pose a menace to human health. Novel methods for the fast detection of these new contaminants are needed. Fluorescence-based detection exploits the use of specific probes to label the MP particles. This method can be environmentally friendly, low-cost, easily scalable but also very sensitive and specific. Here, we present the synthesis and application of a new probe based on perylene-diimide (PDI), which can be prepared in a few minutes by a one-pot reaction using a conventional microwave oven and can be used for the direct detection of MP in water without any further treatment of the sample. The green fluorescence is strongly quenched in water at neutral pH because of the formation dimers. The ability of the probe to label MP was tested for polyvinyl chloride (PVC), polyethylene (PE), polyethylene terephthalate (PET), polypropylene (PP), polystyrene (PS), poly methyl methacrylate (PMMA) and polytetrafluoroethylene (PTFE). The probe showed considerable selectivity to PVC MP, which presented an intense red emission after staining. Interestingly, the fluorescence of the MP after labeling could be detected, under excitation with a blue diode, with a conventional CMOS color camera. Good selectivity was achieved analyzing the red to green fluorescence intensity ratio. UV–Vis absorption, steady-state and time-resolved fluorescence spectroscopy, fluorescence anisotropy, fluorescence wide-field and confocal laser scanning microscopy allowed elucidating the mechanism of the staining in detail.

Keywords: microplastic; ratiometric detection; no-wash fluorescent probe; imaging; one-pot reaction; water remediation; nanoplastic

1. Introduction

In the last years, due to the rising demand of plastic items, plastics that are considered as the greatest invention of the last century, are now one of the main threats for both marine and terrestrial ecosystems [1]. In particular, microplastics (MP), which are plastic particles of micrometric size, are considered emerging contaminants as MP debris are abundantly found in many seas and rivers [2–4], but also in urban environments, reaching even high-level organisms as humans through the food chain due to bioaccumulation [5]. The accumulation of MP in the aquatic environment leads to physical, but also chemical, changes to the ecosystem, as, according to studies, hydrophobic MP are able to concentrate, via partitioning, and eventually re-release organic pollutants [6]. Acting as "Trojan horses", they transport faster and more efficiently the pollutants inside the cells promoting also the bioaccumulation of these organic contaminants in many living organisms such as fishes, crustaceans and mollusks, as well as insects, where their toxicological effects have been demonstrated [5,7]. For example, MP have been reported to affect the homeostasis of dragonfly larvae and zebrafish, causing also REDOX imbalance and increased immunity [8,9].

Indeed, MP are estimated to be more than 50% of the field-retrieved plastic debris found in organisms of all kinds and can be classified as primary and secondary depending on their origin [10]. Primary MP are intentionally manufactured micron-sized plastics, whereas secondary MP are derived by the fragmentation of larger plastic products [11,12]. The main types of MP debris found in the environment are polyethylene (PE), polypropylene (PP), polystyrene (PS), polyvinyl chloride (PVC) and polyethylene terephthalate (PET). Among them, PVC, due to its high density, more easily settles in freshwaters, which in many cases are drinking water sources, being less available during environmental sampling but more available to benthonic organisms [13]. Thus, it can be hypothesized that the real percentage of this polymer in the marine environments is probably underestimated, increasing the need for its detection. Moreover, PVC MP pose a very serious threat to the human organism, as they can circulate in the body and their long exposure can cause immunotoxicity to the cells stimulating the release of interleukin 6 (IL-6) and tumor necrosis factor-α (TNF-α) [14,15]. Large concentrations of PVC can affect freshwater photosynthetic systems, e.g., *Chlorella pyrenoidosa* and *Microcystis flos-aquae algae*, by inhibiting the photosynthetic ability and growth of the organisms [16]. Similar detrimental effects of PVC MP have also been reported in the case of marine phytoplankton, causing dose-dependent toxicity in various diatoms [17]. Thus, it is obvious that methods for the detection of PVC MP in environmental samples, biological tissues or even consumer products are urgently needed to determine the occurrence of this MP in the world.

The detection of MP can be achieved with the use of fluorescent probes [18,19] able to interact and, therefore, stain the polymers. In most cases, Nile Red, able to bind to different polymers leading to colors from yellow to red depending on the plastic's hydrophilicity, has been exploited as a fluorescent probe for the visual inspection and categorization of different MP [20–24]. Nile Red derivatives [25,26] have been presented as possible candidates able to improve the water solubility and selectivity of Nile Red [27]. Nevertheless, there are very few examples of other fluorescent probes. Some examples are the use of commercially available iDyes [28], Rhodamine B [29] and pyrene [30,31], all working optimally in organic solvents or requiring time consuming techniques. Perylene diimides (PDIs) are a class of industrial organic dyes which have been applied for a wide range of applications due to their stability and outstanding optical and electronic properties [32–34]. Focusing on the great need for PVC MP screening, we propose a highly selective, simple and time-saving approach based on a perylene diimide probe (P) shown in Figure 1, which can be carried out in water. This method has the potential to become an outstanding monitoring tool due to its low cost and high throughput.

For the detection of MP in real samples at a global level, fluorescent probes should be easily producible in large scale with environmentally friendly and inexpensive processes. Thus, we propose the use of microwave irradiation to achieve gram-scale production of the probe in a few minutes using a conventional microwave oven.

The main objective of this study was to exploit the effect of aggregation on the fluorescence of PDIs to achieve a no-wash selective probe for MP. In particular, the probe P was designed to be soluble in water at neutral pH, where it can form quenched aggregates giving no background signal and become fluorescent upon absorption on MP. As a result, the fluorescence of the MP can be detected directly in the staining solution without any washing or further treatment.

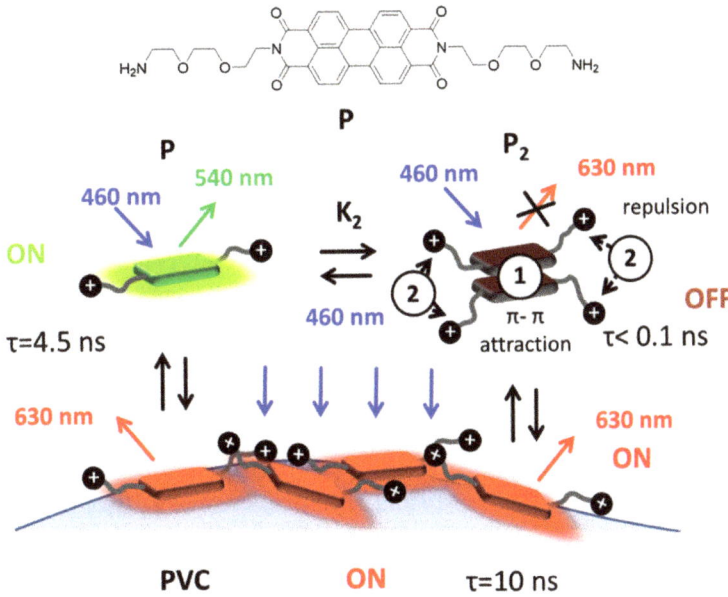

Figure 1. Structure of the fluorescent probe P that includes a central hydrophobic perylene-diimide (PDI) unit and two hydrophilic amino-terminated branches that are protonated in water at pH 7. In neutral water solution, P is, to a large extent, aggregated in the form of very weakly red emitting dimer, P_2, with very short excited state lifetime τ < 0.1 ns. Upon adsorption on polyvinyl chloride (PVC), P becomes intensely red fluorescent and the excited state lifetime increases to 10 ns. The attractive π–π and repulsive electrostatic interactions between P molecules are schematized as 1 and 2. Adsorption of PVC leads to disaggregation and switch-on the red fluorescence (this mechanism is discussed in detail in the main text).

2. Materials and Methods

All reagents, solvents and chemicals were purchased from Sigma-Aldrich (Steinheim, Germany) and used without modification, unless otherwise stated. In particular, the following polymers were purchased (the number in brackets is the Sigma-Aldrich catalogue code): polyvinyl chloride (PVC, 346764) with average Mw = 233,000, medium density polyethylene (PE, 332119), isotactic poly methyl methacrylate (PMMA, 452130), polytetrafluoroethylene (PTFE, 430935), isotactic polypropylene (PP, 428116) with average Mw = 12,000, polyethylene terephthalate (PET, 429252) and polystyrene (PS, 327743) with average Mw 20,000. Note that PP and PET were ground with commercial lab grinder prior to use; their final average grain sizes were 600 ± 300 and 400 ± 200 μm, respectively. Polyaniline (530689) with average Mw = 65,000, poly(acrylic acid) (323667) with average Mw = 1800, poly(acrylic acid) (181285) with average Mw = 450,000, poly(styrene sulfonic acid sodium salt) (81612) with average Mw = 77,000, poly(styrene sulfonic acid sodium salt) (81615) with average Mw = 350,000 and poly(vinyl alcohol) with average Mw = 130,000 were also tested.

Photophysical characterization. The experiments were carried out in air-equilibrated solutions at 25 °C. UV–Vis absorption spectra were recorded with a Perkin-Elmer Lambda 650 (Boston, MA, USA) or Perkin-Elmer Lambda 45 spectrophotometer (Boston, MA, USA) using quartz cells (Hellma, Müllheim, Germany) with path length of 1.0 cm or 2.5 mL macro PMMA or UV disposable cuvettes purchased from BRAND (Wertheim, Germany). The fluorescence spectra were recorded with a Horiba Jobin Yvon Fluoromax-4 (Kyoto, Japan), a Perkin-Elmer LS-55 or an Edinburgh FLS920 (Livingston, UK) equipped with a photomultiplier Hamamatsu R928 phototube (Hamamatsu, Japan). The same instrument connected to a TCC900 card was used for Time Correlated Single Photon Counting (TCSPC) experiments with an LDH-P-C 405 or 635 pulsed diode laser. The fluorescence quantum yields (uncertainty, ±15%) were determined using fluorescein solution in water as a reference with Φ = 1.0 or [Ru(bpy)$_3$]$^{2+}$ in water with Φ = 0.028 [35]. The emission intensities were corrected taking into consideration the inner filter effect according to the standard methods [35]. For the fluorescence anisotropy measurements, an Edinburgh FLS920 equipped with Glan-Thompson polarizers (Livingston, UK) was used. The data were corrected for polarization errors using the G-factor.

Fluorescence microscopy. Polystyrene 24 multi-well costar 3526 plates were used for the preparation of the samples. The set up used for images acquisition was made by an Olympus IX 71 inverted microscope (Shinjuku, Japan) equipped with a Blue diode (470 ± 20 nm, LZ1-10DB00, Led-Engin, Wilmington, MA, USA) for fluorescence excitation and a Basler acA5472-17uc camera with a Sony IMX183 CMOS sensor. The irradiation blue diode was focused on the back focal plane of the objective (4× magnification, Olympus UPLFLN 4X), with a 50 mm focal length after filtering with a band pass filter (469 ± 17.5 nm, Thorlabs, Newton, NJ, USA) and after 90° reflection on a dichroic filter (452–490 nm/505–800 nm, Thorlabs, Newton, NJ, USA). Fluorescence was filtered with a cut-off emission filter (50% transmission at 510 nm, Thorlabs, Newton, NJ, USA).

Confocal microscope. A Nikon A1 (Tokyo, Japan) equipped with a module for fluorescence lifetime image (FLIM) by Picoquant with a pulsed excitation laser at 405 nm with a pulse width of ~70 ps was used.

Dimeric vs. *isodesmic model for the aggregation.*

The equilibrium for the formation of the dimer P$_2$ from P is [36]:

$$P + P \rightleftharpoons P_2$$

Denoting [P] and [P$_2$] as the equilibrium molar concentrations of P and P$_2$, respectively, the equilibrium constant K$_2$ is:

$$K_2 = \frac{[P_2]}{[P]^2}$$

Hence, the molar fraction of monomer, χ_P = [P]/C$_P$, with C$_P$ the total concentration of P, can be expressed as:

$$\chi_P = \frac{-1 + \sqrt{1 + 8K_2C_P}}{4K_2C_P}$$

In the case of the formation of larger aggregates (trimer P$_3$, tetramer P$_4$, etc.), additional equilibria have to be considered:

$$P + P \rightleftharpoons P_2; \; P_2 + P \rightleftharpoons P_3; \; P_3 + P \rightleftharpoons P_4; \ldots$$

Other association constants, K$_3$ for the trimer P$_3$, K$_4$ for the tetramer P$_4$, etc., have to be considered:

$$K_3 = \frac{[P_3]}{[P_2][P]}; \; K_4 = \frac{[P_4]}{[P_3][P]}; \ldots$$

A simple model for the aggregation is the isodesmic one, which considers:

$$K = K_2 = K_3 = K_4, \ldots$$

In this case, a different dependency of the molar fraction of the monomer on the total concentration of P is expected according to the following equation:

$$\chi_P = \frac{2KC_P + 1 - \sqrt{4KC_P + 1}}{2K^2C_P^2}$$

2.1. Preparation and Purification of P Probe

For the synthesis of P (N,N'-Bis(Ethyl-diethoxyethyl-ethylamine)perylene-3,4,9,10-bis(dicarboximide), 200 µL of 2,2-(Ethylenedioxy)bis(ethylamine) (1 mmol) were mixed in 10 mL ethylene glycol in a 100 mL Erlenmeyer flask [37,38]. Afterwards, 20 mg (0.05 mmol) of perylene-3,4,10-tetra-carbocyl dianhydride (PDA) were added to the previous, and the mixture was sonicated for 1 min. The reaction was carried out in a commercial microwave oven (2.45 GHz, 900 W, Benton Harbor, Whirlpool, MI, USA) to achieve the conversion of PDA to PDI-probe, P in 5 min. A color change was observed from red to purple. The product was then precipitated by adding 60 mL of water and isolated by centrifugation at 5000 rpm for 10 min and washed three times with distilled water. The final product was deep purple, and it was dried under vacuum for 24 h.

2.2. Characterization of P

^{1}H-NMR spectra were recorded on Varian 400 (400 MHz) spectrometers (Palo Alto, CA, USA). Chemical shifts are reported in ppm from TMS with the solvent resonance as the internal standard (deuterochloroform: 7.24 ppm). ^{13}C-NMR spectra were recorded on a Varian 400 (100 MHz) spectrometers with complete proton decoupling. Chemical shifts are reported in ppm from TMS with the solvent as the internal standard (deuterochloroform: 77.0 ppm).

^{1}H NMR (400 MHz, DMSO-d6, acetic acid-d4) δ (ppm): 2.97 (t (broad), 4H,), 3.38–3.70 (m (broad), 20 H,), 4.20 (s (broad), 4 H), 8.00–8.32 (m (broad), 8 H)

^{13}C NMR (75.5 MHz, DMSO-d6, acetic acid-d4) δ (ppm): 38.5, 50.8, 59.0, 66.8, 69.6, 69.8, 118.1, 118.7, 124.0, 125.7, 132.8, 138.8, 165.0.

Electrospray ionization (ESI) mass spectra were obtained with Agilent Technologies MSD1100 single-quadrupole mass spectrometer (Santa Clara, CA, USA). Mass spectra were acquired in full-scan mode from m/z 150 to 3000, with a scan time of 0.1 s in the positive ion mode. The parameters of the spray chamber were set as follows: drying gas flow at 10 mL min^{-1}, nebulizer (nitrogen) pressure at 35 psig, temperature at 350 °C and ESI spray voltage at 5000 V. The fragmentor voltage was maintained at 50 V.

Chromatographic analysis with HPLC-MS (Agilent Technologies, Santa Clara, CA, USA) showed a single peak, proving the purity of the product. The masses detected were m/z = 653.4 and m/z = 327.3, corresponding to PH$^+$ (expected m/z = 653.3) and PH$_2^+$ (expected m/z = 327.2), respectively.

High-resolution MS (HRMS) ESI analyses were performed on a Xevo G2-XS QTof mass spectrometer (Waters, Milford, MA, USA). Mass spectrometric detection was performed in the full-scan mode from m/z 50 to 1200, with a scan time of 0.15 s in the positive ion mode and the following settings: cone voltage of 40 V and collision energy of 6.00 eV. ESI was performed with the settings: capillary of 3 kV, cone of 40 V, source temperature of 120 °C, desolvation temperature of 600 °C, cone gas flow of 50 L/h and desolvation gas flow of 1000 L/h.

The exact mass was m/z = 653.259 (expected for PH$^+$ m/z = 652.253).

2.3. Method for Detection of the MP

A water-buffered solution of P (c = 8 µM, phosphate buffer pH = 7) was prepared for the staining of the MP. In particular, PVC, PE, PET, PP, PS, PMMA and PTFE were tested. For the staining, 5 mg of polymer were placed in a 24-well plate and incubated for 2 h with 2 mL of the above-described P-buffered solution. In the next step, the fluorescence images of the polymers were acquired directly in the solution with the fluorescence inverted microscope described above.

3. Results and Discussion

This section describes and discusses the applicability of the probe P for the direct staining of MP in neutral water making them fluorescent and, therefore, easily detectable with a conventional fluorescence microscope. More in detail, two main advantages of this approach are pointed out: (i) the probe P aggregates in neutral-buffered water solution giving almost completely quenched dimers; and (ii) P binds selectively to PVC microplastic, giving a characteristic red fluorescence. As a consequence, it is possible to selectively detect PVC MP by simply mixing the probe in a water suspension of the MP and imaging the suspension without any washing or further treatment, since the fluorescence signal from the background is very weak.

An essential prerequisite of this approach is the use of a water-soluble probe, which still presents a high tendency to aggregate, forming non-fluorescent species. For this reason, we designed the new probe molecule P by functionalizing the fluorophore with two hydrophilic di-ethylene glycolic chains terminated with amino groups that are protonated at neutral pH (pKa = 9.04 ± 0.10), increasing the hydrophilicity of the molecule. These short and linear chains confer to the molecule the required solubility in water, but they are not bulky enough to prevent the aggregation and quenching of the fluorescence. Additionally, the absence of any bulky group directly bound to the chromophore makes possible the interaction of the hydrophobic PDI with the surface of the polymers and, in particular, PVC.

Before testing the ability of P to stain the various MP, we investigated its photophysical properties by UV–Vis absorption spectroscopy and steady-state and time-resolved fluorescence spectroscopy. PDI derivatives are known to aggregate, especially in polar environment, because of strong π–π stacking interactions involving the extended aromatic chromophores. Depending on the solvent and the molecular structure, in particular the presence of bulky substituents, the aggregation can lead to the formation of dimers or oligomers or to formation of large aggregates, up to large nanoaggregates. In these aggregates, the photophysical properties of PDI are widely modified with respect to the starting monomer [32–34].

To investigate the effect of the aggregation of P in water, we first characterized the molecule in dichloromethane (DCM). In fact, for similar molecules, it has been reported that aggregation in this solvent is purely efficient, which allows investigating the properties of PDI in the non-aggregated form [33].The UV–Vis absorption spectrum of P (5.3 µM) in DCM is shown in Figure 2. As shown in the figure, the spectrum exhibited the vibrationally structured bands, typical of the PDI monomer, with a maximum at λ_{max} = 525 nm and a molar absorption coefficient $\varepsilon = 8 \times 10^4$ M^{-1} cm^{-1}. It should be stressed that the presence of the vibrational peaks for the electronic transition (S_0–S_1) in the 400–550 nm range is typical of the monomeric form of PDI, while aggregates (typically H-type) present a much less structured band [32,36].

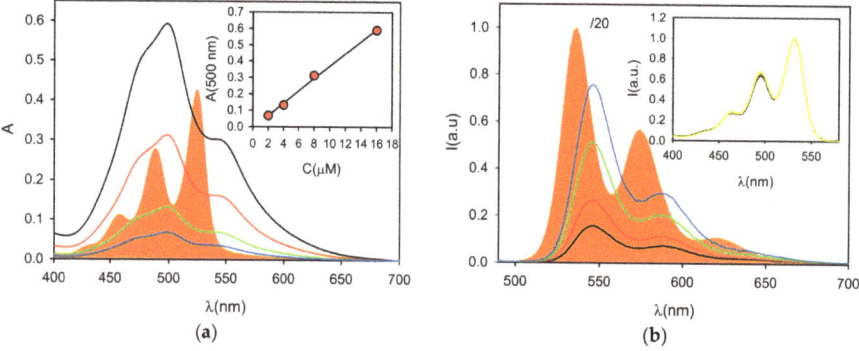

Figure 2. (a) Absorption spectra of P in DCM (filled orange, C = 5.3 µM) and of P in buffered neutral water (pH = 7) at concentrations 16, 8, 4 and 2 µM (black, red, green and blue lines, respectively). The absorbance at 500 nm as a function of the concentration is plotted in the inset. (b) Fluorescence spectra (λ_{exc} = 470 nm) of P in DCM (filled orange, C = 5.3 µM, divided by 20 for scaling) and P in buffered neutral water (pH = 7) at concentrations 16, 8, 4 and 2 µM (black, red, green and blue lines, respectively). The normalized excitation spectra for the four water solutions (λ_{em} = 580 nm) are shown in the inset (same color line). All fluorescence and excitation spectra are corrected for the inner filter effect.

The fluorescence spectrum (λ_{exc} = 480 nm) in DCM, as illustrated in Figure 2, showed a small Stoke shift with respect to the absorption one, and it appeared as the mirror image of the absorption spectrum, maintaining the typical vibrational spectral structure, as also reported for other PDI-based fluorophores. More in detail, the maximum is positioned at λ_{max} = 535 nm, and the fluorescence quantum yield was found to be Φ = 40%, which is lower than reported for similar PDI molecules (~100%) [36]. The excitation spectrum of P in DCM (λ_{em} = 480 nm, not reported) perfectly matched the absorption spectrum. The partial quenching of the fluorescence was confirmed by time resolved fluorescence analysis and in particular by TCSPC that allowed recording a deactivation kinetic of the excited state. The latter could be fitted with a mono-exponential model to give a lifetime of τ = 3.2 ns, while, for similar PDI monomers, a singlet excited state lifetime τ = 3.9 ns has been reported [39,40].

The decrease of the fluorescence quantum yield and of the excited state lifetime of P, with respect to analogous molecules, could be attributed to the presence of the two terminal amino groups which are not protonated in DCM. In fact, amines are known to dynamically quench excited state fluorescence, via electron-transfer process, producing the reduced PDI and the oxidized amine; this process has been widely explored in organic synthetic photocatalysis [41,42]. In the case of dynamic (diffusional) quenching, the back-diffusion of the encounter pair and/or the irreversible degradation of the oxidized amine can lead to the stabilization of the anionic PDI, at least in the absence of oxygen [41]. On the other hand, in the case of P, the quenching is static, and the charge-separated state undergoes fast recombination, producing a partial quenching of the fluorescence. This mechanism was demonstrated by adding trifluoroacetic acid (20 µM) to the P solution in DCM in order to protonate the amines and therefore produce ammonium species which cannot be oxidized by the excited PDI. Indeed, upon protonation, a considerable increase in the lifetime was observed (τ = 4.3 ns).

In the next step, the photophysical behavior of P was investigated in water buffered (pH = 7, phosphate buffer) solution. In particular, since aggregation of P was expected to be strongly concentration-dependent, we prepared a set of different concentrations of the probe in this solvent (0.2×10^{-5}, 0.4×10^{-5}, 0.8×10^{-5} and 1.6×10^{-5} M, referred to as solutions P1, P2, P3 and P4, respectively). The UV–Vis absorption spectra of these water solutions of P are reported in Figure 2. They exhibited broader absorption bands in neutral water with poorly-defined vibrational structures and a maximum at λ_{max} = 498 nm. These spectral changes are characteristic of the H-aggregates formed by PDI dyes that tend to self-assemble and form π–π stacked species, especially in protic solvents such as water.

Since the spectral signature is, to some extent, characteristic for the kind of aggregation and it reflects the ratio between aggregated and non-aggregated molecules, the absorbance spectra after normalization were compared to distinguish any effect of the concentration. All spectra were very well superimposed, and, as shown in Figure 2, the absorbance at 498 nm was found to be directly proportional to the concentration. As a consequence, the molar absorption coefficients, as shown in Table 1, were widely concentration independent. On the other hand, the fluorescence spectra, upon 470 nm excitation, confirmed the presence of a very minor fraction of the free monomer (in all cases, <2%). In fact, looking at the fluorescence spectra shown in Figure 2, the emission profile did not significantly change in water, possessing an emission maximum at 546 nm and the typical structured green fluorescence band of the PBI monomer. The origin of this green emission was confirmed by the excitation spectra of the water solutions at λ_{em} = 580 nm shown in Figure 2, which again presented the typical features of the P monomer, and by the excited state lifetime at λ_{em} = 540 nm.

In fact, the decay of the excited state in neutral water was mono-exponential and the excited state lifetime remained almost unaltered upon concentration change. As shown in Table 1, it was found to be τ ~4.5 ns, which is similar to the value observed for P in DCM after protonation. Further information about the origin of the green fluorescence band was given by fluorescence anisotropy spectroscopy. This technique was chosen for the analysis of the reorientation of the excited state dipole, after excitation, during deactivation. The value of fluorescence anisotropy, r, measured for the green fluorescence was quite low, about 0.03 at all concentrations. This value is compatible with the rotation of monomeric species as P [33].

Table 1. Photophysical properties of P in water (pH = 7) at different concentrations.

C (µM)	ε (M^{-1}cm^{-1}) 500 nm	Φ$_1$ % λ_{exc} = 470 nm	τ$_1$ (ns) λ_{exc} = 405 nm λ_{em} = 540 nm	r$_1$ λ_{exc} = 470 nm	Φ$_2$ % λ_{exc} = 590 nm	τ$_2$ (ns) λ_{exc} = 635 nm λ_{em} = 650 nm	r$_2$ λ_{exc} = 590 nm
2 (P1)	37,000	1.6	4.47	0.026	0.03	<0.1	0.25
4 (P2)	38,000	1.1	4.54	0.023	0.02	<0.1	0.26
6 (P3)	33,000	0.6	4.56	0.023	0.03	<0.1	0.28
16 (P4)	34,000	0.4	4.58	0.024	0.05	<0.1	0.27

Since the absorption spectra clearly revealed the predominant presence of PDI aggregates, we investigated the fluorescence properties of the P solution upon selective excitation at 590 nm (Figure 3), a wavelength at which the monomer cannot be excited (Figure 2). In this excitation condition, all the four P water solutions presented a weak broad emission at 650 nm, typical for PDI aggregates and excimers. As shown in Table 1, for this emission, the fluorescence quantum yields were extremely weak and the excited state lifetimes very short (<0.1 ns). Different from what was reported for the green fluorescence anisotropy, the one for the red emission was very high (r ~ 0.25), considering that the maximum value possible for a randomly oriented system is r = 0.4. Such a high anisotropy is compatible with the presence of aggregates with higher molar mass with respect to the monomer as well as the short excited-state lifetime.

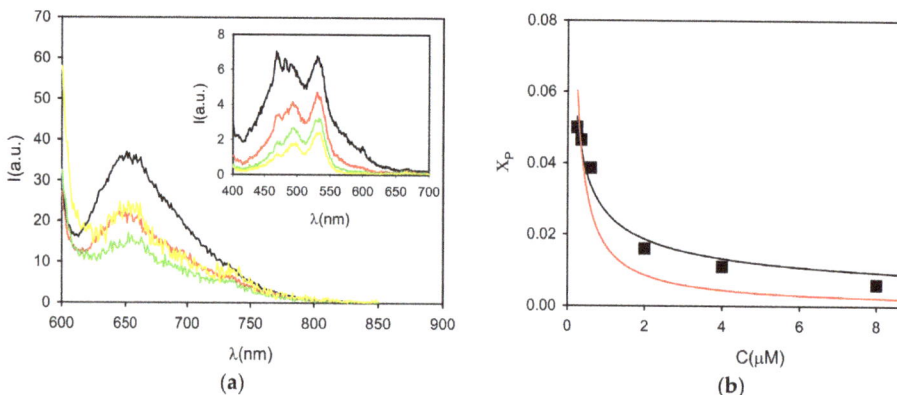

Figure 3. (a) Fluorescence spectra of P in buffered neutral water (pH = 7) at concentrations 16, 8, 4 and 2 µM upon λ_{exc} = 590 nm (black, red, green and yellow lines, respectively). The excitation spectra of the four solutions (λ_{em} = 720 nm) are shown in the inset (same color line). (b) The fluorescence quantum yield is used to calculate the molar fraction of P (χ_P) and it is plotted as a function of concentration (black squares). These data are fitted using either a dimeric (black line $K_2 = 7 \times 10^8$ M^{-1}) or an isodesmic model (red line, K = 5×10^7 M^{-1}).

It must be underlined that, based on their extremely low fluorescence quantum yield, the aggregates can be considered completely quenched with respect to the monomer. Moreover, upon excitation at 470 nm, it is possible to assume that only the fraction of light absorbed by the monomer produces detectable fluorescence. Hence, the average quantum yield in this excitation condition can be used to find the molar fraction of the free monomer. More specifically, as shown in Table 1, the average fluorescence quantum yield in phosphate buffer was found to be much lower (Φ < 2%) than the one measured in DCM, and it was decreasing upon increasing P concentration.

The fraction of free monomer, calculated from the fluorescence quantum yield Φ_1, was plotted as a function of P concentration (Figure 3) to investigate the process of aggregation. In particular, the experimental data were fitted with both dimeric and isodesmic models. As shown in Figure 3, while the dimeric model allowed fitting the experimental points with $K_2 = 7 \times 10^8$ M^{-1}, an optimal fitting with an isodesmic model ($K = 5 \times 10^7$ M^{-1}) failed in describing the experimental results. Thus, we concluded that the aggregation leads to the formation of oligomer and mostly P dimer, but not extended structures.

In the next step, the possibility of P to be applied for the staining of MP was investigated. In particular, the ability of P to selectively stain PVC MP at neutral pH for easy and fast detection of the specific MP debris was verified. Five milligrams of PVC were added to 2 mL of P1–P4 solutions, and they were mixed for 2 h, after which the UV–Vis absorption, fluorescence emission and excitation spectra, as well as excited state lifetime, were acquired again. The absorption spectra of the P1–P4 solutions after 2 h incubation with PVC showed a relevant decrease of the peak at 500 nm without any important change in the spectral profile. On the contrary, from the fluorescence analysis, only a small increase of the average quantum yield of the monomeric (green) fluorescence of P in solution was observed, as shown in Figure 4. These results suggest that the adsorption of P on the PVC surface does not substantially alter the equilibrium of aggregation of P in solution. As before, the interaction of P with PVC was confirmed by excitation spectra, excited state lifetime and fluorescence anisotropy, attributing the remaining green fluorescence to the monomer.

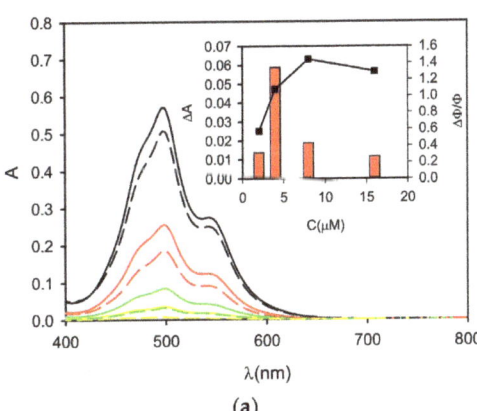
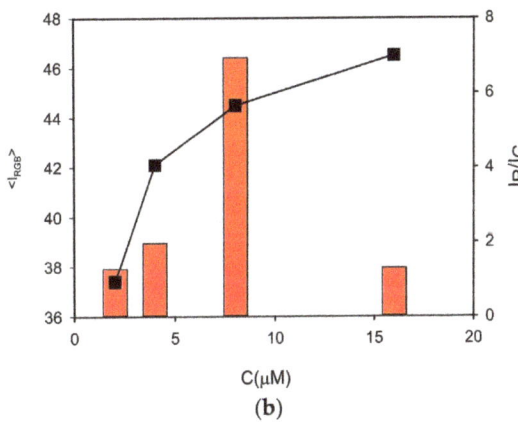

Figure 4. (**a**) Absorption spectra of P in buffered neutral water (pH = 7) at concentrations 16, 8, 4 and 2 µM (black, red, green and yellow lines, respectively) before (continuous line) and after (dashed lines) 2 h incubation with PVC (5 mg in 2 mL solution). The decrease of the absorbance of the solution is plotted in the inset (black squares) as a function of the concentration. The relative increase of the fluorescence intensity of the solution at 540 nm after incubation is plotted as red bars. (**b**) The total intensity measured by wide field microscopy for the PVC particles after the incubation is shown as a function of P concentration (<I_{RGB}>). The ratio of the intensity on the red and green channels (I_R/I_G) is plotted as red bars.

Afterwards, the PVC particles were observed with an inverted fluorescence microscope under excitation at 460 nm, detecting the fluorescence with a conventional color camera after filtering with a cut-off (50% transmittance at 510 nm). The fluorescence images illustrated in Figure 5 show that PVC becomes strongly fluorescent after incubation with P1–P4 solutions and that the intensities and fluorescent colors were, to some extent, concentration dependent. In particular, the total average intensity (I_{RGB}) progressively increased with the P concentration. In fact, when the red component is considered with respect to the green as the I_R/I_B ratio, as plotted in Figure 4, it becomes clear that an optimal red staining was achieved at a concentration of 8 µM.

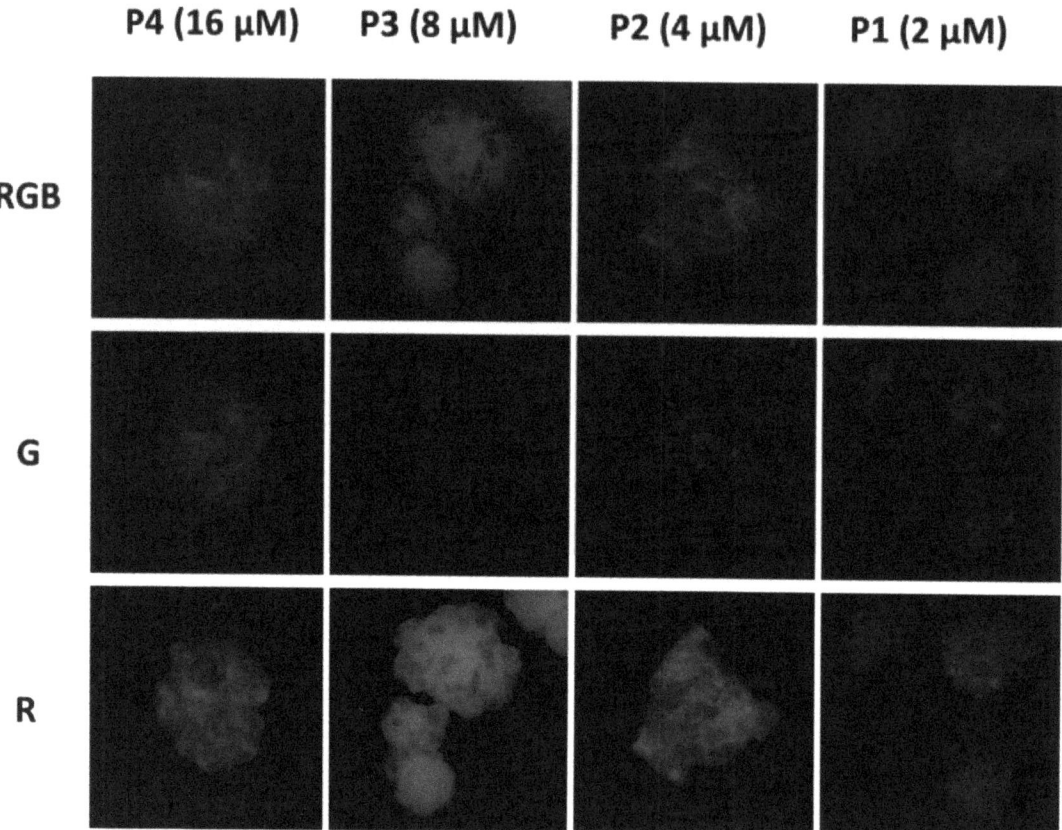

Figure 5. (**Top**) RGB color images of PVC particles after 2 h incubation with P in buffered water solution (pH = 7) at decreasing concentration (16, 8, 4 and 2 μM, solutions P4-P1) obtained with a fluorescence inverted microscope upon illumination with a blue LED diode in epifluorescence mode. The images were acquired with a conventional CMOS camera directly in the P solutions, without any washing. The later size of each image is 450 μm. The images were split into the three color components R, G and B with the software Image J. The G components are shown in the second line of images. The R component are shown at the (**bottom**).

The above results indicate that indeed the adsorption of P onto the PVC surface can lead to an enhancement of the quantum yield. To confirm this hypothesis, we analyzed the red fluorescence of the stained PVC particles by confocal scanning fluorescence microscopy, measuring in particular the local excited state decay with a setup for fluorescence lifetime image microscopy (FLIM). The confocal fluorescence image upon excitation at 489 nm and red channel detection (590/50 nm), as shown in Figure 6, clearly demonstrates that the probe was absorbed only on the surface of the PVC and did not penetrate in the inner layers. On the other hand, the increase of the red emission intensity, with respect to the weak one of the P dimer in solution, is demonstrated by the FLIM image in Figure 6, where an excited state lifetime as long as 10 ns was measured for the red fluorescence on the PVC surface. Considering all these results, and the following relevant known properties of PDI derivatives, we propose for the staining process the mechanism schematized in Figure 1.

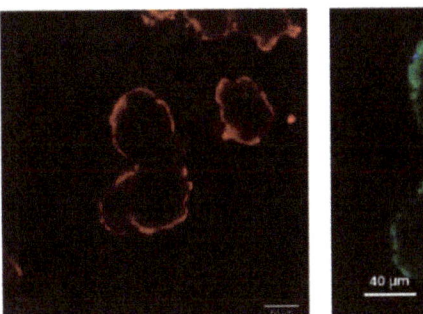
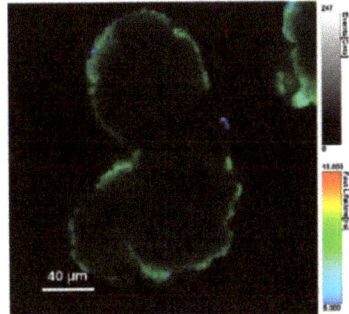

Figure 6. Confocal scanning fluorescence images of PVC particles after 2 h incubation with P (8 μM) in buffered water (pH = 7): (**Left**) excitation at 489 nm and detection at 595/50 nm; and (**Right**) FLIM image showing the local excited state lifetimes.

For PDI in general, it has been reported that strong π–π interactions, such as those in the P_2 dimer, lead to a strong quenching of the typical red fluorescence with a shortening of the excited state lifetime [38]. On the contrary, less compact structures, where the overlap of the π system is less efficient (e.g., those formed in the case of the presence of bulky substituents on the aromatic PDI systems), are more intensely fluorescent and present a much longer excited state lifetime [43]. Regarding the present approach, the efficient formation of the P_2 dimer clearly demonstrates that the π–π stacking interaction (indicated as 1 in Figure 1) prevails over the electrostatic repulsion between the terminal protonated amino groups (indicated as 2 in Figure 1). The previous is possible because the flexible ethylenedioxi-chains can easily bend to increase the distance between the positive charges and, hence, minimize the repulsion. On the contrary, when P is adsorbed on PVC, the PDI unit comes into contact with the polymer surface and the hydrophilic positive ammonium charges are directed toward the solvent creating an accumulation of positive charges on the surface. This results in the repulsion of other protonated P molecule in the solution and a reduced stacking of P on the PVC surface. As a consequence, the less aggregated PDI structures on the PVC surface present more intense fluorescence. Thus, the proposed mechanism is clearly the result of a delicate balance between attractive and repulsive forces that are different in the solution with respect to the PVC surface.

To test the actual selectivity of P towards PVC, we compared the fluorescence of a set of polymers (5 mg) after 2 h incubation with P in buffered solution at the optimal concentration of c = 8 μM. Seven widely used polymer MP were analyzed: PVC, PE, PET, PP, PS, PMMA and PTFE. Figure 7 (top) shows the images of the different polymeric particles after 2 h of incubation with the probe using the inverted fluorescence microscope upon irradiation with blue light. The images were acquired directly in the P solution without any washing. As shown in Figure 7 (bottom), all polymers exhibited a weak green fluorescence when observed in these excitation conditions without the probe. Importantly, after incubation with the probe P, only PVC presents a peculiar red emission, while only minor changes were observed for the other polymers. The bright red color generated in the case of PVC allowed the easy detection and identification of this kind of MP. This different response can be easily observed by analyzing the intensity on the red channel or by representing the red to green intensity ratio I_R/I_G, as shown in Figure 8. Overall, the possibility of using P as probe for PVC MP detection possesses the typical advantages of ratiometric detection, making the identification of the target largely independent of the experimental conditions (e.g., excitation source intensity). Finally, the effects of a series of water-soluble polymers at concentration 2.5 mg/mL on the fluorescence properties of P were investigated. Poly(acrylic) acid, poly(styrene sulfonic acid sodium salt) and poly(vinyl alcohol) were tested. In all cases, a very minor change in the fluorescence intensity was measured with a final quantum yield that was lower than 0.1% for all the aforementioned polymers. Finally, polyaniline, a less common water-insoluble polymer, was tested, where no fluorescence from the polymeric particles was observed after 2 h incubation with P (c = 8 µM).

It should be stressed that the method we present is very effective for the qualitative identification of PVC MP, but, in the present form, it does not allow a precise quantification. The observed selectivity for PVC was in part expected considering the general good solubility of PDI derivatives in chlorinated solvents such as DCM and dichloroethane that reveals a good affinity for chlorinated species and favors the disaggregation of PDI units [40]. As discussed in recent papers, the interaction between organic molecules and MP is the result of several interactions among which the halogen–aromatic interaction can be concluded to be the one playing a major role in the case of PVC–PDI interaction [44,45].

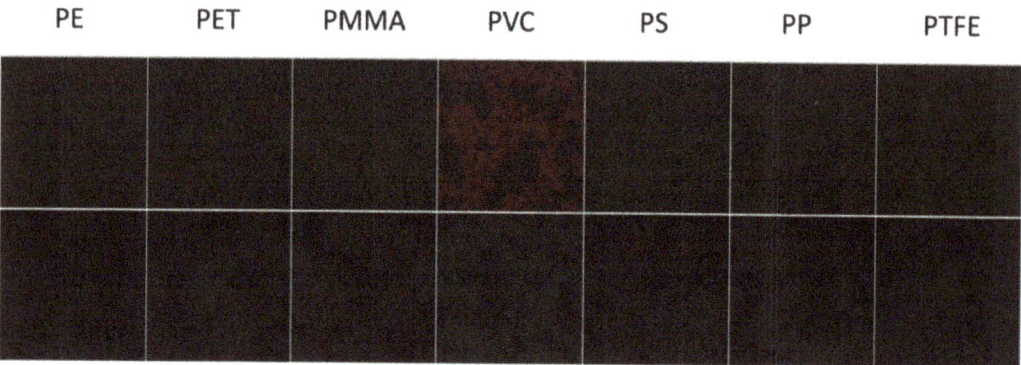

Figure 7. (**Top**) From left to right: RGB color images of PE, PET, PMMA, PVC, PS, PP and PTFE particles after 2 h incubation with P in buffered water solution (pH = 7) at concentration 8 µM obtained with a fluorescence inverted microscope upon illumination with a blue LED diode in epifluorescence mode. Images were acquired with a conventional CMOS camera directly in the P solutions, without any washing. The lateral size of each image is 1.8 mm. (**Bottom**) Images of the fluorescence of the same polymers in the same conditions without staining.

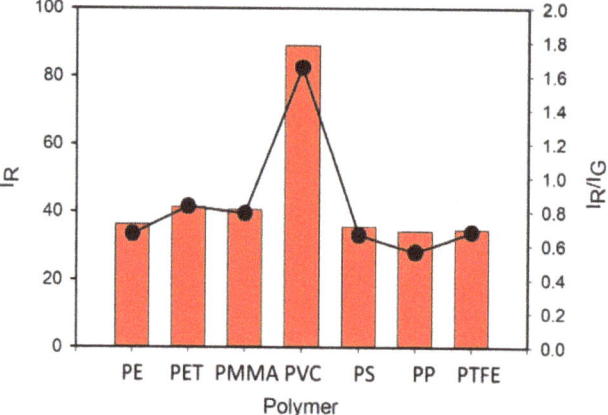

Figure 8. Black dots show the intensity on the red channel (I_R) of the fluorescence images of PE, PET, PMMA, PVC, PS, PP and PTFE particles after 2 h incubation with P in buffered water solution (pH = 7) at concentration 8 µM obtained with a fluorescence inverted microscope upon illumination with a blue LED diode in epifluorescence mode. Images were acquired with a conventional CMOS camera directly in the P solutions, without any washing. The ratio of the intensities on the red and green channels is plotted as red bars.

4. Conclusions

We demonstrated a new approach, established on the design of a PDI-based probe, for the direct labeling and recognition of MP in water (just by mixing). As an advantage, this PDI molecule behaves as a fluorogenic probe, being non-fluorescent in water at neutral pH but becoming intensely red fluorescent upon adsorption on PVC MP. We demonstrated that the labeling process presents great advantages: it can be performed at room temperature and in a short time (2 h). Moreover, the fluorescence of the MP can be detected without any further treatment of the sample (no-washing). The probe exhibited an intense green fluorescence in organic solvent that, in water at neutral pH = 7.0, is almost completely quenched because of the formation of dimers, giving very weak background signal. Interestingly, the probe showed considerable selectivity to PVC particles, which presented an intense red emission after staining that could be detected, under excitation with a blue diode, with a conventional CMOS color camera directly in the probe solution. It is worth noting that the probe was prepared through an environmentally friendly, fast, versatile, cheap and easily scalable method based on microwave heating.

These results pave the way to the development of new specific, fast and convenient probes for the detection of MP in real samples with a simple mixing and no-washing approach using widespread electronic devices.

Author Contributions: V.C. and A.M.-P. synthesized the probe and performed the experiments with the microplastics; M.C. and D.G. performed the measurement at the confocal microscope; E.R. characterized the probe; L.P. and D.F. contributed to the analysis of the results; and M.M. designed the probe and the experiments and wrote the paper. All authors have read and agreed to the published version of the manuscript.

Funding: This research was funded by the Italian Ministry of Education, University and Research (MIUR), (PRIN project: PRIN 2017) 2017E44A9P (BacHound).

Data Availability Statement: The data presented in this study are contained within the article.

Conflicts of Interest: The authors declare no conflict of interest.

References

1. Mammo, F.K.; Amoah, I.D.; Gani, K.M.; Pillay, L.; Ratha, S.K.; Bux, F.; Kumari, S. Microplastics in the environment: Interactions with microbes and chemical contaminants. *Sci. Total Environ.* **2020**, *743*, 140518. [CrossRef]
2. Choong, W.S.; Hadibarata, T.; Tang, D.K.H. Abundance and distribution of microplastics in the water and riverbank sediment in Malaysia—A review. *Biointerface Res. Appl. Chem.* **2021**, *11*, 11700–11712. [CrossRef]
3. Sekudewicz, I.; Dąbrowska, A.M.; Syczewski, M.D. Microplastic pollution in surface water and sediments in the urban section of the Vistula River (Poland). *Sci. Total Environ.* **2021**, *762*, 143111. [CrossRef] [PubMed]
4. Woodall, L.C.; Sanchez-Vidal, A.; Canals, M.; Paterson, G.L.J.; Coppock, R.; Sleight, V.; Calafat, A.; Rogers, A.D.; Narayanaswamy, B.E.; Thompson, R.C. The deep sea is a major sink for microplastic debris. *R. Soc. Open Sci.* **2014**, *1*, 140317. [CrossRef] [PubMed]
5. Rezania, S.; Park, J.; Md Din, M.F.; Mat Taib, S.; Talaiekhozani, A.; Kumar Yadav, K.; Kamyab, H. Microplastics pollution in different aquatic environments and biota: A review of recent studies. *Mar. Pollut. Bull.* **2018**, *133*, 191–208. [CrossRef]
6. Wu, J.; Jiang, R.; Lin, W.; Ouyang, G. Effect of salinity and humic acid on the aggregation and toxicity of polystyrene nanoplastics with different functional groups and charges. *Environ. Pollut.* **2019**, *245*, 836–843. [CrossRef]
7. Pellini, G.; Gomiero, A.; Fortibuoni, T.; Ferrà, C.; Grati, F.; Tassetti, A.N.; Polidori, P.; Fabi, G.; Scarcella, G. Characterization of microplastic litter in the gastrointestinal tract of Solea solea from the Adriatic Sea. *Environ. Pollut.* **2018**, *234*, 943–952. [CrossRef]
8. Chagas, T.Q.; da Costa Araújo, A.P.; Malafaia, G. Biomicroplastics versus conventional microplastics: An insight on the toxicity of these polymers in dragonfly larvae. *Sci. Total Environ.* **2021**, *761*, 143231. [CrossRef]
9. Xu, K.; Zhang, Y.; Huang, Y.; Wang, J. Toxicological effects of microplastics and phenanthrene to zebrafish (Danio rerio). *Sci. Total Environ.* **2021**, *757*, 143730. [CrossRef]
10. Au, S.Y.; Lee, C.M.; Weinstein, J.E.; van den Hurk, P.; Klaine, S.J. Trophic transfer of microplastics in aquatic ecosystems: Identifying critical research needs. *Integr. Environ. Assess. Manag.* **2017**, *13*, 505–509. [CrossRef]
11. Van Wezel, A.; Caris, I.; Kools, S.A.E. Release of primary microplastics from consumer products to wastewater in the Netherlands. *Environ. Toxicol. Chem.* **2016**, *35*, 1627–1631. [CrossRef]
12. Lehtiniemi, M.; Hartikainen, S.; Näkki, P.; Engström-Öst, J.; Koistinen, A.; Setälä, O. Size matters more than shape: Ingestion of primary and secondary microplastics by small predators. *Food Webs* **2018**, *16*, e00097. [CrossRef]
13. Koelmans, A.A.; Mohamed Nor, N.H.; Hermsen, E.; Kooi, M.; Mintenig, S.M.; De France, J. Microplastics in freshwaters and drinking water: Critical review and assessment of data quality. *Water Res.* **2019**, *155*, 410–422. [CrossRef]

14. Mariana, M.; Feiteiro, J.; Verde, I.; Cairrao, E. The effects of phthalates in the cardiovascular and reproductive systems: A review. *Environ. Int.* **2016**, *94*, 758–776. [CrossRef]
15. Han, S.; Bang, J.; Choi, D.; Hwang, J.; Kim, T.; Oh, Y.; Hwang, Y.; Choi, J.; Hong, J. Surface Pattern Analysis of Microplastics and Their Impact on Human-Derived Cells. *ACS Appl. Polym. Mater.* **2020**, *2*, 4541–4550. [CrossRef]
16. Wu, Y.; Guo, P.; Zhang, X.; Zhang, Y.; Xie, S.; Deng, J. Effect of microplastics exposure on the photosynthesis system of freshwater algae. *J. Hazard. Mater.* **2019**, *374*, 219–227. [CrossRef]
17. Wang, S.; Wang, Y.; Liang, Y.; Cao, W.; Sun, C.; Ju, P.; Zheng, L. The interactions between microplastic polyvinyl chloride and marine diatoms: Physiological, morphological, and growth effects. *Ecotoxicol. Environ. Saf.* **2020**, *203*, 111000. [CrossRef]
18. Bonacchi, S.; Rampazzo, E.; Montalti, M.; Prodi, L.; Zaccheroni, N.; Mancin, F.; Teolato, P. Amplified Fluorescence Response of Chemosensors Grafted onto Silica Nanoparticles. *Langmuir* **2008**, *24*, 8387–8392. [CrossRef]
19. Rampazzo, E.; Boschi, F.; Bonacchi, S.; Juris, R.; Montalti, M.; Zaccheroni, N.; Prodi, L.; Calderan, L.; Rossi, B.; Becchi, S.; et al. Multicolor core/shell silica nanoparticles for in vivo and ex vivo imaging. *Nanoscale* **2012**, *4*. [CrossRef]
20. Vermeiren, P.; Munoz, C.; Ikejima, K. Microplastic identification and quantification from organic rich sediments: A validated laboratory protocol *. *Environ. Pollut.* **2020**, *262*, 114298. [CrossRef]
21. Tiwari, M.; Rathod, T.D.; Ajmal, P.Y.; Bhangare, R.C.; Sahu, S.K. Distribution and characterization of microplastics in beach sand from three different Indian coastal environments. *Mar. Pollut. Bull.* **2019**, *140*, 262–273. [CrossRef] [PubMed]
22. Stock, V.; Laurisch, C.; Franke, J.; Dönmez, M.H.; Voss, L.; Böhmert, L.; Braeuning, A.; Sieg, H. Uptake and cellular effects of PE, PP, PET and PVC microplastic particles. *Toxicol. In Vitro* **2021**, *70*, 105021. [CrossRef] [PubMed]
23. Nel, H.A.; Chetwynd, A.J.; Kelleher, L.; Lynch, I.; Mansfield, I.; Margenat, H.; Onoja, S.; Goldberg Oppenheimer, P.; Sambrook Smith, G.H.; Krause, S. Detection limits are central to improve reporting standards when using Nile red for microplastic quantification. *Chemosphere* **2021**, *263*, 127953. [CrossRef] [PubMed]
24. Simmerman, C.B.; Coleman Wasik, J.K. The effect of urban point source contamination on microplastic levels in water and organisms in a cold-water stream. *Limnol. Oceanogr. Lett.* **2020**, 137–146. [CrossRef]
25. Danylchuk, D.I.; Jouard, P.-H.; Klymchenko, A.S. Targeted Solvatochromic Fluorescent Probes for Imaging Lipid Order in Organelles under Oxidative and Mechanical Stress. *J. Am. Chem. Soc.* **2021**, *143*, 912–924. [CrossRef]
26. Khalin, I.; Heimburger, D.; Melnychuk, N.; Collot, M.; Groschup, B.; Hellal, F.; Reisch, A.; Plesnila, N.; Klymchenko, A.S. Ultrabright Fluorescent Polymeric Nanoparticles with a Stealth Pluronic Shell for Live Tracking in the Mouse Brain. *ACS Nano* **2020**, *14*, 9755–9770. [CrossRef]
27. Sturm, M.T.; Horn, H.; Schuhen, K. The potential of fluorescent dyes—Comparative study of Nile red and three derivatives for the detection of microplastics. *Anal. Bioanal. Chem.* **2021**, *413*, 1059–1071. [CrossRef]
28. Karakolis, E.G.; Nguyen, B.; You, J.B.; Rochman, C.M.; Sinton, D. Fluorescent Dyes for Visualizing Microplastic Particles and Fibers in Laboratory-Based Studies. *Environ. Sci. Technol.* **2019**, *6*, 334–340. [CrossRef]
29. Tong, H.; Jiang, Q.; Zhong, X.; Hu, X. Rhodamine B dye staining for visualizing microplastics in laboratory-based studies. *Environ. Sci. Pollut. Res.* **2021**, *28*, 4209–4215. [CrossRef]
30. Costa, C.Q.; Cruz, J.; Martins, J.; Teodósio, M.A.; Jockusch, S.; Ramamurthy, V.; Da Silva, J.P. Fluorescence sensing of microplastics on surfaces. *Environ. Chem. Lett.* **2021**, 1–6. [CrossRef]
31. Battistini, G.; Cozzi, P.G.; Jalkanen, J.-P.; Montalti, M.; Prodi, L.; Zaccheroni, N.; Zerbetto, F. The erratic emission of pyrene on gold nanoparticles. *ACS Nano* **2008**, *2*. [CrossRef]
32. Würthner, F.; Saha-Möller, C.R.; Fimmel, B.; Ogi, S.; Leowanawat, P.; Schmidt, D. Perylene Bisimide Dye Assemblies as Archetype Functional Supramolecular Materials. *Chem. Rev.* **2016**, *116*, 962–1052. [CrossRef]
33. Montalti, M.; Battistelli, G.; Cantelli, A.; Genovese, D. Photo-tunable multicolour fluorescence imaging based on self-assembled fluorogenic nanoparticles. *Chem. Commun.* **2014**, *50*, 5326–5329. [CrossRef]
34. Sun, M.; Müllen, K.; Yin, M. Water-soluble perylenediimides: Design concepts and biological applications. *Chem. Soc. Rev.* **2016**, *45*, 1513–1528. [CrossRef]
35. Montalti, M.; Credi, A.; Prodi, L.; Candolfi, T. *Handbook of Photochemistry*, 3rd ed.; CRC Press: Boca Raton, FL, USA, 2006.
36. Shao, C.; Grüne, M.; Stolte, M.; Würthner, F. Perylene Bisimide Dimer Aggregates: Fundamental Insights into Self-Assembly by NMR and UV/Vis Spectroscopy. *Chem. Eur. J.* **2012**, *18*, 13665–13677. [CrossRef]
37. Liu, K.; Mukhopadhyay, A.; Ashcraft, A.; Liu, C.; Levy, A.; Blackwelder, P.; Olivier, J.-H. Reconfiguration of π-conjugated superstructures enabled by redox-assisted assembly. *Chem. Commun.* **2019**, *55*, 5603–5606. [CrossRef]
38. López-Andarias, J.; Rodriguez, M.J.; Atienza, C.; López, J.L.; Mikie, T.; Casado, S.; Seki, S.; Carrascosa, J.L.; Martín, N. Highly Ordered n/p-Co-assembled Materials with Remarkable Charge Mobilities. *J. Am. Chem. Soc.* **2015**, *137*, 893–897. [CrossRef]
39. Romero, N.; Nicewicz, D.A. Organic Photoredox Catalysis. *Chem. Rev.* **2016**, *116*, 10075–10166. [CrossRef]
40. Ford, W.E.; Kamat, P.V. Photochemistry of 3,4,9,10-perylenetetracarboxylic dianhydride dyes. 3. Singlet and triplet excited-state properties of the bis(2,5-di-tert-butylphenyl)imide derivative. *J. Phys. Chem.* **1987**, *91*, 6373–6380. [CrossRef]
41. Ghosh, I.; Ghosh, T.; Bardagi, J.I.; König, B. Reduction of aryl halides by consecutive visible light-induced electron transfer processes. *Science* **2014**, *346*, 725–728. [CrossRef]
42. Zeman, C.J.; Kim, S.; Zhang, F.; Schanze, K.S. Direct Observation of the Reduction of Aryl Halides by a Photoexcited Perylene Diimide Radical Anion. *J. Am. Chem. Soc.* **2020**, *142*, 2204–2207. [CrossRef] [PubMed]

43. Trofymchuk, K.; Reisch, A.; Shulov, I.; Mély, Y.; Klymchenko, A.S. Tuning the color and photostability of perylene diimides inside polymer nanoparticles: Towards biodegradable substitutes of quantum dots. *Nanoscale* **2014**, *6*, 12934–12942. [CrossRef] [PubMed]
44. Joo, S.H.; Liang, Y.; Kim, M.; Byun, J.; Choi, H. Microplastics with adsorbed contaminants: Mechanisms and Treatment. *Environ. Chall.* **2021**, *3*, 100042. [CrossRef]
45. Mei, W.; Chen, G.; Bao, J.; Song, M.; Li, Y.; Luo, C. Interactions between microplastics and organic compounds in aquatic environments: A mini review. *Sci. Total Environ.* **2020**, *736*, 139472. [CrossRef]

Article

The Role of the Reactive Species Involved in the Photocatalytic Degradation of HDPE Microplastics Using C,N-TiO₂ Powders

Aranza Denisse Vital-Grappin [1], Maria Camila Ariza-Tarazona [2], Valeria Montserrat Luna-Hernández [1], Juan Francisco Villarreal-Chiu [1,3,*], Juan Manuel Hernández-López [1], Cristina Siligardi [2] and Erika Iveth Cedillo-González [1,2,*]

1. Universidad Autónoma de Nuevo León, Facultad de Ciencias Químicas, Av. Universidad S/N Ciudad Universitaria, San Nicolás de los Garza C.P. 66455, Nuevo León, Mexico; aranza.vitalgr@uanl.edu.mx (A.D.V.-G.); valeria.lunahrn@uanl.edu.mx (V.M.L.-H.); juan.hernandezlz@uanl.edu.mx (J.M.H.-L.)
2. Department of Engineering "Enzo Ferrari", University of Modena and Reggio Emilia, Via P. Vivarelli 10/1, 41125 Modena, Italy; mariacamila.arizatarazona@unimore.it (M.C.A.-T.); cristina.siligardi@unimore.it (C.S.)
3. Centro de Investigación en Biotecnología y Nanotecnología (CIByN), Facultad de Ciencias Químicas, Universidad Autónoma de Nuevo León, Parque de Investigación e Innovación Tecnológica, Km. 10 Autopista al Aeropuerto Internacional Mariano Escobedo, Apodaca 66629, Nuevo León, Mexico
* Correspondence: juan.villarrealch@uanl.edu.mx (J.F.V.-C.); ecedillo@unimore.it (E.I.C.-G.)

Citation: Vital-Grappin, A.D.; Ariza-Tarazona, M.C.; Luna-Hernández, V.M.; Villarreal-Chiu, J.F.; Hernández-López, J.M.; Siligardi, C.; Cedillo-González, E.I. The Role of the Reactive Species Involved in the Photocatalytic Degradation of HDPE Microplastics Using C,N-TiO₂ Powders. *Polymers* **2021**, *13*, 999. https://doi.org/10.3390/polym13070999

Academic Editor: Jacopo La Nasa

Received: 28 February 2021
Accepted: 22 March 2021
Published: 24 March 2021

Publisher's Note: MDPI stays neutral with regard to jurisdictional claims in published maps and institutional affiliations.

Copyright: © 2021 by the authors. Licensee MDPI, Basel, Switzerland. This article is an open access article distributed under the terms and conditions of the Creative Commons Attribution (CC BY) license (https://creativecommons.org/licenses/by/4.0/).

Abstract: Microplastics (MPs) are distributed in a wide range of aquatic and terrestrial ecosystems throughout the planet. They are known to adsorb hazardous substances and can transfer them across the trophic web. To eliminate MPs pollution in an environmentally friendly process, we propose using a photocatalytic process that can easily be implemented in wastewater treatment plants (WWTPs). As photocatalysis involves the formation of reactive species such as holes (h^+), electrons (e^-), hydroxyl (OH$^\bullet$), and superoxide ion ($O_2^{\bullet-}$) radicals, it is imperative to determine the role of those species in the degradation process to design an effective photocatalytic system. However, for MPs, this information is limited in the literature. Therefore, we present such reactive species' role in the degradation of high density polyethylene (HDPE) MPs using C,N-TiO₂. Tert-butanol, isopropyl alcohol (IPA), Tiron, and Cu(NO₃)₂ were confirmed as adequate OH$^\bullet$, h^+, $O_2^{\bullet-}$ and e^- scavengers. These results revealed for the first time that the formation of free OH$^\bullet$ through the pathways involving the photogenerated e^- plays an essential role in the MPs' degradation. Furthermore, the degradation behaviors observed when h^+ and $O_2^{\bullet-}$ were removed from the reaction system suggest that these species can also perform the initiating step of degradation.

Keywords: microplastics; HDPE; microbeads; photocatalysis; scavengers; C,N-TiO₂; remediation; nanotechnology; plastic pollution; visible light photodegradation

1. Introduction

Microplastics (MPs) are defined as plastic particles with a diameter ≤ 5 mm and are categorized as primary or secondary according to their origin [1]. Primary MPs are manufactured explicitly to that particle size to be used in cosmetic or personal hygienic products [1]. On the other hand, secondary MPs are derived from large plastic items' fragmentation due to photolytic, mechanical, and biological degradation [1].

MPs pollution has increased so rapidly in recent years that it has become a global concern. Today, MPs can be found in a wide variety of sizes, shapes, and polymers in samples of soil, sand, air, water, wastewater, and snow all around the globe [1,2]. More importantly, due to their high specific surface area, MPs can adsorb persistent organic pollutants (POPs), hydrophobic organic contaminants (HOCs), pharmaceuticals, and heavy metals [1–3]. As MPs are frequently consumed by the biota present in aquatic and terrestrial ecosystems, these hazardous compounds can move easily through the trophic chain up to humans [1,2]. This has been evidenced by the presence of MPs in human stool [4] and even in the placenta [5]. Furthermore, the SARS-CoV-2 pandemic has increased MPs fibers' presence in the environment due to the inadequate disposal of single-use plastic face masks [6].

In this context, there is an urgent need to develop remediation technologies to dissipate MPs pollution. The current methodologies for this include physical removal, biodegradation, and chemical treatments [7,8]. These latter include photocatalysis, a process that has been widely used to mineralize recalcitrant organic pollutants in aqueous and gaseous environments [9].

There are several significant advantages of using photocatalysis as a remediation technology to solve MPs pollution. Firstly, MPs can be entirely mineralized to CO_2 and water. Only a small fraction is known to be converted into innocuous or less toxic chemical substances [10,11]. Secondly, photocatalysis can easily be adapted to wastewater treatment plants (WWTPs), identified as a potential source of MPs to aquatic and terrestrial ecosystems [1,2]. By coupling MPs photocatalysis to WWTPs, not only could the MPs level in discharge waters be significantly reduced, but the harmful problems that these pollutants cause to their facilities could also be avoided (inhibitory effects on activated sludge flocs, reduction of the diversity of biological communities and decline of the abundance of crucial microorganisms [12]). Finally, the photocatalytic process can be regarded as environmentally friendly since it can be performed under visible or solar light. Simultaneously, the photocatalyst can be prepared using renewable raw materials, respecting the sixth, seventh and ninth principles of Green Chemistry.

Photocatalysis has recently been proven to degrade low-density polyethylene (LDPE), high-density polyethylene (HDPE), polystyrene (PS), and polypropylene (PP) MPs in aqueous solutions [10,11,13–19]. However, complete degradation of any of these plastics has not been reported yet. To improve or optimize MPs degradation and be able to achieve their mineralization, different criteria must be considered, such as the photocatalyst's properties, the pollutant's characteristics and the reaction conditions, and the potential role of reactive species such as holes (h^+), electrons (e^-), and reactive oxygen species (ROS) that are formed along the process and could participate in the degradation reaction.

To date, the role of these reactive species in MPs degradation has been lightly investigated. For example, Nabi et al. [18] reported the "solid phase" photodegradation of 400 nm PS MPs using a Triton X-100 derived TiO_2 nanoparticle film with ethylenediaminetetraacetic acid (EDTA), tert-butyl alcohol (TBA), and anaerobic conditions to check the role of holes, hydroxyl radicals ($OH^{\bullet}$), and oxygen (O_2) in the degradation process. By monitoring the change in the PS particle's diameter using scanning electron microscopy (SEM) analysis, their experiments demonstrated that the h^+ was the dominant active specie in their reaction system when submitted under UV light for 6 h [18]. On the other hand, Jiang et al. [11] evaluated the photocatalytic degradation of a 1 g/L aqueous dispersion of HDPE MPs (200–250 µm) using a hydroxy-rich bismuth oxychloride photocatalyst and visible light ($\lambda \geq 420$ nm). The role of h^+, $OH^{\bullet}$, and superoxide ion radical ($O_2^{\bullet -}$) in the degradation process was monitored by Fourier-transform infrared spectroscopy (FTIR) using methanol, isopropyl alcohol (IPA) and p-benzoquinone as scavengers. Results demonstrated that free $OH^{\bullet}$ formed through the reaction of adsorbed water with surface hydroxyl was the main reaction group involved in the degradation process [11].

These studies suggested that different reactive species can promote MPs' degradation depending on the reaction conditions (solid- or aqueous-based reaction). However, there is still a need for a more in-depth analysis regarding the specific role that each one of those species may have on MPs degradation.

The present work is aimed to analyze the specific role of four reactive species in the photocatalytic degradation of primary HDPE MPs using $C,N-TiO_2$ powders. To achieve this, tert-butanol, isopropyl alcohol, Tiron, and copper nitrate were used as $OH^{\bullet}$, h^+, $O_2^{\bullet -}$ and e^- scavengers, respectively. The reactions were performed in an aqueous solution because of its relevance for applying this technology in WWTPs. The $C,N-TiO_2$ powders were chosen because they enable to perform the experiments in visible light and, get a better interaction between the dispersed photocatalyst and MPs [13]. It was found that all four species have an essential role in MPs degradation, suggesting for the first time that the photogenerated e^- plays an essential role in the photocatalytic degradation of HDPE MPs, mainly due to the formation of free $OH^\bullet$. Furthermore, it was demonstrated that the initiating step on MPs degradation is performed by $OH^\bullet$, h^+ or $O_2^{\bullet-}$ species.

2. Materials and Methods

2.1. MPs Obtainment and Characterization

Primary HDPE MPs were obtained from a commercial facial scrub using the extraction methodology proposed by Napper et al. [20]. HDPE MPs particle size and morphology were determined by optical microscopy (OM) using a light microscope (DME, Leica, Wetzlar, Germany) with a microscope camera (ICC50 W, Leica, Wetzlar, Germany). Morphology was also investigated by SEM coupled with energy dispersive X-ray spectroscopy (EDS), using a Thermo-Fisher Scientific FEI E-SEM (Quanta-200, FEI Company, Hillsboro, OR, USA). Before the analysis, the samples were coated with a gold layer of 10 nm. The type of polymer present in the sample was determined by attenuated total reflectance-Fourier-transform infrared spectroscopy (ATR-FTIR) in an infrared spectrometer (ALPHA II, Bruker, Ettlingen, Germany). The spectrum was collected by averaging 32 scans between 400 and 4000 cm^{-1} with a 4 cm^{-1} spectral resolution. The surface area was evaluated by nitrogen adsorption, using a surface area analyzer (Tristar II Plus 3.01, Micromeritics, Norcross, GA, USA) (bath temperature of 77.3 K, equilibration interval of 5 s, and degasification at 30 °C for 23 h).

2.2. C,N-TiO$_2$ Preparation and Characterization

The C,N-TiO$_2$ semiconductor was prepared by the synthesis procedure reported by Zeng et al. [21], with some modifications already described in previous work [13]. The crystalline composition was determined by X-ray diffraction (XRD), using a diffractometer with Cu $K\alpha 1$ radiation, current of 40 mA, accelerating voltage of 40 kV, and 2θ-steps of 0.023° (D5000, SIEMENS, Munich, Germany). The chemical groups adsorbed in the semiconductor's surface were identified by ATR-FTIR (ALPHA II, Bruker, Ettlingen, Germany), using the analysis conditions reported in Section 2.1. The presence of nitrogen and carbon in the TiO$_2$ semiconductor was confirmed by elemental analysis using a CHNS/O analyzer with a combustion temperature of 975 °C and a reducing column with helium at 88 °C (2400 Series II, PerkinElmer, Waltham, MA, USA). The bandgap (E_g) was calculated from the reflectance spectrum (Figure S1, Supplementary Materials) using the Kubelka-Munk equation (Equation S1). The spectrum was measured in an UV-vis/NIR spectrophotometer with BaSO$_4$ (DEQ, Monterrey, Mexico) as reference material (V-670, JASCO, Tokyo, Japan). The microstructure was analyzed by field emission gun scanning electron microscopy (FEG-SEM)(Nova NanoSEM 450, FEI Company, Hillsboro, Oregon, USA). The BET surface area was evaluated by nitrogen adsorption, using a BET analyzer (Gemini 2380, Micromeritics, Norcross, GA, USA) (bath temperature of 77.3 K, equilibration interval of 10 s, and degasification at 110 °C).

2.3. Photocatalytic Experiments in the Absence and Presence of Scavengers

The photocatalytic experiments' reaction conditions were carried out according to our previous work, which demonstrated the photocatalytic degradation of HDPE MPs using the C,N-TiO$_2$ material [13]. Therefore, photocatalysis without scavengers was conducted in a Batch-type glass reactor, using 50 mL of a 0.4 wt./vol % HDPE MPs dispersion (pH 3) and a C,N-TiO$_2$ load of 200 mg. The reactor was placed in a closed reaction chamber equipped with a constant temperature bath (set at 0 ± 2 °C) and irradiated for 50 h with a Slim LED IP65 50 W visible LED lamp (ARE-006, ARTlite, New Delhi, India) placed at 25 cm from the sample (57.2 ± 0.3 W/m^2). MPs were dispersed in the experiments by continuous stirring at 350 rpm. After photocatalysis, the residual (degraded) MPs were separated from the reaction system and washed twice with distilled water to remove potential C,N-TiO$_2$ traces that could remain attached. The photocatalytic experiments with the scavengers were performed according to this methodology, but adding reagent grade tert-butanol (0.2 mol, DEQ, Monterrey, Mexico), IPA (0.5 mol, DEQ, Monterrey, Mexico), 4,5-dihydroxy-1,3-benzenedisulfonic acid disodium salt "tiron" (2.18 × 10^{-5} mol, Sigma Aldrich, Ciudad de México, México) or copper nitrate (5.0 × 10^{-3} mol, DEQ, Monterrey, Mexico) as scavengers [22,23] (Table 1). The quantities of each scavenger were chosen based on previous reports [22,23] and, their concentrations were in excess respect to the HDPE MPs pollutants. Three replicates were performed for all the photocatalytic experiments, and the mean ± S.D. values are reported.

Table 1. Chemical reagents used as OH$^\bullet$, h^+, O$_2^{\bullet-}$ and e^- scavengers.

Reagent	Scavenged Specie	k (M^{-1}s^{-1}) [a]	Reference
Tert-butanol	OH$^\bullet$ in bulk water	6.2 × 10^8	[22]
IPA	h^+	NA	[22]
	OH$^\bullet$	2 × 10^9	[22]
Tiron	O$_2^{\bullet-}$	5 × 10^8	[22]
Copper nitrate	e^-	NA	[23]

[a] Rate constant of the reaction between the reagent and the scavenged species.

To determine the reactive species' role in MPs photocatalysis, three analytical methods were used to monitor and confirm MPs degradation. These methods were particularly used as they are known to be effective independently of the presence of C,N-TiO$_2$ traces. The first method used was gravimetry. This technique is commonly used due to its simplicity and represents a direct way to quantify the polymers' degradation [24]. As expected, no presence of traces of C,N-TiO$_2$ in the residual MPs was detected by this method. All mass measurements were performed in an analytical balance (CY224C, ACZET, Mumbai, India) after 15, 30, and 50 h of photocatalysis. Changes in MPs' concentration were calculated and reported as wt./vol %. After 50 h of photocatalysis, MPs degradation was monitored by FTIR, which can detect polar functional groups, such as ketones and ester carbonyls (typical of oxidative degradation pathways [24]). Nevertheless, as higher intensity of the carbonyl band does not always represent more plastic degradation, this signal has to be compared to second reference band [25]. Consequently, the extent of oxidation during MPs degradation was quantified by calculating their carbonyl index (CI). CI quantifies the absorbance change for the carbonyl stretch relative to the C-H stretching modes [24]. CI was defined as the carbonyl group's absorbance ratio around 1720 cm^{-1} to an internal thickness band as a reference peak at 1380 cm^{-1} [26] (CI = A_{1720}/A_{1380}). As the FTIR adsorption bands of TiO$_2$ are located at the 439–754 cm^{-1} interval [27,28], the presence of C,N-TiO$_2$ traces in the residual MPs does not affect CI calculation. The CI changes of the whole set of samples were compared with the trend in the MPs' concentration. OM and SEM (secondary-electron imaging) were used to analyze the changes in the morphology of the MPs, again, with no interference from the presence of traces of C,N-TiO$_2$. In conjunction with SEM, EDS was used to determine the incorporation of oxygen to the degraded MPs.

3. Results

3.1. HDPE MPs and C,N-TiO$_2$ Characterization

As shown in Figure 1a, the blue microbeads extracted from the commercial facial scrub have an average size of 725 ± 108 μm (50 measurements). The FTIR analysis carried out on the microbeads (Figure 1b) revealed the characteristic vibrational bands of high-density polyethylene (HDPE), these being the bands at 2911 and 2846 cm^{-1} that correspond to the asymmetric and symmetric stretching vibrations of the CH$_2$ group and the bands at 1463 and 719 cm^{-1} that correspond to the scissoring and rocking bending vibrations of the same group [27]. SEM micrographs demonstrated that the surface of the blue microbeads exhibited a homogeneous roughness, which may be related to their application as abrasive in facial scrubs (Figure 1c). Furthermore, some minimal roughness-related porosity was present, which leads to a surface area of 0.41 ± 0.03 m^2/g (Figure 1d). These results confirmed that the blue microbeads extracted from the commercial facial scrub are primary MPs of HDPE.

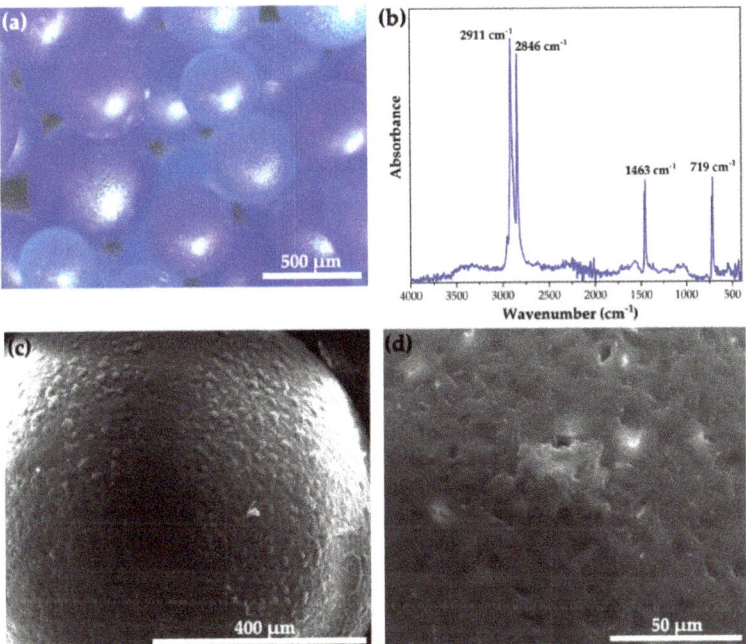

Figure 1. (a) Optical micrography, (b) FTIR spectrum, and (c,d) SEM micrographs of the as−extracted HDPE MPs.

For the C,N-TiO$_2$ semiconductor, the XRD pattern (Figure 2a) showed that the semiconductor was mainly composed of the anatase polymorph of TiO$_2$. However, two small peaks at 36° and 44° attributable to rutile TiO$_2$ were also visible. Although anatase has a higher photocatalytic yield than rutile [29], it is known that the presence of rutile in mixed-phase TiO$_2$ nanocomposites increases the photocatalytic activity due to the interfacial charge separation between anatase and rutile. Thus, inhibiting the recombination process of the photogenerated h^+-e^- pairs [30].

On the other hand, the FTIR analysis (Figure 2b) exhibited the absorption bands corresponding to the Ti–O–Ti bonds' stretching vibrations from the photocatalytic material, located at 439 and 754 cm^{-1} [27,28]. The broad absorption band between 3450 and 3200 cm^{-1} was assigned to the O–H bond's stretching vibration from adsorbed water. The adsorption bands at 1408 cm^{-1} and 1080 cm^{-1} were assigned to the vibrations of the Ti–N bond [31,32], while the band located 1627 cm^{-1} is usually related to the bending vibrations of the N–H and O–H bonds [32]. The additional elemental analysis confirmed that the semiconductor contained 0.41 wt. % of nitrogen, along with 1.37 wt. % of carbon (Table 2). The presence of these elements in the semiconductor's structure is known to promote a reduction of the bandgap of TiO$_2$ from its standard value of 3.2 eV [9] to 2.90 eV (Figure S1), enabling the absorption of visible light of 428 nm (Table 2).

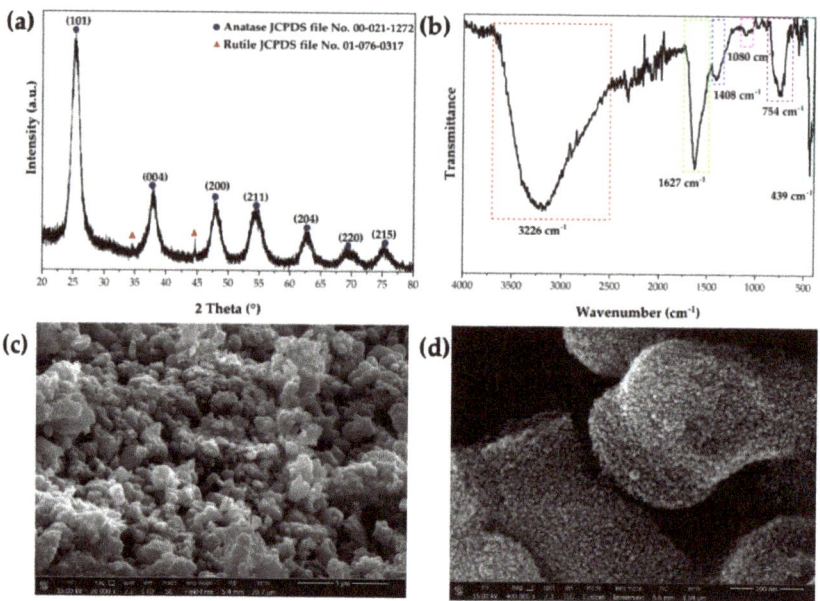

Figure 2. (a) XRD pattern, (b) FTIR spectrum, and (c,d) FEG–SEM micrographs of the C,N–TiO$_2$ semiconductor.

Table 2. Properties of the C,N-TiO$_2$ semiconductor.

Property	Value
Carbon content (wt. %)	1.37
Nitrogen content (wt. %)	0.41
E_g (eV)	2.90
Wavelength of light absorption (nm)	428
BET surface area (m^2/g)	194.0 ± 1.8
Porosity type	Mesoporous

Figure 2c,d show the FEG-SEM micrographs of the C,N-TiO$_2$ photocatalyst at different magnifications. At lower magnifications (Figure 2c), it can be observed that the semiconductor is formed by irregular particles that are ranged between 1 and 3 μm in diameter, which promoted porosity due to their agglomeration. On the other hand, higher magnification (Figure 2d) exposed that these micrometric particles are formed by smaller particles of approximately 10 nm in diameter. Thus, the microstructure of the C,N-TiO$_2$ semiconductor possesses a high surface area. This was demonstrated by BET analysis, which reported that the C,N-TiO$_2$ semiconductor has a surface area of 194.0 ± 1.8 m^2/g (Table 2), being 3.88 times bigger than the surface area of commercial TiO$_2$ (Aeroxide® P25, 50 m^2/g [22]). The isotherm obtained from the nitrogen absorption analysis (Figure S2) belongs to type IV, which corresponds to a mesoporous material according to the IUPAC classification (Table 2). The isotherm presented a hysteresis loop type H1, which is often associated with porous materials that consist of almost uniform agglomerated spheres [33], as observed in Figure 2d.

3.2. Understanding the Practical and Mechanistic Issues of Photocatalytic Degradation of MPs in an Aqueous Medium

Photocatalysis of MPs using aqueous solutions and powder semiconductors should be carefully designed as it presents some inherent practical and mechanistic issues that may negatively affect degradation. As presented in Equation (1), irradiation of the semiconductor with energy higher than its bandgap generates h^+ in its valence band and e^- in the conduction band. If their recombination is avoided, h^+ can oxidize organic pollutants (P) to their radical cation (Equation (2)) or react with adsorbed water or hydroxyl groups (OH^-) to generate $OH^\bullet$ (Equations (3) and (4)), which in turn oxidize the adsorbed pollutants (Equation (5)) [9,22,34]. Photogenerated e^- reduce molecular oxygen to form $O_2^{\bullet-}$ (Equation (6)), which participates in the acid–base equilibrium with its protonated form (Equation (7)) [22]. As the pK_a of $HO_2^\bullet$ is 4.6, the acidic form of $O_2^{\bullet-}$ may be the most abundant species in our reaction system (pH 3) [22,35]. Dismutation reactions of $HO_2^\bullet$ and $O_2^{\bullet-}$ generate H_2O_2 and O_2 (Equations (8) and (9)) [9,22,35]. The generation of H_2O_2 during photocatalysis is a crucial step for further $OH^\bullet$ formation in bulk water. Once formed, H_2O_2 can react with the photogenerated e^- (Equation (10)), $O_2^{\bullet-}$ (Equation (11)) or be decomposed by photolysis (Equation (12)) to form $OH^\bullet$ [9]. The interaction of the hydroxyl radicals with the MPs should lead to the degradation of these tiny plastic pollutants (Equation (13)). It is essential to highlight that the reaction presented in Equation (10) only produces surface-adsorbed $OH^\bullet$, while the reactions presented in Equations (11) and (12) can produce both surface-adsorbed and free $OH^\bullet$ [9]. Additionally, the direct photolysis of H_2O_2 under visible light irradiation can be neglected due to its small molar absorption coefficient ($<1\ M^{-1}cm^{-1}$ above 300 nm) [23].

$$\text{Semiconductor} + h\nu \rightleftharpoons h^+ + e^- \tag{1}$$

$$h^+ + P \rightleftharpoons P^{\bullet+} \tag{2}$$

$$h^+ + H_2O_{ads} \rightarrow OH^\bullet_{ads} \tag{3}$$

$$h^+ + OH^-_{ads} \rightarrow OH^\bullet_{ads} \tag{4}$$

$$P + OH^\bullet_{ads} \rightarrow \text{Oxidation products} \tag{5}$$

$$e^- + O_2 \rightarrow O_2^{\bullet-}\ (k = 1.9 \times 10^{10}\ M^{-1}s^{-1}) \tag{6}$$

$$HO_2^\bullet \rightleftharpoons H^+ + O_2^{\bullet-}\ (pK_a = 4.8) \tag{7}$$

$$2\ HO_2^\bullet \rightarrow H_2O_2 + O_2 \tag{8}$$

$$2O_2^{\bullet-} + 2H^+ \rightarrow H_2O_2 + O_2 \tag{9}$$

$$H_2O_2 + e^- \rightarrow OH^\bullet + OH^- \tag{10}$$

$$H_2O_2 + O_2^{\bullet-} \rightarrow OH^\bullet + O_2 + OH^- \tag{11}$$

$$H_2O_2 + h\nu \rightarrow OH^\bullet + OH^\bullet \tag{12}$$

$$MPs + OH^\bullet \rightarrow \text{Oxidation products} \tag{13}$$

As deduced by the previous reactions, the pollutant's interaction with the generated reactive species (through adsorption) is crucial to the photocatalytic process. When performing heterogeneous photocatalysis in an aqueous medium, the interaction between the pollutant and the semiconductor is usually achieved by the adsorption of the former at the surface of the photocatalyst. However, MPs pollutants cannot be adsorbed at the surface of the photocatalyst. As presented in Figure 3, the size of HDPE MPs limits their interaction with the powder semiconductor.

The MPs used in this work are nearly 241-fold the size of C,N-TiO$_2$ biggest particles, precluding their adsorption on the semiconductor and limiting their oxidation by h^+ and adsorbed $OH^\bullet$. However, degradation may still be promoted whether the semiconductor is used in the form of a powder dispersion. As shown in Figure 3, the C,N-TiO$_2$ semiconductor forms a coating on the HDPE MPs' surface [13]. Nonetheless, the MPs' coverage with C,N-TiO$_2$ may be limited by the low porosity and surface area of the HDPE microbeads.

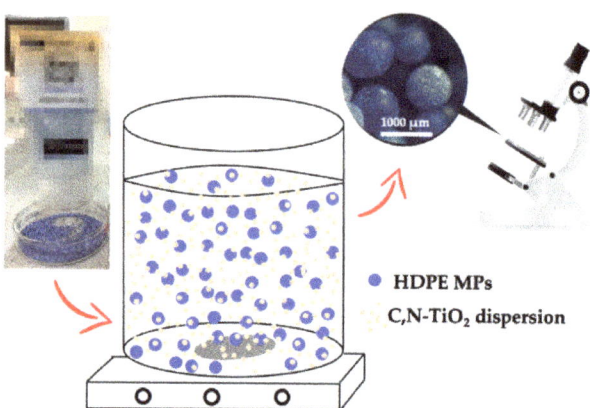

Figure 3. Adsorption of the C,N-TiO$_2$ semiconductor in the surface of HDPE MPs.

3.3. Effect of Free OH$^\bullet$ on HDPE MPs Photocatalytic Degradation

The interaction between the HDPE MPs and the C,N-TiO$_2$ powder semiconductor is limited when photocatalysis is performed in an aqueous medium. Thus, it is very likely that most of the photocatalytic degradation of MPs is mainly conducted by the free OH$^\bullet$ radicals present in the bulk of the medium. To determine the free OH$^\bullet$ contribution in HDPE MPs degradation, photocatalysis was carried out in the absence and presence of tert-butanol as OH$^\bullet$ scavenger.

Figure 4 presents the degradation plot for the visible-light photocatalysis of HDPE MPs by C,N-TiO$_2$ in the absence and presence of tert-butanol. The black curve shows that without scavenger, i.e., when free OH$^\bullet$ are present in the reaction system, the MPs' concentration was reduced by 71.77%. As previously reported, degradation at these conditions was related to the increase of the MPs' surface area due to their fragmentation at 0 °C and the effect of acidic pH in the plastic degradation mechanism and the semiconductor deagglomeration [13]. The red curve of Figure 4 shows that the inhibition of the reaction between HDPE MPs and free OH$^\bullet$ (k for the tert-butanol/OH$^\bullet$ reaction is 6.2 × 10^8 M^{-1}s^{-1}, Table 1) dramatically reduces the degradation of MPs. The MPs concentration reduction passed from 71.77% at normal conditions to 1.98% when tert-butanol is present in the reaction system. This behavior indicates that free OH$^\bullet$-mediated oxidation processes are the predominant pathways leading to the HDPE MPs' degradation. This result is in good agreement with that obtained by Jiang et al. [11] and with the LDPE and PP MPs degradation pathways suggested by the research group of J. Dutta [10,36]. For this work, the reaction systems characteristics discussed in Section 3.2 suggest that oxidation of HDPE MPs by C,N-TiO$_2$ is promoted OH$^\bullet$ formation through the reaction pathways involving the photogenerated e^- (Equations (6)–(11)). The fact that some degradation was still detected at these reaction conditions was related to the deposition of the C,N-TiO$_2$ powders on the MPs' surface (Figure 3), a phenomenon that may facilitate the direct oxidation of the HDPE microbeads with the photogenerated h^+ (Equation (2)). To test this hypothesis, photocatalytic experiments using IPA as an h^+ scavenger were performed.

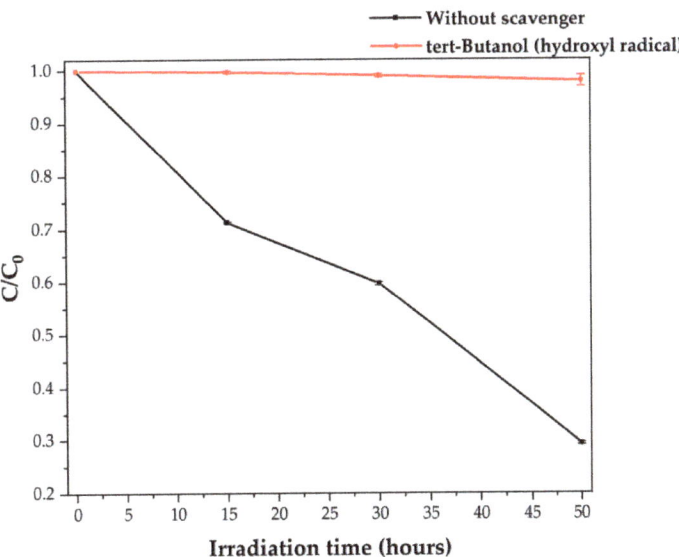

Figure 4. Visible-light photocatalytic degradation of HDPE MPs by C,N-TiO$_2$ in the absence and presence of tert-Butanol as free OH$^\bullet$ scavenger.

3.4. Effect of h^+ on HDPE MPs Photocatalytic Degradation

Holes are the primary oxidizing species in photocatalytic reactions [9]. Although they are not usually related to plastic degradation by photocatalysis, photogenerated h^+ can carry out the direct oxidation of organics when these compounds can act as electron donors [9,37]. For example, Ueckert et al. reported that photoreforming of poly(ethylene terephthalate) (PET) and poly(lactic acid) (PLA) could be carried out by carbon nitride/nickel phosphide (CN$_x$ | Ni$_2$P) photocatalyst [38]. By performing this process in the presence of terephthalic acid (TPA) as an OH$^\bullet$ scavenger, authors found that OH$^\bullet$ plays a minimal role in the oxidation of plastics and that photoreforming instead proceeds via direct h^+ transfer between the catalyst and the substrate [38]. Here, to determine the contribution of h^+ in the degradation of HDPE MPs, photocatalysis was carried out in the presence of IPA as an h^+ scavenger [22]. As tert-butanol, IPA can also act as OH$^\bullet$ scavenger ($k = 2 \times 10^9$ M^{-1}s^{-1}, Table 1). However, due to its smaller size, IPA presents strong dissociative and non-dissociative chemisorption on TiO$_2$ [39]. Thus, adsorbed IPA acts as an electron donor and is oxidized by h^+.

Figure 5 presents the degradation plot for the visible-light photocatalysis of HDPE MPs by C,N-TiO$_2$ in the absence and presence of IPA.

The blue curve of Figure 5 shows that when IPA is added to the reaction system, MPs' photocatalytic removal is inhibited. IPA decreases MPs degradation by adsorbing on the semiconductor surface, reacting with the photogenerated h^+, and avoiding Reaction 2. As IPA also eliminates OH$^\bullet$ from the reaction system, this result suggests that HDPE MPs are oxidized by the free OH$^\bullet$ and directly by the photogenerated h^+. These species can react with the free electron-containing ending groups of the PE chain [40] through the photo-Kolbe reaction [41] (Equation (14)).

$$RCOO^- + h^+ \rightarrow R^\bullet + CO_2 \tag{14}$$

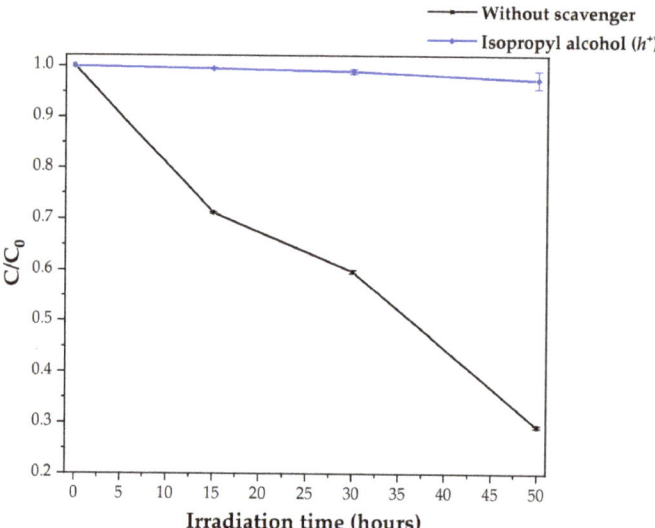

Figure 5. Visible-light photocatalytic degradation of HDPE MPs by C,N-TiO$_2$ in the absence and presence of IPA as h^+ and OH$^\bullet$ scavenger.

At these reaction conditions (similar to those reported in Section 3.3 for tert-butanol scavenger), low degradation of the HDPE MPs was detected (HDPE concentration was reduced by 2.53%). This result was related to the deposition of the C,N-TiO$_2$ powders on the MPs' surface, a phenomenon that in this system (without h^+ and OH$^\bullet$) may facilitate the HDPE oxidation microbeads with $O_2^{\bullet-}$. As recombination of the photogenerated h^+-e^- pairs was avoided by scavenging h^+ with IPA, the reaction presented in Equation (6) is favored, and $O_2^{\bullet-}$ is expected to be present in the reaction system. As OH$^\bullet$, $O_2^{\bullet-}$ is often believed to play an essential role in initiating the polyethylene degradation process by attacking the polymeric chain [17,26]. To test this hypothesis, photocatalytic experiments using Tiron as an $O_2^{\bullet-}$ scavenger were performed.

3.5. Effect of $O_2^{\bullet-}$ on HDPE MPs Photocatalytic Degradation

The role of $O_2^{\bullet-}$ on the photocatalytic degradation organics in water is frequently evaluated using p-benzoquinone scavenger (k for p-benzoquinone/$O_2^{\bullet-}$ reaction is 0.9–1 × 10^9 M^{-1}s^{-1}) [22]. However, p-benzoquinone does not only react with $O_2^{\bullet-}$, but also with e^- ($k = 1.35 \times 10^9$ M^{-1}s^{-1}) and OH$^\bullet$ ($k = 6.6 \times 10^9$ M^{-1}s^{-1}) [22]. The use of p-benzoquinone as an $O_2^{\bullet-}$ scavenger has other drawbacks. When photoexcited with wavelengths above 310 nm, p-benzoquinone produces singlet oxygen ($^1O_2$) [42], which may contribute to the oxidation of organic molecules [9]. On the other hand, Garg et al. have postulated that quinones in natural organic matter are subjected to photolysis under sunlight, generating $O_2^{\bullet-}$, which further evolves to H_2O_2 and OH$^\bullet$ [43]. Tiron was chosen as $O_2^{\bullet-}$ scavenger (k for Tiron/$O_2^{\bullet-}$ reaction is 5×10^8 M^{-1}s^{-1}, Table 1) due to the previous motivations.

Figure 6 presents the degradation plot for the visible-light photocatalysis of HDPE MPs by C,N-TiO$_2$ in the absence and presence of Tiron. The pink curve of Figure 6 shows that when Tiron is added to the reaction system, MPs' photocatalytic removal is reduced but not completely inhibited (MPs' concentration was reduced by 37.98%). This inhibition is in good agreement with the previously proposed photocatalytic degradation pathways of LDPE and PP MPs [10,36]. To explain the partial inhibition of photocatalysis, three phenomena should be taken into account: The absence of (i) $O_2^{\bullet-}$ and (ii) $O_2^{\bullet-}$-derived OH$^\bullet$ in the reaction system and (iii) the presence of adsorbed h^+ on the C,N-TiO$_2$ surface.

First, some reports indicate that $O_2^{\bullet-}$ is involved in the direct oxidations of organic molecules that can act as a proton source [9,44], as $O_2^{\bullet-}$ can react with a proton donor to form $HO_2^{\bullet}$ (Equation (15)) [44]:

$$O_2^{\bullet-} + HX \rightleftharpoons HO_2^{\bullet} + X^- \tag{15}$$

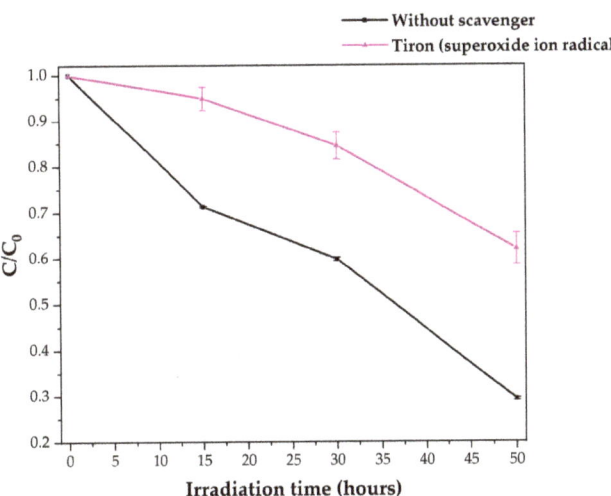

Figure 6. Visible-light photocatalytic degradation of HDPE MPs by C,N-TiO$_2$ in the absence and presence of Tiron as $O_2^{\bullet-}$ scavenger.

Here, $O_2^{\bullet-}$ may promote the initiation step of PE oxidation by the abstraction of a proton of the polymeric chain (see Scheme 1 in Section 3.7). Therefore, degradation is in part inhibited when this species is not present in the reaction system. However, when $O_2^{\bullet-}$ reacts with Tiron, the acid-based equilibrium presented in Equation (7) is perturbed. As a consequence, the formation of H_2O_2 through Equations (8) and (9) is avoided. As H_2O_2 further evolves to $OH^{\bullet}$ through Equations (10) and (11), this specie is not expected to be present in this particular reaction system. As the role of the $OH^{\bullet}$ in the MPs degradation has been ruled out in Section 3.3, the decrease in degradation is explained by its absence from the reaction system. Consequently, this suggests that $O_2^{\bullet-}$ plays an essential role in HDPE MPs photocatalytic degradation by extracting protons from the PE chain and their conversion to H_2O_2 and $OH^{\bullet}$. Lastly, the reduction of the HDPE degradation (instead of the complete inhibition) of Figure 6 confirms MPs' oxidation by the adsorbed h^+ species through Reaction 2. As $^1O_2$ cannot be present in the reaction system because it is formed through oxidation of the absent $O_2^{\bullet-}$ by the adsorbed h^+ [9], the only plausible path for MPs degradation in the presence of Tiron is the direct oxidation by h^+. The lower degradation of MPs concerning the experiment without scavenger (black curve) was related to the limited deposition of C,N-TiO$_2$ powders on the MPs. The crucial role of the $O_2^{\bullet-}$ in the photocatalytic degradation of HDPE MPs presented here differs from the result obtained by Juang et al. [11], where the absence of this reactive species in their system did not affect the degradation of HDPE MPs by BiOCl photocatalyst. This difference can be related to the use of p-benzoquinone as an $O_2^{\bullet-}$ scavenger [11], which can lead to the formation of $^1O_2$ [42], H_2O_2 [43] and $OH^{\bullet}$ [43] in the reaction system. From these, $OH^{\bullet}$ was observed to be the main specie that promotes degradation [11].

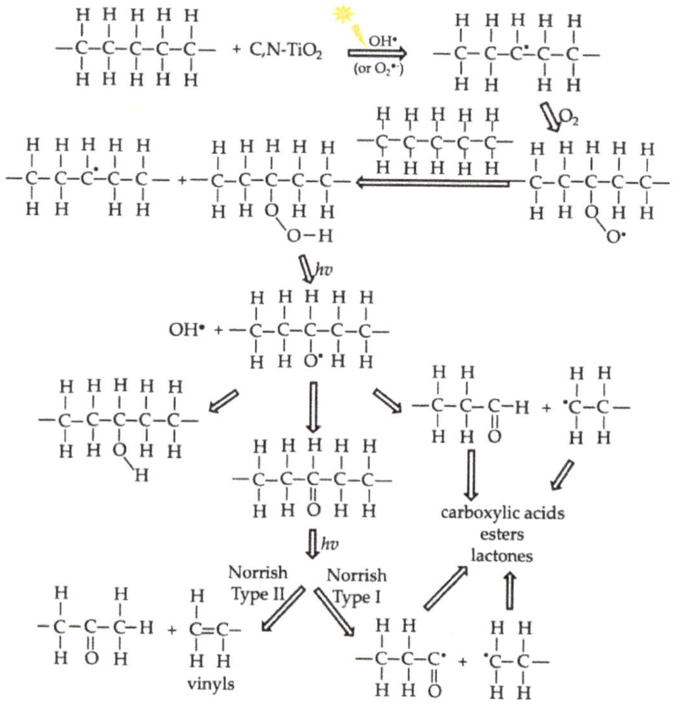

Scheme 1. Proposed mechanism for HDPE MPs photocatalytic degradation with C,N-TiO$_2$.

3.6. Effect of e^- on HDPE MPs Photocatalytic Degradation

Copper nitrate was used to determine the role of the photogenerated e^- in the HDPE degradation. This reagent was used because the adsorption of the NO$_3^-$ anions on the surface of TiO$_2$ is known to be weak [23]. Figure 7 presents the degradation plot for the visible-light photocatalysis of HDPE MPs by C,N-TiO$_2$ in the absence and presence of Cu(NO$_3$)$_2$.

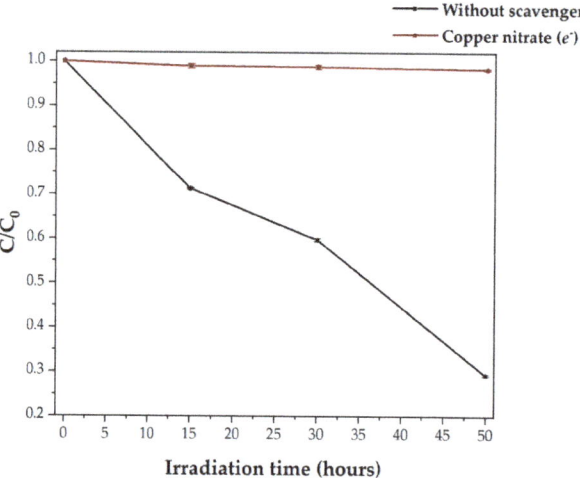

Figure 7. Visible-light photocatalytic degradation of HDPE MPs by C,N-TiO^2 in the absence and presence of copper nitrate as an e^- scavenger.

As observed in the bordeaux colour curve of Figure 7, the presence of copper nitrate in the reaction system inhibits MPs degradation. This result can be explained by analyzing the critical steps of the photocatalytic mechanism where photogenerated e^- participates. As Cu^{2+} cation adsorbs in the surface of C,N-TiO_2, it reacts with the e^-, inhibiting the formation of $O_2^{\bullet-}$ through the reaction presented in Equation (6) [23]. As explained previously, the absence of $O_2^{\bullet-}$ inhibits the generation of H_2O_2 and its further conversion into free $OH^\bullet$ (Equations (6)–(9)). Furthermore, removing e^- from the reaction system inhibits the reduction of H_2O_2 to $OH^\bullet$ (Equation (10)). This latter is a key element for the degradation of HDPE MPs by photocatalysis with C,N-TiO_2. Since the quenching of e^- with copper nitrate avoids h^+-e^- recombination, the low degradation of the MPs at these conditions (1.64%) was attributed to the adsorbed h^+ and their interaction with the MPs.

Figure 8 presents the optical micrographs of the HDPE MPs after visible-light photocatalytic degradation by C,N-TiO_2 in the four scavengers' presence. The MPs subjected to photocatalysis in reaction media without $OH^\bullet$, h^+, and e^- show the same morphology as the as-extracted microplastics of Figure 2a. On the other hand, the photocatalytically degraded MPs in the absence of $O_2^{\bullet-}$ are highly disintegrated (Figure 8c), confirming plastic degradation [45].

Figure 8. Optical micrographs of the HDPE MPs after visible-light photocatalytic degradation for 50 h by C,N-TiO_2 in the presence of (**a**) tert-Butanol, (**b**) isopropyl alcohol (**c**) Tiron, and (**d**) copper nitrate scavengers.

A deeper insight into the changes of the MPs morphology was determined by SEM (Figure 9). The SEM micrographs of the microbeads subjected to photocatalysis without $OH^\bullet$, h^+, and e^- show the same rough surface as the as-extracted microplastics of Figure 2d, while the MPs that were degraded in the absence of $O_2^{\bullet-}$ present a slightly smoother surface. The loss of roughness was related to the degradation of the more external surface of the HDPE MPs by the adsorbed h^+, as most plastics degrade first at the polymer surface because it is exposed and available for the chemical attack [13,45].

EDS chemical microanalysis of the as-extracted and degraded MPs are presented in Figures S3 and S4, respectively. In both cases, the signal from gold corresponds to the conductive layer deposited during sample preparation. For the as-extracted MPs, the EDS spectra were taken from several microbeads and they present the intense signal of carbon from polyethylene polymer. As observed in Figure S3, the oxygen signal is practically absent from this sample, indicating that the as-extracted MPs are not oxidated. The EDS spectra from Figure S4 correspond to the MPs subjected to photocatalysis for 50 h at pH 3 and 0 °C, without scavengers. The oxygen signal is present in all the spectra. However, the EDS analysis interpretation must take into account both oxidation and the presence of C,N-TiO$_2$ on the plastic's surface, as both contribute to the oxygen signal. Figure S5 presents a higher magnification micrograph of the same sample, where the photocatalyst residues are visible. By taking into account the stoichiometry of TiO$_2$, the EDS analysis presented in Table S1 demonstrates that the presence of oxygen in this sample can be related to plastic oxidation (spectrum 6), the semiconductor (spectra 1 and 2) or a combination of both (spectra 3–5).

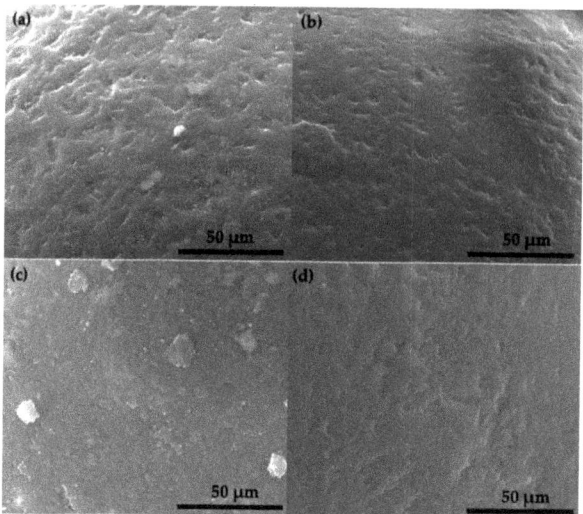

Figure 9. SEM micrographs of the HDPE MPs after visible-light photocatalytic degradation for 50 h by C,N-TiO$_2$ in the (**a**) tert-Butanol, (**b**) isopropyl alcohol (**c**) tiron, and (**d**) copper nitrate scavengers.

The calculation of CI from the FTIR spectra (Figure S6) also confirmed the degradation of plastics. As observed in Table 3, the as-extracted MPs present a CI value of 0.86, which was related to chromophores' presence [26,36]. When photocatalysis is carried out without scavengers, MPs degradation led to an increase in the CI of 29%. The CI values of the microbeads subjected to photocatalysis without OH$^\bullet$, h^+, and e^- are similar to that of the as-extracted microplastics, while the MPs degraded in the absence of O$_2^{\bullet-}$ present an increase of 19% with respect to the pristine MPs. This intermediate value between the original and the most degraded MPs is in good agreement with reducing the HDPE concentration observed for this particular reaction system.

Table 3. Comparison of the CI of the as-extracted HDPE MPs with those subjected to visible-light photocatalytic degradation by C,N-TiO$_2$ in the presence of scavengers.

HDPE MPs	MPs' Degradation (%)	CI (A$_{1720}$/A$_{1380}$)
As-extracted	-	0.86
Without scavengers	71.77 ± 1.88 [13]	1.11
With OH$^\bullet$ scavenger	1.98 ± 0.60	0.85
With h^+ scavenger	2.53 ± 1.00	0.80
With O$_2^{\bullet-}$ scavenger	37.98 ± 1.98	1.03
With e^- scavenger	1.62 ± 0.10	0.70

3.7. Proposed Mechanism for HDPE MPs Photocatalytic Degradation

The proposed mechanism for PE degradation is presented in Scheme 1.

As previously reported in the literature, in the initiation step the OH$^\bullet$ break the C–H bonds of the polymer backbone to produce polyethylene alkyl radicals [36,45]. The results suggest that the initiation step can also be promoted by the abstract of a hydrogen atom from the polymeric chain by $O_2^{\bullet-}$. This reaction's product is $HO_2^\bullet$, which form more OH$^\bullet$ by the dismutation reaction presented in Equation (8). Moreover, the photo-Kolbe type reaction between the free electron-containing ending groups of the PE chain and the h^+ may also generate polyethylene alkyl radicals [41]. During propagation, the alkyl radicals react with O_2 to generate peroxy radicals (–CH_2–CH_2–HCOO$^\bullet$–CH_2–CH_2-), which form the hydroperoxide species (–CH_2–CH_2–HCOOH–CH_2–CH_2-) by extracting a hydrogen atom from another polymer chain. This chemical specie is further divided into two new free oxy and OH$^\bullet$ radicals by the scission of the weak O–O bond [46,47]. After the formation of hydroperoxides, complex radical reactions take place and lead to autoxidation. The propagation step leads to chain scission or crosslinking, promoting a decrease or increase in molecular weight. Finally, termination of the radical reaction takes place when the combination of two radicals forms inert products such as alkenes, aldehydes, and ketones.

The results obtained here confirm that the photocatalysis of MPs is a complex phenomenon that strongly depends on the reaction system's particular characteristics. Due to their bigger size (relative to C,N-TiO_2 particles), MPs do not adsorb on the semiconductor. Instead, the C,N-TiO_2 powders adsorb on the plastic's surface. This phenomenon limits the interaction between MPs, h^+ and OH$^\bullet_{ads}$. Despite this drawback, MPs' oxidation is still possible thanks to the formation of $O_2^{\bullet-}$ and free OH$^\bullet$ (Reactions 6 and 11). Since both species are generated from the reaction pathways involving the photogenerated e^-, the results presented in this work reveal for the first time that the photogenerated e^- plays an essential role in the photocatalytic degradation of HDPE MPs. Furthermore, the degradation behaviors observed when h^+ and $O_2^{\bullet-}$ were absent from the reaction system suggest that the initiating step of MPs degradation can be performed by OH$^\bullet$ and the photogenerated h^+ or $O_2^{\bullet-}$.

4. Conclusions

In this work, an in-depth analysis regarding the participation of OH$^\bullet$, h^+, $O_2^\bullet$ and e^- in the photocatalytic degradation of primary HDPE MPs by C,N-TiO_2 powders was performed using scavengers. The information presented in this work can contribute to the proper design of effective photocatalytic reaction systems that enhance the degradation of HDPE MPs. As the species adsorbed in the photocatalyst's surface have less contribution to degradation due to their limited interaction with the microbeads, degradation should be mainly based on the generation of free OH$^\bullet$ in the bulk of the solution. This can be achieved, for example, by using photocatalysts specifically designed for avoiding h^+-e^- recombination and well-oxygenated reaction mediums that enhance the OH$^\bullet$ generation from the e^--derived reaction pathways. Additionally, photocatalysis can be performed in solar light to further promote the generation of free OH$^\bullet$ from the photolysis of H_2O_2.

Supplementary Materials: The following are available online at https://www.mdpi.com/2073-4360/13/7/999/s1, Figure S1: (a) Reflectance spectra and (b) Plot of $(F(R)*h\nu)^{1/2}$ vs. $h\nu$ of the C,N-TiO_2 photocatalyst. Figure S2. N_2 adsorption–desorption isotherms of the C,N-TiO_2 photocatalyst. Figure S3. SEM-EDS analysis of the as-extracted HDPE MPs. Figure S4 SEM-EDS analysis of the HDPE MPs after photocatalysis at 0 °C and pH 3 for 50 h. Figure S5. SEM micrograph of the HDPE MPs after photocatalysis at 0 °C and pH 3 for 50 h (high magnification). Table S1. EDS chemical microanalysis of the HDPE MPs after photocatalysis at 0 °C and pH 3 for 50 h. Figure S6. FTIR spectra of the HDPE MPs before and after 50 h of photocatalytic degradation with and without scavengers.

Author Contributions: Conceptualization, E.I.C.-G.; Funding acquisition, E.I.C.-G. and J.F.V.-C.; Investigation, A.D.V.-G., M.C.A.-T. and V.M.L.-H.; Methodology, E.I.C.-G. and M.C.A.-T.; Supervision, E.I.C.-G., C.S., J.F.V.-C and J.M.H.-L.; Project administration, E.I.C.-G.; Resources, E.I.C.-G., J.M.H.-L. and C.S.; writing—original draft preparation, A.D.V.-G., M.C.A.-T. and E.I.C.-G.; writing—review and editing, E.I.C.-G., C.S., J.M.H.-L. and J.F.V.-C. All authors have read and agreed to the published version of the manuscript.

Funding: This research was funded by Mexican Council for Science and Technology (CONACYT), grant number PN-2017/5167, Program for Teacher Professional Development for the Superior Type of Mexican Secretary of Public Education, grant number 511-6/18-892 and UANL's PAICYT, grant number IT1277-20. The APC was funded by Universidad Autónoma de Nuevo León.

Institutional Review Board Statement: Not applicable.

Informed Consent Statement: Not applicable.

Data Availability Statement: The data presented in this study are available on request from the corresponding authors.

Acknowledgments: Authors thank the technical support of Paola Miselli from University of Modena and Reggio Emilia for her help with SEM-EDS analysis. Special thanks are given to David Adrian Garza Tamez from the Universidad Autónoma de Nuevo León for his help with performing some experiments.

Conflicts of Interest: The authors declare no conflict of interest. The funders had no role in the design of the study; in the collection, analyses, or interpretation of data; in the writing of the manuscript, or in the decision to publish the results.

References

1. Sharma, S.; Basu, S.; Shetti, N.P.; Nadagouda, M.N.; Aminabhavi, T.M. Microplastics in the environment: Occurrence, perils, and eradication. *Chem. Eng. J.* **2020**, *408*, 127317. [CrossRef]
2. Barceló, D.; Picó, Y. Microplastics in the global aquatic environment: Analysis, effects, remediation and policy solutions. *J. Environ. Chem. Eng.* **2019**, *7*, 103421. [CrossRef]
3. Wang, Y.; Wang, X.; Li, Y.; Li, J.; Liu, Y.; Xia, S.; Zhao, J. Effects of exposure of polyethylene microplastics to air, water and soil on their adsorption behaviors for copper and tetracycline. *Chem. Eng. J.* **2021**, *404*, 126412. [CrossRef]
4. Schwabl, P.; Koppel, S.; Konigshofer, P.; Bucsics, T.; Trauner, M.; Reiberger, T.; Liebmann, B. Detection of Various Microplastics inHuman Stool: A Prospective Case Series. *Ann. Intern. Med.* **2019**, *171*, 453–457. [CrossRef] [PubMed]
5. Ragusa, A.; Svelato, A.; Santacroce, C.; Catalano, P.; Notarstefano, V.; Carnevali, O.; Papa, F.; Rongioletti, M.C.A.; Baiocco, F.; Draghi, S.; et al. Plasticenta: First evidence of microplastics in human placenta. *Environ. Int.* **2021**, *146*, 106274. [CrossRef] [PubMed]
6. Fadare, O.O.; Okoffo, E.D. Covid-19 face masks: A potential source of microplastic fibers in the environment. *Sci. Total Environ.* **2020**, *737*, 140279. [CrossRef] [PubMed]
7. Habib, S.; Iruthayam, A.; Shukor, M.Y.A.; Alias, S.A.; Smykla, J.; Yasid, N.A. Biodeterioration of Untreated Polypropylene Microplastic Particles by Antarctic Bacteria. *Polymers* **2020**, *12*, 2616. [CrossRef]
8. Padervand, M.; Lichtfouse, E.; Robert, D.; Wang, C. Removal of microplastics from the environment. A review. *Environ. Chem. Lett.* **2020**, *18*, 807–828. [CrossRef]
9. Fujishima, A.; Zhang, X.; Tryk, D.A. TiO_2 photocatalysis and related surface phenomena. *Surf. Sci. Rep.* **2008**, *63*, 515–582. [CrossRef]
10. Uheida, A.; Mejía, H.G.; Abdel-Rehim, M.; Hamd, W.; Dutta, J. Visible light photocatalytic degradation of polypropylene microplastics in a continuous water flow system. *J. Hazard. Mater.* **2021**, *406*, 124299. [CrossRef]
11. Jiang, R.; Lu, G.; Yan, Z.; Liu, J.; Wu, D.; Wang, Y. Microplastic degradation by hydroxy-rich bismuth oxychloride. *J. Hazard. Mater.* **2021**, *405*, 124247. [CrossRef]
12. Zhang, Z.; Chen, Y. Effects of microplastics on wastewater and sewage sludge treatment and their removal: A review. *Chem. Eng. J.* **2020**, *382*, 122955. [CrossRef]
13. Ariza-Tarazona, M.C.; Villarreal-Chiu, J.F.; Hernández-López, J.M.; De la Rosa, J.R.; Barbieri, V.; Siligardi, C.; Cedillo-González, E.I. Microplastic pollution reduction by a carbon and nitrogen-doped TiO_2: Effect of pH and temperature in the photocatalytic degradation process. *J. Hazard. Mater.* **2020**, *395*, 122632. [CrossRef]
14. Ariza-Tarazona, M.C.; Villarreal-Chiu, J.F.; Barbieri, V.; Siligardi, C.; Cedillo-González, E.I. New Strategy for Microplastic Degradation: Green Photocatalysis Using a Protein-Based Porous N-TiO_2 Semiconductor. *Ceram. Int.* **2019**, *45*, 9618–9624. [CrossRef]
15. Llorente-García, B.E.; Hernández-López, J.M.; Zaldívar-Cadena, A.A.; Siligardi, C.; Cedillo-González, E.I. First Insights into Photocatalytic Degradation of HDPE and LDPE Microplastics by a Mesoporous N–TiO_2 Coating: Effect of Size and Shape of Microplastics. *Coatings* **2020**, *10*, 658. [CrossRef]
16. Razali, N.; Terengganu, U.M.; Abdullah, W.R.W.; Zikir, N.M. Effect of thermo-photocatalytic process using zinc oxide on degradation of macro/micro-plastic in aqueous environment. *J. Sustain. Sci. Manag.* **2020**, *15*, 1–14. [CrossRef]
17. Tofa, T.S.; Ye, F.; Kunjali, K.L.; Dutta, J. Enhanced Visible Light Photodegradation of Microplastic Fragments with Plasmonic Platinum/Zinc Oxide Nanorod Photocatalysts. *Catalysts* **2019**, *9*, 819. [CrossRef]

18. Nabi, I.; Bacha, A.-U.-R.; Li, K.; Cheng, H.; Wang, T.; Liu, Y.; Ajmal, S.; Yang, Y.; Feng, Y.; Zhang, L. Complete Photocatalytic Mineralization of Microplastic on TiO_2 Nanoparticle Film. *iScience* **2020**, *23*, 101326. [CrossRef]
19. Domínguez-Jaimes, L.P.; Cedillo-González, E.I.; Luévano-Hipólito, E.; Acuña-Bedoya, J.D.; Hernández-López, J.M. Degradation of primary nanoplastics by photocatalysis using different anodized TiO_2 structures. *J. Hazard. Mater.* **2021**, *413*, 125452. [CrossRef]
20. Napper, I.E.; Bakir, A.; Rowland, S.J.; Thompson, R.C. Characterisation, quantity and sorptive properties of microplastics extracted from cosmetics. *Mar. Pollut. Bull.* **2015**, *99*, 178–185. [CrossRef] [PubMed]
21. Zeng, H.; Xie, J.; Xie, H.; Su, B.-L.; Wang, M.; Ping, H.; Wang, W.; Wang, H.; Fu, Z. Bioprocess-inspired synthesis of hierarchically porous nitrogen-doped TiO_2 with high visible-light photocatalytic activity. *J. Mater. Chem. A* **2015**, *3*, 19588–19596. [CrossRef]
22. Rodríguez, E.M.; Márquez, G.; Tena, M.; Álvarez, P.M.; Beltrán, F.J. Determination of main species involved in the first steps of TiO_2 photocatalytic degradation of organics with the use of scavengers: The case of ofloxacin. *Appl. Catal. B Environ.* **2015**, *178*, 44–53. [CrossRef]
23. Pelaez, M.; Falaras, P.; Likodimos, V.; O'Shea, K.; de la Cruz, A.A.; Dunlop, P.S.M.; Byrne, J.A.; Dionysiou, D.D. Use of selected scavengers for the determination of NF-TiO_2 reactive oxygen species during the degradation of microcystin-LA under visible light irradiation. *J. Mol. Catal. A Chem.* **2016**, *425*, 183–189. [CrossRef]
24. Chamas, A.; Moon, H.; Zheng, J.; Qiu, Y.; Tabassum, T.; Jang, J.H.; Abu-Omar, M.M.; Scott, S.L.; Suh, S. Degradation Rates of Plastics in the Environment. *ACS Sustain. Chem. Eng.* **2020**, *8*, 3494–3511. [CrossRef]
25. Davidson, R.; Meek, R. The photodegradation of polyethylene and polypropylene in the presence and absence of added titanium dioxide. *Eur. Polym. J.* **1981**, *17*, 163–167. [CrossRef]
26. Ali, S.S.; Qazi, I.A.; Arshad, M.; Khan, Z.; Voice, T.C.; Mehmood, C.T. Photocatalytic degradation of low density polyethylene (LDPE) films using titania nanotubes. *Environ. Nanotechnol. Monit. Manag.* **2016**, *5*, 44–53. [CrossRef]
27. Socrates, G. *Infrared and Raman Characteristic Group Frequencies: Tables and Charts*, 3rd ed.; John Wiley & Sons: Hoboken, NJ, USA, 2004; ISBN 9780470093078.
28. Yu, J.-G.; Yu, H.-G.; Cheng, B.; Zhao, X.-J.; Yu, J.C.; Ho, W.-K. The Effect of Calcination Temperature on the Surface Microstructure and Photocatalytic Activity of TiO_2 Thin Films Prepared by Liquid Phase Deposition. *J. Phys. Chem. B* **2003**, *107*, 13871–13879. [CrossRef]
29. Zhang, J.; Zhou, P.; Liu, J.; Yu, J. New understanding of the difference of photocatalytic activity among anatase, rutile and brookite TiO_2. *Phys. Chem. Chem. Phys.* **2014**, *16*, 20382–20386. [CrossRef] [PubMed]
30. Li, G.; Richter, C.P.; Milot, R.L.; Cai, L.; Schmuttenmaer, C.A.; Crabtree, R.H.; Brudvig, G.W.; Batista, V.S. Synergistic effect between anatase and rutile TiO_2 nanoparticles in dye-sensitized solar cells. *Dalton Trans.* **2009**, 10078–10085. [CrossRef]
31. Huo, Y.; Jin, Y.; Zhu, J.; Li, H. Highly active TiO_2-x-yNxFyvisible photocatalyst prepared under supercritical conditions in NH_4F/EtOH fluid. *Appl. Catal. B Environ.* **2009**, *89*, 543–550. [CrossRef]
32. Yang, G.; Jiang, Z.; Shi, H.; Xiao, T.; Yan, Z. Preparation of highly visible-light active N doped TiO_2 photocatalyst. *J. Mater. Chem.* **2010**, *20*, 5301–5309. [CrossRef]
33. Yurdakal, S.; Garlisi, C.; Özcan, L.; Bellardita, M.; Palmisano, G. (Photo)catalyst characterization techniques: Adsorption isotherms and BET, SEM, FTIR, UV-Vis, photoluminescence, and electrochemical characterizations. In *Heterogeneous Photocatalysis: Relationships with Heterogeneous Catalysis and Perspectives*, 1st ed.; Marcì, G., Palmisano, L., Eds.; Elsevier: Amsterdam, The Netherlands, 2019; pp. 87–152, ISBN 9780444640154.
34. Bezerra, P.; Cavalcante, R.; Garcia, A.; Wender, H.; Martines, M.; Casagrande, G.; Giménez, J.; Marco, P.; Oliveira, S.; Machulek, A., Jr. Synthesis, Characterization, and Photocatalytic Activity of Pure and N-, B-, or Ag- Doped TiO_2. *J. Braz. Chem. Soc.* **2017**, *28*, 1788–1802. [CrossRef]
35. Du, Y.; Rabani, J. The Measure of TiO_2 Photocatalytic Efficiency and the Comparison of Different Photocatalytic Titania. *J. Phys. Chem. B* **2003**, *107*, 11970–11978. [CrossRef]
36. Tofa, T.S.; Kunjali, K.L.; Paul, S.; Dutta, J. Visible light photocatalytic degradation of microplastic residues with zinc oxide nanorods. *Environ. Chem. Lett.* **2019**, *17*, 1341–1346. [CrossRef]
37. Mills, A.; Lee, S.-K. A web-based overview of semiconductor photochemistry-based current commercial applications. *J. Photochem. Photobiol. A Chem.* **2002**, *152*, 233–247. [CrossRef]
38. Uekert, T.; Kasap, H.; Reisner, E. Photoreforming of Nonrecyclable Plastic Waste over a Carbon Nitride/Nickel Phosphide Catalyst. *J. Am. Chem. Soc.* **2019**, *141*, 15201–15210. [CrossRef] [PubMed]
39. Arsac, F.; Bianchi, D.; Chovelon, J.M.; Ferronato, C.; Herrmann, J.M. Experimental Microkinetic Approach of the Photocatalytic Oxidation of Isopropyl Alcohol on TiO_2. Part 1. Surface Elementary Steps Involving Gaseous and Adsorbed C_3HxO Species. *J. Phys. Chem. A* **2006**, *110*, 4202–4212. [CrossRef]
40. Giles, H.F.; Wagner, J.R.; Mount, E.M. Polymer Overview and Definitions. *Extrusion* **2005**, 165–177. [CrossRef]
41. Yang, R.; Christensen, P.A.; Egerton, T.A.; White, J.R. Degradation products formed during UV exposure of polyethylene–ZnO nano-composites. *Polym. Degrad. Stab.* **2010**, *95*, 1533–1541. [CrossRef]
42. Alegría, A.E.; Ferrer, A.; Santiago, G.; Sepúlveda, E.; Flores, W. Photochemistry of water-soluble quinones. Production of the hydroxyl radical, singlet oxygen and the superoxide ion. *J. Photochem. Photobiol. A Chem.* **1999**, *127*, 57–65. [CrossRef]
43. Garg, S.; Rose, A.L.; Waite, T.D. Production of Reactive Oxygen Species on Photolysis of Dilute Aqueous Quinone Solutions. *Photochem. Photobiol.* **2007**, *83*, 904–913. [CrossRef]

44. Hayyan, M.; Hashim, M.A.; Alnashef, I.M. Superoxide Ion: Generation and Chemical Implications. *Chem. Rev.* **2016**, *116*, 3029–3085. [CrossRef] [PubMed]
45. Gewert, B.; Plassmann, M.M.; MacLeod, M. Pathways for degradation of plastic polymers floating in the marine environment. *Environ. Sci. Process. Impacts* **2015**, *17*, 1513–1521. [CrossRef] [PubMed]
46. Kriston, I. Some Aspects of the Degradation and Stabilization of Phillips Type Polyethylene. Ph.D. Thesis, Budapest University of Technology and Economics, Budapest, Hungary, 26 April 2011.
47. Fotopoulou, K.N.; Karapanagioti, H.K. Degradation of Various Plastics in the Environment. In *Hazardous Chemicals Associated with Plastics in the Marine Environment*, 1st ed.; Takada, H., Karapanagioti, H., Eds.; Springer: Cham, Switzerland, 2019; pp. 71–92.

Article

pH-Stat Titration: A Rapid Assay for Enzymatic Degradability of Bio-Based Polymers

Lukas Miksch *, Lars Gutow and Reinhard Saborowski

Helmholtz Centre for Polar and Marine Research, Alfred Wegener Institute, Am Handelshafen 12, 27570 Bremerhaven, Germany; Lars.Gutow@awi.de (L.G.); Reinhard.Saborowski@awi.de (R.S.)
* Correspondence: Lukas.Miksch@awi.de; Tel.: +49-471-4831-1326

Abstract: Bio-based polymers have been suggested as one possible opportunity to counteract the progressive accumulation of microplastics in the environments. The gradual substitution of conventional plastics by bio-based polymers bears a variety of novel materials. The application of bioplastics is determined by their stability and bio-degradability, respectively. With the increasing implementation of bio-based plastics, there is also a demand for rapid and non-elaborate methods to determine their bio-degradability. Here, we propose an improved pH Stat titration assay optimized for bio-based polymers under environmental conditions and controlled temperature. Exemplarily, suspensions of poly(lactic acid) (PLA) and poly(butylene succinate) (PBS) microparticles were incubated with proteolytic and lipolytic enzymes. The rate of hydrolysis, as determined by counter-titration with a diluted base (NaOH), was recorded for two hours. PLA was hydrolyzed by proteolytic enzymes but not by lipase. PBS, in contrast, showed higher hydrolysis rates with lipase than with proteases. The thermal profile of PLA hydrolysis by protease showed an exponential increase from 4 to 30 °C with a temperature quotient Q_{10} of 5.6. The activation energy was 110 kJ·mol^{-1}. pH-Stat titration proved to be a rapid, sensitive, and reliable procedure supplementing established methods of determining the bio-degradability of polymers under environmental conditions.

Keywords: polymer degradation; microparticles; PLA; PBS; enzymes; specificity; thermal profile; activation energy

Citation: Miksch, L.; Gutow, L.; Saborowski, R. pH-Stat Titration: A Rapid Assay for Enzymatic Degradability of Bio-Based Polymers. *Polymers* **2021**, *13*, 860. https://doi.org/10.3390/polym13060860

Academic Editor: Jacopo La Nasa

Received: 17 February 2021
Accepted: 8 March 2021
Published: 11 March 2021

Publisher's Note: MDPI stays neutral with regard to jurisdictional claims in published maps and institutional affiliations.

Copyright: © 2021 by the authors. Licensee MDPI, Basel, Switzerland. This article is an open access article distributed under the terms and conditions of the Creative Commons Attribution (CC BY) license (https://creativecommons.org/licenses/by/4.0/).

1. Introduction

As an attempt to substitute petroleum-based plastics, policy makers and producers are increasingly promoting the innovation and development of greener so-called "bioplastics" [1]. The term bioplastic is used for plastic material that is either bio-based, biodegradable, or both. The implementation of bio-based polymers as basis for the production of plastic goods is a promising alternative, as common reduce, reuse, and recycle strategies are insufficient to counteract the extensive accumulation of plastic waste in the environment [2]. With the perspective of having sustainably produced, biocompatible, and biodegradable materials, the development of bioplastics especially from waste and renewable resources has become popular [3]. Advances in waste reutilization [4,5] and in the synthesis technology of biopolymers make it possible to convert agricultural and food waste into usable bio-based materials. Natural polymers like poly(lactic acid) (PLA), polyhydroxyalkanoate (PHA), and polyhydroxybutyrate (PHB) have been isolated through different processes from by-products of milk, sugar, biodiesel, and other industrial processes [3,6]. The effective utilization of waste flows, low carbon footprint, and the biodegradability make bioplastics a sustainable option to replace conventional plastics.

Enzymatic biodegradation of synthetic polymers is, basically, a similar process as the degradation of natural polymers, such as cellulose or chitin [7]. Extracellular microbial enzymes attach to the surface of the polymer and form enzyme–substrate complexes. Accordingly, degradation rates are highest in irregular micro-particles with a high surface:volume ratio. These evolve in the environment through fragmentation of larger items by physicochemical and mechanical forcing. Depending on the substrate, specific enzymes hydrolyze the polymer chain and release oligo- and monomers. These degradation products may be finally assimilated and metabolized to water and carbon dioxide [8]. The enzymatic degradation is strongly influenced by environmental factors, such as pH and temperature [9,10], but also by the nature of the polymer [11,12]. Two common biodegradable polymers in commercial use are poly(lactic acid) (PLA) and poly(butylene succinate) (PBS). Both polymers are aliphatic polyesters and exhibit advantageous physical properties resembling those of polypropylene (PP) or low-density polyethylene (LDPE). Their chemical structure consists of monomers linked by ester bonds. Different hydrolytic enzymes, particularly esterases, are able to cleave these linkages. PLA was hydrolyzed in-vitro by serine proteases like proteinase K and subtilisin [13], whereas PBS was primarily hydrolyzed by lipase and cutinase [14,15].

Various standard tests and methods are available for the analytical assessment of the biodegradability of bioplastics [16]. These comprise the carbon dioxide forming test (Organisation for Economic Co-operation and Development, OECD 301b), differential scanning calorimetry (International Organization for Standardization, ISO 11357), weight-loss measurements (ISO 13432), or Fourier transform infrared spectroscopy [17]. Most of these methods are time consuming, laborious, or depend on expensive equipment. Accordingly, there is a demand for effective and accurate procedures that allow for the rapid analysis of the biodegradability of bioplastics under controlled conditions. A promising method is pH-Stat titration. It is based on maintaining a constant pH by the controlled addition of a diluted acid or base in a system where a pH affecting reaction takes place [18]. The hydrolysis of polyesters forms carboxyl groups, which reduce the pH in the solution. The quantity of added base to maintain a constant pH is thus a direct measure of the rate of hydrolysis of the bioplastic and, therefore, of its biodegradability. Although there is no uniform approach described, pH Stat titration has been used in several studies to assay the enzymatic degradation of aliphatic polyesters [19–24]. Most of these studies were conducted under elevated temperatures near enzyme optimum (Table 1). As the oceans are a potential sink of bio-based plastics, it is also of high relevance how natural environmental factors affect the process of degradation. Therefore, investigations of the hydrolysis of biopolymers under realistic environmental conditions are required.

Table 1. Procedural parameters of reported polymer degradation experiments by pH Stat titration.

Polymer (Substrate)	Enzyme [Source]	Temperature	Medium	Period	Reference
Poly-(trimethylene succinate)	Lipase [a]	37 °C	0.9% NaCl	10 h	[19]
Several model polyesters	Lipase [b,c,d,e,f] Hydrolase [g] α-Chrymotrypsin [h] Subtilisin [i] Esterase [j]	25–50 °C	0.9% NaCl	15 min–20 h	[20]
Several model polyesters	Lipase [b,e,k,l,m,n,o] Hydrolase [p] Proteinase K	40 °C	0.9% NaCl	15 min	[21]
Poly(3-hydroxybutyrate) (PHB)	PHB depolymerases	37 °C	1 mmol·L^{-1} Tris-HCl	20 min	[22]
Poly(ethylene terephtalate) (PET)	Cutinase [q,r,s]	30–90 °C	1 mmol·L^{-1} Tris-HCl with 10% glycerol	15 min	[23]
Poly(vinyl acetate) (PVC)	Cutinase [q,r,s]	40 °C, 50 °C, 70 °C	1 mmol·L^{-1} Tris-HCl with 10% glycerol	1–192 h	[24]

[a] Rhizopus delemar, [b] Pseudomonas spp., [c] Chromobacterium viscosum, [d] Rhizomucor miehei, [e] Candida cylindracea, [f] wheat germ, [g] Thermonospora fusca, [h] bovine pancreas, [i] Bacillus subtilis, [j] porcine liver, [k] Aspergillus oryzae, [l] Mucor miehei, [m] porcine pancreas, [n] Pseudomonas cepacia, [o] Pseudomonas fluorescens, [p] Thermobifida fusca, [q] Humilica insolens, [r] Fusarium solani, [s] Pseudomonas mendocrina.

In this study, we adapt the pH-Stat titration method for routine analysis to determine the biodegradability of bio-based plastics under relevant environmental conditions with support of specific proteolytic and lipolytic enzymes. Parameters for an improved performance and reliability of the method were tested and procedural difficulties identified.

2. Materials and Methods

2.1. Chemicals

Proteinase K from *Tritirachium album* (19133) was purchased from Qiagen (Hilden Germany) and the NaOH-standard solution (5564732) from Omnilab (Bremen, Germany). All other chemicals and enzymes were purchased from Sigma-Aldrich (St. Louis, MO, USA): poly(L-lactide) (PLA, 765112), poly(1,4-butylene succinate) (PBS, 448028), protease from *Bacillus licheniformis* (P4860), and lipase from *Candida antarctica* (62288) (Supporting Information, Supplementary Tables S1 and S2).

2.2. Titration Device

We used the automatic titrator TitroLine® 7000 (SI Analytics GmbH, Mainz, Germany) for all pH-Stat applications. The titration unit was equipped with a 20-mL exchangeable head, a magnetic stirrer (TM 235), and a pH-electrode model A 162 2M DIN ID. The reaction vial was a 20-mL glass vial. The opening of the vial (17 mm) was too narrow to insert the electrode and the titration tip. Therefore, the titration tip was equipped with a 1 mm diameter polytetrafluoroethylene (PTFE)-tube. The reaction vial was placed in a custom-made thermostat jacket (Supporting Information, Figure S1). A circulation thermostat (e.g., Lauda, Lauda-Königshofen, Germany) connected to the thermostat jacket maintained a constant temperature in the reaction vial.

2.3. pH-Stat Titration

The titration system was set by default on the following parameters: mode: pH-Stat; endpoint: 8.2; step size: 0.001; measuring interval: 1 min; total time: 150 min; dosing speed: 1.5%; stirring speed: 1200 rpm. The pH was kept constant at 8.2 by titration of NaOH-solution. The concentrations ranged from 5 to 10 mmol·L^{-1}, depending on the velocity of the reaction after the enzyme was added. Substrate (biopolymer) suspensions (3 mg·mL^{-1}) were prepared in a solution of 3.4% sodium chloride in water (referred to as artificial seawater). PLA suspensions were prepared with a commercial product with an average M_n of 10,000. Aggregates of polymer particles were thoroughly crushed with a spatula before the suspension was sonicated in an ultrasonic bath for 8 min. Preliminary experiments showed that 3 mg·mL^{-1} was the highest concentration to ensure homogenous suspension of the particles and, thus, highest saturation. PBS granules were ground with a cryogenic mill (SPEX SamplePrep, 6775 Freezer/Mill) and sieved for fractions smaller than 200 µm. The suspensions were stirred in a glass beaker at 800 rpm for 16 h before aliquots of 10 mL were subjected to pH-Stat titration. Enzyme solutions (10–80 µL) were added to the reaction vial with a 100-µL microsyringe (Model 710 N, Hamilton Bonaduz AG, Bonaduz, Switzerland). The increase of volume due to titrant supply was always less than 1 mL. The electrode was calibrated every day before use. Routine measurements was carried out in triplicate.

2.4. Operational Parameters

To strengthen the quality standards, we separated the pH-Stat titration procedure into specific phases and separated unspecific background reactions from the true hydrolysis of the bioplastic material. To determine the optimal duration of the measurement, incubation experiments with PLA and protease from *Bacillus licheniformis* were run for up to 120 min. A linear regression was fitted to describe the amount of titrant (NaOH) added as a function of time. The coefficient of determination (R^2) of the NaOH consumption curve was progressively calculated from the start to identify the period after which the coefficient of determination of the linear regression was consistently above $R^2 = 0.99$.

2.5. Enzyme Specificity and Enzyme Concentration

Hydrolysis rates of PLA and PBS were determined with protease from *Bacillus licheniformis*, Proteinase K from *Tritirachium album*, and lipase from *Candida antarctica*. The hydrolysis rates were analyzed separately for PLA and PBS. Additionally, hydrolysis of PLA was assayed at five different protease concentrations (three replicates each) of up to 80 µL, which equals 30 to 240 mAU (Anson Units). Additionally, the dependency of the hydrolysis rate on the enzyme concentration was described by the following non-linear regression model:

$$f_{(x)} = a + (b - a) \cdot (1 - e^{(-c \cdot x)}) \quad (1)$$

with a being $f_{(x)}$ when x (concentration) is zero. b is the asymptote to which the curve approaches, which is the maximum reaction velocity. c is the rate constant of the curve.

2.6. Thermal Profiles

The temperature controlling system, consisting of a circulation cooler and a thermostat jacket, allowed maintaining constant and environmentally relevant temperatures during pH-Stat titration. Ten mL of a PLA-suspension (3 mg·mL^{-1}) were incubated with 40 µL protease solution (120 mAU). The hydrolysis of PLA was measured at temperatures between 4 and 30 °C. The Q_{10} temperature coefficient was calculated as:

$$Q_{10} = \left(\frac{V_2}{V_1}\right)^{\frac{10}{T_2 - T_1}} \quad (2)$$

where V_1 and V_2 represent the reaction velocities at the temperatures T_1 and T_2 in °C or K.

The apparent activation energy (E_A) of the reaction was calculated from the slope of the Arrhenius plot:

$$\ln V = -E_A \frac{1}{RT} \quad (3)$$

where V represents the reaction velocity, R is the universal gas constant, and T the temperature in K.

2.7. Statistics

For each of the two polymers (PLA and PHB), the hydrolysis rate was compared between the three enzymes (protease, Proteinase K, and lipase) by a 1-factorial analysis of variance (ANOVA) followed by Tukey's HSD (honestly significant difference) test. Prior to the ANOVA, the data were tested for heteroscedasticity by Levene's test. Similarly, the hydrolysis rates of PLA at five different protease quantities (30, 60, 120, 180, and 240 mAU) were compared by a 1-factorial ANOVA with subsequent Tukey's HSD test for pairwise comparisons. The significance level of all statistical analyses was $\alpha = 0.05$. Analyses and graphs were done with the program GraphPad Prism version 7.05 for Windows, GraphPad Software, La Jolla California USA, www.graphpad.com.

3. Results & Discussion

3.1. Duration and Sequence of pH-Stat Titration

Compared to natural substrates such as proteins, carbohydrates, or lipids, the enzymatic hydrolysis of bioplastics is slow. This is particularly due to the physical characteristics of the bioplastic particles. They are solid and do not dissolve in water. Only the surface of the particles is accessible to enzymatic action. Accordingly, the accurate determination of the in-vitro degradability of bioplastics requires a sensitive and reliable procedure.

The pH-Stat titration assay, as shown for the hydrolysis of PLA by a protease, follows a characteristic temporal profile of pH changes in the reaction vial and titrant (NaOH) supply, which can be separated into four distinct phases (Figure 1).

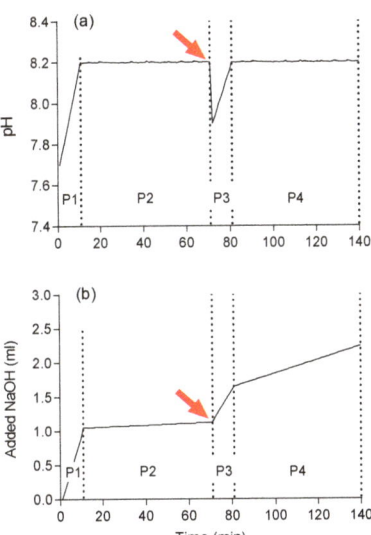

Figure 1. Typical course of (**a**) pH and (**b**) NaOH supply over time in a pH-Stat titration assay of a poly(lactic acid) (PLA) suspension as substrate and a PLA degrading enzyme. Phase 1 (P1): adjustment of the pH to 8.2. Phase 2 (P2): supply of low amounts of titrant to counterbalance an unspecific pH decrease before enzyme addition (substrate blank). Phase 3 (P3): re-adjustment of the pH to 8.2 after enzyme addition (indicated by an arrow). Phase 4 (P4): titration after enzyme addition to counterbalance a pH decrease of substrate hydrolysis (reaction) and an unspecific pH drop (blank).

During Phase 1, the automatic titration system adjusts the pH of the suspension to the starting point of 8.2. Subsequently, the adjusted pH is monitored for 60 min (Phase 2) for conspicuous variations, which may reveal technical errors. During Phase 2, minor quantities of NaOH are continuously added because unspecific reactions in the suspension result in a slight decrease in pH (substrate blank). The addition of the slightly acidic enzyme solution induces a drastic pH drop in the suspension. After re-adjustment to pH 8.2 (Phase 3), NaOH is continuously added to counterbalance the decrease in pH in response to the formation of carboxyl groups from the enzymatic degradation of the substrate (Phase 4). Accordingly, the slope of the curve in Phase 4 is proportional to the hydrolysis rate plus the unspecific reactions of the substrate blank (Phase 2).

To evaluate the rates of unspecific reactions (substrate blank), we tested four substrate (PLA) concentrations (0, 1.5, 3, and 6 g·L^{-1}) and five enzyme volumes of 10, 20, 40, 60, and 80 µL protease in 10 mL artificial seawater, corresponding to enzyme activities of 30, 60, 120, 180, and 240 mAU, respectively. The hydrolysis rate of the substrate blank (Phase 2) ranged between 8 and 26% of the enzyme-catalyzed reaction (Phase 4) for all enzyme concentrations. The blank reaction without enzyme did not correlate with the substrate concentration.

The unspecific reaction may be due to autolysis of biopolymers [25] or diffusion of atmospheric CO_2 into the reaction mixture and dissociation to carbonic acid. We tried to minimize CO_2 diffusion by overlaying the reaction mixture with an inert gas, such as nitrogen or helium. This, however, caused variations in temperature of the reaction solution and was not further considered. Another option to stabilize the initial pH is to use a buffer for suspension [22–24]. However, as the hydrolysis rates at environmental temperatures are expected to be rather low, a buffer could mask potential hydrolysis rates. Furthermore, a buffer does not represent natural conditions. Instead, we used seawater and corrected the slope of Phase 4 for the substrate blank (slope of Phase 2) to obtain the actual enzyme-catalyzed reaction V:

$$V = c \cdot \left[\left(\frac{\delta v_4}{\delta t_4}\right) - \left(\frac{\delta v_2}{\delta t_2}\right)\right] \cdot 10^6 \qquad (4)$$

where c is the concentration of the added NaOH solution (mmol·L^{-1}), v_2 and v_4 the volumes (mL) of NaOH added in phase 2 and 4, and t_2 and t_4 the duration (min) of the phases 2 and 4. The conversion factor 10^6 was used to express the reaction velocity as nmol·min^{-1} under the given conditions. Additionally, the hydrolysis rate of the enzyme alone without substrate in artificial seawater (enzyme blank) was measured and subtracted from the hydrolysis rate of enzyme with substrate, to correct for possible autocatalytic activity of the enzyme solution.

3.2. Titrant Leakage

After the initial adjustment of the pH in Phase 1, further increases in pH beyond the default value of 8.2 occurred during Phase 2 before the enzyme solution was added. This increase in pH was due to leakage of titrant from the tip of the titration device. To evaluate the magnitude of this leakage effect, the increase of the pH over time was analyzed by linear regression. The slope of the linear regression was tested for deviation from zero by an F-test. The increase in pH was more distinct when higher concentrated NaOH solutions were used (Figure 2). In artificial seawater, the increase was always statistically significant (i.e., slope of linear regression always significantly >0) and accounted for 0.01 pH units per hour with 5 mmol·L^{-1} NaOH and 0.02 to 0.06 pH units per hour at higher NaOH concentrations of 10 to 50 mmol·L^{-1}. Upon request, the manufacturer of the titration device confirmed that small amounts of NaOH titrant may constantly diffuse from the titration tip into the reaction solution and raise the pH.

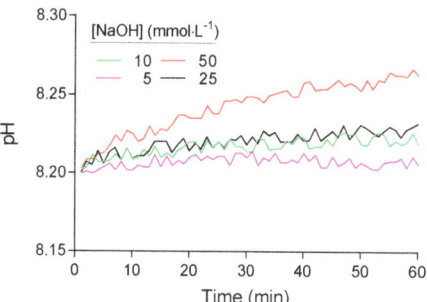

Figure 2. pH drift in artificial seawater over time with different concentrations of NaOH titrant due to diffusion from the titration tip.

To minimize leakage, tips with different opening diameters were tested. However, NaOH diffused from the original tips of the manufacturer, as well as from custom-made tips. Similarly, reducing the opening diameter of the tip below 0.5 mm did not stop leakage of NaOH. To minimize this source of error, the NaOH titrant has to be adjusted to a concentration that does not significantly affect the measurement of the enzymatic activity. Simultaneously, the concentration of the NaOH has to be high enough to avoid excess addition of the titrant that may exceed the capacity of the reaction vial. The use of a buffer would also partially eliminate this increase in pH, but was not considered for reasons described above.

When determining the leakage effect in PLA suspensions of 3 g·L^{-1}, the increase of pH was lower than in artificial seawater (Figure 3). This is probably due to the unspecific acidifying reactions in the PLA suspension as observed in phase 2 of the titration profile (see Section 3.1). This unspecific reaction is stronger in PLA suspensions than in artificial seawater and therefore counterbalances the pH increase due to NaOH leakage. The pH increase in PLA-suspension of 3 mg·mL^{-1} was statistically significant for NaOH concentrations of 25 mmol·L^{-1} ($F_{1,58}$ = 21.96, $p < 0.01$) and 50 mmol·L^{-1} ($F_{1,58}$ = 66.27, $p < 0.01$). At lower NaOH concentrations of 10 and 5 mmol·L^{-1}, the pH of the PLA-suspension dropped below 8.2 because of the unspecific reaction described above (3.1). Here, the titration system added small amounts of NaOH to keep the pH at 8.2. Therefore, NaOH concentrations of 10 mmol·L^{-1} ($F_{1,58}$ = 0.01, p = 0.91) and 5 mmol·L^{-1} ($F_{1,58}$ = 0.82, p = 0.37) showed no significant change of pH in the reaction vial.

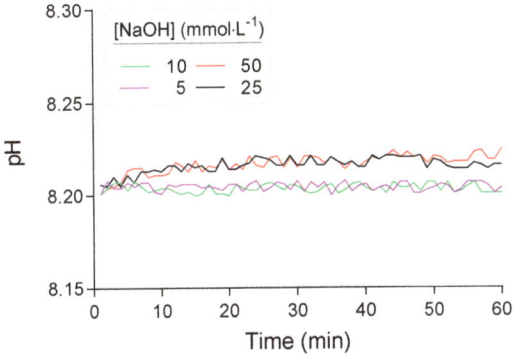

Figure 3. pH drift in substrate (PLA) suspension (3 mg·mL^{-1}) in artificial seawater over time with different concentrations of NaOH titrant. Endpoint of the titration was set to 8.2. Titrant was added at NaOH concentrations of 5 and 10 mmol·L^{-1} (dashed lines) to counterbalance the pH drop (see main text).

3.3. Time Frame of pH-Stat Measurement

With fast reactions due to high temperature (22 °C), the coefficient of determination reached R^2 = 0.99 on average after 17 min in phase 4 of the pH-Stat titration assay. At lower temperatures (12 °C) and slower reactions, R^2 = 0.99 was reached on average after 50 min (Figure 4).

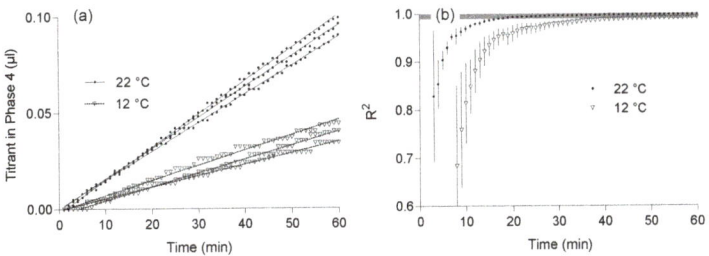

Figure 4. (**a**) Addition of titrant (0.01 mol·L^{-1} NaOH solution) during the 60 min lasting hydrolysis of PLA (3 g·L^{-1}) by protease from *Bacillus licheniformis* (2.4 AU·mL^{-1}). (**b**) Progressive approximation of the coefficient of variation to R^2 = 0.99 (indicated by a grey bar).

Previous studies measuring hydrolysis rates at high temperatures had time frames ranging from 15 min (initial slope) up to several hours (Table 1). From the course and the coefficients of determination of the NaOH consumption curves (Figure 4) we concluded, that 20-min titration duration may be sufficient for fast reactions, whereas slow reactions showed strong pH variations after 30 min. Longer measurement periods of 60 to 120 min further increase the accuracy of the assay. However, depending on the enzyme used and the experimental conditions (temperature, pH) extended exposure times may be disadvantageous for less stable enzymes.

Therefore, prior testing of the enzyme activities under the specific assay conditions is recommended for measurements over longer time periods. Additionally, substrate limitation as a result of continuous substrate conversion will lead to a decrease in hydrolysis rate over time. Therefore, considering time constraints, throughput efficiency, and enzyme stability, we chose 60 min as an appropriate period for the pH-Stat assay of PLA hydrolysis in our study.

3.4. Effect of Enzyme Concentration

The hydrolysis rate varied significantly with the amount of added enzyme (ANOVA: $F_{4,9} = 12.71$, $p < 0.01$–Figure 5). The average reaction velocity continuously increased from 8.0 ± 0.5 nmol·min^{-1} at 30 mAU to 17.9 ± 1.4 nmol·min^{-1} at 120 mAU. No statistically significant increase in reaction velocity was observed at volumes above 120 mAU. The non-linear regression model explained 85% of the variation and identified the maximum reaction velocity at 18.9 nmol·min^{-1}. The calculated maximum velocity is within the range of the 95% confidence intervals of the reaction velocities at added enzyme amounts of 120–240 mAU.

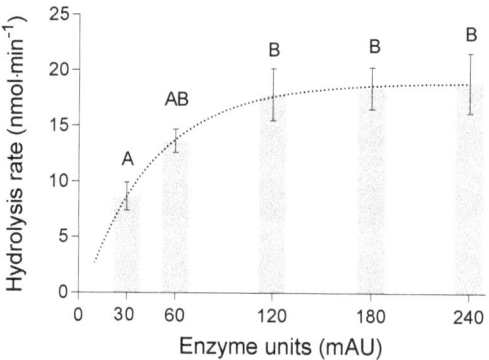

Figure 5. Rate of PLA hydrolysis (3 mg·mL^{-1}) at different quantities (Anson Units) of protease from *Bacillus licheniformis* at 22 °C and pH 8.2 (means ± SD, n = 2–3).

The stagnation of the hydrolysis rate of PLA at higher volumes of the protease solution may be due to the limitation of accessible substrate at higher enzyme concentrations. The surface area of the polymer to which the enzymes can attach and hydrolyze the substrate is limited. If all accessible sites on the polymer are occupied, the cleavage and therefore the hydrolysis rate stagnates even if enzyme is added [26]. Similar results were reported for the hydrolysis of poly-(trimethylene succinate) films by lipase from *Rhizopus delemar* in pH-Stat degradation experiments [19].

3.5. Enzyme Specificity

A set of enzymes was selected to examine their hydrolytic potential for PLA and PBS at 22 °C and pH 8.2. Both polymers were incubated with 100 µL lipase from *Candida antarctica* (18 U), 50 µL proteinase K from *Tritirachium album* (30 mAU), and 10 µL protease from *Bacillus licheniformis* (30 mAU), respectively. The hydrolytic activity for PLA varied significantly between the enzymes (ANOVA: $F_{(2,6)} = 115.8$, $p < 0.0001$–Figure 6a). PLA was efficiently degraded by the proteinase K from *Tritirachium album* with a hydrolysis rate of 13.7 ± 1.0 nmol·min^{-1} and by the protease from *Bacillus licheniformis* at 8.3 ± 1.0 nmol·min^{-1}. The lipase from *Candida antarctica* showed almost no hydrolytic activity with 0.5 ± 0.6 nmol·min^{-1}. Accordingly, both proteases were capable of degrading PLA, whereas the lipase was ineffective in PLA degradation. Efficient degradation of PLA by serine proteases has repeatedly been described [13,27,28]. However, the hydrolytic potential of most lipases is limited to aliphatic polyesters with low melting temperature and a lack of optically active carbon [29], which does not apply to PLA.

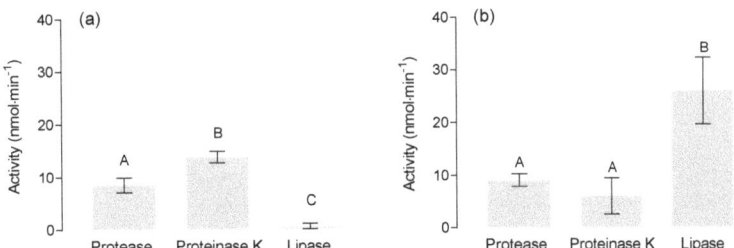

Figure 6. Enzymatic degradation of (**a**) PLA and (**b**) poly(butylene succinate) (PBS) at 22 °C and pH 8.2 (mean ± SD, n = 3). Different letters indicate statistically significant differences (Tukey multiple comparison test).

The hydrolytic activity for PBS also varied between the enzymes (ANOVA: $F_{(2,6)}$ = 19.33, p = 0.0024–Figure 6b). Lipase from *C. antarctica* efficiently degraded PBS with a hydrolysis rate of 25.9 ± 5.3 nmol·min^{-1}. The hydrolysis rates of protease from *B. licheniformis* and the proteinase from *T. album* were significantly lower at 9.1 ± 1.1 and 6.0 ± 1.2 nmol·min^{-1}, respectively, although a potential of proteinase K for degrading PBS has been confirmed previously [22]. PBS is more efficiently hydrolyzed by lipases than proteases because of the conformational similarity of the polymer to tri- and diglycerides and the strong adsorption of lipase to the surface of PBS [30].

The degradation rates obtained by pH-Stat titration were verified with an alternative method. We chose a slightly modified fluorescent assay [31], which is based on the reaction of a fluorophore with liberated carboxyl groups of the PLA. The fluorescent method confirmed that protease from *B. licheniformis* and proteinase K from *T. album* were able to hydrolyze PLA (Supporting Information, Figure S2). However, the fluorescent assay is specific with regard to the polymer type and not applicable to measure the degradation products of PBS. It is also less sensitive than the pH-Stat method. Accordingly, the pH-Stat titration method is applicable to a wider range of polymers and performs better in in-vitro assays on small polymer quantities.

3.6. Thermal Profiles

Temperature controlled pH-Stat titration allows for investigating biodegradability of polymers within a wide range of environmentally relevant temperatures from 4 °C to above 30 °C. However, adjusting the temperature of the reaction mixture below 4 °C was challenging because of heat generation by the stirring unit, which is necessary to dispense the titrated NaOH in the reaction vial efficiently.

The hydrolysis rate of PLA by *B. licheniformis* protease strongly increased with temperature and followed first-order kinetics (Figure 7a). Activities ranged from 1.4 ± 0.2 nmol·min^{-1} at 4 °C to 66.0 ± 10.2 nmol·min^{-1} at 30 °C. The temperature coefficient Q_{10} of 5.6 (calculated over the entire temperature range from 4 °C to 30 °C) reveals that a rise in temperature by 10 °C induces a 5.6-fold increase in hydrolytic activity.

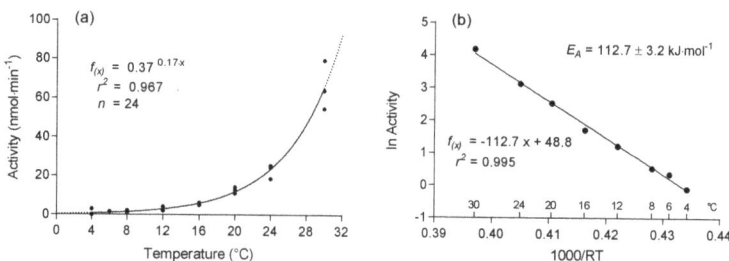

Figure 7. (**a**) Hydrolysis rates of PLA at different temperatures. (**b**) Arrhenius-plot calculated from data of the temperature profile of PLA hydrolysis.

Degradation of PLA by protease (*B. licheniformis*) proceeded more slowly at environmentally relevant seawater temperatures of up to 20 °C compared to seawater temperatures above 25 to 30 °C. Such elevated temperatures occur only in tropical waters [32,33] or in exposed rock pools warmed by solar radiation during low tide [34,35]. On the ground, highest temperatures appear at the surface and depend strongly on irradiation, soil humidity, and topography. At beaches, surface temperatures increase with distance from the shoreline and may exceed 50 °C on dry sand [36,37] but rapidly cool down already a few decimeters below the surface. Accordingly, temperature may become a limiting factor in the decomposition of biodegradable plastics in natural habitats [38]. Consequently, industrial composting processes typically utilize temperatures of 50 to 65 °C to accelerate the degradation process [16,39,40].

The Arrhenius-plot showed a strong linear relation between the ln of the reaction velocity and the reciprocal of the absolute temperature (Figure 7b). The apparent activation energy (E_A) of the hydrolysis of PLA by *B. licheniformis* protease was 112.7 ± 3.2 kJ·mol^{-1}. The activation energy of a chemical reaction is a proxy for the velocity of the reaction. The lower the activation energy, the easier and faster a reaction can proceed.

Thermodynamic data on the hydrolysis of biodegradable polymers at environmentally relevant temperatures are lacking whereas data on thermal degradation at high temperatures of up to several hundred degree Celsius are available. For example, the activation energy of the thermogravimetric degradation of corn starch-based plastic accounts for 120–140 kJ·mol^{-1} [41]. The activation energy of the thermodegradation of hemp-poly lactic acid composites was about 160 kJ·mol^{-1} [42]. Those authors suggested that a high binding energy of the polymers improves the thermal stability of the compound, as it requires a higher activation energy to induce degradation.

Activation energies of thermogravimetric decomposition of natural substrates or their derivatives are in the same range. The E_A of cellulose from different plants are between 150 and 200 kJ·mol^{-1} [43] whereas the apparent E_A of the thermogravimetric decomposition of chitin and chitosan account for 125 and 169 kJ·mol^{-1}, respectively [44].

The E_A of enzymatic degradation of natural polymers is distinctly lower. Cellulolytic activity in soil showed activation energies of 22–28 kJ·mol^{-1} [45]. The E_A of the degradation of cellulose films by cellulose from sac fungus *Trichoderma reesei* was 37 kJ·mol^{-1} [46]. Similarly, the hydrolysis of swollen chitin by chitinase from *Trichoderma harzianum* showed an E_A of 70 and 80 kJ·mol^{-1} [47] and from *Paenibacillus* sp. of 19 kJ·mol^{-1} [48]. Chitinases of the psychrophilic Antarctic bacterium *Arthrobacter* sp. showed E_A of 62–63 kJ·mol^{-1} when hydrolyzing soluble chitin [49].

The capability of our method to establish thermal profiles rapidly and to deduce thermodynamic parameters allows studying material characteristics for various purposes. These may comprise industrial or medical applications, degradation kinetics, and effects of weathering. These results may be important to evaluate the degradability of bio-based polymers under conditions of the natural environment as well as industrial composting facilities.

3.7. Limitations

Despite many advantages, the pH-Stat titration method as described here has also limitations.

(1) Buffering Capacity of Plastic Suspensions

The pH of a polymer suspension depends on the properties of the liquid (water) and the polymer used. This intrinsic pH of the suspension was usually around 8.0 in our experiments with seawater and PLA/PBS. In phase 1 of each titration assay, the pH of the suspension was adjusted to 8.2, at which the hydrolytic reaction was observed.

Seawater or freshwater has a negligible buffer capacity. This has the advantage that even the lowest hydrolysis rates are detectable. However, due to the negligible buffering capacity, we observed that the adjusted pH moved towards the original intrinsic pH of the suspension with time. This may be due to the interaction of the plastic particles or leaching additives with the water.

If the intrinsic pH is lower than the adjusted pH, a decrease in pH appears in phase 2 (substrate blank) and is compensated by the addition of sodium hydroxide. If the intrinsic pH is higher than the adjusted pH, no sodium hydroxide will be added when the pH is increasing and no net change will be measured. This may mask a potential reaction and lead to an underestimation of the hydrolysis rate. To ensure the most accurate measurement of the hydrolysis rate, the adjusted pH should always be slightly higher than the intrinsic pH of the polymer suspension.

(2) Particle Size and Surface Area

The enzymatic reaction takes place in a suspension of very small particles. These particles (powder) have to be prepared first by e.g., cryo-milling and sieving. The enzymes act at the surface of the particles, which, however, cannot be defined in small mg-amounts of sample. Surface characteristics may vary between plastic types.

4. Conclusions

pH-Stat titration is a valuable tool for investigating the biodegradability of polymers, which may ideally complement already existing testing methods or even substitute some of them. When screening bio-based and other polymers for their degradability by various enzymes, the pH-Stat titration provides rapid and reliable results within only a few hours instead of weeks or months. To ensure a smooth and accurate performance of this method under environmental conditions, the tuning of the parameters to the chosen conditions and selected test materials is necessary.

The precise measurement of pH and base consumption, which reflect the formation of carboxyl groups through hydrolysis of the polymer, allows for a high sensitivity of the method and dealing with extremely small polymer quantities. For PLA and PBS microplastics we were able to demonstrate differential enzymatic degradability by proteolytic enzymes and lipase, respectively, and to describe the temperature dependencies of hydrolysis rates. This method is also suitable to monitor the enzymatic biodegradability of any other biopolymer, including future inventions facing the challenge to reduce environmental pollution by persistent plastic materials. No buffer solutions or other chemical reagents, which may interfere with the hydrolytic reaction, are required in the reaction solution to keep the pH constant during the measurement. Accordingly, degradation processes in natural media, such as freshwater or seawater, can be studied accurately. This makes this method an excellent tool for studying polymer degradation under different environmentally relevant conditions in terms of temperature, salinity, or pH. Additionally, the resulting degradation products, i.e., oligomers and monomers, can be isolated and subjected to detailed chemical analysis for studying each step in the process of polymer decomposition, which is also beneficial in the development of innovative future materials.

Supplementary Materials: The following Supplementary Materials are available online at https://www.mdpi.com/2073-4360/13/6/860/s1, Figure S1: Technical illustration of the thermal jacket, Figure S2: Fluorometric assay of PLA degradation products, Table S1: Enzyme specifications, Table S2: Polymer specifications

Author Contributions: L.M.: Conceptualization, Methodology, Validation, Formal analysis, Investigation, Writing-Original draft preparation. L.G.: Conceptualization, Methodology, Formal analysis, Writing-Review & Editing, Supervision. R.S.: Conceptualization, Methodology, Formal analysis, Writing-Review and Editing, Supervision. All authors have read and agreed to the published version of the manuscript.

Funding: This research has received funding from the European Union's Horizon 2020 research and innovation program under grant agreement No. 860407.

Institutional Review Board Statement: Not applicable.

Informed Consent Statement: Not applicable.

Data Availability Statement: The data presented in this study were submitted to PANGAEA to be openly available.

Acknowledgments: We gratefully acknowledge the creativeness and efficiency of our colleagues of the AWI-workshop who realized the jacket for temperature control of the reaction vial.

Conflicts of Interest: The authors declare no conflict of interests.

References

1. Endres, H.-J.; Siebert-Raths, A. *Engineering Biopolymers—Markets, Manufacturing, Properties and Applications*; Carl Hanser Verlag: München, Germany, 2011. [CrossRef]
2. Borrelle, S.B.; Ringma, J.; Law, K.L.; Monnahan, C.C.; Lebreton, L.; McGivern, A.; Murphy, E.; Jambeck, J.; Leonard, G.H.; Hilleary, M.A.; et al. Predicted growth in plastic waste exceeds efforts to mitigate plastic pollution. *Science* **2020**, *369*, 1515–1518. [CrossRef]
3. Rai, P.; Mehrotra, S.; Priya, S.; Gnansounou, E.; Sharma, S.K. Recent advances in the sustainable design and applications of biodegradable polymers. *Bioresour. Technol.* **2021**, *325*, 124739. [CrossRef]
4. Jiang, B.; Na, J.; Wang, L.; Li, D.; Liu, C.; Feng, Z. Reutilization of food waste: One-step extration, purification and characterization of ovalbumin from salted egg white by aqueous two-phase flotation. *Foods* **2019**, *8*, 286. [CrossRef]
5. Jiang, B.; Wang, L.; Wang, M.; Wu, S.; Wang, X.; Li, D.; Liu, C.; Feng, Z.; Chi, Y. Direct separation and purification of α-lactalbumin from cow milk whey by aqueous two-phase flotation of thermo-sensitive polymer/phosphate. *J. Sci. Food Agric.* **2021**. [CrossRef] [PubMed]
6. Saratale, R.G.; Cho, S.-K.; Saratale, G.D.; Kadam, A.A.; Ghodake, G.S.; Kumar, M.; Bharagava, R.N.; Kumar, G.; Kim, D.S.; Mulla, S.I.; et al. A comprehensive overview and recent advances on polyhydroxyalkanoates (PHA) production using various organic waste streams. *Bioresour. Technol.* **2021**, *325*, 124685. [CrossRef] [PubMed]
7. Polman, E.M.; Gruter, G.-J.M.; Parsons, J.R.; Tietema, A. Comparison of the aerobic biodegradation of biopolymers and the corresponding bioplastics: A review. *Sci. Total Environ.* **2021**, *753*, 141953. [CrossRef]
8. Lucas, N.; Bienaime, C.; Belloy, C.; Queneudec, M.; Silvestre, F.; Nava-Saucedo, J.-E. Polymer biodegradation: Mechanisms and estimation techniques—A review. *Chemosphere* **2008**, *73*, 429–442. [CrossRef]
9. Haider, T.P.; Völker, C.; Kramm, J.; Landfester, K.; Wurm, F.R. Plastics of the future? The impact of biodegradable polymers on the environment and on society. *Angew. Chem. Int. Ed.* **2019**, *58*, 50–62. [CrossRef]
10. Azevedo, H.S.; Reis, R.L. Understanding the enzymatic degradation of biodegradable polymers and strategies to control their degradation rate. In *Biodegradable Systems in Tissue Engineering and Regenerative Medicine*; Reis, R.L., San Román, J., Eds.; CRC Press: Boca Raton, FL, USA, 2004; p. 26. [CrossRef]
11. Tokiwa, Y.; Suzuki, T. Hydrolysis of polyesters byrhizopus delemarlipase. *Agric. Biol. Chem.* **1978**, *42*, 1071–1072. [CrossRef]
12. Mochizuki, M.; Hirano, M.; Kanmuri, Y.; Kudo, K.; Tokiwa, Y. Hydrolysis of polycaprolactone fibers by lipase: Effects of draw ratio on enzymatic degradation. *J. Appl. Polym. Sci.* **1995**, *55*, 289–296. [CrossRef]
13. Lim, H.-A.; Raku, T.; Tokiwa, Y. Hydrolysis of polyesters by serine proteases. *Biotechnol. Lett.* **2005**, *27*, 459–464. [CrossRef] [PubMed]
14. Ding, M.; Zhang, M.; Yang, J.; Qiu, J.-H. Study on the enzymatic degradation of PBS and its alcohol acid modified copolymer. *Biodegradation* **2011**, *23*, 127–132. [CrossRef]
15. Hu, X.; Gao, Z.; Wang, Z.; Su, T.; Yang, L.; Li, P. Enzymatic degradation of poly(butylene succinate) by cutinase cloned from Fusarium solani. *Polym. Degrad. Stab.* **2016**, *134*, 211–219. [CrossRef]
16. Ruggero, F.; Gori, R.; Lubello, C. Methodologies to assess biodegradation of bioplastics during aerobic composting and anaerobic digestion: A review. *Waste Manag. Res.* **2019**, *37*, 959–975. [CrossRef]

17. Pamuła, E.; Błażewicz, M.; Paluszkiewicz, C.; Dobrzynski, P. FTIR study of degradation products of aliphatic polyesters-carbon fibres composites. *J. Mol. Struct.* **2001**, *596*, 69–75. [CrossRef]
18. Jacobsen, C.F.; Leonis, J.; Linderstrøm-Lang, K.; Ottesen, M.; Lonis, J.; Linderstrm-Lang, K. The pH-Stat and its use in biochemistry. In *Methods of Biochemical Analysis*; Glick, D., Ed.; Interscience Publishers Inc.: New York, NY, USA, 1957; Volume 4, p. 39. [CrossRef]
19. Walter, T.; Augusta, J.; Müller, R.-J.; Widdecke, H.; Klein, J. Enzymatic degradation of a model polyester by lipase from Rhizopus delemar. *Enzym. Microb. Technol.* **1995**, *17*, 218–224. [CrossRef]
20. Marten, E.; Müller, R.-J.; Deckwer, W.-D. Studies on the enzymatic hydrolysis of polyesters I. Low molecular mass model esters and aliphatic polyesters. *Polym. Degrad. Stab.* **2003**, *80*, 485–501. [CrossRef]
21. Herzog, K.; Müller, R.-J.; Deckwer, W.-D. Mechanism and kinetics of the enzymatic hydrolysis of polyester nanoparticles by lipases. *Polym. Degrad. Stab.* **2006**, *91*, 2486–2498. [CrossRef]
22. Gebauer, B.; Jendrossek, D. Assay of poly(3-hydroxybutyrate) depolymerase activity and product determination. *Appl. Environ. Microbiol.* **2006**, *72*, 6094–6100. [CrossRef]
23. Ronkvist Åsa, M.; Xie, W.; Lu, W.; Gross, R.A. Cutinase-catalyzed hydrolysis of poly(ethylene terephthalate). *Macromolecules* **2009**, *42*, 5128–5138. [CrossRef]
24. Ronkvist Åsa, M.; Lu, W.; Feder, D.; Gross, R.A. Cutinase-catalyzed deacetylation of poly(vinyl acetate). *Macromolecules* **2009**, *42*, 6086–6097. [CrossRef]
25. Cornish-Bowden, A. *Fundamentals of Enzyme Kinetics*, 3rd ed.; Portland Press: London, UK, 2004; ISBN 1-85578-1581.
26. Shih, C. Chain-end scission in acid catalyzed hydrolysis of poly (d,l-lactide) in solution. *J. Control. Release* **1995**, *34*, 9–15. [CrossRef]
27. Williams, D.F. Enzymic hydrolysis of polylactic acid. *Eng. Med.* **1981**, *10*, 5–7. [CrossRef]
28. Lee, S.H.; Kim, I.Y.; Song, W.S. Biodegradation of polylactic acid (PLA) fibers using different enzymes. *Macromol. Res.* **2014**, *22*, 657–663. [CrossRef]
29. Tokiwa, Y.; Jarerat, A. Biodegradation of poly(l-lactide). *Biotechnol. Lett.* **2004**, *26*, 771–777. [CrossRef]
30. Lee, C.W.; Kimura, Y.; Chung, J.-D. Mechanism of enzymatic degradation of poly(butylene succinate). *Macromol. Res.* **2008**, *16*, 651–658. [CrossRef]
31. Vichaibun, V.; Chulavatnatol, M. A new assay for the enzymatic degradation of polylactic acid. *Sci. Asia* **2003**, *29*, 297–300. [CrossRef]
32. Fiedler, P.C.; Talley, L.D. Hydrography of the eastern tropical Pacific: A review. *Prog. Oceanogr.* **2006**, *69*, 143–180. [CrossRef]
33. Deser, C.; Alexander, M.A.; Xie, S.-P.; Phillips, A.S. Sea surface temperature variability: Patterns and mechanisms. *Annu. Rev. Mar. Sci.* **2010**, *2*, 115–143. [CrossRef]
34. Huggett, J.; Griffiths, C. Some relationships between elevation, physico-chemical variables and biota of intertidal rock pools. *Mar. Ecol. Prog. Ser.* **1986**, *29*, 189–197. [CrossRef]
35. Jocque, M.; Vanschoenwinkel, B.; Brendonck, L. Freshwater rock pools: A review of habitat characteristics, faunal diversity and conservation value. *Freshw. Biol.* **2010**, *55*, 1587–1602. [CrossRef]
36. Vohra, F.C. Zonation on a tropical sandy shore. *J. Anim. Ecol.* **1971**, *40*, 679. [CrossRef]
37. Harrison, S.; Morrison, P. Temperatures in a sandy beach under strong solar heating: Patara Beach, Turkey. *Estuar. Coast. Shelf Sci.* **1993**, *37*, 89–97. [CrossRef]
38. Emadian, S.M.; Onay, T.T.; Demirel, B. Biodegradation of bioplastics in natural environments. *Waste Manag.* **2017**, *59*, 526–536. [CrossRef] [PubMed]
39. Kale, G.; Kijchavengkul, T.; Auras, R.; Rubino, M.; Selke, S.E.; Singh, S.P. Compostability of bioplastic packaging materials: An overview. *Macromol. Biosci.* **2007**, *7*, 255–277. [CrossRef]
40. Folino, A.; Karageorgiou, A.; Calabrò, P.S.; Komilis, D. Biodegradation of wasted bioplastics in natural and industrial environments: A review. *Sustainability* **2020**, *12*, 6030. [CrossRef]
41. Patnaik, S.; Panda, A.K.; Kumar, S. Thermal degradation of corn starch based biodegradable plastic plates and determination of kinetic parameters by isoconversional methods using thermogravimetric analyzer. *J. Energy Inst.* **2020**, *93*, 1449–1459. [CrossRef]
42. Oza, S.; Ning, H.; Ferguson, I.; Lu, N. Effect of surface treatment on thermal stability of the hemp-PLA composites: Correlation of activation energy with thermal degradation. *Compos. Part B Eng.* **2014**, *67*, 227–232. [CrossRef]
43. Kaloustian, J.; Pauli, A.M.; Pastor, J. Kinetic study of the thermal decompositions of biopolymers extracted from various plants. *J. Therm. Anal. Calorim.* **2000**, *63*, 7–20. [CrossRef]
44. Moussout, H.; Ahlafi, H.; Aazza, M.; Bourakhouadar, M. Kinetics and mechanism of the thermal degradation of biopolymers chitin and chitosan using thermogravimetric analysis. *Polym. Degrad. Stab.* **2016**, *130*, 1–9. [CrossRef]
45. Deng, S.; Tabatabai, M. Cellulase activity of soils. *Soil Biol. Biochem.* **1994**, *26*, 1347–1354. [CrossRef]
46. Hu, G.; Heitmann, J.A.; Rojas, O.J. In situ monitoring of cellulase activity by microgravimetry with a quartz crystal microbalance. *J. Phys. Chem. B* **2009**, *113*, 14761–14768. [CrossRef] [PubMed]
47. Kapat, A.; Panda, T. pH and thermal stability studies of chitinase from Trichoderma harzianum: A thermodynamic consideration. *Bioprocess Biosyst. Eng.* **1997**, *16*, 269–272. [CrossRef]

48. Singh, A.K.; Chhatpar, H.S. Purification and characterization of chitinase from *Paenibacillus* sp. D1. *Appl. Biochem. Biotechnol.* **2010**, *164*, 77–88. [CrossRef]
49. Lonhienne, T.; Baise, E.; Feller, G.; Bouriotis, V.; Gerday, C. Enzyme activity determination on macromolecular substrates by isothermal titration calorimetry: Application to mesophilic and psychrophilic chitinases. *Biochim. Biophys. Acta* **2001**, *1545*, 349–356. [CrossRef]

Article

Optical Monitoring of Microplastics Filtrated from Wastewater Sludge and Suspended in Ethanol

Benjamin O. Asamoah [1,*], Pauliina Salmi [2], Jukka Räty [3], Kalle Ryymin [4], Julia Talvitie [5], Anna K. Karjalainen [4], Jussi V. K. Kukkonen [6], Matthieu Roussey [1] and Kai-Erik Peiponen [1,*]

1. Department of Physics and Mathematics, University of Eastern Finland, P.O. Box 111, FI-80101 Joensuu, Finland; matthieu.roussey@uef.fi
2. Faculty of Information Technology, University of Jyväskylä, Mattilanniemi 2 (Agora building), P.O. Box 35, FI-40014 Jyväskylä, Finland; pauliina.u.m.salmi@jyu.fi
3. Unit of Measurement Technology, MITY, University of Oulu, Technology Park, P.O. Box 127, FI-87400 Kajaani, Finland; jukka.raty@oulu.fi
4. Department of Biological and Environmental Science, University of Jyväskylä, Survontie 9C (YAC Building), P.O. Box 35, FI-40014 Jyväskylä, Finland; kalle.ryymin@gmail.com (K.R.); anakakarjalainen@gmail.com (A.K.K.)
5. Marine Management, Finnish Environment Institute (SYKE), Latokartanonkaari 11, FI-00790 Helsinki, Finland; julia.talvitie@syke.fi
6. Department of Environmental and Biological Science, University of Eastern Finland, Kuopio Campus, P.O. Box 1627, FI-79211 Kuopio, Finland; jussi.kukkonen@uef.fi
* Correspondence: benjamin.asamoah@uef.fi (B.O.A.); kai.peiponen@uef.fi (K.-E.P.)

Abstract: The abundance of microplastics (MPs) in the atmosphere, on land, and especially in water bodies is well acknowledged. In this study, we establish an optical method based on three different techniques, namely, specular reflection to probe the medium, transmission spectroscopy measurements for the detection and identification, and a speckle pattern for monitoring the sedimentation of MPs filtrated from wastewater sludge and suspended in ethanol. We used first Raman measurements to estimate the presence and types of different MPs in wastewater sludge samples. We also used microscopy to identify the shapes of the main MPs. This allowed us to create a teaching set of samples to be characterized with our optical method. With the developed method, we clearly show that MPs from common plastics, such as polypropylene (PP), polyethylene terephthalate (PET), polystyrene (PS), and polyethylene (PE), are present in wastewater sludge and can be identified. Additionally, the results also indicate that the density of the plastics, which influences the sedimentation, is an essential parameter to consider in optical detection of microplastics in complex natural environments. All of the methods are in good agreement, thus validating the optics-based solution.

Keywords: microplastics; sludge; wastewater; Raman spectroscopy; laser speckle pattern; transmittance; sedimentation

Citation: Asamoah, B.O.; Salmi, P.; Räty, J.; Ryymin, K.; Talvitie, J.; Karjalainen, A.K.; Kukkonen, J.V.K.; Roussey, M.; Peiponen, K.-E. Optical Monitoring of Microplastics Filtrated from Wastewater Sludge and Suspended in Ethanol. *Polymers* **2021**, *13*, 871. https://doi.org/10.3390/polym13060871

Academic Editor: Jacopo La Nasa

Received: 5 February 2021
Accepted: 9 March 2021
Published: 11 March 2021

Publisher's Note: MDPI stays neutral with regard to jurisdictional claims in published maps and institutional affiliations.

Copyright: © 2021 by the authors. Licensee MDPI, Basel, Switzerland. This article is an open access article distributed under the terms and conditions of the Creative Commons Attribution (CC BY) license (https://creativecommons.org/licenses/by/4.0/).

1. Introduction

The recent increase in publications shows that research on microplastics (MPs) is a topic of utmost importance for the environment and health despite the recent emergence and recognition of the problem by the scientific community [1–7]. The multi-disciplinary character of the topic makes the homogenization of the research complex. Although there are still some inconsistencies in the sampling methods [8] and sample processing of the microplastics, as well as in the analysis methods, a trend can be seen in the uniformity of the techniques. Another study emphasized the issue of discrepancies between plastic particle types, size ranges, and concentrations in tests that were carried out either in a laboratory or in real aquatic environments [1]. According to Alexy et al. [1], there is a need to harmonize the methods for better comparability of the results as well as a requirement to have automated sensors for real-time and/or continuous measurements and data processing.

Many analysis methods and techniques have been developed for the identification of MPs [9,10]. Through Fourier-transform-infrared (FTIR) [11] and Raman [12] spectroscopies, optics are widely used, although these two techniques still require sample harvesting and preparation prior to measurements. Regarding the automated and real-time detection of microplastics, optical methods are among the most powerful alternatives and offer many different opportunities based on well-established concepts. However, there is often the remaining problem of sampling microplastics, which are usually required in the form of dry samples for measurements [13].

Plastics can be identified using the spectral fingerprints of synthetic hydrocarbons in the near-infrared (NIR) spectral range. Applying spectral techniques in this wavelength range in the detection of plastics in an aquatic environment is rather complicated due to the strong NIR absorption by water, which interferes with the method of plastic detection and identification. To navigate this problem, dry sample measurements are, therefore, preferred. However, in the vicinity of $\lambda = 1200$ nm, water absorption is weaker, and methods such as airborne detection of floating microplastics have been suggested, but for harvested samples only [14]. An important step forward for understanding the behavior of microplastics in aquatic environments using optical sensing is to realize in situ and real-time measurement methods that require no harvesting or sample treatment of any kind. This is a challenging task, however, due to the relatively low concentration of MPs—with partial or full transparency in some cases—as well as the complex and changing environments. Moreover, most of the current optical techniques for MPs are powerful when used separately. However, for an accurate identification, quantification, and classification of MPs, the development of a multifunctional sensor based on the clever integration of the individual optical methods is necessary for in situ monitoring. Therefore, simulations and tests on different types of optical phenomena in controlled laboratory conditions are needed to progress in the design and development of such in situ optical detectors.

In this paper, we combine spectroscopic methods based on transmittance and laser-based measurements to identify and validate MP detection in a complex matrix. The method is verified with Raman measurements. We demonstrate that this approach is a promising tool in the detection of low concentrations of different MPs in complex environments and for implementation in real-time measurements. Moreover, the study validates the technique, which could serve as a qualitative control in wastewater treatment processes in order to monitor the MPs after treatment [13,15]. The microplastics were obtained from wastewater sludge samples and suspended in ethanol (filtrated and suspended in ethanol MPs: FEMPs). Ethanol, like water, has a spectral fingerprint in the NIR region and, therefore, presents similar challenges for in situ optical detection of MPs in aquatic environment. In addition, ethanol improves the measurement safety, as it kills harmful bacteria and viruses [16], which can exist in wastewaters.

Detecting particles in wastewater sludge is easy; detecting one specific type of particle is the key challenge of current research in this field. To perform such an analysis in the field, we have established a set of samples to be used to learn the behavior of the MP particles coming from such a harsh environment. The outcomes of the investigations on this teaching sample set with our newly developed optical method(s) can then be used in real conditions, i.e., outside the laboratory, and can serve as a basis for the determination of a new detection methodology. The goal is, therefore, to first determine the presence of MPs in wastewater samples through Raman spectroscopy, as well as their shape through microscopy, in order to establish our teaching test set. Further, we used this teaching set to determine whether our optical methods are suitable for detecting the same MPs in the same samples.

2. Materials and Methods

In this section, we give a detailed description of the sample preparation and optical detection schemes that could be an alternative to the time-consuming Raman spectroscopy detection method for MPs. However, as a reference for a comparative study, we use data from a Raman spectrometer that was exploited in the identification of MPs of different plastic types that originated from a municipal wastewater plant. The advantage of the different proposed optical method(s) is the relatively quick measurement time and the ease of combining them in a multi-functional optical device to be used for quality inspections in wastewater treatment.

2.1. Sample Preparation

To obtain the microplastics (20–5000 µm), cake samples of anaerobically digested and dried wastewater sludge from the Nenäinniemi wastewater treatment plant (Jyväskylä, Finland) were collected in aluminum foil and stored at +4 °C in 1 L plastic boxes. Subsamples of 40–42 g wet weight (11–14 mg dry weight) were purified with hydrogen peroxide (30%, WVR International) and enzymes (ASA Special enzyme GmbH, Wolfenbüttel, Germany) in 250 mL glass bottles according to the universal purification protocol developed and validated in [17]. The final step in the protocol was the density separation of microplastics with $ZnCl_2$ solution (ρ = 1.6–1.8 g cm^{-3}). Before the purification protocol, hydrogen peroxide and the enzymes were pre-filtered through GF/A filters (Whatmann, Maidstone, UK) to remove possible microplastic contaminants. The $ZnCl_2$ solution was filtered through a 20 µm stainless-steel filter because it easily blocked the GF/A filters.

The high dry matter content of our samples prevented the direct application of the vacuum filtration described by Löder et al. (2017) [17]. Instead, we washed the samples by centrifugation between each step as follows: Samples were poured in 50 mL centrifuge tubes, covered with foil, and centrifuged at 3000 rcf for 2 min. The supernatant was then filtered using 20 µm mesh-sized stainless-steel filters. The remaining pellet was again centrifuged using ultrapure water to clean the sample, and the supernatant was filtered. The filter and the pellet were transferred to the 250 mL glass bottle and submerged with the reagent of the next step. Another modification introduced to the method of Löder et al. (2017) was the replacement of sodium dodecyl sulfate (SDS) in step 1 with 30% hydrogen peroxide due to the extensive foaming of SDS. Hydrogen peroxide incubations were done at 60 °C.

Finally (step 8, Table 1), the microplastics were rinsed with $ZnCl_2$ (ρ = 1.6–1.8 g·cm^{-3}) from the steel filter into a glass separation funnel. Sand and other relatively heavy materials were separated by incubating for 1–3 days. Microplastics remaining in the $ZnCl_2$ were collected on the steel filter and eluted with ethanol (70% in water) in glass bottles for storage. The resulting mixture was a diluted concentration of different MPs filtrated and preserved in ethanol (FEMPs). Figure 1 shows images of the MPs during the preparation process (left) and contained in the measured samples (in the white circle, middle). One can appreciate their various sizes and irregular shapes, e.g., "hook-like", fibers, fragments, and aggregates. Despite the filtering process, other unidentified organic materials than MPs were clearly present (right) after preparation. Such particles are typical of wastewater and natural environments, and they add to the complexities of the matrix. More information and details are given in the dataset published by P. Salmi et al. [18].

Table 1. Sequence of reagents used to purify the microplastic samples according to Löder et al. (2017) [17].

Step	Reagent	Incubation Temperature and Time
1	Hydrogen peroxide 30%	60 °C, 1 d
2	Protease in Tris-HCl buffer (1:5, pH = 9)	50 °C, 1 d
3	Lipase in Tris-HCl buffer (1:20, pH = 9)	40 °C, 1 d
4	Cellulase in NaOAc buffer (1:4, pH = 5)	40 C, 3 × 1 d
5	Hydrogen peroxide 30%	60 C, 1 d
6	Chitinase in NaOAc (1:20, pH = 5)	37 °C, 1 d
7	Hydrogen peroxide 30%	60 °C, 1 d
8	Density separation with $ZnCl_2$ (ρ = 1.6–1.8 g.cm^{-3})	Room temperature, 1–3 d

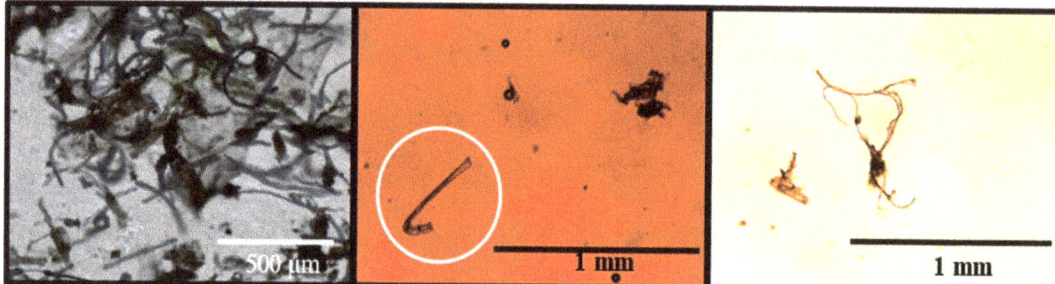

Figure 1. Micrograph of microplastics (MPs) during preparation (left) with different shapes and sizes and those identified in the measured sample (middle). In the white circle, a hook-shaped MP is highlighted, and the other black spots are air bubbles and unidentified organic materials (also in the right image).

2.2. Counting and Characterization of Microplastics with Raman Microscopy

Subsamples of 5 mL of the FEMP samples were filtered on aluminum oxide filters (Anodisc, Whatmann, Maidstone, UK). Microplastics were manually counted on at least 15% of the area of the filter, and the plastic type was analyzed from the field of view of a Raman microscope (Thermo Fischer Raman DXR 2xi, Waltham, MA, USA). Table 2 gives the numbers of particles and subsampling ratios. The Raman spectra of microplastics were obtained with 758 nm laser excitation (20 mW power), an objective with ×10 magnification, a 50 µm slit aperture, and an exposure time of 0.1 s for 20 scans. The acquired spectra were compared to commercial material libraries using the OMNIC xi software (Thermo Fischer Scientific, Waltham, MA, USA). For pristine samples, the match to library spectra was expected to be higher than 70%; however, due to matrix interference, a 50–70% match was obtained based on the Raman peak positions. The particles characterized as plastic were categorized as fragments or fibers based on the deviation from the sting-shaped form of the fibers. Counts were converted to microplastics per gram of dry weight (MP g^{-1} DW) according to Equation (1).

$$MP\ g^{-1}\ DW = [(N \times k) / m] \times df, \quad (1)$$

where N is the number of particles counted, k is the ratio of the filtration area to the area where the particles were counted from, m is the dry mass of the processed sample, and df is the dilution factor because the whole processed sample was not filtered with the aluminum oxide filter.

Table 2. Identified MP types and quantities per dry weight (DW) from Raman microscopy and average densities in pristine form. MPs g^{-1} DW: Microplastics per gram of dry weight. MPs/mL: Microplastics per millilitre ethanol. Particle (P) or fiber (F).

	Sample A		Sample B		Sample C		
Dry Weight [g]	11.5		13.8		13.9		
Proportion of DW analyzed with Raman [%]	0.9		0.9		2		
Fibers counted	6		4		15		
Particles counted	8		7		6		
MPs g^{-1} DW	130		85		85		
MPs/mL	19		15		17		
MP types	F	P	F	P	F	P	Density in pristine form [g cm^{-3}]
PP	0	3	0	1	0	3	0.85
PE	1	0	0	1	3	0	0.94
PET	5	1	4	2	12	1	1.38
PS	0	1	0	0	0	1	1.05
POM	0	3	0	3	0	1	1.42

2.3. Three Optics-Based Alternatives for a New Detection Method

The same samples were also characterized—benchmarking with ethanol—with three other devices to compare with the results of Raman spectroscopy and to determine the dynamic light scattering of the identified MPs. The devices were (1) a spectrophotometer (UV 3600, Shimadzu, Kyoto, Japan) for obtaining the spectral fingerprints, (2) a handheld diffractive optical element (DOE, MGM Devices, Finland)-based optical sensor [19], and, finally, (3) a "basic" optical setup consisting of a Helium–Neon (HeNe) laser light source (5 mW, λ = 632.8 nm) and a charge-coupled device (CCD, Thorlabs, Newton, NJ, USA) camera for probing the samples. Both the basic setup and DOE-based sensors were used for the determination of the dynamic light scattering properties of the FEMP samples.

One of our novel methods is based on the modification of an original handheld glossmeter to create a DOE-based optical sensor. A detailed description of the device is given in [19]. It consists of a handheld glossmeter with a single-cell photodiode for recording the specular reflection and an attached CCD camera—without an objective lens—for recording the forward-scattered light. Moreover, for liquid samples, there is a compartment designed with an 800 µL inner volume to contain the liquid for measurement. The liquid compartment consists of a glass disk base (with either both sides smooth or one side rough) and a black surrounding cup. The use of the rough glass disk enables the generation of a speckle pattern—see Figure 2a—with a coherent laser light source for probing liquid-based samples. The laser speckle pattern is formed due to coherent laser light scattering from microscopic objects, such as MPs in water or in other host liquids. This pattern is sensitive to the local displacement of particles, i.e., sinking or moving in the direction of the beam. The presence of MPs is, therefore, expected to modify this original speckle pattern, leading to the detection of the presence of the MPs. The extent of the speckle pattern modification by the MPs depends on the properties of the MPs, such as the size, volume properties, and surface texture [20,21], as well as the refractive index (RI) mismatch between the present material and the ambient medium. For very small MP sizes, as considered in this study, few modifications to the speckle and interference patterns are expected. Replacing the rough disk with smooth glass leads to the transmission of an interference pattern; see [19]. Connecting the DOE-based sensor, which is equipped with a semiconductor laser light source (0.8 mW, λ = 635 nm), to a computer enables time-dependent measurements [22], which may be essential for studies in the field.

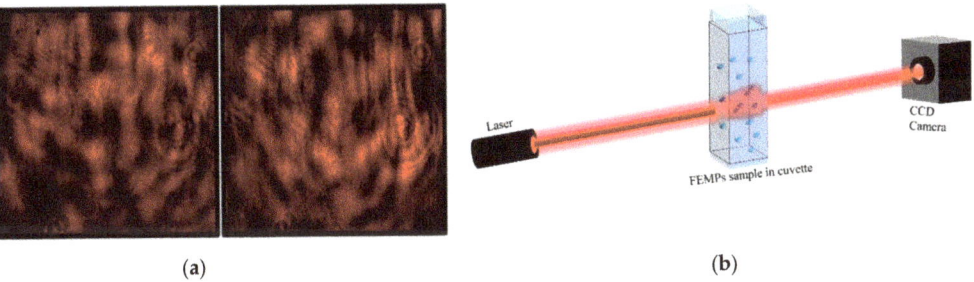

Figure 2. (a) Speckle pattern of samples on a rough glass surface for samples of ethanol (left) and MPs filtrated and preserved in ethanol (FEMPs) (right). The recorded static images (patterns) are very similar to the naked eye for both samples. (b) Basic optical experimental setup for recording the beam profile after propagation through the FEMP sample.

The speckle contrast [23], $C = [<I^2> - <I>]^{1/2} / <I>$, was calculated to obtain information on the dynamics of the samples as a result of the modification to the initially generated speckle pattern with time. I and $<\cdots>$ represent the intensity of the speckle pattern and the ensemble average, respectively. The speckle contrast was calculated using the intensity of the red channel from the red–green–blue (RGB) pattern captured on the CCD camera after correcting with the measured stray light. C has values ranging from 0 to 1 for undeveloped and fully developed speckles, respectively. It has a long history of applications ranging from surface roughness characterization [21,24] to flow rate determination [25]. In this study, we employed it in the qualitative identification of the presence of microplastics. The data obtained with these optical methods, which are presented in Figure 2, will be treated in the last part of the manuscript while describing the detection of the sedimentation of MPs.

2.4. Measurement Procedure with the Spectrophotometer and the New Optical Devices

First, the transmittance of the FEMPs and the reference ethanol samples was measured with the spectrophotometer in the visible-near-infrared (Vis-NIR) spectral range of 500–2500 nm, where most plastics have spectral fingerprints [26–28]. Additionally, a standard spectroscopic cuvette with inner dimensions of 10 mm × 5 mm × 35 mm and an inner volume of ca. 1750 µL was filled with the FEMP sample. The cuvette was set so that light propagated along the 5 mm direction in the sample. The transmittance as a function of time at a fixed wavelength $\lambda = 800$ nm was then measured after shaking the sample in the cuvette. The settling of the MPs after shaking was to simulate the sedimentation process of the different MPs due to variation in densities. In both measurements, an empty cuvette was used for the baseline measurements.

Next, the "basic" laser-based optical setup was used to measure the pure ethanol and the FEMP samples in another standard spectroscopic quartz cuvette with dimensions of 10 mm × 10 mm × 35 mm (ca. 3500 µL). The covered cuvette was shaken and illuminated with a Helium–Neon (HeNe) laser light source with a 0.81 mm beam diameter. The corresponding transmitted beam was also recorded with time on a CCD camera—also without an objective lens—placed 60 cm from the cuvette to determine the dynamic laser-light-scattering properties of the particles in the FEMP sample.

Lastly, using the smooth glass disk base in the modified portable handheld device, we introduced a 15 µL drop of the filtrated sample into the volume compartment of the DOE-based sensor with a glass pipette to measure the specular reflection of the sample with time. Since there was very little probability of having any MPs in such a small volume, we were only able to obtain information on the effective optical medium of the FEMP sample after the preparation process based on the specular reflection. To further probe the dynamic light-scattering properties of the samples in a relatively larger volume, using the same device but with a rough glass disk base, the 800 µL (for an approximate height of 2.8 mm and glass disk diameter of 20 mm) volume compartment was filled with 700 µL of the pure ethanol and the FEMP sample, and the transmitted pattern was again measured with time on the CCD camera. Static images of the recorded speckle patterns are shown in Figure 2a. It is worth mentioning that the images are very similar for both the ethanol and FEMP samples. This optical method simultaneously detected the presence of various types and sizes of mostly irregularly shaped MPs in the liquid matrix.

For comparison, we also performed similar measurements for pure ethanol (96%), in which the FEMPs were preserved. We noted that the speckle pattern measurements with the DOE-based sensor and the HeNe laser setup differed. These gave different transmittance schemes and possibly different numbers of particles in the considered sample volumes. However, both methods showed similar dynamic activity of the FEMP samples. The probing laser light beam was limited by an aperture of 2 mm at the bottom of the glass disk in the DOE-based sensor. In the same device, the FEMP samples were introduced directly on the light beam's path. Hence, the motion of the FEMPs was preferentially falling parallel to the direction of the beam, whereas in the other schemes, the FEMPs were falling orthogonal to it. Therefore, in the case of the recorded speckle pattern, the DOE-based sensor monitored variations in the speckle pattern with time because of local variations in the surface roughness of the glass disk base.

It is to be noted that the use of the spectrophotometer (detection of the transmission spectrum) prevents its use for several minutes or seconds for other purposes, such as monitoring the sedimentation. For this reason, we needed a laser-based sensor that, thanks to coherent radiation, was very sensitive in order to detect any small changes along the path of the laser beam, e.g., an MP passing in the beam.

3. Results and Discussion

3.1. Raman Microscopy

The analysis of the Raman microscopy spectra led to the identification of the following MPs in the FEMP sample: polypropylene (PP), polyethylene (PE), polyethylene terephthalate (PET), polystyrene (PS), and polyoxymethylene (POM). The numbers of MPs counted for each shape of the identified plastic types are presented in Table 2 for the three sets of FEMP samples that were measured (Samples A, B, and C). The corresponding estimated concentrations of the MPs (mean = 17 MPs/mL, SD = 2) from the Raman measurements are indicated for the samples. The average densities of the identified MPs are also presented in the same table to aid in a later discussion of the other results. By comparing the measured Raman spectra to the commercial material library, MPs of five (5) different plastic types could be identified in the FEMP samples, which is in accordance with the most abundant MPs in wastewater [29].

Raman spectroscopy provided a clear picture of the nature and types of MPs in the samples to be investigated with our novel optical method. First, one can remark that PET fibers are predominant. Second, the number of fibers is negligible for PP, PS, and POM. No particles, but only fibers and in a very small amount, were detected for PE.

These results, which were obtained with a recognized method for MP identification, allowed us first to determine the presence and types of MPs in the prepared samples, second, to establish a set of known samples obtained from a wastewater treatment plant, and, third, to further validate our simpler optical methods.

3.2. Specular Reflection Signal

Figure 3 shows the reflection signal detected with the DOE-based sensor after introducing 15 µL of the FEMP sample on the smooth glass disk base. The recorded specular reflection signal had an accuracy of ca. 0.10%. We observe that both the pure ethanol and FEMP samples showed lower reflection signals than with the smooth glass only, with that of pure ethanol being higher than that of the FEMP sample. The clear difference in the magnitudes of the reflection signals between the pure ethanol and FEMP samples was more likely due to the difference in the refractive indexes of the two samples. Pure ethanol has a refractive index (RI) of n = 1.36; by preserving the filtrated sample in ethanol, we changed the effective refractive index of the resultant mixture, which yielded a change in the reflection signal. Obtaining a lower reflection signal than that of pure ethanol indicates that the FEMP sample had a higher effective RI than the pure ethanol. However, we note that, by measuring the specular reflection of such a small volume on the smooth glass, we obtained information on the ambient medium of the FEMP sample rather than on the MPs, the reason being that such a small volume of the FEMP sample has very little probability of containing an MP, although it may contain other particles (see Figure 1) that also affect the effective RI. Additionally, considering the fairly constant nature of the reflection signal with time, one can conclude that both liquids rarely spread over the smooth glass [22,30]. This is especially true for the FEMP sample due to its slightly more viscous nature than that of the pure ethanol.

3.3. Transmission Spectra

The transmission spectra of pure ethanol and the FEMPs are shown in Figure 4a. From the Raman measurements, we know that FEMP samples A, B, and C contained five different plastic types. The measured transmittance is, therefore, a combination of their weighted spectral features. For $\lambda < 900$ nm, both samples showed T > 100%, which is due to the use of air for the baseline correction. Pure ethanol showed a higher transmittance than the FEMP samples across almost all wavelengths, with dominant features for $\lambda > 1100$ nm. The magnitude of MP transmittance in a liquid medium is highly dependent on the concentration [26]. To identify and classify the different MPs in the FEMP samples based on the spectral features, we subtracted the transmittance of pure ethanol from that of the FEMP samples. Figure 4b shows the resulting difference in transmittance, ΔT, where the spectral features, seen as dips, consisting of first-third overtones of C–H vibrations [31], are more visible. In comparison to earlier studies [26–28], these observed spectral features can be attributed to MPs from known plastic types. In the study by Peiponen et al. (2019) [26], where the spectral features of plastics were studied in both air and water, it was observed that the maximum deviation of an identified absorption peak is about 4 nm, despite the relatively large difference in RI mismatch. In some cases, no deviations were observed for some peaks of the same plastic type.

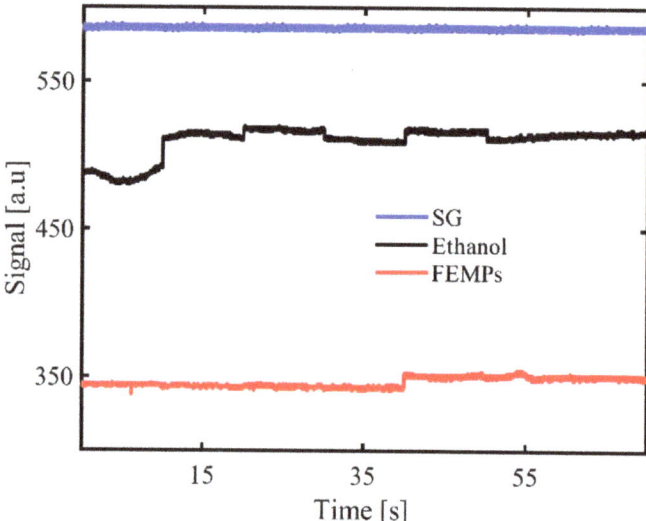

Figure 3. Specular reflection recorded with the diffractive optical element (DOE)-based sensor for the smooth glass (SG) disk base only for the pure ethanol (Ethanol) and FEMP (MPs preserved in ethanol) samples.

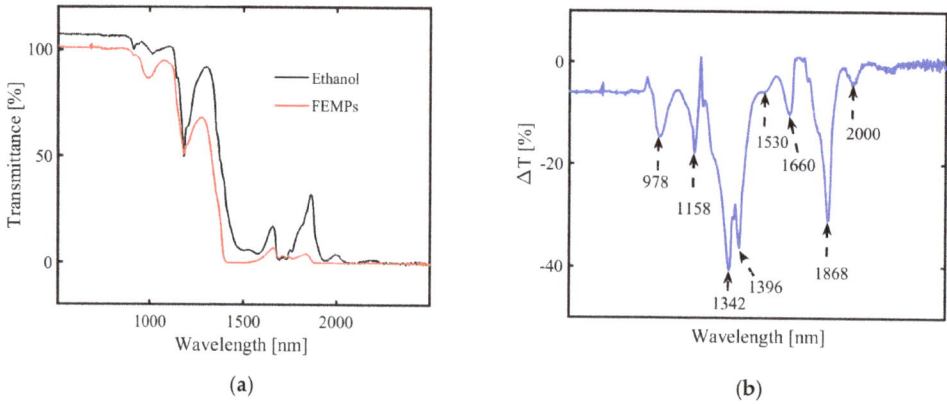

Figure 4. Transmission spectra of (**a**) pure ethanol (black curve) and the FEMP (MPs preserved in ethanol, red curve) samples and (**b**) differences between the two spectra (ΔT).

Imposing similar constraints and considering the small difference in the dispersion of ethanol and water, we can therefore attribute some of the identified spectral dips as follows: 1158 (PS), 1396 (PE), and 1660 nm (PET). This identification and classification, such as in the case of PET, is in close agreement with the results of [28], who performed spectral measurements of plastics in air at NIR and mid-infrared. As is evident from [28], POM shows no spectral features in the Vis-NIR spectral range; hence, none of the fingerprints can be attributed to it, but due to its density, it will contribute to the sedimentation process. Based on these observations, we can classify the mixture of MPs by simply combining transmittance measurements with an a priori library of MP spectral properties, such as that presented in [28]. The creation of such a rich library of intrinsic spectral properties of plastics could also be beneficial in quantifying the concentrations of the different plastic types using chemometric tools, such as principal component analysis [31]. However, we remark that for real-time monitoring and detection of plastics in open water, the problem of suppressing the spectral fingerprints of water absorption could persist [26]. The unattributed dips at 978, 1530, and 1868 nm can be due to the water content present in the preserving ethanol (ca. 30%).

Thus far, based on the transmission and the Raman measurements, similar types of MPs could be identified in the FEMP sample—excluding the POM, which showed no spectral features in the measured range. Although the shapes of the identified MPs could not be distinguished in the transmission measurements, combining both techniques in a sensor promises a more quantitative means of identifying and validating detected MPs in natural environment with higher confidence.

3.4. Sedimentation

We consider the case of MP sedimentation from the three measurement schemes, where the densities of the MPs play a major role. Figure 5a illustrates the evolution of the transmittance with the time measured with the spectrophotometer when the light beam passed through the shaken liquid. Figure 4a shows that the probing wavelength selected for the sedimentation investigations, $\lambda = 800$ nm, ensures that the measured transmittance of the FEMP sample is predominantly due to the scattering and diffraction (depending on the polydisperse particles) rather than absorption. The red dots are the measured transmittance with time, whereas the red curve is the exponential fit of the data points. Similarly, the observed $T > 100\%$, which is due to the use of air for the baseline correction, still exists. It is obvious that the transmittance increases as a function of time. This is attributed to the sedimentation of the MPs in ethanol. Ethanol has a relatively lower density of 0.81 g cm^{-3}, and the densities of the MPs in the FEMPs suggest that they will settle at different depths. However, after shaking the cuvette, the FEMPs are first in a dynamic state before starting to settle, depending on the densities. Based on the densities, most of the MPs will settle at the bottom, leaving fewer particles of PS and some available organic materials with lower density to interact with the probing light after some time has elapsed. Indeed, the curve tends to saturate after 4 min. Thus, the exponential growth of the curve is strongly influenced by the types of particles available.

Next, we consider the case of DOE-based sensor with the full volume (700 µL). The recorded speckle patterns of the pure ethanol and FEMP samples are very similar and indistinguishable to the naked eye, as seen in Figure 2a. However, due to the sensitivity of the speckle pattern to the local variation and the polydisperse particle sizes, the little variations in the pattern may be revealed in the speckle contrast, C.

Figure 5b shows the speckle contrast, with error bars, calculated from the recorded speckle pattern using the DOE-based sensor with a rough glass disk base for the case of 700 µL of pure ethanol and FEMP samples. We observed a fairly constant mean value of C with time in the case of pure ethanol, whilst a similar nonlinear relationship, as seen in Figure 5a, exists between the mean of C and the time for the FEMP sample. The larger variation in C for the pure ethanol could be due to its volatility over longer durations, which is not the case for the FEMP slurry sample. In the absence of any particles in pure ethanol, the scattering is influenced by the RI difference between the (rough) glass and the ethanol, which does not change with time; hence, the horizontal nature of the speckle contrast is as one would expect. However, the FEMP sample is rather complicated, as the scattering, in addition to the RI of the ambient medium, is also influenced by the presence, sizes, and irregular shapes of the MPs and their refractive indices. The different RI values of the MPs cause local changes in the RI for the probing beam, thus modifying the speckle pattern at the detection plane. Further increases in the speckle contrast can also be because of the falling of heavier MPs, such as POM, PS, PET, and PE, and random movement of less dense MPs, such as PP, thus modifying the speckle pattern in similar manner to that found when increasing the surface roughness. In such a small height of 2.8 mm, the speckle contrast tends to be sensitive to such sedimentation dynamics of the particles. Considering the speckle pattern as a transmission component of the DOE-based sensor, the positions of the specular reflection signals and the speckle contrast for the ethanol and the FEMP samples are in good agreement.

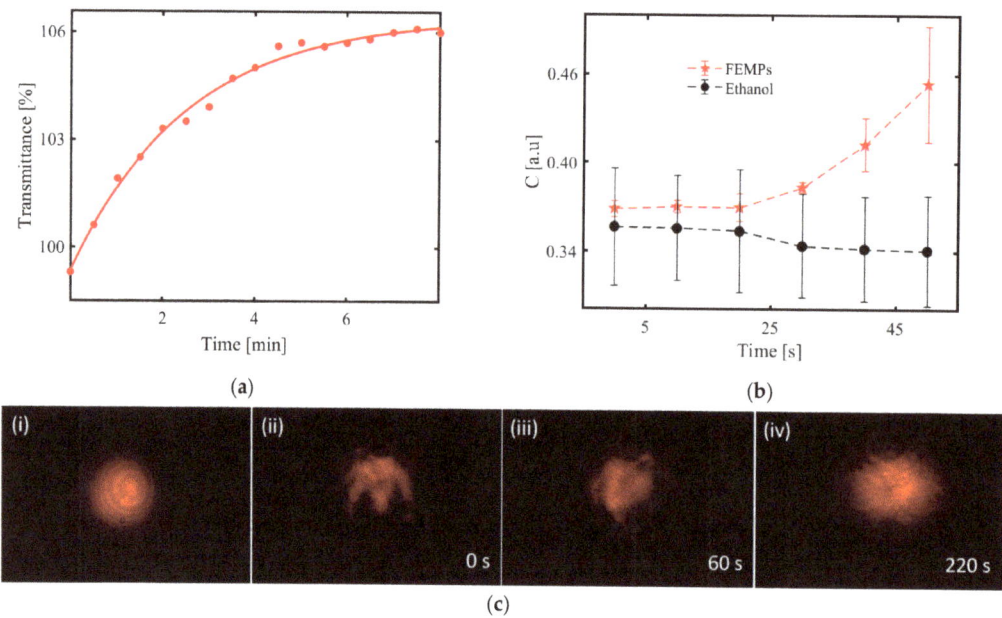

Figure 5. Sedimentation: (**a**) Measured transmittance (red dots) of the FEMP (MPs preserved in ethanol) samples with time at $\lambda = 800$ nm. The red curve is an exponential fit of the data. (**b**) Calculated speckle contrast from the measured speckle pattern using the diffractive optical element (DOE)-based sensor with a rough glass disk base and 700 µL volume of the samples. (**c**) Recorded patterns of the light beam transmitted through (i) pure ethanol and FEMPs after shaking at (ii) 0, (iii) 60, and (iv) 220 s with the charge-coupled device (CCD) camera.

Comparing the integrated transmittance in Figure 5a and the speckle contrast curves in Figure 5b, we observe that both display dynamic activity in the system, which can be attributed to the presence of the MPs in the FEMP sample. However, the integrated transmittance curve saturates at a much longer time (after 4 min), which could be due to the different shapes of the respective containers. In addition to the possible higher number of MPs with the increased sample volume, the height of the cuvette (35 mm) and the area of the sensing beam (3×8 mm^2) contribute to the longer saturation time in the case of the transmittance measurements, unlike in the DOE-based sensor with a 2.8 mm height of the volume compartment. Regarding the simultaneous presence of both regularly and irregularly shaped MPs with different densities, the particles are likely to spend a longer time in the narrow cuvette, leading to a longer saturation time than that for the wider cup of the DOE-based sensor with a much shorter sedimentation height.

The images in Figure 5c were obtained using the optical setup described in Figure 2b. Figure 5c shows static images related to similar measurements, but using the HeNe laser beam and a 10 mm cuvette. In Figure 5c (i), the pattern recorded for pure ethanol, which was stable for various durations, is shown. We observe regular and symmetric annular interference fringes, which could result from the thermo-optic properties of ethanol and the Gaussian intensity profile of the HeNe laser [32–34]. The presence of the annular rings serves as a measure of the extent of distortion of the incident beam by the FEMP sample. Figure 5c (ii) and (iii), on the other hand, show an obviously distorted beam even after 60 s of shaking the cuvette. The distortion of the beam evolves with time. The pattern, however, becomes steady after 220 s, as shown in Figure 5c (iv) with secondary circular rings. Although the macroscopic patterns in Figure 5c (i) and (iv) resemble each other, we assume that the haziness is due to the scattering of light from small particles—possibly micrometer-sized MPs. The patterns taken at 0 and 60 s show that the beam interacted with some particles, which we attribute to the MPs due to the low concentration of the organic matter present in the FEMP sample.

The amplitude of the distortion can be greatly influenced by the properties of the particles, such as the size, number, shape, and surface texture, interacting with the beam. For example, a relatively large particle compared to the beam diameter with some surface roughness can lead to a speckle pattern. In principle, smaller particles can also lead to edge diffraction and interference fringes for thin and long particles [35,36]. Considering the different and complex shapes and sizes of the MPs in the micrograph (Figure 1a), one would expect non-conventional light scattering, resulting in the complex combination of the above-stated optical phenomena, to be at play, with no particularly dominant phenomenon as observed in the 0 and 60 s patterns. In both cases of the transmittance (with time) and beam profile measurements, we observed the effect of the density of the MPs, as well as their relative positions during the sedimentation process with respect to the probing beam's position. In general, the sedimentation process of MPs in a natural aquatic environment can be complex, as the density of plastics is influenced by aging [37,38] and the flow of water, which is constantly changing. It is, therefore, evident that in the design of an optical sensor for the detection of MPs in their natural environments, such as rivers and lakes, where the situation in constantly changing, the density of the particles should be taken into consideration. This can be done by combining the measurements of the intensity and size of the speckle in addition to the contrast [39]. All variations are nonlinear, but they can be detected with optical sensors through a referencing scheme. The movement as well as the nature of the plastic can thus be determined.

3.5. Discussion of the Methods

Raman microscopy allowed us to determine the different plastic types present in the samples with high precision. The first step in our optical method(s) is to examine the optical medium of the sample with specular reflection after the filtration and suspension in ethanol to determine whether we can detect a change in the samples due to the sample preparation. This is confirmed by the results presented in Section 3.2. The results from the transmission spectroscopy of MPs in the liquid are in good agreement with those of the Raman spectroscopy, since most of the plastic types were optically identified by the former. Finally, the most important conclusion to be drawn is the influence of the shape of the MP particles on their sedimentation as a result of their complex light scattering. We clearly see from this set of samples that the time evolution of the speckles is a very good indicator of the plastic shape, which is convoluted with the density of the plastics, i.e., their types. These complementary observations justify the use of several optical methods.

4. Conclusions

Microplastics extracted from wastewater sludge and suspended in ethanol were investigated using Raman microscopy and transmission measurements, as well as novel laser-based techniques. It was demonstrated that combining Raman and transmission techniques is a promising tool in the quantitative identification and validation of common microplastics, such as polypropylene (PP), polyethylene terephthalate (PET), polystyrene (PS), and polyethylene (PE), that are present in processed wastewater sludge samples. Moreover, the results also show that basic laser-based measurement can be a useful technique for qualitatively studying the dynamic properties of microplastics based on their complex light-scattering behavior. Additionally, the results indicate that the density of MPs plays a vital role in the dynamic properties and can affect their detection in natural environments. This study represents a step towards the development of a multifunctional optical sensor for the detection of microplastics and their dynamic properties of light scattering in natural environments based on their spectral features and speckle metrology. Such a device could serve as a qualitative control in wastewater treatment processes by monitoring MPs after treatment. A lack of reliable methods for the detection of complex shapes, which are most common in wastewater, appears in current research on MPs. Most of the optical methods used so far are mainly based on Mie light scattering, which is valid only for spherical particles. Thus, the development of a multifunctional optical sensor that employs both spectroscopic and non-spectroscopic methods, as described in this work, could be useful in the monitoring of micro-or nanoplastics and their essential characteristics, such as complex shapes, sizes, and sedimentation in situ in real field conditions. Studies are ongoing in order to establish further correlations between the proposed techniques and conventional methods and to integrate different optical techniques for the development of a multifunctional optical sensor for the identification of MPs in aquatic environments.

Author Contributions: Conceptualization, J.V.K.K. and K.-E.P.; formal analysis, B.O.A., M.R. and K.-E.P. funding acquisition, M.R.; investigation, B.O.A., P.S., J.R. and K.R.; methodology, B.O.A., P.S., K.R., J.T. and A.K.K.; supervision, J.V.K.K., M.R. and K.-E.P.; writing–original draft, B.O.A. and P.S.; writing–review & editing, B.O.A., P.S., J.T., A.K.K., M.R. and K.-E.P. All authors have read and agreed to the published version of the manuscript

Funding: The work is part of the Academy of Finland Flagship Programme, Photonics Research and Innovation (PREIN), decision 320166.

Institutional Review Board Statement: Not applicable

Informed Consent Statement: Not applicable

Conflicts of Interest: The authors declare no conflict of interest.

References

1. Alexy, P.; Anklam, E.; Emans, T.; Furfari, A.; Galgani, F.; Hanke, G.; Koelmans, A.; Pant, R.; Saveyn, H.; Sokull Kluettgen, B. Managing the Analytical Challenges Related to Micro- and Nanoplastics in the Environment and Food: Filling the Knowledge Gaps. *Food Addit. Contam.-Part A Chem. Anal. Control. Expo. Risk Assess.* **2020**, *37*, 1–10. [CrossRef]
2. Imran, M.; Das, K.R.; Naik, M.M. Co-Selection of Multi-Antibiotic Resistance in Bacterial Pathogens in Metal and Microplastic Contaminated Environments: An Emerging Health Threat. *Chemosphere* **2019**, *215*, 846–857. [CrossRef]
3. Zhou, Q.; Zhang, H.; Fu, C.; Zhou, Y.; Dai, Z.; Li, Y.; Tu, C.; Luo, Y. The Distribution and Morphology of Microplastics in Coastal Soils Adjacent to the Bohai Sea and the Yellow Sea. *Geoderma* **2018**, *322*, 201–208. [CrossRef]
4. Du, C.; Liang, H.; Li, Z.; Gong, J. Pollution Characteristics of Microplastics in Soils in Southeastern Suburbs of Baoding City, China. *Int. J. Environ. Res. Public Health* **2020**, *17*, 845. [CrossRef]
5. Zhang, Y.; Kang, S.; Allen, S.; Allen, D.; Gao, T.; Sillanpää, M. Atmospheric Microplastics: A Review on Current Status and Perspectives. *Earth-Science Rev.* **2020**, *203*, 103118. [CrossRef]
6. Cole, M.; Lindeque, P.; Fileman, E.; Halsband, C.; Goodhead, R.; Moger, J.; Galloway, T.S. Microplastic Ingestion by Zooplankton. *Environ. Sci. Technol.* **2013**, *47*, 6646–6655. [CrossRef] [PubMed]
7. Bordos, G.; Urbanyi, B.; Micsinai, A.; Balazs, K.; Palotai, Z.; Szabo, I.; Hantosi, Z.; Szoboszlay, S. Identification of Microplastics in Fish Ponds and Natural Freshwater Environments of the Carpathian Basin, Europe. *Chemosphere* **2019**, *216*, 110–116. [CrossRef]

8. Yu, Q.; Hu, X.; Yang, B.; Zhang, G.; Wang, J.; Ling, W. Distribution, Abundance and Risks of Microplastics in the Environment. *Chemosphere* **2020**, *249*, 126059. [CrossRef]
9. Klein, S.; Dimzon, I.K.; Eubeler, J.; Knepper, T.P. Analysis, Occurrence, and Degradation of Microplastics in the Aqueous Environment. In *Freshwater Microplastics: Emerging Environmental Contaminants?* Wagner, M., Lambert, S., Eds.; Springer International Publishing: Cham, Switzerland, 2018; pp. 51–67. [CrossRef]
10. Asamoah, B.O.; Uurasjärvi, E.; Räty, J.; Koistinen, A.; Roussey, M. Towards the Development of Portable and In Situ Optical Devices for Detection of Micro and Nanoplastics in Water: A Review on the Current Status. *Polymers (Basel)* **2021**, *13*, 1–30. [CrossRef] [PubMed]
11. Mintenig, S.M.; Int-Veen, I.; Löder, M.G.J.; Primpke, S.; Gerdts, G. Identification of Microplastic in Effluents of Waste Water Treatment Plants Using Focal Plane Array-Based Micro-Fourier-Transform Infrared Imaging. *Water Res.* **2017**, *108*, 365–372. [CrossRef]
12. Araujo, C.F.; Nolasco, M.M.; Ribeiro, A.M.P.; Ribeiro-Claro, P.J.A. Identification of Microplastics Using Raman Spectroscopy: Latest Developments and Future Prospects. *Water Res.* **2018**, *142*, 426–440. [CrossRef]
13. Sun, J.; Dai, X.; Wang, Q.; van Loosdrecht, M.C.M.; Ni, B.J. Microplastics in Wastewater Treatment Plants: Detection, Occurrence and Removal. *Water Res.* **2019**, *152*, 21–37. [CrossRef]
14. Garaba, S.P.; Dierssen, H.M. An Airborne Remote Sensing Case Study of Synthetic Hydrocarbon Detection Using Short Wave Infrared Absorption Features Identified from Marine-Harvested Macro- and Microplastics. *Remote Sens. Environ.* **2018**, *205*, 224–235. [CrossRef]
15. Enfrin, M.; Dumée, L.F.; Lee, J. Nano/Microplastics in Water and Wastewater Treatment Processes–Origin, Impact and Potential Solutions. *Water Res.* **2019**, *161*, 621–638. [CrossRef] [PubMed]
16. Sanjuan-Reyes, S.; Gómez-Oliván, L.M.; Islas-Flores, H. COVID-19 in the Environment. *Chemosphere* **2021**, *263*, 127973. [CrossRef] [PubMed]
17. Löder, M.G.J.; Imhof, H.K.; Ladehoff, M.; Löschel, L.A.; Lorenz, C.; Mintenig, S.; Piehl, S.; Primpke, S.; Schrank, I.; Laforsch, C.; et al. Enzymatic Purification of Microplastics in Environmental Samples. *Environ. Sci. Technol.* **2017**, *51*, 14283–14292. [CrossRef]
18. Salmi, P.; Ryymin, K.; Karjalainen, A.K.; Mikola, A.; Uurasjärvi, E.; Talvitie, J. Particle Balance and Return Loops for Microplastics in a Tertiary-Level Wastewater Treatment Plant. *Archive* **2020**. [CrossRef]
19. Asamoah, B.O.; Kanyathare, B.; Roussey, M.; Peiponen, K.E. A Prototype of a Portable Optical Sensor for the Detection of Transparent and Translucent Microplastics in Freshwater. *Chemosphere* **2019**, *231*, 161–167. [CrossRef] [PubMed]
20. Asamoah, B.O.; Roussey, M.; Peiponen, K. On Optical Sensing of Surface Roughness of Flat and Curved Microplastics in Water. *Chemosphere* **2020**, *254*, 126789. [CrossRef]
21. Asamoah, B.; Amoani, J.; Roussey, M.; Peiponen, K.-E. Laser Beam Scattering for the Detection of Flat, Curved, Smooth, and Rough Microplastics in Water. *Opt. Rev.* **2020**, *27*, 217–224. [CrossRef]
22. Asamoah, B.O.; Kanyathare, B.; Peiponen, K.-E. Table Model and Portable Optical Sensors for the Monitoring of Time-Dependent Liquid Spreading over Rough Surfaces. *J. Eur. Opt. Soc.* **2018**, *14*. [CrossRef]
23. Erf, R.K. *Speckle Metrology*; Erf, R.K., Ed.; Academic Press: Cambridge, MA, USA, 1978; Vol. 1, ISBN 0-12-241360-1. [CrossRef]
24. Fujii, H.; Asakura, T.; Shindo, Y. Measurements of Surface Roughness Properties by Means of Laser Speckle Techniques. *Opt. Commun.* **1976**, *16*. [CrossRef]
25. Kim, J.W.; Jang, H.; Kim, G.H.; Jun, S.W.; Kim, C.-S. Multi-Spectral Laser Speckle Contrast Images Using a Wavelength-Swept Laser. *J. Biomed. Opt.* **2019**, *24*, 1. [CrossRef]
26. Peiponen, K.E.; Räty, J.; Ishaq, U.; Pélisset, S.; Ali, R. Outlook on Optical Identification of Micro- and Nanoplastics in Aquatic Environments. *Chemosphere* **2019**, *214*, 424–429. [CrossRef]
27. Kanyathare, B.; Asamoah, B.O.; Ishaq, U.; Amoani, J. Optical Transmission Spectra Study in Visible and Near-Infrared Spectral Range for Identification of Rough Transparent Plastics in Aquatic Environments. *Chemosphere* **2020**, *248*, 126071. [CrossRef] [PubMed]
28. Vázquez-Guardado, A.; Money, M.; McKinney, N.; Chanda, D. Multi-Spectral Infrared Spectroscopy for Robust Plastic Identification. *Appl. Opt.* **2015**, *54*, 7396. [CrossRef] [PubMed]
29. Zhang, X.; Chen, J.; Li, J. The Removal of Microplastics in the Wastewater Treatment Process and Their Potential Impact on Anaerobic Digestion Due to Pollutants Association. *Chemosphere* **2020**, *251*, 126360. [CrossRef] [PubMed]
30. Yu, Y.S.; Sun, L.; Huang, X.; Zhou, J.Z. Evaporation of Ethanol/Water Mixture Droplets on a Pillar-like PDMS Surface. *Colloids Surfaces A Physicochem. Eng. Asp.* **2019**, *574*, 215–220. [CrossRef]
31. Huth-Fehre, T.; Feldhoff, R.; Kantimm, T.; Quick, L.; Winter, F.; Cammann, K.; van den Broek, W.; Wienke, D.; Melssen, W.; Buydens, L. NIR-Remote Sensing and Artificial Neural Networks for Rapid Identification of Post Consumer Plastics. *J. Mol. Struct.* **1995**, *348*, 143–146. [CrossRef]
32. Durbin, S.D.; Arakelian, S.M.; Shen, Y.R. Laser-Induced Diffraction Rings from a Nematic- Liquid-Crystal Film. *Opt. Lett.* **1981**, *6*, 411–413. [CrossRef]
33. Kim, Y.H.; Park, S.J.; Jeon, S.-W.; Ju, S.; Park, C.-S.; Han, W.-T.; Lee, B.H. Thermo-Optic Coefficient Measurement of Liquids Based on Simultaneous Temperature and Refractive Index Sensing Capability of a Two-Mode Fiber Interferometric Probe. *Opt. Express* **2012**, *20*, 23744. [CrossRef]

34. Arnaud, N.; Georges, J. Investigation of the Thermal Lens Effect in Water-Ethanol Mixtures: Composition Dependence of the Refractive Index Gradient, the Enhancement Factor and the Soret Effect. *Spectrochim. Acta-Part. A Mol. Biomol. Spectrosc.* **2001**, *57*, 1295–1301. [CrossRef]
35. Shen, J.; Jia, X. Diffraction of a Plane Wave by an Infinitely Long Circular Cylinder or a Sphere: Solution from Mie Theory. *Appl. Opt.* **2013**, *52*, 5707–5712. [CrossRef]
36. Tavassoly, M.T.; Hosseini, S.R.; Fard, A.M.; Naraghi, R.R. Applications of Fresnel Diffraction from the Edge of a Transparent Plate in Transmission. *Appl. Opt.* **2012**, *51*, 7170. [CrossRef] [PubMed]
37. Arvidsson, A.; White, J.R. Densification of Polystyrene under Ultraviolet Irradiation. *J. Mater. Sci. Lett.* **2001**, *20*, 2089–2090. [CrossRef]
38. Iacopi, A.V.; White, J.R. Residual Stress, Aging, and Fatigue Fracture in Injection Molded Glassy Polymers, I. Polystyrene. *J. Appl. Polym. Sci.* **1987**, *33*, 577–606. [CrossRef]
39. Chicea, D. Speckle Size, Intensity and Contrast Measurement Application in Micron-Size Particle Concentration Assessment. *Eur. Phys. J. Appl. Phys.* **2007**, *40*, 305–310. [CrossRef]

Review

Towards the Development of Portable and In Situ Optical Devices for Detection of Micro-and Nanoplastics in Water: A Review on the Current Status

Benjamin O. Asamoah [1,*], Emilia Uurasjärvi [2], Jukka Räty [3], Arto Koistinen [2], Matthieu Roussey [1] and Kai-Erik Peiponen [1]

1. Department of Physics and Mathematics, University of Eastern Finland, P.O. Box 111, FI-80101 Joensuu, Finland; matthieu.roussey@uef.fi (M.R); kai.peiponen@uef.fi (K.-E.P.)
2. SIB Labs, University of Eastern Finland, P.O. Box 1627, 70211 Kuopio, Finland; emilia.uurasjarvi@uef.fi (E.U.); arto.koistinen@uef.fi (A.K.)
3. MITY, University of Oulu, Technology Park, P.O. Box 127, FI-87400 Kajaani, Finland; jukka.raty@oulu.fi
* Correspondence: benjamin.asamoah@uef.fi

Citation: Asamoah, B.O.; Uurasjärvi, E.; Räty, J.; Koistinen, A.; Roussey, M.; Peiponen, K.-E. Towards the Development of Portable and In Situ Optical Devices for Detection of Micro-and Nanoplastics in Water: A Review on the Current Status. *Polymers* **2021**, *13*, 730. https://doi.org/10.3390/polym13050730

Academic Editor: Jacopo La Nasa

Received: 9 February 2021
Accepted: 23 February 2021
Published: 27 February 2021

Publisher's Note: MDPI stays neutral with regard to jurisdictional claims in published maps and institutional affiliations.

Copyright: © 2021 by the authors. Licensee MDPI, Basel, Switzerland. This article is an open access article distributed under the terms and conditions of the Creative Commons Attribution (CC BY) license (https://creativecommons.org/licenses/by/4.0/).

Abstract: The prevalent nature of micro and nanoplastics (MP/NPs) on environmental pollution and health-related issues has led to the development of various methods, usually based on Fourier-transform infrared (FTIR) and Raman spectroscopies, for their detection. Unfortunately, most of the developed techniques are laboratory-based with little focus on in situ detection of MPs. In this review, we aim to give an up-to-date report on the different optical measurement methods that have been exploited in the screening of MPs isolated from their natural environments, such as water. The progress and the potential of portable optical sensors for field studies of MPs are described, including remote sensing methods. We also propose other optical methods to be considered for the development of potential in situ integrated optical devices for continuous detection of MPs and NPs. Integrated optical solutions are especially necessary for the development of robust portable and in situ optical sensors for the quantitative detection and classification of water-based MPs.

Keywords: micro and nanoplastics; freshwater; sludge; optical detection; portable devices; in situ detection

1. Introduction

During the past decade, the concern with plastic litter has become a hot topic, both in science and in everyday life [1]. The alarming information on the fragmentation of plastics into smaller particles, e.g., microplastics (MPs; size range of 1 µm–5 mm) and nanoplastics (NPs; 1 nm–100 nm), is a current issue of debate because they can have adverse environmental and health effects [2,3]. Ongoing research [4] on MPs and NPs tries to understand the properties of these particles in the aquatic environment [5,6]. However, one thing is sure: plastic litter in aquatic environments is decaying into smaller particles [7] due to factors such as ultraviolet (UV) light from sunshine and mechanical abrasion [8–10] of both floating and sunken plastic litter. Mechanical abrasion can result from the interaction of plastics with other solid media, such as sand, with water, and wind.

Effort has been put into recycling metals, glass, cardboard, and paper, especially in developed countries. This increase in recycling is due to compliance with the related strict regulations and policies. Unfortunately, due to the current lack of fully efficient suggestions or protocols for managing and recycling, many different kinds of human-induced plastic litter remain untreated. These plastics disperse into natural water bodies, soil, and the atmosphere [11–13]. The presence of plastics in water bodies contributes to raising their toxicity levels. For example, recycled food packing and preservation plastics, although reducing plastic waste, are a potential source for increased toxic chemical additives, which can be released into food or water bodies [14]. The direct and indirect impact of MPs on the environment is now clearly acknowledged. For instance, the interdependence of climate change and microplastics has been recently demonstrated [15]. In reference to climate change, one can talk about the changes in water bodies as a result of the time-dependent increase of MPs and NPs pollution and resuspension, and their cyclical effect on the climate.

MPs are usually characterized in the laboratory rather than in the field. After sampling, they are typically separated from the sample matrix, which can contain other organic and inorganic matter. Then, they are characterized using various types of sophisticated chemical analysis methods [16]. The characterization aims to obtain information not only on the plastic type but also on the particle sizes and concentrations of MPs in environmental samples [17].

Currently, in situ detection and quantification of MPs is difficult or even impossible, because of a lack of applicable methods. Some environments, such as wastewater treatment plants and water-related industrial processes, contain high amounts of organic and inorganic solids, making in situ detection of less abundant MPs complicated compared with other materials, a challenge without sample pretreatments. Moreover, factors such as temperature and pressure variations in natural water bodies and (frozen) lakes complicate the problem. The variations in these parameters impact the local conditions at different heights (e.g., frozen top and unfrozen bottom and top layers) in the water bodies, which consequently affects differently the properties of the local MPs.

Among the different detection methods to meet the harsh and varying measurement conditions, photonics-based sensor solutions are promising in achieving the desired real-time data acquisition from MPs in situ. Demonstrating its potential, many of the existing MP identification methods in the laboratory already utilize photonics. Methods based on Fourier transform infrared spectroscopy (FTIR) and Raman spectroscopy [18–20] are examples. The advantages of photonics are based on the feasibility of non-destructive, label-free, real-time, robust, and inexpensive sensors. Photonics-based methods enable the determination of intrinsic and extrinsic properties of materials. For example, surface roughness, curvature, and transparency of plastics [21–25] can be determined simultaneously with its chemical compositions. These advantages make photonics-based solutions compatible with the detection and discrimination of very complex microplastic types originating from common synthetic plastics such as polyethylene (PE), polypropylene (PP), polyethylene terephthalate (PET), etc. MPs originating from tire wear, however, are hardly identifiable by popular photonics-based methods without further treatment of the samples [26]. Although independent photonics-based solutions, as demonstrated, present a powerful tool for the fight against MPs detection and analysis, their potential as an integrated solution of the various techniques for a robust and portable device for in situ detection of MPs in aquatic environments are yet to be explored due to the complexity of the problem, ipso facto, to be addressed.

In this review, we first briefly consider the physical phenomena behind light and plastic interactions, and the current status of photonics-based spectroscopic methods of MPs as well as their advantages and disadvantages. In particular, we consider the working principles behind the two widely used techniques—FTIR and Raman spectroscopies—to aid the understanding of non-photonics experts. We also consider possible portable optical devices with the potential to detect MPs in field conditions. Moreover, we further propose the development of photonics-based integrated sensors for in situ detection of problematic NPs in water. Finally, we provide hints on how integrated optical systems can be used to enable the development of portable devices for the detection of NPs in the natural complex environment. From these perspectives, we point out the gap between the many current laboratory-based MP- and NP-related studies and the requirements for in situ detection. We also propose and indicate the need for an integrated photonics-based solution towards the development of portable and in situ optical sensors in the fight against aquatic-based MPs. Although multifunctional photonics-based solutions, as proposed, are unavoidable in the tackling of MP identification, we note that specific applications such as in aquaculture will require further considerations in the combination of specific and applicable photonics solutions.

2. Important Optical Properties of Microplastics

Plastic is a general term for the description of synthetic and semi-synthetic materials that are produced from raw materials such as crude oil, coal, cellulose, etc., and are moldable into desired shapes and sizes. Due to their versatility, plastics have a wide range of applications in many areas including packaging, automobile, healthcare, and energy. Plastics can either be classified as thermoplastic—softens upon heating—or thermoset—solidifies when heated—in their response to heat. Our focus is on common and abundant MP and NP pollutants such as PET and PP, which are thermoplastics. These microplastics, with size < 5 mm, can be produced in small sizes (primary sources) or originate from the fragmentation of larger plastics (secondary sources).

For specified applications, plastic materials can be transparent, opaque, or colored. Similar to other materials, when plastics interact with electromagnetic (EM) radiation (light waves), irrespective of the incident wavelength of the light wave, some optical phenomena can occur, as illustrated in Figure 1. These phenomena are reflection (Figure 1a), refraction, absorption (Figure 1e), transmission (Figure 1a), and interference of light (Figure 1b). Upon the interaction of the two—light waves and the plastic—the plastic particle affects the properties of the emitted light wave. For example, the polarization of the electric field of the light wave can be modified, which can be detected by interferometry. The strength of the phenomena, ipso facto, depends on the intrinsic optical properties of the plastic and the wavelength of the incident radiation. The intrinsic optical properties are described by the index of refraction and the absorption coefficient of the medium, which manifests in the optical spectra. Reflection and transmission of the light wave, for example, constitute measurement methods of optical spectra such as Fourier transform infrared (FTIR). Additionally, excitation of MPs with suitably higher energy light waves can lead to subsequent light wave emissions at longer wavelengths. On the other hand, specific light frequencies can also excite the vibrational modes of the molecular constituents of MPs, leading to scattered secondary photons with slightly shifted frequencies.

In addition to the intrinsic optical properties, the geometry of MP particles, when compared to the wavelength of the probing light wave, can influence its optical response (spectra). In such a case, in addition to the five optical phenomena mentioned above, scattering of the light wave can occur. Considering the simple case of spherical primary MPs, classical Mie scattering theory can be used to analyze them. In reality, however, secondary MPs are more typical in aquatic environments and may take complex shapes, including fibers, films, sponges, and fragments [27–29].

In such a case, there is no general theory to describe the scattering of a light wave from the complex-shaped MPs. However, MPs can also diffract light waves from the edges (Figure 1d), which may provide information on the complex structure of the MP. Moreover, although the shape of an MP may be complex, it will exhibit its intrinsic optical properties, such as the index of refraction. The exhibition of intrinsic optical properties also holds for an MP with, due to mechanical wear, a rough surface [24]. Surface roughness is another factor, in addition to the size of an MP, that can affect the strength of light wave scattering [24]. Interacting with a coherent light source, surface roughness leads to the formation of a grainy pattern, called speckles (Figure 1c), that results from the random phase arising from the variation in local surface heights [30]. Below, we also introduce, in the context of Raman spectroscopy, another type of scattering mechanism of the probe wave.

Regarding in situ detection, the natural environment of MPs can also impact their optical characteristics. For example, MPs may adsorb contaminants that can be organic, such as bacteria and viruses, or inorganic, like metals [31,32]. This additional micro-or nanofilm, forming eco-corona, modifies the effective optical properties of MPs. The effect of these contaminants can also be optically detected, however not in situ [33].

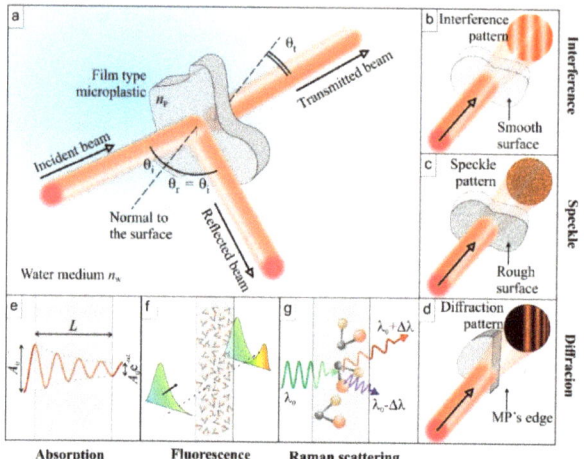

Figure 1. Illustration of some optical phenomena to be exploited for the identification of microplastic (MP) properties. (**a**) Overview of a standard experiment: the incident light is reflected, transmitted, and absorbed. (**b**) Interference: If the MP has two smooth surfaces, multiple reflections inside the film lead to an interferogram. (**c**) Speckle: Rough surfaces will generate a speckle originating from multiple interferences. (**d**) Diffraction: edges of the MP particles can diffract light beams to create organized patterns. (**e**) Absorption: the output amplitude is lower than the incident one. (**f**) Fluorescence: absorbed light energy is re-emitted at other wavelengths when molecules relax. (**g**) Raman scattering: specific light frequencies excite vibrational states of molecules, which lead to the emission of secondary photons at a slightly shifted frequency (Stokes and anti-Stokes).

In the case of aquatic-based NPs, plastics particles with at least one of the dimensions between 1 and 100 nm, other factors such as nano-roughness [34], and the properties of the surrounding matrix can affect the optical response. Unfortunately, current NP detection studies only focus on the commercial nanosphere NPs under controlled laboratory conditions, which differ significantly from the real environmental samples that can be polydispersed and polymorphic [35]. Indeed, for similar MPs harbored in freshwater and saltwater, the samples under the two conditions have shown different or similar optical responses under certain optical characterizations, indicating the need for complementary methods for the realistic detection of MPs and NPs in situ [33].

Similar to the geometry of the MPs, when NPs or their aggregates have a size much smaller than the probe wavelength of the light wave, Rayleigh scattering occurs, which is proportional to the inverse of the fourth power of the wavelength of the incident light wave, whereas Mie scattering dominates when the size is comparable or larger than the wavelength of the probe light wave.

Thus, optical spectroscopies in the frame of the linear optical process, such as reflection, absorption, and transmission of probe light waves, have poor sensitivity to the low number density of NPs and become erroneous in near-infrared (NIR) due to strong absorption of water. Additionally, for very small NPs, spontaneous scattering is very weak, and hence it is difficult to separate inelastically scattered laser light from intense Rayleigh scattering. Thus, efficient measurement methods, to be described in a later section of the paper, unlike the traditional FTIR and Raman spectroscopies, are necessary for better sensitivity and reliability for in situ detection of NPs in complex aquatic environments.

3. Spectroscopic Identification of Microplastics

Spectroscopy is the study of wavelength-dependent light–matter interactions, which has a long history in the characterization of materials, after its discovery. In this section, we examine the different spectroscopic methods used in the identification and characterization of MPs. Being the two most widely used spectroscopic techniques for MP identification, a somewhat detailed description of the process and device is given for Fourier-transform infrared and Raman techniques for readers who are new to the field.

3.1. Fourier Transform Infrared Spectroscopy

Fourier transform infrared spectroscopy (FTIR) is a vibrational spectroscopy, which identifies molecules based on how they absorb infrared (IR) light waves. In molecules, bonds between atoms vibrate in specific energy levels, and IR, interacting with the material, corresponding to this specific energy level are absorbed by the molecules. IR spectra, therefore, represent the absorbance as a function of the wavelength, typically expressed in wavenumbers (cm^{-1}) where the spectral peaks correspond to specific bonds or multi-bonds of the studied chemical structures. As different molecules absorb IR differently, due to the different bonds present, FTIR spectroscopy presents a very powerful method for characterizing known and unknown organic compounds when compared with the spectral library. With its long history of polymer identification, FTIR has consequently gained popularity as an efficient method to study MPs [36]. Different versions of the FTIR devices, namely the transmission, reflection (specular and diffuse), and the attenuated total reflection (ATR)—exponential decay of intensity of evanescent wave—with their corresponding limitations, exist for different applications. In addition to its capabilities, the choice of the FTIR device is largely dependent on the size of the MPs.

The most affordable and easy to use instrumentation is ATR with a benchtop FTIR spectrometer. It has been used for the characterization of MPs, especially when studying larger particle sizes (>1 mm) or verifying the accuracy of other analytical methods, such as light microscopy [37–39]. However, ATR measurements are laborious, requiring a larger particle size for ease of handling and sample pretreatment. Moreover, surface modification of samples can interfere with the quality of the obtained signal [40], perhaps limiting the percentage confidence in identifying environmental MPs from spectra libraries.

To detect MP sizes down to approximately 10 μm, a relatively convenient option is to couple a benchtop FTIR spectrometer with a microscope. The configuration employs one of two detectors, namely a single-pixel or so-called point detector, which measures one spectrum at a time [41–43], and an imaging focal plane array (FPA) detector consisting of a square array of pixels, where each pixel measures one spectrum at a time [44–46]. Practically, a point detector requires manual or automatic [47] pre-selection of particles to measure, whereas FPA is used for larger measurement areas. Although both options of imaging-FTIR allow for the identification of smaller MPs, the efficiency of the pretreatment or separation from the sample matrix becomes very important [16], as inefficient processing leads to increased measurement time in the point detector. On the contrary, the presence of other materials hinders the spectral identification in FPA imaging.

Imaging-FTIR can be operated in reflection or transmission mode [48]. Generally, the transmission mode is more suitable for smaller particles, as larger and thicker particles heavily absorb the incident radiation. Reflection mode, on the other hand, suffers from variations in morphologies [49]. Figure 2 shows examples of FTIR spectra measured with the FPA detector in reflection mode for some common MP samples [50] from a real environment separated from biota samples showing clear distinction among the spectra for PE, PP, and PET, especially at higher wavenumber.

Although FTIR methods provide quantitative information, the classification of the spectra obtained from environmental MPs can be challenging. FTIR-obtained spectral data is usually analyzed by comparing sample and reference spectra [51]. The reference spectra can be commercial, custom-made, or free plastic libraries, including spectra of virgin and/or weathered plastics and MPs [52]. As the sample spectra usually deviate from the characteristics of the collected reference spectra, human interference is needed for classification. This assistance can introduce some bias [53]. Whilst single spectra, despite being time-consuming [53], can be manually searched against spectral libraries, classifying FTIR images comprising millions of spectra is challenging. Therefore, (semi)automatic data analysis methods and software have been developed for quantifying MPs from imaging-FTIR data [54–57]. The Match rate of reference and sample spectra is usually calculated as Pearson's correlation coefficient, but also other algorithms such as hierarchical cluster analysis [51] and curve-fitting Python code [58,59] have been developed for MP research, significantly improving the classification. To improve the efficiency of the data analysis, spectra can be preprocessed to remove noise, correct baseline, and/or normalize the spectra [60].

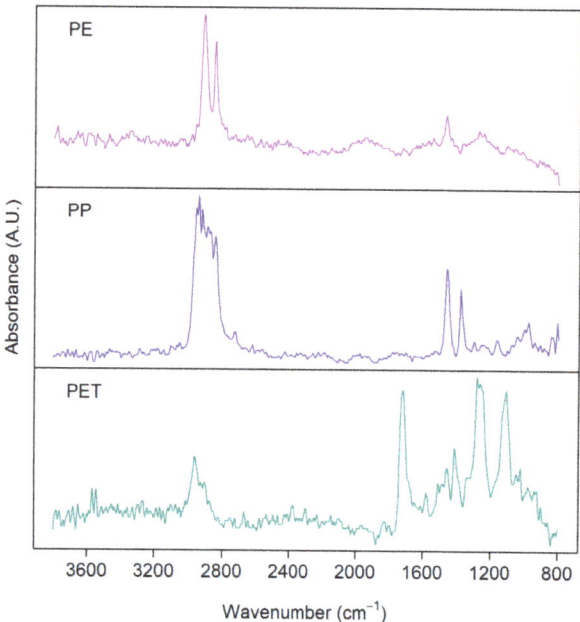

Figure 2. FTIR spectra of microplastics detected from environmental samples: polyethylene (PE), polypropylene (PP), polyethylene terephthalate (PET). Samples were measured with a focal plane array (FPA) detector in reflection mode from silver membrane filters.

Imaging-FTIR with an FPA detector has been proposed to be the standard method for quantifying small MPs from environmental samples [44]. Followed by automatic data analysis methods, it provides the counts, sizes, plastic types, and mass estimations of MPs in samples [54,57]. However, imaging-FTIR has disadvantages: the instrumentation is expensive and it requires special expertise. Additionally, spectral imaging of large areas is time-consuming and produces a large data set, which is also time-consuming to analyze. The major advantage of imaging-FTIR compared to point measurements, however, is the higher degree of automation, which reduces the laboriousness and subjectivity of the analysis.

Various factors affect the quality of FTIR spectra. As mentioned above, the measurement mode can be either reflection or transmission, which dictates the type of substrate to be used. For example, silver membrane filters [54], gold-coated filters [46], or microscope reflection slides [61] are used for reflection measurements, whereas zinc selenide windows [57], aluminum oxide (Anodisc) filters [62], or silicon filters [20] are also used for transmission measurements. In both transmission and reflection modes, the substrate materials should be non-absorbing in the desired wavelength range. The more these prerequisites are met, the higher the integrity of the obtained spectra. In addition to the effects of the substrate, the physical structure of MPs can cause interference and diffraction, as discussed in Chapter 2, which hinders the quality of FTIR spectra. FTIR spectra are usually presented as an average of multiple scans, and the signal-to-noise ratio can be increased by increasing the number of scans. However, for large area-imaging, a compromise between data quality and the single measurement time should be considered [44] to achieve optimum results. Lastly, FTIR techniques are challenged by colored samples, and the presence of water as both conditions leads to increased IR absorption [53].

3.2. Raman Spectroscopy

Raman spectroscopy is also vibrational spectroscopy, similar to FTIR. However, the physical phenomena behind the two non-destructive methods are different. For example, (FT)IR is sensitive to the variation in the dipole moment, whereas Raman is sensitive to the polarizability of the molecule. While FTIR measures the absorbance of light waves, Raman measures the scattering of the incident light waves in the visible light region. Scattering of light occurs when a photon hits a molecule and excites the bonds to the so-called virtual vibrational states. When the bond returns to a lower vibrational energy state, a photon is emitted, and the emission wavelength and intensity are measured. The energy and wavelength difference between absorbed and emitted light is presented as a Raman shift (cm^{-1}). Similar to the FTIR spectrum, the Raman spectrum has peaks in specific wavenumber positions corresponding to bonds in molecules. With a library of Raman spectra, one can identify molecules, such as polymers and other components of MPs [63]. In terms of sample preparation and data analysis, both FTIR and Raman measurements utilize the same methods [64] (see Section 3.1 for FTIR and data analysis).

A Raman spectrometer consists of a monochromatic laser, a grating filter for removing the unscattered Rayleigh radiation (scattered light wave with the same frequency as the incident), and a charge-coupled device (CCD) detector. Raman is inelastic scattering consisting of Stokes and anti-Stokes scattering depending on the initial state of the bond before excitation. Because the ground state (initial state for Stokes) is always more occupied than the first excited state (initial state for anti-Stokes), Stokes scattering is a more common phenomenon than anti-Stokes, and therefore more sensitive to measure. Figure 1g illustrates the Stokes shift and anti-Stokes shift related to the nonlinear optical scattering phenomenon.

For MP studies, Raman microscopy, a spectrometer coupled to a microscope [63], is a popular choice because it enables the characterization of small particles down to 1 µm, lower than the detection limit of FTIR [18], although sample pretreatment is difficult. Another advantage of Raman spectroscopy compared to FTIR is a wider spectral range, which reaches 50 cm^{-1}, enabling the identification of inorganic compounds. Moreover, Raman can provide information about additive compounds of plastics, such as colorants and fillers, in addition to polymer types [65]. Contrary to FTIR, Raman is not sensitive to water, which makes it a feasible option for in situ measurements in water environments. Additionally, contrary to FPA in FTIR, Raman microscopes lack such fast detectors. Therefore, Raman mapping is conducted by measuring areas point by point, and it is generally more time-consuming than FTIR [65].

As opposed to the absorbance in FTIR measurement, Raman scattering (Figure 1f) is rather very weak—only about one photon in a million scatters. Therefore, relatively high laser power is required for an optimum signal, which, on the other hand, can destroy sensitive samples such as very small MP particles. Moreover, the laser light source can induce fluorescence, which causes a strong background in the spectrum and overlaps with the Raman peaks, a problem not encountered in FTIR, complicating the interpretation of the measured data. Laser-induced fluorescence can be avoided using an excitation wavelength in the NIR region of the electromagnetic light wave [66]. Other methods have also been proposed for addressing the fluorescence issue, for example, subtraction of the fluorescence background after simultaneous excitation with two laser light sources with slightly different wavelength [67,68]. Moreover, since fluorescence is a delayed optical response compared to Raman scattering, a short-pulsed laser with gated detectors provides an alternative means to separately collect the suitable Raman signal [69]. In addition to the laser source, fluorescence masking of the Raman signal also arises due to the presence of inorganic and organic materials [18,70] and pigments [71], either limiting signal acquisition or influencing peak position. However, these sources of interference can somewhat be solved by suitable sample pretreatments [72], which may also impose degradation to the samples [73,74]. Additionally, automatic fluorescence correction and improved sensitivity of Raman detectors can address the issue [18].

Besides the fluorescence described above, the mere presence of non-fluorescing materials can also impact the quality of the Raman signal [75]. Although FTIR and Raman are limited by the credibility of the reference library—consisting of spectra from pristine samples—Raman is less sensitive to weathered MPs.

Figure 3 presents examples of Raman spectra measured from the same set of real environmental MP samples shown for the FTIR measurements above. The quality of spectra varies between particles, although they have been measured with the same settings. Consequently, the same measurement parameters, such as focus, objective magnification, laser wavelength, and measurement time, are not suitable for every particle [65]. Thus, an optimal signal-to-noise ratio would require a long time to obtain a representative amount of particles analyzed. As in FTIR, the choice of the measurement parameters is always a trade-off between spectral quality and measurement time.

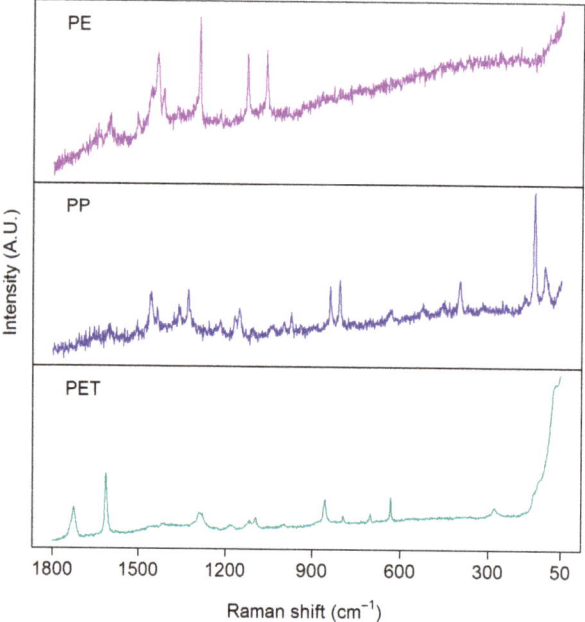

Figure 3. Examples of Raman spectra of microplastics measured from environmental samples: polyethylene (PE), polypropylene (PP), and polyethylene terephthalate (PET). Samples were measured with a 785 nm laser from aluminum oxide (Anodisc) filters.

Raman spectroscopy has also been coupled with other techniques for the identification of NPs. For instance, instead of coupling to microscopy, it is rather coupled to optical tweezers [76], a technique demonstrated by Gilbert et al. [19]. The idea is to tightly focus a laser beam to trap a particle in the region of the focal point of the beam. Essentially, the particle is trapped by the strong electric field at the focal point, a phenomenon that was nicely demonstrated by Arthur Ashkin for which he was granted a Physics Nobel Prize in 2018. The force exerted by the focused light depends on the refractive index of the MP or the NP to be detected. The nice feature of Raman tweezers is that submicron (20 μm down to 50 nm) plastic particles can be identified. Moreover, the method can discriminate between plastic, organic, and mineral particles [19]. Furthermore, thanks to the microscope, information on the size and shape of both MPs and NPs can also be obtained. However, the challenge regarding in situ detection using Raman tweezer is that the force for trapping of a particle is very weak; the typical motion of water in natural water bodies and wastewater filtration systems can disturb the trapping of the MPs.

A well-established measurement method to detect micrometer-sized samples is based on surface-enhanced Raman scattering (SERS) [77]. This technique is based on a nanostructured metal substrate. The metal substrates typically used are silver and gold. The SERS signal in the proximity of the metal is enhanced by several orders of magnitude compared with the conventional Raman signal. The enhancement is due to the so-called plasmonic behavior of metallic nanostructures. Lv et al. [78] successfully exploited SERS in the monitoring of MPs. In a way, this method also requires sample preparation because MPs must be in the proximity of the metal structure. Although the signal from Raman spectroscopy is weaker compared to that of SERS, it has also been used for the detection of MPs in simulated natural water environments [79].

Other Raman-related methods used in the detection of MPs are stimulated Raman scattering microscopy [80] and coherent anti-Stokes Raman scattering microscopy (CARS) [81,82]. CARS belongs to the category of the so-called third-order nonlinear optical phenomena [83]. The CARS spectrum can be detected using two lasers simultaneously—a high-power laser for pumping of the medium, and a wavelength-tunable low-power laser for probing the medium to detect characteristic Raman spectra of the scatterers. However, due to the requirements of expensive instrumentations and expertise, these methods have not gained as wide popularity in MP research as traditional Raman and FTIR spectroscopies.

3.3. Transmission Spectroscopy

In addition to FTIR and Raman, NIR transmission spectroscopy has also been proposed for the identification of MPs in water under laboratory conditions [84]. Transmission spectroscopy, from the sample point of view, works similarly to molecular absorption (Figure 1e) of NIR light waves, leading to spectra that are rich in overlapping overtones and their combinations. However, from the device point of view, transmission spectroscopy is based on the conventional spectrophotometer, which is much slower than the FTIR that is based on Michelson interferometry.

Although not a fully developed identification method for MPs, like in the case of the other two methods, NIR transmission of MPs provides other useful information on MPs. For example, one can monitor the surface roughness and its time-dependent change, which can serve as a measure for the residence time of MPs in aquatic environments [85]. Moreover, transparent, translucent, and MPs with surface roughness can be differentiated from one another in water by measuring their transmittance [24,84]. Figure 4a illustrates an example of transmittance of thin-film PET samples with different roughness in water [24], showing a nonlinear dependence with the average surface roughness [22]. Additionally, for different MP types in a complex matrix, similar identification and classification based on spectra libraries, as in the case of FTIR and Raman, can be obtained via peak attribution of the weighted spectra. Figure 4b illustrates the difference in transmittance (ΔT) spectra for pure ethanol and ethanol with different environmental MPs suspended in it (FEMPs), where some dips in the curve, namely 1158 nm, 1396 nm, and 1660 nm, have been attributed to the PS, PE, and PET, respectively, in comparison to the studies in [84,86]. Thus, the difference in transmittance allows immediate authentication of characteristic peaks of certain plastics in an extremely complex environment, which are otherwise absent in the conventional transmittance of FEMPs. In the case of in situ detection of arbitrary MPs or mesoplastics in water, analytical tools such as Kramers–Kronig light dispersion relation can be convenient for the identification of the plastic type and surface properties [25]. Thus, the existence of a transmission spectra library of MPs under varied and complex conditions can prove very useful for the identification of aquatic MPs using this method. In particular, the use of classification techniques such as PCA [87] could be beneficial to MP identification.

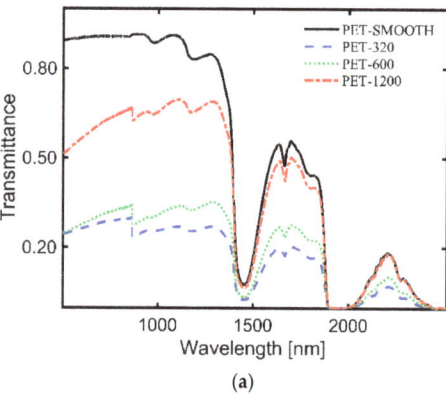
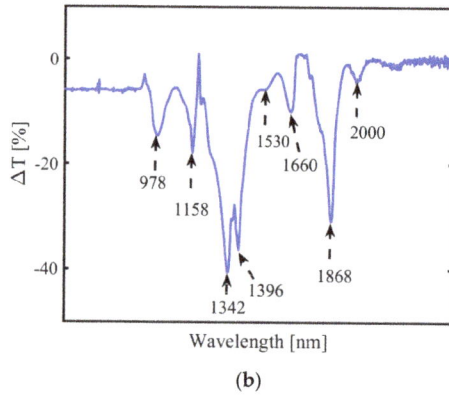

Figure 4. (a) Influence of surface roughness on the transmittance of film-type MPs, with a rough surface on both sides, in water [24]. MPs were prepared from polyethylene terephthalate (PET) plastic roughened with different sandpaper grits, 1200, 600, and 320, to achieve average surface roughness values of 0.34 µm, 0.60 µm, and 1.10 µm, respectively. The transmittance of the MPs nonlinearly decreases with increasing average surface roughness over the whole spectral range. (b) Difference in transmittance (ΔT) between the pure ethanol only and ethanol-containing MPs from a real sludge sample [submitted elsewhere]. It allows the immediate authentication of characteristic peaks of certain plastics in an extremely complex environment: polystyrene (PS, λ = 1158 nm), polyethylene (PE, λ = 1396 nm), and polyethylene terephthalate (PET, λ = 1660 nm).

Despite the potential of NIR transmission spectroscopy in aquatic MP detection, there exist some practical implementation challenges. In the NIR spectra range, water shows strong absorption of the light wave, which overshadows the spectral features of plastics present in the same spectral region. This effect of water absorption can, however, be avoided by decreasing the optical path length of the light radiation. Additionally, since some dense plastics such as polyoxymethylene (POM) have spectral features only in the mid-IR range [86], a much wider spectral range should be considered when using this method.

3.4. Fluorescence Spectroscopy Using Staining of Microplastics

Fluorescence spectroscopy (Figure 1f) is a very sensitive method to monitor the low concentration of typical organic materials. A medium that absorbs UV radiation may emit light at higher wavelengths, resulting in an emission spectrum that can be used for the identification of materials. Not all materials, however, show intrinsic fluorescence. Such materials can, therefore, be detected by labeling with a fluorescence tag.

When the other spectroscopic methods do not suffice, or visual counting preceding identification and characterization of MPs is challenging, fluorescence techniques are employed to enhance the process, as proposed by Andrady [88]. Such techniques, sometimes employing heat-assisted treatment, were successfully used to screen MPs by staining them with Nile Red [89–94]. It has, however, also been considered for NP identification [95]. Recently, Nel et al. [96] have also published a comprehensive study on sample preparation and identification of stained MPs to examine the detection limit when using Nile Red [96]. The use of Nile Red leads to improved selectivity of MPs compared to natural particles [93,97] and provides a relatively low-cost and faster identification [98]. Further, for weathered particles with irregularities, Nile Red fluorescent provides a relatively easier identification than with FTIR.

As with the other techniques, the detection of MPs by staining with Nile Red also has its challenges. For instance, there is the possibility of staining other natural particles, leading to unwanted emissions from these particles and the overestimating of MP concentrations [99]. To mitigate this effect, sample pretreatment and polymer-specific identification threshold should be considered [96]. Additionally, the effectiveness of Nile Red depends on its derivatives, concentration, and the type of solvent, and the excitation source and time [100] should also be considered. Red fluorescence of MPs, as seen in Figure 5 (right), contributes to the background staining, which complicates the actual MP identification, whereas green fluorescence is much more suitable [93,97]. Despite these challenges, a semi-automated method based on Nile Red tagging has been developed for the identification of smaller MPs [89,90], which, however, requires sample pretreatment.

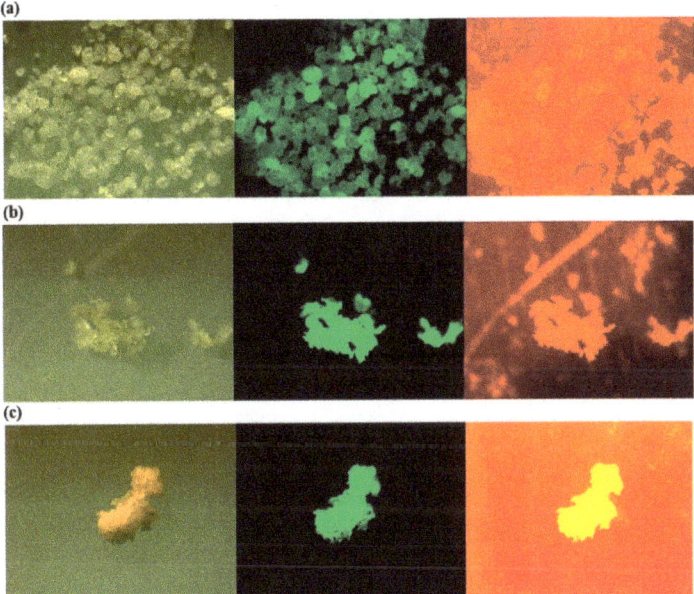

Figure 5. Photos under a microscope (left), a fluorescence microscope with excitation and emission wavelength of 534–558 and 515–565 nm (middle); and 534–558 and >590 nm (right) for the Nile Red stained low density (**a**) polyethylene (LDPE); (**b**) polypropylene (PP); (**c**) expanded polystyrene (EPS) Reprinted from *Marine Pollution Bulletin*, 113, Shim, W.J.; Song, Y.K.; Hong, S.H.; Jang, M., Identification and Quantification of Microplastics Using Nile Red Staining, 469–476. [93], Copyright (2016), with permission from Elsevier.

3.5. Hyperspectral Imaging of Microplastics

Hyperspectral imaging (HI) is another well-established technique that has been exploited in remote air-borne sensing of natural objects such as forests [101]. The goal is to obtain both image and spectral information from an object under study. Different scanning methods for HI exist, namely point, line, and area scanning of samples. However, the core idea is to produce a typical output probe beam in a form of a line. To obtain image information, the line is scanned over the object by either moving the device or the object itself. Similar to the FPA–FTIR method, the detecting camera records the spectrum from each pixel within the defined spectral range, however either in the visible or NIR spectral range. Due to the simultaneous collection of both image and spectral data, it results in a huge data file with some redundant information. Therefore, to obtain useful information, further processing is usually needed. Huge data processing techniques based on supervised and unsupervised algorithms such as principal component analysis (PCA), support vector machine (SVM) in combination with PCA, and partial least squares discriminant analysis (PLS-DA) with PCA, respectively, are some of the popular classification methods [102–104].

HI has been applied under both laboratory and field conditions. Figure 6 shows an image of a laboratory-based hyperspectral imaging system for MP identification. For field applications, HI has been used with unmanned aerial vehicles (UAV), e.g., drones, to monitor coastal MPs [105] and, in combination with photogrammetry, to map coastal transparent plastics such as bottles and bags, which are also primary sources of MPs [106]. The advantages and disadvantages of current MP identification methods, including hyperspectral imaging, have been described in a recent focused review [107]. The method is promising in the detection of reflection spectra from MPs and the provision of information on the size and surface area of a relatively large-size MP. It has also been proven to enable wide spectral range hyperspectral imaging of marine-harvested plastics. [108]. However, there are some issues of strong absorption by water, especially in the NIR range, when the MP is fully or partially covered in it. Additionally, the size of MP detection is currently limited to 100 µm [103] and, therefore, cannot detect NPs. Further, as with the FTIR, environmentally-modified samples show variations in recorded spectra, negatively influencing the accurate identification of MPs [102]. With regards to data processing, issues such as pre-processing of data are usually essential for efficient MP classification [109]. Lastly, since MP identification is based on reflectance spectra from HI, the absolute reflectance should be obtained using, for example, reference panels. However, this objective is hampered by changing the illumination conditions of the object arising from changing the cloudiness of the sky. Nonetheless, calibration issues of HI are well-described in [110].

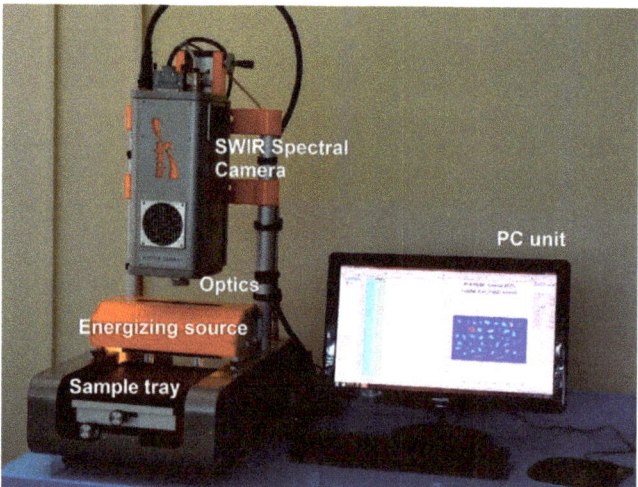

Figure 6. An overview of SISUChema XL™ Chemical Workstation (Specim, Finland) Reprinted from *Waste Management*, 76, Serranti, S.; Palmieri, R.; Bonifazi, G.; Cózar, A. Charac-terization of Microplastic Litter from Oceans by an Innovative Approach Based on Hyperspectral Imaging, 117–125. [104], Copyright (2020), with permission from Elsevier.

4. Handheld Optical Devices

Thanks to the development of small size light sources, detectors, and optical elements and the need for field measurements, various types of portable and handheld commercial spectrometers are available. Some of these miniaturized systems are suitable for environmental monitoring issues [69]. Optical sensors belonging to the category of commercial portable and/or handheld devices are based on FTIR, Raman, their combinations, and HI [69,111] methods (see Figure 7).

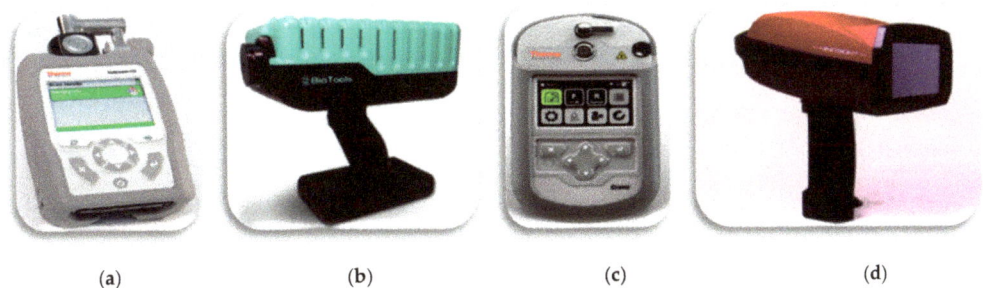

Figure 7. Portable devices: (**a**) FTIR—Thermo Fisher Scientific; (**b**) Raman—BioTools RamTest; (**c**) Gemini-combined Raman and FT-IR handheld analyzer—Thermo Fisher Scientific; (**d**) HI-BaySpec's GoldenEye. Reprinted from *Applied Spectroscopy*, 72, Crocombe, R.A. Portable Spectroscopy, 1701–1751. [69], Copyright (2018), with permission from SAGE Publications.

FTIR- and Raman-based methods have found applications in raw materials quality inspection in pharmacy [112]. In addition to quality inspection, handheld FTIR and Raman methods can be used for the screening of counterfeit pharmaceutical tablets. Typically, the blister material is also made of plastic, through which the tablets are probed. FTIR portable spectrometers are also used for homeland security to detect dangerous gases and for remote monitoring of pollution due to exhaust gases from vehicles, chimneys of factories, and gas emissions from the dump. Gradually, these portable devices are gaining popularity in the detection of MPs.

Recently, a handheld FTIR spectrometer was used in the detection and identification of four different types of meso- and micro-sized plastics that were harvested in the coastal area of Greenland [113]. It is also probable that handheld Raman devices and multi-function handheld sensors, where both FTIR and Raman are coupled together, can be exploited in the monitoring of MPs from the environment. However, the analysis of MPs in situ with portable FTIR/Raman spectrometers is challenging because MP concentrations in the environment are typically low. Additionally, samples for FTIR measurements require the separation of MPs from the matrix before measurement. Moreover, because water absorbs IR, FTIR is less suitable for in situ monitoring of MPs in water environments. The methods for separation and detection of MPs in the field in situ will require major developments. Despite these challenges, Iri et al. [114] developed an inexpensive portable Raman sensor for the detection of micrometer-sized plastic coated magnetic spherical particles in water in a quartz cuvette.

Recently, a prototype of a handheld optical sensor has been developed for the investigation of transparent and translucent a priori MPs (1–5 mm) [21–23]; see Figure 8. Unlike the more quantitative and matured spectroscopic devices like FTIR and Raman, this sensor monitors reflection, transmission, diffraction, and scattering of red laser light waves from MPs using a photodiode and a charge-coupled device (CCD) camera, without an imaging objective, as detectors. The sensor provides information on the smoothness and roughness, flatness/curviness, size, and sedimentation of MPs, parameters that are essential for in situ detection. Samples with surface roughness or volume inhomogeneity, due to porosity, scatter incident radiation of a coherent laser light source to form a speckle pattern on the CCD camera. Figure 9a,b show the interference and speckle patterns of transparent and smooth PET MPs and MPs with surface roughness, respectively, recorded with the handheld sensor. The speckles originate from the rough surface of the MP and the correlation of the magnitude of the surface roughness estimated from analyzing the grainy laser speckle pattern. Thus, the adherence of toxic and non-toxic materials, e.g., heavy metal compounds and organic materials [1,31,115,116], leading to surface modification of MPs can also be monitored. Below, we show some typical results of the device to indicate the scattering characteristics of MPs under different conditions.

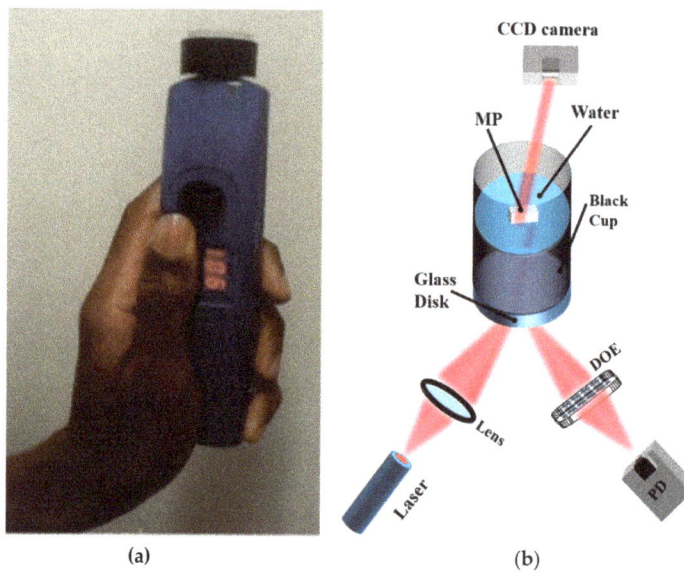

Figure 8. (**a**) Image of a handheld optical sensor for MP detection [117]; (**b**) optical measurement setup. The diffractive optical element (DOE) spatially filters the reflected light signal to obtain the specular component of the light signal on the photodiode (PD). The speckle pattern is measured by the CCD camera Reprinted from *Chemosphere*, 254, Asamoah, B.O.; Roussey, M.; Peiponen, K. On Optical Sensing of Surface Roughness of Flat and Curved Microplastics in Water, 126789. [23], Copyright (2020), with permission from Elsevier.

The speckle pattern can also be projected to monitor moving MPs in water [118]. Figure 9c shows such a result from the handheld sensor. It illustrates the speckle contrast C variation with time, with error bars, calculated from the recorded speckle pattern for 700 µL of pure ethanol and ethanol-containing filtrated MPs (FEMPs). The speckle pattern is first generated by the interaction of the coherent laser source and a rough glass at the volume compartment to probe the liquid volume. When the liquid containing the MPs is introduced to the optical sensor, the particles modify the surface roughness, which correspondingly changes the speckle pattern on the CCD camera. By analyzing the speckle pattern to obtain the speckle contrast [30], one can monitor the time-dependent sedimentation of the MPs in ethanol, which is different from that of pure ethanol.

Although not from the portable device, Figure 9d also shows an exponential growth of transmittance with time for the FEMPs using a spectrophotometer to record the transmittance at a fixed wavelength of 800 nm. It suggests an increase in sedimentation of the particles increases with time, at least for particles across the beam, decreasing the amount of the scattering of the probing beam, and thereby increasing the transmittance. Similar to the handheld measurements, the cuvette containing the FEMPs was shaken before the measurements and probed midway up the cuvette height.

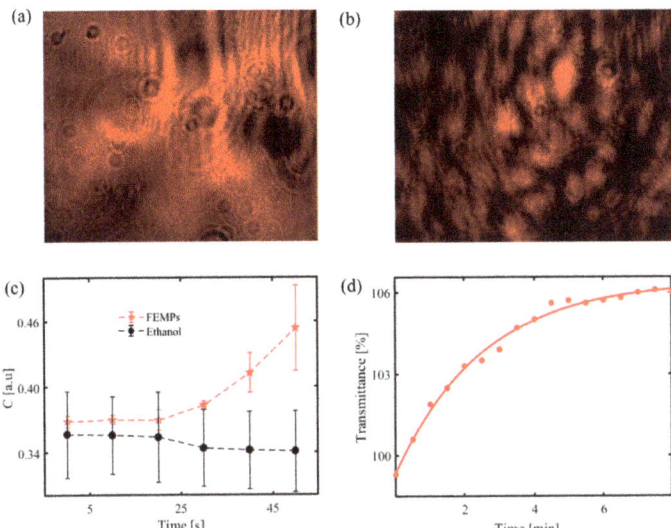

Figure 9. (**a**) Recorded interference pattern for transparent and smooth PET MP [21]; speckle pattern of similar MP with a rough surface [23]; (**b**) using the handheld optical sensor. Identification of MPs in ethanol; (**c**) time-dependent speckle contrast calculated for pure ethanol only and ethanol-containing MPs (FEMPs) on a rough glass disk using a handheld optical sensor; (**d**) time-dependent transmittance signal from ethanol containing MPs at a fixed wavelength (800 nm) using a spectrophotometer. Figure 9a Reprinted from Chemosphere, 231, Asamoah, B.O.; Kanyathare, B.; Roussey, M.; Peiponen, K.E. A Prototype of a Portable Optical Sensor for the Detection of Transparent and Translucent Microplastics in Freshwater, 161–167. [21], Copyright (2019), with permission from Elsevier; Figure 9b Reprinted from *Chemosphere*, 254, Asamoah, B.O.; Roussey, M.; Peiponen, K. On Optical Sensing of Surface Roughness of Flat and Curved Microplastics in Water, 126789. [23], Copyright (2020), with permission from Elsevier.

5. The Gap Between Current Detection and Preferred In Situ Detection

It is well acknowledged that MP and NP pollution is ubiquitous in our ecosystem, and a lot of effort, as described above, is therefore being channeled to address the issue from various fields. However, based on the several studies considered above, it appears that current identification methods are still limited to the laboratory, where sample type and size are a priori known, and measurement conditions are relatively stable. Thus far, only the HI technique in combination with UAV comes close to addressing the issue on the field or in situ. Even with this, it is still limited to the large plastics on the water surface [106], neglecting the in situ challenge. However, most abundant plastics in the environment (aquatic) are small (<500 μm) [89,119–121], requiring the development of more sophisticated techniques for in situ detection and the accurate estimation of MP/NP identification and concentration determination.

The reasons for the gap between current methodology and the required in situ detectors could be summarized as follows: rather than having dedicated solutions to address the issue, we are using conventional methods which are limited in many ways; as evident from MPs from the environment, these particles are usually different in morphology and further show different characteristics from their pristine counterparts. To have a perspective of the two reasons, let us briefly touch on the lifecycle of plastics in the aquatic environment.

It is evident that the plastics released into the environment (aquatic) are severely impacted by the habitat and therefore degrade further depending on their initial conditions. The various types of degradation are nicely summarized in [122]. The effects of these degradations are changes in the physical, optical, mechanical, chemical, etc. properties, leading to, for example, fragmentation, coloration, and variation in optical response [122]. Additionally, since MPs are interacting with other particles in the environment, it leads to the collection and accumulation of foreign materials, in which case the sample becomes heterogeneous. The consequence of these environmental interactions, further complicated by the harsh method of pretreatments, is the manifested variation in the optical response of environmental MPs to methods such as FTIR and HI.

Furthermore, the effect of the morphology and geometry of MPs on other non-spectroscopic techniques is also convoluted. In MP-related studies, information on the size/thickness, shape, refractive index, and concentration are also crucial; therefore, non-spectroscopic methods such as dynamic and multi-angle light scattering (DLS and MALs) [123,124], leading to speckle pattern formation, becomes essential. Such methods have been successfully demonstrated for mono-and poly-dispersive spherical particles under laboratory conditions. However, in situ, the presence of polydispersed MPs of different plastics types and other minerals and irregular organic materials can render this method unreliable. For example, Brownian motion of non-fiber-like MPs in water, and porous MPs with a high probability of multiple scattering, can complicate the scattered pattern, requiring rigorous analysis to extract useful information.

An additional limitation of the current identification techniques is the method of reference library creation and the classification methods. By far, most of the reference libraries are either largely based on pristine samples or are locally created. As we have seen, this approach poses a problem for the classification methods, limiting the accurate identification of MPs. Although fast and automated classifications are being developed in this regard, a consolidated library reference including spectra from different environments could prove beneficial to the integrity of MP identification and quantification, especially when it comes to in situ, paving way for the application of artificial intelligence.

Based on the summary above, it is therefore presupposed that to holistically develop optical sensors for in situ application, the above issues, in addition to the impact of a complex matrix such as water, should be considered, and the current optical methods need to be complemented. Fortunately, other solutions can be adapted. Thus, we consider some optical techniques that may be useful for the development of in situ detectors under the following categories:

1. general (Section 6) and
2. integrated solutions (Section 7).

6. Development of In Situ Detection Techniques for Screening of Microplastics and Nanoplastics

The best option regarding in situ real-time detection of MPs is to eliminate sample treatment. To achieve such a goal, there exist different possibilities. Although one cannot quantify the absolute results of these possibilities, their potential in the development of a sophisticated optical sensor for in situ application is unavoidable.

6.1. Fluorescence, Digital Holography, Dynamic Light Scattering, Optical Fibers, and Immersion Liquid Techniques

One promising option is based on intrinsic fluorescence or phosphorescence, i.e., photoluminescence of MPs, as described in Section 3.4. In such a case, the staining of MPs can be avoided. The potential challenge in in situ detection will be luminescence from organic materials, such as chlorophyll, that disturbs the luminescence from MPs similar to the case of Raman measurements. However, Ornik et al. [70] successfully demonstrated a photoluminescence detection technique (see Figure 10) using a high-power blue light-emitting laser to discriminate between plastic and non-plastic luminescence spectra. Additionally, the disturbance from organic materials may be handled by sophisticated chemometrics.

Moreover, MPs arriving in the aquatic environment may already have coloration due to color print on food and non-food plastic products based on the use of dyes and pigments [125]. The formulation of ink for plastic color print involves different types of resins, oils, and solvents such that ink may have spectral properties similar to those of plastics. Furthermore, each plastic usually requires its specific ink formulation, and thus provides a means to identify such colored MPs using possible fluorescence signals directly from the overlayer print. Moreover, the different spectral properties of the dyes and pigments of the prints, or the coloring agents of plastic without a print, can be exploited for the identification of the MPs.

Another type of optical method that can be used for the identification of both MPs and organic particles to yield information on the size and thickness of the particles could be based on digital holography (DH) [126]. In this method, one uses the imaging principle of holography based on a coherent laser light source. An interference pattern generated by the interaction of light radiation transmitted through an object is captured on a CCD camera together with a reference radiation signal. Thereafter, mathematical data analysis is used to reconstruct the image of the microparticle. Merola et al. [127] studied the identification of MPs and microalgae using digital holography, albeit in a Petri dish. Digital holography can be applied in relatively clear waters for in situ and real-time detection of particulates, including MPs. For example, this has been demonstrated for in situ detection of plankton [128], and an algorithm to analyze the arbitrary shapes of micro objects, including fiber-type, has been given in the literature of DH microscopy [129]. Similarly, toward practical applications, holography, coupled to Raman spectroscopy, has been demonstrated in large water flow volume, although for relatively large MP detection [130]; see Figure 11.

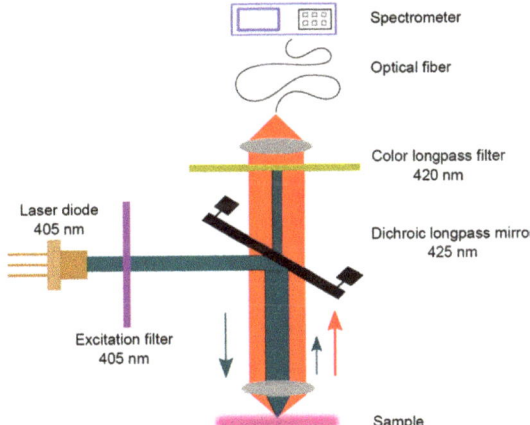

Figure 10. Scheme of the photoluminescence (PL) setup. The blue and red arrows indicate the propagation direction of laser and photoluminescence light, respectively Reprinted from Applied Physics B: Lasers and Optics, 126, Ornik, J.; Sommer, S.; Gies, S.; Weber, M.; Lott, C.; Balzer, J.C.; Koch, M. Could Photoluminescence Spectroscopy Be an Alternative Technique for the Detection of Microplastics? First Experiments Using a 405 Nm Laser for Exci-tation, 1–7. [70], Copyright (2019), with permission from Springer.

While conventional microscopy exploits 2D imaging of the MPs and requires the particle to be in the focal plane of the imaging system, the probability of capturing a single MP in in situ detection is typically very low. On the contrary, DH provides 3D imaging holograms as a function of time and monitors particles out of the focal plane as well. This allows more flexibility because it is possible to monitor, in addition to the shapes of the MPs, their locations and velocities in a water volume, yielding information on the particle thickness and the continuously varying refractive index of the ambient medium [131]. The practical implementation of a DH system can be based on the use of a high-power laser diode as a point light source to create a divergent beam. Such a diverging source could replace the use of imaging lenses, hence simplifying the system. For fast particle tracking purposes, a pulsed laser source, having a pulse width of tens of nanoseconds, may be suitable as the light source.

In in situ measurements, one could also use single-mode optical fibers to guide both the object and reference radiations of the DH system into a measurement location of a suitable detector. The advantage of the single-mode fiber is its stability against mechanical disturbances such as movement in shallow waves and deep-water currents. Furthermore, the output ends will produce strongly diverging light beams and hence enable the monitoring of a large water volume. Already, the potential of DH in MP identification has again been demonstrated by combining it with Raman spectroscopy to identify the size and type of MPs in the laboratory [130].

Apart from using optical fibers in the digital holography, it can also be used for sensing applications or for guiding signals in shallow and deep water. For example, Kenny et al. detected phenols and gasoline in groundwater using laser-induced fluorescence along optical fibers [132]. This rather old technique can also be exploited for the detection of MPs in water. In particular, with the recent development of miniaturized high-power laser sources and sensitive detectors, fiber optic sensor units based on these devices may be convenient for underwater MP detection.

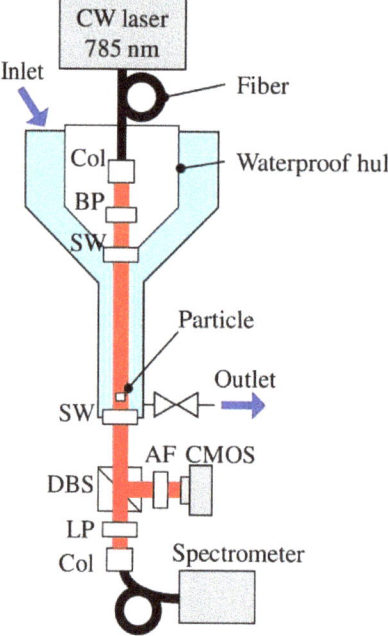

Figure 11. Experimental setup: Col, collimator; BP, bandpass filter; SW, sapphire window; DBS, dichroic beam splitter; AF, attenuation filter; and LP, long pass filter. Trapped particles are probed with holography and Raman techniques [130].

The immersion liquids method [133,134] can be exploited for in situ detection of MPs and NPs. By injecting suitable non-toxic water-soluble liquid into the detection compartment of underwater optical sensors, with controlled water flow, one can detect a change in transmittance both in clear and turbid waters. The transmittance changes as a function of the mismatch between the refractive index of the binary liquid mixture and that of different plastic particles. The nice feature of the immersion liquid method is its validity for the detection of both MPs and NPs with regular or irregular shapes. However, this method may be suitable as a complementary detection technique.

6.2. Hamaker's Constant, Interferometry, Effective Medium, Scattering, and Stoke's Parameters

Regarding the detection of small NPs, light scattering is negligible, and most of the methods described above are questionable. In the case single NP detection, for example, light scattering, in the Rayleigh domain, is weak for NPs of 1–10 nm in size. However, as NPs aggregate, resulting in a relatively large size, the probability of scattering is in the Mie domain and becomes higher; hence, it is easier to detect an aggregate by light scattering. The aggregation of the NPs—homogenous or heterogenous—is due to Van der Waals forces. Homogenous aggregation of NPs can lead to enhanced light scattering beneficial for their detection. Heterogenous aggregation, on the other hand, hinders the identification of the individual plastic types, especially when they show overlapping spectral features. The issue of NP aggregation is likely to be enhanced by the size of the probing source that can lead to a weighted spectrum from the aggregate.

Additionally, in an aquatic environment, one would expect NPs to be irregular, having nanoroughness and other surface coatings that further cause their optical properties to deviate from their macroscopic properties [34]. The effect of Van der Waal's forces causing aggregation can be predicted from UV–Vis–IR spectral properties of macroscopic plastics based on both wavelength-dependent refractive index and the absorption coefficient of the plastics. In nanoplastics aggregation, the so-called Hamaker constant plays an important role. For example, the Derjaguin–Landau–Verwey–Overbeek (DLVO) theory, which exploits the Hamaker's constant and depends on the plastic type of NP, the salinity, as well as pH of water, has been applied for the detection of spherical NPs in water and their aggregation [5,6]. Primarily, the Hamaker's constant can be obtained by using the Kramers–Kronig analysis method and a wide electromagnetic spectrum in the calculation of dielectric properties of the NP–water system. Readers interested in Hamaker's constant and progress in its calculation can find more information in the review article of Rosenholm et al. [135]. Additionally, the detection of sub-diffraction-limit MPs could also benefit from wide-field scanning correlation interferometric microscopy [136]. Indeed, on its own, the background-free interferometric detection method has been used to detect low-index particles in water down to 10 nm in radius [137]; coupled to other spectroscopic methods, this method may be a viable component of the integrated solution for in situ detection of NPs.

Further, to better understand the behavior of NPs in the complex environment, Bruggeman effective medium theory [138] can be exploited in the simulation of such an environment. Bruggeman's model predicts change of spectral fingerprints of the NP–water system as a function of the concentration of NPs in water volume, but under the assumption of negligible scattering of the probe wave. The method is valid for real nanoparticles such as NPs in water. The change in the spectral properties of such a dilute liquid–solid mixture can be effectively predicted by this method.

Further to the developed optical methods in Section 4, the use of speckle analysis is also promising in screening MPs in situ. Analysis of speckle intensity can provide information on the scattering medium such as MPs in water, yielding parameters including the aggregation of MPs [139] in salty water, thickness [140], and surface roughness [141]. Additionally, the change of polarization state of the probe laser light or other light sources can be determined by observing the so-called Stokes parameters [142]. It becomes thus a powerful method to obtain information from the optical properties of MPs. Implementing this method, however, requires typically the use of two light polarizers, one after the light source and the other before the detector, in transmission mode. In reflection mode, both polarizers will be placed on the same side in the paths of the incident and the reflected light waves. The speckle contrast calculation and the fixed-wavelength transmittance measurements can be integrated into a portable sensor to study the sedimentation of MPs in a controlled environment. Similarly, modifying the systems of dynamic and multi-angle light scattering methods to include linear polarizers could lead to a more sensitive device to detect rod-like or other non-spherical particles. Brownian motion can cause rod-like MPs, for example, to rotate, which can be observed in the dynamic change of the scattered light pattern after passing through the polarizer. In principle, such a system with a complex dynamic speckle pattern could be analyzed, exploiting on time-dependent cross-correlation analysis of the speckle pattern to resolve sizes of scattering NPs and MPs, as suggested in [143].

7. Towards Full Integration

Moving a step further in the integration of optical methods requires the development of lab-on-chip [144] or lab-on-fiber [145]. A simplified vision of waveguide sensing provides two options, since the waveguide can be used either as a probe or as a source. In the first case, the sensing or detection mechanism is often evanescent [146,147]. However, specific waveguide configurations may allow interaction between the analyte and the core propagating light, such as in the cases of slot [148], and anti-resonant reflecting optical waveguide (ARROW)-based sensors [149]. In the second case, light emerging from the waveguide can be used to image, illuminate, and interact with the surrounding medium. The two mechanisms are illustrated in Figure 12. An obvious issue with such devices is the typical size of the MPs and NPs to be detected. A waveguide with an output mode diameter ranging from 200 nm (silicon nano-waveguides [150]) to a few micrometers (glass or ceramic waveguides [151]) is common. Fibers can provide modes with very large diameters depending on the core-size. Although the purpose here is not to provide a detailed review of waveguide sensing, we state a few facts that can be used to detect plastic particles at nearly all scales.

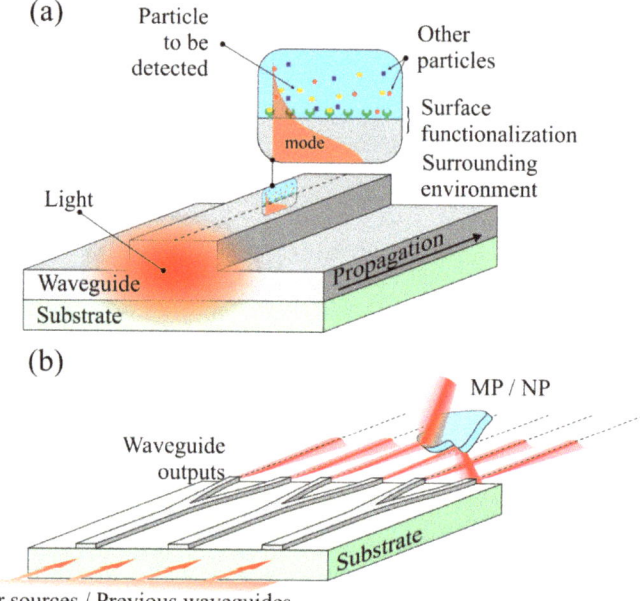

Figure 12. Waveguide-based sensors/detectors. (**a**) The top surface of a waveguide can be functionalized to attract specific molecules. The exponentially decaying portion of the propagating mode is interacting with the trapped particles, which yields a signal modification; (**b**) multiple channel waveguides illuminate a sample (which can be very large), which can reflect, deflect, or diffract the beams.

7.1. Evanescent Sensing

This type of sensing comes from the interaction of the exponentially decaying portion of a guided mode with the environment surrounding the waveguide (see Figure 11a). In such a case, the waveguide can be a channel, a slab, or a fiber. Numerous sensing schemes can be imagined from this configuration, and we refer the reader, for instance, to the following reviews [152–155]. The waveguide (core or cladding) can be chemically functionalized, i.e., labeled, to trap a specific group or molecule on the surface, which will modify the signal [156]. Instead of a particular chemistry, the surface can be patterned onto a nanostructure, leading, for instance, to fiber Bragg gratings that have demonstrated their power in sensing [157]. Such a sensing mechanism is often used for refractive index [158], fluorescence [159], and Raman/SERS [160] measurements. In the particular case of micro and nanoplastics, for which the concentration is extremely small, such a configuration is not suitable, although extreme sensitivity is possible.

7.2. Waveguide-to-Free Space Sensing

Channel waveguides on a chip offers numerous possibilities, especially if we consider a free-space output combined with an imaging lens system allowing the beams to propagate in free space in a controlled manner (focused, collimated). A simplified scheme is illustrated in Figure 11b. It is composed of a source (usually a coherent laser light) coupled to broad channel waveguides that are separated and ramified until they provide the desired amount of outputs. Each output may have a different wavelength or phase. In its simplest form, such a sensor is a Young interferometer, in which the two-point sources are the waveguide outputs from a Y-junction [161]. Arrayed waveguide grating (AWG) [162,163], light detecting and ranging (LiDAR) [164], and phased array (PHASAR) [165] are in constant development and are already in use for the monitoring of very large areas, such as forests. If LiDAR can be used for the detection and positioning in space, similarly to RADAR, the AWG system forms an excellent basis for spectrometry and is already extensively used in telecommunication systems for dense wavelength multiplexing. Fully integrated photonic circuits for LiDAR are used in automotives. Waveguide and fiber sensing are nowadays ubiquitous, since different physics, such as plasmonics [166], for instance, can be merged to raise the sensitivity of the device.

7.3. Waveguide-to-Free Space Sensing

Despite all these tools, so far no application of integrated optics has been devoted to the monitoring, detection, screening, or analysis of MPs and NPs. In the near future, it seems reasonable to expect to see the development of spectrometers, diffractometers, and speckle analyzers combined with artificial intelligence (AI) schemes for data analysis and processing for the detection in situ of micro and nanoplastics. Several key advances are paving the way. It has been recently demonstrated that fluorescence signals from microparticles can be detected even in mixtures [167]. Optical waveguides are ideal devices to be integrated with microfluidic channels, leading to a device capable of detecting antigens, DNA, µRNA, and so on [168,169]. Low power consumption, low weight, mass production compatibility, multi-physics, and chemistry compatibility are some advantages of waveguides over free-space solutions. The price to pay is the extremely challenging fabrication and high sensitivity to fabrication limitations that often make the devices not fully reliable.

In summary, to obtain development of future multi-purpose in situ optical solutions for comprehensive identification and quantification of MPs and NPs beneath water surfaces, a clever way of integrating miniature versions of FTIR, Raman, fluorometer, conventional spectrophotometer, and the novel methods described in this section seem viable options. Based on the similarities between conventional plastics and bioplastics [170], we believe that the methods described above are applicable to MPs and NPs originating from bioplastics. We note that for in situ applications, there are possible technical issues, such as contamination of optical elements of the probing window of the proposed sensor, that can arise. Nevertheless, this problem can be solved by exploiting micro and nanopatterning and ultrasound cleaning techniques. Smart sensors connected to AI for data processing based on integrated optics can be integrated into a water filtration membrane to yield real-time information on the presence of MPs and NPs captured by the membrane.

8. Conclusions

In this review, we have described a variety of different types of useful conventional but also novel optical methods to detect micro-and nanoplastics sampled from aquatic environments at the water surface, beneath the water surface, or embedded within it. The advantages and disadvantages of the individual traditional photonics-based solutions such as FTIR, Raman, and hyperspectral methods and their application to real microplastics obtained from the municipal wastewater system were also discussed. It is evident that FTIR and Raman techniques play a key role in the identification of MPs harvested from the environment. Portable and remote sensing devices based on FTIR, Raman, speckle metrology, and hyperspectral imaging and their potential for MP detection were also highlighted. Furthermore, novel and emerging optical methods such as digital holography, photoluminescence, and optical waveguides were proposed as useful and promising future techniques for in situ identification of micro and nanoplastics in complex aquatic environments. We conclude that the development of a reliable and multifunctional device for the accurate detection and monitoring of MPs/NPs in situ certainly could benefit from the intelligent integration of photonics-based solutions. Indeed, such a study is ongoing where some of the proposed novel methods are coupled together for the identification of MPs in water.

Funding: This work is part of the Academy of Finland Flagship Programme, Photonics Research and Innovation (PREIN), decision 320166.

Institutional Review Board Statement: Not applicable.

Informed Consent Statement: Not applicable.

Data Availability Statement: The data presented in this study are available on request from the corresponding author.

Acknowledgments: We thank Kalle Ryymin for the Raman spectra.

Conflicts of Interest: The authors declare no conflict of interest.

References

1. Campanale, C.; Massarelli, C.; Savino, I.; Locaputo, V.; Uricchio, V.F. A Detailed Review Study on Potential Effects of Microplastics and Additives of Concern on Human Health. *Int. J. Environ. Res. Public Health* **2020**, *17*, 1212. [CrossRef]
2. Wang, J.; Peng, J.; Tan, Z.; Gao, Y.; Zhan, Z.; Chen, Q.; Cai, L. Microplastics in the Surface Sediments from the Beijiang River Littoral Zone: Composition, Abundance, Surface Textures and Interaction with Heavy Metals. *Chemosphere* **2017**, *171*, 248–258. [CrossRef]
3. Gong, J.; Xie, P. Research Progress in Sources, Analytical Methods, Eco-Environmental Effects, and Control Measures of Microplastics. *Chemosphere* **2020**, *254*, 126790. [CrossRef]
4. Derraik, J.G.B. The Pollution of the Marine Environment by Plastic Debris: A Review. *Mar. Pollut. Bull.* **2002**, *44*, 842–852. [CrossRef]
5. Mao, Y.; Li, H.; Huangfu, X.; Liu, Y.; He, Q. Nanoplastics Display Strong Stability in Aqueous Environments: Insights from Aggregation Behaviour and Theoretical Calculations. *Environ. Pollut.* **2020**, *258*, 113760. [CrossRef]
6. Liu, Y.; Hu, Y.; Yang, C.; Chen, C.; Huang, W.; Dang, Z. Aggregation Kinetics of UV Irradiated Nanoplastics in Aquatic Environments. *Water Res.* **2019**, *163*, 114870. [CrossRef]
7. Min, K.; Cuif, J.D.; Mathers, R.T. Ranking Environmental Degradation Trends of Plastic Marine Debris Based on Physical Properties and Molecular Structure. *Nat. Commun.* **2020**, *11*. [CrossRef]
8. Yousif, E.; Haddad, R. Photodegradation and Photostabilization of Polymers, Especially Polystyrene: Review. *Springerplus* **2013**, *2*, 1–32. [CrossRef]
9. Song, Y.K.; Hong, S.H.; Jang, M.; Han, G.M.; Jung, S.W.; Shim, W.J. Combined Effects of UV Exposure Duration and Mechanical Abrasion on Microplastic Fragmentation by Polymer Type. *Environ. Sci. Technol. Fragm.* **2017**, *51*, 4368–4376. [CrossRef]

10. Wahl, A.; Le Juge, C.; Davranche, M.; El Hadri, H.; Grassl, B.; Reynaud, S.; Gigault, J. Nanoplastic Occurrence in a Soil Amended with Plastic Debris. *Chemosphere* **2021**, *262*, 127784. [CrossRef]
11. Zhang, Y.; Kang, S.; Allen, S.; Allen, D.; Gao, T.; Sillanpää, M. Atmospheric Microplastics: A Review on Current Status and Perspectives. *Earth Sci. Rev.* **2020**, *203*, 103118. [CrossRef]
12. Du, C.; Liang, H.; Li, Z.; Gong, J. Pollution Characteristics of Microplastics in Soils in Southeastern Suburbs of Baoding City, China. *Int. J. Environ. Res. Public Health* **2020**, *17*, 845. [CrossRef]
13. Dong, M.; Luo, Z.; Jiang, Q.; Xing, X.; Zhang, Q.; Sun, Y. The Rapid Increases in Microplastics in Urban Lake Sediments. *Sci. Rep.* **2020**, *10*, 1–10. [CrossRef]
14. Geueke, B.; Groh, K.; Muncke, J. Food Packaging in the Circular Economy: Overview of Chemical Safety Aspects for Commonly Used Materials. *J. Clean. Prod.* **2018**, *193*, 491–505. [CrossRef]
15. Zhang, Y.; Liang, J.; Zeng, G.; Tang, W.; Lu, Y.; Luo, Y.; Xing, W.; Tang, N.; Ye, S.; Li, X.; et al. How Climate Change and Eutrophication Interact with Microplastic Pollution and Sediment Resuspension in Shallow Lakes: A Review. *Sci. Total Environ.* **2020**, *705*, 135979. [CrossRef]
16. Löder, M.G.J.; Imhof, H.K.; Ladehoff, M.; Löschel, L.A.; Lorenz, C.; Mintenig, S.; Piehl, S.; Primpke, S.; Schrank, I.; Laforsch, C.; et al. Enzymatic Purification of Microplastics in Environmental Samples. *Environ. Sci. Technol.* **2017**, *51*, 14283–14292. [CrossRef] [PubMed]
17. Weis, J.S. Improving Microplastic Research. *AIMS Environ. Sci.* **2019**, *6*, 326–340. [CrossRef]
18. Araujo, C.F.; Nolasco, M.M.; Ribeiro, A.M.P.; Ribeiro-Claro, P.J.A. Identification of Microplastics Using Raman Spectroscopy: Latest Developments and Future Prospects. *Water Res.* **2018**, *142*, 426–440. [CrossRef]
19. Gillibert, R.; Balakrishnan, G.; Deshoules, Q.; Tardivel, M.; Magazzù, A.; Donato, M.G.; Maragò, O.M.; Lamy De La Chapelle, M.; Colas, F.; Lagarde, F.; et al. Raman Tweezers for Small Microplastics and Nanoplastics Identification in Seawater. *Environ. Sci. Technol.* **2019**, *53*, 9003–9013. [CrossRef]
20. Käppler, A.; Windrich, F.; Löder, M.G.J.; Malanin, M.; Fischer, D.; Labrenz, M.; Eichhorn, K.J.; Voit, B. Identification of Microplastics by FTIR and Raman Microscopy: A Novel Silicon Filter Substrate Opens the Important Spectral Range below 1300 Cm(-1) for FTIR Transmission Measurements. *Anal. Bioanal. Chem.* **2015**, *407*, 6791–6801. [CrossRef] [PubMed]
21. Asamoah, B.O.; Kanyathare, B.; Roussey, M.; Peiponen, K.E. A Prototype of a Portable Optical Sensor for the Detection of Transparent and Translucent Microplastics in Freshwater. *Chemosphere* **2019**, *231*, 161–167. [CrossRef]
22. Asamoah, B.; Amoani, J.; Roussey, M.; Peiponen, K.-E. Laser Beam Scattering for the Detection of Flat, Curved, Smooth, and Rough Microplastics in Water. *Opt. Rev.* **2020**, *27*, 217–224. [CrossRef]
23. Asamoah, B.O.; Roussey, M.; Peiponen, K. On Optical Sensing of Surface Roughness of Flat and Curved Microplastics in Water. *Chemosphere* **2020**, *254*, 126789. [CrossRef]
24. Kanyathare, B.; Asamoah, B.O.; Ishaq, U.; Amoani, J. Optical Transmission Spectra Study in Visible and Near-Infrared Spectral Range for Identification of Rough Transparent Plastics in Aquatic Environments. *Chemosphere* **2020**, *248*, 126071. [CrossRef]
25. Kanyathare, B.E.; Asamoah, B.; Ishaq, M.U.; Amoani, J.; Räty, J.; Peiponen, K. Identification of Plastic Type and Surface Roughness of Film-Type Plastics in Water Using Kramers—Kronig Analysis. *Chemosensors* **2020**, *8*, 88. [CrossRef]
26. Baensch-Baltruschat, B.; Kocher, B.; Stock, F.; Reifferscheid, G. Tyre and Road Wear Particles (TRWP)—A Review of Generation, Properties, Emissions, Human Health Risk, Ecotoxicity, and Fate in the Environment. *Sci. Total Environ.* **2020**, *733*, 137823. [CrossRef]
27. Zhou, Q.; Zhang, H.; Fu, C.; Zhou, Y.; Dai, Z.; Li, Y.; Tu, C.; Luo, Y. The Distribution and Morphology of Microplastics in Coastal Soils Adjacent to the Bohai Sea and the Yellow Sea. *Geoderma* **2018**, *322*, 201–208. [CrossRef]
28. Espiritu, E.Q.; Angeli, S.; Dayrit, S.N.; Coronel, A.S.O.; Paz, N.S.C.; Ronquillo, P.I.L.; Castillo, V.C.G.; Enriquez, E.P. Assessment of Quantity and Quality of Microplastics in the Sediments, Waters, Oysters, and Selected Fish Species in Key Sites Along the Bombong Estuary and the Coastal Waters of Ticalan in San Juan, Batangas. *Philipp. J. Sci.* **2019**, *148*, 789–801.
29. Lares, M.; Ncibi, M.C.; Sillanpää, M.; Sillanpää, M. Occurrence, Identification and Removal of Microplastic Particles and Fibers in Conventional Activated Sludge Process and Advanced MBR Technology. *Water Res.* **2018**, *133*, 236–246. [CrossRef]
30. Pao, Y.; Kelley, P. *Speckle Metrology*; Erf, R.K., Ed.; Academic Press: Cambridge, MA, USA, 2015; Volume 188. [CrossRef]
31. Imran, M.; Das, K.R.; Naik, M.M. Co-Selection of Multi-Antibiotic Resistance in Bacterial Pathogens in Metal and Microplastic Contaminated Environments: An Emerging Health Threat. *Chemosphere* **2018**. [CrossRef]
32. Wang, J.; Qin, X.; Guo, J.; Jia, W.; Wang, Q.; Zhang, M.; Huang, Y. Evidence of Selective Enrichment of Bacterial Assemblages and Antibiotic Resistant Genes by Microplastics in Urban Rivers. *Water Res.* **2020**, *183*, 116113. [CrossRef] [PubMed]
33. Ramsperger, A.F.R.M.; Narayana, V.K.B.; Gross, W.; Mohanraj, J.; Thelakkat, M.; Greiner, A. Environmental Exposure Enhances the Internalization of Microplastic Particles into Cells. *Sci. Adv.* **2020**, *6*, 1–10. [CrossRef]
34. Aspnes, D.E.; Theeten, J. Investigation of Effective-Medium Models of Microscopic Surface Roughness by Spectroscopic Ellipsometry. *Phys. Rev. B* **1979**, *20*, 3292–3302. [CrossRef]
35. El Hadri, H.; Gigault, J.; Maxit, B.; Grassl, B.; Reynaud, S. Nanoplastic from Mechanically Degraded Primary and Secondary Microplastics for Environmental Assessments. *NanoImpact* **2020**, *17*, 100206. [CrossRef]
36. Veerasingam, S.; Ranjani, M.; Venkatachalapathy, R.; Bagaev, A.; Mukhanov, V.; Litvinyuk, D.; Mugilarasan, M.; Gurumoorthi, K.; Guganathan, L.; Aboobacker, V.M.; et al. Contributions of Fourier Transform Infrared Spectroscopy in Microplastic Pollution Research: A Review. *Crit. Rev. Environ. Sci. Technol.* **2020**, 1–63. [CrossRef]

37. Álvarez-Hernández, C.; Cairós, C.; López-Darias, J.; Mazzetti, E.; Hernández-Sánchez, C.; González-Sálamo, J.; Hernández-Borges, J. Microplastic Debris in Beaches of Tenerife (Canary Islands, Spain). *Mar. Pollut. Bull.* **2019**, *146*, 26–32. [CrossRef]
38. Biginagwa, F.J.; Mayoma, B.S.; Shashoua, Y.; Syberg, K.; Khan, F.R. First Evidence of Microplastics in the African Great Lakes: Recovery from Lake Victoria Nile Perch and Nile Tilapia. *J. Great Lakes Res.* **2016**, *42*, 146–149. [CrossRef]
39. Hendrickson, E.; Minor, E.C.; Schreiner, K. Microplastic Abundance and Composition in Western Lake Superior As Determined via Microscopy, Pyr-GC/MS, and FTIR. *Environ. Sci. Technol.* **2018**, *52*, 1787–1796. [CrossRef]
40. Scientific, T. *Guide to the Identification of Microplastics by FTIR and Raman Spectroscopy*; 2018. Available online: https://assets.thermofisher.com/TFS-Assets/MSD/Application-Notes/WP53077-microplastics-identification-ftir-raman-guide.pdf (accessed on 27 February 2021).
41. Bordos, G.; Urbanyi, B.; Micsinai, A.; Balazs, K.; Palotai, Z.; Szabo, I.; Hantosi, Z.; Szoboszlay, S. Identification of Microplastics in Fish Ponds and Natural Freshwater Environments of the Carpathian Basin, Europe. *Chemosphere* **2019**, *216*, 110–116. [CrossRef]
42. De Jesus Negrete Velasco, A.; Rard, L.; Blois, W.; Labrun, D.; Lebrun, F.; Pothe, F.; Stoll, S. Microplastic and Fibre Contamination in a Remote Mountain Lake in Switzerland. *Water* **2020**, *12*, 2410. [CrossRef]
43. Talvitie, J.; Mikola, A.; Setälä, O.; Heinonen, M.; Koistinen, A. How Well Is Microlitter Purified from Wastewater?—A Detailed Study on the Stepwise Removal of Microlitter in a Tertiary Level Wastewater Treatment Plant. *Water Res.* **2017**, *109*, 164–172. [CrossRef]
44. Löder, M.; Kuczera, M.; Lorenz, C.; Gerdts, G. FPA-Based Micro-FTIR Imaging for the Analysis of Microplastics in Environmental Samples. *Environ. Chem.* **2015**, *12*, 563–581. [CrossRef]
45. Rist, S.; Vianello, A.; Winding, M.H.S.; Nielsen, T.G.; Almeda, R.; Torres, R.R.; Vollertsen, J. Quantification of Plankton-Sized Microplastics in a Productive Coastal Arctic Marine Ecosystem. *Environ. Pollut.* **2020**, *266*, 115248. [CrossRef] [PubMed]
46. Uurasjärvi, E.; Pääkkönen, M.; Setälä, O.; Koistinen, A.; Lehtiniemi, M. Microplastics Accumulate to Thin Layers in the Stratified Baltic Sea. *Environ. Pollut.* **2021**, *268*, 115700. [CrossRef]
47. Brandt, J.; Bittrich, L.; Fischer, F.; Kanaki, E.; Tagg, A.; Lenz, R.; Labrenz, M.; Brandes, E.; Fischer, D.; Eichhorn, K.J. High-Throughput Analyses of Microplastic Samples Using Fourier Transform Infrared and Raman Spectrometry. *Appl. Spectrosc.* **2020**, *74*, 1185–1197. [CrossRef]
48. Primpke, S.; Christiansen, S.H.; Cowger, W.; De Frond, H.; Deshpande, A.; Fischer, M.; Holland, E.B.; Meyns, M.; O'Donnell, B.A.; Ossmann, B.E.; et al. Critical Assessment of Analytical Methods for the Harmonized and Cost-Efficient Analysis of Microplastics. *Appl. Spectrosc.* **2020**, *74*, 1012–1047. [CrossRef]
49. Harrison, J.P.; Ojeda, J.J.; Romero-gonzález, M.E. The Applicability of Re Fl Ectance Micro-Fourier-Transform Infrared Spectroscopy for the Detection of Synthetic Microplastics in Marine Sediments. *Sci. Total Environ.* **2012**, *416*, 455–463. [CrossRef] [PubMed]
50. Zhang, X.; Chen, J.; Li, J. The Removal of Microplastics in the Wastewater Treatment Process and Their Potential Impact on Anaerobic Digestion Due to Pollutants Association. *Chemosphere* **2020**, *251*, 126360. [CrossRef]
51. Primpke, S.; Wirth, M.; Lorenz, C.; Gerdts, G. Reference Database Design for the Automated Analysis of Microplastic Samples Based on Fourier Transform Infrared (FTIR) Spectroscopy. *Anal. Bioanal. Chem.* **2018**, *410*, 5131–5141. [CrossRef]
52. Cowger, W.; Gray, A.; Christiansen, S.H.; DeFrond, H.; Deshpande, A.D.; Hemabessiere, L.; Lee, E.; Mill, L.; Munno, K.; Ossmann, B.E.; et al. Critical Review of Processing and Classification Techniques for Images and Spectra in Microplastic Research. *Appl. Spectrosc.* **2020**, *74*, 989–1010. [CrossRef]
53. Löder, M.G.J.; Gerdts, G. Methodology Used for the Detection and Identification of Microplastics—A Critical Appraisal. In *Marine Anthropogenic Litter*; Bergmann, M., Gutow, L., Klages, M., Eds.; Springer Open: Berlin/Heidelberg, Germany, 2015; p. 214.
54. Primpke, S.; Cross, R.K.; Mintenig, S.M.; Simon, M.; Vianello, A.; Gerdts, G.; Vollertsen, J. Toward the Systematic Identification of Microplastics in the Environment: Evaluation of a New Independent Software Tool (SiMPle) for Spectroscopic Analysis. *Appl. Spectrosc.* **2020**, *74*, 1127–1138. [CrossRef]
55. Primpke, S.; Lorenz, C.; Rascher-Friesenhausen, R.; Gerdts, G. An Automated Approach for Microplastics Analysis Using Focal Plane Array (FPA) FTIR Microscopy and Image Analysis. *Anal. Methods* **2017**, *9*, 1499–1511. [CrossRef]
56. Wander, L.; Vianello, A.; Vollertsen, J.; Westad, F.; Braun, U.; Paul, A. Exploratory Analysis of Hyperspectral FTIR Data Obtained from Environmental Microplastics Samples. *Anal. Methods* **2020**, *12*, 781–791. [CrossRef]
57. Liu, F.; Olesen, K.B.; Borregaard, A.R.; Vollertsen, J. Microplastics in Urban and Highway Stormwater Retention Ponds. *Sci. Total Environ.* **2019**, *671*, 992–1000. [CrossRef]
58. Renner, G.; Schmidt, T.C.; Schram, J. Automated Rapid & Intelligent Microplastics Mapping by FTIR Microscopy: A Python—Based Work Flow. *MethodsX* **2020**, *7*, 100742. [CrossRef]
59. Renner, G.; Sauerbier, P.; Schmidt, T.C.; Schram, J. Robust Automatic Identification of Microplastics in Environmental Samples Using FTIR Microscopy. *Anal. Chem.* **2019**, *91*, 9656–9664. [CrossRef] [PubMed]
60. Renner, G.; Nellessen, A.; Schwiers, A.; Wenzel, M.; Schmidt, T.C.; Schram, J. Data Preprocessing & Evaluation Used in the Microplastics Identi Fi Cation Process: A Critical Review & Practical Guide. *Trends Anal. Chem.* **2019**, *111*, 229–238. [CrossRef]
61. Simon, M.; van Alst, N.; Vollertsen, J. Quantification of Microplastic Mass and Removal Rates at Wastewater Treatment Plants Applying Focal Plane Array (FPA)-Based Fourier Transform Infrared (FT-IR) Imaging. *Water Res.* **2018**, *142*, 1–9. [CrossRef] [PubMed]

62. Haave, M.; Lorenz, C.; Primpke, S.; Gerdts, G. Different Stories Told by Small and Large Microplastics in Sediment—First Report of Microplastic Concentrations in an Urban Recipient in Norway. *Mar. Pollut. Bull.* **2019**, *141*, 501–513. [CrossRef]
63. Anger, P.M.; von der Esch, E.; Baumann, T.; Elsner, M.; Niessner, R.; Ivleva, N.P. Raman Microspectroscopy as a Tool for Microplastic Particle Analysis. *TrAC Trends Anal. Chem.* **2018**, *109*, 214–226. [CrossRef]
64. Elert, A.M.; Becker, R.; Duemichen, E.; Eisentraut, P.; Falkenhagen, J.; Sturm, H.; Braun, U. Comparison of Different Methods for MP Detection: What Can We Learn from Them, and Why Asking the Right Question before Measurements Matters? *Environ. Pollut.* **2017**, *231*, 1256–1264. [CrossRef]
65. Käppler, A.; Fischer, D.; Oberbeckmann, S.; Schernewski, G.; Labrenz, M.; Eichhorn, K.J.; Voit, B. Analysis of Environmental Microplastics by Vibrational Microspectroscopy: FTIR, Raman or Both? *Anal. Bioanal. Chem.* **2016**, *408*, 8377–8391. [CrossRef]
66. Yang, S.; Akkus, O. Fluorescence Background Problem in Raman Spectroscopy: Is 1064 Nm Excitation an Improvement of 785 Nm? *Wasatch Photonics* **2013**. Available online: https://www.ccmr.cornell.edu/wp-content/uploads/sites/2/2015/11/FL-Time-1064-nm-vs-785-nm-Shan-9_30.pdf (accessed on 27 February 2021).
67. Shreve, A.P.; Cherepy, N.J.; Mathies, R.A. Effective Rejection of Fluorescence Interference in Raman Spectroscopy Using a Shifted Excitation Difference Technique. *Appl. Spectrosc.* **1992**, *46*, 707–711. [CrossRef]
68. Zhao, J.; Carrabba, M.M.; Allen, F.S. Automated Fluorescence Rejection Using Shifted Excitation Raman Difference Spectroscopy. *Appl. Spectrosc.* **2002**, *56*, 834–845. [CrossRef]
69. Crocombe, R.A. Portable Spectroscopy. *Appl. Spectrosc.* **2018**, *72*, 1701–1751. [CrossRef] [PubMed]
70. Ornik, J.; Sommer, S.; Gies, S.; Weber, M.; Lott, C.; Balzer, J.C.; Koch, M. Could Photoluminescence Spectroscopy Be an Alternative Technique for the Detection of Microplastics? First Experiments Using a 405 Nm Laser for Excitation. *Appl. Phys. B Lasers Opt.* **2020**, *126*, 1–7. [CrossRef]
71. Imhof, H.K.; Laforsch, C.; Wiesheu, A.C.; Schmid, J.; Anger, P.M.; Niessner, R.; Ivleva, N.P. Pigments and Plastic in Limnetic Ecosystems: A Qualitative and Quantitative Study on Microparticles of Different Size Classes. *Water Res.* **2016**, *98*, 64–74. [CrossRef]
72. Imhof, H.K.; Schmid, J.; Niessner, R.; Ivleva, N.P.; Laforsch, C. A Novel, Highly Efficient Method for the Separation and Quantification of Plastic Particles in Sediments of Aquatic Environments. *Limnol. Oceanogr. Methods* **2012**, *10*, 524–537. [CrossRef]
73. Dehaut, A.; Cassone, A.L.; Frère, L.; Hermabessiere, L.; Himber, C.; Rinnert, E.; Rivière, G.; Lambert, C.; Soudant, P.; Huvet, A.; et al. Microplastics in Seafood: Benchmark Protocol for Their Extraction and Characterization. *Environ. Pollut.* **2016**, *215*, 223–233. [CrossRef]
74. Enders, K.; Lenz, R.; Beer, S.; Stedmon, C.A. Extraction of Microplastic from Biota: Recommended Acidic Digestion Destroys Common Plastic Polymers. *ICES J. Mar. Sci.* **2017**, *74*, 326–331. [CrossRef]
75. Lenz, R.; Enders, K.; Stedmon, C.A.; MacKenzie, D.M.A.; Nielsen, T.G. A Critical Assessment of Visual Identification of Marine Microplastic Using Raman Spectroscopy for Analysis Improvement. *Mar. Pollut. Bull.* **2015**, *100*, 82–91. [CrossRef]
76. Ashkin, A. Optical Trapping and Manipulation of Neutral Particles Using Lasers. *Proc. Natl. Acad. Sci. USA* **1997**, *94*, 4853–4860. [CrossRef]
77. Bodelón, G.; Pastoriza-santos, I. Recent Progress in Surface-Enhanced Raman Scattering for the Detection of Chemical Contaminants in Water. *Front. Chem.* **2020**, *8*. [CrossRef]
78. Lv, L.; He, L.; Jiang, S.; Chen, J.; Zhou, X.; Qu, J.; Lu, Y.; Hong, P.; Sun, S.; Li, C. In Situ Surface-Enhanced Raman Spectroscopy for Detecting Microplastics and Nanoplastics in Aquatic Environments. *Sci. Total Environ.* **2020**, *728*, 138449. [CrossRef]
79. Kniggendorf, A.; Wetzel, C.; Roth, B. Microplastics Detection in Streaming Tap Water with Raman Spectroscopy. *Sensors* **2019**, *19*, 1839. [CrossRef]
80. Zada, L.; Leslie, H.A.; Vethaak, A.D.; Tinnevelt, G.H.; Jansen, J.J.; de Boer, J.F.; Ariese, F. Fast Microplastics Identification with Stimulated Raman Scattering Microscopy. *J. Raman Spectrosc.* **2018**, *49*, 1136–1144. [CrossRef]
81. Goodhead, R.M.; Moger, J.; Galloway, T.S.; Tyler, C.R. Tracing Engineered Nanomaterials in Biological Tissues Using Coherent Anti-Stokes Raman Scattering (CARS) Microscopy—A Critical Review. *Nanotoxicology* **2015**, *9*, 928–939. [CrossRef] [PubMed]
82. Cole, M.; Lindeque, P.; Fileman, E.; Halsband, C.; Goodhead, R.; Moger, J.; Galloway, T.S. Microplastic Ingestion by Zooplankton. *Environ. Sci. Technol.* **2013**, *47*, 6646–6655. [CrossRef]
83. Boyd, R.W. *Nonlinear Optics*, 4th ed.; Pitts, T., Ed.; Academic Press: London, UK, 2020.
84. Peiponen, K.E.; Räty, J.; Ishaq, U.; Pélisset, S.; Ali, R. Outlook on Optical Identification of Micro- and Nanoplastics in Aquatic Environments. *Chemosphere* **2019**, *214*, 424–429. [CrossRef] [PubMed]
85. Pan, Z.; Guo, H.; Chen, H.; Wang, S.; Sun, X.; Zou, Q.; Zhang, Y.; Lin, H.; Cai, S.; Huang, J. Microplastics in the Northwestern Paci Fi c: Abundance, Distribution, and Characteristics. *Sci. Total Environ.* **2019**, *650*, 1913–1922. [CrossRef]
86. Vázquez-Guardado, A.; Money, M.; McKinney, N.; Chanda, D. Multi-Spectral Infrared Spectroscopy for Robust Plastic Identification. *Appl. Opt.* **2015**, *54*, 7396. [CrossRef] [PubMed]
87. Andoh, S.S.; Nyave, K.; Asamoah, B.; Kanyathare, B.; Nuutinen, T.; Mingle, C.; Peiponen, K.E.; Roussey, M. Optical Screening for Presence of Banned Sudan III and Sudan IV Dyes in Edible Palm Oils. *Food Addit. Contam. Part A Chem. Anal. Control. Expo. Risk Assess.* **2020**, *37*, 1049–1060. [CrossRef]
88. Andrady, A.L. Using Flow Cytometry to Detect Micro- and Nano-Scale Polymer Particles; Courtney. In *Proceedings of the Second Research Workshop on Microplastic Marine Debris*; Courtney, A., Baker, J., Eds.; 2011; Available online: https://marinedebris.noaa.gov/sites/default/files/publications-files/TM_NOS-ORR_39.pdf (accessed on 27 February 2021).

89. Erni-cassola, G.; Gibson, M.I.; Thompson, R.C. Lost, but Found with Nile Red; a Novel Method to Detect and Quantify Small Microplastics (20 Mm–1 Mm) in Environmental Samples. *Environ. Sci. Technol.* **2017**, *51*, 23. [CrossRef] [PubMed]
90. Prata, J.C.; Reis, V.; Matos, J.T.V.; Costa, J.P.; Duarte, A.C.; Rocha-santos, T. A New Approach for Routine Quantification of Microplastics Using Nile Red and Automated Software (MP-VAT). *Sci. Total Environ.* **2019**, *690*, 1277–1283. [CrossRef]
91. Karakolis, E.G.; Nguyen, B.; You, J.B.; Rochman, C.M.; Sinton, D. Fluorescent Dyes for Visualizing Microplastic Particles and Fibers in Laboratory-Based Studies. *Environ. Sci. Technol. Lett.* **2019**, *6*, 334–340. [CrossRef]
92. Hengstmann, E.; Fischer, E.K. Nile Red Staining in Microplastic Analysis—Proposal for a Reliable and Fast Identification Approach for Large Microplastics. *Environ. Monit Assess.* **2019**, *191*. [CrossRef] [PubMed]
93. Shim, W.J.; Song, Y.K.; Hong, S.H.; Jang, M. Identification and Quantification of Microplastics Using Nile Red Staining. *Mar. Pollut. Bull.* **2016**, *113*, 469–476. [CrossRef]
94. Lv, L.; Qu, J.; Yu, Z.; Chen, D.; Zhou, C.; Hong, P.; Sun, S.; Li, C. A Simple Method for Detecting and Quantifying Microplastics Utilizing Fluorescent Dyes—Safranine T, Fluorescein Isophosphate, Nile Red Based on Thermal Expansion and Contraction Property *. *Environ. Pollut.* **2019**, *255*, 113283. [CrossRef]
95. Gagné, F.; Auclair, J.; Quinn, B. Detection of Polystyrene Nanoplastics in Biological Samples Based on the Solvatochromic Properties of Nile Red: Application in Hydra Attenuata Exposed to Nanoplastics. *Environ. Sci. Pollut. Res.* **2019**, *26*, 33524–33531. [CrossRef]
96. Nel, H.A.; Chetwynd, A.J.; Kelleher, L.; Lynch, I.; Mans, I.; Margenat, H.; Onoja, S.; Goldberg, P.; Sambrook, G.H.; Krause, S. Detection Limits Are Central to Improve Reporting Standards When Using Nile Red for Microplastic Quantification. *Chemosphere* **2021**, *263*, 127953. [CrossRef]
97. Sturm, M.T.; Horn, H.; Schuhen, K. The Potential of Fluorescent Dyes—Comparative Study of Nile Red and Three Derivatives for the Detection of Microplastics. *Anal. Bioanal. Chem.* **2021**, *413*, 1059–1071. [CrossRef]
98. Li, J.; Liu, H.; Chen, J.P. Microplastics in Freshwater Systems: A Review on Occurrence, Environmental Effects, and Methods for Microplastics Detection. *Water Res.* **2018**, *137*, 362–374. [CrossRef]
99. Stanton, T.; Johnson, M.; Nathanail, P.; Gomes, R.L.; Needham, T.; Burson, A. Exploring the Efficacy of Nile Red in Microplastic Quantification: A Costaining Approach. *Environ. Sci. Technol. Lett.* **2019**, *6*, 606–611. [CrossRef]
100. Prata, J.C.; Alves, J.R.; João, P.; Duarte, A.C.; Rocha-santos, T. Major Factors Influencing the Quantification of Nile Red Stained Microplastics and Improved Automatic Quantification (MP-VAT 2.0). *Sci. Total Environ.* **2020**, *719*, 137498. [CrossRef]
101. Hycza, T.; Stereńczak, K.; Bałazy, R. Potential Use of Hyperspectral Data to Classify Forest Tree Species. *N. Z. J. For. Sci.* **2018**, *48*. [CrossRef]
102. Karlsson, T.M.; Grahn, H.; Bavel, V.; Geladi, P. Hyperspectral Imaging and Data Analysis for Detecting and Determining Plastic Contamination in Seawater Filtrates. *J. Near Infrared Spectr.* **2016**, *149*, 141–149. [CrossRef]
103. Shan, J.; Zhao, J.; Zhang, Y.; Liu, L.; Wu, F.; Wang, X. Simple and Rapid Detection of Microplastics in Seawater Using Hyperspectral Imaging Technology. *Anal. Chim. Acta* **2019**, *1050*, 161–168. [CrossRef]
104. Serranti, S.; Palmieri, R.; Bonifazi, G.; Cózar, A. Characterization of Microplastic Litter from Oceans by an Innovative Approach Based on Hyperspectral Imaging. *Waste Manag.* **2020**, *76*, 117–125. [CrossRef]
105. Merlino, S.; Paterni, M.; Berton, A.; Massetti, L. Unmanned Aerial Vehicles for Debris Survey in Coastal Areas: Long-Term Monitoring Programme to Study Spatial and Temporal Accumulation of the Dynamics of Beached Marine Litter. *Remote Sens.* **2020**, *12*, 1260. [CrossRef]
106. Gonçalves, G.; Andriolo, U.; Pinto, L.; Bessa, F. Mapping Marine Litter Using UAS on a Beach-Dune System: A Multidisciplinary Approach. *Sci. Total Environ.* **2020**, *706*, 135742. [CrossRef]
107. Huang, H.; Ullah, J.; Shuchang, Q.; Zehao, L.; Chunfang, S.; Wang, H. Hyperspectral Imaging as a Potential Online Detection Method of Microplastics. *Bull. Environ. Contam. Toxicol.* **2020**. [CrossRef]
108. Garaba, S.P.; Dierssen, H.M. Hyperspectral Ultraviolet to Shortwave Infrared Characteristics of Marine-Harvested, Washed-Ashore and Virgin Plastics. *Earth Syst. Sci. Data* **2020**, *12*, 77–86. [CrossRef]
109. Zhang, Y.; Wang, X.; Shan, J.; Zhao, J.; Zhang, W.; Liu, L.; Wu, F. Hyperspectral Imaging Based Method for Rapid Detection of Microplastics in the Intestinal Tracts of Fish. *Environ. Sci. Technol.* **2019**, *53*, 5151–5158. [CrossRef] [PubMed]
110. Hakala, T.; Markelin, L.; Id, E.H.; Scott, B.; Theocharous, T.; Id, O.N.; Näsi, R.; Suomalainen, J.; Id, N.V.; Greenwell, C. Direct Reflectance Measurements from Drones: Sensor Absolute Radiometric Calibration and System Tests for Forest Reflectance Characterization. *Sensors* **2018**, *18*, 1417. [CrossRef]
111. Scientific, T. *Raman and FTIR Spectroscopy: Complementary Technologies for Chemical and Explosives Identification*; 2015. Available online: http://public.jenck.com/folletos/Gemini%20-%20Complementary%20Technology.pdf (accessed on 27 February 2021).
112. Wartewig, S.; Neubert, R.H.H. Pharmaceutical Applications of Mid-IR and Raman Spectroscopy. *Adv. Drug Deliv. Rev.* **2005**, *57*, 1144–1170. [CrossRef] [PubMed]
113. McCumskay, R.; Tang, P.L.; Mckumskay, R.; Rogerson, M.; Waller, C.; Forster, R. Handheld Portable FTIR Spectroscopy for the Triage of Micro and Meso Sized Plastics in the Marine Environment Incorporating an Accelerated Weathering Study and an Aging Estimation. *Spectroscopy* **2019**, *34*, 54–60.
114. Iri, A.H.; Shahrah, M.H.A.; Ali, A.M.; Qadri, S.A.; Erdem, T.; Ozdur, I.T.; Icoz, K. Optical Detection of Microplastics in Water. *Environ. Sci Pollut Res.* **2021**. [CrossRef]

115. Oz, N.; Kadizade, G.; Yurtsever, M. Investigation of Heavy Metal Adsorption on Microplastics. *Appl. Ecol. Environ. Res.* **2019**, *17*, 7301–7310. [CrossRef]
116. Li, J.; Zhang, K.; Zhang, H. Adsorption of Antibiotics on Microplastics. *Environ. Pollut.* **2018**, *237*, 460–467. [CrossRef] [PubMed]
117. Kanyathare, B.; Kuivalainen, K.; Räty, J.; Silfsten, P.; Bawuah, P.; Peiponen, K.E. A Prototype of an Optical Sensor for the Identification of Diesel Oil Adulterated by Kerosene. *J. Eur. Opt. Soc.* **2018**, *14*. [CrossRef]
118. García, J.; Zalevsky, Z.; García-Martínez, P.; Ferreira, C.; Teicher, M.; Beiderman, Y. Projection of Speckle Patterns for 3D Sensing Projection of Speckle Patterns for 3D Sensing. *J. Phys. Conf. Ser.* **2008**, *139*. [CrossRef]
119. Enders, K.; Lenz, R.; Stedmon, C.A.; Nielsen, T.G. Abundance, Size and Polymer Composition of Marine Microplastics $\geq$10 Mm in the Atlantic Ocean and Their Modelled Vertical Distribution. *Mar. Pollut. Bull.* **2015**, *100*, 70–81. [CrossRef]
120. Cabernard, L.; Roscher, L.; Lorenz, C.; Gerdts, G.; Primpke, S. Comparison of Raman and Fourier Transform Infrared Spectroscopy for the Quantification of Microplastics in the Aquatic Environment. *Environ. Sci. Technol.* **2018**, *52*, 13279–13288. [CrossRef]
121. Pabortsava, K.; Lampitt, R.S. High Concentrations of Plastic Hidden beneath the Surface of the Atlantic Ocean. *Nat. Commun.* **2020**, *11*, 1–11. [CrossRef]
122. Singh, B.; Sharma, N. Mechanistic Implications of Plastic Degradation. *Polym. Degrad. Stab.* **2008**, *93*, 561–584. [CrossRef]
123. Mintenig, S.M.; Bäuerlein, P.S.; Koelmans, A.A.; Dekker, S.C.; Van Wezel, A.P. Closing the Gap between Small and Smaller: Towards a Framework to Analyse Nano- and Microplastics in Aqueous Environmental Samples. *Environ. Sci. Nano* **2018**, *5*, 1640–1649. [CrossRef]
124. Wagner, M.; Holzschuh, S.; Traeger, A.; Fahr, A.; Schubert, U.S. Asymmetric Flow Field-Flow Fractionation in the Field of Nanomedicine. *Anal. Chem.* **2014**, *86*, 5201–5210. [CrossRef]
125. Markarian, J. Back-to-Basics: Adding Colour to Plastics. *Plast. Addit. Compd.* **2009**, *11*, 12–15. [CrossRef]
126. Del Socorro Hernández-Montes, M.; Mendoza-Santoyo, F.; Flores Moreno, M.; de la Torre-Ibarra, M.; Silva Acosta, L.; Palacios-Ortega, N. Macro to Nano Specimen Measurements Using Photons and Electrons with Digital Holographic Interferometry: A Review. *J. Eur. Opt. Soc.* **2020**, *16*. [CrossRef]
127. Merola, F.; Memmolo, P.; Bianco, V.; Paturzo, M.; Mazzocchi, M.G.; Ferraro, P. Searching and Identifying Microplastics in Marine Environment by Digital Holography★. *Eur. Phys. J. Plus* **2018**, *133*. [CrossRef]
128. Sun, H.; Hendry, D.C.; Player, M.A.; Watson, J.; Member, S. In Situ Underwater Electronic Holographic Camera for Studies of Plankton. *IEEE J. Ocean. Eng.* **2007**, *32*, 373–382. [CrossRef]
129. Darakis, E.; Khanam, T.; Rajendran, A.; Kariwala, V.; Naughton, T.J.; Asundi, A.K. Microparticle Characterization Using Digital Holography. *Chem. Eng. Sci.* **2010**, *65*, 1037–1044. [CrossRef]
130. Takahashi, T.; Liu, Z.; Thevar, T.; Burns, N.; Mahajan, S.; Lindsay, D.; Watson, J.; Thornton, B. Identification of Microplastics in a Large Water Volume by Integrated Holography and Raman Spectroscopy. *Appl. Opt.* **2020**, *59*, 5073. [CrossRef]
131. Dardikman, G.; Shaked, N.T. Review on Methods of Solving the Refractive Index—Thickness Coupling Problem in Digital Holographic Microscopy of Biological Cells. *Opt. Commun.* **2018**, *422*, 8–16. [CrossRef]
132. Kenny, J.E.; Jarvis, G.B.; Chudyk, W.A.; Pohlig, K.O. Ground-Water Monitoring Using Remote Laser-Induced Fluorescence. In *Luminescence Applications in Biological, Chemical, Environmental, and Hydrological Sciences*; Goldberg, M.C., Ed.; ACS Publications: Washington, DC, USA, 1989; pp. 233–239.
133. Niskanen, I.; Räty, J.; Peiponen, K.E. Determination of the Refractive Index of Microparticles by Utilizing Light Dispersion Properties of the Particle and an Immersion Liquid. *Talanta* **2013**, *115*, 68–73. [CrossRef]
134. Wang, Z.; Wang, N.; Wang, D.; Tang, Z.; Yu, K.; Wei, W. Method for Refractive Index Measurement of Nanoparticles. *Opt. Lett.* **2014**, *39*, 4251–4254. [CrossRef]
135. Rosenholm, J.B.; Peiponen, K.E.; Gornov, E. Materials Cohesion and Interaction Forces. *Adv. Colloid Interface Sci.* **2008**, *141*, 48–65. [CrossRef]
136. Aygun, U.; Urey, H.; Ozkumur, A.Y. Label-Free Detection of Nanoparticles Using Depth Scanning Correlation Interferometric Microscopy. *Sci. Rep.* **2019**, *9*, 1–8. [CrossRef]
137. Ignatovich, F.V.; Novotny, L. Real-Time and Background-Free Detection of Nanoscale Particles. *Phys. Rev. Lett.* **2006**, *96*, 013901. [CrossRef]
138. Bruggeman, D.A.G. Berechnung Verschiedener Physikalischer Konstanten Von Heterogenen Substanzen. I. Dielektrizitätskonstanten Und Leitfähigkeiten Der Mischkörper Aus Isotropen Substanzen. *Ann. Phys.* **1935**, *416*, 636–678. [CrossRef]
139. Piederriere, Y.; Meur, J.L.; Cariou, J. Particle Aggregation Monitoring by Speckle Size Measurement; Application to Blood Platelets Aggregation. *Opt. Express* **2004**, *12*, 4596–4601. [CrossRef]
140. Sadhwani, A.; Schomacker, K.T.; Tearney, G.J.; Nishioka, N.S. Determination of Teflon Thickness with Laser Speckle. I. Potential for Burn Depth Diagnosis. *Appl. Opt.* **1996**, *35*, 5727–5735. [CrossRef]
141. Lehmann, P. Surface-Roughness Measurement Based on the Intensity Correlation Function of Scattered Light under Speckle-Pattern Illumination. *Appl. Opt.* **1999**, *30*, 1144–1152. [CrossRef]
142. Born, M.; Wolf, E. *Principles of Optics Electromagnetic Theory of Propagation, Interference and Diffraction of Light*, 7th ed.; Cambridge University Press: Cambridge, UK, 1999.
143. Block, I.D.; Scheffold, F. Modulated 3D Cross-Correlation Light Scattering: Improving Turbid Sample Characterization. *Rev. Sci. Instrum.* **2010**, *81*, 123107. [CrossRef]

144. Ozcelik, D.; Cai, H.; Leake, K.D.; Hawkins, A.R.; Schmidt, H. Optofluidic Bioanalysis: Fundamentals and Applications. *Nanophotonics* **2017**, *6*, 647–661. [CrossRef]
145. Pisco, M.; Cusano, A. Lab-On-Fiber Technology: A Roadmap toward Multifunctional Plug and Play Platforms. *Sensors* **2020**, *20*, 4705. [CrossRef]
146. Li, X.; Li, X. Athermal Evanescent Wave Filter Based on Slab Waveguide Coupled Silica Microfiber. *Opt. Commun.* **2020**, *473*, 125950. [CrossRef]
147. Zhao, Y.; Zhao, J.; Zhao, Q. Sensors and Actuators A: Physical Review of No-Core Optical Fiber Sensor and Applications. *Sens. Actuators A Phys.* **2020**, *313*, 112160. [CrossRef]
148. Barrios, C.A.; Gylfason, K.B.; Sánchez, B.; Griol, A.; Sohlström, H.; Holgado, M.; Casquel, R. Slot-Waveguide Biochemical Sensor. *Opt. Lett.* **2007**, *32*, 3080–3082. [CrossRef] [PubMed]
149. Bernini, R.; Campopiano, S.; Zeni, L.; Sarro, P.M. ARROW Optical Waveguides Based Sensors. *Sens. Actuators B* **2004**, *100*, 143–146. [CrossRef]
150. Soref, R. The Past, Present, and Future of Silicon Photonics. *IEEE J. Sel. Top. Quantum Electron.* **2006**, *12*, 1678–1687. [CrossRef]
151. Tervonen, A.; West, B.R.; Honkanen, S. Ion-Exchanged Glass Waveguide Technology: A Review. *Phys. Comput. Sci. Fac. Publ. Opt. Eng.* **2011**, *91*. [CrossRef]
152. Grieshaber, D.; MacKenzie, R.; Vörös1, J.; Reimhult, E. Electrochemical Biosensors—Sensor Principles and Architectures. *Sensors* **2008**, *8*, 1400–1458. [CrossRef]
153. Kozma, P.; Kehl, F.; Ehrentreich-förster, E.; Stamm, C.; Bier, F.F. Biosensors and Bioelectronics Integrated Planar Optical Waveguide Interferometer Biosensors: A Comparative Review. *Biosens. Bioelectron.* **2014**, *58*, 287–307. [CrossRef]
154. Washburn, A.L.; Bailey, R.C. Photonics-on-a-Chip: Recent Advances in Integrated Waveguides as Enabling Detection Elements for Real-World, Lab-on-a-Chip Biosensing Applications. *Analyst* **2011**, *136*, 227–236. [CrossRef]
155. Lee, B. Review of the Present Status of Optical Fiber Sensors. *Opt. Fiber Technol.* **2003**, *9*, 57–79. [CrossRef]
156. Pitois, C.; Wiesmann, D.; Lindgren, M.; Hult, A. Functionalized Fluorinated Hyperbranched Polymers for Optical Waveguide Applications. *Adv. Mater.* **2001**, *13*, 1483. [CrossRef]
157. Kersey, A.D.; Davis, M.A.; Patrick, H.J.; Leblanc, M.; Koo, K.P.; Askins, C.G.; Putnam, M.A.; Friebele, E.J. Fiber Grating Sensors. *J. Light. Technol.* **1997**, *15*, 1442–1463. [CrossRef]
158. Wu, D.K.C.; Kuhlmey, B.T.; Eggleton, B.J. Ultra-Sensitive Photonic Crystal Fiber Refractive Index Sensor. *Opt. Lett.* **2009**, *34*, 322–324. [CrossRef]
159. Park, D.S.; Young, B.M.; You, B.; Singh, V.; Soper, S.A.; Murphy, M.C.; An, A. An Integrated, Optofluidic System With Aligned Optical Waveguides, Microlenses, and Coupling Prisms for Fluorescence Sensing. *J. Microelectromech. Syst.* **2020**, *29*, 600–609. [CrossRef]
160. Liu, L.; Jin, M.; Shi, Y.; Lin, J.; Zhang, Y.; Jiang, L.; Zhou, G.; He, S. Optical Integrated Chips with Micro and Nanostructures for Refractive Index and SERS-Based Optical Label-Free Sensing. *Nanophotonics* **2015**, *4*, 419–436. [CrossRef]
161. Ahmadi, L.; Hiltunen, M.; Stenberg, P.; Hiltunen, J.; Aikio, S.; Roussey, M.; Saarinen, J.; Honkanen, S. Hybrid Layered Polymer Slot Waveguide Young Interferometer. *Opt. Express* **2016**, *24*, 10275–10285. [CrossRef] [PubMed]
162. Bogaerts, W.; Dumon, P.; Member, S.; Van Thourhout, D.; Taillaert, D.; Jaenen, P.; Wouters, J.; Beckx, S.; Wiaux, V.; Baets, R.G.; et al. Compact Wavelength-Selective Functions in Silicon-on-Insulator Photonic Wires. *IEEE J. Sel. Top. Quantum Electron.* **2006**, *12*, 1394–1401. [CrossRef]
163. Pathak, S.; Member, S.; Vanslembrouck, M.; Dumon, P. Optimized Silicon AWG With Flattened Spectral Response Using an MMI Aperture. *J. Light. Technol.* **2013**, *31*, 87–93. [CrossRef]
164. Wulder, M.A.; White, J.C.; Nelson, R.F.; Næsset, E.; Ole, H.; Coops, N.C.; Hilker, T.; Bater, C.W.; Gobakken, T. Remote Sensing of Environment Lidar Sampling for Large-Area Forest Characterization: A Review. *Remote Sens. Environ.* **2012**, *121*, 196–209. [CrossRef]
165. Smit, M.K.; Member, A.; Van Dam, C. PHASAR-Based WDM-Devices: Principles, design and applications. *IEEE J. Sel. Top. Quantum Electron.* **1996**, *2*, 236–250. [CrossRef]
166. Desiatov, B.; Goykhman, I.; Levy, U. Direct Temperature Mapping of Nanoscale Plasmonic Devices. *Nano Lett.* **2014**, *14*, 648–652. [CrossRef]
167. Ohkubo, T. Detection of Fluorescence from a Mixed Specimen Consisting of Micro Particles Attached to Different Fluorescent Substances Using a Light Waveguide Incorporated Optical TAS. *Microsyst. Technol.* **2020**, *26*, 95–109. [CrossRef]
168. Heinze, B.C.; Gamboa, J.R.; Kim, K.; Song, J.; Yoon, J. Microfluidic Immunosensor with Integrated Liquid Core Waveguides for Sensitive Mie Scattering Detection of Avian Influenza Antigens in a Real Biological Matrix. *Anal. Bioanal. Chem.* **2010**, *398*, 2693–2700. [CrossRef]
169. Stimpson, D.I.; Gordon, J. The Utility of Optical Waveguide DNA Array Hybridization and Melting for Rapid Resolution of Mismatches, and for Detection of Minor Mutant Components in the Presence of a Majority of Wild Type Sequence: Statistical Model and Supporting Data. *Genet. Anal.* **1996**, *13*, 73–80. [CrossRef]
170. Fojt, J.; David, J.; Radek, P.; Veronika, Ř. A Critical Review of the Overlooked Challenge of Determining Micro-Bioplastics in Soil. *Sci. Total Environ.* **2020**, *745*. [CrossRef] [PubMed]

Article

Study on the Microstructure of Polyether Ether Ketone Films Irradiated with 170 keV Protons by Grazing Incidence Small Angle X-ray Scattering (GISAXS) Technology

Hongxia Li [1], Jianqun Yang [1], Feng Tian [2], Xingji Li [1,*] and Shangli Dong [1,*]

[1] School of Materials Science and Engineering, Harbin Institute of Technology, Harbin 150001, China; koalalhx@163.com (H.L.); yangjianqun@hit.edu.cn (J.Y.)
[2] Shanghai Synchrotron Radiation Facility, Zhangjiang Lab, Shanghai Advanced Research Institute, Chinese Academy of Sciences, Shanghai 201204, China; tianfeng@zjlab.org.cn
* Correspondence: lxj0218@hit.edu.cn (X.L.); shanglidong@hit.edu.cn (S.D.); Tel.: +86-4518-6412-462 (X.L.)

Received: 31 October 2020; Accepted: 9 November 2020; Published: 17 November 2020

Abstract: Polyether ether ketone (PEEK) films irradiated with 170 keV protons were calculated by the stopping and ranges of ions in matter (SRIM) software. The results showed that the damage caused by 170 keV protons was only several microns of the PEEK surface, and the ionization absorbed dose and displacement absorbed dose were calculated. The surface morphology and roughness of PEEK after proton irradiation were studied by atomic force microscope (AFM). GISAXS was used to analyze the surface structural information of the pristine and irradiated PEEK. The experimental results showed that near the surface of the pristine and irradiated PEEK exists a peak, and the peak gradually disappeared with the increasing of the angles of incidence and the peak changed after irradiation, which implies the 170 keV protons have an effect on PEEK structure. The influences of PEEK irradiated with protons on the melting temperature and crystallization temperature was investigated by differential scanning calorimetry (DSC). The DSC results showed that the crystallinity of the polymer after irradiation decreased. The structure and content of free radicals of pristine and irradiated PEEK were studied by Fourier transform infrared spectroscopy (FTIR) and electron paramagnetic resonance (EPR). The stress and strain test results showed that the yield strength of the PEEK irradiated with 5×10^{15} p/cm^2 and 1×10^{16} p/cm^2 was higher than the pristine, but the elongation at break of the PEEK irradiated with 5×10^{15} p/cm^2 and 1×10^{16} p/cm^2 decreased obviously.

Keywords: PEEK; SIRM; damage mechanisms; GISAXS; irradiation

1. Introduction

Polyether ether ketone (PEEK) is widely used as an electrical insulation material in the aerospace field due to its excellent thermal and mechanical properties. The aromatic rings in the PEEK backbone are responsible for its strength, heat and radiation resistance [1–3]. The pendant ketone group increases the spacing between the molecules, and the ether bond makes the main chain flexible. The excellent mechanical stability and radiation resistance of PEEK has made it a choice material in a number of applications in the space environment [4]. In some areas of application, e.g., in the nuclear industry or space research, the radiation resistance of PEEK is of major importance [5]. The irradiation causes degradation of polymeric chains, breakage of chemical bonds, generation of free radicals and release of gas degradation products [6–8]. Subsequent chemical reactions of transient highly reactive species lead to excessive double bonds [9], low-mass stable degradation products, large cross-linked structures and oxidized structures [10]. So far, although the related research on the PEEK irradiation effect and its damage mechanism have been reported at home and abroad, it is still in the beginning exploration stage. Only a few papers mainly choose the gamma ray, heavy ion and electron irradiation sources to test and evaluate the chemical properties after irradiation, the mechanical behavior and the mechanism of the irradiated PEEK, of which the irradiated damage is only several microns and has not been studied [11–13]. The mechanical behavior of polymer insulating materials, especially a deep understanding of tensile deformation behavior, can help to evaluate the behavior of the material in orbit [14]. In recent years, synchrotron radiation X-ray scattering technology provides an effective method for detecting the microstructure changes of polymer materials [15,16]. The results can provide theoretical basis for the development of light, high performance and low-cost polymer materials, and the technology have important academic value. Our group has studied the structure evolution mechanism of the Low-density Polyethylene (LDPE) and PEEK polymer after electron irradiation by small angle X-ray scattering (SAXS) technology. There are a large number of low-energy charged particles in space, and these particles cause great damage to the structure and performance of the material. The damage caused by low-energy proton irradiation is only on the surface. How the surface structure and properties change after irradiation and the impact on the overall structure and properties has not been studied yet. The changes in the internal crystals of the material will significantly affect the changes in material properties. The effect of surface structure damage on the overall structure and performance degradation of the material is of great significance to the study of material degradation mechanisms.

GISAXS is a powerful tool for studying the surface of the films and interface structures [17–19]. Due to the small incidence angle of grazing incidence, we usually detect the area given by the slender coverage area of X-ray beam on the sample. The horizontal beam width is usually about 0.5 mm, and the coverage area extends the full length of the sample along the beam direction. The typical GISAXS sample size is from 10 mm to 30 mm, so we detect several mm^2 surface macroscopic areas, and the structural period is from 1 nm to 100 nm. In addition, the dispersion signal is proportional to the volume square of the irradiated sample area, and the 100 nm film on the 20 mm base is 10^6 μm^3. In contrast, the area detected by a typical transmitted SAXS beam is about 1 mm × 1 mm, which is less than one-tenth of the amount of scattering, and therefore less than one-hundredth of the intensity of scattering [20]. In addition, the substrate causes attenuation. When the 0.5 mm silicon wafer is irradiated at the speed of 10 keV, the transmittance will be reduced to 3%. And we can't get information about the height of the membrane. So the grazing incidence is used to analyze the microstructure damage. The structure evolution mechanisms of PEEK after low-energy proton irradiation were studied by the advanced GISAXS technology. The damage was only on the surface of the irradiated PEEK.

In this paper, the surface microstructure damage mechanisms of the PEEK irradiated with 170 keV proton were studied. The tensile tests and AFM were used to analyze the stress-strain and surface morphology of PEEK. Synchrotron radiation grazing incidence small angle X-ray scattering (GISAXS), FT-IR and DSC were applied to analyze the structure and crystallinity change of the PEEK after 170 keV proton irradiation. We used Stopping and Range of Ions in Matter (SRIM) to calculate the incident depth and ionization and displacement absorbed dose of the PEEK, combined with the advanced GISAXS technology and other experimental methods to reveal the influence of the material surface damage on the overall structure and performance degradation.

2. Experimental Section

2.1. Materials and Equipment

The density of PEEK material films was 1.3 g/cm^3 and the thickness was 50 μm. They were manufactured by the British Weiges Co., Ltd. (London, United Kingdom), and their molecular formula is $[C_9O_3H_{12}]_n$. The molecular structure diagram is shown in Figure 1. The 170 keV proton irradiation was performed by the experimental device of low-energy charged particles irradiation at Harbin Institute of Technology (Harbin, China).

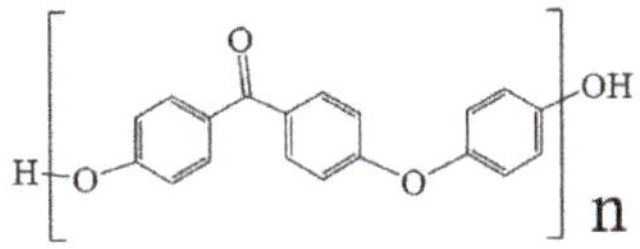

Figure 1. The Schematic of the repeat unit of polyether ether ketone (PEEK).

2.2. Experimental Parameter

The experimental proton energy was 170 keV, the sample chamber was vacuum 10^{-8} Pa, the irradiation area was 5×5 cm^2, the flux was 1×10^{15} p/cm^2·s, and the fluence values were 1×10^{15} p/cm^2, 5×10^{15} p/cm^2 and 1×10^{16} p/cm^2. The sizes of the dumbbell-shaped tensile samples were 12, 4 and 0.5 mm, respectively.

2.3. Microstructural and Mechanical Property Analysis

The AFM images were obtained by atomic force microscope with a multimode scanning probe microscope (Dimension Fastscan) produced by Bruker, Berlin, Germany. The maximum scanning range was 90 µm, the test temperature was −35–250 °C, the elastic modulus range was 1 MPa~100 Gpa, the adhesion force range was 10 pN~10 µN, the surface potential was ±10 V, and the precision was 10 mV. Moving the sample across the x-y plane, a voltage was applied to move the piezoelectric driver along the z-axis to maintain the same detection force, resulting in a three-dimensional image of the height of the sample surface. To evaluate mechanical performances of pristine and irradiated PEEK, elongation at break and tensile strength were tested by the MTS 810 material analysis and testing system of MTS (Shimadzu, Japan). The samples used for the tensile test were the dumbbell-shaped sample with a draw rate of 2 µm/s at room temperature. The length, width and thickness of the dumbbell-shaped tensile test samples were 12, 4, and 0.5 mm, respectively. The Fourier transform infrared (FT-IR) absorption spectra were used to analyze the microstructure of the samples, and the spectra were obtained in wavenumber range from 700 to 4000 cm^{-1} at every 2 cm^{-1} using a Magna-IR 560 spectrometer produced by Thermo scientific, Waltham, MA, USA. The free radicals before and after irradiation were analyzed by electron paramagnetic resonance (EPR) spectrometer (A200) from Bruker company, Berlin, Germany, the magnetic field ranged from 0 to 7000 G at a microwave frequency of 100 kHz. The melting and crystallization behavior were measured by differential scanning calorimetry (DSC) by a calorimeter (204 F1, Netzsch, Selb, Germany), the samples test temperature ranged from 30 to 400 °C at a heating and cooling rate of 10 °C/min. The grazing incident small angle X-ray scattering (GISAXS) measurements were performed on beamline BL16B1 of the Shanghai synchrotron radiation facility (SSRF) located in Shanghai Institute of Applied Physics, China. The Kohzu tilt stage was used as the GISAXS sample stage. The incidence angle of X-ray could be adjusted by the sample stage within an accuracy of 0.001°. In principle, working with an X-ray energy of 10 keV, the beam spot size at the samples was 0.5 mm × 0.5 mm. The sample detector distance was 5000 mm and the wavelength of incident X-ray was 0.124 nm. The schematic diagram of GISAXS test is shown in Figure 2. The GISAXS data was processed by FIT2D (London, UK).

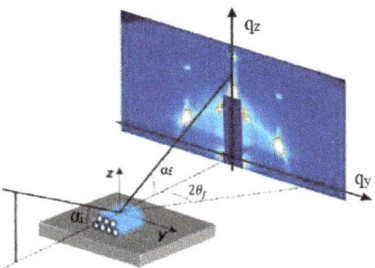

Figure 2. Schematic diagram of GISAXS test.

3. Results

3.1. SIRM Calculation

The ionization and displacement effect are the two aspects of the interaction between high-energy particles and materials. The ionization effect causes the polymer structure to form an excited state and the chemical bonds are broken. The displacement effect causes C, H, O and other atoms to leave the original lattice position to migrate along the polymer chain or in the material, and the atoms will be captured and stabilized in the appropriate position. Understanding the sensitivity of polymer materials to the ionization/displacement effect is of great significance for predicting the performance of polymer materials in an irradiation environment.

The SRIM software simulates the evolution of ionization and displacement dose produced by the PEEK irradiated with 170 keV proton and the content of free radicals changes with the change of ionization and displacement dose, as shown in Figure 3. If it is assumed that the excitation process of electrons and atoms is irrelevant, the energy loss of incident particles can be attributed to the sum of ionization energy loss and displacement energy loss, and the expression is shown in Equation (1):

$$\frac{dE}{dx} = N[S_n(E) + S_e(E)] \tag{1}$$

Among them, $S_n(E)$ is the blocking cross section of nuclear blocking, $S_e(E)$ is the blocking cross section of electron blocking, and N is the atomic density of the material, so the particle range is shown in Equation (2):

$$R = \int_0^R -\frac{dE}{dE/dx} = \frac{1}{N}\int_0^E \frac{dE}{S_n(E) + S_e(E)} \tag{2}$$

It can be seen that the total range of ions only depends on the electron blocking cross section and the nuclear blocking cross section, while the nuclear blocking and electron blocking cross sections depend on the interaction of particle collisions. The absorbed dose irradiated by charged particles refers to the radiant energy absorbed by a unit mass of material. The mathematical expression of the absorbed dose of a single-energy charged particle is shown in Equation (3):

$$D = \frac{1}{\rho} S \cdot \Phi \tag{3}$$

Dose is the absorbed dose, ρ is the density of the irradiated material, S is the stopping power, and Φ is the spectral distribution of particle fluence versus energy. Ionization damage is characterized by linear energy transfer (LET). The absorbed ionization dose can be obtained by multiplying LET by the fluence Φ, and the calculation formula is as follows (4):

$$D_i(\text{rad}) = 1.6 \times 10^{-8} LET \times \Phi \tag{4}$$

1.6×10^{-8} is the unit conversion factor.

Similar to the calculation of ionization damage, displacement damage is characterized by nonionizing energy loss (NIEL), and the calculation formula is as follows (5):

$$D_d(\text{rad}) = 1.6 \times 10^{-8} NIEL \times \Phi \tag{5}$$

According to the above formula, the results of ionization and displacement absorbed dose can be calculated and combined with the results of the EPR test, the result of Figure 3d can be obtained. The content of free radicals increases with the increasing of ionization and displacement absorbed dose.

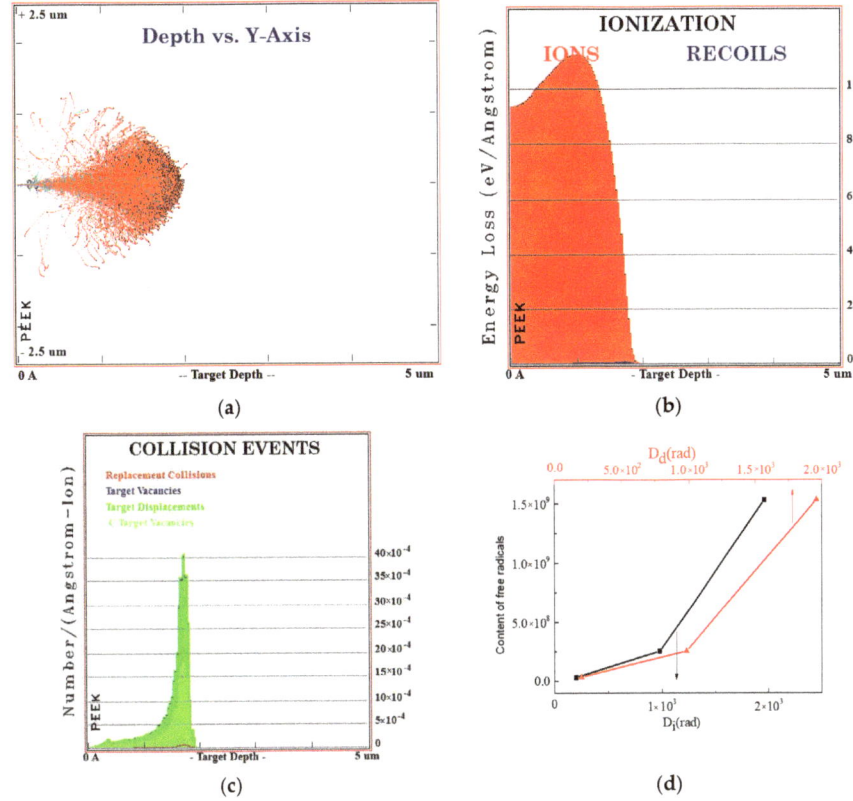

Figure 3. SRIM simulation of PEEK irradiated with 170 keV protons. (**a**) Simulation of particle collision process; (**b**) simulation of ionization energy loss of ion collision process; (**c**) simulation of vacancy distribution after ion irradiation; (**d**) content of free radicals change with ionization and displacement absorbed dose.

3.2. Surface Morphology and Roughness Analysis

The surface roughness values and 3D morphology images of pristine and irradiated PEEK with different fluences analyzed by AFM are presented Figure 4. The Figure 4d shows the roughness variation curves of pristine and irradiated PEEK in real-time. The roughness analysis of these samples was carried out for the chosen scan areas of 20 μm × 20 μm. Mean roughness (Ra) and root mean square (Rq) were calculated and are presented in Table 1. The PEEK film before irradiation (Figure 4a) exhibited an almost flat surface with some small bumps on it and its roughness parameters Ra and Rq were around 18.8 nm and 23.4 nm, respectively. When the PEEK was irradiated at 5×10^{15} p/cm^2, the films (Figure 4b) displayed a moderately creased surface which is revealed by slightly higher roughness values. In this case, Ra as well as Rq were relatively higher, and equal to 20.8 nm and 26.1 nm, respectively. When the irradiation fluences further increased to 1×10^{16} p/cm^2, the surface roughness values (Figure 4c) decreased slightly, the Ra and Rq decreased to 17.5 nm and 22.3 nm, respectively. The reason may be that when the irradiation fluence was 5×10^{15} p/cm^2, the radiation damage on the surface of the PEEK material caused the roughness to increase, and when the irradiation fluence was 1×10^{16} p/cm^2, the irradiation slightly etched the surface of the material to reduce the surface roughness. However, overall, as can be seen from Table 1, the surface roughness of PEEK before and after irradiation changed slightly. Generally speaking, the roughness of the material affected the GISAXS results which ignore the effect of the roughness.

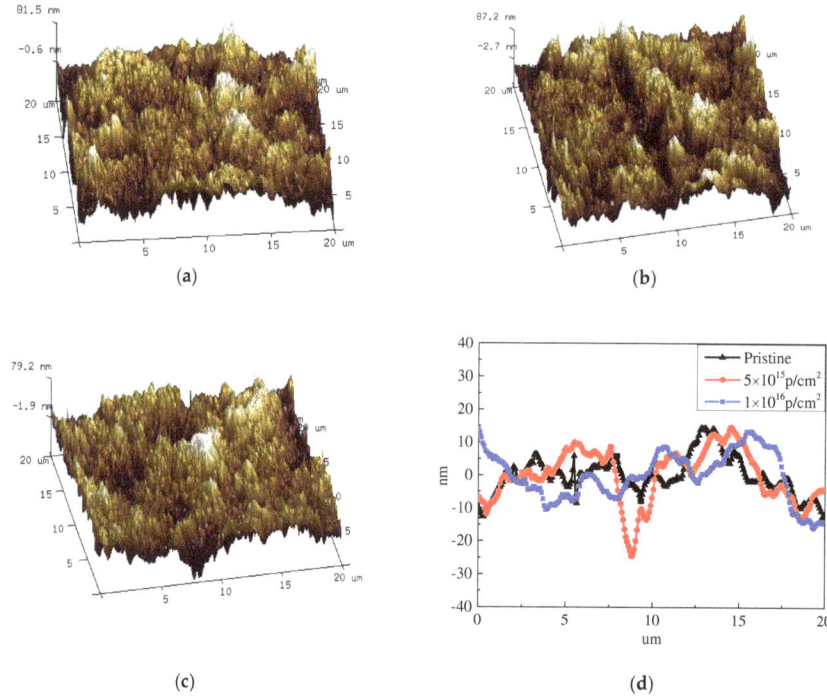

Figure 4. Surface roughness morphology of (**a**) pristine PEEK, (**b**) PEEK irradiated with 5×10^{15} p/cm^2, and (**c**) PEEK irradiated with 1×10^{16} p/cm^2, and (**d**) the surface roughness of the pristine and irradiated PEEK.

Table 1. Values of roughness parameters (Ra and Rq) of the PEEK films irradiated with different fluences.

Sample	Pristine	5×10^{15} p/cm^2	1×10^{16} p/cm^2
Ra(nm)	18.8	20.8	17.5
Rq(nm)	23.4	26.1	22.3

3.3. Micrstructure Analysis Based on GISAXS

The GISAXS scattering patterns and intensity distribution curves with different incident angles for pristine and irradiated PEEK with different fluences are shown in Figure 5. The GISAXS technology provides a powerful tool for studying ordered interface and surface of materials [21,22]. The effect of roughness on the experimental results is very complicated. In this experiment, the AFM experiment was used to prove that the roughness after irradiation was affected slightly, so the effect of roughness on the GISAXS results was ignored. If the layered structure is parallel to the substrate, the dispersion peaks will be obtained along the surface normal in the plane of incidence. For a freely oriented layer materials, we get a ring diffraction peak. Because the scattered X-rays are blocked by the base, the ring peak only can be seen if the angle of departure is greater than zero. If the layered structure is partially oriented, the ring peak will become an arc peak. For the layered structure, we will observe the Bragg reflection in the direction parallel to the substrate surface. The layered structures with substrates and polymer membranes may be observed in disordered systems as rings or arcs. It can be seen from the Figure 5a, there appeared obvious scattering peak and it shows different shapes with different angles of incidence. The horizontal stripe diffraction pattern represents the crystal in the horizontal direction. This peak should be a layered crystal structure. The Figure 5b shows that there are multiple peaks, which may be multiple crystal structures. However, the GISAXS images of the PEEK irradiated by

170 keV protons with different fluences changed significantly, as shown Figure 5c–f. When the angles of incidence increases to a certain degree, the peak gradually changes and eventually disappears. This results combined with the SRIM calculation results, which show that irradiation makes the PEEK materials produce ionization effect and displacement effect, resulting in a large number of vacancies, which affects the crystal structure of the material. It is known that the X-ray incident on crystal materials will cause the Bragg diffraction of the crystals as given by the following formula:

$$2d \sin \theta = n\lambda \tag{6}$$

n is the diffraction order, d is the crystal spacing, θ is the diffraction angle and λ is the incident X-ray wavelength.

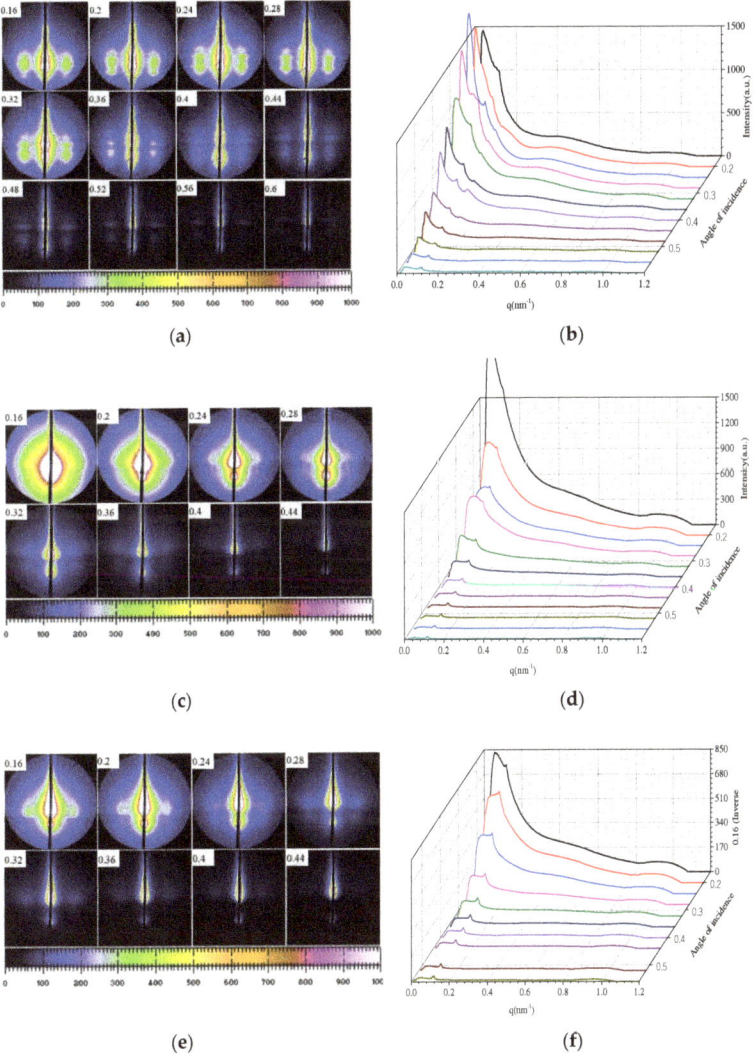

Figure 5. The different GISAXS scattering patterns and intensity distribution curves of pristine and irradiated PEEK with the incident angle, (**a**,**b**) pristine, (**c**,**d**) 5×10^{15} p/cm^2, (**e**,**f**) 1×10^{16} p/cm^2.

Crystalline phases can have their own formula to represent the long period, while giving a larger value L. The interphase distribution of the crystalline and amorphous regions constitutes this structure distance between two adjacent crystal regions. Therefore, when the incident wavelength is constant, the Bragg diffraction can occur at a smaller angle, which corresponds to a smaller scattering vector q. The Bragg diffraction peak generated by the long-period structure in a small angle range can represent the SAXS scattering peak of the pristine PEEK. According to the definition and the formula (6) of the scattering vector, the relationship between the long period and scattering vector can be obtained as follows:

$$q = 2n\pi/L \tag{7}$$

q is the scattering vector, n is the number of diffraction orders, and L is the long period of the crystalline polymer. According to formula (7) and the scattering peak position of the undisturbed PEEK, the long period of the pristine PEEK without stretching was 16.44 nm. The changes in long period of the pristine and the irradiated PEEK with the incident Angle are shown in Table 2. From the table, it can be seen that with the increasing of the incident angle, the long period of the PEEK decreased. Irradiation reduced the long period of the material and changes the internal crystal structure of the material.

Table 2. The changes in long period of the pristine and the irradiated PEEK with the incident angle.

Sample	0.16	0.2	0.24	0.28	0.32	0.36	0.4	0.44	0.48
L(nm)/Pristine	16.44	16.10	16.02	15.90	15.66	15.43	15.24	14.74	14.6
L(nm)/5 × 10^{15} p/cm^2	14.78	14.6	14.37	13.86	13.14	12.41			
L(nm)/1 × 10^{16} p/cm^2	14.4	14.08	13.96	13.36	12.56				

3.4. EPR Analysis

The EPR spectra of pristine and PEEK irradiated with different fluences and the content of free radicals changed with fluences are shown in Figure 6a,b. In polymer materials, the displacement effect will cause some atoms in the molecule chain to change from the original position to bond breakage, resulting in the degradation of the molecular chain. Covalent bond breakage leaves the atoms in an unpaired electronic state and then generates free radicals [23]. The charged particles will interact with atoms in the outer layer of polymer, generating electron excitation and causing ionization, which may also generate free radicals in the molecular chain. It can be seen from Figure 6 that the PEEK after 170 keV proton irradiation produced a large amount of free radicals, and the content of the free radicals increased with the increasing of the fluences. The data were calculated by the Formula (8):

$$N = \iint Sds/m \tag{8}$$

where N is the content of free radicals, S is the measured EPR spectrum, m is the quality of test sample. The g value was 2.0025 in this data, and the g value was stable and it did not change with the change of the irradiated particle energy and fluences. In polymer material, the g value corresponds to two free radicals, one is pyrolytic carbon free radicals, the other is a hydroxy superoxide radicals. The irradiation experiment was under a vacuum environment, there was not enough oxygen reacting with free radicals, so it was impossible to generate such a large number of hydroxyl superoxide radicals. Thus, most of the free radicals should be a pyrolytic carbon free radical. If free radicals were induced abundantly in main chains of PEEK, they would react in different ways to induce oxidation, crosslinking or degradation.

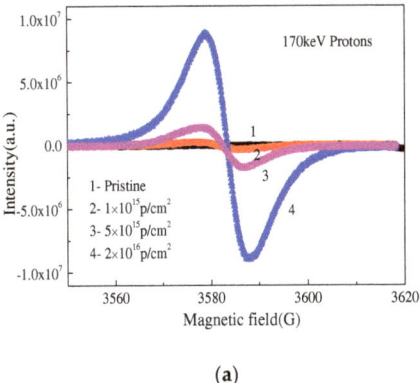 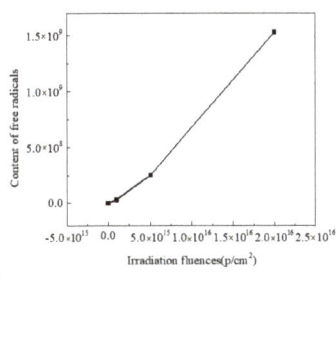

Figure 6. (a) The EPR spectrogram of pristine and irradiated PEEK with different fluences, (b) the content of free radicals changed with fluences.

3.5. FTIR Analysis

The Fourier infrared spectrum of pristine and PEEK irradiated by 170 keV proton with different fluences are shown in Figure 7. With the increasing of the irradiation fluences, the intensity of all the characteristic absorption peaks gradually weakened and even disappeared. The absorbance of the ketone group (C=O) stretching vibration peaked at 1651 cm^{-1}, the aromatic skeletal vibration occurred at 1598, 1490 and 1413 cm^{-1}, the ether bond (C–O–C) was asymmetric stretching at 1280 and 1187 cm^{-1}, aromatic hydrogen in-plane deformation bands occurred at 1157 and 1103 cm^{-1}, the diphenyl ketone band occurred at 927 cm^{-1} and the plane bending modes of the aromatic hydrogens occurred at 860 and 841 cm^{-1}, which was consistent with the previous reports on this material [24,25]. Low-energy proton irradiation has a great effect on the molecular structure of PEEK materials. It has been shown that heavy ion irradiation can destroy benzene rings and generate a large number of aromatic fragments according to the references [26]. It can be inferred that the low energy proton irradiation will destroy the carbon skeleton of the benzene ring in the molecular structure of PEEK, resulting in the decrease of C=O and benzene content in PEEK. By comparing the intensity of the characteristic absorption peaks of the materials before and after the irradiation sources, it can be seen that the intensity of characteristic absorption peaks of each group decreased slightly with the increasing of the irradiation fluences. The reason for this change may be the dipole moment of the material changed due to irradiation, which improved the infrared scattering of the material surface and weakened the absorption intensity. The spectra were obviously changed by 1×10^{16} p/cm^2 irradiation fluences. The transmittance decreased dramatically, in addition to the two oxidation-related peaks at 1730 cm^{-1} and 1100 cm^{-1}. This is also attributed to the scission of C-H bonds and the following oxidation, which in turn destroyed chemically symmetric structures. This structural change would make PEEK hard and brittle, which is a typical mechanical change induced in PEEK aged by irradiation.

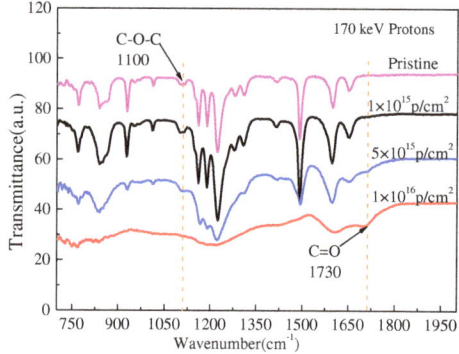

Figure 7. FT-IR spectra of pristine and PEEK irradiated with different fluences.

3.6. DSC Analysis

The DSC spectra of the pristine and PEEK irradiated with different fluences are shown in Figure 8. Table 3 shows the melting and crystallization parameters of PEEK before and after irradiation, including initial melting temperature T_1, ending melting temperature T_2, melting enthalpy ΔH_m, starting crystallization temperature T_3, ending crystallization temperature T_4, crystallization enthalpy ΔH_c and crystallinity X(%). From the figure we can see that the melting peak of the PEEK was about 338 °C, the melting temperature of irradiated PEEK moved towards the higher temperature slightly, and the ΔH_m of irradiated PEEK decreased with the increasing of the fluences. The crystallization peak of PEEK was around 298 °C, the crystallization temperature of irradiated PEEK increased first than decreased slightly with the increasing of the fluences, and the crystallization enthalpy ΔH_c decreased slightly. The crystallinity of the polymer after irradiation reduced from 17% to 13%. The reason is that the crystal structure of PEEK changed after irradiation and the crystallinity decreased, causing the enthalpy of crystallization and melting enthalpy to decrease.

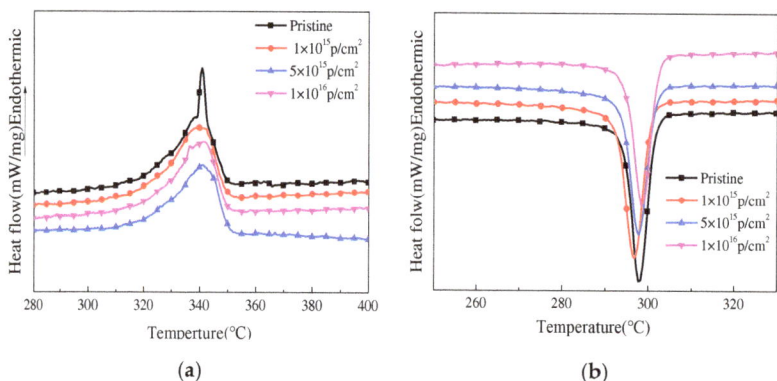

Figure 8. The DSC spectra of the pristine and PEEK irradiated with different fluences, (**a**) The melting temperature; (**b**) The crystallization temperature

Table 3. Melting and crystallization parameters of PEEK before and after irradiation.

Sample	Pristine	5×10^{15} e/cm^2	1×10^{16} e/cm^2
T1 (°C)	301.8 ± 0.5	305.5 ± 0.5	309.2 ± 0.5
T2 (°C)	350.9 ± 0.5	351.3 ± 0.5	352.7 ± 0.5
ΔH_m (J/g)	21.9 ± 5	20.4 ± 5	16.7 ± 5
T3 (°C)	301.4 ± 0.5	302.5 ± 0.5	303.4 ± 0.5
T4 (°C)	294.5 ± 0.5	296.7 ± 0.5	299.9 ± 0.5
ΔH_c (J/g)	40.9 ± 5	43.0 ± 5	39.5 ± 5
X (%)	17	15.8	13

3.7. Stress–Strain Curves Analysis

The tensile stress–strain curves of the pristine and 170 keV proton irradiated PEEK with different fluences at room temperature and the inner figure, which is a locally enlarged view of yield strength, are shown in Figure 9. The tensile curves are divided into three stages including elastic deformation, cold drawing and strain hardening. The first elastic deformation stage was mainly the deformation of the molecular chain in the amorphous region, and the stress increased with the increasing of the strain until it reached the yield point. In the cold drawing stage, the necking cross-sectional area remains largely unchanged, but the necking part increased until the entire sample became thinner. As a result, the stress level at this stage was almost unchanged, and the strain increased gradually. In the strain hardening stage, the stress increased with the increasing of the strain, and the sample was uniformly stretched to the break point. It can be seen from the figure that the basic characteristics of the tensile stress-strain curves were almost the same between the pristine and the irradiated PEEK with 1×10^{15} p/cm^2. The yield strength of the PEEK irradiated with 5×10^{15} p/cm^2 and 1×10^{16} p/cm^2 was higher than that of the pristine, but the elongation at break of the PEEK irradiated with 5×10^{15} p/cm^2 and 1×10^{16} p/cm^2 decreased obviously. The defects of the PEEK films after low energy proton irradiation increased and the molecular weight decreased. It can be seen from the results that, although the irradiation damage was only about 3 μm on the surface of the material, it had a great influence on the elongation at break of the entire material.

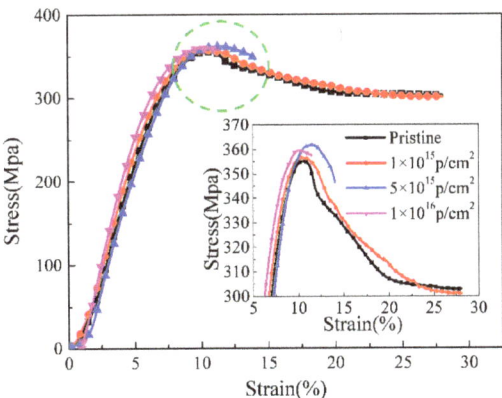

Figure 9. Stress–strain curves of pristine and PEEK films irradiated with different fluences and the inner image which is the local enlarged image.

3.8. Discussion

The schematic of the irradiation process and irradiation degradation process and the EPR results of PEEK repeat unit is shown in Figure 10. The schematic shows that the PEEK films irradiated by the 170 keV proton may produce defects and the free radicals on the surface of the irradiated PEEK films. The accelerated 170 keV proton had enough energy to break all the chemical bonds in organic materials. The irradiation sources may break the benzene rings, ether bonds and C=O bonds of the PEEK films [27]. The most common result of the chemical bonds breakage is the formation of free radicals. From the EPR results, we can know that most of the free radicals were pyrolytic carbon free radicals. The free radicals may have combined with active factors such as oxygen when the irradiated materials were put in the air, which may have led to formation of defects and oxide layer on the surface of PEEK. The irradiation processes can be classified according to the effect of formation of free radicals, including curing, crosslinking, degradation and grafting. It can be seen from the stress-strain curves that the elongation at break of the material decreased after irradiation, so the most likely reason is that the molecular weight of the polymer decreased, and the irradiation caused the PEEK material to degrade. Because the damage was only on the surface of the polymer, the yield strength did not change significantly for the overall material. According to the GISAXS and FTIR results, the radiation had a significant effect on the surface structure damage. The DSC results show that irradiation will cause the crystallinity of PEEK to decrease.

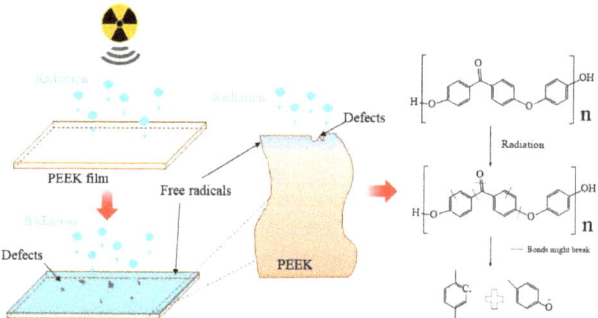

Figure 10. Schematic of the irradiation process and irradiation degradation process and the EPR results of PEEK repeat unit.

4. Conclusions

PEEK films were irradiated by 170 keV protons and the microstructure and the mechanical properties of PEEK before and after proton irradiation were characterized. The SIRM calculation results showed that the irradiation caused vacancies in the material and the damage thickness was about 3 µm. The FTIR results showed that the intensity of all the characteristic absorption peaks gradually weakened and even disappeared with the increasing of the irradiation fluences. The EPR results showed that PEEK produced a lot of free radicals after irradiation, and the content of free radicals increased with the increase of irradiation fluences. The DSC results showed that the crystallinity of the polymer after irradiation decreased. The results of GISXAS showed that the pristine PEEK samples had peaks and the peaks changed with the increasing of angle of incidence. After the irradiation of the 170 keV protons, the scattering patterns and peaks changed gradually. The yield strength of the irradiated PEEK with 5×10^{15} p/cm^2 and 1×10^{16} p/cm^2 increased a little, which compared to that of the pristine, but the elongation at break of the irradiated PEEK with 5×10^{15} p/cm^2 and 1×10^{16} p/cm^2 decreased obviously.

Author Contributions: Conceptualization, H.L.; Data curation, H.L.; Investigation, H.L.; Writing—original draft, H.L.; Supervision, J.Y. and S.D.; Writing—review & editing, J.Y.; Formal analysis, X.L. and S.D.; Validation, S.D.; Visualization, S.D.; Funding acquisition, X.L.; Methodology, F.T.; Project administration, X.L.; Resources, X.L. All authors have read and agreed to the published version of the manuscript.

Funding: This research was funded by Science Challenge Project, grant number NO.TZ2018004.

Conflicts of Interest: The authors declare no conflict of interest.

References

1. Jung, H.; Bae, K.J.; Jin, J.-U.; Oh, Y.; Hong, H.; Youn, S.J.; You, N.-H.; Yu, J. The effect of aqueous polyimide sizing agent on PEEK based carbon fiber composites using experimental techniques and molecular dynamics simulations. *Funct. Compos. Struct.* **2020**, *2*, 025001. [CrossRef]
2. Wu, Y.; Cao, Y.; Wu, Y.; Li, D. Neutron Shielding Performance of 3D-Printed Boron Carbide PEEK Composites. *Materials* **2020**, *13*, 2314. [CrossRef] [PubMed]
3. Zhu, J.; Xie, F.; Dwyer-Joyce, R. PEEK Composites as Self-Lubricating Bush Materials for Articulating Revolute Pin Joints. *Polymers* **2020**, *12*, 665. [CrossRef] [PubMed]
4. Laux, C.J.; Villefort, C.; Ehrbar, S.; Wilke, L.; Guckenberger, M.; Müller, D.A. Carbon Fiber/Polyether Ether Ketone (CF/PEEK) Implants Allow for More Effective Radiation in Long Bones. *Materials* **2020**, *13*, 1754. [CrossRef] [PubMed]
5. Sehyun, K.; Ki-Jun, L.; Yongsok, S. Polyetheretherketone (PEEK) Surface Functionalization by Low-Energy Ion-Beam Irradiation under a Reactive O_2 Environment and Its Effect on the PEEK/Copper Adhesives. *Langmuir* **2004**, *20*, 157–163.
6. Li, H.; Yang, J.-Q.; Dong, S.; Tian, F.; Li, X. Low Dielectric Constant Polyimide Obtained by Four Kinds of Irradiation Sources. *Polymers* **2020**, *12*, 879. [CrossRef] [PubMed]
7. Shyichuk, A.; Stavychna, D.; White, J. Effect of tensile stress on chain scission and crosslinking during photo-oxidation of polypropylene. *Polym. Degrad. Stab.* **2001**, *72*, 279–285. [CrossRef]
8. Liu, B.-X.; Liang, J.; Peng, Z.-J.; Wang, X.-L. Proton irradiation effects on the structural and tribological properties of polytetrafluoroethylene. *Chin. J. Polym. Sci.* **2016**, *34*, 1448–1455. [CrossRef]
9. Dong, L.; Liu, X.-D.; Xiong, Z.-R.; Sheng, D.; Zhou, Y.; Yang, Y. Preparation and Characterization of UV-absorbing PVDF Membranes via Pre-irradiation Induced Graft Polymerization. *Chin. J. Polym. Sci.* **2018**, *37*, 493–499. [CrossRef]
10. Nakamura, T.; Nakamura, H.; Fujita, O.; Noguchi, T.; Imagawa, K. The Space Exposure Experiment of PEEK Sheets under Tensile Stress. *JSME Int. J. Ser. A* **2004**, *47*, 365–370. [CrossRef]
11. Macková, A.; Havránek, V.; Svorcik, V.; Djourelov, N.; Suzuki, T. Degradation of PET, PEEK and PI induced by irradiation with 150keV Ar^+ and 1.76MeV He^+ ions. *Nucl. Instrum. Methods Phys. Res. Sect. B Beam Interact. Mater. Atoms* **2005**, *240*, 245–249. [CrossRef]
12. Narasimha, P.; Sudhir, T.; Henry, C. *Fabrication of Lightweight Radiation Shielding Composite Materials by Field Assisted Sintering Technique (FAST)*; NASA: Washington, DC, USA, 2017.
13. Malinsky, P.A.; Mackova, V.; Hnatowicz, R.I.; Khaibullin, V.F. Properties of polyimide, polyetheretherketone and poly-ethyleneterephthalate implanted by Ni ions to high fluences. *Nucl. Instrum. Methods Phys. Res. Sect. B Beam Interact. Mater. Atoms* **2012**, *272*, 396–399. [CrossRef]
14. Alnajrani, M.N.; Alosime, E.M.; Basfar, A.A. Influence of irradiation crosslinking on the flame-retardant properties of polyolefin blends. *J. Appl. Polym. Sci.* **2019**, *137*, 48649. [CrossRef]
15. Jeun, J.P.; Nho, Y.C.; Kang, P.H. Enhancement of mechanical properties of PEEK using radiation process. *Appl. Chem.* **2008**, *12*, 245–248.
16. Sasuga, T.; Kudoh, H. Ion irradiation effects on thermal and mechanical properties of poly(ether–ether–ketone) (PEEK). *Polymer* **2000**, *41*, 185–194. [CrossRef]
17. Végsö, K.; Siffalovic, P.; Weis, M.; Jergel, M.; Benkovičová, M.; Majkova, E.; Chitu, L.; Halahovets, Y.; Luby, Š.; Capek, I.; et al. In Situ GISAXS monitoring of Langmuir nanoparticle multilayer degradation processes induced by UV photolysis. *Phys. Status Solidi* **2011**, *208*, 2629–2634.
18. Dong, Q.; Brian, D.; Lorcán, B.; Gennaro, S. Rapid surface activation of carbon fibre reinforced PEEK and PPS composites by high-power UV-irradiation for the adhesive joining of dissimilar materials. *Compos. Part A* **2020**, *137*, 105976.
19. Végsö, K.; Siffalovic, P.; Benkovicova, M.; Jergel, M.; Luby, S.; Majkova, E.; Capek, I.; Kocsis, T.; Perlich, J.; Roth, S.V. GISAXS analysis of 3D nanoparticle assemblies—Effect of vertical nanoparticle ordering. *Nanotechnology* **2012**, *23*, 45704–45711. [CrossRef]
20. Xue, F.; Li, H.; You, J.; Lu, C.; Reiter, G.; Jiang, S. The crucial role of cadmium acetate-induced conformational restriction in microscopic structure and stability of polystyrene-block-polyvinyl pyridine thin films. *Polymer* **2014**, *55*, 5801–5810. [CrossRef]

21. Yasin, S.; Shakeel, A.; Ahmad, M.; Ahmad, A.; Iqbal, T. Physico-chemical analysis of semi-crystalline PEEK in aliphatic and aromatic solvents. *Soft Mater.* **2019**, *17*, 143–149. [CrossRef]
22. Susumu, T.; Toshiaki, E.; Hiroaki, I.; Yousuke, O.; Joji, O.; Takuji, K. Highly Efficient Singlet Oxygen Generation and High Oxidation Resistance Enhanced by Arsole-Polymer-Based Photosensitizer: Application as a Recyclable Photooxidation Catalyst. *Macromolecules* **2020**, *53*, 2006–2013.
23. Shaimukhametova, I.F.; Shigabieva, Y.A.; Bogdanova, S.A.; Allayarov, S.R. Effect of γ-Irradiation on the Gel-Forming Capability of a Crosslinked Acrylic Polymer. *High Energy Chem.* **2020**, *54*, 111–114. [CrossRef]
24. Lyu, J.-Z.; Laptev, R.S.; Dubrova, N. Positron Spectroscopy of Free Volume in Poly(vinylidene fluoride) after Helium Ions Irradiation. *Chin. J. Polym. Sci.* **2018**, *37*, 527–534. [CrossRef]
25. Al Lafi, A.G.; Alzier, A.; Allaf, A.W. Two-dimensional FTIR spectroscopic analysis of crystallization in cross-linked poly(ether ether ketone). *Int. J. Plast. Technol.* **2020**, *24*, 1–8. [CrossRef]
26. Kornacka, E.M.; Przybytniak, G.; Nowicki, A. Radical processes initiated by ionizing radiation in PEEK and their macroscopic consequences. *Polym. Adv. Technol.* **2018**, *30*, 79–85. [CrossRef]
27. Claude, B.; Gonon, L.; Duchet, J.; Verney, V.; Gardette, J. Surface cross-linking of polycarbonate under irradiation at long wavelengths. *Polym. Degrad. Stab.* **2004**, *83*, 237–240. [CrossRef]

Publisher's Note: MDPI stays neutral with regard to jurisdictional claims in published maps and institutional affiliations.

© 2020 by the authors. Licensee MDPI, Basel, Switzerland. This article is an open access article distributed under the terms and conditions of the Creative Commons Attribution (CC BY) license (http://creativecommons.org/licenses/by/4.0/).

MDPI
St. Alban-Anlage 66
4052 Basel
Switzerland
Tel. +41 61 683 77 34
Fax +41 61 302 89 18
www.mdpi.com

Polymers Editorial Office
E-mail: polymers@mdpi.com
www.mdpi.com/journal/polymers

www.ingramcontent.com/pod-product-compliance
Lightning Source LLC
LaVergne TN
LVHW070248100526
838202LV00015B/2193